Classical Algebra

for

Undergraduate, honours and Post graduate classes of Calcutta University

DR. SUDHIR KUMAR PUNDIR
M.Sc., M.Phil., N.E.T. (C.S.I.R.) Ph.D.
Associate Professor
Department of Mathematics
S.D. (P.G.) College, Muzaffarnagar (U.P.)

CBS

CBS Publishers & Distributors Pvt. Ltd.

New Delhi • Bengaluru • Chennai • Kochi • Kolkata • Mumbai • Pune
Hyderabad • Nagpur • Patna • Vijayawada

CLASSICAL ALGEBRA

ISBN: 978-81-239-2796-1

First Edition: 2015

Published by:

Satish Kumar Jain for CBS Publishers & Distributors Pvt. Ltd.,
CBS Plaza, 4819/XI Prahlad Street, 24 Ansari Road,
Daryaganj, New Delhi – 110002, India
delhi@cbspd.com, cbspubs@airtelmail.in • www.cbspd.com
Ph.: 23289259, 23266861, 23266867 • Fax: 011-23243014

Branches:
- *Bengaluru:* 2975, 17th Cross, K.R. Road, Bansankari 2nd Stage,
  Bengaluru - 70 • Ph: +91-80-26771678/79 • Fax: +91-80-26771680
  E-mail: cbsbng@gmail.com, bangalore@cbspd.com
- *Chennai:* No. 7, Subbaraya Street, Shenoy Nagar, Chennai - 600030
  Ph: +91-44-26681266, 26680620 • Fax: +91-44-42032115
  E-mail: chennai@cbspd.com
- *Kochi:* Ashana House, 39/1904, A.M. Thomas Road, Valanjambalam,
  Ernakulum, Kochi • Ph: +91-484-4059061-65
  Fax: +91-484-4059065 • E-mail: cochin@cbspd.com
- *Kolkata:* 6-B, Ground Floor, Rameshwar Shaw Road, Kolkata - 700014
  Ph: +91-33-22891126/7/8 • E-mail: kolkata@cbspd.com
- *Mumbai:* 83-C, Dr. E. Moses Road, Worli, Mumbai - 400018
  Ph: +91-9833017933, 022-24902340/41 • E-mail: mumbai@cbspd.com
- *Pune:* Bhuruk Prestige, Sr. No. 52/12/2+1+3/2,
  Narhe, Haveli (Near Katraj-Dehu Road Bypass), Pune - 411041
  Ph: +91-20-64704058/59, 32342277 • E-mail: pune@cbspd.com

Representatives:
- Hyderabad: 0-9885175004
- Patna: 0-9334159340
- Nagpur: 0-9021734563
- Vijayawada: 0-9000660880

Printed at :
Neekunj Print Process, Delhi

Preface

The book entitled 'Classical Algebra' is meant for UG, honours and P.G students of Calcutta University. It will also be very useful for various competitive examinations.

This book provides theoretical background of the subject with well graded set of detailed solved examples. It is a collection and compilation work from various sources and has been endeavoured to include as much as information could be possible. There is plenty of scope in the form of exercise for the reader to try and solve the problem on his own. To make the book self-contained and competition oriented, a chapter review of basic terms, results and questions has been given at the end of each chapter. For better understanding of the subjects a larger number of objective type questions also given at the end of each chapter.

I express my gratitude to the authors and publishers of various books I consulted.

I wish to sincerely thank Sh. S.K. Jain, M.D, CBS publisher and Distributors New Delhi for his encouragement and help in bringing out this book in a present nice form.

My special thank to Sh. Y.N. Arjuna, Sh.B.M. Singh, Sh. Sunil Dutt and entire team of CBS publisher New Delhi whose encouragement and unstinted support enabled to me to complete my book. Mr. Ajay Kumar Malhotra also deserve special mention for nice typesetting.

I must also record my appreciation due to my wife Dr. Rimple, daughter Rijuta and son Shrish for their understanding and love during the long period that I have taken to complete this book.

Above all I am thankful to the Almighty God, without grace nothing is possible for any one.

Readers are welcomed to point out errors, if any and send their valuable suggestions for improving the quality of the book.

Dr. Sudhir Kumar Pundir
email:skpundir05@yahoo.co.in
skpundir05@gmail.com

Contents

CHAPTER-4
Primes and Their Distributions

CHAPTER-5
Congruences

CHAPTER-6
Fermat's Theorems and Its Applications

CHAPTER-7
Number Theoretic Functions

CHAPTER-8
Complex Numbers

CHAPTER-9
Exponential and Trigonometric Functions of a Complex Variable

CHAPTER-10
Logarithm of Complex Quantities

CHAPTER-11
Inverse Circular and Hyperbolic Functions of a Complex Number

CHAPTER-12
Polynomial with Real Coefficients

CHAPTER 1

Well ordering Principle and Mathematical Induction

- ❖ Number System
- ❖ Binary Operations on Integers
- ❖ Ordering of Integers
- ❖ Mathematical Induction
- ❖ Archimedian Property
- ❖ Basic Representation Theorem

1.1 INTRODUCTION

The theory of numbers mainly deals with properties of the natural numbers 1, 2, 3, 4, ... (also called the positive integers). These numbers together with the negative integers and zero, form the set of integers. Properties of these numbers have been studied from earliest time. For example, an integer is divisible by 3, if and only if the sum of its digits is divisible by 3 as in the number 852 with sum of digits 8 + 5 + 2 = 15. The equation $x^2 + y^2 = z^2$ has infinitely many solutions in positive integers, such as $3^4 + 4^4 = 5^4$, whereas $x^3 + y^3 = z^3$ and $x^4 + y^4 = z^4$ have none. There are infinitely many prime numbers, where a prime is a natural number such as 31 that can not be factored into two smaller natural numbers. Thus 33 is not a prime, because 33 = 3 . 11. The fact that the sequence of primes 2, 3, 5, 7, 11, 13, 17, ..., is endless was known to Euclid. Also known to Euclid was the result that $\sqrt{2}$ is an irrational number.

The theory of numbers is closed tied to the other areas of mathematics most especially to abstract algebra, linear algebra, combinatorics, analysis, geometry and even topology.

1.2 NUMBER SYSTEM

The number system plays a key role in mathematics. The real system R is one of the most important and beautiful mathematical system. There are different ways of introducing the real number system, but the most common way is to start with Peano's axioms for the natural numbers.

The axioms for natural numbers, discovered by the Italian Mathematician Peano are

  (i) 1 is a natural number
 (ii) Each natural number n has a successor $(n + 1)$.
(iii) Two natural numbers are equal if their successor are equal.
 (iv) Except 1, each natural number is a successor of natural number.
  (v) Any set of natural numbers which contains 1 and the successor of every natural number k whenever it contains k, in the set N of natural number.

Remark

* Axiom (v) is commonly known as the axiom of induction or principle of finite induction.

(i) **Natural Numbers:** The numbers 1, 2, 3, ... are called natural numbers. We represent the set of natural numbers by N.
   i.e.   N = {1, 2, 3, ...}

(ii) **Integers:** The numbers ..., –3, –2, –1, 0, 1, 2, 3, ... are called integers. We repsent the set of integers by Z.
   i.e.   Z = {..., – 3, – 2, – 1, 0, 1, 2, 3, ...}

(iii) **Rational Numbers:** Any number of the form $\frac{p}{q}(q \neq 0), p, q \in Z$ and p, q have no

common factor (except ± 1) is called a rational number. We represent the set of rational numbers by Q.

   i.e. $Q = \left[\frac{p}{q} : p, q \in Z, q \neq 0 \right]$

(iv) **Irrational Numbers:** Any number, which is not rational, is called an irrational number. For example $\sqrt{2}, \sqrt{3}$ ... *etc.*

Remark

* Every rational number can be expressed as a terminating or recurring decimal whereas every irrational number can be expressed as a non-terminating infinite decimal.

(v) **Real Numbers:** A number which is either rational or irrational is called a real number. We represent the set of real numbers by R.

Remark

* The above discussion of numbers, starting with Peano's axioms for natural numbers was put forward by Karl Weirstrass, Richard Dedekind and George Canter.

1.3 BASIC BINARY OPERATION ON THE SET OF INTEGERS

Generally, there are two binary operations :

(i) addition, denoted by +
(ii) multiplication, denoted by ×

The basic properties of these two operations are as follow :

1.3.1 Addition Axioms

(i) **Closure property,**   i.e. $a + b \in Z, \forall\, a, b \in Z$

(ii) **Commutativity,**   i.e. $a + b = b + a, \forall\, a, b \in Z$

(iii) **Associativity,** i.e. $a + (b + c) = (a + b) + c, \forall\, a, b, c \in Z$

(iv) **Existence of Identity:** There exists a unique integer 0 such that

$$a + 0 = a = 0 + a, \forall\, a \in Z$$

This integer 0 is called the additive identity.

(v) **Existence of Additive Inverse:** If, $a \in Z$ then there exists a unique integer $-a \in Z$ such that

$$-a + a = 0 = a + (-a)$$

The integer $-a$ is called the additive inverse of the integer a.

1.3.2 Multiplication Axioms

(i) **Closure property,**   i.e. $ab \in Z, \forall\, a,b \in Z$

(ii) **Commutativity,**   i.e. $ab = ba, \forall\, a,b \in Z$

(iii) **Associativity,** i.e. $a(bc) = (ab)c, \forall\, a,b,c \in Z$

(iv) **Existence of Identity:** There exists a unique integer 1 such that

$$1 . a = a = a . 1 \ \forall\ a \in Z$$

This integer 1 is called the multiplicative identity.

(v) **Existence of Additive Inverse:** If $a \in Z$, then there exists a unique integer $a^{-1} \in Z$ such that

$$aa^{-1} = 1 = a^{-1}a, \ \forall\ a \in Z$$

The integer a^{-1} is called the additive inverse of the integer a.

1.4 ORDERING OF THE INTEGERS

The set of positive integers, Z^+, having the following two properties:

(1) **The Law of Trichotomy:** If, $a \in Z$ then one and only, one of the following is true.

(i) $a \in Z^+$ (ii) $a = 0$ (iii) $-a \in Z$

(2) If $a,b \in Z^+$, then $a + b \in Z^+$ and $ab \in Z^+$

It is clear that $0 \notin Z^+$ and $1 \in Z^+$

Definition: *If $a,b \in Z$ and $a - b \in Z^+$, then we say that a is greater than b and we write* $a > b$.

If $a < b$ or $a = b$, we write $a \leq b$ and if $a > b$ or $a = b$, we write $a \geq b$.
Clearly, a is positive if and only if $a > 0$ and a is negative if and only if $a < 0$.

Also, if $a \in Z$, then one and only one of the following is true.

$$a \in Z^+, a = 0, -a \in Z$$

i.e. $a > 0, a = 0, a < 0$

1.5 WELL ORDERING PRINCIPLE

Let S be a non-empty subset of Z. If there exists an integer $m \in S$ such that $x \geq m$ for all $x \in S$, then m is said to be the smallest or the least integer in S, i.e., S has a least member. If there exists an integer $n \in S$ such that $x \leq n$ for all $x \in S$, the n is said to be the greatest integer in S.

Statement: *Every non-empty subset of the set of positive integers has a least member.*

1.5.1 Archimedian Property

Statement: *For any two positive integers a and b, there exists a positive integer n such that* $na \geq b$.

Proof: Let if possible there exists no positive integer n such that $na \geq b$. Then, for any positive integer n, we have $na < b$.

Define $A = \{b - na : n \in N\}$

Clearly, A is a non-empty subset of positive integers. Then, by well ordering property, we have a smallest number, say $b - n_1 a$ in A.

Now, $b - (n_1 + 1)a$ also belongs to A and $b - (n_1 + 1)a = b - n_1 a - a < b - n_1 a$.

Therefore, $b - (n_1 + 1)a$ is smaller to $b - n_1 a$ in A, which is a contradiction.

Hence, there exists a positive integer n such that $na \geq b$.

1.6   MATHEMATICAL INDUCTION

'Induction' means the method of inferring a general statement from the validi of particular cases. But in mathematics, this kind of inference is not allowed. If a statement is true for a large number of cases, even then we can not say that the general statement is true for all n unless we establish a relation which is always true. For example, consider

$$f(n) = n^2 + n + 41 \qquad\qquad ...(1)$$

Putting $n = 1, 2, 3, ...$ in (1), we obtain $43, 47, 53, ..., 151$ which are all prime numbers. On the basis of these results, we assert that the substitution of any positive integer for n in $f(n)$ will always yield a prime number. But this reason is fallacious. In fact $f(n)$ yields a prime number for $n = 1, 2, ..., 39$, but for $n = 40$, we have

$f(40) = 40^2 + 40 + 41$

$\qquad = 40^2 + 2 \times 40 + 1 = (40 + 1)^2 = 41 \times 41$, which is composite.

This example shows that we can not make general assertion with respect to any n unless results holds for $n = m + 1$, whenever it holds for $n = m$.

Definition : *Let* $n \in N$ *and P(n) denotes a certain statement or formula or theorem. Then P(n) holds for every natural number n if*

(i) *it holds for n = 1, and*

(ii) *it holds for n = m + 1, whenever it holds for n = m.*

Remarks

❖ In order to prove a certain statement for all natural numbers n, it is essential to establish both the condition (i) and (ii).

❖ As a matter of fact, condition (i) creates the basis for carrying out induction and condition (ii) gives us right of an unlimited automatic extension of this basis.

❖ In some problems, the result does not hold, for n = 1 or 2, ... but the result is true for a natural number > 1. Let us assume that result is true for a natural number m > 1, then we assume that result is true for $n = k$ (a natural number > m). In the next step, we shall show that the result is true for $n = k + 1$ by using the assumption. In this case, this is done, we say that the result is proved for every natural number $\geq m$ by the principle of mathematical induction.

WORKING PROCEDURE

Let $P(n)$ be a statement or a theorem. In order to prove any result by method of induction, we proceed as follows :

Step-1 Verify that result is true for $n = 1$
Step-2 Suppose result is true for $n = m$
Step-3 Show that result is true for $n = m + 1$.

Then $P(n)$ is true for all $n \in N$.

SOLVED EXAMPLES

> ### Based on the following Results
>
> ● Let $n \in N$ and $P(n)$ be a certain statement or formula or theorem. then $P(n)$ holds for every natural number n if
>
> (i) it holds for $n = 1$ and
>
> (ii) it holds for $n = m + 1$ whenever it holds for $n = m$
>
> ● For any two positive integers a and b, there exist a positive integer n such that $na \geq b$.

Example 1. *By mathematical induction, prove that*

$$1 + 2 + 3 + ... + n = \frac{n(n+1)}{2}, \forall n \in N$$

Solution. Here, we have

$$P(n) = 1 + 2 + ... + n = \frac{n(n+1)}{2} \qquad \qquad ...(1)$$

For $\qquad n = 1, P(1) = \dfrac{1(1+1)}{2} = 1$

Also, for $n = 1$, LHS = 1

Therefore, result is true for $n = 1$.

Let us assume result is true for $n = m$.

i.e. $1 + 2 + 3 + \ldots + m = \dfrac{m(m+1)}{2}$ ...(2)

Adding $(m + 1)$ on both sides of (2), we get

$$1 + 2 + 3 + \ldots + m + (m+1) = \frac{m(m+1)}{2} + (m+1) = (m+1)\left(\frac{m}{2} + 1\right)$$

Therefore, $1 + 2 + 3 + \ldots + m + (m+1) = \dfrac{(m+1)(m+2)}{2} = \dfrac{(m+1)[(m+1)+1]}{2}$

Therefore, the result is true for $n = m + 1$.

Hence, by principle of mathematical induction, result (1) is true for all $n \in N$.

Example 2. *Use the principle of mathematical induction to prove that*

$$1 + 4 + 7 + \ldots + (3n - 2) = \frac{n(3n-1)}{2}$$

Solution. We have to show that

$$P(n) = 1 + 4 + 7 + \ldots + (3n-2) = \frac{n(3n-1)}{2}$$ ...(1)

Step (1) For $n = 1$

$$P(1) = 1 = \frac{1(3.1-1)}{2} = \frac{1 \times 2}{2} = 1$$

$\therefore$ Result is true for $n = 1$.

Step (2) Assume that result is true for $n = m$.

i.e. $P(m) = 1 + 4 + 7 + \ldots + (3m-2) = \dfrac{m(3m-1)}{2}$ ...(2)

Step (3) Adding $[3(m + 1) - 2]$, i.e. $3m + 1$ on both sides of (2), we get

$$1 + 4 + 7 + \ldots + (3m-2) + (3m+1) = \frac{m(3m-1)}{2} + 3m + 1$$

$$= \frac{m(3m-1) + 6m + 2}{2} = \frac{3m^2 + 5m + 2}{2}$$

$$= \frac{(m+1)(3m+2)}{2}$$

$$= \frac{(m+1)[3(m+1)-1]}{2}$$

which shows that result is true for $n = m + 1$.

Hence, by the principle of mathematical induction, result (1) is true for all $n \in N$.

Example 3. *Show by induction that*

$$\frac{a^n + b^n}{2} \geq \left(\frac{a+b}{2}\right)^n, \forall\, n \in N$$

where a, b are positive real numbers.

Solution. Let $P(n): \dfrac{a^n + b^n}{2} \geq \left(\dfrac{a+b}{2}\right)^n$, $\forall \ n \in N$

Step (1) $P(1) = \dfrac{a+b}{2} \geq \left(\dfrac{a+b}{2}\right)$, which is true.

Step (2) Let $P(n)$ is true for $n = m$.

i.e., $\qquad \dfrac{a^m + b^m}{n} \geq \left(\dfrac{a+b}{2}\right)^m$ $\qquad\qquad$ ...(1)

Step (3) Multiplying both sides of (1) by $\left(\dfrac{a+b}{2}\right)$, we get

$$\left(\dfrac{a^m + b^m}{2}\right)\left(\dfrac{a+b}{2}\right) \geq \left(\dfrac{a+b}{2}\right)^{m+1}$$

or $\qquad \left(\dfrac{a+b}{2}\right)^{m+1} \leq \left(\dfrac{a^m + b^m}{2}\right)\left(\dfrac{a+b}{2}\right)$

$\Rightarrow \qquad \left(\dfrac{a+b}{2}\right)^{m+1} \leq \dfrac{a^{m+1} + ab^m + a^m b + b^{m+1}}{4}$

$\Rightarrow \qquad \left(\dfrac{a+b}{2}\right)^{m+1} \leq \dfrac{a^{m+1} + b^{m+1}}{4} + \dfrac{ab^m + a^m b}{4}$ $\qquad$ ...(2)

Now, we have the following two cases :

Case (I) if $a \geq b$

Then, $a^m \geq b^m \Rightarrow a - b \geq 0$ and $a^m - b^m \geq 0$

$\Rightarrow \qquad (a - b)(a^m - b^m) \geq 0$

$\Rightarrow \qquad a^{m+1} + b^{m+1} - ab^m - a^m b \geq 0$

$\Rightarrow \qquad ab^m + a^m b \leq a^{m+1} + b^{m+1}$

Case (II) if $a \leq b$

Then, $\qquad b^m \geq a^m \Rightarrow b - a \geq 0$ and $b^m - a^m \geq 0$

$\qquad\qquad \Rightarrow (b - a)(b^m - a^m) \geq 0$

$\qquad\qquad \Rightarrow b^{m+1} + a^{m+1} - ab^m - ba^m \geq 0$

$\qquad\qquad \Rightarrow ab^m + ba^m \leq a^{m+1} + b^{m+1}$

We observe that in both cases

$\qquad ab^m + ba^m \leq a^{m+1} + b^{m+1}$ $\qquad\qquad$ ...(3)

Using (2) and (3), we get

$$\left(\dfrac{a+b}{2}\right)^{m+1} \leq \dfrac{a^{m+1} + b^{m+1}}{4} + \dfrac{a^{m+1} + b^{m+1}}{4}$$

or
$$\left(\frac{a+b}{2}\right)^{m+1} \le \frac{a^{m+1}+b^{m+1}}{2}$$

or
$$\frac{a^{m+1}+b^{m+1}}{2} \ge \left(\frac{a+b}{2}\right)^{m+1}$$

Therefore, $P(m+1)$ is true.

Hence, by the princple of mathematical induction, $P(n)$ is true for all $n \in N$.

Example 4. *Let $x, y \in Z$ such that $x \neq y$, show by the method of induction that $(x^n - y^n)$ is divisible by $(x - y)$ for all $n \in N$.*

Solution. Let $P(n)$: $x^n - y^n$ is divisible by $(x - y)$.

Step (1) $(x^1 - y^1)$ is divisible by $(x - y)$.

$\Rightarrow$ $P(1)$ is true.

Step (2) Let $P(m)$ be true for $n = m$ such that

$P(m)$: $x^m - y^m$ is divisible by $x-y$,

i.e. $x^m - y^m = k(x - y)$ for some k.

Step (3) consider

$$\begin{aligned}
x^{m+1} - y^{m+1} &= x^{m+1} - x^m y + x^m y - y^{m+1} \\
&= x^m(x - y) + y(x^m - y^m) = x^m(x - y) + ky(x - y) \\
&= (x - y)(x^m + ky)
\end{aligned}$$

which is divisible by $(x - y)$

Thus, $P(m + 1)$: $(x^{m+1} - y^{m+1})$ is divisible by $(x - y)$ is true.

Hence, by the principle of mathematical induction, $P(n)$ is true for all $n \in N$.

Example 5. *Use principle of mathematical induction to prove that*

$$\frac{1}{2.3} + \frac{1}{3.4} + \frac{1}{4.5} + \dots + \frac{1}{(n+1)(n+2)} = \frac{n}{2(n+2)}, \forall n \in N$$

Solution. Let $P(n): \dfrac{1}{2.3} + \dfrac{1}{3.4} + \dfrac{1}{4.5} + \dots + \dfrac{1}{(n+1)(n+2)} = \dfrac{n}{2(n+2)}, \quad \forall n \in N$

Step (1) For $n = 1$

$$P(1) = \frac{1}{2.3} = \frac{1}{2(1+2)} = \frac{1}{6}$$

$\Rightarrow$ $P(1)$ is true.

Step (2) Let result be true for $n = m$.

i.e. $P(m): \dfrac{1}{2.3} + \dfrac{1}{3.4} + \dfrac{1}{4.5} + \dots + \dfrac{1}{(m+1)(m+2)} = \dfrac{m}{2(m+2)}$ ....(1)

Step (3) Adding $\dfrac{1}{(m+2)(m+3)}$ on both sides of (1), we get

$$\frac{1}{2.3}+\frac{1}{3.4}+\frac{1}{4.5}+...+\frac{1}{(m+1)(m+2)}+\frac{1}{(m+2)(m+3)} = \frac{m}{2(m+2)}+\frac{1}{(m+2)(m+3)}$$

$$= \frac{1}{m+2}\left[\frac{m}{2}+\frac{1}{m+3}\right]$$

$$= \frac{1}{m+2}\left[\frac{m^2+3m+2}{2(m+3)}\right]$$

$$= \frac{1}{(m+2)}\left[\frac{(m+1)(m+2)}{2(m+3)}\right]$$

$$\therefore \frac{1}{2.3}+\frac{1}{3.4}+\frac{1}{4.5}+...+\frac{1}{(m+1)(m+2)}+\frac{1}{(m+2)(m+3)} = \frac{m+1}{2(m+3)}$$

$\therefore$ Result is true for $n = m + 1$, i.e. $P(m + 1)$ is true.

Hence, by the principle of mathematical induction, result is true for all $n \in N$.

Example 6. *Using principle of mathematical induction, show that for all positive integer n*

$$\frac{n^7}{7}+\frac{n^5}{5}+\frac{2}{3}n^3 - \frac{n}{105} \text{ is an integer.}$$

Solution. Let $P(n)$: $\dfrac{n^7}{7}+\dfrac{n^5}{5}+\dfrac{2}{3}n^3-\dfrac{n}{105}$ be an integer.

Step (1) $P(1)$: $\dfrac{1}{7}+\dfrac{1}{5}+\dfrac{2}{3}-\dfrac{1}{105}$ is an integer.

Since, $\dfrac{1}{7}+\dfrac{1}{5}+\dfrac{1}{3}-\dfrac{1}{105} = \dfrac{15+21+70-1}{105} = 1$ is an integer.

$\Rightarrow$ $P(1)$ is true.

Step (2) Let $P(m)$ be true.

i.e. $P(m): \dfrac{m^7}{7}+\dfrac{m^5}{5}+\dfrac{2}{3}m^3-\dfrac{m}{105}$ is an integer

$= k$ (say)

Step (3) Now, $\dfrac{(m+1)^7}{7}+\dfrac{(m+1)^5}{5}+\dfrac{2(m+1)^3}{3}-\dfrac{(m+1)}{105}$

$$= \frac{1}{7}(m^7+7m^6+21m^5+35m^4+35m^3+21m^2+7m+1)$$

$$+\frac{1}{5}(m^5+5m^4+10m^3+10m^2+5m+1)$$

$$+\frac{2}{3}(m^3+3m^2+3m+1)-\frac{m}{105}-\frac{1}{105}$$

$$= \left(\frac{m^7}{7} + \frac{m^5}{5} + \frac{2}{3}m^3 - \frac{m}{105} \right) + m^6 + 3m^5 + 6m^4 + 7m^3 + 7m^2 + 4m + 1$$

$$= k + m^6 + 3m^5 + 6m^4 + 7m^3 + 7m^2 + 4m + 1 \qquad \text{[Using (1)]}$$

$$= \text{an integer}$$

$\therefore \quad P(m+1): \dfrac{(m+1)^7}{7} + \dfrac{(m+1)^5}{5} + \dfrac{2(m+1)^3}{3} - \dfrac{(m+1)}{105}$ is an integer, is true.

Hence, by the principle of mathematical induction, result is true for all $n \in N$.

Example 7. *Using principle of mathematical induction, show that*

$$(1+x)^n > 1 + nx \text{ for } n \geq 2 \text{ and } x > -1, x \neq 0$$

Solution. Let $P(n):(1+x)^n > 1 + nx$ for $n \geq 2$ and $x > -1, x \neq 0$.

Step (1) For $n = 2$

$$P(2):(1+x)^2 > 1 + 2x$$

or $\qquad (1 + x^2 + 2x) > 1 + 2x$, which is always true.

Step (2) Let us assume, result is true for $n = m$.

i.e. $\qquad P(m):(1+x)^m > 1 + mx$ $\qquad\qquad$ ...(1)

Step (3) Since, $x > -1 \Rightarrow 1 + x > 0$

Multiplying (1) by $(1 + x)$ on both sides, we get

$\Rightarrow \qquad (1+x)^{m+1} > (1+mx)(1+x)$

$\Rightarrow \qquad (1+x)^{m+1} > 1 + x + mx + mx^2$

$\Rightarrow \qquad (1+x)^{m+1} > 1 + (m+1)x + mx^2$

$\Rightarrow \qquad (1+x)^{m+1} > 1 + (m+1)x$ $\qquad\qquad (\because mx^2 > 0)$

$\therefore \qquad$ Result is true for $n = m + 1$.

Hence, by the principle of mathemtical induction, result is true for all $n \in N$.

Example 8. *For n > 1, show that*

$$\text{(i) } n! < \left(\frac{n+1}{2} \right)^n \qquad\qquad \text{(ii) } \frac{(2n!)}{2^{2n}(n!)^2} \leq \frac{1}{(3n+1)^{1/2}}$$

Solution. Let $P(n):n! < \left(\dfrac{n+1}{2} \right)^n$

Step (1) For $n = 2$, the inequality is valid since

$$2! < \left(\frac{2+1}{2} \right)^2$$

i.e. $\qquad 2 < 9 / 4$

Therefore, $P(2)$ is true.

Step (2) Let result be true for $n = m$.

i.e. $\qquad P(m) : m! < \left(\dfrac{m+1}{2}\right)^m$ is true $\qquad$ ...(1)

Step (3) Consider

$$(m+1)! = (m+1)m! < (m+1)\left(\frac{m+1}{2}\right)^m \qquad \text{[Using (1)]} \qquad \text{...(2)}$$

We now prove that

$$(m+1)\left(\frac{m+1}{2}\right)^m < \left(\frac{m+2}{2}\right)^{m+1} \qquad \text{...(3)}$$

Inequality (3) can clearly be written as

$$\frac{2^{m+1}}{2^m} < \left(\frac{m+2}{m+1}\right)^{m+1}$$

or $\qquad 2 < \left(1 + \dfrac{1}{m+1}\right)^{m+1}$

But, by binomial theorem

$$\left(1 + \frac{1}{m+1}\right)^{m+1} = 1 + (m+1)\frac{1}{m+1} + \ldots > 2$$

So, the inequality (3) holds. It now follows from (2) and (3) that

$$(m+1)! < \left(\frac{m+2}{2}\right)^{m+1}$$

Therefore, $P(n)$ is true for $n = m + 1$.

Hence, by the principle of mathematical induction, result is true for all $n \in N$.

(ii) Let $P(n) : \dfrac{(2n!)}{2^{2n}(n!)^2} \le \dfrac{1}{(3n+1)^{1/2}} \ (n \ge 1)$

Step (1) For $n = 1$, both LHS and RHS are equal to ½.

Therefore, $P(1)$ is true.

Step (2) Let us assume the inequality holds for $n = m$.

i.e. $\qquad \dfrac{(2m!)}{2^{2m}(m!)^2} \le \dfrac{1}{(3m+1)^{1/2}} (m \ge 1)$ $\qquad$ ...(1)

Step (3) Now, we shall prove that inequality holds for $n = m + 1$, for which, we will show that

$$\frac{(2m+2)!}{2^{2m+2}\{(m+1!)\}^2} \le \frac{1}{(3m+4)^{1/2}} \qquad \text{...(2)}$$

$$\text{LHS} = \frac{(2m+2)(2m+1)(2m)!}{4(m+1)^2.2^{2m}.(m!)^2} \le \frac{(2m+2)(2m+1)}{4(m+1)^2}.\frac{1}{(3m+1)^{1/2}} \qquad \text{(Using (1))}$$

$$= \frac{2m+1}{2(m+1)}.\frac{1}{(3m+1)^{1/2}} \le \frac{1}{(3m+4)^{1/2}} \qquad \text{(Using (2))}$$

Above will be proved if on squaring, we establish that

$$(3m+4)(2m+1)^2 \le 4(m+1)^2(3m+1) \qquad \qquad ...(3)$$

It should be noted that all the factors are positive

$$3m\{(2m+1)^2 - 4(m+1)^2\} + 4\{(2m+1)^2 - (m+1)^2\} \le 0$$

or $\qquad 3m\{-4m-3\} + 4\{3m^2 + 2m\} \le 0 \quad \text{or} \quad -m \le 0$

Above is true since $m \ge 1$. Therefore, (2) is proved.

Hence, by the principle of mathematical induction, result is true for all $n \in$ N.

Example 9. *Show by induction that the sum of the cubes of three consecutive natural numbers is divisible by 9.*

Solution. Let $P(n)$: sum of cubes of three consecutive natural numbers starting from n and divisible by 9.

Step (1) Here, we have

$P(1)$: sum of cubes of first three consecutive natural numbers divisible by 9.

$$\because 1^3 + 2^3 + 3^3 = 36, \text{ which is divisible by } 9.$$

Therefore, $P(1)$ is true.

Step (2) Let $P(m)$ be true, i.e. sum of the cubes of three consecutive natural numbers starting with m is divisible by 9.

$\Rightarrow \qquad\qquad m^3 + (m+1)^3 + (m+2)^3$ is divisible by 9.

$\Rightarrow \qquad\qquad m^3 + (m+1)^3 + (m+2)^3 = 9k, k \in$ N $\qquad\qquad ...(1)$

Step (3) Now, we shall prove that $P(m+1)$ is true, i.e. $(m+1)^3 + (m+2)^3 + (m+3)^3$ is divisible by 9.

Consider

$$(m+1)^3 + (m+2)^3 + (m+3)^3 = (m+1)^3 + (m+2)^3 + m^3 + 9m^2 + 27m + 27$$

$$= m^3 + (m+1)^3 + (m+2)^3 + 9(m^2 + 3m + 3)$$

$$= 9k + 9(m^2 + 3m + 3) \qquad \text{(Using (1))}$$

$$= 9(k + m^2 + 3m + 3)$$

$$= \text{divisible by } 9.$$

Therefore, $P(m+1)$ is true.

Hence, by the principle of mathematical induction, result is true for all $n \in$ N.

Example 10. *Prove by the method of induction that every even power of every odd number greater than 1 when divided by 8 leaves 1 as a remainder.*

Solution.

Step (1) We first prove that the square of every odd number greater than 1 when divided by 8 leaves 1 as a remainder.

First odd integer greater than 1 is 3 and $3^2 = 9 = 8 \times 1 + 1$

Thus the square of 3, when divided by 8 leaves 1 as a remainder.

Now, assume $(2m+1)^2 = 8k + 1$ where, $k \in Z^+$ ...(1)

Then, we have

$$(2m+3)^2 - (2m+1)^2 = 8(m+1)$$

So that $(2m+3)^2 = (2m+1)^2 + 8(m+1)$

$$= 8k + 1 + 8(m+1) \hspace{2cm} \text{[Using (1)]}$$

$$= 8(k+m+1) + 1$$

Therefore, $(2m+3)^2$ when divided by 8 leaves 1 as a remainder. Then, by mathematical induction, it follows that for all n, $(2n+1)^2$ when divided by 8 leaves 1 as a remainder, i.e. we have proved that :

$$(2n+1)^2 = 8k + 1 \hspace{4cm} ...(2)$$

Step (2) Let us assume that result is true for $m = n$.

i.e. $(2n+1)^{2m}$, where $m \in Z^+$, when divided by 8 leaves 1 as remainder.

i.e. assume that

$$(2n+1)^{2m} = 8p + 1 \hspace{4cm} ...(3)$$

where p is a positive integer

Then, $(2n+1)^{2m+2} = (2n+1)^{2n}(2n+1)^2$

$$= (8p+1)(8k+1) \hspace{2cm} \text{[Using (2) and (3)]}$$

$$= 8(8pk + p + k) + 1$$

This shows that $(2n+1)^{2m+2}$, when divided by 8 leaves 1 as remainder.

Hence, by the principle of mathematical induction, result is true for all n.

Example 11. *Let $u_1 = 1$, $u_2 = 1$ and $u_{n+2} = u_{n+1} + u_n$ for $n \geq 1$. Use mathematical induction to show that*

$$u_n = \frac{1}{\sqrt{5}}\left[\left(\frac{1+\sqrt{5}}{2}\right)^n - \left(\frac{1-\sqrt{5}}{2}\right)^n\right] \text{ for all } n \geq 1.$$

Solution. We have to prove that

$$u_n = \frac{1}{\sqrt{5}}\left[\left(\frac{1+\sqrt{5}}{2}\right)^n - \left(\frac{1-\sqrt{5}}{2}\right)^n\right] \text{ for all } n \geq 1.$$

Step (1) We obviously have

$$u_1 = 1 = \frac{1}{\sqrt{5}}\left[\left(\frac{1+\sqrt{5}}{2}\right) - \left(\frac{1-\sqrt{5}}{2}\right)\right]$$

$$\text{and } u_2 = 1 = \frac{1}{\sqrt{5}}\left[\left(\frac{1+\sqrt{5}}{2}\right)^2 - \left(\frac{1-\sqrt{5}}{2}\right)^2\right]$$

Therefore, result is true for $n = 1$ and $n = 2$.

Step (2) Let result is true for $n = k$.

i.e. $\qquad u_k = \dfrac{1}{\sqrt{5}}\left[\left(\dfrac{1+\sqrt{5}}{2}\right)^k - \left(\dfrac{1-\sqrt{5}}{2}\right)^k\right]$ $\qquad\qquad (k = 1, 2, 3, ..., m)$

Now, $\qquad u_{m+2} = u_{m+1} + u_m$ for $m \geq 1$

$\Rightarrow \qquad u_{m+1} = u_m + u_{m-1}$ for $m \geq 2$

Threfore, by induction hypothesis on u_k, we have

$$u_{m+1} = u_m + u_{m-1}$$

$$= \frac{1}{\sqrt{5}}\left[\left(\frac{1+\sqrt{5}}{2}\right)^m - \left(\frac{1-\sqrt{5}}{2}\right)^m\right] + \frac{1}{\sqrt{5}}\left[\left(\frac{1+\sqrt{5}}{2}\right)^{m-1} - \left(\frac{1-\sqrt{5}}{2}\right)^{m-1}\right]$$

$$= \frac{1}{\sqrt{5}}\left[\left(\frac{1+\sqrt{5}}{2}\right)^{m-1}\left\{\frac{1+\sqrt{5}}{2}+1\right\} - \left(\frac{1-\sqrt{5}}{2}\right)^{m-1}\left\{\frac{1-\sqrt{5}}{2}+1\right\}\right]$$

$$= \frac{1}{\sqrt{5}}\left[\left(\frac{1+\sqrt{5}}{2}\right)^{m-1}\left(\frac{6+2\sqrt{5}}{4}\right) - \left(\frac{1-\sqrt{5}}{2}\right)^{m-1}\left(\frac{6-2\sqrt{5}}{4}\right)\right]$$

$$= \frac{1}{\sqrt{5}}\left[\left(\frac{1+\sqrt{5}}{2}\right)^{m-1}\left(\frac{1+\sqrt{5}}{2}\right)^2 - \left(\frac{1-\sqrt{5}}{2}\right)^{m-1}\left(\frac{1-\sqrt{5}}{2}\right)^2\right]$$

$$= \frac{1}{\sqrt{5}}\left[\left(\frac{1+\sqrt{5}}{2}\right)^{m+1} - \left(\frac{1-\sqrt{5}}{2}\right)^{m+1}\right]$$

Therefore, result is true for $n = m + 1$.

Hence, by the principle of mathematical induction, result is true for all n.

Example 12. *Show that at any time, the total number of persons on the earth who shake hands an odd number of times is even.*

Solution. To prove this result, we first assign to each hand shake a number in natural order. Then, our assertion is equivalent to the following for every n, after a hand shake with number n, the number of people who have made an odd number of hand shaken is even. This statement depends on n and will be proved by induction. For convenience, we call the people who have made an odd number of hand shakes type A and the rest type B, that is of type B are those people who had made an even number of hand Shake.

After the hand shake with number 1, we have two people of type A, an even number. After the mth hand shake, let the number of people of type A be even and let the hand shake number $(m + 1)$ take place. Now, there are following three cases, when the hand shake number $(m + 1)$ will occur between

  (i) Two persons of type A
  (ii) Two persons of type B

(iii) A person of type A and a person of type B

In case (i) two persons of type A and one handshake to their odd number of handshake and becomes of type B; In case (ii), two persons of type B becomes of type A and in case (iii), a person of type A becomes of type B and C person oftype B is changed into type A. thus the number of persons of type A, either decreases by two or increases by two or remains unchaged. In any cases, the number remain even and proof is complete.

Example 13. *Use mathematical induction to show that*

$$2 + 4 + 6 + ... + 2n = n^2 + n$$

Solution.   Here, we have n terms in LHS

and   RHS $= n^2 + n$

Step (1) :   For $n = 1$

$$2.1 = 1 + 1 \Rightarrow 2 = 2$$

$\Rightarrow$ Result is true for n = 1.

Step (2)   Let result be true for $n = k$, i.e.

$$2 + 4 + 6 + ... + 2k = k^2 + k$$

Step (3)   To show result is true for $n = k + 1$, i.e.

$$2 + 4 + 6 + ... + 2(k + 1) = (k + 1)^2 + (k + 1) \qquad ...(1)$$

$$\text{LHS} = 2 + 4 + 6 + ... + 2k + 2(k + 1)$$

$$= k^2 + k + 2k + 2 \qquad \text{[Using Step (I)]}$$

$$= k^2 + 3k + 2$$

Now, $$\text{RHS} = (k + 1)^2 + (k + 1)$$

$$= k^2 + 1 + 2k + k + 1$$

$$= k^2 + 3k + 2$$

$\Rightarrow$ $$\text{LHS} = \text{RHS}$$

Therefore, (1) is satisfied.

$\Rightarrow$ Result is true for $n = k + 1$.

Hence, by the principle of mathematical induction, result is true for all n.

1.6.1. Principles of Finite Induction

First Principle: Let A be a subset of positive integers N with the properties

(i) $1 \in A$ and

(ii) whenever $m \in A \Rightarrow m + 1 \in A$,   Then, $A = N$.

Second Principle: Let A be a subset of positive integers N with the properties

(i) $1 \in A$ and

(ii) if m is a positive integer such that 1, 2, ..., $m \in A \Rightarrow (m + 1) \in A$

Then, $A = N$.

Remark

❖ The well ordering principle is equivalent to the principle of finite induction.

SOLVED EXAMPLES

Example 1. *Show that*

$$^nC_r + {}^nC_{r-1} = {}^{n+1}C_r, 1 \le r \le n \qquad \text{(Pascal's rule)}$$

Solution. We know that

$$\frac{1}{r} + \frac{1}{n-r+1} = \frac{n-r+1+r}{r(n-r+1)} = \frac{n+1}{r(n-r+1)}$$

Multiplying both sides by $\dfrac{n!}{(r-1)!(n-r)!}$, we get

$$\frac{n!}{r(r-1)!(n-r)!} + \frac{n!}{(r-1)!(n-r+1)(n-r)!} = \frac{(n+1)n!}{(r-1)!r(n-r+1)(n-r)!}$$

which implies

$$\frac{n!}{r!(n-r)!} + \frac{n!}{(r-1)!(n-(r-1))} = \frac{(n+1)!}{r!(n+1-r)!}$$

Hence, $^nC_r + {}^nC_{r-1} = {}^{n+1}C_r$

Example 2. *Show that each binomial coefficient* $^nC_r, \forall\, n \ge 1$ *and for all r satisfying* $0 \le r \le n$ *is an integer.*

Solution. For $n = 1$, we have $r = 0$ or 1.

and $\qquad ^nC_r = {}^1C_0 = 1;\ ^nC_r = {}^1C_1 = 1$

Clearly, both are integers.

Now, suppose that the result is true for $n-1 \ge 1$ and for all r satisfying $0 \le r \le n-1$, i.e. $^{n-1}C_r$, is an integer.

Now, we shall show the result for n.

Let $r \le n$ be an integer.

If $r = n$ then, $^nC_r = {}^nC_n = 1$

If $r < n$, then $r \le n-1$ and $^nC_r = {}^{n-1}C_{r-1} + {}^{n-1}C_r$.

But according to our assumption, $^{n-1}C_{r-1} = {}^{n-1}C_{r-1}$ are integers. Thus, ^nC_r is an integer.

Hence, by the principle of mathematical induction, the coefficients ^nC_r are integers.

Example 3. *Show that product of any r consecutive integers is divisible by r !*

Solution. Let $P = (n-r)(n-r+1) \dots (n-1)$ be the product of r consecutive integers. Then

$$P = \frac{(n-r)(n-r+1)\dots(n-1)(n-r-1)\dots 2.1}{(n-r-1)\dots 2.1}$$

$$= \frac{1.2\dots(n-r-1)(n-r)\dots(n-1)}{2.1\dots(n-1-r)}$$

$$= \frac{(n-1)!}{(n-1-r)!} = \frac{(n-1)!\,r!}{r!(n-1-r)!} = {}^{n-1}C_r \cdot r!$$

Now, since ${}^{n-1}C_r$ is an integer, then clearly P is divisible by $r!$.

SOME INTERESTING FACTS ABOUT NUMBERS

1. Every natural number upto 1000 can be expressed as a sum of four squares of natural numbers.
2. No n^{th} power is a sum of fewer than n^{th} powers.
3. Leonard Euler conjecture asserts that no n^{th} power is a sum of fewer than n, n^{th} powers. In particular, for $n = 3$, this would assert that no cube is the sum of two smaller cubes.
4. The Goldback's conjecture asserts that every even integer greater than 2 is the sum of two primes. For example,
   $$4 = 2 + 2, 6 = 3 + 3, 20 = 7 + 13, 50 = 3 + 47, 100 = 29 + 71$$
5. Every prime p is a divisor of $(p-1)! + 1$.
6. Every prime number of the form $(4n + 1)$ is a sum of two squares.

1.7 BASIC REPRESENTATION THEOREM

Statement: *For a fixed integer k, every positive integer n has unique representation of the form*

$$n = a_0 k^r + a_1 k^{r-1} + \dots + a_r,\, a_0 \neq 0, 0 \leq a_i < n, i = 0,1,2,\dots,r$$

Proof: Existence of such representation:

Here, we have the following cases

Case (1) : $n < k$, then n is the unique representation of n.

Case (2) : $n = k$, then $n = 1 . k + 0$. This is again a unique representation.

Case (3) : $n > k$, then there exists unique integers q and r such that $n = qk + r$, $0 \leq r < k$. If $q < k$, we take $q = a_0$ and $r = a$. Then

$$n = a_0 k + a_1,\text{ which is a unique representation of } n.$$

If $q \geq k$, then there exist integers q_1 and r_1 such that $q = q_1 k + r_1, 0 \leq r_1 < k$. If $q_1 < k$, then

$$n = (q_1 k + r_1)k + r$$

$$= q_1 k^2 + r_1 k + r$$

Taking $q_1 = a_0, r_1 = a_1$ and $r = a_2$.

Then, $n = a_0 k^2 + a_1 k + a_2$ is the required representation.

Further, if $q_1 \geq k$, we continue the process and this process will terminate after a finite number of steps and we get

$$n = a_0 k^r + a_1 k^{r-1} + \ldots + a_r$$

Uniqueness

Let, if possible, n has two distinct representations given by

$$n = a_0 k^{r_1} + a_1 k^{r_1-1} + \ldots + a_{r-1}, a_0 \neq 0, 0 \leq a_i < k, 0 \leq i \leq r_1 - 1 \qquad \ldots(1)$$

and $$n = b_0 k^{r_2} + b_1 k^{r_2-1} + \ldots + b_{r_2-1}, b_0 \neq 0, 0 \leq b_i < k, 0 \leq i \leq r_2 - 1 \qquad \ldots(2)$$

Since r_1 and r_2 are integers then by law of trichotomy, we have

either $r_1 > r_2$ or $r_1 = r_2$ or $r_1 < r_2$

If $r_1 > r_2$, then

$$k^{r_1} = (k-1) \sum_{i=0}^{r_1-1} k^i + 1$$

Thus, $$n = k^n - \left[(k-1) \sum_{i=0}^{r_1} n^i + 1 \right] + b_0 k^{r_2} + \ldots\ldots + b_{r_2}$$

$$= c_0 k^{r_1} + c_1 k^{r_1-1} + \ldots + c_{r_1}, 0 \leq c_i < k, c_0 \neq 0, 0 \leq i \leq r_1 - 1 \qquad \ldots(3)$$

Therefore, we may assume two different representations of n with same number of terms are as follows

$$x = a_0 k^{r_1} + a_1 k^{r_1-1} + \ldots + a_{r_1}, 0 \leq a_i \leq k, a_0 \neq 0, 0 \leq i \leq r_1 - 1 \qquad \ldots(4)$$

and $$x = b_0 k^{r_1} + b_1 k^{r_1-1} + \ldots + b_{r_1}, 0 \leq b_i \leq k, b_0 \neq 0, 0 \leq i \leq r_1 - 1 \qquad \ldots(5)$$

Let i be the largest integer such that $a_i \neq b_i$, then

$$a_{i+1} = b_{i+1}, \ldots, a_{r_1} = b_{r_1}$$

Then, from (4) and (5), we conclude that

$$(a_0 - b_0)k^{r_1} + (a_1 - b_1)k^{r_1-1} + \ldots k(a_i - b_i)k^i = 0$$

Since, $k > 1$ and $i < r_1$, we have

$$a_i - b_i = -k[(a_0 - b_0)k^{r_1-i-1} + \ldots(a_{i-1} - b_{i-1})]$$

This shows that $k \mid (a_i - b_i)$, which is not true because a_i and b_i both are less than k. This contradiction shows that representation must be unique.

SOLVED EXAMPLES

Example 1. *Write 506 in decimal system.*

Solution. We have $506 = 50 \times 10 + 6$

Also, $50 = 5 \times 10$

Thus, we can write

$$506 = 5 \times 10 \times 10 + 6 = 5 \times 10^2 + 0 \times 10 + 6$$

Hence, $(506) = (506)_{10}$.

Example 2. *Write 25 with base 2.*

Solution. We have the following division scheme

$$
\begin{array}{rl}
2\underline{)25} & \\
2\underline{)12} & \text{remainder} \\
2\underline{)\ 6} & 1 \\
2\underline{)\ 3} & 0 \\
2\underline{)\ 1} & 0 \\
\ \ 0 & 1 \\
 & 1
\end{array}
$$

Hence, $(25)_{10} = (11001)_2$

Example 3. *Write 347 with base 8.*

Solution. We have the following division scheme

$$
\begin{array}{rl}
8\underline{)347} & \\
8\underline{)\ 43} & \text{remainder} \\
8\underline{)\ 5} & 3 \\
\ \ 0\ \cdot & 3 \\
 & 5
\end{array}
$$

Hence, $(347)_{10} = (533)_8$

Example 4. *Show that if $a_r k^r + a_{r-1}k^{r-1} + ... + a_0$ is a representation of n to the base k, then*
$0 < n \le k^{r+1} - 1$

Solution. Here, we have

$$0 < n = a_r k^r + a_{r-1}k^{r-1} + ... + a_0$$

$$\le 1.k^r + 1.k^{r-1} + ... + 1$$

$$= k^r + k^{r-1} + ... + 1 = \frac{k^{r+1} - 1}{k - 1}$$

$$\le \frac{k^{r+1} - 1}{2 - 1} \qquad [\because k \ge 2]$$

$$= k^{r+1} - 1$$

EXERCISE 1.1

Prove the following by the principle of mathematical induction : **(Qus. 1-9)**

1. $n(n+1)(2n+1)$ is divisible by 6, $\forall\, n \in N$.

2. $1 + 4 + 7 + ... + (3n - 2) = \dfrac{1}{2}n(3n - 1), \forall\, n \in N.$

3. $\dfrac{1}{1.2} + \dfrac{1}{2.3} + \dfrac{1}{3.4} + ... + \dfrac{1}{n(n+1)} = \dfrac{n}{n+1}, \forall\, n \in N.$

4. $(2^{3n} - 1)$ is divisible by 7, $\forall\, n \in N$.

5. $(10^{2n-1} + 1)$ is divisible by 11.

6. $n < 2^n, \forall\, n \in N$

7. For $n \in N, 10^n + 3(4)^{n+2} + 5$ is divisible by 9.

8. $1^3 + 2^3 + 3^3 + ... + n^3 = \left(\dfrac{n(n+1)}{2}\right)^2$

9. (i) $\underset{n\ \text{digit}}{7 + 77 + 777 + ... + 777...7} = \dfrac{7}{81}(10^{n+1} - 9n - 10)$

   (ii) $\dfrac{1}{3.7} + \dfrac{1}{7.11} + \dfrac{1}{11.15} + ... + \dfrac{1}{(4n-1)(4n+3)} = \dfrac{n}{3(4n+3)}$

   (iii) $1.6 + 2.9 + 3.12 + ... + n(3n + 3) = n(n+1)(n+2)$

10. Show that $7^{2n} + (2^{3n-3})3^{n-1}$ is divisible by 25, $n \in N$.

11. (i) Show that $5^{2n+2} - 24n - 25$ is divisible by 576.

    (ii) Show that $2.7^n + 3.5^n - 5$ is divisible by 24 for all $n \geq 0$.

12. If p is a natural number, then show that $p^{n+1} + (p+1)^{2n-1}$ is divisible by $p^2 + p + 1$ for every positive integer n.

13. Prove the binomial theorem.
    $$(x + a)^n = x^n + {}^nc_1 x^{n-1}a + {}^nc_2 x^{n-2}a^2 + ... + {}^nc_r x^{n-r}a^r + ... + a^n$$

14. If $x^3 = x + 1$, then show that $x^{3n} = a_n x + b_n - c_n x^{-1}$,

    where, $a_{n+1} = a_n + b_n; b_{n+1} = a_n + b_n + c_n, c_{n+1} = a_n + b_n$

15. Use the principle of mathematical induction to show that

    (i) $5^{2n} - 1$ is divisible by 24.

    (ii) $4^n - 3n - 1$ is divisible by 9.

    (iii) $10^{2n-1} + 1$ is divisible by 11.

    (iv) $a^{2n-1} - 1$ is divisible by $a - 1$

    (v) $n^2 - n + 41$ is prime

16. Show that

    (i) $1.4.7 + 2.5.8 + 3.6.9 + ... + n(n+3)(n+6) = \dfrac{n}{4}(n+1)(n+6)(n+7)$

    (ii) $1 + 3 + 5 + ... + (2n - 1) = n^2$

17. If n straight lines in a plane are such that no two of them are parallel and no three of them are concurrent, show that they intersect each other in $\dfrac{n(n-1)}{2}$ points.

18. Let P(n) be the statement "The arithmetic mean of n and $(n + 2)$ is the same as their geometric mean". Show that $P(1)$ is not true. Also, show that if $P(n)$ is true, then $P(n + 1)$ is also true. How does this contradict the principle of induction ?

19. Show that $(x + a_1)(x + a_2)(x + a_3)...(x + a_n) = x^n + P_1 x^{n-1} + P_2 x^{n-2} + ... + P_{n-1} x + P_n$

where, $P_1 = \Sigma a_i, P_2 = \Sigma a_i a_j, P_3 = \Sigma a_i a_j a_k ; 1 \le i \le n, 1 \le i \le j \le n, 1 \le i \le j \le k \le n$

$P_n = a_1 a_2 ...... a_n$

20. Let P(n) be the statement : $2^n \ge 3n$. If $P(r)$ is true, show that $P(r+1)$ is true. Do you conclude that $P(n)$ is true for all $n \in N$.

21. Show that

(i) $\dfrac{n^7}{7} + \dfrac{n^5}{5} + \dfrac{n^3}{3} + \dfrac{n^2}{2} - \dfrac{37}{210} n$, is a positive integer, $\forall n \in N$

(ii) $\dfrac{n^{11}}{11} + \dfrac{n^5}{5} + \dfrac{n^3}{3} + \dfrac{62}{165}$ is a positive integer, $\forall n \in N$

(iii) $\dfrac{n^5}{5} + \dfrac{n^3}{3} + \dfrac{7n}{15} n$, is a natural number.

22. Show by the method of induction that for all $n \in N, 3^{2n}$, when divided by 8, the remainder is always 1.

23. Show that the product of any r consecutive natural numbers is always divisible by $r!$.

24. Show that $\displaystyle\sum_{k=0}^{n} k^2 C_k = n(n+1)2^{n-2}$ for $n \ge 1$.

25. (i) For what natural number n, the inequality $2^n > 2n + 1$ is valid?

(ii) For what natural number n, the inequality $2^n > n^2$ is valid ?

26. For all integers $n \ge 2$, prove that $\displaystyle\sum_{r=1}^{n-1} i(i+1) = n(n-1)(n+1))/3$.

27. Prove that if $a_r k^r + a_{r-1} k^{r-1} + ... + a_0$ is a representation of n to the base k, then

$$0 < n \le k^{r+1} - 1$$

28. Show that $\dfrac{(2n)!}{n!(n+1)!}$ is an integer.

ANSWERS

(25) (i) for $n \ge 3$; (ii) for $n = 1$ and for all natural numbers $x \ge 5$.

CHAPTER REVIEW : A COMPETITIVE APPROACH

Selected terms and Results

TERMS

- **Natural Numbers (N) :** The numbers 1, 2, 3, ... are called natural numbers.
- **Integers (Z) :** The numbers ..., –3, –2, –1, 0, 1, 2, 3, ... are called integers.
- **Rational numbers (Q) :** Any number of the form $\frac{p}{q}(q \neq 0)$, $p, q \in Z$ and p, q have no common factor except ± 1, is called rational number.
- **Irrational number (I) :** Any number which is not rational is called irrational number.
- **Real numbers (R) :** A number which is either rational or irrational is called a real number.
- **Induction :** Induction means the method of inferring a general statement from the validity of particular cases.
- **Mathematical Induction :** Let $n \in N$ and $P(n)$ denotes a certain statement then $P(n)$ holds for every natural number n if it holds for
  (i) $n = 1$ and
  (ii) $n = m + 1$ whenever it holds for $n = m$.

RESULTS

- For any two positive integers a and b, there exists a positive integer n such that $na \geq b$ (Archimedean property)
- Every non-empty subset of the set of positive integers has a least member (Well-ordering property)
- The well-ordering principle is equiv-alent to the principle of finite induction.
- Every natural number upto 1000 can be expressed as a sum of four squares of natural numbers.
- No n^{th} power is a sum of fewer than n^{th} powers.
- No cube is the sum of two smaller cubes.
- Every even integer greater than 2 is the sum of two primes.
- Every prime p is a divisor of $(p - 1)! + 1$.
- Every prime numbr of the form $(4n + 1)$ is a sum of two squares.

Review Questions and Project Work

1. Prove the second principle of mathematical induction using first principle.
2. Prove the first principle of mathematical induction, using the well ordering principle.
3. Prove that every non-empty set of negative integers has a largest element.
4. Let $n_0 \in Z$ and S be a non empty subset of $T = \{n \in Z : n \geq n_0\}$ and l be a least element of the set $T^* = \{n - n_0 + 1 : n \in S\}$ then, prove that $n_0 + l - 1$ is a least element of S.
5. Let S_n be the sum of elements in the nth set of the sequence of sets of squares $\{1\}, \{4, 9\}, \{16, 25, 36\}, ...$ Find a formula for S_n.
6. Prove that one more than four times the product of any two consecutive integers is a perfect square.
7. Find a positive integer that can be ex-

pressed as the sum of two cubes in two different ways.

8. Find three consecutive positive integers such that the sum of their cubes is also a cube.

9. Find four consecutive positive integers such that the sum of their cubes is also a cube.

10. Let S denote the sum of the elements in the nth set in the sequence of positive integers $\{1\}, \{2,3, ..., 8\}, \{9, 10, ..., 21\} \{22, 23, ..., 40\}$. Find a formula for S.

Objective Type Questions

Fill in the blanks:

1. The fact that the sequence of primes 2, 3, 5, 7, 11, 13, 17, ... is endless was known to ...

2. A number which is either rational or irrational is called ... numbers.

3. If $a \in Z$, then one and only one of the following is true

   $a \in Z^+, a = 0, -a \in Z$ and

   $a, b \in Z^+ \Rightarrow a + b \in Z^+, \ ab \in Z^+$

   this law is called ...

4. 'Every non-empty subset of the set of positive integers has a least member'. This principle is called ...

5. Archimedian property states that for any two positive integers a and b there exist a positive integer n such that .

True/False: *Write 'T' for true and 'F' for false statement.*

1. Induction means the method of inferring a general statement from the validity of particular case.   (T/F)

2. Every non-empty subset of the set of positive integers has a least member. (T/F)

3. For any two positive integers a and b there exist a positive integer n such that $na < b$.   (T/F)

4. Sum of the cubes of three consecutive natural numbers is not necessarily divisible by 9.   (T/F)

5. Every even integer greater than 2 is the sum of two primes.   (T/F)

Multiple Choice Questions : *Choose the most appropriate one :*

1. Which one of the following is not a statement of mathematical induction:
   (a) Verify the result for $n = 1$
   (b) Suppose that result is true for $n = k$
   (c) Show that result is true for $n = k + 1$
   (d) none of these

2. For any two positive integers a and b, there exists a positive integers n such that $na \geq b$. This property is called
   (a) Archimedian property
   (b) Well ordering property
   (c) Both (a) and (b) are true
   (d) none of these

3. Every non-empty subset of the set of positive integer has a least number. This property is called
   (a) Archimedian property
   (b) Well ordering property
   (c) Both (a) and (b) are true
   (d) none of these

4. Which one of the following is not true
   (a) no n^{th} power is a sum of fewer than n^{th} powers
   (b) every even integer greater than 2 is the sum of two primes
   (c) every prime p is a divisor of $(p - 1)! + 1$
   (d) all are true

5. Each binomial coefficient $^nC_r, \forall n \geq 1$ and for all r satisfying $0 \leq r \leq n$ is a/an
   (a) integer
   (b) rational
   (c) natural number
   (d) none of these

ANSWERS

Fill in the blanks

(1) Euclid   (2) Real
(3) Law of trichotomy   (4) Well ordering principle
(5) $> b$

True/False

(1) T   (2) T
(3) F   (4) F
(5) T

Multiple choice questions

(1) *d*   (2) *a*
(3) *b*   (4) *d*
(5) *a*

CHAPTER 2

Divisibility Theory

> ❖ Divisibility
>
> ❖ Greatest Common Divisor
>
> ❖ Least Common Multiple
>
> ❖ Lame's Theorem
>
> ❖ Division Algorithm
>
> ❖ Euclid's Algorithm
>
> ❖ Fibonacci Sequence
>
> ❖ Kronecker's Theorem

2.1 INTRODUCTION

An algorithm is a step by step process, complete in a finite number of steps, for solving a given problem. By the division algorithm, we mean that process with which the student became familiar in arithmetic. Divisors, multiples, prime and composite numbers are concepts that have been known and studied at least since the time of Euclid, about 350 BC. In this chapter, we shall discuss all these topics with divisibility theory of integers.

2.2 DIVISION ALGORITHM

Definition : *An interger b is divisible by an integer a, not zero, if there is an integer x such that b = ax and we write a | b. In case, b is not divisible by a, we write a∤b.*

THEOREM-1

(a) *For any given integer a (the dividend) and any given non-zero integer b (the divisor) there exist integers q (the quotient) and r (the remainder) such that*

$$a = qb + r \qquad \qquad ...(1)$$
$$and \quad 0 \le r < |b| \qquad \qquad ...(2)$$

(b) *q and r are unique.*

Proof. Since a lies between two consecutive integers of the sequence

$$... -2|b|, -|b|, 0, |b|, 2|b|, ...$$

we may assume

$$q|b| \le a < (q+1)|b|$$

Then, $a - q|b| \ge 0, a - q|b| < |b|$. Let $a - q|b| = r$. Then, $0 \le r < |b|$ and therefore, we have

$$a = qb + r, \text{ when } b > 0.$$

and $a = (-q)b + r$, when $b < 0$.

Hence, the existence of q and r is proved. Now, we shall prove the uniqueness, as follows:

If there exists another representation given by

$$a = q_1 b + r_1, 0 \le r_1 < |b|, \text{ then}$$

$$(q - q_1)b = r_1 - r, 0 \le |r_1 - r| < |b|$$

i.e. $|q - q_1| |b| < |b|$. Thus $|q - q_1| < 1$. Since q, q_1, are both integers, therefore, $q = q_1$ and consequently $r = r_1$ which implies that q and r are unique.

Remarks

❖ The integer r is called the least non-negative remainder or briefly the remainder of a divided by b. If $r = 0$, then $a = qb$ and hence a is a multiple of b.

❖ If $a | b$ and $0 < a < b$, then a is called a proper divisor of b.

❖ It is understood that we never use 0 as the left member of the pair of integers in $a | b$. On the other hand, not only may 0 occur as the right member of the pair, but also in such instances we always have divisibility. Thus, $a | 0$ for every integer a not zero.

❖ The notation $a^k | | b$ is sometimes used to indicate that $a^k | b$ but $a^{k+1} \nmid b$.

THEOREM-1

(1) $a | b$ implies $a | bc$ for any integer c.

(2) $a | b$ and $b | c$ imply $a | c$.   *(Transitive property)*

(3) $a | b$ and $a | c$ imply $a | (bx + cy)$ for any integers x and y.   *(Linearity property)*

(4) $a | b$ and $b | a$ imply $a = \pm b$.

(5) $a | b, a > 0, b > 0$ imply $a \le b$.

(6) If $m \ne 0$, $a | b$ implied by $ma | mb$.   *(Multiplication property)*

(7) $1 | a$ *(1 divides every integer)*

(8) $a | 0$.   *(Every integers divides 0)*

Proof. The proof of these results follow at once from the definition of divisibility. To give a sample proof, consider (3). Since $a | b$ and $a | c$ are given, this implies that there are integers r and s such that $b = ar$ and $c = as$. Therefore, $bx + cy$ can be written as $a(rx + sy)$ which proves that a is a divisor of $bx + cy$.

Remark

❖ The property (3) can be extended to any finite set as follows :

$$a | b_1, a | b_2, ..., a | b_n \text{ imply } a | \sum_{j=1}^{n} b_j x_j \text{ for any integers } x_j.$$

THEOREM-2

If $c = ax + by$ and $d | a$ but $d \nmid c$, then $d \nmid b$.

Proof. Here, we have

$d | a$ implies there exists an integer q_1 such that
$$a = dq_1$$
Therefore $c = ax + by = dq_1 x + by$

Let if possible $d \mid b$, then there exists an integer q_2 such that $b = dq_2$.

$\therefore \qquad c = dq_1 x + dq_2 y = d(q_1 x + q_2 y)$

$\Rightarrow \qquad \qquad d \mid c$, which is a contradiction.

Hence, contrapositively

$$d \nmid c \Rightarrow d \nmid b$$

THEOREM-3

For any two integers a and $b > 0$ there exists integers q_1 and r_1 such that $a = bq_1 + cr_1$, $0 \le r_1 < b/2, c = +1$ or -1.

Proof. By division algorithm, we have

$$a = bq + r, 0 \le r < b$$

Now, there are following cases.

Case 1 : If $r < b/2$. If we take $q = q_1, r = r_1, c = 1$, then (1) gives

$$a = bq_1 + cr_1, 0 \le r_1 < b/2$$

Case 2 : If $r > b/2$. Then $0 < b - r < b/2$. If we take $q = q_1 + 1, r = b - r_1$ and $c = -1$, then (1) gives

$$a = b(q_1 + 1) - (b - r_1)$$
$$= bq_1 + cr_1, 0 \le r_1 < b/2$$

Case 3 : If $r = b/2$. If $q = q_1, r = r_1$ and $c = 1$, then from (1) we have

$$a = bq_1 + cr_1, r_1 = b/2$$

Further, if we replace q_1 by $q + 1$, r_1 by $b - r$ and c by -1, we get

$$a = b(q+1) - (b - r)$$
$$= bq_1 + cr_1, r_1 = b/2$$

DEDUCTIONS :

(1) Every integer having one of the following form :

    (a) $3q$ or $(3q \pm 1)$ (Taking $b = 3$ in the above theorem)

    (b) $4q$, $(4q \pm 1)$ or $(4q \pm 2)$ (Taking $b = 4$ in the above theorem)

    (c) $5q$, $(5q \pm 1)$ or $(5q \pm 2)$ (Taking $b = 5$ in the above theorem)

(2) Every old integer having one of the following form :

    (a) $2q + 1$ (b) $2q - 1$ (c) $4q \pm 1$ (d) $\pm (4q + 1)$

THEOREM-4

Every square number is of the form $9k$ or $3k + 1$ where k is an integer.

Proof. Since any interger can be written in the form $3q$ or $3q \pm 1$,

$$(3q)^2 = 9q^2 = 9k$$

$$\text{and} (3q \pm 1)^2 = 3(3q^2 \pm 2q) + 1 = 3k + 1.$$

DEDUCTION : If n is a positive odd integer and $n = ab$, then

$$n = ab = \left(\frac{a+b}{2}\right)^2 - \left(\frac{a-b}{2}\right)^2$$

Since a and b are both odd, then $\dfrac{a+b}{2}$ and $\dfrac{a-b}{2}$ are integers. Therefore, we can say that if a positive odd integer a can be decomposed into a product of two divisors, then a can be written as the difference of two square numbers.

THEOREM-5

The square of an odd integer is of the form $8q + 1$.

Proof. Let k be any odd integer. Then, we have

$$k = 4q_1 + 1 \text{ or } k = -(4q_1 + 1) \text{ for integer } q_1.$$

Then, $\begin{aligned} k^2 = [\pm(4q_1 + 1)]^2 &= 16q_1^2 + 8q_1 + 1 \\ &= 8(2q_1^2 + q_1) + 1 \\ &= 8q + 1, \text{ where } q = 2q_1^2 + q_1, \text{ again an integer.} \end{aligned}$

Hence, square of an odd integer is of the form $8q + 1$.

THEOREM-6

The product of any three consecutive integers is a multiple of 3.

Proof. Since any integer can be written in the form of $3n$ or $3n \pm 1$, the difference of two integers is of the same form is a multiple of 3 and therefore, not less than 3. But the difference of any two of three consecutive integers is less than 3, so that the three consecutive integers are respectively of the above three forms, among which one of the form $3n$, i.e. a multiple of 3.

DEDUCTION : From above theorem, it follows that the product of three consecutive integers is a multiple of 3.2. A generalization of this property gives the following result:

"The product of any three consecutive integers is divisible by 3!"

THEOREM-7

Let p be a positive integer greater than 1. Then every positive integer a can be written uniquely in the form

$$a = c_n p^n + \ldots + c_1 p + c_0 \qquad \qquad \ldots(1)$$

where, $n \geq 0$, c_i is an integer $0 \leq c_i < p$, $c_n \neq 0$. p is called the base of a, which is denoted by $(c_n c_{n-1} \ldots c_1 c_0)_p$.

Proof. We shall prove this theorem by induction on a.

When $a = 1$, we have $n = 0$ and $c_0 = 1$. Then (1) is true for $a = 1$. Now assume that theorem is true for any integer less than a. Since $p > 1$, $a > 0$, therefore, a must lie between two certain consecutive numbers of the following sequence.

$$p^0, p^1, p^2, \ldots p^n \ldots$$

i.e. there exists a unique integer n, such that

$$p^n \le a < p^{n+1}$$

Then, by division algorithm, we have

$$a = c_n p^n + r, 0 \le r < p^n$$

Clearly, $p > c_n > 0$, if $r = 0$, then

$$a = c_n p^n + 0.\ p + \dots + 0.p + 0$$

if $r \ne 0$, then by induction hypothesis

$$r = b_t p^t + \dots + b_1 p + b_0, \ t < n, where \ 0 \le b_i < p$$

Therefore,

$$a = c_n p^n + b_t p^t + \dots b_1 p + b_0$$

and (1) is true.

Uniqueness : To prove uniqueness, let us assume that there is another representation

$$a = d_m p^m + \dots d_1 p + d_0 \qquad \dots(2)$$

with $m \ge 0$, $0 \le d_i < p$. If c_i and d_i are not equal, by subtracting (1) from (2), we get

$$0 = e_s p^s + \dots + e_1 p + e_0$$

where, s is the largest value of i for which, $c_i \ne d_i$ so that $e_s \ne 0$. If $s = 0$, then $c_1 = c_0 = 0$, which is a contradiction.

If $s > 0$ we have

$$|e_i| = |c_i - d_i| < p - 1, i = 0, 1, \dots, s - 1$$

and $e_s p^s = -(e_{s-1} p^{s-1} + \dots + e_0)$

Therefore

$$p^s < |e_s p^s| = |e_{s-1} p^{s-1} + \dots e_0|$$
$$< (p-1)(p^{s-1} + \dots + p + 1) = p^s - 1$$

Which is again a contradiction. Hence, we conclude that c_i and d_i are all equal, *i.e.*,

$$n = m, c_i = d_i, i = 0,1,2, \dots, n.$$

Hence, the representation is unique.

DEDUCTION

If we take $p = 2$, then every positive integer may be represented as the sum of distinct powers of 2, i.e.

$$a = c_n 2^n + \dots + c_1.2 + c_0$$

where, each c_i is either 0 or 1.

For example : We consider $a = 2107$. If $p = 10$, then

$$2107 = 2(10)^3 + (10)^2 + 7 = (2107)_{10}$$

If $p = 12$, since

$$2107 = 175 \times 12 + 7, \quad 175 = 14 \times 12 + 7, \quad 14 = 12 + 2$$

then $\qquad (2107) = (1277)_{12}$

Similarly, if $p = 2$, then

$$(2107) = (100000011111)_2$$

SOLVED EXAMPLES

> **Based on the following Results :**
> - An integer b is divisible by an integer a (not zero) if there is an integer x such that $b = ax$
> - Every square number is of the form $9k$ or $3k + 1$, where k is an integer.
> - The square of an odd integer is of the form $8k+1$.
> - The product of three consecutive integers is a multiple of 3!

Example 1. *If $(a - s) \mid (ab + st)$, then show that $(a - s) \mid (at + bs)$.*

Solution. Since $(ab + st) - (at + bs) = (a - s)(b - t)$ and the hypothesis is that $ab + st$ is a multiple of $a - s$. Thus, $at + bs$ is a multiple of $a - s$.

Example 2. *Show that, one of every three consecutive integers is divisible by 3.*

Solution. Let $n, n + 1, n + 2$ be any three consecutive integers. Then, we know that n is of the form $3q, (3q + 1)$ or $(3q - 1)$

If $n = 3q$, then clearly it is divisible by 3. If $n = 3q + 1$, then

$$n + 2 = 3q + 1 + 2 = 3q + 3 = 3(q + 1)$$

which is again divisible by 3.

Finally, if $n = 3q - 1$, then $n + 1 = 3q - 1 + 1 = 3q$, which is also divisible by 3.

Hence, one of every three consecutive integers is divisible by 3.

Example 3. *Show that if n is an even number, then $3^n + 1$ is divisible by 2; if n is odd number, then $3^n + 1$ is divisible by 2^2, if n is any number, whether even or odd, then $3^n + 1$ is not divisible by 2^m with $m \geq 3$.*

Solution. We know that the square of an odd number minus 1 is a multiple of 8.

Now, when $n = 2m$, we have

$$3^n = 3^{2m} = (3^m)^2 = 8a + 1$$

and therefore $3^n + 1 = 2(4a + 1)$

when $n = 2m + 1$, we have

$$3^n + 1 = 3^{2m + 1} + 1 = 3(8a + 1) + 1 = 4(6a + 1)$$

Since $4a + 1$ and $6a + 1$ are odd, the statement is true.

Example 4. *If a and b are any two odd integers, then one of the two numbers $\dfrac{a+b}{2}$ and $\dfrac{a-b}{2}$ is odd and the other is even.*

Solution. Let us assume that

$$a = 2k_1 + 1 \quad \text{and} \quad b = 2k_2 + 1 \quad \text{where, } k_1 \text{ and } k_2 \text{ are any two integers.}$$

Then $\qquad \dfrac{a+b}{2} = \dfrac{2k_1 + 1 + 2k_2 + 1}{2} = (k_1 + k_2) + 1$ $\qquad\qquad$ ...(1)

also, $\qquad \dfrac{a-b}{2} = \dfrac{(2k_1 + 1) - (2k_2 + 1)}{2} = (k_1 - k_2)$ $\qquad\qquad$ ...(2)

From (1) and (2), we conclude that if k_1 and k_2 both are even (or both odd), then

$\dfrac{a+b}{2}$ is an even integer and $\dfrac{a-b}{2}$ is an odd integer.

Hence, if a and b are any two integers, then one of the two numbers $\dfrac{a+b}{2}$ and $\dfrac{a-b}{2}$ is odd and the other is even.

Example 5. *Show that if a is any positive integer, then $a^2 + a + 1$ is not a square number.*

Solution. Since, we have

$$a^2 < a^2 + a + 1 < a^2 + 2a + 1 = (a+1)^2$$

The next square number greater than a^2 is $(a + 1)^2$. Hence, $a^2 + a + 1$ is not a square number.

Remark

❖ If an integer a is a square of some other integer, then a is called a square number.

Example 6. *If a is an odd integer, then show that $\dfrac{a^4 + 4a^2 + 11}{16}$ is an integer.*

Solution. Let us suppose $a = 2n + 1$

Then, consider

$$\begin{aligned}
\frac{a^4 + 4a^2 + 11}{16} &= \frac{(2n+1)^4 + 4(2n+1)^2 + 11}{16}\\
&= \frac{16n^4 + 32n^3 + 40n^2 + 24n + 16}{16}\\
&= n^4 + 2n^3 + 1 + \frac{n(5n+3)}{2}
\end{aligned}$$

Clearly, $\dfrac{n(5n+3)}{2}$ is in integer, if n is even integer. If n is an odd integer, then $5n + 3$ is an even integer and therefore, $\dfrac{n(5n+3)}{2}$ is again an even integer.

Hence, the given quantity is an integer.

Example 7. *Show that if $1 < a_1 < a_2 < ... < a_{n-1} < a_n$, then there exist i and j with $i < j$ such that $a_i \mid a_j$.*

Solution. Let $a_i = 2^n b_i, n_i \geq 0, b_i$ is odd. Since, among $1, 2, ..., 2n$, there are only n distinct odd numbers $b_1, b_2, ..., b_{n+1}$ are not all distinct. In other words, among them there are some equal odd numbers. Let $b_i = b_j$. Then, $a_i \mid a_j$.

Example 8. *Show that the number of the form $\dfrac{a(a^2+2)}{3}$ is an integer where a is an integer greater than or equal to 1.*

Solution. Since, we know that every integer a is of the form $3n$, $3n + 1$, or $3n +2$. If $a = 3n$, then

$$\frac{a(a^2+2)}{3} = \frac{3n[(3n)^2+2]}{3} = n(9n^2+1), \text{ which is an integer.}$$

If $a = 3n + 1$, then

$$\frac{a(a^2+2)}{3} = \frac{(3n+1)[(3n+1)^2+2]}{3} = (3n+1)(3n^2+2n+1),$$

which is again an integer.

If $a = 3n + 2$ then

$$\frac{a(a^2+2)}{3} = \frac{(3n+2)[(3n+2)^2+2]}{3} = (3n+2)(3n^2+4n+2),$$

which is also an interger. Hence, a number of the form $\dfrac{a(a^2+2)}{3}$ is an integer, where $a \geq 1$.

EXERCISE 2.1

1. If integer b divides a positive integer a then show that b is not necessarily greater than a.

2. Show that we can choose two integers from any given three integers such that their sum and difference are even numbers.

3. If a is odd integer, then show that $\dfrac{a^4+(a+2)^3+(a+4)^2+1}{12}$ is an integer.

4. Show that if $d_1, ..., d_k$ are all positive divisors of n, then $(d_1 ... d_k)^2 = n^k$

5. Show that $n^2 - n$ is divisible by 2 for every integer n; $n^2 - n$ is divisible by 6; and $n^5 - n$ is divisible by 30.

6. If $a > 0, n \geq 2$, then show that a^n can be expressed as the sum of n consecutive positive integers.

7. Show that if x and y are odd, then $x^2 + y^2$ is even but not divisible by 4.

8. Show that $a^{n+1} - (a-1)n - a$ can be divided by $(a-1)^2$.

9. Show that for any positive integer n, there exists at least n consecutive integers such that each of them has a divisor which is a square number.

10. Show that there are no positive integers $a, b, n > 1$ such that $(a^n - b^n)|(a^n + b^n)$.

11. Find 9 integers such that they form an arithmetic progression and the sum of the squares of each of them is a square integer.

12. Show that $(n-1)^2|(n^k-1)$ if and only if $(n-1)|k$.

2.3 GREATEST COMMON DIVISOR

If c is a divisor of a and a divisor of b simultaneously, i.e. $c \mid a$ and $c \mid b$, then c is called a common divisor of a and b. For example, 1 is a common divisor of a and b. Since any non-zero integer has only a finite number of divisors, then a and b (both not zero) also have a finite number of common divisors, the largest integers among which is called the greatest common divisor of a and b and is written as (a, b).

Definition : *Let a and b be any two given integers (both not zero), then the greatest common divisor of a and b denoted by (a, b) is the positive integer d such that*

(i) $d \mid a$ and $d \mid b$

(ii) *if $c \mid a$ and $c \mid b$ then $c \mid d$.*

The greatest common divisor of any two integers, both not zero can be found by following the Euclidean algorithm.

Euclid's Algorithm

Let a and b be any two positive integers, then we obtain an integer $k \geq 1$ such that

$$\left.\begin{aligned}
a &= q_1 b + r_1; 0 \leq r_1 < b \\
b &= q_2 r_1 + r_2; 0 \leq r_2 < r_1 \\
&\cdots \quad \cdots \quad \cdots \\
r_{k-2} &= q_k r_{k-1} + r_k; 0 < r_k < r_{k-1} \\
r_{k-1} &= q_{k+1} r_k
\end{aligned}\right\} \qquad \ldots(1)$$

From the first equation, we have $(a, b) = (b, r)$ and therefore

$$(a, b) = (b, r) = (r_1, r_2) = \ldots = (r_{k-1}, r_k) = r_k.$$

Hence, r_k is the required g.c.d (a, b) that is to say we can find the greatest common divisor by using Educlidean algorithm.

2.4 ABSOLUTELY LEAST REMAINDER ALGORITHM OR MINIMAL ALGORITHM

Let a and b be two positive integers such that $a > b$. Then, there exist integers q_1 and r_1 such that

$$a = bq_1 + e_1 r_1, 0 < r_1 \leq \frac{b}{2} \qquad \ldots(1)$$

Again, there exists integers q_2 and r_2 such that

$$b = r_1 q_2 + e_2 r_2, 0 < r_2 \leq \frac{r_1}{2} \qquad \ldots(2)$$

Continuing this process, we get

$$r_1 = r_2 q_3 + e_3 r_3, 0 < r_3 \leq \frac{r_2}{2} \qquad \ldots(3)$$

$$\vdots$$

$$r_{n-3} = r_{n-2} q_{n-1} + e_{n-1} r_{n-1}, 0 < r_{n-1} \leq \frac{r_{n-2}}{2} - (n-1)$$

$$r_{n-2} = r_{n-1}q_n + e_n r_n, \, r_n = 0 \qquad \qquad \ldots(4)$$

where $e_1, e_2, \ldots, e_n$ all are either $+1$ or -1.

Finally, since $a > b > r_1 > r_2 \ldots > r_n$ form a decreasing sequence of non-negative integers. Therefore $r_n = 0$ for some integer n. The g.c.d of a and b will be r_{n-1} as in Euclid's algorithm.

THEOREM-1

Let a and b be positive integers such that $a > b$ and let $r_k = 0$ in Euclid's algrothm. Then, r_{k-1} is the g.c.d. of a and b.

Proof. We know that

$$r_{k-2} = r_{k-1}q_k$$

which implies $r_{k-1} \mid r_{k-2}$

Further, we have

$$\begin{aligned} r_{k-3} &= r_{k-2}q_{k-1} + r_{k-1} \\ &= r_{k-1}q_k q_{k-1} + r_{k-1} \\ &= r_{k-1}[q_k q_{k-1} + 1] \end{aligned}$$

which implies

$$r_{k-1} \mid r_{k-3}$$

Continuing this process, finally we get

$$r_{k-1} \mid a \text{ and } r_{k-1} \mid b$$

Now, let c divides a and b. Since, $a = bq_1 + r_1$, then c divides b and r_1. Also, $b = r_1 q_2 + r_2$, which implies c divides r_1 and r_2. Continuing the process, we get c divides r_{k-1}.

Hence, g.c.d. $(a, b) = r_{k-1}$.

THEOREM-2

Any common divisor of a and b is a divisor of their greatest common divisor (a, b).

Proof. If $c \mid a$, $c \mid b$, then, we have $c \mid r_1$. Also, since $c \mid b$ and $c \mid r_1$, then $c \mid r_2$. Continuing this process, at least we obtain $c \mid r_k$. Hence $c \mid (a, b)$.

THEOREM-3

$$(a, b)c = (ac, bc), \, c > 0$$

Proof. By multiplying each equation of (1) by c, the integers a, b and r_1 become ac, bc and $r_1 c$ respectively and hence, $(ac, bc) = (a, b)c$.

THEOREM-4

If $(a, b) = 1$, then $(ac, b) = (c, b)$.

Proof. Since $(ac, b) \mid ac$ and $(ac, b) \mid bc$, we have

$$(ac, b) \mid (ac, bc) = (a, b) \mid |c| = |c|$$

But, $(ac, b) \mid b$, therefore $(ac, b) \mid (c, b)$

Further, since $(c, b) \mid ac$ and $(c, b) \mid b$, then $(c, b) \mid (ac, b)$. Hence, $(ac, b) = (c, b)$.

DEDUCTIONS :

From above theorem, we can easily obtain the following useful results :

(1) If $b \mid ac$ and $(a, b) = 1$, then $b \mid c$ and therefore $(ac, b) = b$ which together with theorem (4) gives $(c, b) = b$. Therefore $b \mid c$. On the other hand, if $(a, b) = 1$, then $(a, b^n) = 1$.

(2) If $a \mid c$, $b \mid c$ and $(a, b) = 1$, then $ab \mid c$. In fact, from $a \mid c$, it follows that $c = ac_1$ so that $b \mid ac_1$. Consequently $b \mid ac_1$ and $ab \mid ac_1$ that is $ab \mid c$.

(3) If $(a, c) = 1$ and $(b, c) = 1$, then $(ab, c) = 1$. For, by theorem (4), we have $(ab, c) = (b, c) = 1$.

(4) If $(a, b) = 1$, then $(ab, a + b) = 1$. For, from $(a, a + b) = 1$, $(b, a + b) = 1$, it follows that $(ab, a + b) = 1$.

THEOREM-5

If a and b are any two integers not both zero, then (a, b) uniquely exists.

Proof. We know that (a, b) is not affected by the sign a and b. Therefore, we assume that both a and b are positive and $a \geq b$. By division algorithm, we have

$$a = bq_1 + r_1, \quad 0 \leq r_1 < b \qquad \text{...(1)}$$

Now, there are following cases :

If, $r_1 = 0$, then $b \mid a$ and $(a, b) = b$. Therefore, (a, b) exists.

If $r_1 \neq 0$, then by division algorithm, we have $b = r_1 q_2 + r_2, 0 \leq r_2 < r_1$ $\qquad$ ...(2)

If $r_2 = 0$, then $r_1 \mid b$ and so from (1)

$$a = (r_1 q_2)q_1 + r_1$$
$$= r_1(q_2 q_1 + 1)$$

Which implies $r_1 \mid a$

Further, let $s \mid a, s \mid b \Rightarrow s \mid a - bq_1$

$$\Rightarrow s \mid r_1 \qquad \text{[Using (1)]}$$

$\therefore$ $(a, b) = r_1$, which shows the existence of (a, b) at this stage.

If $r_2 \neq 0$ we again apply the same process.
After n steps, we get zero remainder.
Thus, we get a sequence of integers r_i such that

$$0 \leq r_n < r_{n-1} \ldots < r_2 < r_1 < b$$
$$r_{n-2} = r_{n-1}q_n + r_n, \quad n \geq 3$$

and $r_{n-1} = q_{n+1}r_n$.

Therefore, $r_n \mid r_{n-1}, r_n \mid r_{n-2} \ldots r_n \mid b$ and $r_n \mid a$.

Further, if s is a common factor of a and b, then $s \mid a$ and $s \mid b$, which implies

$$s \mid a - bq$$

$\Rightarrow \qquad s \mid r_1$

$\Rightarrow \qquad s \mid r_2$

...

$\Rightarrow \qquad s \mid r_n$

Hence, $(a, b) = r_n$

Now, we shall prove the uniqueness of (a, b).

Let, if possible d_1 and d_2 are two g.c.d's of a and b. Then, by definition

$$d_1 \geq d_2 \quad \text{and} \quad d_2 \geq d_1$$

i.e. $d_1 = d_2$

Hence, (a, b) is unique.

THEOREM-6

If g is the greatest common divisor of a and b, then there exists integers x and y such that $g = (a, b) = ax + by$.

Proof. Consider the linear combination $ax_0 + by_0$, where x_0 and y_0 range over all integers. This set of integers $\{ax_0 + by_0\}$ includes positive and negative values, and also 0 by the choice $x_0 = y_0 = 0$. Choose x and y such that $ax + by$ is the least positive integer l in the set, so $l = ax + by$.

Next, we prove that $l \mid a$ and $l \mid b$, let if possible $l \nmid a$. then, $l \nmid a$ implies there exist integers q and r such that

$a = lq + r$

with $0 < r < l$. Therefore, we have $r \quad = a - lq = a - q(ax + by)$

$\qquad \qquad \qquad \qquad \qquad = a(1 - qx) + b(-qy)$

and thus r is the set $\{ax_0 + by_0\}$. This contradicts the fact that l is the least positive integer in the set $\{ax_0 + by_0\}$.

Now, since g is the greatest common divisor of a and b, we may write

$a = g\mathrm{A}, b = g\mathrm{B}$ and $l = ax + by = g(\mathrm{A}x + \mathrm{B}y)$.

Thus $g \mid l$ which gives $g \leq l$. Now $g < l$ is not possible, since g is the greatest common divisor, therefore $g = l = ax + by$.

DEDUCTIONS :

(1) The greatest common divisor g of a and b can be characterized in the following two ways:

   (a) It is the least positive value of $ax + by$ where x and y range over all integer.

   (b) It is the positive common divisor of a and b that is divisible by every common divisor.

(2) If an integer d is expressible in the form of $d = ax + by$, then d is not necessarily the g.c.d. of (a, b). However, it does not follow from such an equation that (a, b) is a divisor of d. In particular, if $ax + by = 1$ for some integers x and y, then $(a, b) = 1$.

THEOREM-7

For any positive integer m $(ma, mb) = m(a, b)$

Proof. By theorem-6 we have

$$(ma, mb) = \text{least positive value of } max + mby$$
$$= m. [\text{least positive value of } ax + by]$$
$$= m(a, b)$$

DEDUCTIONS :

(1) If $d \mid a$ and $d \mid b$ and $d > 0$, then $\left(\dfrac{a}{d}, \dfrac{b}{d}\right) = \dfrac{1}{d} (a, b)$

(2) If $(a, b) = g$, then $\left(\dfrac{a}{g}, \dfrac{b}{g}\right) = 1$

Proof. If $g = (a, b)$, then, we have $g \mid a$ and $g \mid b$

Therefore, $\dfrac{a}{g}$ and $\dfrac{b}{g}$ both are integers.

Also, $\qquad g = (a, b)$

$$= \left(g \cdot \frac{a}{g}, g \frac{b}{g}\right) = g\left(\frac{a}{g}, \frac{b}{g}\right)$$

$$\Rightarrow \qquad \left(\frac{a}{g}, \frac{b}{g}\right) = 1$$

THEOREM-8

Let $a > 1$, and m, n be positive integers. Then
$$(a^m - 1, a^n - 1) = a^{(m, n)} - 1$$

Proof. When $m = n$, then result is obvious.
Now, suppose that $m > n$, $m = qn + r$, then

$$a^m - 1 = (a^n - 1)a^{m-n} + a^{m-n} - 1$$
$$= (a^n - 1)a^{m-n} + (a^n - 1)a^{m-2n} + a^{m-2n} - 1$$
$$= (a^n - 1)(a^{m-n} + a^{m-2n} + ... + a^{m-qn}) + a^r - 1$$

Hence, $(a^m - 1, a^n - 1) = (a^n - 1, a^r - 1) = (a^r - 1, a^{r_1} - 1) = ...$
$$= (a^d - 1, a^0 - 1) = a^d - 1$$

where, $d = (m, n)$

2.5 RELATIVELY PRIME INTEGERS

If the greatest common divisors of a and b is 1, then a and b are said to be relatively prime.

Also, $a_1, a_2, ... a_n$ are said to be relatively prime in pairs, if $(a_i, a_j) = 1$ for all $i = 1, 2, ..., n$ and $= 1, 2, ..., n$ with $i \neq j$.

Remarks

❖ The fact that $(a, b) = 1$ is sometimes expressed by saying that a and b are coprime or by saying that a is prime to b.

❖ If $(a_i a_j) > 1$, whenever $i \neq j$, the numbers $a_1, ..., a_n$ are said to be relatively prime in pairs. If $(a_1, ..., a_n)$ are relatively prime in pairs, then $(a_1, ..., a_n) = 1$.

THEOREM-1

The integers a and b are relatively prime if and only if there exists integers x and y such that $ax + by = 1$

Proof. Let us first suppose there exist integers x and y such that $ax + by = 1$. To show a and b are relatively prime.

Let d be the common divisor of a and b, such that $a = p.d$ and $b = qd$.

Then, $(px + qy)d = 1$, shows that d must be a unit.

∴ $g = (a, b) = \pm 1$, where g is the g.c.d. of (a, b)

Hence, a and b are relatively prime.

Conversely, if $(a, b) = 1$, then Euclid algorithm guarentees the existence of x and y such that $ax + by = 1$

DEDUCTIONS

(1) *If* $(a, b) = g$, *and* $a = Ag$, $b = Bg$ *then* $(A, B) = 1$

Proof. By the Euclid algorithm, there exist integers x and y such that
$g = ax + by$
then, from $g = (Ax + By)g$, we get $Ax + By = 1$
Then, by above theorem, we get $(A, B) = 1$

(2) *If* $(a, b) = 1$ *and* $(a, c) = 1$ *then* $(a, bc) = 1$

Proof. Using above theorem, x_1 and y_1 exist so that $ax_1 + by_1 = 1$ and x_2 and y_2 exist so that $ax_2 + cy_2 = 1$.

Then, $1 = 1.1 = (ax_1 + by_1)(ax_2 + cy_2)$

$$= a(x_1ax_2 + x_1cy_2 + by_1x_2) + bc(y_1y_2)$$

Therefore, integers $x_3 = x_1ax_2 + x_1cy_2 + by_1y_2$ and $y_3 = y_1y_2$ exist so that $ax_3 + bcy_3 = 1$.

Hence, again by above theorem – 1, we have

$(a, bc) = 1$

THEOREM-2

If a and b are relatively prime and if a divides bc, then a must divide c.

Proof. Since a and b are relatively prime, therefore, by definition, we have $(a, b) = 1$. Therefore, there exists integers x and y such that $1 = ax + by$. Hence,

$c = c(ax + by) = acx + bcy$

but, a divides bc, therefore $bc = aP$, which shows that

$c = a(cx + Py)$

Hence, a divides c.

THEOREM-3 (EUCLID'S LEMMA)

If $a \mid bc$ and $(a, b) = 1$, then $a \mid c$.

Proof. We have $(a, b) = 1$.

Therefore, there exists integers x and y such that $ax + by = 1$

$\Rightarrow \quad c(ax + by) = c.1$

$\Rightarrow \quad c = cax + cby$

Now, $a \mid ac$ and $a \mid bc$

$\Rightarrow \quad a \mid acx + bcy$

$\Rightarrow \quad a \mid c$

DEDUCTIONS :

(1) If a and b are integers, p is a prime such that $p \mid ab$ and $p \nmid a$, then $p \mid b$.

(2) If $p \mid a_1 a_2 \dots a_n$, then there exists some i such that $p \mid a_i$.

Remark

❖ If a and b are not relatively prime, then Euclid's Lemma does not holds good.

THEOREM-4

If a and b are any two integers, not both zero. Then there exists a positive integer $g = (a, b)$ if and only if

(i) *$g \mid a$ and $g \mid b$*

(ii) *whenever $c \mid a$ and $c \mid b$, then $c \mid g$.*

Proof. Let, $g = (a, b)$. Then, clearly, $g \mid a$ and $g \mid b$. Therefore, condition (i) is satisfied. Now, since $g = (a, b)$, therefore, there exist integers x and y, such that $g = ax + by$. Now, $c \mid a$ and $a \mid b$ implies $c \mid ax + by$. Thus, condition (ii) is satisfied.

Conversely, let g be any positive integer satisfying conditions (i) and (ii). To show $g = (a, b)$. Using condition (ii), we have, if c is a common divisior of a and b, then $c \mid g$, therefore $g \geq c$ and hence, g is the greatest common divisor of a and b, i.e.

$(a, b) = g$

THEOREM-5

For any integer x, $(a, b) = (b, a) = (a, -b) = (a, b + ax)$.

Proof. Denote (a, b) by d and $(a, b + ax)$ by g. Here, it is clear that $(a, -b) = d$.

Now, we know that there exist integers x_0 and y_0 such that

$d = ax_0 + by_0$

Therefore, we can write

$d = a(x_0 - xy_0) + (b + ax)y_0$

which shows that the greatest common divisors of a and $b + ax$ is a divisor of d, i.e.,

$g \mid d$. Now, we shall prove that $d \mid g$. Since $d \mid a$ and $d \mid b$, we have $d \mid (b + ax)$. Since, we know that every common divisor of a and $b + ax$ is a divisor of their g.c.d., i.e. a divisor of g. Thus $d \mid g$. Hence, we conclude that $d = \pm g$. However, d and g are both positive, by definition so $d = g$.

SOLVED EXAMPLES

Based on the following results

- The greatest common divisors of two given integers a and b (both not zero) denoted by (a, b) is the positive integer d such that

  (i) $d \mid a$ and $d \mid b$

  (ii) if $c \mid a$ and $c \mid b$ then $c \mid d$.

- If a and b are any two integers not both zero then (a, b) exist uniquely.

- If $g = (a, b)$ then $\exists$ integers x and y such that $g = ax + by$

- If $(a, b) = 1$ then a and b are relatively prime.

Example 1. *Find the greatest common divisor of 525 and 231.*

Solution. We have

$$
\begin{aligned}
525 &= 2 \times 231 + 63 \\
231 &= 3 \times 63 + 42 \\
63 &= 1 \times 42 + 21 \\
42 &= 2 \times 21
\end{aligned}
$$

Hence, $(525, 231) = 21$.

Example 2. *Find g.c.d of 396 and 671.*

Solution. We can write

$$
\begin{aligned}
671 &= 2 \times 396 - 121 \\
396 &= 3 \times 121 + 33 \\
121 &= 4 \times 33 - 11 \\
33 &= 3 \times 11 + 0
\end{aligned}
$$

Hence, $(396, 671) = 11$.

Aliter :

$$
\begin{aligned}
671 &= 1(396) + 275 \\
396 &= 1(275) + 121 \\
275 &= 2(121) + 33 \\
121 &= 3(33) + 22 \\
33 &= 1(22) + 11 \\
22 &= 2(11) = 11
\end{aligned}
$$

$\Rightarrow$ g.c.d $(396, 671) = 11$

2.6 ALGORITHM TO FIND G.C.D.

The investigation of the set of integers $\{bx + cy\}$ to find a smallest positive element is not practical for large values of b and c. If b and c are small, value of g, x_0 and y_0 such that $g = bx_0$

+ cy_0 can be found by inspection

For example, if $b = 10$ and $c = 6$, it is clear that $g = 2$ and one pair of values for x_0, y_0 is 2 and -3. But if b and c are large, we can not use inspection method. Theorem-5 can be used to calculate g effectively and also to get values of x_0 and y_0. We now discuss an example to show how Theorem-5 can be used to calculate the greatest common divisor.

Consider the case $b = 963$, $c = 657$. If we divide c into b, we get a quotient $q = 1$ and remainder $r = 306$. Therefore, $b = cq + r$ or $r = b - cq$, in particular $306 = 963 - 1$. (657).

Now, $(b, c) = (b - cq, c)$ by replacing a and x by c and $-q$ in theorem-5, therefore, we have

$$(963, 657) = (963 - 1(657), 657) = (306, 657)$$

The integer 963 has been replaced by the smallest integer 306 and we repeat the procedure. Therefore, we divide 306 into 657 to get a quotient 2 and a remainder 45 and

$$(306, 657) = (306, 657 - 2(306)) = (306, 45)$$

Next, 45 is divided into 306 with quotient 6 and remainder 36. The 36 is divided into 45 with quotient 1 and remainder 9. Thus, we conclude that

$$(963, 657) = (306, 657) = (306, 45) = (36, 45) = (36, 9)$$
$$\Rightarrow (963, 657) = 9$$

and we can express 9 as a linear combination of 963 and 657 by sequentially writing each remainder as a linear combination of the two original numbers.

$$
\begin{aligned}
306 &= 963 - 657 \\
45 &= 657 - 2(306) \\
&= 657 - 2(963 - 657) \\
&= 3(657) - 2(963) \\
36 &= 306 - 6(45) \\
&= (963 - 657) - 6(3 \times 657 - 2 \times 963) \\
&= 13(963) - 19(657) \\
9 &= 45 - 36 = 3(657) - 2(963) - [13(963) - 19(657)] \\
&= 22(657) - 15(963)
\end{aligned}
$$

Thus, we can find $g = 9$, $x = -15$, $y = 22$

These values of x_0 and y_0 are not unique.

Remarks

❖ To find the greatest common divisor of any two integers b and c, we now generalize what is done in the special case above. The process will also give integers x_0 and y_0 satisfying the equation $bx_0 + cy_0 = (b, c)$.

❖ The case $c = 0$ is special because $(b, 0) = |b|$. For $c \neq 0$, we have $(b, c) = (b, -c)$ and hence, we presume that c is positive.

SOLVED EXAMPLES

Example 1. *Find the greatest common divisor of* 42823 *and* 6409.

Solution. We have

$$42823 = 6(6409) + 4369 \qquad = (42823, 6409)$$
$$6409 = 1(4369) + 2040 \qquad = (6409, 4369)$$
$$4369 = 2(2040) + 289 \qquad = (4369, 2040)$$
$$2040 = 7(289) + 17 \qquad = (2040, 289)$$
$$289 = 17(17) \qquad = (289, 17) = 17$$

Example 2. *Find g.c.d. of 256 and 1166 and express g.c.d as linear combination of 256 and 1166.*

Solution. Here, we have
$$1166 = 256(4) + 142$$
$$256 = 142(1) + 114$$
$$142 = 114(1) + 28$$
$$114 = 28(4) + 2$$
$$28 = 2(14) + 0$$

Thus, $(256, 1166) = 2$

Also,
$$2 = 114 - 28(4)$$
$$= 114 - 4(142 - 114(1))$$
$$= 114 - 4(142) + 4(114)$$
$$= 5(114) - 4(142)$$
$$= 5(256 - 142(1)) - 4(142)$$
$$= 5(256 - 5(142) - 4(142)$$
$$= 5(256) - 9(142)$$
$$= 5(256) - 9(1166) + 36(256)$$
$$= 41(256) - 9(1166)$$
$$= 256x + 1166y$$

where, $x = 41$, and $y = -9$. Hence, g.c.d. 2 has been expressed as linear combination of 256 and 1166.

Example 3. *Find integer x and y to satisfy 42823x + 6409y = 17*

Solution. We find integers x_i and y_i such that
$$42823x_i + 6409y_i = r_i$$

Here, it is natural to consider $i = 1, 2, ...$, but to initiate the process, we also consider $i = 0$ and $i = -1$. We put $r_{-1} = 42823$ and write
$$42823(1) + 6409(0) = 42823$$

Similarly, we put $r_0 = 6409$ and write
$$42823(0) + 6409(1) = 6409$$

We multiply the second of these equation by $q_1 = 6$ and subtract the result from the first equation to obtain.
$$42823(1) + 6409(-6) = 4369$$

Multiply this equation by $q_2 = 1$ and subtract it from the preceding equation we find
$$42823(-1) + 6409(7) = 2040$$

We multiply this by $q_3 = 2$ and subtract the result from the preceding equation to find that

$$42823(3) + 6409(-20) = 289$$

Next, multiply this by $q_4 = 7$ and subtract the result from the preceding equation to find that

$$42823.(-22) + 6409.(147) = 17$$

On dividing 17 into 289, we find that $q_5 = 17$ and that $289 = 17 \times 17$. Therefore, r_4 is the last positive remainder so that $g = 17$ and we may take $x = -22$, $y = 147$.

Example 4. *Find g.c.d. of 28 and 49. Express it as linear combination of these numbers.*

Solution. We have

$$49 = 28(1) + 21$$
$$28 = 21(1) + 7$$
$$21 = 7(3) + 0$$

Which implies that $r_2 = 7$ is the required greatest common divisor. Also,

$$\begin{aligned}
7 &= 28 - 21.(1) \\
&= 28 - (49 - 28(1)) \\
&= 28.(2) - 49(1) \\
&= 28x + 49y
\end{aligned}$$

when $x = 2$, $y = -1$

Example 5. *Find $g = (b, c)$ where $b = 5033464705$ and $c = 3137640337$ and determine x and y such that $bx + cy = g$.*

Solution. Proceeding same as above, we get the following calculation scheme :

	5033464705	1	0
1	3137640337	0	1
1	1895824368	1	-1
1	1241815699	-1	2
1	654008399	2	-3
1	587807570	-3	5
8	66200829	5	-8
1	58200938	-43	69
7	7999891	48	-77
3	2201701	-379	608
1	1394788	1185	-1901
1	806913	-1564	2509
1	587875	2749	-4410
2	219038	-4313	6919
1	149799	11375	-18248
2	69239	-15688	25167
6	11321	42751	-68582
8	1313	-272194	436659
1	817	2220303	-3561854
1	496	-2492497	3998513

1	321	4712800	– 7560367
1	175	– 7205297	11558880
1	146	11918097	– 19119247
5	29	– 19123394	30678127
29	1	107535067	– 1752509882

Hence, $g = 1$ and we may take $x = 107535067$, $y = – 172509882$.

2.7   GREATEST COMMON DIVISOR OF MORE THAN TWO INTEGERS

In the previous section, we introduced the concept of the greatest common divisor of two integers. Now, in the same way, we can also define greatest common divisor of more than two integers, not all zeroes. The g.c.d. of integers $a_1, a_2, ..., a_n$ not all zero, is the largest integer which is divisor of each of these integers, it exists uniquely and is denoted by $(a_1, a_2, ..., a_n)$

If $(a_1, a_2, ..., a_n) = 1$, then we say that the integers $a_1, a_2, ..., a_n$ are mutually relatively prime. If each pair of integers a_i and a_j from the set is relatively prime, then these integers are called pairwise relatively prime.

Clearly, if integers are pairwise relatively prime, then they must be mutually relatively prime. The converse is not true. For example :

Since, $(16, 10, 15) = 1$, therefore, 6, 10 and 15 are mutually relatively prime, but any two of these integers are not relatively prime. Hence, they are not pairwise relatively prime.

WORKING PROCEDURE

To find the greatest common divisor of (a, b, c) of three integers a, b, and c, we shall first find the g.c.d. (a, b) of a and b and then find the g.c.d. $((a, b), c)$ of (a, b) and c; the rsult is the required (a, b, c).

Remark

❖ Similarly, the infinite set of integers $a_1, a_2, ..., a_n, ...$ also has the greatest common divisor $(a_1, a_2, ..., a_n, ...)$ which can be obtained by using the same procedure.

SOLVED EXAMPLES

Example 1. *Find the greatest common divisor of 136, 221, 391.*

Solution. We have

$$(136, 221, 391) = (136, (221, 391)$$
$$= (136, 17)$$
$$= 17$$

$\because$ $391 = 1(221) + 170$
$221 = 1(170) + 51$
$170 = 3(51) + 17$
$51 = 3(17)$

We can also compute it as follows :

$$(136, 221, 391) = (136, 221 – 136, 391 – 2(136))$$
$$= (138, 85, 119) = (51, 85, 34)$$
$$= (17, 17, 34) = 17$$

Moreover, since

$$5(136) – 3(221) = 17$$
$$(– 22).17 + 1(391) = 17$$

We have

$136.(-110) + 221(66) + 391(1) = 17$

Example 2. Let $a = qc + r$, $b = q_1c + r_1$, show that $(a, b, c) = (r, r_1, c)$.

Solution. We have

$$\begin{aligned}
(a, b, c) &= ((a, c), b) = ((c, r), b) \\
&= (r, (b, c)) \\
&= (r, (c, r_1)) = (r, r_1, c)
\end{aligned}$$

2.8 LEAST COMMON MULTIPLE

Let $m \neq 0$ be a multiple of a and b. Then, m is called a common multiple of a and b. Clearly ab is a common multiple of a and b. Among the common multiple of a and b there is no greatest integer, but there is a unique positive least integer, which is called the least common multiple of a and b and denoted by $[a, b]$.

Definition : *If a and b are two non-zero integers, then a positive integer m is called their least common multiple if*

(i) $a \mid m$ and $b \mid m$

and (ii) *there exists a positive integer n such that if $a \mid n$ and $b \mid n$, then $m \leq n$, equivalently $m \mid n$.*

THEOREM-1

A common multiple of a and b is a multiple of the least common multiple $[a, b]$.

Proof. Let k be a common multiple of a and b. Divide k by $[a, b] = m$, then we get

$k = qm + r, 0 \leq r < m$

Since, $a \mid k$ and $a \mid m$, so $a \mid r$. Similarly, we can show that $b \mid r$. If $r \neq 0$, then r is a common multiple of a and b. This contradicts the assumption that m is the least common multiple. Therefore, $r = 0$ and $k = qm$, i.e. k is a multiple of m.

THEOREM-2

If $a > b > 0$ be two integers, then $[a, b] (a, b) = a.b$

Proof. Let $[a, b] = m$ and $(a, b) = d$. Since, $a \mid m$ and $b \mid m$, we have $ab \mid ma$, $ab \mid mb$ and hence $ab \mid (ma, mb)$, that is $ab \mid md$.

Also, since $a \mid ab/d$ and $b \mid ab/d$, that is ab/d is a common multiple of a and b.

By previous theorem, $m \mid ab/d$, therefore md/ab and $ab = md$.

Verification : For example, $[6, 9] = 18$ $(6, 9) = 3$, here $18.3 = 6.9$.

DEDUCTION :

If $(a, b) = 1$, then, we have $[a, b] = ab$, so that ab is the least common multiple of a and b. Conversely, if the least common multiple of a and b is ab, then $(a, b) = 1$. Hence, a necessary and sufficient condition for $[a, b] = ab$ is $(a, b) = 1$.

THEOREM-3

If $k > 0$ is a common multiple of a and b, then $\left(\dfrac{k}{a}, \dfrac{k}{b}\right) = \left(\dfrac{k}{[a, b]}\right).$

Proof. Since, we have

$$\left(\frac{k}{a},\frac{k}{b}\right)|ab| = (kb, ka) = k(a, b)$$

$$= k.\frac{a.b}{[a,b]} \qquad \left(\because (a,b)=\frac{a.b}{[a,b]}\right)$$

Hence, $\qquad \left(\dfrac{k}{a},\dfrac{k}{b}\right) = \dfrac{k}{[a,b]}$

DEDUCTIONS :

(1) If $k = [a, b]$, then $\left(\dfrac{k}{a},\dfrac{k}{b}\right) = 1$. Conversely, if $\left(\dfrac{k}{a},\dfrac{k}{b}\right) = 1$, then $k = [a, b]$.

(2) If $k > 0$, then a necessary and sufficient condition for $k = [a, b]$ is $\left(\dfrac{k}{a},\dfrac{k}{b}\right) = 1$.

Remark

❖   We have $\qquad [a, b] = \dfrac{ab}{(a,b)}$

Therefore, we first find (a, b) and then the formula to get the required $[a, b]$.

THEOREM-4

Let a and b be two positive integers, then $(a + b)[a, b] = b[a, a + b]$

Proof. We have

$$b[a, a + b] = \frac{ba(a+b)}{(a, a+b)}$$

$$= \frac{(a+b)ab}{(a,b)}$$

$$= (a, b)[a, b]$$

THEOREM-5

The least common multiple of two non-zero integers is unique.

Proof. Let a and b be any two non-zero integers. Let, if possible, there are two least common multiples m_1 and m_2 of a and b. By definition, a and b divide both m_1 and m_2.
Again, by definition of *l.c.m.*,

$\qquad\qquad m_1$ divides m_2

and m_2 divides m_1.

Hence, $\qquad m_1 = m_2$

i.e. least common multiplier is unique.

2.9   LEAST COMMON MULTIPLE OF *n* INTEGERS

The least common multiple of $[a, b, c]$ of three integers, a, b and c can be found by using the following formula

$$[a, b, c] = [[a, b], c] \qquad \qquad ...(1)$$

i.e. we first find the least common multiple $[a, b]$ of a and b, and then the least common multiple $[[a, b], c]$ of $[a, b]$ and c, which is the required $[a, b, c]$.

Generalization : The above formula (1) can be generalized to the case of n integers a_1, a_2, ... a_n as follows :

If $[a_1, a_2] = m_2$, $[m_2, a_3] = m_3$, ..., $[m_{n-1}, a_n] = m_n$

Then, $[a_1, a_2, ... a_n] = m_n$

If $a_1, a_2, ... a_n$ are mutually relatively prime, then

$$[a_1, a_2, ... a_n] = a_1 a_2 ... a_n$$

THEOREM-1

Let a, b, c be three positive integers. Then $[a, b, c] = \dfrac{abc}{(ab, bc, ca)}$

Proof. We have

$$[a, b, c] = [[a, b], c] = \frac{[a, b].c}{([a, b], c)}$$

$$= \frac{abc}{(a, b)([a, b], c)} = \frac{abc}{(a, b, (a, b)c)}$$

$$= \frac{abc}{(ab, bc, ca)}$$

Remark

❖ The above formula is also true for any number of integers.

SOLVED EXAMPLES

> **Based on the following results**
>
> - A positive integer m is called *l.c.m* of two non-zero integers a and b denoted by $[a, b]$ if
>
>   (i) $a \mid m, b \mid m$
>
>   and (ii) $\exists$ a positive integer n such that if $a \mid n$ and $b \mid n$ then $m \leq n$ or $m \mid n$.
>
> - $[a, b].(a.b) = a.b$
>
> - L.c.m of two non-zero integers is unique.

Example 1. *Find the g.c.d. and l.c.m. of 119 and 272.*

Solution. Here, we have

$$272 = 119(2) + 34$$
$$119 = 34(3) + 17$$
$$34 = 17(2) + 0$$

which implies that 17 is the g.c.d. of 119 and 272.

Also, $[119, 272] = \dfrac{119 \times 272}{(119, 272)}$ $\left[\because [a,b] = \dfrac{a.b}{(a,b)}\right]$

$= \dfrac{119 \times 272}{17} = 1904$

Example 2. *Find the l.c.m of 136, 221 and 391.*

Solution. We have

$[136, 221, 391] = [[136, 221], 391]$

$= \left[\dfrac{136 \times 221}{17}, 391\right] = [1768, 391]$ $(\because$ g.c.d. of 136 and 221 is 17)

$= \dfrac{1768 \times 391}{17} = 40664$

Example 3. *Find the l.c.m. of 8, 12, 15, 20 and 25.*

Solution. Using $[a, b] = \dfrac{ab}{(a,b)}$

We have

$[8, 12] = \dfrac{8 \times 12}{(8,12)} = \dfrac{8 \times 12}{4} = 24$

$[24, 15] = \dfrac{24 \times 15}{(24,15)} = \dfrac{24 \times 15}{3} = 120$

$[120, 20] = \dfrac{120 \times 20}{(120,20)} = \dfrac{120 \times 20}{20} = 120$

$[120, 25] = \dfrac{120 \times 25}{(120,25)} = \dfrac{120 \times 25}{5} = 600$

Hence, l.c.m. of 8, 12, 15, 20 and 25 is given by $[8, 12, 15, 20, 25] = 600$

2.10   FIBONACCI SEQUENCE

The sequence $a_1, a_2, a_3, \ldots$ *in which* $a_1 = 1, a_2 = 1$ *and* $a_n = a_{n-1} + a_{n-2}$ *for every* $n > 2$ *is called a Fibonacci sequence.*

The first few Fibonacci numbers are

1, 1, 2, 3, 5, 8, 13, 21, 34, 55, 89, 144, ...

It is surprising to discover that the Fibonacci numbers can be extracted from Pascal's tiangle by adding the numbers along the north-east diagonals as follows :

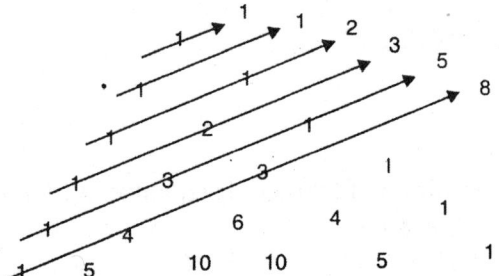

Interestingly enough, Fibonacci numbers appear in nature, music, geography and geometry. They can be found in the spiral arrangement of seeds in sunflowers, the scale patterns of pine cones, the numbers of petals in flowers and the arrangement of leaves on trees.

Fibonacci number can be obtained explicitly using Binet's formula

$$F_n = \frac{\alpha^n - \beta^n}{\alpha - \beta}, \text{ where } \alpha = \frac{1+\sqrt{5}}{2}, \beta = \frac{1-\sqrt{5}}{2}$$

which are root of quadratic equation $x^2 = x + 1$.

2.10.1 Lucas Number : The Lucas number L_n are determined by relation $L_1 = 1$, $L_2 = 3$ and $L_n = L_{n-1} + L_{n-2}$ for $n > 2$.

We have $L_n = \left(\frac{1+\sqrt{5}}{2}\right)^n + \left(\frac{1-\sqrt{5}}{2}\right)^n$

Lucas numbers can also be obtained by Binet's formula

$$L_n = \alpha^n + \beta^n, \text{ where } \alpha = \frac{1+\sqrt{5}}{2}, \beta = \frac{1-\sqrt{5}}{2}$$

Remark

❖ The Fibonacci number a_n and Lucas number L_n satisfy the same recurrence relation but with different initial conditions.

THEOREM-1

If a_n is the nth form of the Fibonacci sequence and $\alpha = \dfrac{1+\sqrt{5}}{2}$, then

$a_n > \alpha^{n-1}$ *for $n > 1$.*

Proof. We can easily verify that

$$\alpha^2 = \left(\frac{1+\sqrt{5}}{2}\right)^2 = \frac{1+\sqrt{5}}{2} + 1 = \alpha + 1$$

Thus, for any $k > 1$, we have

$$\alpha^k = \alpha^{k-2}.\alpha^2 = \alpha^{k-2}[\alpha + 1] \qquad [\because \alpha^2 = \alpha + 1]$$
$$= \alpha^{k-1} + \alpha^{k-2}$$

Now, $a_2 = 2$ and $\alpha = \dfrac{1+\sqrt{5}}{2}$

$\therefore \quad a_2 > \alpha = \alpha^{2-1}$

i.e. result is true for $n = 2$.

Further, from the definition of Fibonacci sequence, we have

$$\begin{aligned} a_3 &= a_2 + a_1 > \alpha + 1 \\ &= \alpha^2 = \alpha^{3-1} \end{aligned}$$

i.e. result is true for $n = 3$.

Now, suppose that result is true for $n = 2, 3, ..., k$, then

$$a_k > \alpha^{k-1} \text{ and } a_{k-1} > \alpha^{k-2}$$

Therefore,

$$\begin{aligned} a_{k+1} &= a_k + a_{k-1} \\ &> \alpha^{k-1} + \alpha^{k-2} \\ &= \alpha^{k-2}[\alpha + 1] \\ &= \alpha^{k-2} \cdot \alpha^2 \\ &= \alpha^k \\ &= \alpha^{k+1} - 1 \end{aligned}$$

i.e. result is true for $n = k + 1$.

Hence, by principle of mathematical induction, the result is true for all values of $n > 1$.

THEOREM-2 (LAME'S THEOREM)

If a and b are any two positive integers with a > b and n is the number of divisions in Euclid's algorithm, then $n \le 5p$, where p is the number of digits in b.

Proof. Suppose that a and b be two positive integers and n be the number of divisions in Euclid's algorithm. Then, we get a positive number $r_{n-1}, r_{n-2}, ..., r_2, r_1, b, a$ in increasing form. Comparing this sequence with Fibonacci sequence, we have

$$r_{n-1} > 1 = a_1 \qquad \qquad \qquad \text{...(1)}$$

Also $\quad r_{n-2} = r_{n-1} \cdot q_1$ and $q_2 \ge 2$

$\therefore \qquad r_{n-2} \ge a_2 \qquad \qquad \qquad \text{...(2)}$

Again for $k > 2$, we have

$$\begin{aligned} r_{n-k} &= r_{n-(k-1)} q_{n-(k-2)} + r_{n-(k-2)} \qquad \text{...(3)} \\ &\ge r_{n-(k-1)} + r_{n-(k-2)} \end{aligned}$$

Now, putting $k = 3, 4, ..., n-1$ in (3), we get

$$r_{n-3} \ge r_{n-2} + r_{n-1} \ge a_2 + a_1 = a_3$$
$$r_{n-4} \ge r_{n-3} + r_{n-2} \ge a_3 + a_2 = a_4$$

$$\vdots \qquad \vdots \qquad \vdots \qquad \vdots$$

$$r_2 \ge r_3 + r_4 \ge a_{n-2}$$
$$r_1 \ge r_2 + r_3 \ge a_{n-1}$$

Also, $\qquad b = r_1 q_2 + r_2$

$$\ge r_1 + r_2$$

$$\geq a_{n-1} + a_{n-2}$$
$$= a_n$$
$$> \alpha^{n-1}, \text{ where } \alpha = \frac{1+\sqrt{5}}{2}$$

Thus, $\log_{10} b > (n-1) \log_{10} \alpha$

$$= (n-1) \log 10 \left(\frac{1+\sqrt{5}}{2} \right) = \frac{1}{5} \qquad \qquad \text{...(4)}$$

Since, p is the number of digits in b, therefore,
$$p = \log_{10} b \qquad \qquad \text{...(5)}$$
Using (4) and (5), we conclude that
$$p > (n-1)/5$$
$$\Rightarrow \qquad n < 5p + 1$$
$$\Rightarrow \qquad n \leq 5p$$

DEDUCTION :

$$\bullet \ n \leq \frac{1}{\log_{10} \alpha}, \text{ where } \alpha = \frac{1+\sqrt{5}}{2}$$

THEOREM-3

If a and b are any two positive integers with a > b and n is the number of divisions in minimal algorithm for a and b, then $(n-1) < \dfrac{10}{3} p$, *where p is the number of digits in b.*

Proof. Let a and b be two positive integers such that $a > b$. Then we have

$$r_1 \leq \frac{b}{2}$$

$$r_2 \leq \frac{r_1}{2} \leq \frac{b}{2^2}$$

$$r_3 \leq \frac{r_2}{2} \leq \frac{b}{2^3}$$

$$\vdots \qquad \vdots$$

$$r_{n-1} \leq \frac{r_{n-2}}{2} \leq \frac{b}{2^{n-1}}$$

Also, $r_{n-1} \geq 1$
$$\therefore \qquad b \geq 2^{n-1}.r_{n-1}$$
$$\geq 2^{n-1}$$
Taking log of both the sides, we get
$$\log_{10} b \geq (n-1)\log_{10} 2$$

$$\geq (n-1).\frac{3}{10} \qquad \qquad \left(\because \log_{10} 2 > \frac{3}{10} \right)$$

Also, $\qquad p > \log_{10} b$

which conclude that

$$p > (n-1)\frac{3}{10}$$

i.e. $(n-1) < \dfrac{3}{10p}$

THEOREM-4 (KRONECKER'S THEOREM)

The number of divisions in minimal algorithm for two positive integers a and b is not greater than the number of divisions in Euclid's algrothm for the same integers, i.e. $M(a, b) \le E(a, b)$, where M(a, b) is the number of division in minimal algorithm and E(a, b) is the number of divisions in Euclid's algorithm.

Proof. To prove this theorem, we shall use the principle of mathematical induction.

Let, $a = 2, b = 1$. Then, we have

$\quad\quad 2 = 2 . 1 + 0$

$\Rightarrow\quad\quad E(a, b) = 1$ and $M(a, b) = 1$

$\Rightarrow\quad\quad M[a, b] \le E(a, b)$

$\Rightarrow\quad\quad$ Result is true for $a = 2$

Now, for $a = 3, b = 1$, we have

$\quad\quad 3 = 3. 1 + 0$

$\Rightarrow\quad\quad E(a, b) = 1, M(a, b) = 1$

$\Rightarrow\quad\quad M(a, b) \le E(a, b)$

For $a = 3, b = 2$, we have

$\quad\quad 3 = 2 . 1 + 1$

$\quad\quad 2 = 1 . 2 + 0$

$\Rightarrow\quad\quad E(a, b) = 2, M(a, b) = 1$

$\Rightarrow\quad\quad M(a, b) \le E(a, b)$

$\Rightarrow\quad\quad$ Result is true for $a = 3$.

Now, suppose that theorem is true for $a = 2, 3, ..., k - 1$. To show that result is true for $a = k$, let b be a positive integer less than k. Then, we have

$\quad\quad k = bQ + eR$, where $e = \pm 1$ and $0 \le R \le b/2$

Thus $M(k, b) = 1 + M(b, R)$ ...(1)

$\quad\quad k = bq + r, \quad 0 \le r < b$

$\Rightarrow\quad\quad (k, b) = 1 + E(b, r)$ ...(2)

Now, there are following three possibilities :

(i) $0 = R = r$: In this case, we have

$\quad\quad M(k, b) = 1$ and $E(k, b) = 1$

$\Rightarrow\quad\quad$ Theorem holds for $a = k$.

(ii) $0 < R = r$: In this case, we have

$\quad\quad E(k, b) = 1 = 1 + E(b, r)$

but, $b < k$ and $R < b$, then by our assumption

$\quad\quad E(b, R) \ge M(b, R)$ ...(3)

From (2), (3) and (1), we have

$$E(k, b) = 1 + E(b, R)$$
$$\geq 1 + M(b, R) = M(k, B)$$

$\Rightarrow$ Theorem holds for $a = k$.

(iii) $0 < R < r$: Let $r = b - R$

Put this value in (2), we get

$$E(k, b) = 1 + E(b, b - R) \qquad \qquad ...(4)$$

Further, $b < k$ and $b - R < b$, therefore, by induction

$$E(b, b - k) \geq M(b, b - k)$$
$$E(k, b) = 1 + E(b, b - R) \qquad \qquad ...(5)$$

From (4) and (5), we have

$$E(k, b) = 1 + E(b, b - R)$$
$$\geq 1 + M(b, b - R)$$
$$\geq 1 + M(b, R) = M(k, b)$$

$\Rightarrow$ Theorem holds for $a = k$.

Hence, by principle of mathematical induction, we have, the result is true for $a > 1$.

2.11 THE PIGEONHOLE PRINCIPLE AND DIVISION ALGORITHM

The pigeonhole principle is known as the Dirichlet box principle after the German mathematician Gustav Peter Lejeune Dirichlet who used it in his work on number theory.

Pigeon Hole Principle : *If m pigeons are assigned n pigeon holes, where m > n, then at least two pigeons must occupy the same pigeon hole.*

SOLVED EXAMPLES

Example. *Let b be an integer ≥ 2. Suppose $(b + 1)$ integers are randomly selected. Prove that the difference of two of them is divisibly by b.*

Solution. Let a be an integer and it leaves quotient q and remainder r when it divide by b. Then by division algorithm, $a = bq + r$, where $0 \leq r < b$. Now $b + 1$ integer gives $b + 1$ remainders (pigeons); but there are only b possible remainders (pigeonhole). Therefore by pigeonhole principle, two of the remainders must be equal.

Let x and y be such integers which leave same remainder after dividing by b,

i.e. $x = bq_1 + r$, $y = bq_2 + r$, where q_1 and q_2 are integers.

$\therefore \qquad x - y = (bq_1 + r) - (bq_2 + r)$
$$= b(q_1 - q_2)$$

Thus, $x - y$ is divisible by b.

ADDITIONAL SOLVED EXAMPLES

Example 1. *Show that $([a, b], c) = [(a, c), (b, c)]$, $[(a, b), c] = ([a, c], [b, c])$.*

Solution. We have

$$([a, b], c) = \frac{[a, b]\, c}{[[a, b], c]} = \frac{[a, b].c}{[a, b, c]}$$

$$= \frac{[ab, bc, ca]}{(a, b)}$$

Also, $\quad [(a, c)], (b, c)] = \dfrac{(a, c)(b, c)}{((a, c), (b, c))} = \dfrac{(a, c)(b, c)}{(a, b, c)}$

as $\qquad \dfrac{(ab, bc, ca)}{(a, b)} = \dfrac{(a, c)(b, c)}{(a, b, c)}$

i.e. $\quad (a, b)\,(b, c)(c, a) = (a, b, c)(ab, bc, ca)$

Example 2. *If $0 < b \le a/3$, then show that $M(a, b) < M(a, a - b)$.*

Solution. Given that $0 < b \le a/3$

The given condition can also be written as

$$0 < b \le \frac{a - a/3}{2} \le \frac{a - b}{2} \qquad\qquad \text{...(1)}$$

Now, there are following two cases :

(i) **b divides a :** In this case, we have

$$a = bq_1 + 0 \qquad\qquad \text{...(2)}$$

for some integers q_1. Therefore $M(a, b) = 1$.

Also, we have

$$a = (a - b) + b$$

Here, the dividend is $(a - b)$ and divisor is b.

$\therefore \qquad a - b = bq_1 - b \qquad\qquad\qquad\qquad\qquad\qquad$ [Using (2)]

$$= b(q_1 - 1) + 0$$

Thus, $\quad M(a, a - b) = 2$. In this case

$$M(a, b) < M(a, a - b)$$

(ii) **a is not divisible by b**

Here, the given condition implies that

$$0 < b < \frac{b - a/3}{2} < \frac{a - b}{2} \qquad\qquad \text{...(3)}$$

Now, in the first step of minimal algorithm, we have

$$a = bq_1 + e_1 r_1 \qquad\qquad \text{...(4)}$$

for some integers $q_1, e_1 = \pm 1$ and $0 < r_1 < b/2$.

Thus, $\quad M(a, b) = 1 + M(b, r_1) \qquad\qquad \text{...(5)}$

For the second step, the dividend is $a - b$ and the divisor is b. In this step, we have

$$a - b = bq_1 + e_1 r_1 - b$$
$$= b(q_1 - 1) + e_1 r_1, \quad 0 < r_1 \le b/2$$

Therefore,

$$M(a, b - b) = 2 + M(b, r_1) \qquad\qquad \text{...(6)}$$

Using (5) and (6), we conclude that

$$M(a, b) - 1 = M(a, a - b) - 2$$
$$\Rightarrow \qquad M(a, b) = M(a, a - b) - 1$$

Hence, $M(a, b) < M(a, a - b)$

Example 3. *Show that if $5 \nmid (n - 1)$, $5 \nmid n$ and $5 \nmid (n + 1)$, then $5 \mid n^2 + 1$.*

Solution. Since, we know that a positive integer is divisible by 5, if the digit of the unit place of the given number is either 0 or 5. Also, the digit of unit place of a number not divisible by 5 must be 1, 2, 3, 4, 6, 7, 8, 9. Since, it is given that $5 \nmid (n - 1)$, $5 \nmid n$ and $5 \nmid (n + 1)$. Therefore, unit digit of n can not be 4 or 9 otherwise $(n + 1)$ will be divisible by 5. Also, if the unit digit of n is 2 or 8, then the unit digit of $n^2 + 1$ will be 4 + 1 = 5, i.e. $5 \mid n^2 + 1$.

Similarly, if the unit digit of n is 3 or 7, then 0 will be the unit digit of $n^2 + 1$, which is again divisible by 5.

Hence, if $5 \nmid (n - 1)$, $5 \nmid n$ and $5 \nmid (n - 1)$, then $5 \mid n^2 + 1$.

EXERCISE 2.2

1. Show that if $ac \mid bc$, then $a \mid b$.

2. Let n be an odd integer greater than 1, show that $n^3 - n$ is a multiple of 24.

3. Prove that if a and b are positive integers satisfying $(a, b) = [a, b]$, then $a = b$.

4. Show that common divisor of integers $a_1, a_2, ..., a_n, ...$ is a divisor of their greatest common divisor.

5. Let $n \geq 2$ and k be any positive integers. Prove that $(n - 1)^2 \mid (n^k - 1)$ if and only if $(n - 1) \mid k$.

6. If $m > 0$, $n > 0$ and m is an odd integer, show that $(2^m - 1)(2^n + 1) = 1$.

7. If $(a, b) = 1$, show that $(a - b, a + b) = 1$ or 2.

8. If a and b are non-zero integers, then there exist four integers, h, k, r, s such that $hs - kr = 1$, $ak + bs = 0$

9. Show that if $(a, b) = 1$, then $(a + b, a^2 - ab + b^2) = 1$ or 3.

10. Show that $M(a, b) = M(a \pm b, b)$

11. If $\dfrac{a}{3} < b < \dfrac{2}{5} a$, then show that $M(a, -a - b) = 1 + M[b, b - (3b - a)]$.

12. If $(a, b) = d$, then d is the least positive integer among all the integers of the form $ax + by$, i.e. $d = ax + by$, where x and y are any integers.

13. For Fibonacci sequence, show that $a_1 + a_2 + a_3 + ... + a_n = a_{n+2} - 1$.

14. For Fibonacci sequence, show that $a_{n+1} a_{n-1} - a_n^2 = (-1)^n$.

15. For Fibonacci sequence, show that $a_{m+n} = a_{m-1} a_n + a_m a_{n+1}$ for any positive integer m and n and hence prove that $a_m \mid a_n$ if $m \mid n$.

16. If the sum of two reduced fractions is an integer, say $\left(\dfrac{a}{b}\right) + \left(\dfrac{c}{d}\right) = n$, then show that $|b| = |d|$

17. Show that for every $n \geq 1$, there exists uniquely determined $a > 0$, $b > 0$ such that $n = a^2 b$, where b is a square free integer.

ANSWERS (WITH HINTS)

(2) Since $n^3 - n = (n - 1) n (n + 1)$ is the product of three consecutive integers, it is a multiple of 3. If n is an odd integer, $n - 1$ and $n + 1$ both must be even and one of them is a multiple of 4. Hence, $(n - 1)(n + 1)$ is a multiple of 8. Thus, $n^3 - n$ is the multiple of $3 \cdot 8 = 24$.

(4) Let b be the common divisor and d is the greatest common divisor. As $[b, d] = c$, then $d \le c$. Again as $b \mid a_i, d \mid a_i$, then $c \mid a_i$ and hence c is the common divisor of $a_1, ..., a_m$, but d is the greatest common divisor, hence, $c \le d$. Hence, $c = d$ and $b \mid d$.

(5) Use $n^k = ((n - 1) + 1)^k$.

(6) Let $(2^m - 1, 2^m + 1) = d$. Then $2^m = kd + 1, 2^n = ld - 1, kl > 0$. Therefore, $2^{mn} = (kd + 1)^n = td + 1, 2^{mn} = (ld - 1)^m = sd - 1$, where $(s - t)d = 2$. Thus $d \mid 2$, i.e. $d = 1$ or 2, but as $2^m - 1, 2^n + 1$ are both odd integers, hence, $d = 1$

(7) Let $(a - ba + b) = d$. Then, $a - b = kd, a + b = ld$. Therefore, $2a = (k + 1)d, n/lb = (k - l)d$. Hence, $2 = (2a, 2b) = d(k + 1, k - l)$ or $d \mid 2$.

(8) Let a and b be two positive integers. Since $[a, b] (a, b) = ab$ letting $[a, b] = ak$, $s = - a$, which is divisible by (a, b), we get $ak + bs = 0$. Also, since $l = ab/a(a, b) = b/(a, b)$, then $(k, s) = 1$ and hence $hs + (-k)r = 1$.

CHAPTER REVIEW : A COMPETITIVE APPROACH

Selected terms and Results

TERMS

- **Divisibility :** An integer b is divisible by an integer $a(\neq 0)$ if there is an integer x such that $b = ax$.

- **Greatest Common Divisor :** Let a and b be any two given integers (both, not zero) then the greatest common divisor of a and b, denoted by (a, b) is the positive integer d such that (i) $d \mid a$ and $d \mid b$ (ii) if $c \mid a$ and $c \mid b$ then $c \mid d$.

- **Relatively Prime Integers :** If greatest common divisors of a and b is 1 then a and b are said to be relatively prime.

- **Least common multiple :** If a and b are two non-zero integers then a positive integer m is called their least common multiple if

    (i) $a \mid m$ and $b \mid m$.

    and (ii) $\exists$ a positive integer n such that if $a \mid n, b \mid n$ then $m \mid n$.

- **Fibonacci sequence :** A sequence a_1, a_2, ... in which $a_1 = 1$, $a_2 = 1$ and $a_n = a_{n-1} + a_{n-2}$ for every $n > 2$ is called Fibonacci sequence.

- **Lucas Number :** The Lukas number L_n are determined by the relation $L_1 = 1, L_2 = 3$ and $L_n = L_{n-1} + L_{n-2}$ for $n > 2$.

RESULTS

- The sum, difference and product of two integers are obviously integers, but the quotient of two integers may or may not be in an integer.

- The product of three consecutive integers is a multiple of 3!.

- The sum of square of two odd integers is not a square integer.

- Every positive integer greater than 1 may be used as base.

- Any non-zero integer has only a finite number of divisors.

- Any common divisor of a and b is a divisor of their greatest common divisor (a, b).

- A common multiple of a and b is a multiple of the least common multiple $[a, b]$.

- A necessary and sufficient condition for $[a, b] = ab$ is $(a, b) = 1$.

- If $ab > 0$, then $[a, b] (a, b) = ab$.

- The infinite set of integers $a_1, ..., a_n, ...$ also has the greatest common divisor $(a_1, a_2, ..., a_n, ...)$.

- The greatest common divisor of two numbers is always unique.

- The least common multiple of two numbers is always unique.

- The formula $[a, b, c] = \dfrac{abc}{(ab, bc, ca)}$ is true for any number of integers.

Review Questions and Project Work

1. For any prime p and positive integers a, b and n prove the following
    (i) if $p \mid a^n \Rightarrow p \mid a$
    (ii) if $p \mid a^2 \Rightarrow p \mid a$
    (iii) the product of any n integers of the form $4k + 1$ is also of the same form.
    (iv) $(a, b) = 1$ if and only if $(a^n, b^n) = 1$

2. Prove the following results :
    $(a, - b) = (a, b) = (- a, b) = (- a, - b)$
    $= (a, a + b)$

3. If a certain integer a has the remainder 11 when divided by $b = 16$, de-

termined the remainder r with $0 \leq r < b$ when $a + k$ is divided by b for each of the cases $k = 1, 3, 5, 7, 9, 105, -17$.

4. If $b = 2c + 1 > 0$, prove that for any integer a there exist integers q and r such that $a = qb + r$ with $|r| \leq c$. Also show that q and r are unique.

5. With respect to divisibility by 6, show that the set of all integers is divided into six classes, where members of the same class have the same remainder.

6. If $d = (a, b) = ax + by$, prove that x and y are relatively prime and not unique.

7. Show that for all values of k, the integers $a = 22k + 7$ and $b = 33k + 5$ are relatively prime.

8. For fixed integers a and b (not both zero) show that least positive integer $d = ax + by$ of the set S' of all integers of the form $as + bt$ divides every member to the set S'. Show that S' coincides with the set of all multiples of d.

Objective Type Questions

Fill in the blanks:

1. $a \mid b \Rightarrow a \mid bc$ for ....

2. If $c = ax + by$ and $d \mid a$ then d ....

3. The product of any three consecutive integers is a multiple of ....

4. If a is any integer then $a^2 + a + 1$ is .... square number.

5. If a is an odd integer then $\dfrac{a^4 + 4a^2 + 11}{16}$ is .... an integer.

6. Any common divisor of a and b is a divisor of their ....

7. If a and b are any two integers (not both zero) then (a, b) exist ....

8. If $(a, b) = 1$ then a and b are called ......

9. The g.c.d. of 525 and 231 is ....

10. The g.c.d. of 256 and 1166 is ....

True/False: *Write 'T' for true and 'F' for false statement.*

1. There are no positive integers a, b, $n > 1$ such that $(a^n - b^n)/(a^n + b^n)$. (T/F)

2. If $a > 0$, $n \geq 2$ then a^n can be expressed as the sum of consecutive positive integers. (T/F)

3. If a is any positive integer then $a^2 + a + 1$ is a square number. (T/F)

4. The product of any three consecutive integers is divisible by 3!. (T/F)

5. Every square number is of the form $9k$ or $3k + 1$. (T/F)

6. $a \mid b$ and $a \mid c \Rightarrow a \mid (bx + cy)$ (T/F)

7. $a \mid b$ and $b \mid a \Rightarrow a = \pm b$ (T/F)

8. If $c = ax + by$ and $d \mid a$ but $d \mid c$ then $d \mid b$ (T/F)

Multiple Choice Questions : *Choose the most appropriate one :*

1. Let a and b be any two integers ($b > 0$) then there exist unique integers q and r such that $a = bq + r$; $0 \leq r < b$. This is called :
   (a) Division algorithm
   (b) Euler's algorithm
   (c) Euclid's algorithm
   (d) none of these

2. The product of any three consecutive integers is divisible by
   (a) 2!   (b) 3!
   (c) $n!$ (d) none of these

3. The number $a^{n+1} - (a - 1) - a$ is divisible by :
   (a) $(a + 1)$
   (b) $(a - 1)^2$
   (c) both (a) and (b) are true
   (d) none of these

4. If g.c.d $(28, 49) = 7$ and $7 = 28x + 49y$ then values of x and y are respectively given by
   (a) 2, 1   (b) 2, -1
   (c) 1, -2 (d) none of these

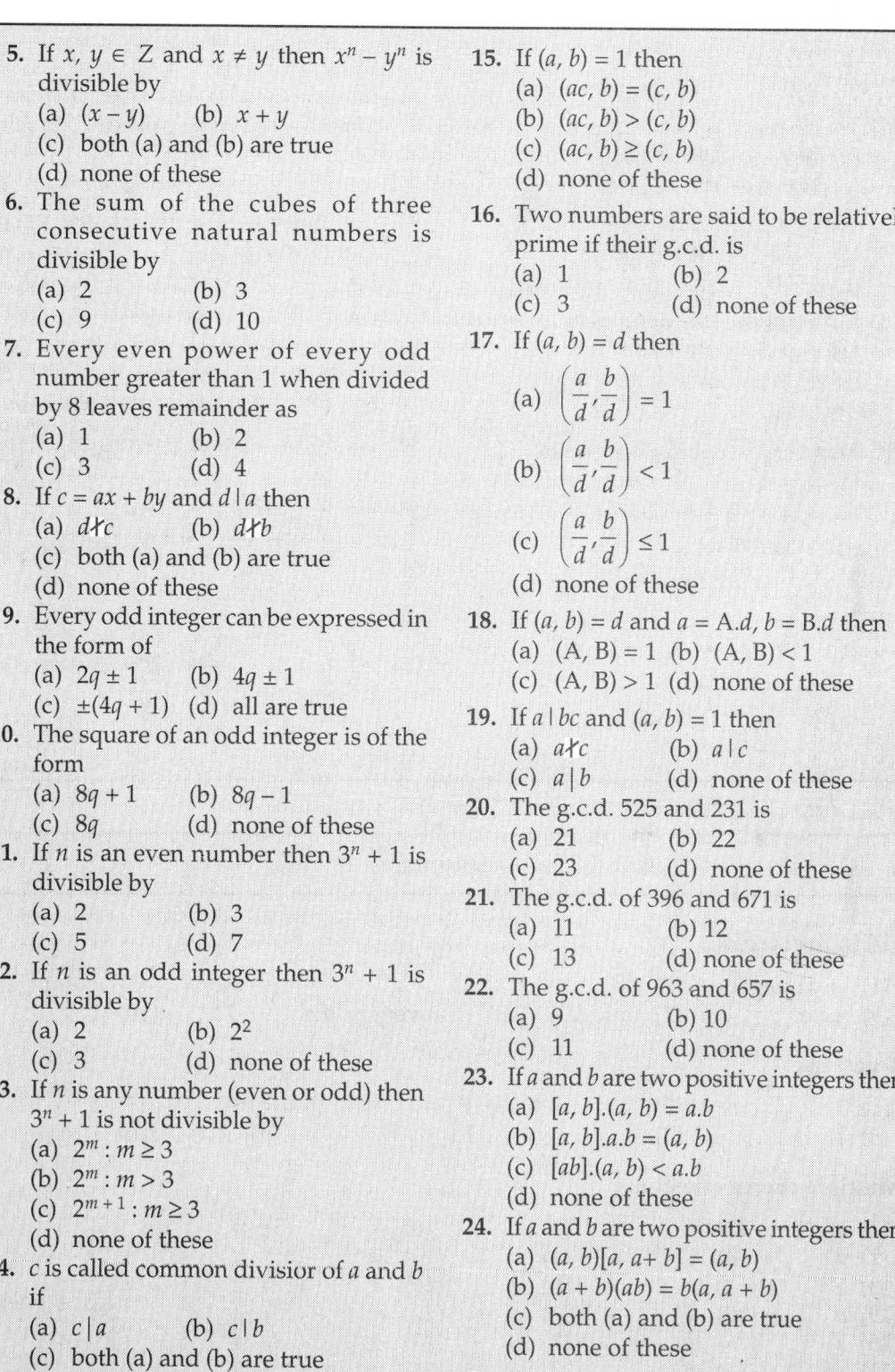

5. If $x, y \in Z$ and $x \neq y$ then $x^n - y^n$ is divisible by
   (a) $(x - y)$ (b) $x + y$
   (c) both (a) and (b) are true
   (d) none of these

6. The sum of the cubes of three consecutive natural numbers is divisible by
   (a) 2   (b) 3
   (c) 9   (d) 10

7. Every even power of every odd number greater than 1 when divided by 8 leaves remainder as
   (a) 1   (b) 2
   (c) 3   (d) 4

8. If $c = ax + by$ and $d \mid a$ then
   (a) $d \nmid c$ (b) $d \nmid b$
   (c) both (a) and (b) are true
   (d) none of these

9. Every odd integer can be expressed in the form of
   (a) $2q \pm 1$ (b) $4q \pm 1$
   (c) $\pm(4q + 1)$ (d) all are true

10. The square of an odd integer is of the form
    (a) $8q + 1$ (b) $8q - 1$
    (c) $8q$ (d) none of these

11. If n is an even number then $3^n + 1$ is divisible by
    (a) 2   (b) 3
    (c) 5   (d) 7

12. If n is an odd integer then $3^n + 1$ is divisible by
    (a) 2   (b) 2^2
    (c) 3   (d) none of these

13. If n is any number (even or odd) then $3^n + 1$ is not divisible by
    (a) $2^m : m \geq 3$
    (b) $2^m : m > 3$
    (c) $2^{m+1} : m \geq 3$
    (d) none of these

14. c is called common divisior of a and b if
    (a) $c \mid a$ (b) $c \mid b$
    (c) both (a) and (b) are true
    (d) none of these

15. If $(a, b) = 1$ then
    (a) $(ac, b) = (c, b)$
    (b) $(ac, b) > (c, b)$
    (c) $(ac, b) \geq (c, b)$
    (d) none of these

16. Two numbers are said to be relatively prime if their g.c.d. is
    (a) 1   (b) 2
    (c) 3   (d) none of these

17. If $(a, b) = d$ then
    (a) $\left(\dfrac{a}{d}, \dfrac{b}{d}\right) = 1$
    (b) $\left(\dfrac{a}{d}, \dfrac{b}{d}\right) < 1$
    (c) $\left(\dfrac{a}{d}, \dfrac{b}{d}\right) \leq 1$
    (d) none of these

18. If $(a, b) = d$ and $a = A.d, b = B.d$ then
    (a) $(A, B) = 1$ (b) $(A, B) < 1$
    (c) $(A, B) > 1$ (d) none of these

19. If $a \mid bc$ and $(a, b) = 1$ then
    (a) $a \nmid c$ (b) $a \mid c$
    (c) $a \mid b$ (d) none of these

20. The g.c.d. 525 and 231 is
    (a) 21   (b) 22
    (c) 23   (d) none of these

21. The g.c.d. of 396 and 671 is
    (a) 11   (b) 12
    (c) 13   (d) none of these

22. The g.c.d. of 963 and 657 is
    (a) 9   (b) 10
    (c) 11   (d) none of these

23. If a and b are two positive integers then
    (a) $[a, b].(a, b) = a.b$
    (b) $[a, b].a.b = (a, b)$
    (c) $[ab].(a, b) < a.b$
    (d) none of these

24. If a and b are two positive integers then
    (a) $(a, b)[a, a+ b] = (a, b)$
    (b) $(a + b)(ab) = b(a, a + b)$
    (c) both (a) and (b) are true
    (d) none of these

25. The l.c.m. of 992 and 550 is

(a) 272800   (b) 2728

(c) 272700   (d) none of these

26. The l.c.m. of 172 and 20 is

(a) 865   (b) 860

(c) 680   (d) none of these

27. The l.c.m. of 16, 10, 15 is

(a) 2450   (b) 2400

(c) 2500   (d) none of these

28. Luca's number is defined by

(a) $L_1 = 1$ (b) $L_2 = 3$

(c) $L_n = L_{n-1} + L_n - 2, \ \forall \ n > 2$

(d) all are true

29. Fibonacci's series is defined by

(a) $a_1 = 1$ (b) $a_2 = 1$

(c) $a_n = a_{n-1} + a_{n-2} \ \forall \ n > 2$

(d) all are true

30. If a_n is the nth term of the fibonacci's

sequence and $\alpha = \dfrac{1+\sqrt{5}}{2}$ then

(a) $a_n > \alpha^{n-1} \ \forall \ n > 1$

(b) $a_n < \alpha^{n-1} \ \forall \ n > 1$

(c) $a_n > \alpha \ \forall \ n > 1$

(d) none of these

31. If n is an odd integer greater than 1 then $n^3 - n$ is a multiple of

(a) 24   (b) 26

(c) 25   (d) none of these

32. If a and b are positive integers such that $(a, b) = [a, b]$

(a) $a \neq b$ (b) $a = b$

(c) $a > b$ (d) $a < b$

33. If $(a, b) = 1$ then $(a - b, a + b)$

(a) 2   (b) 3

(c) 1   (d) 0

34. If $m > 0, n > 0$ and m is an odd integer, then

(a) $(2^m - 1)(2^n + 1) = 1$

(b) $(2^m - 1)(2^n + 1) = 0$

(c) $(2^m - 1)(2^{n+1}) = 1$

(d) none of these

35. If $(a, b) = 1$ then $(a + b, a^2 - ab + b^2) = $

(a) 1   (b) 2

(c) 3   (d) 0

ANSWERS

Fill in the blanks

(1) any integer c   (2) ∤b   (3) 3   (4) not a   (5) an integer

(6) g.c.d.   (7) uniquely (8) relatively prime (9) 21   (10) 2

True/False

(1) T   (2) T   (3) F   (4) T   (5) T

(6) T   (7) T   (8) F   (9) T   (10) T

Multiple choice questions

(1) a   (2) b   (3) c   (4) b   (5) a   (6) b

(7) a   (8) c   (9) d   (10) a   (11) a   (12) b

(13) a   (14) c   (15) a   (16) a   (17) a   (18) a

(19) b   (20) a   (21) a   (22) a   (23) a   (24) b

(25) a   (26) b   (27) b   (28) d   (29) a   (30) a

(31) a   (32) b   (33) c   (34) a   (35) a

CHAPTER 3

Linear Diophantine Equations

- ❖ Diophantine Equations of Second Degree
- ❖ Diophantine Equation of Three or More Unknowns
- ❖ General Solution of Pythagores Equation

3.1 INTRODUCTION

An equation which has two or more than two unknowns is called an indeterminate or Diophantine equation, after the name of Greek mathematician Diophantine. Gener-ally, a system of equation is called indeterminate or Diophantine, if the number of equations is less than that of the unknown. For such type of equation, we only look for the solutions in a restricted class of numbers such as positive integers, negative integers, or integers. In this chapter, we shall discuss the simplest Diopantine equations; when do they have solutions? When do we describe them explicitly.

3.2 LINEAR DIOPHANTINE EQUATION

Definition : *An equation of the form*

$$ax + by = c$$

with $a \neq 0, b \neq 0$ and c integers, is called a linear diophantine equation in two unknowns x and y.

3.3 SOLUTION OF THE EQUATION $ax + by = c$

Here, we want to find all pairs of integers x, y which satisfy the equation

$$ax + by = c \qquad \qquad ...(1)$$

in which a, b and c are given integers with $a \neq 0, b \neq 0$

If $a = b = c = 0$, then every pair (x, y) of integers is a solution of (1), whereas if $a = b = 0$ and $c \neq 0$, then (1) has no solution. The following is a fundamental theorem which lets us know when an indeterminate equation has solutions and when it does not.

THEOREM-1

The linear diophantine equation

$$ax + by = c \qquad \qquad \text{...(1)}$$

If a, b, c are integers, has integer solutions if and only if $d \mid c$ where d = g.c.d. of a and b.
Moreover if $x = x_0$, $y = y_0$ is a particular solution, then any solution can be written as

$$x = x_0 + \frac{b}{d}.t, \ y = y_0 - \frac{a}{d}.t, \text{ where } t \text{ is any integer.}$$

Proof. Let us first suppose $d \mid c$, then we have

$$c = rd, \ \text{ where } r \text{ is any integer.}$$

Now, since $(a, b) = d$, then by definition, there exist integers x_1 and y_1 such that

$$ax_1 + by_1 = d \qquad \qquad \text{...(1)}$$

Multiplying both sides of (1) by $\dfrac{c}{d}$, we have

$$\frac{c}{d}.ax_1 + \frac{c}{d}.by_1 = d.\frac{c}{d} = c$$

$$\Rightarrow \qquad c = a\left(\frac{c}{d}x_1\right) + b\left(\frac{c}{d}y_1\right) = ax + by$$

$$\Rightarrow \qquad \left(\frac{c}{d}x_1\right) \text{ and } \left(\frac{c}{d}y_1\right) \text{ satisfy the equation (1)}$$

Thus, linear diopantine equation has a solution.

Conversely, let us suppose that the equation $ax + by = c$ has a solution, say (x_0, y_0).

Then, $\quad ax_0 + by_0 = c$

But, $\quad ax_0 + by_0$ must be a multiple of d, i.e.

$$ax_0 + by_0 = rd, \text{ where } r \text{ is any integer.}$$

Therefore,

$$c = rd$$

$$\Rightarrow \qquad d \mid c$$

Further, if $x = x_0$, $y = y_0$ is solution of (1), then

$$ax_0 + by_0 = c$$

Subtracting (1) from this equation, we get

$$a(x_0 - x_1) + b(y_0 - y_1) = 0, \ \text{i.e. } a(x_0 - x_1) = -b(y_0 - y_1)$$

$$\Rightarrow \qquad a(x_0 - x_1) = b(y_1 - y_0) \qquad \qquad \text{...(2)}$$

for $\qquad (x, y) = (x_1, y_1)$

Now, since $(a, b) = d$; there exist integers r_1 and r_2 such that $a = r_1 d$, $b = r_2 d$

Putting these values in (2), we get

$$r_1 d[x_1 - x_0] = -r_2 d(y_1 - y_0)$$

$$\Rightarrow \qquad r_1(x_1 - x_0) = -r_2(y_1 - y_0)$$

$$\Rightarrow \qquad \frac{x_1 - x_0}{r_2} = -\frac{y_1 - y_0}{r_1} = t \text{ (some integer)} \qquad \qquad \text{...(3)}$$

Therefore, by division algorithm, we can write

$$y_1 = y_0 - tr_1, \text{ for some integer } t$$

(or) $\quad y_1 = y_0 - \dfrac{a}{d} t$

Now, from (3), we have

$$r_1(x_1 - x_0) = r_2 r_1 t$$
$$\Rightarrow \qquad x_1 - x_0 = + r_2 t$$
$$\Rightarrow \qquad x_1 = x_0 + r_2 t$$
$$= x_0 + \dfrac{b}{d} t$$

Hence, $x_1 = x_0 + \dfrac{b}{d} t$ and $\quad y_1 = y_0 - \dfrac{a}{d} t$ is the general solution of (1).

DEDUCTIONS :

(1) If $(a, b) = 1$, the solution (1) can be written as $x = x_0 + bt$, $y = y_0 - at$ where, $x = x_0$, $y = y_0$ is a solution of (1)

(2) If (x_0, y_0) is one solution of $ax + by = c$, $(a, b) = d$, then

$$x_1 = x_0 + \dfrac{b}{d} . t, \ y_1 = y_0 - \dfrac{a}{d} . t$$

SOLVED EXAMPLES

Example 1. *Determine if the linear diophantine equation* (i) $12x + 18y = 30$, (ii) $2x + 3y = 4$ *and* (iii) $6x + 8y = 25$ *are solvable.*

Solution. Comparing the given equation with $ax + by = c$, we have

(i) $a = 12$, $b = 18$, $c = 30$ and $(12, 18) = 6$ and $6 \mid 30$, so the linear diophantine equation $12x + 18y = 30$ has a solution.

(ii) $a = 2$, $b = 3$, $c = 4$ and $(2, 3) = 1$ and $1 \mid 4$ so the linear diophantine equation $2x + 3y = 4$ also have a solution.

(iii) $a = 6$, $b = 8$, $c = 25$ and $(6, 8) = 2$ but $2 \nmid 25$, so the linear diophantine equation $6x + 8y = 25$ is not solvable.

Example 2 (*Mahavira Puzzle*). *Twenty three weary travelers entered the outskirts of a lush green and beautiful forest. They found 63 equal heaps of plantains (fruit) and seven single fruits. They divided them equally. Find the number of fruits in each heap.*

Solution. Let x denote the number of plantains in a heap and y the number of plantains received by a traveler. Then according to the given problem, we get the diophantine equation

$$63x + 7 = 23y$$

The linear equation in Mahavira's puzzle is

$63x - 23y = -7$, g.c.d. of 63 and 23 is 1, i.e. $(63, 23) = 1$, so diophantine equation has the solution of type $x = x_0 + bt$ and $y = y_0 - at$.

To find a particular solution x_0, y_0, first we express the g.c.d. 1 as a linear combination of 63 and 23. To do this, we apply Euclidean algorithm.

$$63 = 2(23) + 17$$
$$23 = 1(17) + 6$$
$$17 = 2(6) + 5$$
$$6 = 1(5) + 1$$
$$5 = 5(1) + 0$$

Now, use the first four equations in reverse order

$$1 = 6 - 1(5)$$
$$= 6 - 1[(17 - 2(6)]$$
$$= 3(6) - 1(17)$$
$$= 3(23 - 1(17) - 1(17)$$
$$= 3 . 23 - 4 . 17$$
$$= 3 . 23 - 4(63 - 2 . 23)$$
$$= -4 . 63 + 11 . 23$$

Multiply both side by -7 to get -7×1 for R.H.S. of Diophantine equations

$$-7 = (-7)(-4)63 + (-7) . 11 . 23$$
$$= 63 . 28 - 23 . 11 \text{ which shows that}$$
$$x_0 = 28 \text{ and } y_0 = 77$$

Hence $x = x_0 + bt = 28 - 23t$ and $y = y_0 - at = 77 - 63t$ are general solution of given diophantine equation where t is an artibtrary integers.

Example 3. *Find the general solution of* $70x + 112y = 168$.

Solution. Here, the given equation is

$$70x + 112y = 168 \qquad \qquad \qquad ...(1)$$

Firstly, we shall find the g.c.d of 70 and 112 in the following manner,

$$112 = 70(1) + 42$$
$$70 = 42(1) + 28$$
$$42 = 28(1) + 14$$
$$28 = 14(2) + 0$$
$$\Rightarrow \qquad (70, 112) = 14$$

Since, $14 \mid 168$, therefore, equation (1) has a solution.

Dividing (1) by 14, we get

$$5x + 8y = 12$$

We can easily see that $x = -4$ and $y = 4$ satisfy the above equation. Its general solution is given by

$$x_1 = x_0 + \frac{112}{14}t = -4 + 8t$$

and $\qquad y_1 = y_0 - \frac{70}{14}t = 4 - 5t$

Example 4 . *Solve the diophantine equation*

$$525x + 231y = 42$$

Solution. We can easily find that

$$(525, 231) = 21$$

Therefore, dividing the given equation by 21, we get

$$25x + 11y = 2$$

Again, as $(25, 11) = 1$, by the Euclid's algorithm, we have
$$25(4) + 11(-9) = 1$$
Hence, $x = 2 \cdot 4 = 8, y = 2(-9) = -18$ is a solution of the given equation.
Therefore, the required general solution is given by
$$x = 8 + 11t, \quad y = -18 - 25\,t$$
Clearly, there are no positive integer solutions.

3.4 EULER'S METHOD FOR SOLVING LINEAR DIOPHANTINE EQUATIONS :

There are so many ways of obtaining a particular solution. When the coefficient of (1) are not large, it can sometimes be found by inspection. Beside this, we use the process of successively diminishing the coefficients. To make it more clear, see the following examples.

SOLVED EXAMPLES

Example 1. *Find the positive integer solution of*
$$7x + 19y = 213$$

Solution. Dividing the given equation by the smaller coefficient 7, we get
$$x = \frac{213 - 19y}{7} = 30 - 2y + \frac{3 - 5y}{7}$$
Since x is an integer, y is also an integer. Therefore
$$\frac{3 - 5y}{7} = u \text{ is also an integer. Now, we have}$$
$$5y + 7u = 3$$
Dividing it by 5, we have
$$y = \frac{3 - 7u}{5} = -u + \frac{3 - 2u}{5}$$
or $\qquad 2u + 5v = 3$

Clearly, $u = -1, v = 1$ is a solution. Hence, $x = 25, y = 2$. Thus, the general solution of the given equation is
$$x = 25 + 19t, \quad y = 2 - 7t$$
Since, we require the solution to be positive, i.e.
$$25 + 19t > 0, \quad 2 - 7t > 0$$
We require
$$-\frac{25}{19} < t < \frac{2}{7}$$
Thus, $t = 0$ or $t = -1$. Hence, the required positive integer solutions are
$$x = 25, y = 2, \quad x = 6, y = 9$$

Example 2. *Solve the linear diopantine equation* $1076x + 2076y = 3076$ *by Euler's method.*

Solution. Clearly, $(1076, 2076) = 4$ and $4 | 3076$

$\Rightarrow \qquad$ given equation has infinitely many solutions

Divide the given equation by smallest coeffiecient, i.e. 1076, we can write
$$x = \frac{-2076y + 3076}{1076}$$

$$= -y + 2 + \frac{-1000y + 924}{1076} \quad \text{(By division algorithm)} \qquad ...(1)\backslash$$

Let $u = \dfrac{-1000y + 924}{1076}$ then we have $1076u + 1000y = 924$

$$\Rightarrow \qquad y = \frac{-1076u + 924}{1000} = -u + \frac{-764 + 924}{1000} \quad \text{(By division algorithm)} \qquad ...(2)$$

Let $v = \dfrac{-76u + 924}{1000} \Rightarrow 76u + 1000\,v = 924 \Rightarrow u = \dfrac{-1000v + 924}{76} = -13v + 12 + \dfrac{(-12)v + 12}{76}$

$$\text{(By division algorithm)} \qquad ...(3)$$

Now let $w = \dfrac{-12v + 12}{76} \Rightarrow 12v + 76w = 12 \Rightarrow v = \dfrac{76w + 12}{12} = -6w + 1 - \dfrac{w}{3}$

To find a particular solution, we let $t = 0$; then $w = 0$ and work through the equation (1), (2) and (3) in reverse order, we get

$$v = -6w + 1 - \frac{w}{3} = -6(0) + 1 - 0 = 1; u = \frac{-1000v + 924}{76} = \frac{-1000 + 924}{76} = -1$$

$$y = -\frac{-1076u + 924}{1000} = \frac{1076 + 924}{1000} = 2 \; ; x = \frac{-2076y + 3076}{1076}$$

$$= \frac{-4152 + 3076}{1076} - 1$$

We may verify that $x_0 = -1$, $y_0 = 2$ is the solution of given linear diophantine equation
To find the general solution, we use successive substitution in reverse order

We get $w = 3t$, $u = -6w + 1 - \dfrac{w}{3} = -19t + 1$

$$u = -13v + 12 + w = 250t - 1$$

$$y = -4 + v = -269t + 2$$

$$x = -y + 2 + u = 519t - 1$$

Hence, the general solution is given by

$$x = 519t - 1, \; y = -269t + 2$$

THEOREM-2

If $ax + by = c$, $(a, b) = 1$ and b is numerically smaller of the two coefficients a and b and a_1 and c_1 are the minimal remainders of a and c respectively with respect to $|b|$. Then, (1) can be written in the form ...(1)

$$a_1 x + |b| x_1 = c_1$$

where $|a_1| \le \dfrac{|b|}{2}$ and $|c| \le \dfrac{|b|}{2}$

Proof. As per given, a_1 and c_1 are minimal remainders of a and c with respect to $|b|$, we have

$$a = |b| q_1 + a_1 \qquad 0 < |a_1| \le \frac{|b|}{2}$$

$$c = |b| q_2 + c_1 \qquad 0 < |c_1| \le \frac{|b|}{2}$$

Therefore, (1) reduces to

$$(|b|q_1 + a_1)x + by = |b|q_2 + c_1$$

or $\quad a_1 x + |b|\left(q_1 x + \dfrac{b}{|b|}y - q_2\right) = c_1$

Putting $x_1 = q_1 x + \dfrac{b}{|b|}y - q_2$, the above equation reduces to

$$a_1 x + |b|x_1 = c_1$$

SOLVED EXAMPLES

Example 1. *Find the general solution of* $21x + 13y = 1791$ *(By Euler method.)*

Solution. Here, we have

$$21x + 13y = 1791 \qquad \qquad \text{...(1)}$$

$$\Rightarrow \qquad y = \frac{1791 - 21x}{13} \qquad \qquad [\because 13 < 21]$$

$$= \frac{138 \times 13 - 3 - 26x + 5x}{13}$$

$$= 138 - 2x + \frac{5x - 3}{13} \qquad \qquad \text{...(2)}$$

Putting $y_1 = \dfrac{5x - 3}{13}$, we get

$$13y_1 - 5x = -3$$

$$\Rightarrow \qquad x = \frac{3 + 13y_1}{5} \qquad \qquad \text{...(3)}$$

$$= \frac{5 - 2 + 15y_1 - 2y_1}{5} = 1 + 3y_1 - \frac{(2 + 2y_1)}{5}$$

Now, putting

$$x_1 = \frac{2 + 2y_1}{5}$$

$$\Rightarrow \qquad 5x_1 = 2 + 2y_1, \text{ i.e. } y_1 = \frac{5x_1 - 2}{2}.$$

Putting $x_1 = 0$, we get $y_1 = -1$. Therefore, from (3), we have

$$x = \frac{3 + 13(-1)}{5} = \frac{3 - 13}{5} = -\frac{10}{5} = -2$$

Again from (2), we get

$$y = \frac{1791 - 21(2)}{13} = \frac{1791 + 42}{13} = \frac{1833}{13} = 141$$

Hence, the general solution of (1) is given by

$$x = -2t, \ y = 141 + 13t$$

SOLVED EXAMPLES

Example 1. *Find the possible solution of* $11x + 5y = 79$.

Solution. Clearly, we have $(11, 5) = 1$.

Now, $11 = 5 . 2 + 1$
$\qquad$ $79 = 5 . 16 - 1$

Then, (1) can be written as
$\qquad (5 . 2 + 1) x + 5y = 5 \times 16 - 1$
$\Rightarrow \qquad 5[2x + y - 16] + x = -1$
$\Rightarrow \qquad 5u + x = -1$, where $u = 2x + y - 16$

Putting $u = 0$, we get $x = -1$, then from (1), we get $y = 18$.
$\Rightarrow \qquad x = -1, y = 18$ is one solution.

The general solution is given by
$\qquad x = -1 + 5t, \ y = 18 - 11t$

Since, we require the solution to be positive, therefore, we have to find the value of t for which x and y are positive.

Putting $t = 1$, we get
$\qquad x = -1 + 5 \times 1 = 4$
$\qquad y = 18 - 11 \times 1 = 7$

Further, for $t \geq 2$, y will be positive.

Hence, the only positive solution is $x = 4$ and $y = 7$.

Example 2. *Find the general solution of* $311x - 112 y = 73$.

Solution. Here the given equation is
$\qquad 311x - 112y = 73$ $\hspace{4cm}$ ...(1)
Also, $(311, 112) = 1$
Here, 112 is numerically smaller.

Consider 311 and 73. Then
$\qquad 311 = 112(3) - 25$
$\qquad 73 = 112(1) - 39$

Then, (1) reduces to
$\qquad (112 \times 3 - 25)x - 112y = 112(1) - 39$
$\Rightarrow \qquad -25x + 112(3x - y - 1) = -39$
$\Rightarrow \qquad -25x + 112u = -39$ $\hspace{4cm}$ ...(2)
where, $u = 3x - y - 1$

Equation (2) is satisfied by $u = 3, x = 15$.
Thus, we have
$\qquad -25x + 112u = -39, \Rightarrow u = 3, \ x = 15$ $\hspace{2cm}$ ...(3)

In this equation neither of the two coeffiecient is unity. Therefore (2) can be treated in the same way as (1).

Consider 112 and 39.
$\qquad 112 = 25(4) + 12$
$\qquad 39 = 25(2) - 11$

Then, (2) can be written as
$\qquad -25x + (25 \times 4 + 12)u = -(25(2) - 11)$
$\Rightarrow \qquad 25(-x + 4u + 2) + 12u = 11$
$\Rightarrow \qquad 25v + 12u = 11$ $\hspace{4cm}$ ...(4)
where, $v = -x + 4u + 2$

Equation (4) is satisfied by $v = -1$ and and $u = 3$.

Therefore, we have

$$25v + 12u = 11$$
$$v = -1$$
$$u = 3 \qquad \qquad ...(5)$$

Again, neither of the two coefficient is unity. Therefore, consider 25 and 11 such that

$$25 = 12 \times 2 + 1$$
$$11 = 12 \times 1 - 1$$

Then, (4) can be written as

$$(12 \times 2 + 1)v + 12u = 12(1) - 1$$
$$\Rightarrow \qquad 12[2v + u - 1] + v = -1$$
$$\Rightarrow \qquad 12w + v = -1$$

where, $w = 2v + u - 1$ \qquad \qquad ...(6)

Here, one coefficient is unity. Putting $w = 0$, we get

$$v = -1$$
$$u = 3$$
$$x = 15$$
$$y = 41$$
$$\Rightarrow \qquad x = 15 \text{ and } y = 41 \text{ is one solution.}$$

Hence, the general solution is given by

$$x = 15 - 112t, \quad y = 41 - 311t, \quad t \text{ is integer.}$$

Example 3. *Find the solution of linear Diophantine equation* $172x + 20y = 1000$.

Solution. Applying the Euclidean algorithm to find the g.c.d. (172, 20), we find that

$$172 = 8 \,.\, (20) + 12$$
$$20 = 1 \,.\, (12) + 8$$
$$12 = 1 \,.\, (8) + 4$$
$$8 = 2 \,.\, (4)$$

so g.c.d. (172, 20) = 4.

Also $4 \mid 1000$, so the solution of given problem exist.

To obtain the integer 4 as a linear combination of 172 and 20, we proceed backward through the previous calculation as follows :

$$4 = 12 - 8$$
$$= 12 - (20 - 12)$$
$$= 2 \,.\, 12 - 20$$
$$= 2(172 - 8 \,.\, 20) - 20$$
$$= 2 \,.\, 172 + (-17) \,.\, 20$$

Multiply by 250 on both sides, we get

$$1000 = 250[\, 2 \,.\, 172 + (-17) \,.\, 20]$$
$$= 500 \,.\, 172 + (-4250) \,.\, 20$$

so that $x = 500$ and $y = -4250$ provide one solution of the given Diophantine equation. All other solutions are given by

$$x = 500 + \left(\frac{20}{4}\right)t = 500 + 5t$$

$$y = -4250 - \left(\frac{172}{4}\right)t = -4250 - 43t, \text{ for some integer } t.$$

To find the positive integers solution, we must choose the value of *t* such that it satisfy the inequalities

$$500 + 5t > 0; \ -43t - 4250 > 0$$

Solving these, we get

$$-98\frac{36}{43} > t > -100$$

so integer value of $t = -99$

Thus, only positive solution is
$$x = 500 + 5(-99) = 5$$
$$y = -4250 - 43(-99) = 7$$

3.5   DIOPHANTINE EQUATION IN THREE OR MORE UNKNOWNS

We can solve the Diophantine equation in three or more unknowns in a similar manner. The whole process can be understood by the following example.

Example 1. *Solve* $50x + 45y + 36z = 10$.

Solution. The given equation can be decomposed into two equations given by
$$50x + 45y = 5t, \ \ 5t + 36z = 10$$

As $50t + 45(-t) = 5t, \ 5(-70) + 36(10) = 10$, the solutions of the above two equations are respectively given by

$$\begin{cases} x = t + 9k_1, & t = -70 + 36k_2 \\ y = -t - 10k_1, & z = 10 - 5k_2 \end{cases}$$

On eliminating t, we get the required solutions
$$x = -70 + 9k_1 + 36k_2$$
$$y = 70 - 10k_1 - 36k_2$$
$$z = 10$$
or
$$x = 2 + 9k_1 + 36k_2$$
$$y = -2 - 10k_1 - 36k_2$$
$$z = -5k_2, \text{ where } k_1 \text{ and } k_2 \text{ are any integers.}$$

Aliter. The above equation can be solved by the process of successively eliminating the coefficients.

Since 36 is the smallest coefficient, therefore we can write the given equation as
$$36(x + y + z) + 14x + 9y = 10$$
Let
$$x + y + z = k_1, \text{ then}$$
$$14x + 9y + 36k_1 = 10$$
i.e.
$$9(x + y + 4k_1) + 5x = 10$$
Again, let
$$x + y + 4k_1 = 5k_2, \text{ we get}$$
$$5x + 45k_2 = 10, \text{ i.e. } x + 9k_2 = 2$$

Hence, the required soution is given by
$$x = 2 - 9k_2$$
$$y = -2 - 4k_1 + 14k_2$$
$$z = 5k_1 - 5k_2$$
where k_1 and k_2 are any integers.

THEOREM-1

The linear Diophantine equation
$$a_1 x_1 + \ldots + a_n x_n = c, \ a_i, \ c \ being \ all \ integers$$
has integer solution if and only if
$$d = (a_1, \ldots, a_n) \mid c.$$

Proof. The necessary part of the theorem is obvious. Now, we prove the sufficient part.

Let, $b_1, b_2, \ldots, b_n$ be such that
$$a_1 b_1 + \ldots + a_n b_n = d$$
Putting $c = dc_1$, we have
$$a_1(b_1 c_1) + \ldots + a_n(b_n c_1) = dc_1 = c$$
$$\Rightarrow \qquad x_1 = b_1 c_1, \ldots, x_n = b_n c_1 \text{ is the solution of the given equation.}$$

THEOREM-2

Let a and b be two relatively positive primes, then the necessary and sufficient condition that the equation
$$ax + by = n, \ 0 \le n < ab \qquad\qquad \ldots(1)$$
has no non-negative integer solution.
$$n = ab - ka - la; \ k, \ l \ being \ integers.$$

Proof. Let $n = ab - ka - la, \ 0 \le n < ab, \ k \ge 1, l > 1$. If $x \ge 0, y \ge 0$ is the integer solution of (1), i.e. $ax + by = ab - ka - lb$, then
$$a(x + k) + b(y + l) = ab$$
Now, $\qquad b \mid (x + k), a \mid (y + l)$

$\therefore \qquad x + k \ge b, y + l \ge a$

or $a(x + k) + b \ (y + l) > 2ab$, which is a contradiction. Hence, the sufficient condition holds. Now, we shall prove the necessary condition.

Consider the following three properties :

  (1) If (1) has non-negative integer solution, the solution is unique.

  (2) Find the number of n such that (1) has non-negative integer solution. If (1) has solution, the solution is unique, therefore, the solution (x, y) are in one-one correspondence with n. Therefore, the number of n is equal to the number of integer points which are in the region as shown in Fig. 1.

Clearly, this number is $\dfrac{(a+1)(b+1)-2}{2}$.

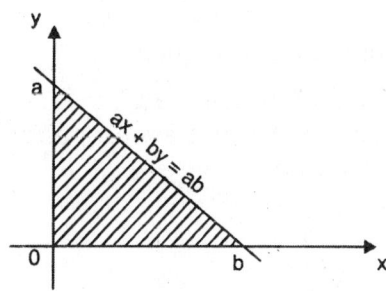

Fig. 1

(3) Find the number n such that
$$n = ab - a - lb, \ 0 \le n < ab, \ k, l \ge 1$$

This num1ber is not greater than the number of equations of the form (1) which have no non-negative integer solution. By the same reason, the number of n equals that for the integer points (k, l) in the region as shown in Fig. 2.

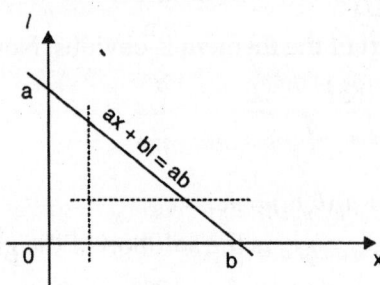

Fig. 2.

Clearly, it is $\dfrac{(a-1)(b-1)}{2}$.

Hence, the total number of n such that (1) has non-negative integer solution, (whether the positive or not) is
$$\frac{(a+1)(b+1)-2}{2} + \frac{(a-1)(b-1)}{2} = ab$$

It is just the total number of integers in the interval $[0, ab[$, i.e. (1) has no non-negative integer solution only when n is of the form as in 3.

Hence, the necessary condition holds.

AN IMPORTANT RESULT :

Let a, b and c be integers with not both a and b equal to 0 and let $g = $ g.c.d. (a, b).

If $g \nmid c$, then the equation
$$ax + by = c \qquad\qquad\qquad \text{...(1)}$$

has no solution in integers. If $g \mid c$, then this equation has infinitely many solutions. If the pair (x_1, y_1) is one integral solution, then all others are of the form $x = x_1 + kb \mid g$, $y = y - ka \mid g$, where k is any integer.

3.6   DIOPHANTINE EQUATION OF THE SECOND DEGREE

The Diophantine equation of the second degree
$$x^2 + y^2 = z^2 \qquad\qquad\qquad \text{...(1)}$$
is called the Shang-gao or Pythagores equation.

If one of x, y, z is zero, say $x = 0$, then $y = \pm z$.

Therefore, we may assume that x, y, z are all positive. Also, we may assume that $(x, y) = 1$, for if $(x, y) = d$, then $d \mid z$ and
$$\left(\frac{x}{d}\right)^2 + \left(\frac{y}{d}\right)^2 = \left(\frac{z}{d}\right)^2$$

where $\left(\dfrac{x}{d}, \dfrac{y}{d}\right) = 1$. Hence, if we can find the solution of (1), with $(x, y) = 1$. its general solution is easily obtained.

Also, if $(x, y) = 1$, then one of x and y must be even and other must be odd.

If x, y are both odd, then $x^2 + y^2 \equiv 1 \pmod 4$ and $z^2 = x^2 + y^2 \equiv 2 \pmod 4$ which is not possible, since $z^2 = 1$ or $0 \pmod 4$. Hence, we may assume y to be even.

3.7 GENERAL INTEGER SOLUTION OF PYTHAGORES EQUATION

The general integer solution of Pythagores equation
$$x^2 + y^2 = z^2 \qquad \qquad \text{...(1)}$$
with $x > 0$, $y > 0$, $z > 0$, $(x, y) = 1$ and y is even, is given by
$$x = a^2 - b^2,\ y = 2ab,\ z = a^2 + b^2 \qquad \qquad \text{...(2)}$$
where, $a > b > 0$, $(a, b) = 1$ and one of a, b is odd, and other even.

Proof. Putting the values of (2) in (1), we can easily verify that (2) is the solution of (1). Now, it remains to prove that every solution that exist from suitably chosen a and b satisfies the conditions of the theorem.

Let x, y, z be a solution satisfying the given conditions. Since, y is even, from (1), we have

$$\frac{z+x}{2} \cdot \frac{z-x}{2} = \left(\frac{y}{2}\right)^2 \qquad \qquad \text{...(3)}$$

But as
$$\left(\frac{z+x}{2}, \frac{z-x}{2}\right)\left(\frac{z+x}{2} + \frac{z-x}{2}\right) = z$$

$$\left(\frac{z+x}{2}, \frac{z-x}{2}\right)\left(\frac{z+x}{2} - \frac{z-x}{2}\right) = x$$

and
$$(x, y) = 1$$
$$\left(\frac{z+x}{2}, \frac{z-x}{2}\right) = 1$$

Let us assume
$$\frac{z+x}{2} = a^2,\ \frac{z-x}{2} = b^2$$

Clearly, $a > b$, $(a, b) = 1$. Thus
$$x = a^2 - b^2,\ z = a^2 + b^2,\ y = 2ab$$

Since z is odd, one of a, b is even and other is odd. Therefore, a, b satisfy the condition given in theorem.

Remarks

❖ In general, if $(x, y) = d$, then $(x, y, z) = d$. Thus, the general solution of (1) is given by $x = \pm d(a^2 - b^2)$, $y = \pm 2abd$, $z = \pm d(a^2 + b^2)$.

❖ All rational points on the unit circle can be written as
$$\left(\pm \frac{a^2 - b^2}{a^2 + b^2}, \pm \frac{2ab}{a^2 + b^2}\right)$$

or
$$\left(\pm \frac{2ab}{a^2 + b^2}, \pm \frac{a^2 - b^2}{a^2 + b^2}\right),\ \text{where } (a, b) = 1.$$

3.8   GENERAL INTEGER SOLUTION OF THE EQUATION
$x^2 + y^2 + z^2 = w^2$, $(x, y, z, w) = 1$

THEOREM-1

The general integer solution of the equation
$$x^2 + y^2 + z^2 = w^2, \ (x, y, z, w) = 1$$
is given by
$$x = a^2 - b^2 + c^2 - d^2, \ y = 2ab + 2cd$$
$$z = 2ab - 2bc, \qquad w = a^2 + b^2 + c^2 + d^2$$

Proof. Without loss of any generality, we may assume that y, z are both even and x, w both are odd. Let $\left(\dfrac{y}{2}, \dfrac{z}{2}\right) = r$, $(x, w) = s$ and $y = 2ry_1$, $z = 2rz_1$, $x = sx_1$, $w = sw_1$. As x_1, w_1 are both odd, the given equation reduces to

$$r^2(y_1^2 + z_1^2) = s^2 \frac{w_1 + x_1}{2} \cdot \frac{w_1 - x_1}{2}$$

Since, $(r, s) = 1$, $\left(\dfrac{w_1 + x_1}{2}, \dfrac{w_1 - x_1}{2}\right) = (w_1 - x_1) = 1$, we have

$$r = r_1 \cdot r_2 \cdot (r_1, r_2) = 1$$

$$r_1^2 \left| \frac{w_1 + x_1}{2} \right., \ r_2^2 \left| \frac{w_1 - x_1}{2} \right.$$

Now, $$y_1^2 + z_1^2 = s^2 \frac{w_1 + x_1}{2r_1^2} \cdot \frac{w_1 - x_1}{2r_2^2}$$

We know that odd prime factors of $y_1^2 + z_1^2$ are one of the form $4k + 1$. Further, prime factor $4k + 1$ can be expressed as the sum of two square numbers. Thus, there are integers, a_1', b_1', c_1', d_1' such that

$$\frac{w_1 + x_1}{2r_1^2} = a_1'^2 + b_1'^2 = c_1'^2 + d_1'^2$$

$\Rightarrow$ $$y_1^2 + z_1^2 = (a_1^2 + b_1^2)(c_1^2 + d_1^2)$$

Now, $$y_1^2 + z_1^2 = (y_1 + (\sqrt{-1})z_1)(y_1 - (\sqrt{-1})z_1)$$

and $$y_1 + \sqrt{-1}.z_1 = (a_1 + \sqrt{-1}.c_1)(b_1 + \sqrt{-1}.d_1)$$

$$y_1 - \sqrt{-1}.z_1 = (a_1 - \sqrt{-1}.c_1)(b_1 - \sqrt{-1}.d_1)$$

Then, we get
$$y_1 = a_1 b_1 - c_1 d_1, \qquad z_1 = a_1 d_1 + b_1 c_1$$

$$x_1 = \frac{1}{s}\left\{r_1^2(a_1^2 + c_1^2) - r_2^2(b_1^2 + d_1^2)\right\}$$

$$w_1 = \frac{1}{s}\left\{r_1^2(a_1^2 + c_1^2) + r_2^2(b_1^2 + d_1^2)\right\}$$

Hence, we have

$$x = a^2 - b^2 + c^2 - d^2, \qquad y = 2ab + 2cd$$
$$z = 2ad + 2bc, \qquad w = a^2 + b^2 + c^2 + d^2$$

where, $a = r_1 a_1$, $b = r_2 b_1$, $c = r_1 c_1$, $d = r_2 d_1$ are all non-negative integers.

THEOREM-2

If x, y, z are solution of $x^2 + y^2 = z^2$ and $(x, y, z) = 1$, then of x, y, there is a multiple of 3 and a multiple of 4. Also, one of x, y, z is a multiple of 5. Hence, xyz is a multiple of 60.

Proof. Let us suppose that both x, y are not multiple of 3. Further, since an integer not a multiple of 3 is of the form $3n \pm 1$, with square
$$(3n \pm 1)^2 = 3(3n^2 \pm 2n) + 1 = 3k + 1$$

therefore, $x^2 + y^2$ is of the form $3k - 1$, which is not a square. It is impossible since z^2 is a square. Therefore, one of the x, y is a multiple of 3.

Further, suppose that x, y both are not multiple of 4. But we know that an integer not a multiple of 4 of the form $4n + 1$ or $4n \div 2$ with squares
$$(4n \pm 1)^2 = 8(2n^2 \pm n) + 1 = 8k + 1$$
$$(4n \pm 2)^2 = 8k + 4$$

If x, y are both of the form $4n \pm 1$ or one of the form $4n \pm 1$ and the other $4n + 2$, then $x^2 + y^2$ is of the form $8k + 2$ or $8k + 5$, which are not squares.

Hence, one of x, y must be a multiple of 4.

Again if none of x, y, z is a multiple of 5, then they must have the forms $5n \pm 1$, $5n \pm 2$. But as
$$(5n \pm 1)^2 = 5k + 1, \quad (5n \pm 2)^2 = 5k - 1$$
$x^2 + y^2$ is of the form $5k - 2, 5k - 2$ or $5k$. In the former, $x^2 + y^2$ is not a square, in the latter, z is a multiple of 5, these contradict the assumption and therefore, one of x, y, z must be a multiple of 5.

3.9 SOLUTION OF GENERAL DIOPHANTINE EQUATION OF SECOND DEGREE

Here, we will discuss, how to solve the general Diophantine equation of the second degree, by illustrating the method as in following example.

SOLVED EXAMPLES

Example 1. *Find the integer solution of*
$$x^2 + xy + y^2 - 3x + 4y - 6 = 0.$$

Solution. We can treat the given equation as a quadratic equation in x. Then, we have

$$x = \frac{1}{2}\left\{-(y-3) \pm \sqrt{5y^2 - 22y + 33}\right\}$$

Since, x is an integer, $5y^2 - 22y + 33$ must be a square. We can easily see that when $y = 1$, it is a square number 16. Thus,

$$x = \frac{1}{2}(2 \pm 4) = 3, -1$$

Therefore, the required solutions are given by

$$x = 3, \quad x = -1,$$
$$y = 1, \quad y = 1$$

Similarly, when $y = 2$, it is again a square number 9, and when $y = 4$, it is a square number 25. Thus, we also obtain the required solution as

$$x = 2, y = 2, \quad x = -1, y = 2 \quad x = 2, y = 4 \text{ and } x = -3, y = 4$$

3.10   EQUATION $x^n + y^n = z^n$

The equation $x^n + y^n = z^n$ has no solution with non-zero integers x, y and z if $n > 2$. This is the well known Fermat's last or Fermat's great theorem. We know that any integer greater than 2 is divisible by 4 or by an odd prime. Thus, if we can prove that when $n = 4$ and when n is any prime, the equation has no solution, then Fermat's last theorem is true.

THEREOM-1

The diophantine equation $x^4 + y^4 = z^2$ has no solution with non-zero positive integers x, y and z.

Proof. Let if possible

$$(x^2)^2 + (y^2)^2 = z^2$$

has positive integer solution. Suppose $(x, y) = 1$, y is even and that z is the smallest of the z in all solutions. We can easily prove that $x^2 = a^2 - b^2$, $y^2 = 2ab$, $z = a^2 + b^2$.

where, $(a, b) = 1$ and exactly one of a, b is even. If a is even, then b is odd.

Therefore,

$$1 = x^2 = a^2 - b^2 \equiv -1 \pmod 4, \text{ which is not possible.}$$

Therefore, a must be odd and b even.

Again, we obtain

$$x = p^2 - q^2, \quad b = 2pq, \quad a = p^2 + q^2, \text{ where } (p, q) = 1, p > q > 0$$

and both p and q are not odd.

For $y^2 = 2ab$, we have

$$y^2 = 2pq(p^2 + q^2)$$

Since, any two of p, q and $p^2 + q^2$ are relatively prime and hence each must be a square. i.e.

$$p = r^2, \quad q = s^2, \quad p^2 + q^2 = t^2$$

We then have

$$r^4 + s^4 = t^2$$

Now, $\quad z = a^2 = b^2 > a = t^2 > t$

i.e. z is greater than t.

which is a contradiction. Therefore, there is no non-zero solution.

Remark

* Similarly, we can prove that for any positive integer n, the Diophantine equation $x^n + y^n = z^n$ has no positive solution larger than n.

SOME MORE SOLVED EXAMPLES

Example 1. *Show that the positive integer solution of* $x^{-1} + y^{-1} = z^{-1}$, $(x, y, z) = 1$ *must have the form*

$$x = a\,(a + b), \quad y = b(a + b), \quad z = ab.$$

where, a, b > 0, (a, b) = 1

Solution. If (x, y, z) is a solution of the given equation $(x, y) = c$, i.e. $x = ca$, $y = cb$,

$(a, b) = 1$. Then $z^{-1} = x^{-1} + y^{-1} = \dfrac{a+b}{cab}$, thus

$$z = \frac{cab}{a+b}$$

Since, $(a, b) = 1$ is given, therefore $(ab, a + b) = 1$. Thus,

$$(a + b) \mid c$$

Setting $c = c'\,(a + b)$, then $z = c'ab$

Now, since $(x, y, z) = 1$, we have

$$(ca, cb, c'ab) + (c(a, b), c'ab) = (c'(a + b), c'(ab)) = c' = 1$$

i.e. $\quad c = (a + b), c' = 1.$

Example 2. *Show that the positive integer solution of* $x^2 + 2y^2 = z^2$, $(x, y) = 1$, *can be expressed as*

$$x = \pm (a^2 - 2b^2), \quad y = 2ab, \quad z = a^2 + 2b^2.$$

Solution. From the given equation, we can easily find

$$2y^2 = z^2 - x^2 = (z - x)(z + x)$$

Since, $(x, z) = 1$, then $(z + x, z - x) = 1$ or 2.

Now, since x, y can only be both odd, hence $(z + x, z - x) = 2$.

Therefore,

$$y^2 = \frac{z+x}{2}\,(z - x) \qquad\qquad ...(1)$$

But, $z + x, z - x$ can not both be multiple of 4. If $z + x$ is not so, then $\dfrac{z+x}{2}$ is odd, so that

$\left(\dfrac{z+x}{2}, z-x\right) = 1$. Therefore, odd $\dfrac{z+x}{2}$ and even $z-x$ are both square numbers. Suppose,

$\dfrac{z+x}{2} = a^2, z-x = 2b^2$.

Then, $x = a^2 - 2b^2, \ y = 2ab, \ z = a^2 + 2b^2$

If $z - x$ is not a multiple of 4, then from (1)

Letting $z + x = (2b)^2, \ \dfrac{z+x}{2} = a^2$, we get

$\qquad x = -(a^2 - 2b^2)$

$\qquad y = 2ab$

$\qquad z = a^2 + 2b^2$

Example 3. *Show that the positive integer solution of the equation* $x^{-2} + y^{-2} = z^{-2}, (x, y, z) = 1$ *is given by*

$$x = a^4 - b^4, \ \ y = 2ab(a^2 + b^2), \ \ z = 2ab(a^2 - b^2)$$

such that $a > b > 0 \ (a, b) = 1 \ a, b$ *can not be both even or both odd.*

Solution. As per given, we have $(x, y, z) = 1$, so that $(x^2, y^2, z^2) = 1$.

Then, proceeding same as example (1), we get

$$x^2 = r(r + s), \ y^2 = s(r + s), \ z^2 = rs$$

where, $r, s > 0, \ (r, s) = 1$

Now, from $z^2 = rs$, we know that r, s are both square numbers and from $x^2 = r(r + s)$, we find that $r + s$ is also a square number. Putting

$$r = r_1^2, \ z = s_1^2, \ r + s = t_1^2$$

we have

$$r_1^2 + s_1^2 = t_1^2, \ r, s > 0, \ (r, s) = 1$$

Then, using Theorem-1 of Pythagores equation, we get

$$r_1 = a^2 - b^2, \ s_1 = 2ab, t_1^2 = a^2 + b^2$$

where $a > b > 0, (a, b) = 1$, one of a, b is odd and the other is even.

Therefore,

$$x = r_1 t_1 = a^4 - b^4$$
$$y = s_1 t_1 = 2ab(a^2 + b^2)$$
$$z = r_1 s_1 = 2ab(a^2 - b^2)$$

EXERCISE 3.1

1. Find the general solution of $170x - 455y = 625$

2. Find the general solution of $39x - 56y = 11$.

3. Find all positive solution of $5x + 14y = 620$.

4. Solve the following equations :

(a) $2072x + 1813y = 2849$

(b) $8x - 18y + 10z = 16$

(c) $4x + 10y + 14z + 6t = 20$

5. Find the sum of all positive integers each of which has 2 digits and has remainder 4 when divided by 4.

6. Find the value of a if $x^2 + y^2 = axy$ has positive integer solution.

7. Show that the equation $213x + 441y = 10002$
has solution in integers but none where both x and y are positive integers.

8. Show that if $ax + by = n$ has any solution in integers we may assume the problem reduced to the form $Ax + By = N$, where $(A, B) = 1$ and $B > 0$.

9. Show that the equation $x^2 + y^2 + z^2 = x^2y^2$ has no integer solution, except $x = y = z = 0$

10. If n is any positive integer, show that $x^2 + y^2 = z^n$ always has positive integer solution.

ANSWERS (With Hints)

(1) $x_1 = 1 + 91t$, $y_1 = -1 + 34t$.

(2) $x_1 = 29 + 56t$, $y_1 = 20 + 34t$.

(3) $x_1 = 110 - 4t$, $y_1 = 5 + 5t$, $t = 1, 2, ..., 7$.

(4) (a) $x_1 = -3 + 7t$, $y_1 = 5 - 8t$.

(b) $x_1 = 4 - 9t_1 - 10t_2$, $y_1 = 2 - 4t_1 - 5t_2$, $z = 2$.

(c) $x_1 = -18 - 54c + 21b + 5a$, $y_1 = 6 + 18c - 7b - 2a$, $z_1 = 1 + 3c - b$, $t_1 = 3 - c$.

(5) Let $4k + 1$ be a number containing 2 digits. Since $10 \le 4k + 1 < 100$ or

$2\dfrac{1}{4} < k < 24\dfrac{3}{4}$, there are 22 integer values between 3 and 24 that k can take.

Hence, we get 22 integers which all contain 2 digits and form an arithmetic progression whose sum is 1210.

(6) Let $(x, y) = d$, $x = dx'$, $y = dy'$

Then $x'^2 + y'^2 = ax'y'$.

Hence, $x' | y'^2$, $y' | x'^2$ or $x' = 1$, $y' = 1 \Rightarrow a = 2$.

(11) Let a, b, c be any system of solutions of the Pythagores equation, then
$(ac^{n-1})^2 + (bc^{n-1})^2 = (c^2)^n$.

CHAPTER REVIEW : A COMPETITIVE APPROACH

Selected terms and Results

TERMS

- **Linear diophantine equation :** An equation of the form $ax + by = c$ with $a \neq 0$, $b \neq 0$ and c integers, is called linear diophantine equation.

- **Shang - gao or Pythorgores equation:** The diophentine equation of the second degree $x^2 + y^2 = z^2$ is called the Shang-gao or Pythogores equation.

RESULTS

- The linear diophantine equation $ax + by = c$ where a, b, c are integers, has integer solution if and only if $d \mid c$ where $d = $ g.c.d. of a and b.

- If $d = (a, b) = 1$, the solution of above equation (1) can be written as $x = x_0 + bt, y = y_0 - at$, where $x = x_0$ and $y = y_0$ is a solution of (1).

- If (x_0, y_0) is one solution of $ax + by = c$,

  $(a, b) = d$ then $x_1 = x_0 + \dfrac{b}{d} .t; \ y_1 = y_0 - \dfrac{a}{d} .t.$

- The linear diophantine equation $a_1 x_1 + a_2 x_2 + ... + a_n x_n = c$ a_i, c being all integers has integer solution if and only if $d = (a_1, a_2, ... a_n) \mid c$.

- Let a and b be two relatively positive primes, then the necessary and sufficient condition that the equation $ax + by = n$, $0 \leq n < ab$ has no non-negative integer solution.

- Let a, b and c be integers with not both a and b equal to zero and let $q = $ g.c.d.(a, b). If $q \nmid c$ then the equation $ax + by = c$ ...(1) has no solution in integer. If $q \mid c$ then this equation has infinitely many solution.

- The diophantine equation $x^4 + y^4 = z^2$ has no solution with non-zero positive integers x, y and z.

- For any positive integer n, the diophantine equation $x^n + y^n = z^n$ has no positive solution larger than n.

Review Questions and Project Work

1. Check, which of the following diophantine equation is solvable:
   (i) $14x + 16y = 15$
   (ii) $28x + 91y = 119$
   (iii) $1076x + 2076y = 1155$
   (iv) $2x + 3y + 4z = 5$
   (v) $12x + 30y - 42z = 66$

   Ans : (iii), (iv) and (v)

2. Find the general solution of each linear diophantine equation
   (i) $12x + 16y = 20$
   (Ans. $x = 3 + 4t; \ y = -1 - 3t$)
   (ii) $15x + 21y = 39$
   (Ans. $x = -3 + 7t; \ y = 4 - 5t$)
   (iii) $1776x + 1976y = 4152$
   (Ans. $x = 41001 - 247t; \ y = -36849 + 222t$)

3. A six digit positive integer is cut up in the middle into two three-digits numbers. If the square of their sum yields the original number, find the number.

4. If a and b are both positive, discuss the number of pairs x, y solving $ax + by = n$, in which x and y are both positive integers.

5. Show that $213x + 441y = 10002$ has solution in integers, but none where both x and y are positive integers.

6. Show that all solutions of $ax + by = n$ in positive integers x, y if there are any can be found by solving $x > 0, y > 0$.

Objective Type Questions

Fill in the blanks :

1. An equation of the form $ax + by = c$ $a \neq 0$, $b \neq 0$ and c are integers is called a .... diophantine equation.

2. Linear diophantine equation $ax + by = c$ has integer solution if and only if .... when $d = (a, b)$

3. If a, b, c are integers ($a \neq 0$, $b \neq 0$) and $d = (a, b)$. If $d \nmid c$ then the equation $ax + by = c$ has .... in integer.

4. The diophantine equation of the second degree is also called .... equation.

5. The diophantine equation $x^n + y^n = z^n$ has no positive solution .... than n.

True/False: *Write 'T' for true and 'F' for false statement.*

1. The diophantine equation $x^4 + y^4 = z^2$ has no solution with non-zero positive integers x, y and z. (T/F)

2. The diophantine equation $x^n + y^n = z^n$ has one positive solution larger than n. (T/F)

3. The equation $x^n + y^n = z^n$ has no solution with non-zero integers x, y and z if $n > 2$. (T/F)

4. The diophantine equation of second degree $x^2 + y^2 = z^2$ is called Shang-gao equation (T/F)

5. The diophantine equation of second degree $x^2 + y^2 = z^2$ is also called Pythogores equation. (T/F)

Multiple Choice Questions : *Choose the most appropriate one :*

1. An equation of the form $ax + by = c$, a, b, $c \in Z$; $a \neq 0$, $b \neq 0$ is called:
   (a) diophantine equation
   (b) indeterminate equation
   (c) both (a) and (b) are true
   (d) none of these

2. The diophantine equation $ax + by = c$ has an integer solution iff (d = g.c.d. (a, b))
   (a) $c \mid d$
   (b) $d \mid c$
   (c) Both (a) and (b) are true
   (d) none of these

3. If $x = x_0$, $y = y_0$ is a particular solution of $ax + by = c$ then its general solution can be written as
   (a) $x = x_0 + \dfrac{b}{d}t$, $y = y_0 - \dfrac{a}{d}t$

   (b) $x = x_0 - \dfrac{b}{d}t$, $y = y_0 - \dfrac{a}{d}t$

   (c) both (a) and (b) are true
   (d) none of these

4. Which one of the following diophantine equation is not solvable.
   (a) $12x + 18y = 30$
   (b) $2x + 3y = 4$
   (c) $6x + 8y = 25$
   (d) none of these

5. Which one of the following diophantine equation is not solvable
   (a) $63x + 23y = 7$
   (b) $6x + 5y = 8$
   (c) $70x + 112y = 168$
   (d) none of these

6. Which one of the following diophantine equation is solvable
   (a) $11x + 5y = 79$
   (b) $311x - 112y = 73$
   (c) $7x + 19y = 213$
   (d) all are true

7. The general solution of the equation $70x + 112y = 168$ is
   (a) $x = -4 + 8t$, $y = 4 - 5t$
   (b) $x = 4 - 8t$, $y = -4 + 5t$
   (c) $x = 5 - 6t$, $y = 7 - 8t$
   (d) none of these

8. The general solution of the equation $2x + 3y = 4$ is
   (a) $x = 8 + 3t$, $y = -4 - 2t$

    (b) $x = 8 - 3t, y = -4 + 2t$
    (c) both (a) and (b) are true
    (d) none of these

9. The general solution of the equation
    $6x + 5y = 8$ is given by
    (a) $x = 3 + 5t, y = -2 - 6t$
    (b) $x = 3 - 5t, y = -2 - 6t$
    (c) both (a) and (b) are true
    (d) none of these

10. The general solution of the equation
    $7x + 19y = 213$ is
    (a) $x = 25 + 9t, y = 2 - 7t$
    (b) $x = 25 + 19t, y = 2 - 7t$
    (c) both (a) and (b) are true
    (d) none of these

ANSWERS

Fill in the blanks

(1) linear
(4) Sang-gao or Pythogores

(2) $d \mid c$
(5) larger

(3) no solution

True/False

(1) T
(4) T

(2) F
(5) T

(3) T

Multiple Choice Questions

(1) *c*
(4) *c*
(7) *a*
(10) *b*

(2) *b*
(5) *d*
(8) *a*

(3) *a*
(6) *d*
(9) *a*

CHAPTER 4

Primes and Their Distributions

❖ Prime and Composite Numbers	❖ Fundamental Theorem of Arithmetic
❖ The Sieve of Eratosthenes	❖ Positive Divisors of a Positive Integer
❖ The Goldbach's Conjecture	❖ Bonse's Inequality

4.1 INTRODUCTION

The idea of prime numbers is very simple. Among all the positive integers 1, 2, 3, 4, ..., we can find that some integers have only two positive divisors and the other have more than two positive divisors, expect the integer 1, which has just one positive divisor, namely itself. A positive integer which is greater than 1 and has only two positive divisors 1 and itself is called a prime number. An integer which is greater than 1, but is not a prime, is called a composite number.

For example 2, 3, 5, 7, 11, ... are prime but 4, 6, 8, 9, 10, ... are composite.

Remarks

❖ 1 is neither prime nor composite number.

❖ Among all even integers, only 2 is prime and all other composite.

4.2 PRIME NUMBERS

(i) **Prime Number :** Any integer $p > 1$ which has no divisor except 1 and the number itself is called a prime number. It is denoted by p. Also, the nth prime is denoted by p_n.

(ii) **Composite Number :** Any integer greater than 1, which is not prime is called composite number.

(iii) **Twin Primes :** Two successive odd integers p and $p + 2$ which are primes, are called twin primes. For example 3, 5; 5, 7; 11, 13; 17, 19 etc.

(iv) **Palindromic Primes :** A Palindrom prime is a prime number that reads the same backward as well as forward. There are 19 such prime < 1000 : 2, 3, 5, 7, 11, 101, 131, 151, 181, 313, 353, 373, 383, 727, 757, 787, 797, 919 and 929. The largest known palindromic prime is $10^{11310} + 4661644 \cdot 10^{56752} + 1$.

THEOREM-1

The least divisor, other than 1 of a composite number is a prime.

Proof. Let if possible the least divisor $q(> 0)$ of an integer a, is not prime, then it has a divisor q_1, $1 < q_1 < q$. Clearly, q_1 is a divisor of a which contradicts the fact that q is the least divisor, since if a is not prime, then its least positive divisor is a prime divisor.

THEOREM-2

If a is a composite divisor, and q is its least positive divisor, then $q \leq \sqrt{a}$.

Proof. As per given, q is a divisor of a, i.e. $q \mid a$, then we have

$$a = qa_1$$
$$\Rightarrow \qquad a_1 \geq q$$

Hence, $\quad a \geq q^2$, i.e. $q \leq \sqrt{a}$

THEOREM-3

If p is a prime and a is any integer, then either (a, p) = 1 or a is a multiple of p.

Proof. As per given, p is prime, then it has two divisors 1 and p. Therefore,
$$(a, p) = 1 \qquad \text{or} \qquad (a, p) = p$$
In case of $(a, p) = 1$, result is obvious.

If $(a, p) = p$, then by definition of g.c.d., a is a multiple of p.

THEOREM-4 (EUCLID'S THEOREM)

The number of primes is infinite, i.e. there is no end to the sequence of primes.

Proof. Let p be any prime and q is a prime divisor of $a = p! + 1$. Now, since $q \nmid p!$, $q > p$. Therefore, for any given prime number, there is a greater prime number, and hence, there are infinitely many primes.

THEOREM-5

The number of primes of the form $(4n - 1)$ is infinite.

Proof. Let $4n_1 - 1, 4n_2 - 1, ..., 4n_k - 1$ be all the primes not greater than $4n - 1$ with same form. Using the argument of Theorem-4, we can write
$$a = 4(4n_1 - 1). (4n_2 - 1), ..., (4n_k - 1) - 1$$
has a prime divisor, different from $4n_i - 1$. Clearly, the prime divisor of a is an odd number. Since an odd number can be written as $4n + 1$ or $4n - 1$ and
$$(4m + 1) \mid (4l + 1) = 4(4lm + l + m) + 1$$
the prime divisor of a can not be all of the form $4n + 1$ and therefore among them, there exist an integer of the form $4n - 1$.

Remark

❖ Similarly, we can prove the number of primes of the form $4n + 1$ is infinite.

THEOREM-6

If p is prime and $p \mid ab$ then either $p \mid a$ or $p \mid b$.

Proof. If $p \mid a$, then $(a, p) = 1$, then theorem is proved.

Now, suppose that $p \nmid a$. The only divisor of p are 1 and p, therefore

$$(p, a) = p$$

or $\qquad (p, a) = 1$

Therefore, there exist integers x and y such that $1 = abx + py$.

Multiply by b, we get

$$b = abx + bpy$$

Since, by hypothesis p divides ab and since obviously p divides p, it follows that p divides b.

DEDUCTIONS :

If a prime p divides a product $a_1 a_2, ..., a_n$, then p must divide at least one of the factors $a_1, a_2, ..., a_n$.

THEOREM-7 (FUNDAMENTAL THEOREM OF ARITHMETIC)

Every positive integer greater than 1 can be expressed uniquely as a product of primes upto the order of the factors. More precisely, any positive integer a can be expressed as

$$a = p_1 p_2 \cdots p_r, \text{ all } p_i \text{ being primes.}$$

Further, if a is also given by

$$a = q_1 q_2 \cdots q_s, \text{ all } q_j \text{ being primes.}$$

then $r = s$ and $p_1, p_2, \cdots p_r$ and $q_1 q_2, ..., q_s$ differ only in their orders.

Proof. Let $a > 1$ be an integer. If a is prime, then result is obvious. If n is composite number, then there exists a prime p_1 such that

$$a = p_1 a_1 \text{ for some integer } a_1$$

If a_1 is prime, then a can be expressed as the product of prime factors. But if a_1 is a composite number, then there exists a prime p_2 such that

$$a = p_1 a_1 = p_1 p_2 a_2, \text{ for some integer } a_2.$$

If a_2 is prime, then again, it can be expressed as the product of prime factors. If a_2 is a composite number, then we proceed continuously as above.

Now, since $a > a_1 > a_2 \cdots$

the process can not continue infinitely. Thus, after a finite number of steps, we get

$$a = p_1 p_2 \cdots p_k \text{ where all } p_i's \text{ are primes.}$$

Now, we shall show the uniqueness of factorization.

Let, if possible, a can be represented as a product of primes in two ways, as follows

$$a = p_1 p_2 \cdots p_r = q_1 q_2 \cdots q_s, \quad r < s \cdots \qquad ...(1)$$

where p_i and q_i are primes in the non-decreasing order, i.e.,

$$p_1 \le p_2 \le p_3 \cdots \le p_r \text{ and } q_1 \le q_2 \le q_3 \cdots \le q_s$$

Since, $p_1 \mid q_1 q_2 \cdots q_n$ there exists some q_i such that $p_1 \mid q_k$. But p_1 and q_k are both primes.

Thus, $p_1 = q_k$. Rearrange q_i's such that $p_1 = q_1$. On cancelling p_1 and q_1 in (1), we get

$$p_2 \cdot p_3 \cdots p_r = q_2 q_3 \cdots q_s$$

We continue this process till all p_i's are exhausted. Now, since $r < s$, we get

$$1 = q_{r+1} q_{r+2} \cdots q_s$$

which is not possible because q_i's are primes.

Therefore, r can not be less than s. Similarly we can show that s can not be less than r. Hence, $r = s$ and

$$p_i = q_i, \ \forall \ i$$

$\Rightarrow$ Factorization is unique.

Remarks

❖ If we take 1 as a prime, then above theorem does not hold. This is also a reason, why 1 can not be prime.

❖ The above theorem is known as unique factorization theorem for integers.

DEDUCTIONS :

(1) Collecting the same primes in (1), we get

$$a = p_1^{a_1} p_2^{a_2} \cdots p_k^{a_k}$$

where $a_i \geq 1, i = 1, 2, \ldots k$.

Each p_i is prime and $p_i \neq p_j$ for $i \neq j$.

This representation is called standard factorization or standard representation of a; where a is the highest exponent of p_i in a and is denoted by $p_i(a)$ so that $a_i \mid p_i(a)$.

Clearly, $p_i(ab) = p_i(a) + p_i(b)$

(2) The necessary and sufficient condition that d be a positive divisor of a is

$$a = p_1^{d_1} p_2^{d_2} \cdots p_k^{d_k}, \qquad\qquad 0 \leq d_i \leq a_i$$

Also, if $a = p_1^{a_1} p_2^{a_2} \cdots p_m^{a_m},$ $b = p_1^{b_1} p_2^{b_2} \cdots p_m^{b_m}$

where a_j, b_j are non-negative integers and some of them may be zero. Then

$$(a, b) = p_1^{c_1} p_2^{c_2} \cdots p_m^{c_m}$$

$$[a, b] = p_1^{d_1} p_2^{d_2} \cdots p_m^{d_m}$$

where c_i and d_i are respectively the smallest and the greatest of a_i and b_i, i.e.

$$c_i = \min \{a_i, b_i\}$$

$$d_i = \max \{a_i, b_i\}$$

(3) If $a \mid c$ and $b \mid c$, then $[a, b] \mid c$ and $ab = (a, b)[a, b]$.

(4) When a and b are relatively prime, i.e. $[a, b] = 1$, if

$$a = p_1^{a_1} p_2^{a_2} \cdots p_r^{a_r}, \qquad\qquad b = q_1^{b_1} q_2^{b_2} \cdots q_s^{b_s}, \qquad\qquad a_i, b_i > 0$$

where, $p_1, \ldots p_r, \ q_1, \ldots q_s$ are distinct primes, then

$$ab = p_1^{a_1} \cdots p_r^{a_r} q_1^{b_1} \cdots q_s^{b_s}$$

is the standard factorization.

4.3 THE SIEVE OF ERATOSTHENES

The sieves of Eratosthenes of ancient origin is the device of preparing a list of prime numbers less than a given limit by writing down all the integers upto that limit and then in a systematic way eliminating all the composite integers. One such device is given by Eratosthenes.

For example, with a limit of $n = 100$, we first set down a list of the integers from 2 to 100. Recognizing that 2 is a prime but that all proper multiples of 2 are composite, we cross out 4, 6, 8, ..., 100. The next number not crossed out is 3, which must be a prime for the only possible proper factor is 2 and 3 is not a multiple of 2, else it would have been crossed out. Since all proper multiples of 3 are composite, we cross out 6, 9, 12, ... 99, although is not a actually necessary to cross out 6, 12, ..., 18, 96 again, since they are already crossed out being the multiples of 2. The next number not crossed out is 5; this number must be a prime, for if it were composite, it would have to have as a proper factor a prime less than 5, namely either 2 or 3; but since 5 is not crossed out, it is not a multiple of 2 or 3. Crossing out all the multiples of 5, not previously crossed out namely 25, 35, 55, 65, 85, 95, we find as the same reasoning as before that the next number not crossed out must be a prime, it is 7. The only multiples of 7, not previously crossed out are 49, 77, 91 and these we now cancel. Now, unless we have been analysing the sieve process carefully, we get all the remaining numbers which have survived the sieve are primes. The sieve appears as follows :

2	3	4	5	6	7	8	9	10	11	12	13	14	15	16
17	18	19	20	21	22	23	24	25	26	27	28	29	30	31	32
33	34	35	36	37	38	39	40	41	42	43	44	45	46	47	48
49	50	51	52	53	54	55	56	57	58	59	60	61	62	63	64
65	66	67	68	69	70	71	72	73	74	75	76	77	78	79	80
81	82	83	84	85	86	87	88	89	90	91	92	93	94	95	96
97	98	99	100												

Remark

❖ We are always sure to reach the end of the sieve process when we have crossed out the proper multiples of p', where p' is the largest prime such that $p' \le \sqrt{n}$.

THEOREM-1

If the smallest prime factor of n is greater than $n^{1/3}$, then n has only two factors which are prime.

Proof. We can write

$$n = p_1^{a_1} p_2^{a_2} \cdots p_r^{a_r}$$

Let, if possible, p_1, p_2, p_3, are prime greater than $n^{1/3}$ and $n > p_1 p_2 p_3$. Thus

$$n = n > n^{1/3} n^{1/3} n^{1/3} = n, \text{ which is not possible.}$$

Hence, $\quad n = p_1 p_2$

THEOREM-2

If $m = p_1^{a_1} p_2^{a_2} \cdots p_r^{a_r} = \prod p^a$

                  $n = p_1^{b_1} p_2^{b_2} \cdots p_r^{b_r} = \prod p^b$

Then, $(m, n) = \prod p^{\min (a, b)}$

Proof. Let us write

$$\prod p^{\min (a, b)} = d$$

To show $(m, n) = d$

Since, $\min (a, b) \le a$, we have $\prod p^{\min (a, b)} \mid \prod p^a$

$\Rightarrow$ $d \mid m$

Similarly, $\min (a, b) \le b$ gives $\prod p^{\min (a, b)} \mid \prod p^b$

$\Rightarrow$ $d \mid n$

$\Rightarrow$ d divides both m and n.

$\Rightarrow$ d is the divisor of m and n.

It remains to prove that d is the greatest divisor.

Let c be any integer which divides both m and n.

Then, c is the product of primes from $p_1, p_2, \ldots, p_r$.

Let, $c = p_1^{c_1} p_2^{c_2} \cdots p_r^{c_r} = \prod p^c$

$\therefore$ $\prod p^c \mid \prod p^a$ and $\prod p^c \mid \prod p^b$

$\Rightarrow$ $c \le a$ and $c \le b$

$\Rightarrow$ $c \le \min (a, b)$

$\Rightarrow$ $\prod p^c$ divides $\prod p^{\min (a, b)}$

$\Rightarrow$ $c \mid d$

$\Rightarrow$ d is the greatest divisor of m and n.

Hence, $d = (m, n)$

Remark

❖ In a similar manner, we can prove that

"If $m = \prod p^a$ and $n = \prod p^b$, then

$[m, n] = \prod p^{\max(a, b)}$.

THEOREM-3

If p is the rth prime number, then $p_r \le 2^{2r-1}$.

Proof. We prove this result by the principle of mathematical induction.

For $r = 1$, we have

                  LHS $= p_1 = 2$, RHS $= 2^{2-1} = 2$

$\Rightarrow$ $p_1 \le 2$, i.e. result is true for $r = 1$.

Now, assume that result is true for $r = m$, i.e. $p_m \le 2^{2^{m-1}}$.

To show result is true for $r = m + 1$

i.e. $p_{m+1} \le 2^{2^{m+1}-1}$.

we have

$$p_{m+1} \le p_1 p_2 \cdots p_{r+1}$$
$$\le 2 \cdot 2^2 \dots 2^{2^m-1} + 1$$
$$= 2^{1+2+2^2+2^{m-1}\dots} + 1$$
$$= 2^{2^m-1} + 1$$

But $1 \le 2^{2^m-1}$

Thus, $p_{m+1} \le 2^{2^m-1} + 2^{2^m-1}$

$$= 2 \cdot 2^{2^m-1} = 2^{2^{m+1}+1}$$

i.e. the result is true for $r = m + 1$.

Hence, by the principle of mathematical induction, the result will be true for all n.

DEDUCTION :

For $n > 1$, there are at least $(n + 1)$ primes less than 2^{2^n}

THEOREM-4

If $n(> 2)$ terms of an A.P.
$$p, p + d, p + 2d, \dots p + (n - 1)d$$
are all prime numbers then the common difference d is divisible by every prime $q > n$.

Proof. Let if possible $q \nmid d$.

The first q terms of A.P. are given by
$$p, p + d, p + 2d, \dots p + (q - 1)d$$

Divide these terms by q, then we get non-zero divisors. If there are two integers i and j with $0 \le i \le j < q$ such that the numbers $p + id$ and $p + jd$ give the same remainder on division by q.

Then, clearly q divides their difference $(j - i)a$.

But, $(q, d) = 1 \Rightarrow q | (j - i)$, which is not possible because of $j - i < q$. Thus, all the remainder will be different. The q remainder will be from $0, 1, 2, \dots, q - 1$.

Hence, one remainder must be zero.

This gives $q | (p + kd)$ for $0 \le k < q$.

Now, $q < n \le p \le p + kd$

We conclude that $p + kd$ is composite, which is a contradiction. Hence, $q | d$.

THEOREM-5 (BONSE'S INEQUALITY)

If $n > 3$, then $p^2_{n+1} < p_1 p_2 \cdots p_n$.

Proof. For $n = 4$, $11^2 < 2 \cdot 3 \cdot 5 \cdot 7$

For $n = 5$, $13^2 < 2 \cdot 3 \cdot 5 \cdot 7 \cdot 11$

For $n = 6$, $17^2 < 2 \cdot 3 \cdot 5 \cdot 7 \cdot 11 \cdot 13$

For $n = 7$, $19^2 < 2 \cdot 3 \cdot 5 \cdot 7 \cdot 11 \cdot 13 \cdot 17$

$\Rightarrow$ Result is true for $n = 4, 5, 6, 7$.

For $n = 8$, we have $m = \left\lfloor \dfrac{8}{2} \right\rfloor = 4 > 3$.

Then, we have the following result

"If $m = \left[\dfrac{n}{2}\right]$, then $n - (m - 1) < p_m$ for $n \geq 8$ and $n \geq 2m$.   ...(1)

Let $k = p_1 p_2 \cdots p_{m-1}$ and let

$$S = \{k - 1, 2k - 1, \ldots p_m k - 1\}$$

Now, suppose that $r_1 k - 1$ and $r_2 k - 1$ of S are divisible by a prime p_{m+i}, where $i = 0, 1, 2, \ldots, n - m$

Then,

$$p_{m+1} \mid [(r_1 k - 1) - (r_2 k - 1) \Rightarrow p_{m+1} \mid k(r_1 - r_2) \qquad [\because (p_{m+i}, k) = 1]$$

and $(r_1 - r_2) < p_{m+i}$

Therefore, none of the primes $p_m, p_{m+1}, \ldots, p_n$ divides more than one element of S.

Now, since $p_m > n - (m - 1)$, there are more integers in S than the number of primes. Thus, there are one or more elements of S which are not divisible by any of the primes $p_m p_{m+1} \cdots p_n$.

Hence, there exist at least one integer in S which is not divisible by any of the primes $p_1, p_2, \ldots p_n$. Suppose this integer is $rk - 1$, for some r such that $1 \leq r \leq p_m$.

$\Rightarrow$ $rk - 1$ is a prime greater than p_m.

$\therefore$ $p_{n+1} < p_1 p_2 \cdots p_m$

$\qquad\qquad p^2_{n+1} < (p_1 p_2 \cdots p_m)(p_1 p_2 \cdots p_m)$

$\qquad\qquad\qquad < (p_1 p_2 \cdots p_m)(p_{m+1} p_{m+2} \cdots p_{2m})$

$\qquad\qquad\qquad < (p_1 p_2 \cdots p_n)$ $(\because n \geq 2m)$

SOLVED EXAMPLES

Example 1. *Express 560 as the product of prime factors.*

Solution. We have

$$\begin{aligned}
560 &= 2 \times 280 \\
&= 2 \times 2 \times 140 \\
&= 2 \times 2 \times 2 \times 70 \\
&= 2 \times 2 \times 2 \times 2 \times 35 \\
&= 2 \times 2 \times 2 \times 2 \times 5 \times 7 \\
&= 2^4 \times 5 \times 7
\end{aligned}$$

Example 2. *Show that every prime of the form $3k + 1$ is necessarily of the form $6m + 1$.*

Solution. Let us suppose $n = 3k + 1$ be a prime number for any positive integer k. Further k may be even or odd.

i.e. $k = 2m$ or $k = 2m + 1$

Case 1 : If $k = 2m$, then

$$n = 3k + 1 = 3 \times 2m + 1 = 6m + 1$$

Case 2 : If $k = 2m + 1$, then

$$n = 3(2m + 1) + 1 = 6m + 3 + 1$$

$$= 2(3m + 2)$$

which is not possibe, because $2(3m + 2)$ is not prime.

Hence, prime n of the form $3k + 1$ will also be of the form $6m + 1$.

Example 3. *For each positive integer n, show that there exist n consecutive numbers each one of which is composite.*

Solution. Consider the following sequence of n integers.

$$(n + 1)! + 2, (n + 1)! + 3, \ldots (n + 1)! + (n + 1)$$

The general term of this sequence is given by

$$(n + 1)! + k = (n + 1) n(n - 1) \ldots k (k - 1) \ldots 2.1 + k$$
$$= k[(n + 1)n \ldots (k + 1).(k - 1) \ldots 2.1 + 1]$$

which is a composite number.

Hence, for each positive integer n, there exist n consecutive numbers each one of which is composite.

Example 4. *If a prime number $p > 3$, then show that $2p + 1$ and $4p + 1$ can not be prime simultaneously.*

Solution. We have already proved that a prime number p can only take the form $3k + 1$ or $3k - 1$.

If $p = 3k + 1$, then

$$2p + 1 = 2(3k + 1) + 1 = 3(2k + 1)$$
$\Rightarrow$ $2p + 1$ is composite.

If $p = 3k - 1$, then

$$4p + 1 = 4(3k - 1) + 1 = 3(4k - 1)$$
$\Rightarrow$ $4p + 1$ is composite.

Example 5. *If p is prime, show that there exist no positive integers a and b such that $a^2 = pb^2$.*

Solution. Let if possible, there exist two positive integers a and b such that

$$a^2 = pb^2 \qquad \qquad \ldots(1)$$

Denote $d = (a, b)$

Then, by definition of g.c.d.

$$a = da_1, \ b = db_1, (a_1, b_1) = 1$$

Putting these values in (1), we get

$$a_1^2 = pb_1^2$$
$\Rightarrow$ $p \,|\, a_1^2$ and then $p \,|\, a_1$

Putting $a_1 = pa_2$, we get

$$a_2^2 p = b_1^2$$

Clearly, $p \,|\, b_1^2$ and then $p \,|\, b_2$.

i.e. p is a common divisor of a_1 and b_1, which is a contradiction because $(a_1, b_1) = 1$. Hence, there exist no positive integers a and b such that $a^2 = pb^2$.

Example 6. *If a, b are relatively primes and $d \,|\, ab$, then exists unique $d_1 \,|\, a$, $d_2 \,|\, b$ such that $d = d_1 d_2$.*

Solution. We know that

$$ab = p_1^{a_1} \ldots p_r^{a_r} q_1^{b_1} \ldots q_s^{b_s} \qquad \qquad \ldots(1)$$

Thus, we have

$$d = p_1^{l_1} \dots p_r^{l_r} q_1^{k_1} \dots q_s^{k_s}, \qquad 0 < d_i \le a, \qquad 0 \le k_i \le b_i$$

Let $\qquad d_1 = p_1^{l_1} \dots p_r^{l_r}, \quad d_2 = q^{k_1} \dots q_s^{k_s}$

Then, $\qquad d = d_1 d_2$

and clearly $d_1 d_2$ are unique.

4.4 POSITIVE DIVISORS OF A POSITIVE INTEGER

Let a be a positive integer. Let $T(a)$, $S(a)$ and $P(a)$ respectively denote the total number, the sum and the product of all positive divisors of a (including 1 and a).

i.e. $\qquad T(a) = \sum_{d|a} 1, \qquad S(a) = \sum_{d|a} d, \qquad P(a) = \prod_{d|a} d$

For example :

a	1	2	3	4	5	6	7	8	9	10
$T(a)$	1	2	2	3	2	4	2	4	3	4
$S(a)$	1	3	4	7	6	12	8	15	13	18
$P(a)$	1	2	3	8	5	36	7	64	27	100

If a is prime, i.e. $a = p$

Then, $\qquad T(p) = 2, \quad S(p) = p + 1, \quad P(p) = p$

We can easily find first all the possible divisor of positive integer a, and then their total number and sum respectively.

$$T(a) = (a_1 + 1)(a_2 + 1) \dots (a_k + 1) = \prod_{i=1}^{k} (a_i + 1)$$

$$S(a) = \sum_{t_1=0}^{a_1} \dots \sum_{t_k=0}^{a_k} p_1^{t_1} \dots p_k^{t_k} = \left(\sum_{t_1=0}^{a_1} p_1^{t_1} \right) \dots \left(\sum_{t_k=0}^{a_k} p_k^{t_k} \right)$$

$$= \frac{p_1^{a_1+1} - 1}{p_1 - 1} \dots \frac{p_k^{a_k+1} - 1}{p_k - 1} = \prod_{i=1}^{k} \frac{p_i^{a_i+1} - 1}{p_i - 1}$$

Suppose $a_1, \dots a_{T(a)}$ are all positive divisors of a. If we set $a = a_i . a_i'$, then $a_1', \dots a'_{T(a)}$ are also all the positive divisors of a and hence, their product is

$$p(a) = a_1 a_2 \dots a_{T(a)} = \sqrt{(a_1 a_1')(a_2 a_2') \dots (a_{T(a)} a_{T(a)}')}$$

$$= \underbrace{\sqrt{a \dots a}}_{T(a) \text{times}} = a^{T(a)/2}$$

For example : If $a = 60 = 2 . 3 . 5$, then

$$T(a) = (2 + 1)(1 + 1)(1 + 1) = 3 . 2 . 2 = 12$$

$$S(a) = \frac{2^3 - 1}{2 - 1} . \frac{3^2 - 1}{3 - 1} . \frac{5^2 - 1}{5 - 1} = 7 . 4 . 6 = 168$$

$$P(a) = 60^6 = 46656000000$$

DEDUCTIONS :

If a, b are relatively prime, then we have

$$T(a, b) = T(a) . T(b), \quad S(ab) = S(a)S(b)$$

$$P(a, b) = (ab)^{T(ab)/2} = (ab)^{T(a)T(b)/2}$$

$$= P(a)^{T(a)} . P(b)^{T(b)}$$

Remarks

* ❖ If $(a, b) \neq 1$, then above three formulae do not hold.

* ❖ **Lerch Formula :** $T(a) = a - \sum_{k-1}^{a-1} T(a-k, k)$.

4.5 THE GOLDBACH'S CONJECTURE

The Goldbach's conjecture asserts that every even integer greater than 2 is the sum of two primes.

For example :

$4 = 2 + 2$	$14 = 3 + 1$
$6 = 3 + 3$	$16 = 5 + 11$
$8 = 3 + 5$	$18 = 7 + 11$
$10 = 3 + 7$	$20 = 7 + 13$
$12 = 5 + 7$	$22 = 5 + 17$

Goldbach's conjecture has been shown to be true for all even integers less than 10^{10}. Goldbach's conjecture is one of most difficult unsolved problem in mathematics. It is stated by Christian Goldbach in 1742 and verified upto 100,000 at least.

4.5.1 Bertrand's Conjecture : Joseph Bertrand conjectured in 1845 that there is a prime between n and $2n$ for every integer $n \geq 2$.

For example,

3 is prime between 2 and 4.

5 is prime between 3 and 6.

7 is prime between 4 and 8 and so on.

Bertrand was able to verify it for all positive integers ≤ 3 million although he could not establish the validity of his conjecture.

THEOREM-1

There are at least $3[n/2]$ primes between the number n and $n!$ where $n \geq 4$.

Proof: This statement is true for $4 \leq n \leq 9$.

Now, we prove this in two cases for $n \geq 10$.

Case (i) : When n is even, say $n = 2k$, where $k \geq 5$, then

$$n! = 1 . 2 . 3 \ldots (2k - 2)(2k - 1)n$$

$$= 2^{k-1}[1 . 2 . 3 \ldots (k - 1)][1 . 3 . 5 \ldots (2k - 1)]n$$

$$> 2^{k-1}(k-1)!2^{k+2}n$$
$$> 2^{k-1}.2^{k-1}.2^{k+2}n, \text{ since } k \geq 5$$
$$> 2^{3k}n.$$

A repeated application of Bertrand's conjecture shows there are at least $3k = 3(n/2)$ primes in the range from n to $n!$ through $2^{3k}.n$.

Case (ii) : Suppose n is odd, say $n > 2k + 1$, where $k \geq 5$, then

$$n! = 1.2.3 \ldots (2k-1)(2k)n$$
$$= 2^k k!1.2.3 \ldots (2k-1)]n$$
$$> 2^k.2^k 2^{k+2}n \quad \text{ since } k \geq 5$$
$$> 2^{3k}n.$$

Thus, as before, there are at least $3k = 3\left(\dfrac{n-1}{2}\right) = 3[n/2]$ primes in the range from n to $n!$ through $2^{3k}n$.

Thus, in both cases, result is true.

SOLVED EXAMPLES

Example 1. *If a is square number, then show that S(a) is an odd integer, if a is not a square number, but an odd integer, then S(a) is an even integer.*

Solution. Let $a = p_1^{2a_1} \ldots p_k^{2a_k}$, where $a_i > 0$ and p_i are prime. Then

$$S(a) = (1 + p_1 + \ldots + p_1^{2a_1}) \ldots (1 + p_k + \ldots + p_k^{2a_k})$$

Since any factor $1 + p_i + \ldots + p_i^{2a_i}$ is the sum of an odd number of integers, whether p_i is odd or even it must be odd and therefore, $S(a)$ is odd.

If $a = p_1^{a_1} \ldots + p_k^{a_k}$ and a_1 is odd, then $S(a)$ is even, since $1 + p_1 + \ldots + p_1^{a_1}$ is even.

Example 2. *If an integer is greater than 2, show that $S(a) < a\sqrt{a}$.*

Solution. Let $a = 2^k$, $k \geq 2$.

Then, we have

$$S(a) = 2^{k+1} - 1 < 2^{k+1} = a - 2^{k/2} = a\sqrt{a}.$$

If $a = p^k$ and p is an odd prime, then

$$S(a) = \frac{p^{k+1}-1}{p-1} = \frac{p^k - 1/p}{1 - 1/p} = \frac{a}{1 - 1/p} \leq \frac{a}{1 - 1/3}$$

$$= a.\frac{3}{2} < a\sqrt{3} \leq a\sqrt{a}.$$

Example 3. *Show that $n^4 + 4^n$ is composite for all integers, n greater than 1.*

Solution. We shall prove this result by the principle of mathematical induction on n. If $n = 2$, we have

$$2^4 + 4^2 = 16 + 16 = 32, \text{ which is a composite number.}$$

$\Rightarrow$ Result is true for $n = 2$.

Assume that result is true for $n = m$, i.e. $m^4 + 4^m$ is a composite number. To show result is true for $n = m + 1$, i.e.

$(m + 1)^4 + (4)^{m+1}$ is a composite number.

Now, $(m + 1)^4 + 4^{m+1} = m^4 + 4m^3 + 6m^2 + 4m + 1 + 1 + 4.4^m$

$$= (m^4 + 4^m) + 2m[2m^2 + m + 2] + (2m + 1) + 3.4^m$$

Here, every term of RHS is a composite number, therefore sum is also composite. Hence, by the principle of mathematical induction, the result is true for all n.

Example 4. *Show that every odd number greater than equal to 9 is the sum of three odd primes.*

Solution. We know that, the smallest number, which is the sum of three odd primes is $3 + 3 + 3 = 9$

Let n be the given odd integer greater than or equal to 9. Also, let p_1 be an odd prime such that $p_1 < n - 9$. Then $n - p_1$ is even and greater than 4. Thus, by Goldbach's conjecture, it can be expressed as the sum of two odd primes p_2 and p_3, i.e.

$$n - p_1 = p_2 + p_3$$
$$\Rightarrow \qquad n = p_1 + p_2 + p_3$$

Hence, the given odd integer n can be expressed as the sum of three odd primes.

Example 5. *If p and q are twin primes, then show that $12 \mid (p + q)$, when $p > 3$.*

Solution. Since p and q are twin primes, therefore

$$q = p + 2$$

Then, $p + q = p + p + 2$

$$\begin{aligned}
&= 2(p + 1) \\
&= 2(3k + 2 + 1) \qquad &\text{(Taking } p = 3k + 2) \\
&= 2(3k + 3) \\
&= 6(k + 1) \qquad &[\because k \text{ can't be even}] \\
&= 6 .2m \\
&= 12 . m
\end{aligned}$$

$\Rightarrow 12 \mid (p + q)$

EXERCISE 4.1

1. If $n > 2$, show that between n and $n!$ there exist at least one prime number.

2. Show that there are infinitely many primes of the form $3n + 2$.

3. If $n > 1$, show that $n!$ is never a perfect square.

4. Show that $n^4 + 4$ is a composite number, when n is any integer greater than 1.

5. Show that the necessary and sufficient condition that a positive integer equals the product of all its positive divisors (excluding itself) is that this positive integer be a cube of prime number or a product of two distinct primes.

6. If p, q are primes and $p - 1 = q + 1$, show that $p^q + q^q$ is a multiple of $p + q$.

7. If a and b are any two odd integers then show that $a^2 + b^2$ is not a perfect square.

8. If n is not divisible by any prime less than or equal to $\sqrt{n}$, then show that n is a prime.

9. If $(a, b) = 1$, then show that there are infinitely many primes of the form $aq + b$ (Dirichlets's theorem).

10. If $m = \left(\dfrac{n}{2}\right)$, then show that

   (i) $n \geq 2m$, $\forall n \in N$

   (ii) $n - (m - 1) < p_m$ for $n \geq 8$, where p_m is the mth prime.

11. If $n \geq 9$, then show that $p_n > 2n + 3$.

12. If p and q are twin primes, then show that $(pq + 1)$ is a perfect square.

13. Determine, whether the integer 1009 is a prime.

14. Determine, whether the integers 701 is a prime by testing all primes $p \leq \sqrt{701}$ as possible divisors.

ANSWERS (With hints)

(1) If $p_1, p_2, \dots p_k$ be primes not greater than n and $q = p_1 p_2 \dots p_{k-1}$. Clearly, q has a prime factor p different from p_i. Thus $p > n$ and $p < q < n! - 1 < n!$.

(2) Let us suppose $p_1, p_2, \dots, p_r$ are primes of the form $3n + 2$.

Put $a = 3p_1 p_2 \dots p_{r-1} = q_1 q_2 \dots q_t$, where q_i are primes.

Then, q_i are different from 3, $p_1, p_2, \dots p_r$.

If all q_i are of the form $3n + 1$, then their product a is also of the form $3n + 1$, which is a contradiction, because $a = 3p_1 p_2 \dots p_{r-1}$.

Hence, at least one q_i is of the form $3n + 2$.

(4) $(n^4 + 4) = (n^2 + 2)^2 - (2n)^2 = (n^2 + 2n + 2)(n^2 - 2n + 2)$. If $n > t$, then $n^2 - 2n + 2 \geq 2$. Hence, $n^4 + 4$ is a composite number.

(5) Since $a^{\frac{1}{2}T(a)-1} = a$ or $a^{\frac{1}{2}T(a)} = a^2$. Then $T(a) = 4$. Therefore, $(a_1 + 1)(a_2 + 1) = 4$ and hence $a_1 = 3$, $a_2 = 0$ or $a_1 = 1$, i.e. $a = p^3$ or $a = pq$

(6) Give $p - 1 = q + 1$, $p^p + q^q = (p^p - 1) + (q^q + 1)$, $p^p - 1 = (p - 1)(p^{p-1} + \dots + p + 1) = (p - 1)(2m - 1)$

$$q^q + 1 = (q + 1)(2n + 1) = (p - 1)(2n + 1)$$

$$\Rightarrow \quad p^p + q^q = 2(p - 1)(m + n + 1) = (p + q)(m + n + 1)$$

(11) Use the principle of mathematical induction.

(12) If p and q are twins prime, then we take $q = p + 2$. Then put this value of q in $pq + 1$

(13) $31 < \sqrt{1009} < 32$. The primes less than 31 are 2, 3, 5, 7, 11, 13, 17, 19, 23, 29, 31. None of these primes divide 1009. Hence, 1009 must be prime.

(14) Do same as in (13).

CHAPTER REVIEW : A COMPETITIVE APPROACH

Selected terms and Results

TERMS

- **Prime Number :** Any integer $p > 1$ which has no divisors except 1 and the number itself is called a prime number.
- **Composite Number :** Any integer greater than 1, which is not prime is called composite number.
- **Twin Prime :** Two successive odd integers p and $p + 2$ which are prime are called twin prime.
- **Palindromic Prime :** A prime number that reads the same backward as well as forward is called palindromic prime.
- **The Sieve of Eratosthenes :** A device of preparing a list of prime numbers less than a given limit by writing down all the integers upto that limit and then in a systematic way eliminating all the composite integers.

RESULTS

- If a is a composite integers, and q is its least positive divisor, then $q \le \sqrt{a}$.
- For any integer n (> 1), there are n consecutive composite numbers.
- The sequence of primes does not come to an end, i.e. number of primes is infinite.
- If p is prime and $p \mid ab$ then $p \mid a$ or $p \mid b$.
- 2 is only even number, which is prime.
- 1 can not be prime.
- Every odd number greater than or equal to 9 is the sum of three odd primes.
- The Goldbach's conjecture states that every even integer can be written as sum of two numbers that are either prime or 1.
- If $(a, b) = 1$, then there are infinitely many primes of the form $aq + b$.
- For each positive integer n, there exist n consecutive numbers each of which is composite.

- If a and b are two odd integers, then $a^2 + b^2$ can not be a perfect square.
- Fundamental theorem of arithmetic states that, every positive integer greater than 1 can be uniquely expressed as the product of prime factors.

Review Questions and Project Work

1. Show that there is no polynomial $f(n)$ with integral coefficients that will produce primes for all integer k.
2. Show that the number of primes $\le$ 100 is 25.
3. Show that for every positive integer n, there are n consecutive integers that are composite numbers.
4. Find six consecutive integers that are composites. (**Ans.** 90, 91, 92, 93, 94, 95)
5. Find the primes such that the digits in their decimal values alternate between 0's and 1's beginning and ending with 1. (**Ans.** 101)
6. If p and $p^2 + 8$ are primes, then show that $p^3 + 4$ is also a prime number.
7. Show that every integer greater than or equal to 9 has a prime factor.
8. If p_r is the rth prime, show that
$$p_{n+1} \le p_1 \cdot p_2 \cdots p_n + 1 \qquad n \ge 1.$$
9. Show that one more than the product of two twin primes is a perfect square.
10. Show that 1601 is a prime number.

Objective Type Questions

Fill in the blanks:

1. Any integer $p > 1$ which has no divisor except 1 and itself is called .... number
2. Any integer greater than 1, which is not prime is called .... number.
3. A prime number that reads the same backward as well as forward is called .... primes.

4. The number of primes of the form $(4n - 1)$ is ....

5. If the smallest prime factor of n is greater than $n^{1/3}$, then n has only .... factors which are prime.

True/False: *Write 'T' for true and 'F' for false statement.*

1. If p is the rth prime number then $p_r \le 2^{2^r - 1}$. (T/F)

2. If $n > 3$ then $p^2_{n+1} < p_1 p_2 \cdots p_n$ (T/F)

3. For each positive integer n, there exist n consecutive numbers each one of which is prime. (T/F)

4. If p is prime then there exist no positive integers a and b such that $a^2 = pb^2$ (T/F)

5. Every odd number greater than equal to 9 is the sum of three odd primes. (T/F)

Multiple Choice Questions : *Choose the most appropriate one :*

1. If p and q are twin prime then $12/(p + q)$ when
   (a) $p = 1$ (b) $p = 2$
   (c) $p > 3$ (d) none of these

2. The number $n^4 + 4^n$ is composite for all n
   (a) less than 1
   (b) greater than 1
   (c) greater than equal to 1
   (d) none of these

3. If n is not divisible by any prime $\le \sqrt{n}$ then n is
   (a) prime   (b) composite
   (c) may or may not be prime

(d) none of these

4. If p is prime and $p \,|\, a_1.a_2 \ldots a_n$ then $p \,|\, a_k$ where
   (a) $1 = k$ (b) $n = k$
   (c) $1 \le k \le n$ (d) none of these

5. Every prime of the form $3m + 1$ is necessarily of the form (where m is a positive integer)
   (a) $6k + 1$ (b) $6k + 2$
   (c) $6k + 3$ (d) none of these

6. If a and b are relatively prime positive integers, then the arithmetic progression $a, a + b, a + 2b, \ldots$ contains infinitely many prime. This is called
   (a) Dirichlet's theorem
   (b) Euler's theorem
   (c) Goldbach's conjecture
   (d) none of these

7. For each prime $p \ge 5$, $p^2 + 2$ is a
   (a) prime number
   (b) composite number
   (c) may or may not be prime
   (d) none of these

8. The prime which can be expressed as $x^7 - 1$, (where x is an integer) is
   (a) 128   (b) 127
   (c) 126   (d) none of these

9. If a and b are any two odd integers then $a^2 + b^2$ is
   (a) a perfect square
   (b) not a perfect square
   (c) may be a perfect square
   (d) none of these

10. If $a \,|\, c$ and $b \,|\, c$ then $(a, b)[a, b] = \ldots$
    (a) a (b) b
    (c) ab (d) none of these

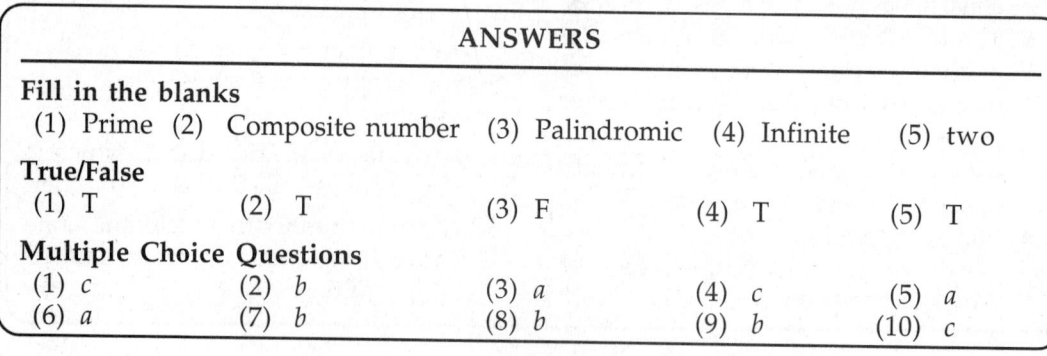

ANSWERS

Fill in the blanks

(1) Prime (2) Composite number (3) Palindromic (4) Infinite (5) two

True/False

(1) T   (2) T   (3) F   (4) T   (5) T

Multiple Choice Questions

(1) c (2) b (3) a (4) c (5) a
(6) a (7) b (8) b (9) b (10) c

CHAPTER 5

Congruences

❖ Congruence	❖ Least and Minimal Residue
❖ Complete and Reduced Residue Systems	❖ Divisibility Tests
❖ Linear Congruence	❖ Chinese Remainder Theorem
❖ System of Congruences	❖ Congruence of Higher Degree

5.1 INTRODUCTION

The theory of congruences was introduced by Carl Friedrich Gauss (1777-1855), one of the greatest mathematician of all times. Gauss contributed to the theory of numbers in many outstanding ways, including the basic idea of this chapter. Although, Pierre de Fermat (1601-1665) has earlier studied number theory in a somewhat systematic way, Gauss was first to develop the subject as a branch of mathematics, rather than just a scattered collection of interesting problems. In his book 'Disquisitioner Arithmetic' written at age 24, Gauss introduced the theory of congruences, which gained ready acceptance as a fundamental tool for the study of number theory.

5.2 CONGRUENCE

Let m be a positive integer. Two integers a and b leave remainders when divided by m. If the remainders are the same, we say that a is congruent to b modulo m and write

$a \equiv b \pmod{m}$ or $a \equiv b \ (m)$

Clearly, if a and b are congruent modulo m, then $m \mid a - b$; if a and b are incongruent modulo m, then $m \nmid ab$.

Hence, the necessary and sufficient condition that $a \equiv b \pmod{m}$ is $m \mid (a - b)$.

Remark

❖ Congruence is an equivalence relation.

5.3 PROPERTIES OF CONGRUENCES

Congruences have many properties in common with inequalities. Some properties are listed in the following theorems.

THEOREM-1

Congruence relations satisfy the following properties of equivalences.
(1) **Reflexive law** : *If a is any integer, than* $a \equiv a \pmod{m}$
(2) **Symmetric law** : *If* $a \equiv b \pmod{m}$, *then* $b \equiv a \pmod{m}$
(3) **Transitive law** : *if* $a \equiv b \pmod{m}$, $b \equiv c \pmod{m}$, *then* $a \equiv c \pmod{m}$

Proof. (1) We have

$$a - a = 0 . m \text{ which gives } m \mid (a - a)$$
$$\Rightarrow \qquad a \equiv a \pmod{m}$$

(2) $\qquad a \equiv b \pmod{m} \Rightarrow m \mid (a - b)$
$\Rightarrow \qquad m \mid (b - a) \quad \Rightarrow b \equiv a \pmod{m}$

(3) $\qquad a \equiv b \pmod{m}, \ b \equiv c \pmod{m}$
$\Rightarrow \qquad m \mid (a - b), \ m \mid (b - c)$
$\Rightarrow \qquad m \mid (a - b + b - c)$
$\Rightarrow \qquad m \mid (a - c)$
$\Rightarrow \qquad a \equiv c \pmod{m}$

THEOREM-2

If $a_1 \equiv b_1 \pmod{m}$, $a_2 \equiv b_2 \pmod{m}$, *then*
$$a_1 \pm a_2 \equiv b_1 \pm b_2 \pmod{m}$$
$$a_1 \, a_2 \equiv b_1 \, b_2 \pmod{m}$$
$$ca_1 \equiv cb_1 \pmod{m}, \ \textit{for any integer c.}$$

Proof. Here, we have
$$a_1 \equiv b_1 \pmod{m} \quad \Rightarrow m \mid (a_1 - b_1)$$
Similarly, $\quad a_2 \equiv b_2 \pmod{m} \quad \Rightarrow m \mid (a_2 - b_2)$
Then, $\qquad m \mid \{(a_1 - b_1) \pm (a_2 - b_2)\} = (a_1 + a_2) - (b_1 + b_2)\}$
i.e. $\qquad a_1 + a_2 \equiv b_1 + b_2 \pmod{m}$
Now, since $m \mid (a_1 - b_1), m \mid (ca_1 - cb_1)$
and therefore,
$$ca_1 \equiv cb_1 \pmod{m}$$
Further, $\quad a_1 \equiv b_1 \pmod{m}$ gives $a_1 a_2 \equiv b_1 a_2 \pmod{m}$
Similarly, $\quad b_1 a_2 \equiv b_1 b_2 \pmod{m}$
$$a_1 a_2 \equiv b_1 b_2 \pmod{m}$$

DEDUCTION :

In particular, if $a \equiv b \pmod{m}$, then for any positive integer n, we have
$$a^n \equiv b^n \pmod{m}$$

Remark

❖ In case of congruence, we can not divide both sides by an integer directly.

THEOREM-3 (CANCELLATION LAW)

If $ca \equiv cb \pmod{m}$, $(c, m) = 1$, *then*
$\quad a \equiv b \pmod{m}$

Proof. Since $ca \equiv cb \pmod{m}$
Therefore, $m \mid c (a - b)$ and $(c, m) = 1$

Then, $m \mid (a - b)$

$\Rightarrow$ $a \equiv b \pmod{m}$

DEDUCTION :

If $ca \equiv cb \pmod{m}$, $(c, m) = d$, then $a \equiv b \pmod{\dfrac{m}{d}}$

Proof. From $m \mid c (a - b)$, we have $\dfrac{m}{d} \mid \dfrac{c}{d} (a - b)$

But, $\left(\dfrac{c}{d}, \dfrac{m}{d} \right) = 1.$ Hence, $\dfrac{m}{d} \mid (a - b)$

THEOREM-4

If $a \equiv b \pmod{m}$, $a \equiv b \pmod{n}$ and $k = [m, n]$, then $a \equiv b \pmod{k}$.

Proof. As per given,

$a \equiv b \pmod{m}$ $\Rightarrow$ $m \mid (a - b)$

Also $a \equiv b \pmod{n}$ $\Rightarrow$ $n \mid (a - b)$

Therefore, $a - b$ is a common multiple of m and n. Then, it is also the multiple of their least common multiple k, thus $k \mid (a - b)$.

$\Rightarrow$ $a \equiv b \pmod{k}$

DEDUCTION :

If $a \equiv b \pmod{m}$, $a \equiv b \pmod{n}$ and $(m, n) = 1$ Then, $a \equiv b \pmod{mn}$

THEOREM-5

The necessary and sufficient condition that a positive integer n can be divided by 3 is that the sum of its digits is divisible by 3.

Proof. Any number n can be written as

$n = a_0 + a_1(10) + a_2(10)^2 + \ldots a_{k-1}(10)^{k-1}, 0 \le a_i < 1$

Since $10 \equiv 1 \pmod 3$, we have

$a_0 + a_1.10 + a_2.10^2 + \ldots a_{k-1}.10^{k-1} = a_0 + a_1 + a_2 + \ldots a_{k-1} \pmod 3$

Hence, $n \equiv 0 \pmod 3$ if and only if

$a_0 + a_1 + a_2 + \ldots a_{k-1} \equiv 0 \pmod 3$

Remarks

❖ Since $10 \equiv 1 \pmod 9$, the necessary and sufficient condition that n can be divided by 9 is that $a_0 + a_1 + \ldots + a_{k-1} \equiv 0 \pmod 9$

SOLVED EXAMPLES

Example 1. *Show that any integer satisfies at least one of the following five congruences*

$x \equiv 0 \pmod 2$, $x \equiv 0 \pmod 3$, $x \equiv 1 \pmod 4$

$x \equiv 5 \pmod 6$, $x \equiv 7 \pmod{12}$

Solution. We know that any even integer satisfies

$x \equiv 0 \pmod 2$

For modulo 12, all odd integers can be classified into six classes i.e.

$$12k + 1, 12k + 3, 12k + 5, 12k + 7, 12k + 9, 12k + 11$$

where k is any integer. Obviously, the classes $12k + 3, 12k + 9$ satisfy

$$x \equiv 0 \ (\text{mod } 3)$$

The classes $12k + 1, 12k + 3$ satisfy $x \equiv 1 \ (\text{mod } 4)$ and the classes $12k + 7, 12k + 11$ satisfy $x \equiv 7 \ (\text{mod } 12)$, $x \equiv 5 \ (\text{mod } 6)$, respectively.

Example 2. *Show that $x^2 + y^2 = z^2$ has no solution consisting of only primes, i.e., no prime solution.*

Solution. Let, if possible, $x = a, y = b, z = c$ form a prime solution.

If $a = 2$, from $c^2 - b^2 = a^2$, we get

$$(c - b)(c + b) = 4$$

$\Rightarrow \qquad c + b = 4, c - b = 1$

i.e. $\qquad c = 5/2$, which is not possible.

Thus, a and b are both odd.

Then, $\qquad a^2 \equiv 1 \ (\text{mod } 4)$

$\qquad\qquad b^2 \equiv 1 \ (\text{mod } 4)$

Hence, $a^2 + b^2 \equiv 2 \equiv c^2 \ (\text{mod } 4)$ and $2 \mid c$, which is a contradiction. This contradiction shows that a, b, c all are not prime.

Example 3. *Obtain the necessary and sufficient condition that a positive integer n can be divided by 7.*

Solution. Since, we have

$$1000 \equiv -1 \ (\text{mod } 7)$$

We can write $n = a_0 + a_1(1000) + a_2(1000)^2 + \dots + a_{k-1}(1000)^{k-1}, 0 \leq a_i < 1000$

Then, we get required condition

$$(a_0 + a_2 + \dots) - (a_1 + a_3 + \dots) = \sum_{i=0}^{k-1} (-1)^i a_i \equiv 0 \ (\text{mod } 7)$$

For example :

$$n = 637693 = 693 + 637 \ (1000)$$

$\Rightarrow \qquad a_0 = 693, a_1 = 637$

Since, $\qquad a_0 - a_1 = 693 - 637 = 56 \equiv 0 \ (\text{mod } 7)$

Hence, 637693 is a multiple of 7.

5.4   LEAST AND MINIMAL RESIDUE

Definition (1) : *If $a \equiv b \ (\text{mod } m), \ 0 \leq b < m$, then b is known as the least residue of a (mod m).*

Definition (2) : *If $a \equiv b \ (\text{mod } m), \ 0 \leq |b| \leq m/2$, then m is called the minimal residue of a (mod m).*

Remarks

❖ Minimal residue is also called the absolutely least residue of a (mod m).

SOLVED EXAMPLES

Example 1. *Find the remainder when 5^{48} is divided by 24.*

Solution. Here, we have

$$5^{48} = (5^2)^{24}$$
$$= (25)^{24}$$
$$= (1)^{24} \text{ (mod 24)}$$
$$= 1 \text{ (mod 24)}$$

Hence, the required remainder is 1.

Example 2. *Find the remainder when 2^{340} is divided by 341.*

Solution. Here, we have

$$341 = 11 \times 31$$
$$340 = 68 \times 5$$

Further, $2^5 = 32 = -1 \text{ (mod 11)}$

Thus, $(2)^{340} = (2^5)^{68} = 32^{68}$
$$= [(-1 \text{ (mod 11)}]^{68}$$
$$= (-1)^{68} \text{ (mod (11))}$$
$$= 1 \text{ (mod 11)}$$

Similarly,

$$2^5 = 1 \text{ (mod 31)}$$
$$\therefore \quad (2^5)^{68} = 1 \text{ (mod 31)}$$
$$\Rightarrow \quad 2^{340} = [\text{(mod 11} \times \text{31)}]$$
$$= 1 \text{ (mod 341)}$$

Example 3. *Find the remainder when the sum*
$S = 1! + 2! + 3! + \dots + 1000!$ *is divisible by 8.*

Solution. We know that $k!$ is divisible by 8 for all $k \geq 4$, we have

$$S = 1! + 2! + 3! + \dots + 1000!$$
$$= 1! + 2! + 3! \text{ (mod 8)}$$
$$= 1 + 2 + 6$$
$$= 9$$
$$= 1 \text{ (mod 8)}$$

Hence, the required remainder is 1.

Example 4. *Find the positive integer n for which $\sum_{1}^{n} k!$ is a square.*

Solution. We know that when $k \geq 5$, $k! \equiv 0 \text{ (mod 10)}$ [Since each $k!$ for $k \geq 5$ have 5 and 2 as it's factor] so let $n \geq 5$. If S denote the given sum. Then

$$S = \text{ones digit in } \sum_{1}^{n} k! \text{ (mod 10)}$$
$$= (1! + 2! + 3! + 4!) \text{ (mod 10)}$$
$$= (1 + 2 + 6 + 24) \text{ (mod 10)}$$
$$= 3 \text{ (mod 10)}$$

Thus, the ones digit in S is 3 if $n \geq 5$.

But square of every integer must end in 0, 1, 4, 5, 6 or 9. Thus, if $n \geq 5$, S can be a square.

When $n = 1$, S = 1 and when $n = 3$, S = 9 so the perfect square but S is not a square when $n = 2$ or 4.

Thus, there are exactly two positive integer n for which S is a square namely 1 and 3.

Example 5. *Find the remainder when* $1! + 2! + ... 100!$ *is divided by* $15!$.

Solution. When $k \geq 5$, $k!$ is divisible by 15 since there is 5 and 3 as factor in each $k!$ for $k \geq 5$.

So, $1! + 2! + ... 100! \equiv 1! + 2! + 3! + 4! + 0 + ... 0 \pmod{15}$
$$\equiv 1 + 2 + 6 + 24 \pmod{15}$$
$$\equiv 1 + 2 + 0 \pmod{15}$$
$$\equiv 3 \pmod{15}$$

Thus the remainder is 3 when the given sum is divided by 15.

Example 6. *Find the remainder when* 16^{53} *is divided by* 7.

Solution. First reduce the base to the least residue
$$16 \equiv 2 \pmod 7 \text{ so}$$
$$16^{53} \equiv 2^{53} \pmod 7$$

Now, express a suitable power of 2 congruent modulo 7 to a number less than $7 : 2^3 \equiv 1 \pmod 7$.

$\therefore$ $2^{53} = 2^{3 \times 17 + 2} = (2^3)^{17+2}$
$$\equiv 1^{17} . 4 \pmod 7$$
$$= 4 \pmod 7$$

So, $16^{53} = 4 \pmod 7$. Thus, when 16^{53} is divided by 7, the remainder is 4.

Example 7. *Find the remainder when* 3^{247} *is divided by* 17.

Solution. We have
$$3^3 = 27 \equiv 10 \pmod{17}$$

Squaring both sides, we get
$$3^6 \equiv 100 \pmod{17}$$
$$\equiv -2 \pmod{17}$$

Raise 4th power on both sides
$$3^{24} \equiv (-2)^4 \pmod{17}$$
$$\equiv -1 \pmod{17}$$

Now, $3^{247} = 3^{24 \times 10 + 7} = (3^{24})^{10} . 3^6 3$
$$\equiv (-1)^{10} . (-2) 3 \pmod{17}$$
$$\equiv -6 \pmod{17}$$

Change -6 to positive least residue
$$\equiv 11 \pmod{17}$$

Thus remainder is 11.

EXERCISE 5.1

1. Show that 41 divides $2^{20} - 1$.

2. Show that if $a \equiv b \pmod m$, then $(a, m) = (b, m)$.

3. Find the remainder when

(i) 2^{14} is divided by 17.

(ii) 3^{287} is divided by 23.

(iii) 11^{35} is divided by 13.

4. Show that $53^{103} + 103^{53}$ is divisible by 39.

5. Show that $504 \mid (n^9 - n^3)$, where n is any integer.

6. If a, b are any integers, p is prime, show that $(a + b)^p \equiv a^p + b^p \pmod{p}$

7. Show that every number containing more than 3 digits can be divided by 8 if and only if the number formed by its last 3 digits is a multiple of 8.

8. Show that $13 \mid (3^{n + 2} + 4^{2n + 1})$.

9. If n is odd, then show that $a^{2n} \equiv 1 \pmod{2^{n + 2}}$, $n \geq 1$.

10. Show that, if a is any integer, then

$$a^2 + a + 1 \equiv 1 \pmod{3} \quad \text{or} \quad a^2 + a + 1 \equiv 0 \pmod{3}$$

ANSWERS

(3) (*i*) 13   (*ii*) 3   (*iii*) 6

5.5 COMPLETE AND REDUCED RESIDUE SYSTEMS

In dealing with integers modulo m, we are performing the ordinary operations of arithmetic but are disregarding multiple of m. Therefore, we are not distinguishing between a and $a + mx$, where x is any integer. For any integer a, let q and r be the quotient and remainder on division by m, so $a = qm + r$. Now, $a \equiv r \pmod{m}$ and since r satisfies the inequalities $0 \leq r < m$, we see that every integer is congruent modulo m to one of the values 0, 1, 2, ..., $m - 1$. Also, no two of these m integers are congruent modulo m. These m values constitute a complete residue system modulo m.

Definition (1) : *If $x \equiv y \pmod{m}$, then y is called a residue of x modulo m. A set $x_1, x_2, ..., x_m$ is called a complete residue system (CRS) modulo m if for every integer y there is one and only one x_i such that $y \equiv x_i \pmod{m}$.*

Definition (2) : *A reduced residue system (RRS) modulo m is a set of integers r_i such that $(r_i, m) = 1$, $r_i \not\equiv r_j \pmod{m}$ if $i \neq j$ and such that every x prime to x is congruent modulo m to some number r_i of the set.*

Remarks

❖ These are infinitely many complete residue system modulo m.

❖ A set of m integers form a complete residue system modulo m if and only if no two integers in the set are congruent modulo m.

❖ For fixed integers a and $m > 0$, the set of all integers x satisfy $x \equiv a \pmod{m}$ is the arithmetic progression ..., $a - 3m, a - 2m, a - m, a, a + m, a + 2m, a + 3m, ...$ This set is called a residue class or congruence class, modulo m.

THEOREM-1

A set of k integers $a_1, a_2, ... a_k$ is a complete residue system modulo m if and only if
(i) $k = m$

(ii) $a_i \neq a_j \pmod{m}, i \neq j$

Proof. Let $a_1, a_2, ..., a_k$ be a complete residue system modulo m, then they must have been taken out one each from m different classes. Therefore, the number of a_i's is m, i.e. $k = m$.

Since any two integers $a_i, a_j, (i \neq j)$ belong to different classes
$$a_i \equiv a_j \pmod{m}$$

Conversely, if $a_1, a_2, ..., a_m$ are m integers such that $a_i \neq a_j \pmod{m}, i \neq j$, then a_i, a_j do not belong to the same class. Hence, these m integers distribute in m different classes, such that every class has just an integer, therefore $a_1, a_2, ..., a_m$ form a complete residue system modulo m.

Remark

❖ Any m incongruent integers modulo m form the complete residue system modulo m.

THEOREM-2

If $a_1, a_2, ..., a_m$ is a complete residue system modulo m and (a, m) = 1, then
$$aa_1 + b, aa_2 + b, ..., aa_m + b, b \text{ is any integer.}$$
are also a complete residue system modulo m.

Proof. In view of theorem-1, we have only to prove that
$$aa_i + b \neq aa_j + b \pmod{m}, i \neq j$$

Let, if possible
$$au_i + b \equiv aa_j + b \pmod{m}$$

Then, $aa_i \equiv aa_j \pmod{m}$. But $(a, m) = 1$

Therefore, $a_i = a_j \pmod{m}$, which is a contradiction, because $a_k ... a_m$ is a complete residue system of m. Hence, $aa_i + b$ is a complete residue system modulo m.

5.6   EULER'S FUNCTION

If a is prime to m, any integer in the class to which a belongs is prime to m; if a is not prime to m, any integer in that class are not prime to m. Thus, the number of integers in the reduced residue is independent of the complete residue system.

It is uniquely determined by m. Euler denoted it by $\phi(m)$ and it is called Euler's function defined as follows :

"*$\phi(m)$ expresses the number of positive integers not exceeding m which are relatively prime to m.*"

For Euler's function, we have to following useful table

m	1	2	3	4	5	6	7	8	9	10
$\phi(m)$	1	1	2	2	4	2	6	4	6	4

Definition : *Let m be a positive integer. If m is prime, then $\phi(m) = m - 1$. If m is composite, then $\phi(m) < m - 1$. Thus, for any integer m (> 1), we have $\phi(m) \le m - 1$.*

Remarks

❖ m is prime if and only if $\phi(m) = m - 1$

SOLVED EXAMPLES

Example 1. *Show that the set of integers $\{-19, -1, 22, 43, 46, 113, 452\}$ is a RRS (mod 15).*

Solution. The least residue (mod 15) of the integers of the given set are

$$11, 14, 7, 13, 1, 4, 8, 2 \qquad \qquad ...(1)$$

Further, the integers less than 15 and prime to 15 are

$$1, 2, 4, 7, 8, 11, 13, 14 \qquad \qquad ...(2)$$

Clearly, (1) is a permutation of (2). Hence, the given set of integers is RRS (mod 15).

Example 2. *Show that the set $\{-3, 14, 3, 12, 37, 57, -1\}$ (mod 7) is a complete system of residue.*

Solution. Since, the set consists of seven integers whose least residue (mod 7) are 4, 0, 3, 5, 2, 1, 6 which is a permutation of the numbers 0, 1, 2, 3, 4, 5, 6. Hence, the given set is a complete system of residue.

Example 3. *If $m \equiv 0$ (mod 2) $a_1, ..., a_m$ and $b_1, ... b_m$ are two complete residue systems, modulo m, then show that $a_1 + b_1, ... a_m + b_m$ is not a complete residue system modulo m.*

Solution. As per given, $a_1, ..., a_m$ form a complete residue system modulo m.

$$\sum_{i=1}^{m} a_i \equiv \sum_{i=1}^{m} i = \frac{m(m+1)}{2} = \frac{m}{2} \ (\text{mod } m).$$

Similarly, $\sum b_i \equiv \frac{m}{2} \ (\text{mod } m)$

If $a_1 + b_1, ..., a_m + b_m$ also form a complete residue system modulo m, then

$$\sum (a_i + b_i) \equiv \frac{m}{2} \ (\text{mod } m)$$

but $\qquad \sum (a_i + b_i) \equiv \frac{m}{2} + \frac{m}{2} \equiv m \equiv 0 \ (\text{mod } m)$

Hence, $\dfrac{m}{2} \equiv 0$ (mod m) is not allowed.

Remark

❖ When $m > 2$, if $a_1, ..., a_m$ and $b_1, ..., b_m$ are two complete residue system of m, then $a_1 b_1, ..., a_m b_m$ is not a complete residue of m.

5.7 SPECIAL DIVISIBILITY TESTS

We can use congruence to test whether a given integer is divisible by another integer or not.

THEOREM-1

Let n be a positive integer such that
$$n = a_m 10^m + a_{m-1} 10^{m-1} + \ldots a_1.10 + a_0,\ q_m \neq 0, 0 \leq a_i < 10, 0 \leq i < m - 1$$

Then,   (1) $2 \mid n$ if and only if $2 \mid a_0$.

(2) $3 \mid n$ if and only if $3 \mid \sum\limits_{i=1}^{m} a_i$

(3) $4 \mid n$ if and only if $4 \mid a_0 + 2a_1$

(4) $5 \mid n$ if and only if $5 \mid a_0$

(5) $9 \mid n$ if and only if $9 \mid \sum\limits_{i=1}^{m} a_i$

(6) $11 \mid n$ if and only if $11 \mid \sum\limits_{i=0}^{m} (-1)^i a_i$

Proof. Since

$$10 = \begin{bmatrix} 0\ (\bmod\ 2)\ and\ (\bmod\ 5) \\ 1\ (\bmod\ 3)\ and\ (\bmod\ 9) \\ 2\ (\bmod\ 4) \\ -1\ (\bmod\ 11) \end{bmatrix}$$

Therefore,

$$n = \begin{bmatrix} a_0\ (\bmod\ 2)\ and\ (\bmod\ 5) \\ \Sigma a_i\ (\bmod\ 3)\ and\ (\bmod\ 9) \\ 2a_1 + a_0 (\bmod\ 4) \\ \Sigma (-1)^i\ a_i (\bmod\ 11) \end{bmatrix} \qquad \ldots(1)$$

Using (1), we can easily deduce the required results.

THEOREM-2 (DIVISIBILITY BY 7 OR 13)

A number $n = (a_{k-1}, a_{k-2}, \ldots, a_1 a_0)$ is divisible by 7 or 13 if and only if
$(a_2 a_1 a_0) - (a_5 a_4 a_3) + (a_8 a_7 a_6)$ is divisible by 7 or 13.

Proof. Let $n = (a_{k-1}, a_{k-2}, \ldots, a_1 a_0)$. Then, we can write
$$n = (a_2 a_1 a_0) + (a_5 a_4 a_3) \times 1000 + (a_8 a_7 a_6) \times 1000^2 + \ldots$$
Also, $1000 \equiv -1 \ (\bmod\ 7\ or\ 13)$
Therefore
$$n \equiv (a_2 a_1 a_0) + (a_5 a_4 a_3)(-1) + (a_8 a_7 a_6)(-1)^2 \ldots (\bmod\ 7\ or\ 13)$$
Hence, we conclude that n is divisible by 7 or 13 if and only if the right hand side of above congruence is divisible by 7 or 13.

WORKING PROCEDURE (A Test for Divisibility by 7)

Starting with any positive integer n, subtract double the units digits from the integer obtained from n by removing the unit digits, giving a smaller integer r. For example, if $n = 41283$ with unit digit 3, we subtract 6 from 4128 to get 4122. Now unit digit is 2 so subtract 4 (= 2 × 2) from 412 to get 408 and then to 24 by subtracting 16 from 40. Since 24 is not divisible by 7. Hence 41283 is not divisible by 7.

THEOREM-3 (DIVISIBILITY BY 37)

A number $n = (a_{k-1}, a_{k-2}, ..., a_1a_0)$ *is divisible by 37 if*

$$(a_2a_1a_0) + (a_5a_4a_3) + ... \text{ is divisible } 37.$$

Proof. As per given

$$n = (a_{k-1}, a_{k-2}, ..., a_1a_0).$$

The given number n can be written as

$$n = (a_2a_1a_0) + (a_5a_4a_3)\ 1000 + (a_8a_7a_6) \times 1000^2 + ...$$

Since, $1000 \equiv 1 \pmod{37}$, therefore

$$n = (a_2a_1a_0) + (a_5a_4a_3) + ... \pmod{37}$$

Hence, we conclude that n is divisible by 37 if RHS is divisible by 37.

SOLVED EXAMPLES

Example 1. *Show that the number 5117247 is divisible by 9.*

Solution. We can write the given number as

$$5117247 = 5 \times 10^6 + 1 \times 10^5 + 1 \times 10^4 + 7 \times 10^3 + 2 \times 10^2 + 4 \times 10 + 7$$

The sum of coefficients is given by

$$= 5 + 1 + 1 + 7 + 2 + 4 + 7 = 27$$

Further, since 9 | 27, hence 9 | 5117247.

Example 2. *Show that the number 1571427 is divisible by 11.*

Solution. The alternating sum of the digits is given by

$$1 - 5 + 7 - 1 + 4 - 2 + 7 = 11$$

Which is divisible by 11. Hence, given number is divisible by 11.

Example 3. *Show that the number 2587322568103 is divisible by 7 and 13.*

Solution. We have

$$103 - 568 + 322 - 587 + 2 = -728$$

which is divisible by 7 and 13. Hence, the given number is divisible by 7 and 13.

1.8 LINEAR CONGRUENCE

Definition (1) : *An expression of the form*

$$ax \equiv b \pmod{m}, \ a \not\equiv 0 \pmod{m} \qquad ...(1)$$

is called a linear congruence **mod** *m.*

Solution : Equation (1) can also be written as

$$ax - b \equiv 0 \ (\text{mod } m) \qquad \qquad ...(2)$$

Consider $ax \equiv b \ (\text{mod } m)$. Then $ax = my + b$ for some integer y. Clearly $ax \equiv b \ (\text{mod } m)$ is solvable if and only if linear diophantine equation $ax - my = b$ is solvable and we have already observed that linear diophantine equation may have infinitely many solutions. Thus, if the congruence $ax \equiv b \ (\text{mod } m)$ is solvable, it has infinitely many solutions. For instance, since 3 is solution of linear congruence $3x \equiv 4 \ (\text{mod } 5)$ every member of the congruence class $[3] = \{..., -7, -2, 3, 8, 13, ...\}$ is a solution. They are given by $x = 3 + 5t$. Consequently, we are interested in the incongruent solutions only.

For example, the congruence $9x \equiv 6 \ (\text{mod } 12)$ has three incongruent solution only namely 2, 6 and 10 as $9 \cdot 2 \equiv 6 \ (\text{mod } 12)$, $9 \cdot 6 \equiv 6 \ (\text{mod } 12)$ and $9 \cdot 10 \equiv 6 \ (\text{mod } 12)$.

THEOREM-1

The linear congruence $ax \equiv b \ (\text{mod } m)$ is solvable if and only if $d \mid b$ where $d = (a, m)$. If $d \nmid b$, then it has d incongruent solutions.

Proof. The linear congruence $ax \equiv b \ (\text{mod } m)$ is equivalent to the linear diophantine equation $ax - my = b$; so the congruence is solvable if and only if the linear diophantine equation is solvable. But the linear diophantine equation is solvable if and only if $d \mid b$. Thus, $ax \equiv b \ (\text{mod } m)$ is solvable if and only if $d \mid b$. When $d \nmid b$, the linear diophantine equation has infinitely solution, given by $x = x_0 + \left(\dfrac{m}{d}\right)t$, $y = y_0 + \left(\dfrac{a}{d}\right)t$. So the congruence has infinitely many solutions $x = x_0 + \left(\dfrac{m}{d}\right)t$, where x_0 is a particular solution.

To find the number of incongruent solutions, when congruence is solvable.

Suppose $x_1 = x_0 + \left(\dfrac{m}{d}\right)t_1$, $x_2 = x_0 + \left(\dfrac{m}{d}\right)t_2$ are two congruent solutions. Then

$$x_0 + \left(\frac{m}{d}\right)t_1 \equiv x_0 + \left(\frac{m}{d}\right)t_2 \ (\text{mod } m) \quad \Rightarrow \quad \left(\frac{m}{d}\right)t_1 \equiv \left(\frac{m}{d}\right)t_2 \ (\text{mod } m)$$

Since, $\dfrac{m}{d} \mid m$, then $t_1 \equiv t_2 \ (\text{mod } d)$. Thus, the solutions x_1 and x_2 are congruent if and only if $t_1 \equiv t_2 \ (\text{mod } d)$ that is, if and only if t_1 and t_2 belong to the same congruence class modulo d. In other words, they are incongruent solution if and only if they belong to distinct congruence class, but there are exactly d incongruent class modulo d. Therefore, the linear congruence, when solvable has exactly d incongruent solution given by

$$x = x_0 + \left(\frac{m}{d}\right)t, \text{ where } 0 \le t < d$$

Remark

The linear congruence $ax \equiv b \ (\text{mod } m)$ has unique solution if and only if $(a, m) = 1$.

Example.1 *Determine if the congruences* $8x \equiv 10 \pmod 6$, $2x \equiv 3 \pmod 4$ *and* $4x \equiv 7$ *(mod 5) are solvable. Find the number of solutions when a congruence is solvable.*

Solution.

From $8x \equiv 10 \pmod 6$, we have $(8, 6) = 2$ and $2 \mid 10$ so the congruence $8x \equiv 10 \pmod 6$ is solvable and it has two incongruent solutions modulo 6.

From $2x \equiv 3 \pmod 4$, we have $(2, 4) = 2$ but $2 \nmid 3$, so the congruence has no solution.

From $4x \equiv 7 \pmod 5$, we have $(4, 7) = 1$, so the congruence has only one incongruent solution.

5.9 CHINESE REMAINDER THEOREM

Let $m_1, m_2, ..., m_r$, be pairwise relatively prime numbers such that g.c.d. $(m_i, m_j) = 1$ for $i \neq j$. Then, the equation of linear congruences

$$x \equiv a_1 \pmod{m_1}$$
$$x \equiv a_2 \pmod{m_2}$$
$$\vdots$$
$$x \equiv a_r \pmod{m_r}$$

has a simultaneous solution which is unique modulo $m_1, m_2, ... m_r$.

Proof. Let us write

$$m = m_1 m_2 ... m_r$$

Define $M_k = \dfrac{m}{m_k} = m_1 m_2 ..., m_{k-1} \cdot m_{k+1}, ..., m_k$ for each $k = 1, 2, ..., r$.

As per given, m_i are relatively prime in pairs, i.e. $(M_k, m_k) = 1$. Therefore, it is possible to solve the linear congruence $M_k x \equiv 1 \pmod{m_k}$. Let it be have a unique solution x_k.

To show

$$\bar{x} = a_1 M_1 x_1 + a_2 M_2 x_2 + ... a_r M_r x_r$$

is a simultaneous solution of the given system.

We have $M_i \equiv 0 \pmod{m_k}$, for $i \neq k$ as $m_k \mid m_i$.

Thus $\bar{x} = a_1 M_1 x_1 + ... + a_r M_r x_r \equiv a_k M_k x_k \pmod{m_k}$

But, x_k satisfies the congruence $M_k x \equiv 1 \pmod{m_k}$

$\therefore$ $\bar{x} = a_k \cdot 1 \equiv a_k \pmod{m_k}$

Now, we shall prove the uniqueness of this solution. Let if possible, $\bar{x}_1$ and $\bar{x}_2$ are two solutions of the given system. Then

$$\bar{x}_1 = a_k \equiv \bar{x}_2 \pmod{m_k}, k = 1, 2, ... r.$$

$\Rightarrow$ $m_k \mid (\bar{x}_1 - \bar{x}_2)$, for each k.

Since, $(m_i, m_j) = 1$, for $i \neq j$, we have $m_1 m_2 ... m_r \mid (\bar{x}_1 - \bar{x}_2)$

Hence, $\bar{x}_1 \equiv \bar{x}_2 \pmod m$.

Remarks

❖ In the Chinese remainder theorem, the hypothesis that the moduli m_j should be pairwise relatively prime is absolutely essential when this hypothesis fails, the existence of a solution x of the simultaneous system is no longer guaranteed and when such an x does not exist, it is unique modulo $(m_1, m_2, ... m_r)$ not modulo m.

❖ In case there is no solution, we call the system inconsistent.

IMPORTANT RESULTS

(1) The necessary and sufficient condition that the congruence
$$ax + by + c \equiv 0 \pmod{m} \qquad \qquad ...(1)$$
has solution is $d \mid c$, where $d = (a, b, m)$. If this condition holds, then (1) has $m \mid d$ solutions

(2) The system of congruences
$$\left. \begin{array}{l} x \equiv a \pmod{m} \\ x \equiv b \pmod{n} \end{array} \right\} \qquad \qquad ...(2)$$
has solution if and only if
$$a \equiv b \pmod{(m, n)}$$
If this condition is satisfied, then (2) has only one unique solution modulo $[m, n]$.

(3) A system of linear congruence $x \equiv a_i \pmod{m_i}$ is solvable if and only if (m_i, m_j) divides $(a_i - a_j)$

SOLVED EXAMPLES

Example 1. *Solve $12x + 15 \equiv 0 \pmod{45}$.*

Solution. Since $(12, 45) = 3$ and $3 \mid 15$.

Therefore, the given congruence has solution. Actually, it will have 3 solutions.

By inspection, the congruence
$$4x + 5 \equiv 0 \pmod{15}$$
has solution

∴ $\qquad x \equiv -5 \equiv 10 \pmod{15}$

Hence, the required solutions are
$$x \equiv 10, 10 + 15, 10 + 30 \pmod{45}$$
i.e. $\qquad x = 10, 25, 40 \bmod (45)$

Example 2. *Solve $9x \equiv 21 \pmod{30}$.*

Solution. Since $(9, 30) = 3$. Thus, there are three solutions.

To solve the given congruence, we proceed as follows :

Corresponding Diophantine equation is given by
$$9x + 30y = 21 \qquad \qquad ...(1)$$

$\Rightarrow \qquad 9x + (9 \times 3 + 3)y = 9 \times 2 + 3$

or $\qquad 9(x + 3y - 2) + 3y = 3$

or $\qquad 9u + 3y = 3$, where $u = x + 3y - 2$.

$\Rightarrow \qquad 3u + y = 1$

If $u = 0$, then $y = 1$.

Putting $y = 1$ in (1), we have $x = -1$. Further, the least residue -1 (mod 30) is 29.

The other two solutions will be $29 - \dfrac{30}{3}$ and $-29 - 2\dfrac{30}{3}$, i.e. -1 and -21.

Example 3. *Solve $103x \equiv 57$ (mod 211).*

Solution. Since 103 and 211 both are primes, the given congruence has only one solution.

Now, $\quad 211 = 2 \times 103 + 5$

Multiplying by 2, we get

$\qquad 206\,x \equiv 114$ (mod 211)

Now, since $211x \equiv 0$ (mod 211), subtracting it from the above congruence, we get

$\qquad 5x \equiv -114 \equiv 97$ (mod 211)

Now, since $211 = 42 \times 5 + 1$

$\qquad 42 . 5x \equiv 42 \times 97 \equiv 65$ (mod 211)

which gives $x \equiv -65$ (mod 211), which is the required solution.

Example 4. *Solve the linear congruence $12x \equiv 44$ (mod 59).*

Solution. Compare the given congruence with

$\qquad ax \equiv b(\text{mod } m),\ (a, m) = 1$ ...(1)

we get $a = 12, b = 44, m = 59$.

Also, $\qquad (12, 59) = 1$

From (1), we have

$$x \equiv \frac{b}{a} \ (\text{mod } m)$$

$$\equiv \frac{b + mh}{a} \ (\text{mod } m) \quad (\text{Remember})$$

$$\equiv \frac{44}{12} \ (\text{mod } 59)$$

$$\equiv 3 + \frac{2 + 2 \times 59}{3} \ (\text{mod } 59)$$

$$\equiv 43 \ (\text{mod } 59), \text{ which is the required solution.}$$

Example 5. *Solve $863\,x \equiv 880$ (mod 2151).*

Solution. The given congruence can also be written as

$\qquad 2151y \equiv -880$ (mod 863)

Therefore

$\qquad 425y \equiv -880$ (mod 863)

On cancelling the common factor, we have

$\qquad 85y \equiv -176$ (mod 863)

$\Rightarrow \qquad 863z \equiv 176$ (mod 85)

$\Rightarrow$ $13z \equiv 6 \pmod{85}$

Also, $85\,w \equiv -6 \pmod{13}$

i.e. $7w \equiv -6 \pmod{13}$

Hence, $w_0 \equiv 1$. Thus

$$z_0 \equiv \frac{85+6}{13} = 7$$

$$y_0 \equiv \frac{863 \times 7 - 176}{85} = 69$$

$$x_0 \equiv \frac{2151 \times 69 + 880}{863} = 173$$

Hence, the required solution is $x \equiv 173 \pmod{2151}$.

Since $(863, 2151) = 1$, the given congruence has only this unique solution.

Example 6. *Find the solution of* $47x \equiv 11 \pmod{249}$.

Solution. Since $(47, 249) = 1$.

Thus, $x \equiv \dfrac{11}{47} \pmod{249}$ $\left[\text{Using } x \equiv \dfrac{b}{a} \pmod{m} = \dfrac{b+mh}{a} \pmod{m}\right]$

$$\equiv \frac{11 + 16 \times 249 \pmod{249}}{47}$$

$$\equiv 85 \pmod{249}$$

Example 7. *Solve* $2x + 7y \equiv 5 \pmod{12}$.

Solution. The given congruence can also be written as

$$2x \equiv 5 - 7y \pmod{12}$$

Since $(2, 12) \equiv 2$, we have

$$7y \equiv 5 \pmod{12}$$

$\therefore$ $y \equiv 1 \pmod 2$

or $y = 1 + 2t$

Putting these values in the given congruence, we get

$$2x \equiv -2 - 14t \pmod{12}$$

i.e. $x \equiv -1 - 7t \pmod 6$

$\Rightarrow$ $x = -1 - 7t + 6s$

Therefore, all required solutions are given by

$$y \equiv 1 + 2t, \quad x \equiv -1 - 7t + 6s \pmod{12}$$

where, $t = 0, 1, 2, ..., 5,\ s = 0, 1$. There are following twelve sets of solutions in all, namely

$$\begin{cases} x \equiv 5 \\ y \equiv 1 \end{cases} \quad \begin{cases} x \equiv 11 \\ y \equiv 1 \end{cases} \quad \begin{cases} x \equiv 4 \\ y \equiv 3 \end{cases} \quad \begin{cases} x \equiv 10 \\ y \equiv 3 \end{cases}$$

$$\begin{cases} x \equiv 3 \\ y \equiv 5 \end{cases} \quad \begin{cases} x \equiv 9 \\ y \equiv 5 \end{cases} \quad \begin{cases} x \equiv 2 \\ y \equiv 7 \end{cases} \quad \begin{cases} x \equiv 8 \\ y \equiv 7 \end{cases}$$

$$\begin{cases} x \equiv 1 \\ y \equiv 9 \end{cases} \quad \begin{cases} x \equiv 7 \\ y \equiv 9 \end{cases} \quad \begin{cases} x \equiv 0 \\ y \equiv 11 \end{cases} \quad \begin{cases} x \equiv 6 \\ y \equiv 11 \end{cases}$$

Example 8. *Solve* $49x \equiv 47 \pmod{81}$.

Solution. Since $(49, 81) = 1$. Therefore, there is only one solution.

The corresponding Diophantine equation is

$$49x + 81y = 47 \qquad \qquad ...(1)$$

$$\Rightarrow 49x + (49 \times 2 - 17)y = 49 \times 1 - 2$$

$$\Rightarrow 49[x + 2y - 1] - 17y = -2$$

$$\Rightarrow \qquad 49u - 17y = -2, \text{ where } u = x + 2y - 1 \qquad \qquad ...(2)$$

$$\Rightarrow (17 \times 3 - 2)u - 17y = -2$$

$$\Rightarrow \qquad 17(3u - y) - 2u = -2$$

$$\Rightarrow \qquad \qquad 17v - 2u = -2, \text{ where } v = 3u - y \qquad \qquad ...(3)$$

$$\Rightarrow \qquad u = 1, \quad v = 0$$

Putting $u = 1$ in (2), we get $y = 3$.

Putting $y = 3$ in (1), we get $x = -4$.

Also, the least residue of $-4 \pmod{81}$ is 77.

Hence, the required solution of given congruence is 77 (mod 81).

SOLVED EXAMPLES (Based on System of Congruences)

Example 1. *Solve the system of congruences*

$$x \equiv 2 \pmod{3}; \ x \equiv 3 \pmod{5}; \ x \equiv 2 \pmod{7}$$

Solution. We have $m = 3 \cdot 5 \cdot 7 = 105$, $x_1 = 2$, $x_2 = 3$, $x_3 = 2$, $m_1 = 3$, $m_2 = 5$ and $m_3 = 7$.

Now, $\qquad M_1 = \dfrac{m}{m_1} = \dfrac{105}{3} = 35$

$$M_2 = \dfrac{m}{m_2} = \dfrac{105}{5} = 21$$

$$M_3 = \dfrac{m}{m_3} = \dfrac{105}{7} = 15$$

We solve the linear congruences

$$35x \equiv 1 \pmod{3}$$
$$21x \equiv 1 \pmod{5}$$
$$15x \equiv 1 \pmod{7}$$

Solving the above congruences, we get respectively the following congruences

$$x \equiv 2 \pmod{3}$$
$$x \equiv 1 \pmod{5}$$
$$x \equiv 1 \pmod{7}$$

Hence, the required solution is given by

$$\bar{x} = a_1 M_1 x_1 + a_2 M_2 x_2 + a_3 M_3 x_3$$
$$\bar{x} = 35 \cdot 2 \cdot 2 + 21 \cdot 1 \cdot 3 + 15 \cdot 1 \cdot 2 \equiv 233 \pmod{105}$$

Example 2. *Solve*

$$2x \equiv 1 \pmod{5}$$
$$3x \equiv 4 \pmod{7}$$

Solution. Solving $2x \equiv 1 \pmod{5}$, we get $x \equiv 3 \pmod{5}$. Also, from $3x \equiv 4 \pmod{7}$, we get $x \equiv 6 \pmod{7}$. Thus, the given system is equivalent to the system

$$x \equiv 3 \pmod 5$$
$$x \equiv 6 \pmod 7$$
...(1)

Let $x = 3 + 5y$, then $3 + 5y \equiv 6 \pmod 7$, thus $5y \equiv 3 \pmod 7$
Clearly, $y \equiv 2 \pmod 7$.
Hence, the required solution is given by
$$x = 3 + 5y \equiv 3 + 10 \equiv 13 \pmod{35}$$

Aliter. We can also solve (1) as follows
$$x = 3 + 5y = 6 + 7z, \text{ i.e. } 5y - 7z = 3$$
Clearly, $y = 2, \; z = 1$. Therefore
$$x = 3 + 10 = 6 + 7 = 13$$

Example 3. *Solve the system of linear congruence*
$$x \equiv 3 \pmod{11}$$
$$x \equiv 5 \pmod{19}$$
$$x \equiv 10 \pmod{29}$$

Solution. We have
$$m = 11 . 19 . 29 = 6061$$
$$x_1 = 3, x_2 = 5, x_3 = 10, m_1 = 11, m_2 = 19, m_3 = 29$$
$$M_1 = \frac{m}{m_1} = \frac{6061}{11} = 551$$
$$M_2 = \frac{m}{m_2} = \frac{6061}{19} = 319$$
$$M_3 = \frac{m}{m_3} = \frac{6061}{29} = 209$$

Thus, the given system reduces to
$$551x \equiv 1 \pmod{11}$$
$$319x \equiv 1 \pmod{19}$$
$$209x \equiv 1 \pmod{29}$$
which are satisfied by $x_1 = 1, x_2 = 14$ and $x_3 = 5$, respectively.
Further, a solution of the system is given by
$$\bar{x} = a_1 M_1 x_1 + a_2 M_2 x_2 + a_3 M_3 x_3$$
$$= 3 \times 551 \times 1 + 5 \times 319 \times 14 + 10 \times 209 \times 5$$
$$= 1653 + 22330 + 10450$$
$$= 34433$$
$$= 4128 \pmod{6061}$$
Hence, the unique solution is given by
$$\bar{x} \equiv 4128 \pmod{6061}$$

Example 4. *Find the last two digits of the number 9^{9^9}*
Solution. Here, we have
$$9^2 \equiv 1 \pmod{10}$$

$$\Rightarrow \qquad (9^2)^4 \equiv 1 \ (\text{mod } 10)$$
$$\Rightarrow \qquad 9^8 \equiv 1 \ (\text{mod } 10)$$

i.e., $\qquad 9^9 \equiv 9 \ (\text{mod } 10)$

Therefore,
$$9^9 \equiv 10k + 9 \qquad\qquad\qquad ...(1)$$

Again $\quad 999 = 99 + 10k$
$$= 9^9 \times 9^{10k} \qquad\qquad\qquad ...(2)$$

But $\qquad 9^{10} \equiv 1 \ (\text{mod } 100)$
$$\Rightarrow \qquad 9^{1-k} \equiv 1 \ (\text{mod } 100) \qquad\qquad\qquad ...(3)$$

From (2) and (3), we have
$$9^{9^9} \equiv 9^9 \times 1 \ (\text{mod } 100)$$
$$\equiv 9 \times 9^8 \ (\text{mod } 100) \equiv 9 \times 21 \ (\text{mod } 100)$$
$$\equiv 89 \ (\text{mod } 100)$$

5.10 CONGRUENCES OF HIGHER DEGREE

We have the general form of a congruence with prime p modulo as
$$f(x) = a_0 x^n + ... + a_n = 0 \ (\text{mod } p), \ p \nmid a_0 \qquad\qquad ...(1)$$
To find the solution of (1), we proceed as follows :

If some coefficients of $f(x)$ in (1) are greater than p, we reduce them to less than p. Also, if the degree of $f(x)$ is not less than p, we get remainder $r(x)$ such that
$$f(x) = (x^p - x) \, q(x) + r(x)$$
where, the degree of $r(x)$ is less than p, by using $x^p - x$ to divide $f(x)$, i.e. we can reduce $f(x)$ to $r(x)$ whose degree is less than p such that
$$f(x) \equiv r(x) \ (\text{mod } p)$$
by using $x^p \equiv x \ (\text{mod } p)$. Now, the problem of solving (1) becomes that of solving $r(x) \equiv 0 \ (\text{mod } p)$

Since the degree of $r(x)$ is less than that of $f(x)$, then calculation is easy. Further, if $f(x) \equiv f_1(x).f_2(x) \ (\text{mod } p)$, i.e. $f_i(x)$ is a factor of $f(x)$ modulo p, then to solve (1), is to solve
$$f_1(x) \equiv 0 \ (\text{mod } p) \quad \text{and} \, f_2(x) \equiv 0 \ (\text{mod } p)$$
Again, if $x \equiv a \ (\text{mod } p)$ is a solution of (1), then from $f(x) \equiv (x - a) \, g(x) + r$
we have $r \equiv 0 \ (\text{mod } p)$

Therefore,
$$(x - a) \, g(x) \ (\text{mod } p), \text{ i.e. } x - a \text{ is a factor of } f(x) \text{ modulo } p.$$
Hence, the problem of solving (1) becomes that of solving $g(x) \equiv 0 \ (\text{mod } p)$.

SOLVED EXAMPLES

Example 1. *Solve* $f(x) \equiv x^7 - 2x^6 - 7x^5 + x + 2 \equiv 0 \ (\text{mod } 5)$.

Solution. After simplification, we get
$$r(x) = x^3 - 2x^2 - x + 2 \equiv 0 \ (\text{mod } 5)$$
Putting the complete residue systems $- 2, - 1, 0, 1, 2$ mod 5 for x, we get the following three solutions
$$x \equiv - 1, 1, 2 \ (\text{mod } 5)$$

Example 2. *Show that $x^2 \equiv 1$ (mod 5) has only two solutions, 1 and $p - 1$, where p is prime.*

Solution. If a is a solution of the given equation, i.e.

$$a^2 - 1 \equiv 0 \text{ (mod 5)}$$

then, $p \mid (a + 1)(a - 1)$. Since p is prime, therefore,

$$p \mid (a + 1) \quad \text{or} \quad p \mid (a - 1)$$

Thus, $a + 1 \equiv 0 \text{ (mod } p) \text{ or } a - 1 \equiv 0 \text{ (mod } p)$

Taking a for the least positive residues, we have

$$a \equiv p - 1 \quad \text{or} \quad a \equiv 1.$$

EXERCISE 5.2

1. Show that 41 divides $2^{20} - 1$.

2. Find the remainder when 2^{24} is divided by 17.

3. Find the remainder when 3^{287} is divided by 23.

4. Show that the set of integers $\{1, 5, 7, 11\}$ is a reduced residue system (mod k).

5. Show that the set $\{49, 20, 10, 17, -18, -27\}$ (mod 6) is a complete system of residue.

6. Solve the following linear congruences :

   (i) $9x \equiv 12$ (mod 15)

   (ii) $6x \equiv 15$ (mod 21)

   (iii) $30x \equiv 52$ (mod 49)

   (iv) $17x \equiv 9$ (mod 276)

7. Solve the following system of linear congruences

   (i) $x \equiv 1$ (mod 6)   (ii) $x \equiv 5$ (mod 6)
         $x \equiv 5$ (mod 7) $x \equiv 4$ (mod 11)
         $x \equiv 2$ (mod 11) $x \equiv 3$ (mod 17)

8. Solve the following Diophantine equation by finding the solution of congruences

   (i) $37x + 49y = 1$ (ii) $4x + 51y = 9$

9. Show that the congruence $f(x) = x^3 + x^2 + 4x + 29 \equiv 0$ (mod 125) is satisfied by $x = 109$.

10. Solve the following congruence
    $$x^4 + 3x^3 + 3x^2 + 3x + 2 \equiv 0 \text{ (mod 30)}.$$

ANSWERS (with hints)

(1) Write $2^{20} - 1 = 81 \times 81 - 1 = 0$ (mod 41).

(2) $2^{24} = (2^4)^6 = (16)^6 = [-1(\text{mod } 17)]^6 = (-1)^6$ (mod 17) $= 1$ (mod 17). Remainder is 1.

(3) 3

(6) (i) 3, 8, 13, 3   (ii) 6, 13, 20   (iii) 5   (iv) $x \equiv 33$ (mod 276)

(7) (i) 453 (mod 462)   (ii) $x = 785$ (mod 1122)

(8) (i) $x = 4 + 49t, y = -3 + 37t$ (ii) $x = 15 + 15t, y = -1, + 4t, t \in Z$

CHAPTER REVIEW : A COMPETITIVE APPROACH

Selected terms and Results

TERMS

- **Congruence :** Let n be a positive integer. Two integers a and b leave remainders when divided by m. If remainders are the same, we say that a is congruent to b modulo m and write $a \equiv b$ (mod m).

- **Least Residue :** If $a \equiv b$ (mod m), $0 \le b < m$ then b is known as the least residue of a (mod m)

- **Minimal Residue :** If $a \equiv b$ (mod m) and $0 \le |b| \le \dfrac{m}{2}$ then m is called the minimal residue of a (mod m).

- **Complete Residue System :** If $x \equiv y$ (mod m) then y is called a residue of x modulo m. A set $x_1 x_2, \ldots x_m$ is called a complete residue system (CRS) modulo if for every integer y there is one and only one x_i such that $y \equiv x_i$ (mod m)

- **Reduced Residue System :** A reduced residue system (RRS) modulo m is a set of integers r_i such that $(r_i, m) = 1$, $r_i \not\equiv r_j$ (mod m) if $i \ne j$ and such that every x prime to x is congruent modulo m to same number r_i of the set.

RESULTS

- An integer a is said to be another integer b modulo m, m is any fixed integer if $m | (a - b)$. It is written as $a \equiv b$ (mod m).

- All usual algebraic law hold for congruence.

- $a \equiv b$ (mod m) if and only if a and b have the same remainders with respect to m.

- If $a \equiv b$ (mod m), $0 \le b < m$, then b is called the least residue of a (mod m).

- If $a \equiv b$ (mod m) and $0 \le |b| \le (m/2)$, then m is called the minimal residue of a(mod m).

- Minimal residue is also called absolutely least residue of a (mod m).

- An expression of the form $ax \equiv b$ (mod m), $a \ne 0$ (mod m) is called a linear congruence mod m.

- The linear congruence $ax \equiv b$(mod m) has a solution if and only if $d | b$ where $d = (a, m)$.

- A system of linear congruence $x \equiv a_i$ (mod m_i) is solvable if and only if (m_i, m_j) divides $(a_i - a_j)$.

- The necessary and sufficient condition that a positive integer n can be divided by 3 is that the sum of its digits is divisible by 3.

- Every number containing more than two digits can be divided by 4 if and only if the number formed by its least two digits can be divided by 4.

Review Questions and Project Work

1. If $a \equiv b$ (mod m) and $c \equiv d$ (mod m) then show that $a - c \equiv b - d$(mod m)

2. Show that no integer of the form $8n + 7$ can be expressed as a sum of three squares.

3. Let $f(x)$ be a polynomial with integral coefficients and $a \equiv b$ (mod m) then $f(a) \equiv f(b)$ (mod m)

4. Show that every odd prime except 3 is congruent to 1 or -1 modulo 6.

5. If $a \equiv b$ (mod m) and c is any integer. Then prove the following
   (i) $a + c \equiv b + c$ (mod m)
   (ii) $a - c \equiv b - c$ (mod m)

(iii) $ac \equiv bc \pmod{m}$
(iv) $a^2 \equiv b^2 \pmod{m}$

Objective Type Questions

Fill in the blanks:

1. No integer of the form $8n + 7$ can be expressed as a sum of .... squares.

2. The positive integers n for which
$$\sum_{1}^{n} k!$$
is a square are .... and ....

3. No prime of the form $4n + 3$ can be expressed as the sum of .... squares.

4. $a \equiv b \pmod{m}$ if and only if a and b leave the same remainder when divided by ....

5. $a \equiv b \pmod{m}$ if and only if $a = b + ...$ for some integer k.

True/False: *Write 'T' for true and 'F' for false statement.*

1. We use congruence modulo 12 to tell the time of the day.   (T/F)

2. We use congruence modulo 7 to tell the day of the week.   (T/F)

3. Every integer is congruent to exactly one of the least residue 0, 1, 2, ... $(m - 1)$ modulo m.   (T/F)

4. When $1! + 2! + ... + 100!$ is divided by 15, remainder is 2.   (T/F)

5. When 3^{247} is divided by 17, the remainder, is 11.   (T/F)

Multiple Choice Questions : *Choose the most appropriate one :*

1. When 11^{35} is divided by 13, the remainder is :
   (a) 4   (b) 5
   (c) 6   (d) none of these

2. When $3^{12} + 5^{12}$ is divided by 13, the remainder is
   (a) 2   (b) 3
   (c) 4   (d) none of these

3. If $(a, m) = d$ then the linear congruence $ax \equiv b \pmod{m}$ has a solution if and only if
   (a) $b \mid d$ (b) $d \mid b$
   (c) $d \mid m$ (d) none of these

4. The last two digits of the number 9^{9^9} is
   (a) 89   (b) 99
   (c) 98   (d) none of these

5. The remainder, when 3^{247} is divided by 25 is
   (a) 3   (b) 4
   (c) 12   (d) none of these

ANSWERS

Fill in the blanks

(1) 3   (2) 1 and 3   (3) 2   (4) m (5) km

True/False

(1) T   (2) T   (3) T   (4) F   (5) T

Multiple choice questions

(1) c (2) a (3) b (4) a (5) c

CHAPTER 6

Fermat's Theorems and Its Applications

❖ Fermat's Factorization Method	❖ Fermat's Numbers
❖ Fermat's Little Theorem	❖ Wilson's Theorem
❖ Fermat's Last Theorem	❖ Euler's Factorization Method
❖ Mersenne's Factorization Method	❖ Absolute Pseudo Prime

6.1 INTRODUCTION

Fermat wrote that the Diophantine equation

$$x^n + y^n = z^n$$

is impossible of solution in positive integers x, y, z for $n = 3, 4, 5...$ That he had found a truly remarkable way of proving the statement but that, unfortunately, the margin was not large enough to permit his writing out the proof. This unsolved problem is known as Fermat's last theorem. It is easy to see that the problem will be completely solved if it can be shown that Fermat's conjecture is true for every odd prime $n = p$ and for $n = 4$. Supposed that n is composite of the form $n = kp$, where p is an odd prime for which Fermat's proposed theorem is known to be true, then the theorem is also true for n. In this chapter, we shall discuss Fermat's factorization method, Fermat's little theorem, Wilson's theorem and Euler's factorization method.

6.2 FERMAT'S FACTORIZATION METHOD

(i) **Fermat Numbers :** *The numbers* $F_n = 2^{2^n} + 1$ *are called the Fermat's numbers*, after Pierre Fermat who thought they might all be primes. Here, we can easily show that F_n is prime for $n = 0, 1, ..., 4$, these are the only n for which F_n is known to be prime. For $n = 5, 6, ..., 21$, F_n is composite.

(ii) **Fermat's Factorization Method :** Let n be any odd positive integer such that it can be written as the difference of square of two integers, i.e.

$$n = x^2 - y^2 = (x - y)(x + y)$$

If $n = ab$, $a \geq b \geq 1$, then we can write

$$n = \left(\frac{a+b}{2}\right)^2 - \left(\frac{a-b}{2}\right)^2$$

We want to find the integral value of x and y which satisfy

$$n = x^2 - y^2$$

or $$x^2 - n = y^2$$

By inspection, we observe the following numbers

$$k^2 - n, \ (k + 1)^2 - n, \ (k + 2)^2 - n, \ ...$$

Let $m \geq \sqrt{n}$ such that $m^2 - n$ is a square. If n can be factorized, then it is not expressible as difference of squares of two numbers and then n has no factor other than n and 1. Hence, n is prime.

6.3   FERMAT'S LITTLE THEOREM

Statement : *If p is prime and $(a, p) = 1$, then $a^{p-1} - 1$ is divisible by p, i.e.* $a^{p-1} \equiv 1 \pmod{p}$.

Proof: Using binomial expansion, we have

$$(x_1 + x_2)^p = x_1^p + {}^pC_1 x_1^{p-1}.x_2 + {}^pC_2 x_1^{p-2}.x_2^2 + ... + {}^pC_{p-1}x_1 x_2^{p-1} + x_2^p$$

$$= (x_1^p + x_2^p) + f(p), \text{ where } f(p) \text{ contains those terms which are divisible by } p.$$

$$= (x_1^p + x_2^p) \pmod{p}$$

Similarly, $(x_1 + x_2 + ... + x_a)^p \equiv (x_1^p + x_2^p ... x_a^p) \pmod{p}$ ...(1)

Now, putting $x_1 = x_2 = ... x_a = 1$ in (1), we get

$$a^{p-1} \equiv a \pmod{p}$$ ...(2)

It is given that $(a, p) = 1$. Therefore, we cancel the common factor a in (2). Then, we have

$$a^{p-1} \equiv 1 \pmod{p}$$

$\Rightarrow \quad a^{p-1} - 1 \equiv 0 \pmod{p}$

$\Rightarrow \qquad (a^{p-1} - 1)$ is divisible by p.

THEOREM-1

Let p be a prime and a be any integer such that $p \nmid a$. Then a^{p-2} is an inverse of a mod p.

Proof. By Fermat little theorem, $a^{p-1} \equiv 1 \pmod{p}$. Thus $a.a^{p-2} \equiv 1 \pmod{p}$ so a^{p-2} is an inverse of a mod p.

For example. Let $p = 7$ and $a = 12$. Then, by this theorem, 12^5 is an inverse of 12 mod 7. But $12 \equiv -2 \pmod{7}$. So $12^5 \equiv (-2)^5 \equiv (-2^2 . 2^3) \equiv -4 . 1 = 3 \pmod{7}$.

Thus 3 is an inverse of 12 module 7. Since $12 . 3 \equiv 1 \pmod{7}$.

Example. *Prove that $2^{340} \equiv 1 \pmod{341}$.*

Solution. We have $341 = 11 . 31$.

Also, $2^{10} = 1024 = 31 . 33 + 1$.

Thus, $2^{11} = 2 . 2^{10} \equiv 2 . 1 \equiv 2 \pmod{31}$
$$2^{31} = 2(2^{10})^3 = 2 . 1^3 \equiv 2 \pmod{11}$$

So, $2^{11 . 31} = 2 \bmod (11 . 31)$

If p and q are distinct primes with $a^p \equiv a \pmod{q}$ and $a^q \equiv a \pmod{p}$, then $a^{pq} \equiv a \pmod{pq}$}

Therefore, $2^{341} \equiv 2 \pmod{341}$

or $\qquad 2^{340} \equiv 1 \pmod{341}$ $\qquad$ (after cancelling the factor 2)

THEOREM-2

Let p be a prime and a be any integer such that $p \nmid a$. Then the solution of linear congruence $ax \equiv b \pmod{p}$ is given by $x \equiv a^{p-2} \pmod{p}$.

Proof. Since $p \nmid a$, so congruence $ax \equiv b \pmod{p}$ has a unique solution. Now a^{p-2} is an inverse of a modulo p. Multiplying both the sides of the congruence by a^{p-2}, we have

$a^{p-2} (ax) \equiv a^{p-2} b \pmod{p}$

$a^{p-1} x \equiv a^{p-2} b \pmod{p}$

$x \equiv a^{p-2} b \pmod{p}$, by Fermat's little theorem.

SOLVED EXAMPLES

Example 1. *Solve the linear congruence $12x \equiv 6 \pmod{7}$.*

Solution. We have $12^5 \equiv 3 \pmod{7}$ is an inverse of 12 modulo 7. Multiplying both sides of the congruence by 3, we have

$\qquad 3(12x) = 3.6 \pmod{7}$

$\qquad x = 4 \pmod{7}$ $\qquad$ Since $3 . 12 \equiv 1 \pmod{7}$

Example. 2. *Solve the congruence $24x \equiv 11 \pmod{17}$.*

Solution. $24x \equiv 11 \pmod{17}$

$\qquad 7x = 11 \pmod{17}$

By above theorem, $x = 7^{15} .11 \pmod{17}$.

Now, we need to find the least residue of $7^{15}. 11 \pmod{17}$. Because, $7^2 \equiv -2 \pmod{17}$; $7^4 \equiv 4 \pmod{17}$; $7^8 \equiv -1 \pmod{17}$. Therefore, $7^{15} \equiv 7^8. 7^4 . 7^2 . 7 \equiv (-1) . 4 . (-2) . 7 \equiv 5 \pmod{17}$. Thus $x = 5 . 11 \equiv 4 \pmod{17}$.

Remarks

❖ The converse of the above theorem need not be true, i.e. if $a^{p-1} \equiv a \pmod{p}$ for some integers a and p, then p need not be a prime.

❖ If p is prime, then $a^p \equiv a \pmod{p}$ for any integer a.

GENERAL DEFINITIONS

(i) Order of a modulo m : If $(a, m) = 1$ and λ is the smallest positive integer such that $a^\lambda \equiv 1 \pmod{m}$ that is

$\qquad a^\lambda \equiv 1 \pmod{m}$ $a^k \not\equiv 1 \pmod{m}$, $0 < k < \lambda$.

Then λ is called the order of a modulo m.

The order of a modulo m tell us that some power of any integer relatively prime to m must be conguent to 1 modulo m.

(ii) Pseudo Prime: Any integer n is called a pseudo prime if $2^n \equiv 2 \pmod{n}$. For example, The number 341 is a pseudo prime because

$\qquad 2^{341} \equiv 2 \pmod{341}$

(iii) **Absolutely Pseudo Primes :** A composite number n is called an absolute pseudo prime if

$$a^n \equiv a \ (\text{mod } n), \ \forall a$$

Remarks

❖ The smallest pseudo prime is 341.

❖ The absolutely pesudo prime number is also know as Chermichael number.

❖ The least absolutely pseudo prime number is 561.

THEOREM-1

If p and q are distinct primes such that $a^p \equiv a$ (mod p) and $a^q \equiv a$ (mod q). Then
$$a^{pq} \equiv a \ (\text{mod } pq)$$

Proof. We know that if p is prime, then

$$a^p \equiv a \ (\text{mod } p)$$

Therefore, $(a^p)^q \equiv a^p$ (mod q)
which implies $a^{pq} \equiv a$ (mod q)

$\Rightarrow \qquad q$ divides $a^{pq} - a$

Similarly $(a^q)^p = a^q$ (mod p) and $a^q \equiv a$ (mod q)
$\Rightarrow \qquad a^{pq} = a$ (mod p)

Thus, p divides $a^{pq} - a$
which conclude that pq divides $a^{pq} - a$
Hence, $a^{pq} \equiv a$ (mod pq).

THEOREM-2

Let n be a composite square free integer. Also, let $n = p_1 . p_2 ... p_r$, p_i are distinct prime.
if $(p_i - 1) | (n - 1)$ for $i = 1, 2, ... r$, then n is an absolutely pseudo prime.

Proof. Consider an integer a such that $(a, n) = 1$
Then, $(a, p_i) = 1$ for each i.

By Fermat's theorem, we have $p_i | (a^{p_i - 1} - 1)$

Since $(p_i - 1) | (n - 1)$, we have $p_i | (a^{n-1} - 1)$

Thus, $p_i | (a^n - a)$, $\forall a$ and $i = 1, 2, ..., r$
Hence, $n | (a^n - a)$, which implies $a^n \equiv a$ (mod n)
Therefore, n is an absolute pseudo prime.

THEOREM-3

The odd prime factor of $a^{2^n} + 1$ $(a > 1)$ is of the form $2^{n+1} t + 1$
Proof. We shall prove this result by mathematical induction.

If $n = 0$, then result is trivially true.

Now, assume that result is true for $n - 1$, i.e.

$$a^{2^{n-1}} + 1 \equiv 0 \pmod{q_1}, \quad q_1 = 2^n t_1 + 1$$

Let $\qquad a^{2^n} + 1 \equiv 0 \pmod{q}$

or $\qquad (a^2)^{2^{n-1}} + 1 \equiv 0 \pmod{q}$

By induction, we have $q = 2^n t + 1$.

Since $\qquad (a^{2^n})^t \equiv (-1)^t \pmod{q}$

i.e. $\qquad a^{q-1} \equiv (-1)^t \pmod{q}$

Then, from Fermat's theorem, we have

$$(-1)^t \equiv 1 \pmod{q}$$

Since, $\qquad q > 2$, t is even, so $t = 2t_2$

Hence, $q = 2^{n+1}.t_2 + 1$

Hence by the principle of mathematical induction, result is true for all n.

THEOREM-4

Every prime factor of the Fermat number F_n $(n > 2)$ is of the from $2^{n+2}.t + 1$.

Proof : Suppose that q is a prime factor of F_n. If we can show $F_{n-1}^{2^{n+1}} + 1 \equiv 0 \pmod{F_n}$, then q is also a prime factor of $F_{n-1}^{2^{n+1}} + 1$.

By previous theorem, we have

$$q = 2^{n+2}.t + 1$$

because

$$F_{n-1}^{2^{n+1}} + (2^{2^{n-1}} + 1)^{2^{n+1}} = (2^{2^n} + 2^{2^{n-1}} + 1)^{2^n} \equiv (2^{2^{n-1}+1})^{2^n}$$

$$= (2^{2^n})^{2^{n-1}+1} = (-1)^{2^{n-1}+1} \equiv -1 \pmod{F_n}$$

That is $F_{n-1}^{2^{n+1}} + 1 \equiv 0 \pmod{F_n}$.

6.4 FERMAT'S LAST THEOREM

Statement : *The Diophantine equation*

$$x^n + y^n = z^n$$

has no intergral solution for $n > 2$, other than the trivial solution in which x or y is zero.

Fermat himself proved it for $n = 4$. In 1760, Euler proved it for $n = 3$. In 1823, Legendre prove it for $n = 5$. Gauss attempted to solve it for $n = 7$, but did not succeed. In 1839, Lame proved it for $n = 7$. In 1978, Wagstall proved it that theorem is true when $2 < n < 125000$ by using a super computer. In 1985 Rother proved that this theorem is correct for $2 < n < 41000000$. Even if n is odd prime and x, y, z are relatively prime to n, i.e. $(x, n) = (y, n) = (z, n) = 1$. At this time, we only know that when $n < 253747889$, the theorem is correct, i.e. $x^n + y^n = z^n$ has no integer solution. In 1983, Falltings established that for $n \geq 3$, $x^n + y^n = 2^n$ has atmost a finite number of solutions. This was a great breakthrough.

SOLVED EXAMPLES

Example 1. *Factorize 10541 by Fermat's factorization method.*

Solution. We can easily see that

$$102^2 < 10541 < 103^2$$

Thus, we have considered the values of $k^2 - 10541$ for $k \geq 103$
We have

$$103^2 - 10541 = 10609 - 10541 = 68$$
$$104^2 - 10541 = 10816 - 10541 = 275$$
$$105^2 - 10541 = 11025 - 10541 = 484 = 22^2$$
$$10541 = 105^2 - 22^2$$
$$\Rightarrow \qquad = (105 + 22)(105 - 22) = 127 \times 83$$

which are the required factors.

Example 2. *Factorize $2^{11} - 1$.*

Solution. We have $2^{11} - 1 = 2047$

which can be written as

$$45^2 < 2047 < 46^2$$

Thus, we consider the values $k^2 - 2047$ for $k \geq 46$
Taking $k = 46, 47, \ldots$, we have

$$46^2 - 2047 = 69$$
$$47^2 - 2047 = 162$$
$$48^2 - 2047 = 257$$
$$49^2 - 2047 = 354$$
$$50^2 - 2047 = 453$$
$$51^2 - 2047 = 554$$
$$52^2 - 2047 = 657$$
$$53^2 - 2047 = 762$$
$$54^2 - 2047 = 869$$
$$55^2 - 2047 = 978$$
$$56^2 - 2047 = 1089 = 33^2$$
$$\Rightarrow \qquad 2047 = 56^2 - 33^2$$
$$= (56 + 33)(56 - 33) = 89 \times 23.$$

Example 3. *Find the remainder when 5^{11} is divided by 7.*

Solution. We have $(5, 7) = 1$ and 7 is prime.

Thus, $5^6 \equiv 1 \pmod 7$

$$5^4 \equiv 2 \pmod 7$$

$\therefore$ $5^{11} = 5^6 \times 5^4 \times 5 \pmod 7$

$$= 1 \times 2 \times 5 \pmod 7$$

$$= 10 \ (\text{mod } 7)$$
$$= 3 \ (\text{mod } 7)$$

Hence, the required remainder is 3.

Example 4. *Show that* $5^{38} \equiv 4 \ (\text{mod } 11)$.

Solution. Since $(5, 11) = 1$ and 11 is prime.

$$\therefore \qquad 5^{11} \equiv 5 \ (\text{mod } 11)$$
$$5^{33} \equiv 5^3 (\text{mod } 11)$$

Now, $\quad 5^{38} = 5^{33} \times 5^5$

$$\equiv 5^3 \times 5^5 \ (\text{mod } 11)$$
$$\equiv 5^8 (\text{mod } 11)$$
$$\equiv (5^2)^4 \ (\text{mod } 11)$$
$$\equiv 3^4 \ (\text{mod } 11)$$
$$\equiv 81 \ (\text{mod } 11) \equiv 4 \ (\text{mod } 11)$$

Example 5. *Find the remainder when* 72^{1001} *is divided by* 31.

Solution. Here, we have

$$72 \equiv 10 \ (\text{mod } 31)$$
$$\therefore \qquad 72^{1001} \equiv 10^{1001} \ (\text{mod } 31)$$

Now, $(10, 31) = 1$ and 31 is prime. Thus, by **Fermat's** theorem, we have

$$10^{30} \equiv 1 \ (\text{mod } 31)$$
$$\Rightarrow \qquad 10^{990} \equiv 1 \ (\text{mod } 31)$$

Further, $10^2 \equiv 7 \ (\text{mod } 31)$
$$10^4 \equiv -13 \ (\text{mod } 31)$$
$$10^8 \equiv 14 \ (\text{mod } 31)$$

Again $72^{1001} \equiv 10^{1001} \ (\text{mod } 31)$
$$\equiv 10^{990} \times 10^8 \times 10^2 \times 10 \ (\text{mod } 31)$$
$$\equiv 1 \times 14 \times 7 \times 10 \ (\text{mod } 31)$$
$$\equiv 5 \times 10 \ (\text{mod } 31)$$
$$\equiv 19 \ (\text{mod } 31)$$

Hence, the required remainder is 19.

Example 6. *Show that 1729 is an absolute pseduo prime.*

Solution. Here, we have

$$1729 = 7 \times 13 \times 19$$

If $(a, 1729) = 1$, then

$$(a, 7) = 1, (a, 13) = 1, (a, 19) = 1$$

Then, by Fermat's theorem

$$a^6 \equiv 1 \ (\text{mod } 7), \quad a^{12} \equiv 1 \ (\text{mod } 13), \quad a^{18} \equiv 1 \ (\text{mod } 17)$$

which gives

$$a^{1728} \equiv (a^6)^{288} \equiv 1 \ (\text{mod } 7)$$
$$\equiv (a^{12})^{144} \equiv 1 \ (\text{mod } 13)$$
$$\equiv (a^{18})^{96} \equiv 1 \ (\text{mod } 19)$$
$$\equiv 1 \ (\text{mod } 7 . 13 . 19) = 1 \ (\text{mod } 1729)$$

Hence, 1729 is an absolute pseduo prime.

Example 7. *Find the remainder when 24^{1947} is divided by 17.*

Solution. We have

$$24 \equiv 7 \ (\text{mod } 17)$$
$$24^{1947} = 7^{1947} \ (\text{mod } 17)$$

But by Fermat's little theorem, $7^{16} \equiv 1 \ (\text{mod } 17)$

So, $7^{1947} = 7^{16.121 + 11} = (7^{16})^{121} . \ 7^{11}$

$$\equiv 1^{121}. \ 7^{11} \ (\text{mod } 17) = 7^{11} \ (\text{mod } 17)$$

But $7^2 \equiv -2 \ (\text{mod } 17)$, so $7^{11} = (7^2)^5 . \ 7 = (-2)^5 . \ 7 \equiv -32.7$

$$= 2 . \ 7 \equiv 14 \ (\text{mod } 17)$$

Thus, the remainder is 14 when 24^{1947} is divided by 17.

Example 8. *The square of every odd integer is congruent to 1 modulo 8.*

Proof. Let a be an odd integer, say $a = 2i + 1$ for some integer i. Then, $a^2 = 4i^2 + 4i + 1 = 4i(i + 1) + 1$. Since $2/i(i + 1)$, $8/4i(i + 1)$, so $a^2 \equiv 1 \ (\text{mod } 8)$.

6.5   WILSON'S THEOREM

Wilson's theorem gives the necessary and sufficient condition for a number p to be prime.

Statement : *An integer p is prime if and only if*

$$(p - 1)! + 1 \equiv 0 \ (\text{mod } p) \hspace{4cm} ...(1)$$

Solution. If p is prime, since $x \equiv 1, 2, ..., p - 1 \ (\text{mod } p)$ are the solution of $x^{p-1} - 1 \equiv 0 \ (\text{mod } p)$, we have

$$x^{p-1} - 1 \equiv (x - 1)(x - 2) ..... (x - p + 1) \ (\text{mod } p)$$

Putting $x = 0$, we get

$$(-1) \equiv (-1)^{p-1}(p - 1)! \ (\text{mod } p)$$

Therefore, when $p = 2$, equation (1) is true.

When p is an odd prime, $p - 1$ is even and (1) is also true.

Conversely, if (1) holds. To show p is prime. Let if possible, p is composite. Let q be the proper factor of p.

Since $1 < q < p$, we have $(p - 1)! \equiv (\text{mod } q)$
and therefore,

$$(p - 1)! + 1 \not\equiv 0 \ (\text{mod } q)$$

which is a contradiction.

Hence, p is prime.

The following example shows an application of Wilson's theorem.

Example. *Let p be a prime and n be any positive integer. Prove that*

$$\frac{(np)!}{n!p^n} \equiv (-1)^n \ (\text{mod } p)$$

Solution. Let a be any positive integer congruent modulo to 1 and p. Then by Wilson's theorem, $a(a + 1) ... [a + (p - 2)] \equiv (p - 1)! \equiv -1 \ (\text{mod } p)$.

i.e. the product of $(p - 1)$ integers between any two consecutive multiples of p is congruent to -1 modulo p. Then

$$\frac{(np)!}{n!p^n} = \frac{(np)!}{p.2p.3p\ldots(np)}$$

$$= \prod_{r=1}^{n} [(r-1)\,p+1]\ldots[(r-1)p+(p-1)]$$

$$= \prod_{r=1}^{n} (p-1)! \pmod p \equiv \prod_{1}^{n} (-1) \pmod p \equiv (-1)^n \pmod p$$

Remarks

❖ The above theorem can be restated as follows : "p is prime iff $(p-1)! \equiv -1 \pmod p$"

❖ In theory, the problem of determining whether a given number is prime, is completely solved. But for large integers, there are great computational difficulties.

5.5.1 General Form of Wilson's Theorem

If $x_1, x_2, \ldots x_{\phi(m)}$ is the reduced residue system modulo prime p. Then, we get
$$x_1, \ldots, x_{\phi(m)} + 1 \equiv 0 \pmod p, \text{ where } p \text{ is prime.}$$

THEOREM-1

The quadratic congruence $x^2 \equiv -1 \pmod p$, p is prime, has a solution if and only if $p \equiv 1 \pmod 4$.

Proof. Let a be any solution of the given quadratic congruence $x^2 \equiv -1 \pmod p$.

Then, $a^2 \equiv -1 \pmod p$ and $(a, p) = 1$.

Then, by Fermat's theorem, we have

$$a^{p-1} \equiv 1 \pmod p$$
$$1 \equiv a^{p-1} \equiv (a^2)^{(p-1)/2} \pmod p$$

Since p is odd, $\left(\dfrac{p-1}{2}\right)$ is an integer and $a^2 \equiv -1 \pmod p$

Therefore, $1 \equiv (-1)^{(p-1)/2} \pmod p$

Which is true only when $\dfrac{p-1}{2} = 2k$ (even)

i.e. $p = 4k + 1 \implies p \equiv 1 \pmod 4$

Conversely, suppose $p \equiv 1 \pmod 4$, then $p = 4k + 1$, k is any integer.

Then, By Wilson's theorem, we have $(p-1)! \equiv -1 \pmod p$

which implies

$$1 \cdot 2 \left(\frac{p-1}{2}\right) \cdot \left(\frac{p+1}{2}\right) \ldots (p-1) \equiv -1 \pmod p \qquad \ldots(1)$$

But
$$\frac{p+1}{2} \equiv -\left(\frac{p-1}{2}\right)(\bmod\ p)$$
$$\frac{p+3}{2} \equiv -\left(\frac{p-3}{2}\right)(\bmod\ p)$$
$$\vdots \qquad\qquad \vdots$$
$$p-1 \equiv -1\,(\bmod\ p)$$

...(2

From (1) and (2), we get

$$(-1)^{\left(\frac{p-1}{2}\right)}\left[1.2\ldots\frac{(p-1)}{2}\right]^2 = -1\ (\bmod\ p)$$

i.e.
$$(-1)^{2k}\left[\left(\frac{p-1}{2}\right)!\right]^2 \equiv -1\ (\bmod\ p)$$

$$\Rightarrow \qquad \left[\left(\frac{p-1}{2}\right)!\right]^2 \equiv -1\ (\bmod\ p)$$

Hence, $\left(\dfrac{p-1}{2}\right)!$ is a solution of the congruence $x^2 \equiv -1\ (\bmod\ p)$

6.6 EULER'S FACTORIZATION METHOD

Let n be an odd integer. We want to factorize n.
Let us suppose
$$n = a^2 + b^2, \quad \text{where } a \text{ is even and } b \text{ is odd.}$$
and $\qquad n = c^2 + d^2, \quad \text{where } c \text{ is even and } d \text{ is odd.}$
Therefore, we can write
$$a^2 + b^2 = c^2 + d^2, \quad \text{i.e.,} \quad a^2 - c^2 = d^2 - b^2$$
which gives
$$(a - c)(a + c) = (d - b)(d + b) \qquad \qquad \text{...(1)}$$
Let k be the g.c.d of $a - c$ and $d - b$
i.e. $\qquad (a - c, d - b) = k$
Then, k is an even integer greater than 2.
If $a - c = kr_1$ and $d - b = kr_2$, then $(r_1, r_2) = 1$
Also, $\qquad (a - c)(a + c) = kr_1\,(a + c)$
and $\qquad (d - b)(d + b) = kr_2(d + b) \qquad \qquad \text{...(2)}$
Using (1) and (2), we have
$$r_1(a + c) = r_2(d + b)$$
$$\Rightarrow \qquad r \mid (d + b) \qquad \qquad \text{...(3)}$$
Let $\qquad d + b = r_1 t$, then from (3), we have
$$a + c = r_2 t$$
Clearly, $t = (a + c, d + b)$ is an even integer.

Therefore, we can write
$$a - c = kr_1, \quad a + c = r_2 t, \quad d - b = kr_2 \quad \text{and} \quad d + b = tr_1$$
From these values, we can easily find

$$a = \frac{kr_1 + tr_2}{2}$$

$$c = \frac{tr_2 - kr_1}{2}$$

$$d = \frac{kr_2 + tr_1}{2}$$

and $\qquad b = \dfrac{tr_1 - kr_2}{2}$

Thus, $\qquad n = a^2 + b^2$ gives

$$n = \left(\frac{kr_1 + tr_2}{2}\right)^2 + \left(\frac{tr_1 - kr_2}{2}\right)^2$$

$$= \frac{k^2 r_1^2 + t^2 r_2^2 + t^2 r_1^2 + k^2 r_2^2}{4} = (r_1^2 + r_2^2)\left(\frac{k^2}{4} + \frac{t^2}{4}\right)$$

which are the required factor of n.

.7 MERSENNE'S FACTORIZATION METHOD

Let n be any integer which is to be factorized. To factorize n by Mersenne's method, we use the following formula:
If $n = a^2 + b^2 = c^2 + d^2$, then

$$n = \frac{(ac + bd)(ac - bd)}{(a + d)(a - d)}$$

SOLVED EXAMPLES

Example 1. *Find the remainder when 2(28!) is divided by 31.*

Solution. By alternative form of Wilson's theorem, we have
$$(p - 1)! \equiv -1 \pmod{p}$$
Therefore,
$$(p - 2)! \equiv 1 \pmod{p}$$
$\therefore \qquad 2(p - 2)! \equiv 2 \pmod{p}$

or $\qquad 2(p - 2)(p - 3)! \equiv 2 \pmod{p}$

i.e. $\qquad 2(p - 3)! \equiv -1 \pmod{p}$

Putting $p = 31$, we get
$$2(28)! \equiv -1 \pmod{31}$$
$$\equiv 30 \pmod{31}$$
Hence, the required remainder is 30.

Example 2. *Show that the integer p (> 2) is prime if and only if*
$$(p-2)! - 1 \equiv 0 \pmod{p}$$

Solution. Using Wilson's theorem, we may get
$$(p-2)!(p-1) - (p-1) \equiv 0 \pmod{p} \qquad \ldots(1)$$

Then, $(p-1)\,[(p-2)! - 1] \equiv 0 \pmod{p}$

But, since $(p-1, p) = 1$. Therefore, $(p-2)! - 1 \equiv 0 \pmod{p}$

Example 3. *For any prime p, show that*
$$p \mid (a^p + (p-1)!\,a), \text{ a is any integer.}$$

Solution. Using Fermat's theorem, we have
$$p \mid (a^{p-1} - 1)$$

Also, by Wilson's theorem
$$p \mid (p-1)! + 1$$

On combining both the above results, we get
$$p \mid (a^{p-1} - 1 + (p-1)! + 1)$$

i.e. $p \mid (a^{p-1} + (p-1)!)$

$\Rightarrow$ $p \mid (a^p + (p-1)!\,a)$

Example 4. *If p, n be positive integers, p > n > 0, show that p is prime if and only if*
$$(n-1)!(p-n)! \equiv (-1)^n \pmod{p}$$

Solution. We know that
$$(-1)^k \equiv p - k \pmod{p}$$

and $(n-1)! = (-1)^{n-1}(p-1) \ldots (p-n+1) \pmod{p}$

Then, from Wilson's theorem, we have
$$(n-1)!(p-n)! \equiv (-1)^{n-1}(p-1)! \equiv (-1)^{n-1}(-1)$$
$$\equiv (-1)^n \pmod{p}$$

Example 5. *Factorize 38025 by Euler's factorization method.*

Solution. We can write
$$38025 = 168^2 + 99^2$$

and $38025 = 156^2 + 117^2$

Therefore,
$$168^2 + 99^2 = 156^2 + 117^2$$

$\Rightarrow$ $117^2 - 99^2 = 168^2 - 156^2$

i.e. $(117 - 99)(117 + 99) = (168 - 156)(168 + 156)$

$\Rightarrow$ $18(216) = 12(324)$

Also, $(216, 324) = 108$, $(18, 12) = 6$

and $216 = 2 \times 108$
$$324 = 3 \times 108$$

$$18 = 3 \times 6$$
$$12 = 2 \times 6$$

Therefore,

$$r_1 = 3, r_2 = 2, k = 6 \text{ and } t = 108$$

Hence,

$$38025 = (r_1^2 + r_2^2)\left(\frac{k^2 + t^2}{4}\right)$$

$$= (3^2 + 2^2)\left(\frac{6^2 + 108^2}{4}\right)$$

$$= \frac{13 \times 11700}{4}$$

$$= 13 \times 2925$$

ADDITIONAL SOLVED EXAMPLES

Example 1. *Show that 8th power of any number is of the form $17n$ or $17n \pm 1$, where n is any integer.*

Solution. Let x be any integer. If 17 is a divisor of x, then we can write $x = 17q$, $q \in Z$. Therefore, $x^8 = 17n$ for some $n \in Z$.

If 17 is not a divisor of x, then since 17 is a prime, we have $(17, x) = 1$. Then, by Fermat's theorem

$$x^{17} - 1 \equiv 1 \pmod{17}$$
$$\Rightarrow \quad x^{16} - 1 \equiv 0 \pmod{17}$$
i.e. $\quad (x^8 - 1)(x^8 + 1) \equiv 0 \pmod{17}$

Therefore,

$$17 \mid (x^8 - 1)(x^8 + 1)$$
$$17 \mid (x^8 - 1) \quad \text{or} \quad 17 \mid (x^8 + 1), \quad \text{because 17 is a prime.}$$

Now, $17 \mid (x^8 - 1)$ implies $x^8 - 1$ is a multiple of 17.

i.e. $\quad x^8 - 1 = 17n \Rightarrow x^8 = 17n + 1$

Further

$$17 \mid (x^8 + 1) \Rightarrow x^8 + 1 \text{ is a multiple of 17.}$$
$$\Rightarrow \quad x^8 + 1 = 17n$$
i.e. $\quad x^8 = 17n - 1$

Hence, if $x \in Z$, then x^8 is of the form $17n$ or $17n \pm 1$, where n is any integer.

Example 2. *If m, n are primes, then show that $m^{n-1} + n^{m-1} - 1$ is a multiple of mn.*

Solution. We know that $m \mid m^{n-1}$ ...(1)

Now, since, m and n are primes, therefore $(m, n) = 1$.

Then, by Fermat's theorem

$$n^{m-1} \equiv 1 \pmod{m}, \quad \text{i.e.} \quad m \mid (n^{m-1} - 1) \qquad \text{...(2)}$$

From (1) and (2), we conclude that

$$m \mid (m^{n-1} + n^{m-1} - 1) \qquad \qquad ...(3)$$

Again $n \mid n^{m-1}$, then by Fermat's theorem

$$m^{n-1} = 1 (\bmod n), \quad \text{i.e.} \quad n \mid (m^{n-1} - 1)$$

$$\Rightarrow \qquad n \mid (n^{m-1} + m^{n-1} - 1) \qquad \qquad ....(4)$$

Further, from (3) and (4), we have

$$m^{n-1} + n^{m-1} - 1 \equiv 0 \ (\bmod \ m)$$

and $\qquad \qquad m^{n-1} + n^{m-1} - 1 \equiv 0 \ (\bmod \ n)$

Therefore,

$$m^{n-1} + n^{m-1} - 1 \equiv 0 \ (\bmod \ (m, n))$$

$$\equiv 0 \ (\bmod \ (mn))$$

Hence, $mn \mid (m^{n-1} + n^{m-1} - 1)$, i.e. $m^{n-1} + n^{m-1} - 1$ is a multiple of mn.

Example 3. *If p is prime, show that $2(p - 3)! + 1$ is a multiple of p.*

Solution. As per given, p is prime. Then by Wilson's theorem, we have

$$(p - 1)! + 1 \equiv 0 \ (\bmod \ p)$$

$$\Rightarrow \qquad (p - 1)(p - 2)(p - 3)! + 1 \equiv 0 \ (\bmod \ p)$$

$$\Rightarrow \qquad (p^2 - 3p + 2)(p - 3)! + 1 \equiv 0 \ (\bmod \ p)$$

$$\Rightarrow \qquad 2(p - 3)! + 1 + (p^2 - 3p)(p - 3)! \equiv 0 \ (\bmod \ p)$$

$$\Rightarrow \qquad 2(p - 3)! + 1 + p(p - 3)(p - 3)! \equiv 0 \ (\bmod \ p)$$

$$\Rightarrow \qquad 2(p - 3)! + 1 \equiv 0 \ (\bmod \ p) \qquad \qquad [\because p(p - 3)(p - 3)! \equiv 0 \ (\bmod \ p)]$$

Example 4. *If p is an odd prime and $(a, b) = 1$, then show that either $a^{(p-1)/2} \equiv 1 \ (\bmod \ p)$ or $a^{(p-1)/2} \equiv -1 \ (\bmod \ p)$.*

Solution. Since p is an odd integer, therefore $(p - 1)$ is even and so $\dfrac{p-1}{2}$ is positive integer.

Now $\qquad a^{p-1} - 1 = \left[a^{(p-1)/2}\right]^2 - 1^2$

$$= \left[a^{(p-1)/2} - 1\right]\left[a^{(p-1)/2} + 1\right] \qquad \qquad ...(1)$$

Since, p is prime and $(a, p) = 1$, therefore, by Fermat's theorem, we have

$$a^{p-1} \equiv 1 \ (\bmod \ p)$$

$$\Rightarrow \qquad p \mid (a^{p-1} - 1)$$

$$\Rightarrow \qquad p \mid \left[a^{(p-1)/2} - 1\right]\left[a^{(p-1)/2} + 1\right]$$

$$\Rightarrow \qquad p \mid \left[a^{(p-1)/2} - 1\right] \ \text{or} \ \ p \mid \left[a^{(p-1)/2} + 1\right] \qquad \qquad (\because p \text{ is prime})$$

Hence, $a^{(p-1)/2} \equiv 1 \ (\bmod \ p)$ or $a^{(p-1)/2} \equiv -1 \ (\bmod \ p)$

Example 5. *If p is prime and a, b ∈ Z, then show that*

$$a^p - b^p = a - b + M(p), \text{ where } M(p) \text{ denotes the numbers of multiple of } p.$$

Solution. As per given, p is prime and a, b are any integers. Therefore

$$a^p \equiv a \pmod{p} \qquad \qquad ...(1)$$

and $\qquad b^p \equiv b \pmod{p} \qquad \qquad ...(2)$

which implies

$$a^p - b^p \equiv a - b \pmod{p}$$

$\Rightarrow \qquad p \mid (a^p - b^p) - (a - b)$

$\Rightarrow \qquad (a^p - b^p) - (a - b) = M(p), \text{ i.e. multiple of } p$

Hence, $\qquad a^p - b^p = a - b + M(p)$

Example 6. *Apply Wilson's theorem to show that*

$\qquad$ (i) $18! + 1 \equiv 0 \pmod{19}$ $\qquad$ (ii) $18! + 1 \equiv 0 \pmod{23}$

Then, show that $18! + 1 \equiv 0 \pmod{437}$

Solution. Since 19 is prime, therefore, by Wilson's theorem, we have

$$(19 - 1)! + 1 \equiv 0 \pmod{19}$$

i.e. $\qquad 18! + 1 \equiv 0 \pmod{19}$

Further, since 23 is prime, then by Wilson's theorem, we have

$$(23 - 1)! + 1 \equiv 0 \pmod{23}$$

$\Rightarrow \qquad 22! + 1 \equiv 0 \pmod{23}$

$\Rightarrow \qquad 22 . 21 . 20 . 19 . 18! + 1 \equiv 0 \pmod{23}$

$\Rightarrow \qquad (23 - 1)(23 - 2)(23 - 3)(23 - 4)18! + 1 \equiv 0 \pmod{23}$

$\Rightarrow \qquad (-1)(-2)(-3)(-4)18! + 1 \equiv 0 \pmod{23}$

$\Rightarrow \qquad 24 . 18! + 1 \equiv 0 \pmod{23}$

$\Rightarrow \qquad (23 + 1) . 18! + 1 \equiv 0 \pmod{23}$

$\Rightarrow \qquad (23)18! + 18! + 1 \equiv 0 \pmod{23}$

$\Rightarrow \qquad 18! + 1 \equiv 0 \pmod{23}$

Now, $\qquad 18! + 1 \equiv 0 \pmod{19} \quad$ and $18! + 1 \equiv 0 \pmod{23}$

$\Rightarrow \qquad 18! + 1 \equiv 0 \pmod{19, 23}$

$\hookrightarrow \qquad 18! + 1 \equiv 0 \pmod{19 \times 23}$

Hence, $\qquad 18! + 1 \equiv 0 \pmod{437}$

Example 7. *Show that $28! + 233 \equiv 0 \pmod{899}$*

Solution. We have $899 = 29 \times 31$, where both 29 and 31 are primes. Now, since 29 is prime, then by Wilson's theorem, we have

$$(29 - 1)! + 1 \equiv 0 \pmod{29}$$

$\Rightarrow \qquad 28! + 1 \equiv 0 \pmod{29}$

$\Rightarrow \qquad 28! + 1 + 8 . 29 \equiv 0 \pmod{29}$

$\Rightarrow \qquad 28! + 1 + 232 \equiv 0 \pmod{29}$

$\Rightarrow \qquad 28! + 233 \equiv 0 \pmod{29}$

Further, since 31 is prime, then by Wilson's theorem, we have

$$(31 - 1)! + 1 \equiv 0 \;(\text{mod } 31) \implies 30! + 1 \equiv 0 \;(\text{mod } 31)$$

$\implies \qquad 30 \cdot 29 \cdot 28! + 1 \equiv 0 \;(\text{mod } 31)$

$\implies \qquad (31 - 1)(31 - 2)28! + 1 \equiv 0 \;(\text{mod } 31)$

$\implies \qquad (- 1)(- 2)28! + 1 \equiv 0 \;(\text{mod } 31)$

$\implies \qquad 2 \cdot 28! + 1 \equiv 0 \;(\text{mod } 31)$

$\implies \qquad 2 \cdot 28! + 31.15 + 1 \equiv 0 \;(\text{mod } 31)$

$\implies \qquad 2 \cdot 28! + 466 \equiv 0 \;(\text{mod } 31)$

$\implies \qquad 28! + 233 \equiv 0 \;(\text{mod } 31) \qquad\qquad\qquad [\because (2, 31) = 1]$

Now, $28! + 233 \equiv 0 \;(\text{mod } 29)$ and $28! + 233 \equiv 0 \;(\text{mod } 31)$

Hence, $28! + 233 \equiv 0 \;(\text{mod } 31)$

$\qquad\qquad\qquad\qquad = 0 \;(\text{mod } 899)$

Example 8. *If p is a prime, show that*

$$1^{p-1} + 2^{p-1} + 3^{p-1} + ... + (p - 1)^{p-1} + 1 = M(p)$$

Solution. Since, p is prime, then by Fermat's theorem for any integer a, we have

$$a^{p-1} \equiv 1 \;(\text{mod } p)$$

Taking $a = 1, 2, ..., p - 1$, we have

$$1^{p-1} \equiv 1 \;(\text{mod } p), \;\; 2^{p-1} \equiv 1 \;(\text{mod } p), \;..., (p - 1)^{p-1} \equiv 1 \;(\text{mod } p)$$

$\implies \qquad 1^{p-1} + 2^{p-1} + 3^{p-1} + ... + (p - 1)^{p-1} \equiv 1 + 1 + 1 ... + 1 \;(\text{mod } p) \; (p - 1)$ times

Therefore,

$$1^{p-1} + 2^{p-1} + 3^{p-1} + ... + (p - 1)^{p-1} \equiv p - 1 \;(\text{mod } p)$$

$\implies \qquad 1^{p-1} + 2^{p-1} + 3^{p-1} + ... + (p - 1)^{p-1} + 1 - p \equiv 0 \;(\text{mod } p)$

$\implies \qquad 1^{p-1} + 2^{p-1} + ... + (p - 1)^{p-1} + 1 \equiv 0 \;(\text{mod } p)$

$\implies \qquad 1^{p-1} + 2^{p-1} + 3^{p-1} + ... + (p - 1)^{p-1} + 1 \equiv M(p)$, a multiple of p.

Example 9. *Show that $n^{13} - n$ is divisible by 2, 3, 5, 7 and 13 for any integer n.*

Solution. Using Fermat's theorem, if p is a positive prime and n is any integer, then

$$n^p \equiv n \;(\text{mod } p), \;\; \text{i.e. } n^p - n \text{ is divisible by } p.$$

Since 13 is prime, thus $n^{13} - n$ is divisible by 13.

We can write $n^{13} - n = n(n^{12} - 1)$

$$= n(n^6 - 1)(n^6 + 1) = (n^7 - n)(n^6 + 1)$$

Since 7 is prime, therefore, $(n^7 - n)$ is divisible by 7 and so $(n^{13} - n)$ is also divisible by 7.

Again, we can write

$$n^{13} - n = n(n^{12} - 1) = n((n^4)^3 - 1^3)$$

$$= n(n^4 - 1)(n^8 + n^4 + 1) = (n^5 - n)(n^8 + n^4 + 1)$$

Since, 5 is prime, therefore $(n^5 - n)$ is divisible by 5 and therefore, $(n^{13} - n)$ is also divisible by 5.

Again, we can write

$$n^{13} - n = n(n^4 - 1)(n^8 + n^4 + 1) = n(n^2 - 1)(n^2 + 1)(n^8 + n^4 + 1)$$
$$= (n^3 - n)(n^2 + 1)(n^8 + n^4 + 1)$$

Since, 3 is prime therefore, $(n^3 - n)$ is divisible by 3 and therefore $(n^{13} - n)$ is also divisible by 3. Finally, we can write

$$n^{13} - n = n(n^2 - 1)(n^2 + 1)(n^8 + n^4 + 1) = n(n - 1)(n + 1)(n^2 + 1)(n^8 + n^4 + 1)$$
$$= (n^2 - 1)(n + 1)(n^2 + 1)(n^8 + n^4 + 1)$$

Since 2 is prime, therefore, $n^2 - n$ is divisible by 2 and hence $n^{13} - n$ is also divisible by 2.

EXERCISE 6.1

1. Use Fermat's method to factorize 23449.
2. Show that $8^{30} - 1$ is divisible by 31.
3. If $7 \nmid a$, show that either $a^3 + 1$ or $a^3 - 1$ is divisible by 7.
4. Find the remainder when 41^{75} is divided by 3.
5. Find the remainder when $2^{1000000}$ is divided by 7.
6. Find the remainder when 3^{287} is divided by 23.
7. Show that 561 is an absolute pseudo prime.
8. Show that 17 is a prime by proving $16! = -1 \pmod{17}$.
9. For any prime, show that $p \mid (p - 1)! \, a^p + a$.
10. Using Euler's factorization method, factorize 493.
11. If p and q are distinct primes, show that $pq \mid a^{pq} - a^p - a^q + a$, for any integer a.
12. If $(a, 42) = 1$, show that $a^6 \equiv 1 \pmod{168}$.
13. Show that $18! \equiv -1 \pmod{437}$
14. Show that $4(29)! + 5!$ is divisible by 17.

ANSWERS (with hints)

(1) Write $153^2 < 23449 < 154^2$ and consider the values of $k^2 - 23449$ for $k > 154$. The required factors are 179 and 131.

(2) Since $(8, 31) = 1$ and 31 is prime, then use Fermat's theorem.

(3) Since 7 is prime and $(7, a) = 1$, then use Fermat's theorem to get $a^{7-1} \equiv 1 \pmod 7$.

(4) We have $41 \equiv 2 \pmod 3$ $\Rightarrow$ $41^{75} \equiv 2^{75} \pmod 3$
     $(2, 3) = 1$ and 3 is prime. Then use Fermat's theorem.

(5) 2. Do same as (4).

(6) 3.

(10) Write $493 = 13^2 + 18^2$ and $493 = 3^2 + 22^2$.

CHAPTER REVIEW : A COMPETITIVE APPROACH

Selected terms and Results

TERMS

- **Fermat Number :** The number $F_n = 2^{2^n} + 1$ is called the Fermat's number.
- **Order of a modulo m :** If $(a, m) = 1$ and λ is the least positive integer such that $a^\lambda \equiv 1 \pmod{m}$ that is $a^\lambda \equiv 1 \pmod{m}$, $a^k \not\equiv 1 \pmod{m}$, $0 < k < 1$ Then λ is called the order of $á$ modulo m.
- **Pseudo Prime :** Any integer n is called a pseduo prime if $2^n \equiv 2 \pmod{n}$.
- **Absolutely pseudo prime :** A composite number n is called an absolute pseudo prime if $a^n \equiv a \pmod{n}$ $\forall$ a.

RESULTS

- Fermat's little theorem states that "If p is prime and $(a, p) = 1$, then a^{p-1} is divisible by p^n.
- If p is prime, then $a^p \equiv a \pmod{p}$ for any integer a.
- If p and q are distinct primes such that $a^p \equiv a \pmod{q}$ and $a^q \equiv a \pmod{p}$ then $a^{pq} \equiv a \pmod{pq}$.
- A composite number n is called absolute pseudo prime if $a^n = a \pmod{n}$ for all integers a.
- The least absolute pseudo prime is 561.
- The least pseudo prime number is 341.
- If p is prime, then $(p-1)! \equiv -1 \pmod{p}$
- If $(m-1)! + 1$ is divisible by m, then m is prime.

- If $n = a^2 + b^2 = c^2 + d^2$, then
$$n = \frac{(ac+bd)(ac-bd)}{(a+d)(a-d)}$$

- If a is an integer such that $p \nmid a$ then a^{p-2} is an inverse of a modulo p.
- If a is an integer such that $p \nmid a$ then $a^{p-1} \equiv 1 \pmod{p}$
$$\approx a^p \equiv a \pmod{p}$$

- There is an infinite number odd pseudo primes.

Review Questions and Project Work

1. Find the primes p for which $\dfrac{2^{p-1}-1}{p}$ is a square. (**Ans.** 3 or 7)

2. Solve the following linear congruence.
   (i) $12x \equiv 6 \pmod{7}$

   (**Ans.** $x \equiv 4 \pmod{7}$)
   (ii) $24x \equiv 11 \pmod{17}$

   (**Ans.** $x \equiv 4 \pmod{17}$)

3. (i) Find the remainder when 7^{1001} is divided by 17   (**Ans.** 10)
   (ii) Find the remainder when 15^{1976} is divided by 12   (**Ans.** 12)

4. For any distinct primes p and q, and arbitrary positive integers a and b prove the following
   (i) $a^{pq} - a^p - a^q + a \equiv 0 \pmod{pq}$
   (ii) If $a^p \equiv b^p \pmod{p}$ then $a \equiv b \pmod{p}$
   (iii) If $a^p \equiv b^p \pmod{p}$ then $a^p \equiv b^p \pmod{p^2}$
   (iv) $p^q + q^p \equiv p + q \pmod{pq}$

5. Verify the following
   (i) $5^{123} \not\equiv 4 \pmod{124}$
   (ii) $12^{64} \not\equiv 2 \pmod{65}$
   (iii) $4^{14} \equiv 1 \pmod{15}$
   (iv) $2^{340} \not\equiv 2 \pmod{340}$

6. Verify Wilson's theorem for the prime 19 and 23.

7. Use Fermat's theorem to show that every prime except 2 and 5 divides infinitely many of the integers 9, 99, 999, 9999, ...

8. Show that for every integer n, the number $n^{13} - n$ is divisible by 2730.

Objective Type Questions

Fill in the blanks:

1. If p is prime then $(p - 1)! \equiv$ .... $\pmod{p}$

2. If n is a positive integer such that $(n - 1)! \equiv -1 \pmod{n}$ then n is ....

3. A composite number n such that $2^n \equiv 2 \pmod{n}$ is called ....

4. If m and n are positive integers such that $m \mid n$ then $2^m - 1 \mid$ ....

5. There is .... number of odd pseudo primes.

6. If n is an odd pseudo prime then $2^n - 1$ is ....

7. $a^p \equiv a \pmod{p}$ is true for any integer a and any prime p greater than ....

8. Let a be a solution of the congruence $x^2 \equiv 1 \pmod{m}$ then .... is also a solution of this congruence.

True/False: *Write 'T' for true and 'F' for false statement.*

1. If the congruence $x^2 \equiv 1 \pmod{p}$ has exactly two solutions then p is prime. (T/F)

2. If p is an odd prime, then $2(p - 3)! \equiv -1 \pmod{p}$. (T/F)

3. A positive integer $n \geq 2$ is prime if and only if $(n - 2)! \equiv 1 \pmod{n}$ (T/F)

4. Let p be prime and n is any integer then $\dfrac{(np)!}{n!p^n} \not\equiv (-1)^n \pmod{p}$. (T/F)

5. Let p be an odd prime, then $(p - 1)(p - 2) \dots (p - k)$ $\equiv (-1)^k . k! \pmod{p}$ where $1 \leq k \leq p$ (T/F)

Multiple Choice Questions : *Choose the most appropriate one :*

1. When 3^{287} is divided by 23, the remainder is
   (a) 1   (b) 2
   (c) 3   (d) none of these

2. When 5^{48} is divided by 24, the remainder is
   (a) 1   (b) 2
   (c) 3   (d) none of these

3. When the sum $1^5 + 2^5 + 3^5 + ... + 100^5$ is divided by 4, the remainder is
   (a) 0   (b) 1
   (c) 2   (d) 3

4. When $3^{12} + 5^{12}$ is divided by 13, the remainder is
   (a) 1   (b) 2
   (c) 3   (d) none of these

5. Using Fermat factorization method, the factor of $2^{11} - 1$ are
   (a) 88, 23   (b) 87, 22
   (c) 89, 23   (d) none of these

6. If p is prime then $(p - 1)! \equiv -1 \pmod{p}$. This is called
   (a) Fermat's theorem
   (b) Wilson's theorem
   (c) Dirichlet's Theorem
   (d) none of these

7. If p is prime such that $p \nmid a$ then $a^{p-1} \equiv -1 \pmod p$. This is
   (a) Fermat's theorem
   (b) Wilson's theorem
   (c) Dirichlet's Theorem
   (d) none of these

8. The value of prime p for which
   $$\frac{2^{p-1}-1}{p}$$
   is a square is
   (a) 3   (b) 7
   (c) both (a) and (b) are true
   (d) none of these

9. Let p be prime and a is any integer such that $p \nmid a$. Then the solution of the linear congruence $ax \equiv b \pmod p$ is
   (a) $x \equiv a^{p-2}b \pmod p$
   (b) $x \equiv a^p b \pmod p$
   (c) both (a) and (b) are true
   (d) none of these

10. If a is an integer such that $p \nmid a$ then
   (a) $a^{p-1} \equiv -1 \pmod p$
   (b) $a^p \equiv a \pmod p$
   (c) a^{p-2} is an inverse of a modulo p
   (d) all are true.

ANSWERS

Fill in the blanks

(1) -1

(3) Pseudo prime

(5) infinite

(7) 3

(2) Prime

(4) $2^n - 1$

(6) Pseudo prime

(8) $m - a$

True/False

(1) F

(3) T

(5) T

(2) T

(4) F

Multiple choice questions

(1) c

(3) a

(5) c

(7) a

(9) a

(2) a

(4) b

(6) b

(8) c

(10) d

CHAPTER 7

Number Theoretic Functions

* ❖ Functions τ and σ
* ❖ Mobius Function
* ❖ Function $T(n)$, $S(n)$, $\overline{\phi}(n)$
* ❖ Greatest integer Function or Bracket Function
* ❖ Euler's Function
* ❖ Square free integer

7.1   INTRODUCTION

A Function, whose domain is the set of natural numbers is called a number theoretic function. These type of functions having special importance in the theory of numbers. In this chapter, we shall discuss some important number theoretic function with their important properties.

7.2   THE FUNCTION τ AND σ

Let us define a function $\tau(n)$ to give the number of positive integer divisors of any given positive integer n. Such a function must be of a very different character from the functions usually studied in algebra or analysis, for it depends in a critical way not only upon the value of n, but also upon the standard representation of n.

Definition: *For each positive integer n, $\tau(n)$ is the number of positive divisors of n including 1 and n, i.e.*

$$\tau : N \to N, \text{ such that}$$

$$\tau(n) = \sum_{\substack{d \mid n \\ d \geq 1}} 1,$$

where, $\Sigma 1$ denotes the sum of as many 1's as there are positive divisors of n.

Example. *Evaluate $\tau(18)$ and $\tau(23)$.*

Solution. The positive divisors of 18 are 1, 2, 3, 6, 9 and 18, and there are six in number.

So, $\tau(18) = 6$

23, being a prime, has exactly two positive divisors i.e 1 and 23, so $\tau(23) = 2$.

Alternative Method

If n is written in standard form as $n = p_1^{a_1} p_2^{a_2} \dots p_k^{a_k}$ then all the positive integer divisors of n are given, without repetition of the form

$$d = p_1^{b_1} p_2^{b_2} \cdots p_k^{b_k}$$

where, each value of i, the b_i runs independently, through the following range of values : $b_i : 0, 1, 2, ..., a_i$.

Now, by combinatorial principal, it follows that if b_1 can be chosen in $a_1 + 1$ ways, if b_2 can be chosen in $a_2 + 1$ ways ... and if b_k can be chosen in $a_k + 1$ ways, then $b_1, b_2 ..., b_k$ all together can be selected in a number of ways given by the product.

$$(a_1 + 1) (a_2 + 1) ... (a_k + 1)$$

Hence, we find that the number $\tau(n)$ of positive integer divisors of n is exactly given by

$$\tau(n) \;=\; (a_1 + 1) (a_2 + 1) ... (a_k + 1)$$

For Example : $2520 = 2^3 . 3^2 . 5 . 7$, hence

$$\tau(2520) \;=\; (3 + 1) (2 + 1) (1+ 1) (1 + 1) = 48$$

so 2520 has exactly 48 distinct positive integer divisors.

$\sigma(n)$ **: The sum of divisors of n :** It is clear from the definition of $\tau(n)$, the sum $\sigma(n)$ of all the distinct positive integer divisors of a given positive integer $n > 1$ *is* given by the following product

$$\sigma(n) \;=\; (1 + p_1 + p_1^2 + ... + p_1^{a_1}) (1 + p_2 + ... + p_2^{a_2}) ... (1 + p_k + ... + p_k^{a_k})$$

because, in this product, each of the divisors d of n appears once and only once as a summend, when the product has been expanded.

Definition : *For each positive integer n, $\sigma(n)$ is the sum of positive divisors of n including 1 and n. It is defined as follows :*

$$\sigma : N \to N \; such \; that$$

$$\sigma(n) = \sum_{\substack{d|n \\ d \geq 1}} d \,, where \sum_{d|n} d \; is \; the \; sum \; of \; positive \; divisors \; of \; n.$$

SOLVED EXAMPLES

Example 1. *Evaluate* $\sigma(12)$ *and* $\sigma(28)$

Solution. The positive divisors of 12 are 1, 2 , 3, 4, 6 and 12.

so, $\sigma(12) \;=\; 1 + 2 + 3 + 4 + 6 + 12 = 28$

The positive divisors of 28 are 1, 2, 4, 7, 14 and 28.

so, $\sigma(28) \;=\; 1 + 2 + 4 + 7 + 14 + 28 = 56$

Remarks

❖ For $p_i > 1$, $\sigma(n)$ can also be written in the following form

$$\sigma(n) \;=\; \frac{p_1^{a_1+1} -1}{p_1 -1} . \frac{p_2^{a_2+1} -1}{p_2 -1} ... \frac{p_k^{a_k+1} -1}{p_k -1}$$

❖ If p is prime, then $\tau(p) = 2$ and $\sigma(p) = p + 1$.

Example 2. *Calculate* $\tau(n)$ *and* $\sigma(n)$ *for the integer* $n = 180$.

Solution. The prime factor of number $180 = 2^2 . 3^2 .5$.

So, $\tau(180) \;=\; (2 + 1) (2 + 1) (1 + 1) = 18$

i.e. Number of positive divisors of number 180 are 18 and sum of these divisors

$$\sigma(180) = \frac{2^3-1}{2-1} \cdot \frac{3^2-1}{3-1} \cdot \frac{5^2-1}{5-1} = \frac{7}{1} \cdot \frac{26}{2} \cdot \frac{24}{4} = 7.13.6 = 546$$

THEOREM-1

If $n = p_1^{a_1} p_2^{a_2} \dots p_k^{a_k}$ is the prime factorization of $n > 1$, then the divisors of n are the various terms of the expansion of

$$P = (1 + p_1 + p_1^2 + \dots + p_1^{a_1}).(1 + p_2 + p_2^2 + \dots + p_2^{a_2}) \dots (1 + p_k + \dots + p_k^{a_k})$$

Proof. We know that every divisor d of P is of the form $p_1^{b_1} p_2^{b_2} \dots p_k^{b_k}$, where $0 \le b_1 \le a_1$, $0 \le b_2 \le a_2, \dots 0 \le b_k \le a_k$. Therefore, d appears among the terms of the expansion of P. Also, we have that every term in the expansion of p is of the form $p_1^{b_1} p_2^{b_2} \dots p_k^{b_k}$ and therefore, it is a divisor of n. Moreover, all the terms in the expansion of P are distinct. Hence, divisors of n are the various terms of the expansion of P.

THEOREM-2

For any integer $n > 1$, $\tau(n)$ is odd if and only if n is a perfect square.

Proof. Let $n = p_1^{a_1} p_2^{a_2} \dots p_k^{a_k}$ be the square of an integer. Clearly, $a_1, a_2, \dots a_k$ are all even, i.e., $(a_1 + 1), (a_2 + 1), \dots (a_k + 1)$ are all odd.

Therefore,
$$\tau(n) = (a_1 + 1)(a_2 + 1) \dots (a_k + 1)$$

Clearly $\tau(n)$ is an odd integer, because product of odd integer is an add integer.

Conversely, let $\tau(n) = (a_1 + 1).(a_2 + 1) \dots (a_k + 1)$ is an odd integer. Then, clearly, each $(a_i + 1)$ is an odd integer, which gives each a_i is even.

Then, we can write
$$a_i = 2m_i$$
Therefore,
$$n = p_1^{2m_1} . p_2^{2m_2} \dots p_k^{2m_k}$$
$$= \left[p_1^{m_1} . p_2^{m_2} \dots p_k^{m_k} \right]^2$$

which show that n is a perfect square.

THEOREM-3

For any integer n, $\sigma(n)$ is odd if and only if n is a perfect square or twice of a perfect square.

Proof. We can write n and $\sigma(n)$ as follows:

$$n = p_1^{a_1} p_2^{a_2} \dots p_k^{a_k}$$

$$\sigma(n) = \left(\frac{p_1^{a_1+1}}{p_1-1} \right) \left(\frac{p_2^{a_2+1}}{p_2-1} \right) \dots \left(\frac{p_k^{a_k+1}}{p_k-1} \right)$$

$$= (1 + p_1 + \dots + p_1^{a_1})(1 + p_2 + \dots + p_2^{a_2}) \dots (1 + p_k + \dots + p_k^{a_k})$$

Further, $\sigma(n)$ is odd if and only if

$$(1 + p_1 + \ldots + p_1^{a_1})(1 + p_2 + \ldots + p_2^{a_2}) \ldots (1 + p_k + \ldots + p_k^{a_k})$$

iff $(1 + p_i + p_i^{a_i})$ is odd for all $i = 1, 2, \ldots r$

iff a_i is even for all $i = 1, 2, \ldots r$ and a_1 and if one p_i say $p_1 = 2$, then a_1 is any integer.

iff $a_i = 2m_i,\ i = 1, 2, \ldots, r.$

iff $n = p_1^{2m_1} \cdot p_2^{2m_2} \ldots p_r^{2m_r}$

$$= 2^{a_1}(p_2^{2m_2} \cdot p_3^{2m_3} \ldots p_r^{2m_r}) \quad \text{if } p_1 = 2$$

$$= 2^{a_1}(p_2^{m_2} \cdot p_3^{m_3} \ldots p_r^{m_r})^2$$

$$= (2^{a_1/2} p_2^{m_2} \cdot p_3^{m_3} \ldots p_r^{m_r})^2$$

Now, if a_1 is even, then

$$n = (2^{a_1/2} p_2^{m_2} \ldots p_r^{m_r})^2$$

and if a_1 is odd, say $a_1 = 2m + 1$, then $n = 2 \ (2^{m_1} p_2^{m_2} \ldots p_r^{m_r})^2$.

Hence, $\sigma(n)$ is odd if and only if n is a perfect square or twice of a perfect square.

THEOREM-4

If n is an integer greater than 1, then $\displaystyle\prod_{d|n} d = n^{\left(\frac{\tau(n)}{2}\right)}$

Proof. Let d be any divisor of n; then $\dfrac{n}{d}$ is also a divisor of n. Therefore, we always get

the divisors of n in pairs $\left(d, \dfrac{n}{d}\right)$.

Thus, the product of all divisors of n

$\displaystyle\prod_{d|n} d = n^{\left(\frac{\tau(n)}{2}\right)}$, where $\tau(n)$ is the number of divisors.

THEOREM-5

For each positive integer n, $\displaystyle\tau(n) = \prod_{p^a|n} (\alpha + 1)$

Proof. Let $n = \prod p^\alpha$ be the canonical factorization of n. A positive integer $d = \prod p^\beta$ divides n if and only if $0 \le \beta(p) \le \alpha(p)$ for all prime numbers p. Since $\beta(p)$ may take on any one of the values $0, 1, \ldots \alpha(p)$, there are $\alpha(p) + 1$ possible values for $\beta(p)$.

Hence, the number of divisors is $\displaystyle\prod_{p^a|n} (\alpha + 1)$

Remark

❖ From above theorem, it is clear that if $(m, n) = 1$, then $\tau(mn) = \tau(m)\,\tau(n)$.

THEOREM-6

Let x and y by real numbers. Then, we have the following properties:

(1) $[x] \le x < [x] + 1, x - 1 < [x] \le x, 0 \le x - [x] < 1$

(2) $[x] = \displaystyle\sum_{1 \le i \le x} 1 \ if \ x \ge 0.$

(3) $[x + m] = [x] + m;\ m$ *being any integer.*

(4) $[x] + [y] \le [x + y] \le [x] + [y] + 1$

(5) $[x] + [-x] = \begin{cases} 0;\ if\ x\ is\ any\ integer \\ -1;\ otherwise \end{cases}$

(6) $\left[\dfrac{[x]}{m}\right] = \left[\dfrac{x}{m}\right];\ m$ *is any positive integer.*

(7) $-[-x]$ *is the least integer greater than x.*

(8) $\left[x + \dfrac{1}{2}\right]$ *is the nearest integer to x. If two integers are equally near to x, it is the larger of the two.*

(9) $-\left[-x + \dfrac{1}{2}\right]$ *is the nearest integer to x. If two integers are equally near to x, it is the smaller of the two.*

(10) *If n and a are positive integers, $[n/a]$ is the number of integers among $1, 2, 3, ..., n$ that are divisible by a.*

Proof. (1) The first part of (1) is just the definition of $[x]$ in algebraic form. The two other parts are rearrangements of the first part.

(2) In (2), for $x \ge 0$, the sum counts the number of positive integers i that are less than or equal to x. This number is clearly just $[x]$.

(3) Result is obvious, by definition.

(4) We can write $x = n + v,\ y = m + \mu$, where n and m are integers and $0 \le v < 1$, $0 \le \mu \le 1$. Then

$$
\begin{aligned}
[x] + [y] \ &= \ n + m \le [n + v + m + \mu] = [x + y] \\
&= \ n + m + [v + \mu] \le n + m + 1 \\
&= \ [x] + [y] + 1
\end{aligned}
$$

(5) Writing $x = n + v$, we also have $-x = -n - 1 + 1 - v, 0 < 1 - v \le 1$. Then

$$[x] + [-x] = n + [-n - 1 + 1 - v]$$

$$= n - n - 1 + (1 - v) = \begin{cases} 0;\ if\ v = 0 \\ -1;\ if\ v > 0 \end{cases}$$

(6) Let us write $x = n + v$, $n = qm + r$, $0 \le v + 1$, $0 \le r \le m - 1$ and hence

$$\left[\frac{x}{m}\right] = \left[\frac{qm + r + v}{m}\right] = q + \left[\frac{r + v}{m}\right] = q, \text{ since } 0 \le r + v < m$$

Then $\left[\frac{[x]}{m}\right] = \left[\frac{n}{m}\right] = \left[q + \frac{r}{m}\right] = q$

(7) Replacing x by $-x$ in (1), we get $-x - 1 < (-x) \le -x$ *and hence* $x \le -[-x] < x + 1$.

(8) Let n be the nearest integer to x, taking the larger one if two are equally distant.

Then, $n = x + 8$, $-\frac{1}{2} < \theta \le \frac{1}{2}$ and $\left[x + \frac{1}{2}\right] = n + \left[-8 + \frac{1}{2}\right]$, since $0 \le -8 + \frac{1}{2} < 1$.

(9) The proof is similar to that of (8).

(10) If a, $2a$, $3a$, ... ja are all positive integers, less than r equal to n, that are divisible by a, then we must prove that $[x/a] = j$. But we have

$(j + 1)$ a exceeds n, so

$ja \le n < (j + 1)a$,

$\Rightarrow j \le n / a < (j + 1)$

Hence, $[n/a] = j$

7.3 DE POLIGNAC'S FORMULA

THEOREM-7

Let p be a prime. Then the largest exponent e such that $p^e \mid n!$ is $e = \sum\limits_{i=1}^{\infty}\left[\dfrac{n}{p^i}\right]$.

Proof. If $p^i > n$, then $[n/p^i] = 0$ Therefore, The sum terminates, it is not really an infinite series. We prove this theorem by mathematical induction.

Obviously, result is true for $1!$.

Assume that result is true for $(n - 1)!$ and let j denote the largest integer such that $p^j \mid n$. Since $n! = n \cdot (n - 1)!$, we must prove that $\Sigma[n/p^i] - \Sigma[(n - 1)/p^i] = j$,

but

$$\left[\frac{n}{p^i}\right] - \left[\frac{n-1}{p^i}\right] = \begin{cases} 1, \text{if } p^i \mid n \\ 0, \text{if } p^i \nmid n \end{cases}$$

and hence,

$$\sum\left[\frac{n}{p^i}\right] - \sum\left[\frac{n-1}{p^i}\right] = j.$$

Aliter. Let $a_1, a_2, ..., a_n$ be non-negative integers. Let $f(1)$ denote the number of them that are greater than or equal to 1, $f(2)$ the number greater than or equal to 2, and so on.

Then, $a_1 + a_2 + ... + a_n = f(1) + f(2) + ... + f(n)$

Since a_i contributes 1 to each of the numbers $f(1), f(2), ..., f(a_i)$. For $1 \leq j \leq n$, let a_j be the largest integer such that $p^{a_j} \mid j$. Then, we have $e = a_1 + a_2 + ... + a_n$. Also, $f(1)$ counts the number of integers $\leq n$ that are divisible by p, $f(2)$ the number divisible by p^2, and so on. Hence, $f(k)$ counts the integers $p^k, 2p^k, 3p^k, ..., [n/p^k]$, so that

$$f(k) = [n/p^k]$$

Therefore, we have

$$e = a_1 + a_2 + ... + a_n = \sum_{i=1}^{\infty} f(i) = \sum_{i=1}^{\infty} \left[\frac{n}{p^i} \right]$$

7.3.1 Applications of De Polignac's Formula :

(1) The above theorem is used to computing e. For example, if we wish to find the highest power of 7 that divides 1000 !, we find

$$[1000/7] = 142, [142/7] = 20, [20/7] = 2, [2/7] = 0$$

on adding, we find that

$$7^{164} \mid 1000!, \quad 7^{165} \nmid 1000!$$

(2) The day of the week from the date : The problem is to verify a given formula for calculating the day of the week for any given date. Any date, such as January 1, 2001 defines four integers N, M, C, Y as follows. Let N be the number of days in the month, so that N = 1. Let M be the number of month counting from march, so that M = 1 for March, M = 2 for April, ..., M = 10 for December and M = 11 for January. Let C denote the hundreds in the year and Y the rest, so that C = 20 and y = 01 for 2001. If d denotes the day of the week where $d = 0$ for Sunday, $d = 1$ for Monday,....., $d = 6$ for Saturday, then

$$d = N + [2. 6M - 0.2] + Y + [Y / 4] + [C/4] - 2C + (1 + L) [M / 11] \pmod{7}$$

where L = 1 for a leap year and L = 0 for a non-leap year. For example, in the case of January 01, 2001, we have L = 0, so

$$d = 1 + [28.4] + 1 + [1/4] + [20/4] - 40 - [11/11] \equiv 1 \pmod{7}$$

and hence, the first day of 2001 falls on a Monday.

Remark

❖ The above formula holds for any date after   ?, following the adoption of the Gregorian Calender at that time.

THEOREM-8

The highest power of a prime p which divides $n!$ is $\displaystyle\sum_{i=1}^{k} \left[\frac{n}{p} \right]$ *where* $p^k \leq k \leq p^{k+1}$.

Proof. Let p be a prime. Define $N = \{1, 2. ..., n\}$.

Then, N has

$$\left[\frac{n}{p} \right]$$ integers divisible by p.   ...(1)

$$\left[\frac{n}{p^2}\right] \text{ integers divisible by } p^2. \qquad \qquad ...(2)$$

$$\left[\frac{n}{p^3}\right] \text{ integers divisible by } p^3. \qquad \qquad ...(3)$$

$$\vdots$$

$$\left[\frac{n}{p^k}\right] \text{ integers divisible by } p^k. \qquad \qquad ...(4)$$

Now, since $p^k \le n \le p^{k+1}$, there will be no integer in N, divisible by higher powers of p than p^k. Also, since p is prime, there will be no more integers of N, the product of which is divisible by p.

Further, every number of (1) gives a factor p of n !. Every number of (2) is included in (1) and gives an additional factor p to $n!$.

Proceeding in the same way, we can say that every number of (3) is included in (2) and gives one additional factor p to n ! and so on.

Hence, the total number of factors p in n ! will be given by

$$\left[\frac{n}{p}\right] + \left[\frac{n}{p^2}\right] + ... \left[\frac{n}{p^k}\right]$$

DEDUCTIONS :

(1) Since $\left[\dfrac{n}{p^{k+r}}\right] = 0$, for $r > 0$, we have the highest power of a prime p dividing

$n!$ is $\displaystyle\sum_{k=1}^{\infty}\left[\frac{n}{p^k}\right]$.

(2) The number of integers of $S = \{1, 2,..., n\}$ divisible by a positive integer a is

$$\left[\frac{n}{a}\right].$$

THEOREM-9

The highest power of a prime contained in n ! is $h_1 + h_2 + ... + h_k$, where

$$h_1 = \left[\frac{n}{p}\right], \ h_2 = \left[\frac{h_1}{p}\right], \ ...., h_k = \left[\frac{h_{k-1}}{p}\right]$$

Proof. We know that, the highest power of p in $n!$ is

$$\left[\frac{n}{p}\right] + \left[\frac{n}{p^2}\right] + \left[\frac{n}{p^3}\right] + ... + \left[\frac{n}{p^k}\right]$$

Now, $$\left[\frac{n}{p}\right] = h_1$$

$$\left[\frac{n}{p^2}\right] = \left[\frac{1}{p}\left[\frac{n}{p}\right]\right] = \left[\frac{h_1}{p}\right] = h_2$$

$$\left[\frac{n}{p^3}\right] = \left[\frac{1}{p}\left[\frac{n}{p^2}\right]\right] = \left[\frac{h_2}{p}\right] = h_3$$

$$\vdots \qquad \vdots \qquad \vdots$$

$$\left[\frac{n}{p^k}\right] = \left[\frac{1}{p}\left[\frac{n}{p^{k-1}}\right]\right] = \left[\frac{h_{k-1}}{p}\right] = h_k$$

Hence, the highest power of a prime p contained in $n!$ is given by
$$h_1 + h_2 + \dots + h_k$$

THEOREM-10

The number $\dfrac{n!}{a_1!a_2!a_3!}$ *is an integer, when* $a_1, a_2, a_3 \geq 0$ *and* $n = a_1 + a_2 + a_3$

Proof. Let p be a prime such that p divides $n!$. Then, the highest power of p, which divides $n!$ is given by

$$P_H = \sum_{k=1}^{\infty} \frac{n}{p^k}$$

The highest power of p dividing $a_i! : i = 1, 2, 3$ is given by

$$P_{H_i} = \sum_{k=1}^{\infty} \frac{a_i}{p^k}$$

Now, $$\left[\frac{a_1}{p^k}\right] + \left[\frac{a_2}{p^k}\right] + \left[\frac{a_3}{p^k}\right] \leq \left[\frac{a_1 + a_2 + a_3}{p^k}\right] = \left[\frac{n}{p^k}\right], \ \forall \, k$$

Therefore,

$$\sum_{k=1}^{\infty}\left[\left[\frac{a_1}{p^k}\right] + \left[\frac{a_2}{p^k}\right] + \left[\frac{a_3}{p^k}\right]\right] \leq \left[\frac{n}{p^k}\right]$$

which clearly shows that the highest power of a prime dividing the numerator of the given expression is greater than or equal to the highest power of that prime dividing the numerator. Hence,

$$\frac{n!}{a_1!a_2!a_3!} \text{ is an integer.}$$

7.4   MULTIPLICATIVE FUNCTION

If $f(n)$ is an arithmetic function not identically zero such that

$$f(mn) = f(m). f(n)$$

for every pair of positive integers m, n satisfying $(m, n) = 1$

Then, $f(n)$ is said to be multiplicative function.

If $f(mn) = f(m) f(n)$, whether m and n are relatively prime or not, i.e. (m, n) may or may not equal to 1, then $f(n)$ is said to be totally multiplicative.

Remarks

❖ If f is a multiplicative function, then $f(1) = 1$.

❖ $f(m_1, m_2, \dots m_r) = f(m_1) \, f(m_2) \dots f(m_r)$. This result would hold if the integers m_1. $m_2, \dots m_r$, are prime powers of distinct primes.

❖ If f and g are multiplicative functions such that $f(p^\alpha) = g(p^\alpha)$ for all primes p and all positive integers α, then $f(n) = g(n)$ for all positive integers α.

❖ If f and g are totally multiplicative function such that $f(p) = g(p)$ for all primes p, then $f = g$.

THEOREM-1

Let $f(n)$ be a multiplicative function and let $F(n) = \displaystyle\sum_{d|n} f(d)$. Then $F(n)$ is multiplicative.

Proof. Let us suppose that $m = m_1 m_2$ with $(m_1, m_2) = 1$, i.e. m_1 and m_2 are relatively prime.

If $d \mid m$, then we set

$$d_1 = (d, m_1)$$
$$d_2 = (d, m_2)$$

Therefore, $d = d_1 d_2, d_1 \mid m_1, d_2 \mid m_2$

Conversely, if a pair d_1, d_2, of divisors of m_1 and m_2 are given, then

$$d = d_1 d_2$$

is a divisor of m and

$$d_1 = (d, m_1), d_2 = (d, m_2)$$

Therefore, there exists a one-to-one correspondence between the positive divisors d of m and pairs d_1, d_2, of positive divisors of m_1 and m_2. Thus

$$F(m) = \sum_{d|m} f(d) = \sum_{d_1|m_1} \sum_{d_2|m_2} f(d_1, d_2), \text{ for any arithmetic function } f.$$

Since $(d_1, d_2) = 1$, it follows from the hypothesis that f is multiplicative that the RHS is

$$\sum_{d_1|m_1} \sum_{d_2|m_2} f(d_1)f(d_2) = \left(\sum_{d_1|m_1} f(d_1)\right) \left(\sum_{d_2|m_2} f(d_2)\right) = F(m_1)F(m_2)$$

THEOREM-2

The arithmetic functions τ and σ are multiplicative functions.

Proof. Consider the case when $m = n = 1$, then in this case, result is trivially true.

Now, consider the case when $m > 1$, $n > 1$ be two relatively prime integers, i.e. $(m, n) = 1$

Let $\qquad m = p_1^{a_1} p_2^{a_2} \dots p_r^{a_r}$

and $\qquad n = q_1^{b_1} q_2^{b_2} \dots q_s^{b_s}$

be prime factorization of m and n.

Since, $(m, n) = 1$, therefore, no p_i will be among q_j and no q_j will be among p_i.

Thus the prime factorization of mn is given by

$$mn = p_1^{a_1} p_2^{a_2} \dots p_r^{a_r} \cdot q_1^{b_1} q_2^{b_2} \dots q_s^{b_s}$$

$$\therefore \quad \tau(mn) = [(a_1 + 1)(a_2 + 1) \dots (a_r + 1)][(b_1 + 1)(b_2 + 1) \dots (b_s + 1)]$$

$$= \tau(m)\tau(n) \qquad \qquad \dots(1)$$

Further,

$$\sigma(mn) = \left[\left(\frac{p_1^{a_1+1}-1}{p_1-1}\right) \dots \left(\frac{p_r^{a_r+1}-1}{p_r-1}\right)\right]\left[\left(\frac{q_1^{b_1+1}-1}{q_1-1}\right) \dots \left(\frac{q_s^{b_s+1}-1}{q_s-1}\right)\right]$$

$$= \sigma(m)\sigma(n)$$

DEDUCTION :

If $(m, n) = 1$ and $d_1 \mid m$ and $d_2 \mid n$, then the set of positive divisors of mn has all products $d_1 d_2$.

7.5 MOBIUS FUNCTION

Definition : *For positive integer n, we can define*

$$\mu(n) = \begin{cases} (-1)^{\omega(n)}, & \text{if } n \text{ is square free} \\ 0, & \text{otherwise} \end{cases}$$

Then $\mu(n)$ is the Mobius mu (μ) function.

Here, $\omega(n)$ is the number of distinct primes dividing n.

An alternative definition of Mobius function μ is given by

$$\mu(n) = \begin{cases} 1, & \text{if } n = 1 \\ 0, & \text{if } p^2 \mid n \text{ for some prime } p \\ (-1)^k, & \text{if } n = p_1 p_2 \dots p_k \text{ is the product of } k \text{ distinct primes} \end{cases}$$

THEOREM-1

The Mobius function is multiplicative.

Proof. If $a = 1$, then $\mu(ab) = \mu(b) = 1$. $\mu(b) = \mu(a)\,\mu(b)$

If $a > 1$ and $b > 1$ and if $(a, b) = 1$. Then, we can write

$$ab = p_1^{a_1} \dots p_k^{a_k} \cdot q_1^{b_1} \dots q_s^{b_s} \text{ with the } p\text{'s distinct from } q\text{'s.}$$

If $\mu(ab) = 0$, either some $a_i > 1$ or some $b_i > 1$.

In the first case, say $\mu(a) = 0$, therefore

$$\mu(a)\,\mu(b) = 0 = \mu(ab)$$

If $\mu(b) = 0$ Then $\mu(a)\,\mu(b) = 0 = \mu(ab)$

If $\mu(ab) \neq 0$, then every $a_i = 1$ and every $b_i = 1$. Thus

$$\mu(a) = (-1)^k \text{ and } \mu(b) = (-1)^s$$

Therefore,

$$\mu(a)\,\mu(b) = (-1)^k(-1)^s = (-1)^{k+s} = \mu(ab)$$

Since $(a, b) = 1$ implies the p's are distinct from the q's. In every case for which $(a, b) = 1$, it has been shown that $\mu(ab) = \mu(a)\,\mu(b)$.

Hence, $\mu(n)$ is multiplicative function.

THEOREM-2   (F. MORTEN'S LEMMA)

For each integer $n \geq 1$

$$\sum_{d|n} \mu(d) = \begin{cases} 1, & \text{if } n = 1 \\ 0, & \text{if } n > 1 \end{cases}$$

Proof. If $n = 1$, then

$$\sum_{d|n} \mu(d) = \mu(1) = 1$$

For $n > 1$, we can write

$$n = p_1^{k_1} p_2^{k_2} \dots p_r^{k_r}, \ p_i \text{ are distinct primes.}$$

Therefore,

$$\sum_{d|n} \mu(d) = \mu(1) + \sum_{i=1} \cdot \sum_{\substack{i<j \\ i<j=1}} \mu(p_i p_j) + \mu(p_1 p_2 \dots p_r) + 0 + \dots$$

$$(\because \text{ The terms containing square factors are zero)}$$

$$= 1 + r(-1) + (^rC_2)(-1)^2 + \dots + (-1)^r$$
$$= 1 + {}^rC_1(-1) + {}^rC_2(-1)^2 + \dots + (-1)^r$$
$$= (1-1)^r = 0$$

THEOREM-3   (MOBIUS INVERSION FORMULA)

Let $F(n) = \sum\limits_{d|n} f(d)$

then $f(n) = \sum\limits_{d|n} \mu(d)F(n|d) = \sum\limits_{d|n} \mu(n|d)F(d)$

Proof. Here, we have

$$\sum_{d|n} \mu(d)F(n|d) = \sum_{d|n}\left[\mu(d)\sum_{c|(n|d)} f(c)\right]$$

$$= \sum_{d|n}\left[\sum_{c|(n|d)}\mu(d)\,f(c)\right] = \sum_{c|n}\left[\sum_{d|(n|c)}f(c)\,\mu(d)\right]$$

[Using $d \mid n$ and $c \mid (n \mid d) \Leftrightarrow c \mid n$ and $d \mid (n \mid c)$]

$$= \sum_{c|n}\left[f(c)\sum_{d|(n|c)}\mu(d)\right]$$

$$= \sum_{c=n}f(c).1 = f(n)$$

Now, replacing $n \mid d$ by d' we get the second result.

THEOREM-4

If $F(n)$ is a multiplicative function and $F(n) = \sum_{d|n} f(d)$; then f is also multiplicative.

Proof. Consider two relatively prime integers m and n. Then, we have, any divisor d of mn can be written uniquely as $d = d_1 d_2$ where $d_1 \mid m$ and $d_2 \mid n$ such that $(d_1, d_2) = 1$

Using above theorem, we have

$$f(mn) = \sum_{d|m}\mu(d)F\left(\frac{mn}{d}\right)$$

$$= \sum_{\substack{d_1|m \\ d_2|m}}\mu(d_1 d_2)F\left(\frac{mn}{d_1 d_2}\right)$$

$$= \sum_{\substack{d_1|m \\ d_2|m}}\mu(d_1)\mu(d_2)F\left(\frac{m}{d_1}\right)F\left(\frac{n}{d_2}\right)$$

[$\because \mu$ and F both are multiplicative functions]

$$= \sum_{d_1|m}\mu(d_1)F\left(\frac{m}{d_1}\right)\sum_{d_2|n}\mu(d_2)F\left(\frac{n}{d_2}\right)$$

$$= f(m)f(n)$$

Hence, f is a multiplicative function.

THEOREM-5

If $n = p_1^{a_1} p_1^{a_2} \ldots p_k^{a_k}$, then $\sum_{d|n}|\mu(d)| = 2^k$.

Proof. We know that, for every divisor d for which $\mu(d) \neq 0$, $|\mu(d)| = 1$. Using definition of Mobius function, the number of divisors is given by

$$1 + {}^kC_1 + {}^kC_2 + \ldots + {}^kC_k = (1 + 1)^k = 2^k$$

Hence, $\sum_{d|n}|\mu(d)| = 2^k.1 = 2^k$

THEOREM-6

If $n = p_1^{a_1} p_2^{a_2} \ldots p_k^{a_k}$, then

$$\sum_{d|n} \frac{\mu(d)}{d} = \left(1 - \frac{1}{p_1}\right)\left(1 - \frac{1}{p_2}\right)\ldots\left(1 - \frac{1}{p^k}\right).$$

Proof. Write

$$f(n) = \frac{1}{n}$$

Then, $\sum_{d|n} \mu(d) f(d) = \sum_{d|n} \frac{\mu(d)}{d}$

Since $f(n) = \dfrac{1}{n}$ is a multiplicative function $\left[\because f(nm) = \dfrac{1}{mn} = \dfrac{1}{m} \cdot \dfrac{1}{n} = f(m) f(n)\right]$

Therefore,

$$\sum_{d|n} \mu(d) f(d) = [1 - f(p_1)][1 - f(p_2)] \ldots [1 - f(p_k)]$$

$$= \left(1 - \frac{1}{p_1}\right)\left(1 - \frac{1}{p_2}\right)\ldots\left(1 - \frac{1}{p^k}\right)$$

$\Rightarrow \qquad \sum_{d|n} \mu(d) . \frac{1}{d} = \left(1 - \frac{1}{p_1}\right)\left(1 - \frac{1}{p_2}\right)\ldots\left(1 - \frac{1}{p^k}\right)$

Hence, $\sum_{d|n} \dfrac{\mu(d)}{d} = \left(1 - \dfrac{1}{p_1}\right)\left(1 - \dfrac{1}{p_2}\right)\ldots\left(1 - \dfrac{1}{p^k}\right)$

SOLVED EXAMPLES

Based on the following Results

- $\tau(n)$ = Number of positive divisors of n including 1 and n.
- $\sigma(n)$ = The sum of positive divisors of n including 1 and n.
- $f(n)$ is said to be multiplicative function if $f(mn) = f(m).f(n)$; $(m, n) = 1$
- Mobius function $\mu(n) = \begin{cases} (-1)^{\omega(n)}; \text{ if } n \text{ is square free} \\ 0; \text{ otherwise} \end{cases}$

  Here $\omega(n)$ is the number of distinct primes dividing n.

Example 1. *Evaluate* τ *and* σ *for* $n = 3000$.

Solution. We can write

$$3000 = 2^3 . 3^1 . 5^3$$

Therefore,

$$\tau(3000) = (3 + 1)(1 + 1)(3 + 1) = 4 . 2 . 4 = 32$$

Again, $\sigma(3000) = \dfrac{(2^4-1)}{2-1}.\dfrac{(3^2-1)}{3-1}.\dfrac{(5^4-1)}{5-1} = \dfrac{(16-1)}{1}.\dfrac{(9-1)}{2}.\dfrac{(625-1)}{4}$

$= 15.4.156 = 9360$

Example 2. *Show that* $\tau(n) = \tau(n+1) = \tau(n+2)$ *for* $n = 4503$.

Solution. We can write

$$n = 4503 = 3 \times 19 \times 79$$
$$\tau(n) = \tau(4503) = (1+1)(1+1)(1+1) = 8 \qquad ...(1)$$

Again, $n+1 = 4504 = 2^3.563$

Thus, $\tau(n+1) = (3+1)(1+1) = 8 \qquad ...(2)$

Also, $n+2 = 4505 = 5 \times 17 \times 53$

Thus, $\tau(n+2) = (1+1)(1+1)(1+1) = 8 \qquad ...(3)$

Hence, from (1), (2) and (3), we conclude that
$$\tau(n) = \tau(n+1) = \tau(n+2)$$

Example 3. *If* $f(n) = n^2 + 2$ *and* $n = 6$, *then show that*

$$\sum_{d|6} f(d) = \sum_{d|6} f\left(\frac{6}{d}\right)$$

Solution. The divisors of 6 are given by

$$d = 1, 2, 3, 6$$

$$\therefore \qquad \frac{6}{d} = 6, 3, 2, 1$$

$$\Rightarrow \quad \sum_{d|6} f(d) = (1^2+2) + (2^2+2) + (3^2+2) + (6^2+2) = 58$$

Also $\displaystyle\sum_{d|6} f\left(\frac{6}{d}\right) = (6^2+2) + (3^2+2) + (2^2+2) + (1^2+2) = 58$

Hence, $\displaystyle\sum_{d|6} f(d) = \sum_{d|6} f\left(\frac{6}{d}\right)$

Example 4. *Show that* $\tau(n)$ *is equal to the lattice points on the hyperbola* $xy = n$, *lying in the first quadrant.*

Solution. Let $d_1, d_2, ..., d_r$ be the divisors of n.

Then, we have the lattice points $\left(d_1, \dfrac{n}{d_1}\right), \left(d_2, \dfrac{n}{d_2}\right), ..., \left(d_r, \dfrac{n}{d_r}\right)$ lies on the hyperbola

$xy = n$ in the first quadrant.

Again, if $\left(d, \dfrac{n}{d}\right)$ be any lattice point on the hyperbola $xy = n$, then it belongs to the

above lattice points, because $\dfrac{n}{d}$ is an integer. Hence, $\tau(n)$ is equal to the number of lattice points on the hyperbola $xy = n$.

Example 5. *Find the highest power of* 13 *contained in* 20000!.

Solution. We can find the highest power of 13 contained in 20000! by using\

$$H = h_1 + h_2 + h_3$$

where, $\quad h_1 = \left(\dfrac{20000}{13}\right) = 1538$

$$h_2 = \left(\dfrac{1538}{13}\right) = 118$$

$$h_3 = \left(\dfrac{118}{13}\right) = 9$$

Hence, $H = 1538 + 118 + 9 = 1665$.

Example 6. *Find the highest power of* 3 *contained in* 40!.

Solution. We know that, the highest power of a prime p dividing $n!$ is given by

$$\left[\dfrac{n}{p}\right] + \left[\dfrac{n}{p^2}\right] + \dots + \left[\dfrac{n}{p^k}\right], \text{ where } p^k \le n < p^{k+1}$$

Hence, $\left[\dfrac{40}{3}\right] + \left[\dfrac{40}{3^2}\right] + \left[\dfrac{40}{3^3}\right] = 13 + 4 + 118$, is the highest power of 3 which divides 40!.

Example 7. *Find the highest power of* 18 *contained in* 500!.

Solution. The composite number 18 can be decomposed into a product of prime, as

$$18 = 2 \times 3^2$$

Thus, the primes contained in 18 are 2 and 3.

Further, the highest powers of 2 in 500! is given by

$$H_1 = \left[\dfrac{500}{2}\right] + \left[\dfrac{250}{2}\right] + \left[\dfrac{125}{2}\right] + \left[\dfrac{62}{2}\right] + \left[\dfrac{31}{2}\right] + \left[\dfrac{15}{2}\right] + \left[\dfrac{7}{2}\right] + \left[\dfrac{3}{2}\right]$$

$$= 250 + 125 + 62 + 31 + 15 + 7 + 3 + 1 = 494$$

Also, the highest power of 3, contained in 500! is given by

$$H_2 = \left[\dfrac{500}{3}\right] + \left[\dfrac{166}{3}\right] + \left[\dfrac{55}{3}\right] + \left[\dfrac{18}{3}\right] + \left[\dfrac{6}{3}\right]$$

$$= 166 + 55 + 18 + 6 + 2 = 247$$

$\therefore \quad 500! = 3^{494} \times 3^{247} \times p, \quad$ for some p such that $(p, 18) = 1$

$$= (2^2 \times 3)^{247} \times p$$

Hence, the highest power of 18 in 500! is 247.

Example 8. *Show that* $\dfrac{(2n)!}{(n!)^2}$ *is an even integer.*

Solution. Let $f(n) = \dfrac{(2n)!}{(n!)^2}$

We prove this result by the principle of mathematical induction.

For $n = 1$

$$f(1) = \frac{(2.1)!}{(1!)^2} = 2, \text{ which is even.}$$

$\Rightarrow$ Result is true for $n = 1$.

Let us suppose that result is true for $n = m$.

i.e. $$f(m) = \frac{(2m)!}{(m!)^2} = 2k, \text{ an even integer.}$$

Then, $$f(m + 1) = \frac{[2(m+1)]!}{((m+1)!)^2} = \frac{(2m+2)(2m+1)(2m)!}{(m+1)^2(m!)^2}$$

$$= \frac{2[2m+1]2k}{(m+1)}, \text{ which is even for any value of } m.$$

Hence, by the principle of mathematical induction, we have

$\dfrac{(2n)!}{(n!)^2}$ is an even integer.

Example 9. *Find a positive integer n such that*
$$\mu(n) + \mu(n+1) + \mu(n+2) = 3$$

Solution. Here, we have

$$\mu(33) = \mu(3.11) = (-1)^2 = 1$$
$$\mu(34) = \mu(2.17) = (-1)^2 = 1$$
$$\mu(35) = \mu(5.7) = (-1)^2 = 1$$

Therefore, $\mu(33) + \mu(34) + \mu(35) = 3$

which implies n should be equal to 33.

Remark

❖ Similarly, we can find the integers $n = 85$, $n = 93$, which satisfy the given relation.

Example 10. *If n is a positive integer such that $n \geq 3$, then show that*

$$\sum_{k=1}^{n} (k!) = 1$$

Solution. We prove this result by using the principle of mathematical induction.

For $n = 3$, we have

$$\sum_{k=1}^{3} \mu(k!) = \mu(1!) + \mu(2!) + \mu(3!)$$

$$= \mu(1) + \mu(2) + \mu(6)$$
$$= 1 + (-1) + \mu(2.3)$$
$$= (-1)^2 = 1 \Rightarrow \text{ Result is true for } n = 3.$$

Suppose, result is true for $n = m$, i.e.,

$$\sum_{k=1}^{m} \mu(k!) = 1$$

To show, result is true for $n = m + 1$. For this, we shall show that

$$\sum_{k=1}^{m+1} \mu(k!) = 1$$

Consider $\sum_{k=1}^{m+1} \mu(k!) = \sum_{k=1}^{m} \mu(k!) + \mu[(m+1)]!]$

$$= 1 + 0 \qquad\qquad [\because \text{ in } (m+1)!, \text{ there will be one factor } 2^2]$$

$$= 1$$

$\Rightarrow$ Result is true for $n = m + 1$

Hence, by principle of mathematical induction, the result is true for all $n \geq 3$.

EXERCISE 7.1

1. Show that for $n = 3655$, $\tau = 8$ and $\sigma = 4752$.

2. Discuss the behaviour of $\tau(n)$ and $\sigma(n)$ when $n \to \infty$.

3. For each positive integer n, show that $\mu(n)\, \mu(n + 1)\mu(n + 2)\mu(n + 3) = 0$.

4. Show that the highest power of 7 contained in 2000! is 330.

5. Show that $\displaystyle\sum_{d|n} [\tau(n)]^3 = \left[\sum_{d|n} \tau(n)\right]^2$, for $n \geq 1$.

6. Show that $\sigma(n) = \sigma(n + 1)$ for $n = 957$.

7. If $2^k - 1$ is a prime and $n = 2^{k-1}(2^k - 1)$, then show that $\tau(n) = 2n$.

8. Show that $\sigma(2^{k-1}) = 2^k - 1$.

9. Show that for every positive integer n, $\displaystyle\sum_{d|n} |\mu(d)| = 2^{\omega(n)}$

10. What is the highest power of 2 dividing 533!? The highest power of 3? The highest power of 6? The highest power of 12? The highest power of 70?

11. For what real number x is it true that

    (1) $[x] + [x] = [2x]$ (2) $[x + 3] = 3 + [x]$ (3) $[x + 3] = 3 + x$

    (4) $\left[x + \dfrac{1}{2}\right] + \left[x - \dfrac{1}{2}\right] = [2x]$ (5) $[9x] = 9$

12. For every positive integer n, show that $n!\,(n-1)!$ is a divisor of $(2n-2)!$

13. If $(m, n) = 1$, show that

    $$\sum_{x=1}^{m-1}\left[\frac{mx}{n}\right] = \frac{(m-1)(n-1)}{2}$$

14. If $m \geq 1$, show that $[(1 + \sqrt{3}\,)^{2m+1}]$ is divisible by 2^{m+1} but not 2^{m+2}.

15. If n is any positive integer and x is any real number, show that

$$[x] + \left[x + \frac{1}{x}\right] + ... \left[x + \frac{n-1}{x}\right] = [nx]$$

16. Given any positive integer k, show that there exists infinitely many integers n such that

$$\mu(n + 1) = \mu(n + 2) = \mu(n + 3) = ... = \mu(n + k).$$

17. Show that for each positive integer n, $\displaystyle\sum_{\substack{a=1 \\ (a,n)=1}}^{n} e^{2\pi i a/n} = \mu(n)$

18. Suppose that $f(n)$ is an arithmetic function whose values are all non-zero and if

$$F(n) = \prod_{d|n} f(d) \text{, show that } f(n) = \prod_{d|n} F\left(\frac{n}{d}\right)^{\mu(d)}$$

7.6 EULER'S FUNCTION

The Euler function or Euler phi function is a widely used number theoretic function, represented by $\phi(n)$. For $n = 1$, we define $\phi(1) = 1$ and when $n > 1$, we define $\phi(n)$ to be the number of positive integers less then n and relatively prime to n.

For example : $\phi(12) = 4$ because the only positive integers less than 12 and relatively prime to 12 are 1, 5, 7, 11, i.e. 4.

THEOREM-1

For any positive integer n, $n = \Sigma\phi(d)$, where the summation extends over all the positive divisors d of n.

Proof. For $n = 1$, result is obvious, because $\phi(1) = 1$.

Let $n > 1$. For every positive integer $x \leq n$, $(x, n) = d$, where d is a uniquely determined divisor of n. on this basis alone the n numbers 1, 2,..., n are divided into mutually exclusive d-classes. From $(x, n) = d$, we have $a = kd$, $n = d'd$, with $(k, d') = 1$ and with $k \leq d'$, since $x \leq n$. The case $k = d'$ is exceptional for from the condition $(k, d') = 1$, this case can arise only when $d' = 1$. Hence, in all cases we find that there are exactly $\phi(d')$ choices for k and hence $\phi(d')$ integers x which belong to the d-classes. Therefore, by the use of d-classes, we have found $n = \Sigma\phi(d')$ where $d'd = n$ and the summation is overall divisors d of n. However, the set of numbers $\{d'\}$ is simply the set $\{d\}$ in another order. Hence, we can written $n = \Sigma\phi(d') = \Sigma\phi(d)$.

7.6.1 General Formula for Calculating $\phi(m)$

First consider its value at prime and then at prime powers. We know that if p is prime, $\phi(p) = p - 1$. In complete residue system modulo p^k, only multiples of p, i.e. p, $2p$,..., $p^{k-1}.p$ are integers not relatively prime to p. There are p^{k-1} integers in all.

The other

$$p^k - p^{k-1} = p^k\left(1 - \frac{1}{p}\right)$$ integers are relatively prime to p. Hence, the reduced residue

system modulo p contains $p^k\left(1 - \frac{1}{p}\right)$ integers, i.e.

$$\phi(p^k) = p^k\left(1 - \frac{1}{p}\right)$$

Proof : To find $\phi(p^k)$, we have to find the number of positive integers that are less than p^k and relatively prime to p^k.

The set S of positive integers less than p^k contains $p^k - 1$ elements since for every integer a

$$(a, p^k) = 1 \quad \Leftrightarrow \quad (a, p) = 1 \quad \text{or} \quad (a, p^n) \neq 1 \quad \Leftrightarrow \quad [a, p] \neq 1.$$

Therefore, the positive integers in S that are not relatively prime to p^k are only those which are multiple of p.

Now the integers in S which are multiple of p are.

$$p, 2p, ..., (p-1), p, pp, (p+1)p, ..., (p^{n-1} - 1)p$$

Thus, the number of positive integers in S that are not relatively prime to p^k is equal to $p^{k-1} - 1$.

Hence, $\phi(p^k) = (p^k - 1) - (p^{k-1} - 1) = p^k - p^{k-1} = p^k\left(1 - \frac{1}{p}\right)$

For example : Since 2 is a prime, therefore by the above theorem

$$\phi(2^k) = 2^k - 2^{k-1} = 2^{k-1}(2-1) = 2^{k-1}$$

DEDUCTION :

If p is a positive prime and n is any positive integer, then
$$\phi(1) + \phi(p) + \phi(p^2) + ... + \phi(p^{n-1}) + \phi(p^n) = p^n$$

Proof. We know that $\phi(1) = 1$.

Also, by using $\phi(p^k) = p^k - p^{k-1}$, we get

$$\phi(p) = p^1 - p^0 = p - 1$$
$$\phi(p^2) = p^2 - p$$
$$\phi(p^3) = p^3 - p^2$$
$$...\qquad ...$$
$$...\qquad ...$$
$$\phi(p^{k-1}) = p^{k-1} - p^{k-2}$$
$$\phi(p^k) = p^k - p^{k-1}$$

Adding all the above relations, we get

$$\phi(1) + \phi(p) + \phi(p^2) + ... + \phi(p^n) = p^n$$

THEOREM-1

If m and n are relatively prime positive integers, i.e. (m, n) = 1, then

$$\phi(mn) = \phi(m)\phi(n)$$

In other words, we can say that the Euler's function ϕ is multiplicative.

Proof. To find the positive integers less than mn and relatively prime to mn, we arrange the integers $1, 2, 3, \ldots, mn$ in a rectangular array having n rows and m columns in the following manner:

1	2	...	k	...	m
$m + 1$	$m + 2$	...	$m + k$	...	$2m$
$2m + 1$	$2m + 2$	...	$2m + k$	...	$3m$
...	...	...	...	...	...
...	...	...	...	...	...
$(n-1)m + 1$	$(n-1)m + 2$	...	$(n-1)m + k$	...	nm

In the above arrangement of integers, we have to find the integers a such that $1 \le a < mn$ and $(a, mn) = 1$.

But, we know that if $a \in Z$, then $(a, mn) = 1$ if and only if $(a, m) = 1$ and $(a, n) = 1$.

Thus, in the above arrangement of integers, we have to find the integer a such that

$1 \le a < mn$ and $(a, m) = 1$ as well as $(a, n) = 1$.

For this, we first consider the m columns that begin with $1, 2, \ldots, k, \ldots, m$.

If b is any integer, then we have

$$(m, k) = 1 \text{ iff } (m, bm + k) = 1.$$

Thus, if the first term of a column is relatively prime to m, then, every other term of that column is also relatively prime to m. But the number of integers $1, 2, \ldots, m$ that are relatively prime to m is $\phi(m)$. Therefore, there are $\phi(m)$ columns in each of which every integer is relatively prime to m.

Now, we have to find the number of integers in these $\phi(m)$ columns that are also relatively prime to n.

Out of these $\phi(m)$ columns, let us consider the columns that begins with k. The terms in this column are

$$k, m + k, 2m + k, \ldots, (n-1)m + k$$

which are m in number. These n integers when divided by n yield remainders $0, 1, 2, \ldots, n - 1$, though not necessarily in this order.

If r is the remainder when $bm + k$ is divided by n, then

$$bm + k = nq + r, \text{ where } q \in Z \text{ and } 0 \le r < n$$

Now, $$(bm + k, n) = (nq + r, n) = (r, n)$$

$\therefore$ $$(bm + k, n) = 1 \text{ if and only if } (r, n) = 1.$$

But the number of positive integers among the n remainders $0, 1, 2, \ldots, n - 1$ that are relatively prime to n is $\phi(m)$. Therefore, the number of integers in the kth column that are relatively prime to n is $\phi(n)$. This is true for each of the $\phi(m)$ columns. Thus, each of the $\phi(m)$ columns in which every term is relatively prime to m contains $\phi(n)$ integers which are also relatively prime to n. Hence, in the whole arrangement, there are $\phi(m)$

integers that are relatively prime to m as well as to n and consequently relatively prime to mn.

Hence, $\phi(mn) = \phi(m)\phi(n)$

DEDUCTION (1) : If $m_1, m_2, ..., m_r$ are r positive integers which are relatively prime to each other in pairs, then

$$\phi(m_1, m_2, ..., m_r) = \phi(m_1) \, \phi \, (m_2) ... \, \phi(m_r)$$

(2) : If $m = p_1^{m_1} p_2^{m_2} ... p_k^{m_k}$, then using theorem–1, we can derive easily

$$\phi(m) = \phi(p_1^{m_1})\phi(p_2^{m_2}) ... \phi(p_k^{m_k})$$

which is the general formula for $f(m)$

Remarks

❖ The formula $\phi(mn) = \phi(m).\phi(n)$ is applicable only when $(m, n) = 1$. For example, we have $\phi(20) = 8$ because 1, 3, 7, 9, 11, 13, 17, 19 are the only eight positive integers which are less than 20 and relatively prime to 20. Now $20 = 10 \times 2$ and 10 and 2 are not relatively prime, we have $\phi(10) = 4$ and $\phi(2) = 1$. Therefore

$$\phi(20) = \phi(10.2) \neq \phi(10). \phi(2).$$

❖ Also, we can write $20 = 4. 5$ and 4 and 5 are relatively prime, we have $\phi(4) = 2$ and $\phi(5) = 4$. Therefore

$$\phi(20) = \phi(4. 5) = \phi(4) \, \phi(5)$$

THEOREM-2

If $n > 1$ and $p_1, p_2, ..., p_m$ are the distinct prime factors of n, then

$$\phi(n) = n\left(1-\frac{1}{p_1}\right)\left(1-\frac{1}{p_2}\right)...\left(1-\frac{1}{p_m}\right).$$

Proof. Using fundamental theorem of arithmetic, we have

$$n = p_1^{\alpha_1} p_2^{\alpha_2} ... p_m^{\alpha_m} ,$$

where $\alpha_1, ..., \alpha_m$ are positive integers. Since, $p_1, p_2, ..., p_m$ are relatively prime to each other in pairs, thus

$$\phi(m) = \phi(p_1^{\alpha_1})\phi(p_2^{\alpha_2}) ... \phi(p_m^{\alpha_m})$$

$$= p_1^{\alpha_1}\left(1-\frac{1}{p_1}\right).p_2^{\alpha_2}\left(1-\frac{1}{p_2}\right)...p_m^{\alpha_m}\left(1-\frac{1}{p_m}\right)$$

$$= p_1^{\alpha_1} p_2^{\alpha_2} ... p_m^{\alpha_m} \left(1-\frac{1}{p_1}\right)\left(1-\frac{1}{p_2}\right)...\left(1-\frac{1}{p_m}\right)$$

$$= n\left(1-\frac{1}{p_1}\right)\left(1-\frac{1}{p_2}\right)...\left(1-\frac{1}{p_m}\right)$$

Example. Calculate (i) $\phi(360)$, (ii) $\phi(1001)$ and (iii) $\phi(36,000)$.

Solution. (i) $\phi(360)$. The prime power decomposition of 360 is $2^3 . 3^2 . 5$ so

$$\phi(360) = 360\left(1-\frac{1}{2}\right)\left(1-\frac{1}{3}\right)\left(1-\frac{1}{5}\right)$$

$$= 360 . \frac{1}{2}.\frac{2}{3}.\frac{4}{5} = 96$$

(ii) $\phi(1001)$. Here

$$1001 = 7 \times 11 \times 13$$

So, $\phi(1001) = 1001 \left(1-\frac{1}{7}\right)\left(1-\frac{1}{11}\right)\left(1-\frac{1}{13}\right)$

$$= 1001 . \frac{6}{7}.\frac{10}{11}.\frac{12}{13}$$

$$= 720$$

(iii) $\phi(36000)$. Here

$$36000 = 2^5 . 3^2 . 5^2$$

So, $\phi(36000) = 36000\left(1-\frac{1}{2}\right)\left(1-\frac{1}{3}\right)\left(1-\frac{1}{5}\right)$

$$= 36000.\frac{1}{2}.\frac{2}{3}.\frac{4}{5}$$

$$= 9600.$$

Remark

❖ Clearly $\phi(m) \le m - 1$. If $m \ge 2$, then $\phi(m)$ is even. If $n \mid m$, then $\phi(n) \mid \phi(m)$.

THEOREM-3

If $(a, b) = d$, then

$$\phi(ab) = \phi(a)\phi(b)\frac{d}{\phi(d)}.$$

Proof. Using above theorem, we can easily obtain

$$\frac{\phi(ab)}{ab} = \prod_{p\mid ab}\left(1-\frac{1}{p}\right) = \frac{\prod\limits_{p\mid a}\left(1-\frac{1}{p}\right)\prod\limits_{p\mid b}\left(1-\frac{1}{p}\right)}{\prod\limits_{p\mid(a,b)}\left(1-\frac{1}{p}\right)}$$

$$= \frac{\dfrac{\phi(a)}{a}.\dfrac{\phi(b)}{b}}{\dfrac{\phi(d)}{d}} = \frac{1}{ab}\phi(a)\phi(b).\frac{d}{\phi(d)}$$

THEOREM-4

The sum of $\phi(m)$, positive integers less than m (> 1) and relatively prime to m is $\dfrac{m}{2}\,\phi(m)$.

Proof. Let $x_1, x_2, ..., x_{\phi(m)}$ be the given $\phi(m)$ positive integers. Clearly, they are a reduced residue system of m. Then, $m - x_1, ..., m - x_{\phi(m)}$ are also a reduced system of m.

Thus, $(m - x_1) + ... + (m - x_{\phi(m)}) = x_1 + ...\, x_{\phi(m)}$

i.e. $m\phi(m) = 2(x_1 + ... + x_{\phi(m)})$

Hence, $x_1 + x_2 + ... + x_{\phi(m)} = \dfrac{m\phi(m)}{2}$

THEOREM-5

If the integer of an A.P.

$$r, r + m, r + 2m, ... r + (n - 1)m \qquad ...(1)$$

with common difference m relatively prime to n, then the remainders obtained are 0, 1, 2, ..., ($n - 1$) in some order.

Proof. If the terms in (1) are divided by n, we get n remainders

$\{0, 1, 2, ..., n - 1\}$, in some order.

These remainders will be distinct. Suppose, two terms of (1) say $r + xm$ and $r + ym$ have the same remainders where x and y are integers such that

$0 \leq x < y \leq n - 1$. Then we have

$(r + ym) - (r + xm) = (y - x)m$

is divisible by n, which is not possible, because $(m, n) = 1$ and $y - x < n$. Hence, when we divide the terms of (1) by n, then we have n distinct remainders 0, 1, 2,..., ($n - 1$) in some order.

THEOREM-6

If S = $\{r, r + m, r + 2m, ..., r + (n - 1)m\}$ are n terms of an A.P. whose common difference m relatively prime to n, then there are exactly $\phi(n)$ integers prime to n.

Proof. If we divide the integers given in S by n, then we get remainders $S_1 = \{0, 1, 2, ..., n - 1\}$ in some order. Also, there are as many integers prime to n in S as there are in S_1. But there are $\phi(n)$ integers prime to n in S_1. Hence, the number of integers prime to n in S is $\phi(n)$.

THEOREM-7

If d is a positive divisor of a positive integer m, then the number of integers in the complete residue system modulo m, with m have the greatest common divisor d, is $\phi\left(\dfrac{m}{a}\right)$.

Proof. We know that in the complete residue system 1, 2, ..., m modulo m, the multiples of d are of the form $kd \leq k \leq \dfrac{m}{d}$. If $(kd, m) = d$, then $\left(k, \dfrac{m}{d}\right) = 1$.

Thus, only if k is one of the positive integers not exceeding $\dfrac{m}{d}$ and are relatively prime

to $\dfrac{m}{d}$, then the greatest common divisor of kd and m is d.

Hence, the number of k is $\phi\left(\dfrac{m}{d}\right)$.

THEOREM-8

If $d_1, d_2, ..., d_r$ be the distinct positive divisors of a positive integer n, then
$$\phi(d_1) + \phi(d_2) + ... + \phi(d_r) = n$$

Proof. If $n = 1$, then we know that 1 is the only positive divisor of 1 and we have $\phi(1) = 1$.

i.e. result is true for $n = 1$

Let $n > 1$. Then, by fundamental theorem of arithmetic, on being expressed in canonical form, we have
$$n = p_1^{\alpha_1} p_2^{\alpha_2} ... p_m^{\alpha_m}$$

Any positive divisor d of n is of the form $p_1^{\beta_1} p_2^{\beta_2} ... p_m^{\beta_m}$ where $0 \leq \beta_1 \leq \alpha_1$, $0 \leq \beta_2 \leq \alpha_2, ..., 0 \leq \beta_m \leq \alpha_m$

Now, consider the product

$$P = [1 + \phi(p_1) + \phi(p_1^2) + ... \phi(p_1^{\alpha_1})][1 + \phi(p_2) + \phi(p_2^2) + ... \phi(p_2^{\alpha_2})]$$
$$... [1 + \phi(p_m) + \phi(p_m^2) + ... \phi(p_m^{\alpha_m})]$$

whose general term is

$$= \phi(p_1^{\beta_1})\phi(p_2^{\beta_2}) ... \phi(p_m^{\beta_m})$$
$$= \phi(p_1^{\beta_1} p_2^{\beta_2} ... p_m^{\beta_m}) = \phi(d) \qquad [\because \phi \text{ is a multiplicative function}]$$

Therefore, if d is any positive divisor of n, then $\phi(d)$ is equal to the value of one and only one term in the product P.

Thus, if $d_1, d_2, ... d_r$ be the positive distinct divisors of n, then
$$\Sigma\phi(d) = \phi(d_1) + \phi(d_2) + ... + \phi(d_r)$$
$$= \text{sum of all the terms in the product } P.$$
$$= P$$

Also, we have
$$1 + \phi(p_1) + \phi(p_1^2) + ... + \phi(p_1^{\alpha_1}) = p_1^{\alpha_1}, \text{ etc.}$$

Hence, $\Sigma\phi(d) = P = p_1^{\alpha_1} p_2^{\alpha_2} \ldots p_m^{\alpha_m} = n$

Remark

❖ The above theorem is also known as Gauss theorem.

7.7   SOME OTHER IMPORTANT NUMBER THEORETIC FUNCTIONS

(1) Function T(n)

We can define the function $T(n)$ as follows :

$$T(n) = \tau(1) + \tau(2) + \tau(3) + \ldots + \tau(n)$$

$$= \sum_{r=1}^{n} \tau(r)$$

(2) Function S(n)

We can define the function $S(n)$ as follows :

$$S(n) = \sigma(1) + \sigma(2) + \ldots \sigma(n)$$

$$= \sum_{r=1}^{n} \sigma(r)$$

(3) Function ζ(n)

If n is any real number greater than 1, then the function ζ is defined as follows

$$\zeta(n) = \sum_{m=1}^{\infty} \frac{1}{m^n}, \; n > 1$$

$$= \frac{1}{n-1} + c, \;\; \text{where } c \text{ is any constant less than or equal to 1.}$$

For example : $\zeta(2) = 1 + \dfrac{1}{2^2} + \dfrac{1}{3^2} + \ldots = \dfrac{\pi^2}{6}$

(4) Function $\bar{\phi}(n)$

We can define the function $\bar{\phi}(n)$ as follows

$$\bar{\phi}(n) = \phi(1) + \phi(2) + \ldots \phi(n) = \sum_{r=1}^{n} \phi(r)$$

It is known as symmetry function of $\phi(n)$.

For example : $\bar{\phi}(8) = \phi(1) + \phi(2) + \ldots \phi(8)$

$$= 1 + 1 + 2 + 2 + 4 + 2 + 6 + 4 = 22$$

(5) Recurrence Function

We say that the arithmetic function $f(n)$ satisfies a linear recurrence (or recursion) if

$$f(n) = af(n-1) + bf(n-2) \quad \text{for } n = 2, 3, \ldots$$

where, a and b are fixed number, which may be real or complex.

(6) Square Free Integers

A number a is said to be square free if 1 is the largest square dividing a. Thus, a is square free if and only if the exponents take only the values 0 and 1.

In other words, an integer (> 0) is called square free if it is not divisible by any square greater than 1.

General Form : The general form of a square free integer is $p_1 p_2 \ldots p_n$ where each p_i are distinct primes.

THEOREM-1

If n is any given positive integer, then

$$\sum_{x=n+1}^{\infty} \frac{1}{x^2} = 0\left(\frac{1}{n}\right)$$

Proof. Since we know that $\dfrac{1}{x^2}$ is a strictly decreasing function, therefore, we have

$$0 < \frac{1}{(n+1)^2} < \int_n^{n+1}\left(\frac{dx}{x^2}\right)$$

$$0 < \frac{1}{(n+2)^2} < \int_{n+1}^{n+2}\left(\frac{dx}{x^2}\right)$$

$$\vdots \qquad \vdots \qquad \vdots$$

$$0 < \frac{1}{(n+r)^2} < \int_{n+r-1}^{n+r}\left(\frac{dx}{x^2}\right)$$

On adding, we get

$$0 < \sum_{x=n+1}^{n+r} \frac{1}{x^2} < \int_n^{n+r}\left(\frac{dx}{x^2}\right) = \frac{1}{n} - \frac{1}{n+r}$$

Finally, taking limit as $r \to \infty$, i.e. $k + r \to \infty$, we get

$$\sum_{x=n+1}^{\infty} \frac{1}{x^2} < \frac{1}{n} = 0\left(\frac{1}{k}\right)$$

DEDUCTION : For any positive integer n, $\displaystyle\sum_{x=1}^{n} \frac{1}{x^2} = \frac{\pi^2}{6} + 0\left(\frac{1}{n}\right)$

Proof. Using above theorem, we have

$$\sum_{x=1}^{n} \frac{1}{x^2} = \sum_{x=1}^{\infty} \frac{1}{x^2} - \sum_{x=n+1}^{\infty} \frac{1}{x^2}$$

$$= \frac{\pi^2}{6} + 0\left(\frac{1}{n}\right) \qquad\qquad \left[\because \sum_{x=1}^{\infty} \frac{1}{x^2} = \frac{\pi^2}{6}\right]$$

THEOREM-2

If n is greater than 1, then

$$\zeta(n) = \prod_{p}\left(1-\frac{1}{p^n}\right)^{-1}$$

Proof. Let $f(x) = \frac{1}{x^n}$

Clearly, this function is multiplicative and $\sum_{n=1}^{\infty}\frac{1}{x^n}$ is absolutely convergent.

Thus $\sum_{x=1}^{\infty}\frac{1}{x^n} = \prod_{p}\left[1+f(p)+f(p^2)+...\right]$

which gives

$$\zeta(n) = \prod_{p}\left[1+\frac{1}{p^n}+\frac{1}{p^{2n}}+...\right]$$

$$= \prod_{p}\left[1-\frac{1}{p^n}\right]^{-1}$$

THEOREM-3

If $n > 1$, then $\sum\frac{\mu(x)}{x^n} = \prod_{p}\left(1-\frac{1}{p^n}\right)=\frac{1}{\zeta(n)}$

Proof. We know that $\mu(x)$ and $\frac{1}{x^n}$ both are multiplicative functions. Also product of

two multiplicative functions is again multiplicative. Further, $\frac{\mu(x)}{x^n}$ is absolute convergent.

Hence, $\sum\frac{\mu(x)}{x^n} = \prod\left[1+\frac{\mu(p)}{p^n}+\frac{\mu(p^2)}{p^{2n}}+...\right]$

$$= \prod_{p}\left(1-\frac{1}{p^n}\right) \qquad \text{(By using } \mu(p)=-1, \mu(p^2)=\mu(p^3)=...=0)$$

$$= \frac{1}{\zeta(n)}$$

THEOREM-4

$$\bar{\phi}(n) = \frac{1}{2}\sum_{x=1}^{n}\mu(x)\left[\left[\frac{n}{x}\right]^2 + \left[\frac{n}{x}\right]\right]$$

Proof. Here, we have

$$\bar{\phi}(n) = \phi(1) + \phi(2) + \dots + \phi(n) \qquad \dots(1)$$

$$= 1.\sum_{d|1}\frac{\mu(d)}{d} + 2\sum_{d|2}\frac{\mu(d)}{d} + \dots + n\sum_{d|n}\frac{\mu(d)}{n}$$

The above terms can be rearranged as

$$\bar{\phi}(n) = a_1\mu(1) + a_1\mu(2) + \dots + a_n\mu(n)$$

$$= \sum_{x=1}^{n}a_n\mu(x)$$

Since, $\mu(x)$ appears once in the following terms of (1)

$$x\sum_{d|x}\frac{\mu(d)}{d}, 2x\sum_{d|2x}\frac{\mu(d)}{d}, \dots, \left[\frac{n}{x}\right]x\sum_{d|\left[\frac{n}{x}\right]x}\frac{\mu(d)}{d}$$

Thus, $\quad a_x = x\frac{1}{x} + 2x.\frac{1}{x} + \dots + \left[\frac{n}{x}\right].x.\frac{1}{x}$

$$= 1 + 2 + \dots + \left[\frac{n}{x}\right] = \frac{1}{2}\left[\left[\frac{n}{x}\right]^2 + \left[\frac{n}{x}\right]\right]$$

Hence, $\bar{\phi}(n) = \frac{1}{2}\sum_{x=1}^{n}\mu(x)\left[\left[\frac{n}{x}\right]^2 + \left[\frac{n}{x}\right]\right]$

SOLVED EXAMPLES

Based on the following results

- **Euler's function** $\quad \phi(n) = \begin{cases} 1 & \text{if } n = 1 \\ \text{number of positive integers less than } n \text{ and} \\ \text{relatively prime to } n, \text{if } n > 1 \end{cases}$

- $T(n) = \sum_{r=1}^{n}t(r)$

- $S(n) = \sum_{r=1}^{n}s(r)$

- A number a is said to be square free if 1 is the largest square dividing a.

Example 1. *Show that* $\phi(n) = \phi(n + 1) = \phi(n + 2)$ *for* $n = 5186$.

Solution. We can clearly find that

$$\phi(n) = \phi(5186) = 2592$$

Further
$$\phi(n + 1) = \phi(5187)$$
$$= \phi(3.7.13.19) = \phi(3)\phi(7)\phi(13)\phi(19)$$
$$= (3 - 1)(7 - 1)(13 - 1)(19 - 1) = 2592$$

Again $\phi(n + 2) = \phi(5188)$
$$= \phi(2^2 . 1297)$$
$$= \phi(2^2)\phi(1297)= (2^2 - 2)(1297 - 1) = 2 \times 1296 = 2592$$

Hence, the result.

Example 2. *If n and n + 2 both are primes, then show that*
$$\phi(n + 2) = \phi(n) + 2.$$

Solution. Consider LHS $= \phi(n + 2) = (p + 2 - 1) = (p + 1)$

Now RHS $= \phi(n) + 2 = p - 1 + 2 = p + 1$

Hence $\phi(n + 2) = \phi(n) + 2$

Example 3. *Show that* $\phi(n) = \phi(n + 2)$ *is satisfied by* $n = 2(2p - 1)$ *whenever p and* $2p - 1$ *are both odd primes.*

Solution: Consider
$$\text{LHS} = \phi(n)$$
$$= \phi(2(2p - 1)) = \phi(2) \phi (2p - 1)$$
$$= (2 - 1)(2p - 1 - 1) = 2(p - 1)$$

Further, $\phi(n + 2) = \phi(2(2p - 1) + 2)= \phi(4p) = \phi(2^2p)$
$$= \phi(2^2)\phi(p)$$
$$= (2^2 - 2)(p - 1)$$
$$= 2(p - 1)$$

Hence, $\phi(n) = \phi(n + 2)$

Example 4. *If n is an odd integer, then* $\phi(2n) = \phi(n)$

Solution. As per given, n is an odd integer, therefore, it is relatively prime to 2 i.e. $(2, n) = 1$.

Thus, $\phi(2n) = \phi(2)\phi(n)$
$$= (2 - 1)\phi(n)$$
$$= \phi(2n)$$

Example 5. *If every prime that divides n, also divides m, then*
$$\phi(mn) = n\phi(m)$$

Solution. We know that every prime divisor of n is also a prime divisor of m, therefore, we can write
$$n = p_1^{k_1} p_2^{k_2} \dots p_r^{k_r}$$
and $m = p_1^{l_1} p_2^{l_2} \dots p_r^{l_r} p_{r+1}^{l_{r+1}} \dots p_t^{l_t}$
where all p_i are prime k_i. l_i are integers greater than or equal to 1 and $t \geq r$.

Now, $\phi(mn) = \phi(p_1^{k_1+l_1} \cdot p_2^{k_2+l_2} \dots p_r^{k_r+l_r} \ p_{r+1}^{l_{r+1}} \dots p_t^{l_t})$

$\qquad\qquad = \phi(p_1^{k_1+l_1}).\phi(p_2^{k_2+l_2}) \dots \phi(p_r^{k_r+l_r})\phi(p_{r+1}^{l_{r+1}}) \dots \phi(p_t^{l_t})$

$\qquad\qquad = (p_1^{k_1+l_1} - p_1^{k_1+l_1-1})(p_2^{k_2+l_2} - p_2^{k_2+l_2-1}) \dots (p_r^{k_r+l_r} - p_r^{k_r+l_r-1}).\phi(p_{r+1}^{l_{r+1}}) \dots \phi(p_t^{l_t})$

$\qquad\qquad = p_1^{k_1} \cdot p_2^{k_2} \dots p_r^{k_r} (p_1^{l_1} - p_1^{l_1-1})(p_2^{l_2} - p_2^{l_2-1}) \dots (p_r^{l_r} - p_r^{l_r-1}).\phi(p_{r+1}^{l_{r+1}}) - \phi(p_t^{l_t})$

$\qquad\qquad = n\phi(p_1^{l_1})\phi(p_2^{l_2}) \dots \phi(p_t^{l_t})$

$\qquad\qquad = n\phi(m).$

Remark

❖ In particular, if we take $m = n$, then

$\qquad \phi(n^2) = n\phi(n)$

Example 6. *If n is a composite number, then $\phi(n) \leq n - \sqrt{n}$.*

Solution. Let $n = p_1^{k_1} p_2^{k_2} \dots p_r^{k_r}$ where p_1 is the smallest prime.

Then, $\quad \phi(n) = n\left(1 - \dfrac{1}{p_1}\right)\left(1 - \dfrac{1}{p_2}\right) \dots \left(1 - \dfrac{1}{p_r}\right)$

$\qquad\qquad \leq n\left(1 - \dfrac{1}{p_1}\right)$

$\qquad\qquad \leq n\left(1 - \dfrac{1}{\sqrt{n}}\right) = n - \sqrt{n}$

Hence, $\phi(n) \leq n - \sqrt{n}$

Example 7. *If p and $2p + 1$ are both primes and $n = 4p$, then show that*
$$\phi(n + 2) = \phi(n) + 2$$
Solution: Consider

$\qquad\qquad$ LHS $= \phi(n + 2)$

$\qquad\qquad\qquad = \phi(4p + 2)$

$\qquad\qquad\qquad = \phi(2(2p + 1)) = \phi(2)\phi(2p + 1)$

$\qquad\qquad\qquad = (2 - 1)(2p + 1 - 1) = 2p$

Further, $\qquad$ RHS $= \phi(n) + 2$

$\qquad\qquad\qquad = \phi(4p) + 2$

$\qquad\qquad\qquad = \phi(2^2p) + 2$

$\qquad\qquad\qquad = \phi(2^2)\phi(p) + 2$

$\qquad\qquad\qquad = (2^2 - 2)(p - 1) + 2$

$\qquad\qquad\qquad = 2p - 2 + 2 = 2p$

Hence, $\phi(n + 2) = \phi(n) + 2$

Example 8. *Show that if n is the product of twin primes, i.e.* $n = p\,(p+2)$, *where p and p + 2 are primes, then*

$$\phi(n)\sigma(n) = (n+1)(n-3)$$

Solution. Consider

$$
\begin{aligned}
\phi(n) &= \phi(p(p+2)) \\
&= \phi(p)\phi(p+2) \\
&= (p-1)(p+2-1) \\
&= (p-1)(p+1) = p^2 - 1
\end{aligned}
$$

Further

$$
\begin{aligned}
\sigma(n) &= \sigma[p(p+2)] \\
&= \sigma(p)\sigma(p+2) \\
&= (p+1)(p+2+1) \\
&= (p+1)(p+3) = p^2 + 4p + 3
\end{aligned}
$$

Consider LHS.

$$
\begin{aligned}
\phi(n)\sigma(n) &= (p^2-1)(p^2+4p+3) \\
&= p^4 + 4p^3 + 2p^2 - 4p - 3 \\
&= p^4 + 2p^3 + 2p^3 + 4p^2 - 2p^2 - 4p - 3 \\
&= (p^2+2p)(p^2+2p) - 2(p^2+2p) - 3 \\
&= n^2 - 2n - 3 \\
&= (n+1)(n-3)
\end{aligned}
$$

Example 9. *Show that*

$$T(n) = \left[\frac{n}{1}\right] + \left[\frac{n}{2}\right] + \dots + \left[\frac{n}{x}\right] = \sum_{x=1}^{n}\left[\frac{n}{x}\right]$$

Solution. Since, we know that function $T(n)$ is equal to the lattice points in the region given by $x > 0$, $y > 0$, $xy \le n$. This will be equal to the lattice points on the vertices at $x = 1$, $x = 2$, ..., $x = n$ on $xy = n$. The lengths of these vertices are $\dfrac{n}{1}, \dfrac{n}{2}, \dots, \dfrac{n}{n}$ respectively.

Thus, the number of lattice points are respectively given by $\left[\dfrac{n}{1}\right], \left[\dfrac{n}{2}\right], \dots, \left[\dfrac{n}{n}\right]$

Therefore,

$$T(n) = \left[\frac{n}{1}\right] + \left[\frac{n}{2}\right] + \dots + \left[\frac{n}{n}\right]$$

Example 10. *If n is a positive integer, then show that*

$$\sum_{m=1}^{n}\frac{1}{m} = \log n + \gamma + \frac{\theta}{n},\ \text{where } \gamma \text{ is Euler's constant and } 0 < \theta < 1.$$

Solution. Let $f(x) = \dfrac{1}{x}$, clearly for $x > 0$, $f(x)$ is a decreasing function. By property of definite integral, we have

$$\frac{1}{m+1} < \int\limits_m^{m+1} \frac{dx}{x} < \frac{1}{m}, \quad m > 0$$

$$\Rightarrow \qquad 0 < \frac{1}{m} - \int\limits_m^{m+1} \frac{dx}{x} < \frac{1}{m} - \frac{1}{m+1} \qquad \qquad ...(1)$$

Define
$$F(m) = \frac{1}{m} - \int\limits_m^{m+1} \frac{dx}{x} \qquad \qquad ...(2)$$

Then, from (1)

$$0 < F(m) < \frac{1}{m} - \frac{1}{m+1} \qquad \qquad ...(3)$$

Consider a fixed integer a such that $r > a$.

Now, putting $m = a, a + 1, a + 2, ..., r$ in (3) successively, we get

$$0 < F(a) < \frac{1}{a} - \frac{1}{a+1}$$

$$0 < F(a+1) < \frac{1}{a+1} - \frac{1}{a+2}$$

$$\vdots \qquad \vdots \qquad \qquad \vdots \qquad \vdots$$

$$0 < F(r) < \frac{1}{r} - \frac{1}{r+1}$$

On adding, we get

$$0 < \sum_{m=a}^{r} F(m) < \frac{1}{a} - \frac{1}{r+1} \qquad \qquad ...(4)$$

Obviously, the sum function $\sum\limits_{m=1}^{r} F(m)$ is an increasing function of r and bounded above by a. Also, $F(m)$ tends to a finite limit as $r \to \infty$. Therefore,

$$\sum_{m=1}^{\infty} F(m) = \text{a finite quantity}$$

$$\le \frac{1}{a} \qquad \qquad ...(5)$$

Let
$$\gamma = \sum_{m=1}^{\infty} F(m), \text{ be a constant.}$$

Then,
$$\phi = \sum_{m=1}^{n} F(m) + \sum_{m=n+1}^{\infty} F(m)$$

$$= \sum_{m=1}^{n} \frac{1}{m} - \int_1^{n+1} \frac{dx}{x} + \sum_{m=n+1}^{\infty} F(m)$$

$$= \sum_{m=1}^{n} \frac{1}{m} - \int_1^{n} \frac{dx}{x} - \int_n^{n+1} \frac{dx}{x} + \sum_{m=n+1}^{\infty} F(m)$$

$$= \sum_{m=1}^{n} \frac{1}{m} - \log n - \left[\int_{n}^{n+1} \frac{dx}{x} - \sum_{m=n+1}^{\infty} F(m) \right] \qquad ...(6)$$

But $\dfrac{1}{n+1} < \int_{n}^{n+1} \dfrac{dx}{x} \le \dfrac{1}{n}$

Using (5), we get

$$0 < \sum_{m=n+1}^{\infty} (m) \le \frac{1}{n+1}$$

Thus, $\quad 0 < \int_{n}^{n+1} \dfrac{dx}{x} - \sum_{m=n+1}^{\infty} F(m) < \dfrac{1}{n}$

$$\Rightarrow \qquad \int_{n}^{n+1} \frac{dx}{x} - \sum_{m=n+1}^{\infty} F(m) = \frac{\theta}{n} \qquad ...(7)$$

where θ is a function of n such that $0 < \theta < 1$.
From (6) and (7), we conclude that

$$\gamma = \sum_{m=1}^{n} \frac{1}{m} - \log n - \frac{\theta}{n}$$

i.e. $\sum_{m=1}^{n} \dfrac{1}{m} = \gamma + \log n + \dfrac{\theta}{n}$

Example 11. *If α is any real number greater than or equal to 1, then*

$$\gamma = \sum_{m=1}^{[\alpha]} \frac{1}{m} - \log \alpha - \frac{\theta_1}{\alpha}$$

Solution. We know that

$$\gamma = \sum_{m=1}^{\infty} F(m)$$

$$= \sum_{m=1}^{[\alpha]} F(m) + \sum_{m=[\alpha]+1}^{\infty} F(m)$$

$$= \sum_{m=1}^{[\alpha]} \frac{1}{m} - \int_{1}^{\alpha} \frac{dx}{x} - \int_{1}^{[\alpha]+1} \frac{dx}{x} + \sum_{m=[\alpha]+1}^{\infty} F(m)$$

$$= \sum_{m=1}^{[\alpha]} \frac{1}{m} - \log \alpha - \left[\int_{1}^{[\alpha]+1} \frac{dx}{x} + \sum_{m=[\alpha]+1}^{\infty} F(m) \right] \qquad ...(1)$$

From example (10), we have

$$\theta < \sum_{m=[\alpha]+1}^{\infty} F(m) < \frac{1}{[\alpha]+1} < \frac{1}{\alpha}.$$

Again, by the property of definite integral, we have

$$0 \le \left| \int_{\alpha}^{[\alpha]+1} \frac{dx}{x} - \sum_{m=[\alpha]+1}^{\infty} F(m) \right| < \frac{1}{\alpha}$$

$$\int_\alpha^{[\alpha]+1} \frac{dx}{x} - \sum_{m=[\alpha]+1}^\infty F(m) = \frac{\theta_1}{\alpha}, \quad -1 < \theta_1 < 1 \qquad \qquad ...(2)$$

Hence, from (1) and (2), we conclude that

$$\gamma = \sum_{m=1}^{[\alpha]} \frac{1}{m} - \log \alpha - \frac{\theta_1}{\alpha}$$

Example 12. Show that $\bar{\phi}(n) = \dfrac{3n^2}{\pi^2} + 0\,(n \log n)$

Solution. We know that

$$\bar{\phi}(n) = \frac{1}{2} \sum_{x=1}^n \mu(x) \left[\left[\frac{n}{x} \right]^2 + \left[\frac{n}{x} \right] \right]$$

$$= \frac{1}{2} \sum_{x=1}^n \mu(x) \left[\left(\frac{n}{x} + 0(1) \right)^2 + \left(\frac{n}{x} + 0(1) \right) \right]$$

$$= \frac{1}{2} \sum_{x=1}^n \mu(x) \left[\frac{n^2}{x^2} + \frac{n}{x}(2.0\,(1) + 1) + 0(1)^2 + 0(1) \right]$$

$$= \frac{1}{2} \sum_{x=1}^n \mu(x) \left[\frac{n^2}{x^2} + \frac{n}{x}.0(1) + 0(1)^2 \right]$$

$$= \frac{1}{2} \sum_{x=1}^n \mu(x) \left[\frac{n^2}{x^2} + \frac{0(n)}{x} \right]$$

$$= \frac{n^2}{2} \sum_{x=1}^n \frac{\mu(x)}{x^2} + 0(n) \sum_{x=1}^n \frac{\mu(x)}{x} \qquad \qquad ...(1)$$

Further, $\dfrac{n^2}{2} \displaystyle\sum_{x=1}^n \dfrac{\mu(x)}{x^2} = \dfrac{n^2}{2} \left[\dfrac{6}{\pi^2} + 0\left(\dfrac{1}{n} \right) \right]$

$$= \frac{3n^2}{\pi^2} + 0(n) \qquad \qquad ...(2)$$

Further, $\left| \displaystyle\sum_{x=1}^n \dfrac{\mu(x)}{x} \right| < \displaystyle\sum_{x=1}^n \dfrac{1}{x} = \log n + 0(n)$

Therefore,

$$\sum_{x=1}^n \frac{\mu(x)}{x} = 0[\log n + 0(1)]$$

$$= 0\,(\log n)$$

$$0(n) \sum_{x=1}^n \frac{\mu(x)}{x} = 0\,(n)0(\log n) = 0\,(n \log n) \qquad \qquad ...(3)$$

Hence, from (1), (2) and (3), we conclude that

$$\bar{\phi}(n) = \frac{3n^2}{\pi^2} + 0(n) + 0\,(n \log n)$$

$$= \frac{3n^2}{\pi^2} + 0 \, (n \log n)$$

Example 13. *Find all positive integers m and n which satisfy the expression given by*

$$\phi(mn) = \phi(m) + \phi(n)$$

Solution. We know that

$$\phi(mn) = \phi(m) \, \phi(n). \frac{d}{\phi(d)}, \quad \text{where } d = (m, n)$$

Then, given expression can be written as

$$\frac{1}{a} + \frac{1}{b} = b, \, a = \frac{\phi(m)}{\phi(d)}, \, b = \frac{\phi(n)}{\phi(d)}$$

Since a, b and d are positive integers, either $d = 2$, $a = b = 1$ or $d = 1$, $a = b = 2$. For the former, $\phi(m) = \phi(n) = 1$ then $m = n = 2$. For the latter $\phi(m) = \phi(n) = 2$, then one of m, n is 3 and the other is 4. These are the required integers.

Example 14. *Show that* $\phi(m) \geq \dfrac{m}{T(m)}$.

Solution. Let $m = p_1^{m_1} p_2^{m_2} \dots p_k^{m_k}$. Then

$$\phi(m)T(m) = \left(1 - \frac{1}{p_1}\right) \dots \left(1 - \frac{1}{p_k}\right)(m_1 + 1) \dots (m_k + 1)$$

$$\geq m \left(\frac{1}{2}\right)^k .2^k = m$$

Example 15. *Show that* $\displaystyle\sum_{d \mid m} \phi(d).\mu(d) = 0$ *if and only if m is even.*

Solution. Suppose that

$$F(m) = \sum_{d \mid m} \phi(d).\mu(d)$$

Since, if $(a, b) = 1$, then

$$\phi(ab) = \phi(a)\phi(b)$$

and $\mu(ab) = \mu(a)\mu(b)$

Then, we can easily prove that

$$F(ab) = F(a) \, F(b)$$

Further, since $F(p^i) = 1 - (p - 1) = 2 - p$, where p is prime, in general, we have

$$F(m) = \prod_{p \mid m} (2 - p)$$

$\therefore$ If m is even, then $F(m) = 0$.

Conversely, if $F(m) = 0$, then clearly $m = 0$.

Hence, $F(m) = 0$ if and only if $m = 0$.

EXERCISE 7.2

1. Show that an integer a is relatively prime to n if and only if its remainder with respect to n is relatively prime to n.

2. Find the last two digits in the decimal representation of 3^{100}.

3. Show that $\phi(n) = n \sum \dfrac{\mu(d)}{d}$.

4. If $(r, n) = (r - 1, n) = 1$, show that $1 + r + r^2 + ... + r^{\phi(n)-1} \equiv 0 \pmod{n}$.

5. Show that there is no integer m such that $\phi(m) = 14$.

6. If m is a positive integer, show that

$$\mu(m) = \begin{cases} \phi(m) & ; \ if \ m \ is \ odd \\ 2\phi(m); & if \ m \ is \ even \end{cases}$$

7. Show that the number of irreducible proper fractions whose denominators are not greater than n, equals

$$\phi(2) + ... + \phi(n)$$

8. Show that the solution x of $\phi(x) = 4n - 2$, $n > 1$ is of the form p^α or $2p^\alpha$, where p is a prime of the form $4k - 1$.

9. Find all the positive integers x, y such that $x^{\phi(y)} = y$.

10. Show that the number of irreducible proper fractions whose denominators are positive integer n equals $\phi(n)$.

ANSWERS (with Hints)

(7) $\dfrac{1}{2}, \dfrac{1}{3}, \dfrac{2}{3}, ... \dfrac{m_1}{n}, ... \dfrac{m_{\phi(n)}}{n}$ are $\phi(2) + \phi(3) + ... + \phi(n)$ in number.

(8) Let $x = p_1^{\alpha_1} ... p_r^{\alpha_r}$. Then $\phi(m) = p_1^{m_1-1} ... p_r^{m_r-1}(p_1 - 1) ... (p_r - 1)$. When $\phi(m)$ is odd, then p_i is not odd and hence $p_1 = 2$, $m_1 = 1$ or 2.

(9) Let $\phi(y) = r$, when $x > 2$, then $r = \phi(x^{p(y)}) > x^{\phi(y \cdot 1)} = x^{r-1}$

$\Rightarrow \quad r > x^{r-1}$, $x > 2$ which is not possible.

Thus, $x \le 2$. Hence, the required solutions are $x = 1$, $y = 1$; $x = 2$, $y = 2$; $x = 2$, $y = 4$.

(10) If $\dfrac{m}{n}$ is reduced proper fraction, then m is a number in the reduced system of n.

CHAPTER REVIEW : A COMPETITIVE APPROACH

Selected terms and Results

TERMS

- **τ-function :** $\tau(n)$ is the number of positive divisors of n including 1 and n.

- **σ-function :** $\sigma(n)$ is the sum of positive divisors of n including 1 and n.

- **Multiplicative function :** If $f(n)$ is an arithmetic function not identically zero such that $f(mn) = f(m).f(n)$ for every pair of positive integers m, n satisfying $(m, n) = 1$. Then, $f(n)$ is said to be multiplicative function.

- **Mobius function :**
$$\mu(n) = \begin{cases} 1 \; ; \text{if } n = 1 \\ 0 \; ; \text{if } p^2 \mid n \text{ for some prime } p \\ (-1)^k \; ; \text{if } n = p_1 . p_2 ..... p_k \text{ is the product of } k \text{ distinct primes.} \end{cases}$$

- **Euler's function :**
$$\phi(n) = \begin{cases} 1, \text{if } n = 1 \\ \text{number of positive integers less than } n \text{ and relatively prime to } n; \text{ for } n > 1. \end{cases}$$

- **T(n) function :** $T(n) = \displaystyle\sum_{r=1}^{n} t(r)$

- **S(n) :** $S(n) = \displaystyle\sum_{r=1}^{n} s(r)$

- **Square free integers :** A number a is said to be square free if 1 is the largest square dividing a.

RESULTS

- A function whose domain is the set of positive integers is called a number theoretic function.

- An arithmetic function f is said to be multiplicative if $f(mn) = f(m)f(n)$ and $(m, n) = 1$.

- σ and τ both are multiplicative functions.

- The Mobius function $\mu(n)$ is multiplicative.

- If $F(n)$ is multiplicative function, and $F(n) = \displaystyle\sum_{d \mid n} f(d)$ then f is also multiplicative.

- $\phi(n) = n - 1$ if and only if n is prime.

- An integer a is relatively prime to n if its remainder with respect to n is relatively prime to n.

Review Questions and Project Work

1. Let f be a number theoretic function such that $f(n) = 0$ for every positive integer n. Show that $f(n)$ is a multiplicative function.

2. Prove that if $(m, n) = p$ then

$$\phi(mn) = \frac{p}{p-1} \phi(m)\phi(n) = \frac{p}{\phi(p)} \phi(m)\phi(n)$$

3. If f is a multiplicative function then show that $f(1) = 1$

4. If n is an odd integer, show that
$$\phi(4n) = 2\phi(n)$$

5. If $\sigma(n)$ is odd, then n is a perfect square or twice a perfect square (n is odd).

6. Show that the product of the positive divisors of a positive integer n is $n^{\tau(n)2}$.

Objective Type Questions

Fill in the blanks:

1. $\displaystyle\sum_{d|18} \phi(d)$ = ...

2. If $n = 2^j$ where $j \geq 1$ then $n =$ ....

3. If $n \geq 3$, then $\phi(n)$ is ....

4. The value of $\sigma(12) =$ ....

5. The value of $\sigma(28) =$ ....

True/False: *Write 'T' for true and 'F' for false statement.*

1. If f is a multiplicative function then

   $F(n) = \displaystyle\sum_{d|n} f(d)$ is also multiplicative.

   (T/F)

2. The τ(tau) and σ(sigma) functions are multiplicative. (T/F)

3. $\tau(36) = 91$. (T/F)

4. $\sigma(36) = 9$ (T/F)

5. The mobius function is multiplicative. (T/F)

Multiple Choice Questions : *Choose the most appropriate one :*

1. The value of $\tau(6120) =$
   (a) 84   (b) 48
   (c) 408   (d) none of these

2. The highest power of 3 contained in 40! is
   (a) 18   (b) 8
   (c) 81   (d) none of these

3. For $n = 4503$ the value of $\tau(n)$ equals
   (a) $\tau(n + 1)$
   (b) $\tau(n + 2)$
   (c) both (a) and (b) are true
   (d) none of these

4. Which of the following is not a multiplicative function
   (a) Mobius function
   (b) τ-function

   (c) σ function
   (d) none of these

5. If $f(n) = n^2 + 2$ and $n = 6$ then $\displaystyle\sum_{d|6} f(d)$
   is equal to
   (a) 38   (b) 48
   (c) 58   (d) none of these

6. The value of mobius function $\mu(49)$ is given by
   (a) 0   (b) 1
   (c) 2   (d) none of these

7. Let $F(n) = \displaystyle\sum_{d|n} f(d)$ then $f(n)$ equals to

   (a) $\displaystyle\sum_{d|n} \mu(d)F(n|d)$

   (b) $\displaystyle\sum_{d|n} \mu(n|d)F(d)$

   (c) both (a) and (b) are true
   (d) none of these

8. If $p_1^{d_1} \cdot p_2^{d_2} \ldots p_k^{d_k}$ then $\displaystyle\sum_{d|n} |\mu(d)|$ is equal to
   (a) 2   (b) 2^k
   (c) 2^{k+1} (d) none of these

9. The value of $\mu(93) + \mu(94) + \mu(95) =$
   (a) 1   (b) 2
   (c) 3   (d) none of these

10. The value of $[-\pi]$ is equal to ([.] denote the greatest integer function)
    (a) 3   (b) -3
    (c) 1   (d) 0

11. The value of $[x] + [-x] = \ldots.$, where x is an integer
    (a) 0   (b) 1
    (c) 2   (d) 3

12. If n is a natural number greater than or equal to 3 then $\displaystyle\sum_{k=1}^{n} \mu(k!) =$
    (a) 1   (b) 2
    (c) 0   (d) none of these

13. For each integer $n \geq 1$ the value of

$$\sum_{d|n} \mu(d) =$$

  (a) 1 if $n = 1$
  (b) 0 of $n > 1$
  (c) both (a) and (b) are true
  (d) none of these

14. The least integer greater than x is
  (a) $[-x]$ (b) $[x]$
  (c) $-[-x]$ (d) none of these

15. The value of $[x] + [-x]$ is equal to, when x is not an integer
  (a) 1   (b) 0
  (c) -1 (d) none of these

ANSWERS

Fill in the blanks

  (1) 18   (2) $2\,\phi(n)$ (3) even   (4) 28   (5) 56

True/False

  (1) T   (2) T   (3) F   (4) F   (5) T

Multiple choice questions

  (1) *b*   (2) *a*   (3) *c*   (4) *d*   (5) *c*
  (6) *a*   (7) *c*   (8) *b*   (9) *c*   (10) *b*
  (11) *a*   (12) *a*   (13) *c*   (14) *c*   (15) *c*

CHAPTER 8

Complex Numbers

❖ Complex Numbers	❖ Modulus and Arguments
❖ De'moivre's Theorem	❖ Conjugate of a Complex Number

8.1 INTRODUCTION

The necessity of studying the system of complex number arise due to the fact that equations of type.

$$x^2 = -1$$

has no solution in the set of real numbers. There is no real numbers, positive or negative, integral or fractional, rational or irrational which when squared gives − 1. The solutions of such equations introduce the concept of imaginary or complex numbers.

Remarks

❖ Leonhard Euler (1707-1783) was the first to use the symbol, i for $\sqrt{(-1)}$ having the property $i^2 = -1$. He called the symbol i imaginary (also known as iota). With the introduction of this symbol he was able to discover the unknown roots of the equation $x^2 + 1 = 0$

❖ This discovery of Euler is an important landmark in the history of mathematical progress, for it enabled the number system to be extended. The numbers developed before the creation of the symbol 'i' came to known as real numbers.

8.2 COMPLEX NUMBERS

Let x and y be two real numbers; then the set of ordered pairs (x, y) are called the system of complex numbers if it satisfies the following definitions of equality, addition, subtraction, multiplication and division.

(i) **Equality.** Two complex numbers (x, y) and (α, β) are equal if and only if $x = \alpha, y = \beta$.

(ii) **Addition.** The sum of two complex numbers is defined by the equation

$$(x, y) + (\alpha, \beta) = (x + \alpha, y + \beta).$$

(iii) **Subtraction** is defined by the equation
$$(x, y) - (\alpha, \beta) = (x - \alpha, y - \beta).$$

(iv) **Multiplication** is defined by the equation
$$(x, y) \times (\alpha, \beta) = (x\alpha - y\beta, x\beta + y\alpha).$$

(v) **Division** $(x, y) / (\alpha, \beta)$ is defined as the complex number (c, d) given by
$$(\alpha, \beta) \times (c, d) = (x, y)$$
provided that such a number exists.

(vi) We also define $(x, 0)$ to be the real number x and $(0, 0)$ to be the real number 0.

It is easy to verify that definitions (i) to (v) given above to reduce the ordinary rules of algebra when the numbers concerned are of the form $(x, 0)$.

8.3 NOTATIONS

It is convenient to denote the complex number (x, y) by the compound symbol $(x + iy)$, where $i^2 = -1$, for then we can obtain the sum, difference, product etc. of complex numbers by the usual rules of algebra (such as per applicable to real numbers). In this notation
$$x + iy + \alpha + i\beta = (x + \alpha) + i(y + \beta)$$
$$x + iy - (\alpha + i\beta) = (x - \alpha) + i(y - \beta)$$
$$(x + iy)(\alpha + i\beta) = (x\alpha - y\beta) + i(x\beta + y\alpha) \qquad \text{(since } i^2 = -1)$$
and these results are in agreement with the definitions given above for sum, difference and product of two complex numbers.

The complex number $x + iy$ is also denoted by z, x is called the real part of z, i.e. $R(z)$ and y the imaginary part of z, i.e. $I(z)$.

In the case of real number, $a \times a \times a$ .p.. to n factors is denoted by a^n; similarly $z \times z \times z \times \ldots$ to n factors is denoted by z^n, where n is any positive integer.

8.4 REDUCTION OF COMPLEX NUMBERS TO THE STANDARD FORM $r(\cos \theta + i \sin \theta)$

Let $z = x + iy$ by the given complex number.

If (x, y) be taken as the cartesian co-ordinates of a point P and (r, θ) the polar co-ordinate of the same point, we have

$$x = r \cos \theta \qquad \qquad \ldots(A)$$
$$\left. \begin{array}{l} y = r \sin \theta \end{array} \right\}$$
$$\therefore \qquad z = x + iy = r(\cos \theta + i \sin \theta), \qquad \ldots(1)$$

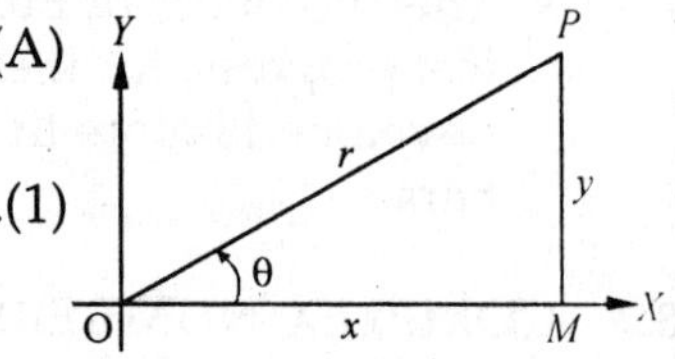

where
$$r = +\sqrt{(x^2 + y^2)} \qquad \tan \theta = \frac{y}{x}$$

or
$$\theta = \tan^{-1} \frac{y}{x} \qquad \qquad \ldots(2)$$

FIG. 1

We notice that all complex number can be put in the form $r(\cos \theta + i \sin \theta)$, Here r or the positive square root of $x^2 + y^2$ is called the **absolute value is modulus** of the complex number $x + iy$, and is designated by $|z|$ or $|x + iy|$. The circular measure of θ which satisfies conditions (A), *viz.*

$$\cos \theta = -\frac{x}{\sqrt{(x^2 + y^2)}}, \sin \theta = \frac{y}{\sqrt{(x^2 + y^2)}}$$

is called the argument or amplitude of $x + iy$. Its value must be taken between $-\pi$ and $+\pi$. The factor $\cos \theta + i \sin \theta$ sometimes abbreviated as cis θ is called the direction factor of the complex number. It modulus is unity. The form $r(\cos \theta + i \sin \theta)$ is called modulus-amplitudes form or polar form or trigonometrical form.

8.5 GEOMETRICAL REPRESENTATION OF COMPLEX NUMBERS

The complex number $z = x + iy$ is represented by a point P whose Cartesian co-ordinates are (x, y) referred to rectangular axes OX and OY, usually called real and imaginary axes respectively. If r is the modulus and θ the argument of the complex number z, then clearly the polar co-ordinates of the point P are (r, θ). The plane whose points are represented by complex numbers is called Argand's plane or Argand's diagram after Argand. (This is also called Complex plane or Gaussian plane).

The complex number z is called the **affix** of the point (x, y) which represent it.

8.5.1 Vector Representation of Complex Number

Let P be a point (x, y) on the Argand's plane corresponding to the complex number $z = x + iy$ referred to OX and OY as co-ordinate axes. The modulus and argument of z are represented by the length (or modulus) and direction of the vector $\overrightarrow{OP}$ respectively and *vice versa*.

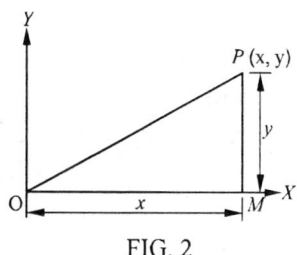

FIG. 2

8.5.2 The Sum and Difference of Two Complex Numbers Represented by a Point on The Argand's Plane

(A) Sum. Let the two complex numbers $z_1 = x_1 + iy_1$ and $z_2 = x_2 + iy_2$ be represented by the points $P(x_1, y_1)$ and $Q(x_2, y_2)$ respectively on the Argand's plane.

Complete the parallelogram OPRQ. The middle points of the diagonals PQ and OR are same. But the mid-point of PQ is

$$\left(\frac{1}{2}(x_1 + x_2), \frac{1}{2}(y_1 + y_2) \right)$$

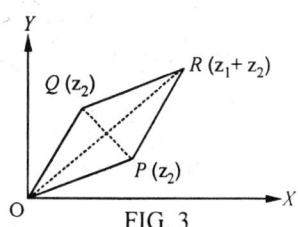

FIG. 3

Since the point O is $(0, 0)$ and therefore the co-ordinates of R are $(x_1 + x_2, y_1 + y_2)$. Hence, the sum of the two complex number z_1 and z_2 is represented by the point R. In vectorial notation $z_1 + z_2 = \overrightarrow{OP} + \overrightarrow{OQ} = \overrightarrow{OP} + \overrightarrow{PR} = \overrightarrow{OR}$

(B) Difference. We represent $-z_2$ by Q so that O is the middle point of QQ'. Complete the parallelogram OPRQ. The middle points of PQ and OR are the same. But the middle point of PQ is

$$\left(\frac{1}{2}(x_1 - x_2), \frac{1}{2}(y_1 - y_2) \right)$$

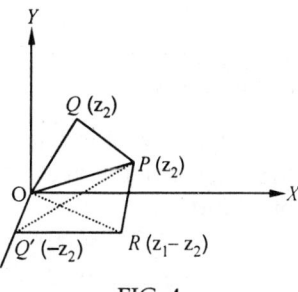

FIG. 4

Therefore, the co-ordinates of R are $(x_1 - x_2, y_1 - y_2)$. Hence the difference $z_1 - z_2$ of the two complex numbers z_1 and z_2 is represented by the point R.

Since OQ is equal and parallel to RP, so ORPQ is parallelogram and hence $\vec{OR} = \vec{QP}$. Thus, in vectorial notation, we have

$$z_1 - z_2 = z_1 + (-z_2) = \vec{OP} + \vec{OQ}' = \vec{OP} + \vec{QO} = \vec{OP} + \vec{PR} = \vec{OR} = \vec{OP}$$

Therefore, it follows that the complex number $z_1 - z_2$ is represented by the vector $\vec{QP}$.

8.6 PRODUCT AND QUOTIENT OF TWO COMPLEX NUMBERS

(A) *If z_1 and z_2 be two complex numbers then*
(i) $|z_1 z_2| = |z_1| . |z_2|$ *and* (ii) $arg(z_1 z_2) = arg\ z_1 + arg\ z_2$

| Let | $z_1 = x_1 + iy_1$ | and | $z_2 = x_2 + iy_2$. |

Then $z_1 z_2 = (x_1 + iy_1)(x_2 + iy_2) = (x_1 x_2 - y_1 y_2) + i(x_1 y_2 + x_2 y_1)$

Further, let $x_1 = r_1 \cos \theta_1$, $y_1 = r_1 \sin \theta_1$

and $x_2 = r_2 \cos \theta_2$, $y_2 = r_2 \sin \theta_2$

where $|z_1| = r_1$, $|z_2| = r_2$

and $arg\ z_1 = \theta_1$, $arg\ z_2 = \theta_2$.

∴ $z_1 z_2 = (r_1 \cos \theta_1 + ir_1 \sin \theta_1).(r_2 \cos \theta_2 + ir_2 \sin \theta_2)$

$= r_1 r_2 [(\cos \theta_1 \cos \theta_2 - \sin \theta_1 \sin \theta_2) + i(\sin \theta_1 \cos \theta_2 + \cos \theta_1 \sin \theta_2)]$

$= r_1 r_2 [\cos(\theta_1 + \theta_2) + i \sin(\theta_1 + \theta_2)]$.   ...(1)

Thus, from (1), we clearly have

$$|z_1 z_2| = r_1 r_2 = |z_1| . |z_2|$$

and $arg(z_1 z_2) = \theta_1 + \theta_2 = arg\ z_1 + arg\ z_2$.

In a similar manner as above, we have in general

$$|z_1 z_2 \ldots z_n| = |z_1| |z_2| \ldots |z_n|$$

and $arg|z_1 z_2 \ldots z_n| = arg\ z_1 + arg\ z_2 + \ldots + arg\ z_n$.

Thus, we can state that the modulus of the product of complex numbers is equal to the product of the moduli of these complex numbers and the arguments of the product of the complex numbers is equal to the sum of their arguments.

8.6.1 Geometrical Representation

Let the points P and Q represent the complex numbers z_1

and z_2 so that the vectors $\vec{OP}$ and $\vec{OQ}$ represent z_1 and z_2 respectively.

FIG. 5

Draw a vector $\vec{OR}$ whose length is equal to the product of the lengths of the vectors z_1 and z_2 and the arg of $z_1 z_2$ is equal to the sum of the arguments of z_1 and z_2.

i.e. $OR = r_1 r_2 = |z_1| |z_2|$

$arg\ z_1 z_2 = \angle ROX = \angle ROQ + \angle QOX = \angle POX + \angle QOX$ $[\because \angle ROX = \angle POX]$

$= \theta_1 + \theta_2 = arg\ z_1 + arg\ z_2$.

(ii) *If z_1 and z_2 be two complex numbers then*

$$\left|\frac{z_1}{z_2}\right| = \frac{|z_1|}{|z_2|} \text{ and } \arg\left(\frac{z_1}{z_2}\right) = \arg z_1 - \arg z_2.$$

Considering two complex numbers z_1 and z_2 as above we have

$$\frac{z_1}{z_2} = \frac{r_1(\cos\theta_1 + i\sin\theta_1)}{r_2(\cos\theta_2 + i\sin\theta_2)} = \frac{r_1(\cos\theta_1 + i\sin\theta_1)(\cos\theta_2 - i\sin\theta_2)}{r_2(\cos\theta_2 + i\sin\theta_2)(\cos\theta_2 - i\sin\theta_2)}$$

$$= \frac{r_1[\cos\theta_1\cos\theta_2 + \sin\theta_1\sin\theta_2) + i(\sin\theta_1\cos\theta_2 - \cos\theta_1\sin\theta_2)]}{r_2(\cos^2\theta_2 + \sin^2\theta_2)}$$

$$= \frac{r_1}{r_2}[\cos(\theta_1 - \theta_2) + i\sin(\theta_1 - \theta_2)] \qquad \ldots(1)$$

Thus from (1), we clearly have

$$\left|\frac{z_1}{z_2}\right| = \frac{r_1}{r_2} = \frac{|z_1|}{|z_2|} \qquad \ldots(2)$$

and $\qquad \arg\left(\dfrac{z_1}{z_2}\right) = \theta_1 - \theta_2 = \arg z_1 - \arg z_2.$ $\qquad \ldots(3)$.

8.6.2. Geometrical Representation

With the help of (2) and (3), we can express z_1/z_2 geometrically.

Let the vectors $\overrightarrow{OP}$ and $\overrightarrow{OQ}$ represents the complex numbers z_1 and z_2 respectively on the argand plane, so that

$$|z_1| = OP = r_1, \; |z_2| = OQ = r_2$$
$$\arg z_1 = \angle POX = \theta_1,$$
$$\arg z_2 = \angle OQX = \theta_2,$$

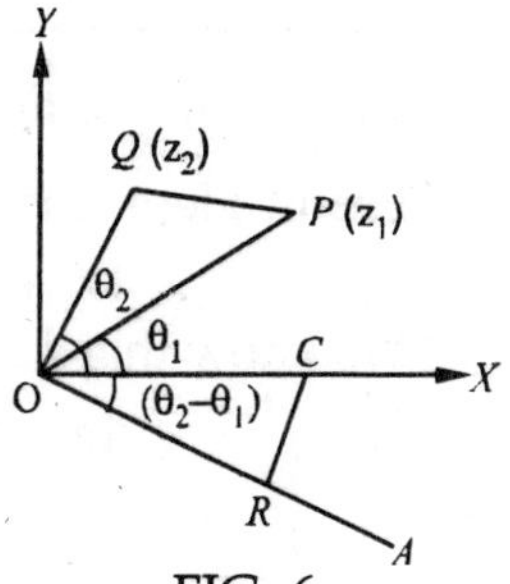

Rotate the line OP in the clockwise direction in such a way that $\angle POA = \arg z_2 = \theta_2$ and let OA be its new position. Cut off $OC = 1$ unit length from OX. Now draw a line CR through C to meet OA in R, so that

$$\angle OCR = \angle OQP$$

FIG. 6

$\therefore$ The point R represents the complex number z_1/z_2. It is shown as follows:

The triangles OCR and OQP are similar, so that we have

$$\frac{OR}{OC} = \frac{OP}{OQ} \quad \text{or } OR = \frac{OP}{OQ} \qquad\qquad [\because OC = 1]$$

or $\qquad\qquad OR = \dfrac{|z_1|}{|z_2|} = \left|\dfrac{z_1}{z_2}\right|$

Also $\qquad\qquad \angle COR = \angle POR - \angle POX = \theta_1 - \theta_2.$

$\therefore$ Vectorial angle of $R = \angle COR = \theta_1 - \theta_2.$

Thus the polar co-ordinates of R are $\left(\left|\dfrac{z_1}{z_2}\right|, \theta_1 - \theta_2\right)$ and hence the complex number $\dfrac{z_1}{z_2}$ is represent by the point R.

8.7 CONJUGATE OF A COMPLEX NUMBER

If $z = x + iy$ is any complex number, then the complex number $x - iy$ is called the conjugate to z and is written as $\bar{z}$. Clearly conjugate to $z_1 + z_2$ is $\bar{z}_1 + \bar{z}_2$ and conjugate of z_1z_2 is $\bar{z}_1\bar{z}_2$. we have

$$|z|^2 = z\bar{z} = |\bar{z}|^2, \qquad \therefore \quad |z| = |\bar{z}|$$

$$2R(z) = z + \bar{z}$$

$$2iI(z) = z - \bar{z}.$$

Also $(\bar{\bar{z}}) = z.$

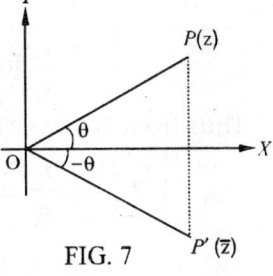

FIG. 7

Geometrically, the reflection (or image) of z in the real axis is the conjugate of z. Thus if (r, θ) are the polar co-ordinates of P, then the polar co-ordinates of its reflection P' are $(r, -\theta)$.

Clearly $|z| = |\bar{z}| = r = \sqrt{(x^2 + y^2)},$

$$\text{amp}(z) = \theta, \text{ amp}(\bar{z}) = -\theta$$

$$\text{amp}(z) = -\text{amp}(\bar{z})$$

8.8 PROPERTIES OF MODULI

THEOREM-1

The modulus of the sum of two complex numbers can never exceed the sum of their moduli.

Proof. Here it is required to prove that

$$|z_1 + z_2| \le |z_1| + |z_2|$$

We know that $|z^2| = z\bar{z}$.

$\therefore$ $|z_1 + z_2|^2 = (z_1 + z_2)(\bar{z}_1 + \bar{z}_2)$

$$= (z_1 + z_2)(\bar{z}_1 + \bar{z}_2) = z_1\bar{z}_1 + z_1\bar{z}_2 + z_2\bar{z}_1 + z_2\bar{z}_2 \qquad \text{...(1)}$$

But $z_1\bar{z}_1 = |z_1|^2, \; z_2\bar{z}_2 = |z_2|^2$

$$z_1\bar{z}_2 + z_2\bar{z}_1 = (x_1 + iy_1)(x_2 - iy_2) + (x_2 + iy_2)(x_1 - iy_1)$$

$$= 2(x_1x_2 + y_1y_2) = 2R(z_1\bar{z}_2).$$

Substituting the values in (1), we have

$$|z_1 + z_2|^2 = |z_1|^2 + |z_2|^2 + 2R(z_1\bar{z}_2) \le |z_1|^2 + |z_2|^2 + 2|z_1||z_2|$$

$$[\because |z_1\bar{z}_2| \ge R(z_1\bar{z}_2)]$$

(since real part of a complex number can never exceed its modulus)

or $|z_1 + z_2|^2 \le |z_1|^2 + |z_2|^2 + 2|z_1||z_2|^2$

$$[\because |z_1\bar{z}_2| = |z_1| \cdot |\bar{z}_2| = |z_1| \cdot |z_2|)]$$

or $|z_1 + z_2|^2 \le (|z_1|^2 + |z_2|)^2$

or $|z_1 + z_2| \le |z_1| + |z_2|$

GEOMETRICAL INTERPRETATION

Let the complex numbers z_1 and z_2 be represented by the points P and Q respectively and R is the point of affix $(z_1 + z_2)$.

$\therefore$ $\quad |z_1| = OP, \ |z_2| = OQ = PR \quad$ and $|z_1 + z_2| = OR$.

Now we know that in any triangle the sum of two sides is greater than the third side.

$\therefore$ $\quad OP + PR > OR$

or $\quad |z_1| + |z_2| > |z_1 + z_2|$ or $|z_1 + z_2| < |z_1| + |z_2|$

If the points O, P and Q are in a straight line, then equality will hold. Hence, we have

$$|z_1 + z_2| \le |z_1| + |z_2|$$

Remarks

❖ By the use of above theorem, we have
$$|z_1 + z_2 + z_3| \le |z_1 + z_2| + |z_3| \le |z_1| + |z_2| + |z_3|$$

❖ In general, by induction, we have

$$\left| \sum_{m=1}^{n} z_m \right| \le \sum_{m=1}^{n} |z_m|$$

THEOREM-2

The modulus of the difference of two complex numbers can never be less than the difference of their moduli :

i.e. $\quad |z_1 - z_2| \ge |z_1| - |z_2|.$

Proof. Consider $|z_1 - z_2|^2 = (z_1 - z_2)\overline{(z_1 - z_2)}$

$\qquad = (z_1 - z_2)(\overline{z}_1 - \overline{z}_2) = |z_1|^2 + |z_2|^2 - 2R(z_2 \overline{z}_2)$ $\qquad$ [See theorem 1]

$\qquad \ge |z_1|^2 + |z_2|^2 - 2 \cdot |z_2 \overline{z}_2|$

$\qquad = |z_1|^2 + |z_2|^2 - 2 |z_1| |\overline{z}_2| = (|z_1| - |z_2|)^2$

$\therefore \qquad |z_1 - z_2| \ge |z_1| - |z_2|.$

GEOMETRICAL INTERPRETATION :

Let $\qquad OP = |z_1|, OQ = |z_2|, \ OP = |z_1 - z_2|$

We know that in a triangle the difference of any two sides is less than the third side.

So from $\triangle OPQ$,

$\qquad OP - OQ < QP$

or $\qquad |z_1| - |z_2| < |z_1 - z_2|$

or $\qquad |z_1 - z_2| > |z_1| - |z_2|$

If the points O, P, Q are in a straight line, then equality will hold. Hence we have

$$|z_1 - z_2| \ge |z_1| - |z_2|$$

Aliter. Theorem 2, can be derived from theorem 1 as follows

$$|z_1| = |(z_1 - z_2) + z_2| \le |z_1 - z_2| + |z_2|$$

or $$|z_1 - z_2| \le |z_1| - |z_2|$$

THEOREM-3

The modulus of the sum of two complex numbers is always greater than or equal to the difference of their moduli.

i.e. $$|z_1 + z_2| \ge |z_1| - |z_2|.$$

Proof. Using Theorem 1 above, we have

$$|z_1 + z_2|^2 = |z_1|^2 + |z_2|^2 + 2R(z_1\bar{z}_2) \qquad \qquad ...(1)$$

But $$|z_1 z_2| \le R(z_1\bar{z}_2)$$

∴ $$- |z_1 z_2| \le R(z_1\bar{z}_2) \quad \text{or} \quad -2|z_1 z_2| \le R(z_1\bar{z}_2) \qquad \qquad ...(2)$$

Adding $|z_1|^2 + |z_2|^2$ to both sides of (2), we get

$$|z_1|^2 + |z_2|^2 - 2|z_1 z_2| \le |z_1|^2 + |z_2|^2 + 2R(z_2\bar{z}_2)$$

or $$(|z_1|^2 - |z_2|)^2 \le |z_1| + |z_2|^2 \qquad \qquad \text{[using (1)]}$$

or $$|z_1 + z_2| \ge |z_1| - |z_2|.$$

SOLVED EXAMPLES

Based on the following Results

- $z = x + iy \quad x = r \cos \theta, y = r \sin \theta, \quad \theta = \tan^{-1} \dfrac{y}{x} \text{ and } r = + \sqrt{x^2 + y^2}$
- $|z_1 z_2| = |z_1|.|z_2|$
- $\arg (z_1.z_2.z_3 ... z_n) = \arg z_1 + \arg z_2 + ... + \arg z_n$
- $|z_1 + z_2| \le |z_1| + |z_2|$
- $|z_1 - z_2| \ge |z_1| - |z_2|$
- $|z_1 + z_2| \ge |z_1| - |z_2|$

Example 1. *Express* $\dfrac{5+7i}{2-3i}$ *in the form of x + iy.*

Solution. Multiplying the denominator and numerator by 2 + 3i, the conjugate complex of denominator, we have

$$\frac{5+7i}{2-3i} = \frac{5+7i}{2-3i} \times \frac{2+3i}{2+3i} = \frac{10-21+i(14+15)}{4+9} = -\frac{11}{13} + i\frac{29}{13}.$$

Example 2. *Express the following complex numbers in the modulus-amplitude form:*

(i) $1 - i.$ (ii) $- \sqrt{3} + i$ (iii) $\dfrac{(1+i)(2-i)}{3+i}$

Solution. (i) Let $$1 - i = r (\cos \theta + \sin \theta)$$

$\therefore$ $\qquad$ $r \cos \theta = 1$ and $r \sin \theta = -1$.

This gives $r = \sqrt{(1+1)} = \sqrt{2}$ and $\tan \theta = -1 = \tan \dfrac{3\pi}{4} \Rightarrow \theta = \dfrac{3\pi}{4}$

$\therefore$ $\qquad$ $1 - i = \sqrt{2}\left(\cos \dfrac{3\pi}{4} + i \sin \dfrac{3\pi}{4}\right)$

(ii) Let $\qquad$ $-\sqrt{3} + i = r(\cos \theta + \sin \theta)$

$\therefore$ $\qquad$ $r \cos \theta = -\sqrt{3}$ and $r \sin \theta = 1$.

This gives $r = \sqrt{(3+1)} = 2$ and $\tan \theta = \dfrac{-1}{\sqrt{3}} = \tan \dfrac{5\pi}{6}$,

$\Rightarrow$ $\qquad$ $\theta = \dfrac{5\pi}{6}$

$\therefore$ $\qquad$ $-\sqrt{3} + i = 2\left(\cos \dfrac{5\pi}{6} + i \sin \dfrac{5\pi}{6}\right)$.

(iii) We have $\dfrac{(1+i)(2-i)}{3+i} = \dfrac{2 - i + 2i - i^2}{3+i} = \dfrac{3+i}{3+i} = 1 = i(\cos 0 + i \sin 0)$

Example 3. *If $(1 + i)(1 + 2i)(1 + 3i) \dots (1 + ni) = A + iB$. Show that*
$$2 . 5. 10 \dots (1 + n^2) = A^2 + B^2$$

Solution. We have
$$(1 + i)(1 + 2i)(1 + 3i) \dots (1 + ni) = A + iB \qquad \dots(1)$$

Taking conjugate of (1), we have
$$(1 - i)(1 - 2i)(1 - 3i) \dots (1 - ni) = A - iB$$

Multiplying (1) and (2) pair-wise, we have
$$(1 - i^2)(1 - 2^2i^2)(1 - 3^2i^2) \dots (1 - n^2i^2) = A^2 - i^2B^2$$
or $\qquad$ $(1 + 1)(1 + 4)(1 + 9) \dots (1 + n^2) = A^2 + B^2$
or $\qquad$ $2 . 5 . 10 \dots (1 + n^2) = A^2 + B^2$.

Example 4. *Using argand-diagram find the product of $(2 + i)$ and $(2 - i)$.*

Solution. Let $\qquad$ $z_1 = 2 + i = r_1[\cos \theta_1 + \sin \theta_1]$
and $\qquad$ $z_2 = 2 - i = r_2[\cos \theta_2 + \sin \theta_2]$

Then, clearly $\qquad$ $r_1 = \sqrt{5}, \theta_1 = \tan^{-1}\left(\dfrac{1}{2}\right);$

$\qquad\qquad$ $r_2 = \sqrt{5}, \theta_2 = \tan^{-1}\left(-\dfrac{1}{2}\right).$

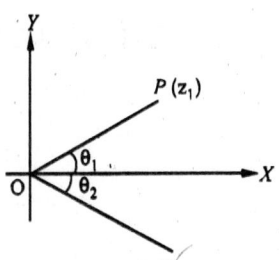

FIG. 8

Now $z_1.z_2 = r_1r_2[\cos(\theta_1 + \theta_2) + i \sin(\theta_1 + \theta_2)]$

$\therefore$ $\qquad$ $|z| = |z_1|.|z_2| = \sqrt{5} \times \sqrt{5} = 5$

and $\quad$ $\arg z = \theta_1 + \theta_2 = 0$.

Example 5. *Prove that* $|z_1 + z_2|^2 + |z_1 - z_2|^2 = 2|z_1|^2 + 2|z_2|^2$ *and interpret the result geometrically.*

Deduce that $|\alpha + \sqrt{(\alpha^2 + \beta^2)}| + |\alpha - \sqrt{(\alpha^2 - \beta^2)}| = |\alpha + \beta| + |\alpha - \beta|$ *all numbers involved, being complex.*

Solution. We have

$$|z_1 + z_2|^2 = (z_1 + z_2)\overline{(z_1 + z_2)}$$

$$= (z_1 + z_2)(\bar{z}_1 + \bar{z}_2) = z_1\bar{z}_1 + z_2\bar{z}_2 + z_1\bar{z}_2 + z_2\bar{z}_1 \qquad \qquad ...(1)$$

and $\qquad |z_1 - z_2|^2 = (z_1 - z_2)(\bar{z}_1 - \bar{z}_2) = z_1\bar{z}_2 + z_2\bar{z}_2 - z_1\bar{z}_2 - z_2\bar{z}_1 \qquad ...(2)$

Adding (1) and (2), we have

$$|z_1 + z_2|^2 + |z_1 - z_2|^2 = 2z_1\bar{z}_1 + 2z_2\bar{z}_2 = 2|z_1|^2 + 2|z_2|^2. \qquad ...(3)$$

GEOMETRICAL INTERPRETATION :

Let P and Q be the points of affix z_1 and z_2 respectively. Complete the parallelogram OPRQ. Then

$$|z_1| = OP, \ |z_2| = OQ, \ |z_1 + z_2| = OR, \ |z_1 - z_2| = QP.$$

Using these values in the relation (3), we have

$$OR^2 + QP^2 = 2OP^2 + 2OQ^2 = OP^2 + OP^2 + OQ^2 + OQ^2$$

$$= OP^2 + QR^2 + OQ^2 + PR^2 \qquad \qquad [\because OP = QR, OQ = PR]$$

This shows that the sum of the squares of their diagonals is equal to the sum of the squares of the sides of a parallelogram.

Deduction : Let $\qquad z_1 = \alpha + \sqrt{(\alpha^2 - \beta^2)}, z_2 = \alpha - \sqrt{(\alpha^2 - \beta^2)}.$

Now from equation (3), we have

$$|z_1|^2 + |z_2|^2 = \frac{1}{2}|z_1 + z_2|^2 + \frac{1}{2}|z_1 - z_2|^2$$

$$= \frac{1}{2}|2\alpha^2| + \frac{1}{2}|2\sqrt{(\alpha^2 - \beta^2)}|^2 = 2|\alpha|^2 + 2|\alpha^2 - \beta^2| \qquad ...(4)$$

Again $\{|z_1| + |z_2|\}^2 = |z_1|^2 + |z_2|^2 + |2|z_1| |z_2|$

$$= 2|\alpha|^2 + |2|\alpha^2 - \beta^2| + 2|\alpha^2 - (\alpha^2 - \beta^2)| \qquad \qquad [\text{using (4)}]$$

$$= 2|\alpha|^2 + 2|\alpha^2 - \beta^2| + 2|\beta^2)|$$

$$= 2|\alpha|^2 + 2|\alpha^2 - \beta^2|^2 + 2|\alpha + \beta| |\alpha - \beta|$$

$$= |\alpha + \beta|^2 + |\alpha - \beta|^2 + 2|\alpha + \beta| |\alpha - \beta| \qquad \qquad [\text{using (3)}]$$

$$= \{|\alpha + \beta| + |\alpha - \beta|\}^2$$

$\therefore \qquad |z_1| + |z_2| = |\alpha + \beta| + |\alpha - \beta|$

or $\qquad |\alpha + \sqrt{(\alpha^2 - \beta^2)}| + |\alpha - \sqrt{(\alpha^2 - \beta^2)}| = |\alpha + \beta| + |\alpha - \beta|$

Example 6. *Prove that the area of the triangle whose vertices are the points represented by the complex numbers z_1, z_2, z_3 on the argand plane.*

Solution. $\quad$ Area $= \dfrac{1}{2}\begin{vmatrix} x_1 & y_1 & 1 \\ x_2 & y_2 & 1 \\ x_3 & y_3 & 1 \end{vmatrix} = \dfrac{1}{2i}\begin{vmatrix} x_1 & iy_1 & 1 \\ x_2 & iy_2 & 1 \\ x_3 & iy_3 & 1 \end{vmatrix}$

Now applying $C_1 \to C_2 + C_1$

$$= \frac{1}{2i}\begin{vmatrix} x_1 & x_1 + iy_1 & 1 \\ x_2 & x_2 + iy_2 & 1 \\ x_3 & x_3 + iy_3 & 1 \end{vmatrix} = \frac{1}{2i}\begin{vmatrix} x_1 & z_1 & 1 \\ x_2 & z_2 & 1 \\ x_3 & z_3 & 1 \end{vmatrix}$$

Expanding by the column, we get

$$\text{Area} = \frac{1}{2i}\Sigma[x_1(z_1 - z_3)] = \frac{1}{2i}\Sigma\left[\frac{1}{2}(z_1 + \bar{z}_1)(z_2 - z_3)\right] \qquad \left[\because z_1 = \frac{1}{2}(z_1 + \bar{z}_1)\right]$$

$$= \frac{1}{4i}\Sigma[(z_1(z_2 - z_3) + \Sigma\{\bar{z}_1(z_2 - z_3)\}]$$

$$= \frac{1}{4i}\left[\Sigma(z_1 z_2) - \Sigma(z_1 z_3) + \Sigma\left\{\frac{z_1\bar{z}_1(z_2 - z_3)}{z_1}\right\}\right]$$

$$= \frac{1}{4i}\left[\Sigma(z_1 z_2) - \Sigma(z_1 z_2) + \Sigma\left\{\frac{|z_1|^2 (z_2 - z_3)}{z_1}\right\}\right] \qquad [\because \Sigma z_1 z_2 = \Sigma z_1 z_3]$$

$$= \Sigma\left[\frac{|z_1|^2 (z_2 - z_3)}{4iz_1}\right]$$

Example 7. *Show that the triangle whose vertices are the points represented by the complex numbers z_1, z_2, z_3, on the argand plane is equilateral if and only if*

$$\frac{1}{z_2 - z_3} + \frac{1}{z_3 - z_1} + \frac{1}{z_1 - z_2} = 0$$

that is iff $\quad z_1^2 + z_2^2 + z_3^2 - z_1 z_2 - z_2 z_3 - z_3 z_1 = 0$

Solution. Let the vertices A, B, C of a $\Delta\,ABC$ be represented by the complex numbers z_1, z_2, z_3, respectively, then

$$BC = |z_2 - z_3|, CA = |z_3 - z_1|, AB = |z_1 - z_2|$$

Now suppose

$$z_2 - z_3 = \alpha, \ z_3 - z_1 = \beta, \ z_1 - z_1 = \gamma$$

$\therefore \qquad\qquad \alpha + \beta + \gamma = 0 \qquad\qquad\qquad\qquad\qquad ...(1)$

$\qquad\qquad\qquad \overline{\alpha + \beta + \gamma} = 0 \ \text{ or } \ \bar{\alpha} + \bar{\beta} + \bar{\gamma} = 0 \qquad ...(2)$

The condition is necessary i.e. if part. Let the $\Delta\,ABC$ be equilateral.

Then $BC = CA = AB$ or $|z_2 - z_3| = |z_3 - z_1| = |z_1 - z_2|$

or $|\alpha| = |\beta| = |\gamma|$ or $|\alpha|^2 = |\alpha|^2 = |\beta|^2 = |\gamma|^2$

or $\alpha\bar{\alpha} = \beta\bar{\beta} = \gamma\bar{\gamma} = k$, (say).   ...(3)

Using (3) in (2), we have

$$\frac{k}{\alpha} + \frac{k}{\beta} + \frac{k}{\gamma} = 0 \quad \text{or} \quad \frac{1}{\alpha} + \frac{1}{\beta} + \frac{1}{\gamma} = 0$$

or
$$\frac{1}{z_2 - z_3} + \frac{1}{z_3 - z_1} + \frac{1}{z_1 - z_2} = 0 \tag{...(4)}$$

or $z_1^2 + z_2^2 + z_3^2 - z_1z_2 - z_2z_3 - z_3z_1 = 0$ ...(5)

This proves the necessary condition.

The condition is sufficient i.e. only if part. Let the condition (4) hold, then to prove that the $\triangle ABC$ is equilateral.

From (4), we have

$$\frac{1}{\alpha} + \frac{1}{\beta} + \frac{1}{\gamma} = 0 \qquad\qquad [\because z_2 - z_3 = \alpha \text{ etc.}]$$

or $\beta\gamma + \alpha\gamma + \alpha\beta = 0$ or $\beta\gamma + (\gamma + \beta)\alpha = 0$

or $\beta\gamma + (-\alpha)\alpha = 0$ [using (1)]

$\alpha^2 = \beta\gamma$.   ...(5)

Conjugate of (5) is,

$$\overline{(\alpha^2)} = \overline{(\beta\gamma)} \quad \text{or} \quad \bar{\alpha}^2 = \bar{\beta}\bar{\gamma} \tag{...(6)}$$

Multiplying (5) and (6), we get

$$\alpha^2\bar{\alpha}^2 = \beta\gamma\bar{\beta}\bar{\gamma} \quad \text{or} \quad (\alpha\bar{\alpha})^2 = \beta\bar{\beta}\gamma\bar{\gamma}$$

or $(\alpha\bar{\alpha})^3 = \alpha\bar{\alpha}\beta\bar{\beta}\gamma\bar{\gamma}$ ...(7)

Similarly, $(\beta\bar{\beta})^3 = \alpha\bar{\alpha}\beta\bar{\beta}\gamma\bar{\gamma} = (\gamma\bar{\gamma})^3$ ...(8)

From (7) and (8), we get

$(\alpha\bar{\alpha})^3 + (\beta\bar{\beta})^3 = (\gamma\bar{\gamma})^3$ or $\alpha\bar{\alpha} = \beta\bar{\beta} = \gamma\bar{\gamma}$

or $|\alpha|^2 = |\beta|^2 = |\gamma|^2$ or $|z_2 - z_3|^2 = |z_3 - z_1|^2 = |z_1 - z_2|^2$

or $BC^2 = CA^2 = AB^2$ or $BC = CA = AB$

Therefore the $\triangle ABC$ is equilateral.

Example 8. *If* $|z_1| = |z_2| = ... |z_n| = 1$, *prove that*

$$|z_1 + z_2 + ... + z_n| = \left| \frac{1}{z_1} + \frac{1}{z_2} + ... \frac{1}{z_n} \right|$$

Solution. We have $|z_i| = 1 \Rightarrow |z_i|^2 = 1$ for $i = 1, 2, 3, ... n$

$\Rightarrow$ $z_i\bar{z}_i = 1$ ...(1)

We know that $|z| = |\bar{z}|$

Hence
$$|z_1 + z_2 + ... + z_n| = |\overline{z_1 + z_2 + ... z_n}| = |\overline{z}_1 + \overline{z}_2 + ... \overline{z}_n|$$

$$= \left| \frac{1}{z_1} + \frac{1}{z_2} + ... \frac{1}{z_n} \right| \qquad \text{[using (1)]}$$

Example 9. *If* $x + iy = \dfrac{3}{2 + \cos\theta + i\sin\theta}$ *prove that* $(x-1)(x-3) + y^2 = 0$.

Solution. We have

$$x + iy = \frac{3}{2 + \cos\theta + i\sin\theta} \qquad \qquad \text{...(1)}$$

$$x - iy = \frac{3}{2 + \cos\theta + i\sin\theta} \qquad \qquad \text{...(2)}$$

Multiplying (1) and (2), we have

$$x^2 + y^2 = \frac{9}{5 + 4\cos\theta} \qquad \qquad \text{...(3)}$$

Adding (1) and (2), we have

$$2x = \frac{6[2 + \cos\theta]}{5 + 4\cos\theta} \qquad \qquad \text{...(4)}$$

Now $(x - 1)(x - 3) + y^2 = x^2 - 4x + 3 + y^2 = x^2 + y^2 - 2 \times 2x + 3$

$$= \frac{9}{5 + 4\cos\theta} - \frac{2 \times 6(2 + \cos\theta)}{5 + 4} + 3 = 0$$

Example 10. *A variable complex number* $z = x + iy$ *is such that the amplitude of fraction*
$\dfrac{z-1}{z+1}$ *is always* $\dfrac{\pi}{3}$. *Show that* $x^2 + y^2 - \dfrac{2y}{\sqrt{3}} - 1 = 0$

Solution. Let
$$w = \frac{z-1}{z+1} = \frac{x + iy - 1}{x + iy + 1} = u + iv \text{ (say)}$$

Then
$$u = \frac{(x-1)(x+1) + y^2}{(x+1)^2 + y^2} \quad \text{and} \quad v = \frac{2y}{(x+1)^2 + y^2}$$

Now
$$\text{amp } (w) = \frac{\pi}{3}$$

$$\therefore \qquad \tan\frac{\pi}{3} = \frac{v}{u} = \frac{2y}{x^2 + y^2 - 1} \quad \text{or} \quad \frac{2y}{x^2 + y^2 - 1} = \sqrt{3}$$

or $x^2 + y^2 - \dfrac{2}{\sqrt{3}} y - 1 = 0$

Example 11. *If* z_1 *and* z_2 *are two complex numbers such that* z_1/z_2 *is purely imaginary, prove that*

$$|z_1 + z_2|^2 = |z_1|^2 + |z_2|^2.$$

Solution. We have $\dfrac{z_1}{z_2}$ purely imaginary

$$\therefore \qquad \frac{z_1}{z_2} + \frac{\overline{z}_1}{\overline{z}_2} = 0 = \frac{z_1\overline{z}_2 + \overline{z}_1 z_2}{|z_2|^2} = 0$$

$\Rightarrow \qquad z_1\bar{z}_2 + \bar{z}_1 z_2 = 0$ $\qquad\qquad\qquad\qquad\qquad\qquad\qquad\qquad\qquad$ ...(1)

Now, $\qquad |z_1 + z_2|^2 = (z_1 + z_2)\overline{(z_1 + z_2)} = (z_1 + z_2)(\bar{z}_1 + \bar{z}_2)$

$\qquad\qquad\qquad\qquad = |z_1|^2 + |z_2|^2 + (z_1\bar{z}_2 + \bar{z}_1 z_2) = |z_1|^2 + |z_2|^2$ $\qquad$ [using (1)]

EXERCISE 8.1

1. Express the following complex numbers in the form of $r(\cos\theta + i\sin\theta)$

   (i) $\sqrt{3} - \sqrt{-1}$ $\qquad\qquad$ (ii) $2 + \sqrt{3} + i$ $\qquad\qquad$ (iii) $-5 - 12i$

   (iv) $1 - i$ $\qquad\qquad\qquad$ (v) $-\sqrt{3} + i$ $\qquad\qquad$ (vi) $\dfrac{(1+i)(2-i)}{3+i}$

2. Put the following in $A + iB$ form :

   (i) $(2 + 3i)^2$ $\qquad\qquad$ (ii) $(5 - 6i)^2$ $\qquad\qquad$ (iii) $\dfrac{2 + 5i}{7 - 7i}$

3. Find the numbers A and B if

   (i) $A + iB = \dfrac{1}{(1 - 2i)(2 + 3i)}$ $\qquad$ (ii) $A + iB = \dfrac{3 - 2i}{7 + 4i}$

4. If $\left|\dfrac{z-1}{z+1}\right| = 2$ prove that the locus if z on the argand plane is a circle whose centre has

   affix $\left(-\dfrac{5}{3}, 0\right)$ and whose radius is $\dfrac{4}{3}$.

5. If $\left|\dfrac{z-i}{z+i}\right| = 5$, prove that the locus of z on the argand diagram is a circle, whose centre

   has affix $\left(0, -\dfrac{13}{12}\right)$ and radius is $\dfrac{5}{12}$.

6. If $\left|\dfrac{2x+5}{z-3}\right| = 1$, show that the locus of z on the Argand plane is $3x^2 + 3y^2 + 26x + 16 = 0$.

7. A variable complex number $z = x + iy$ is such that the amplitude of the fraction

   $\left|\dfrac{z-1}{z+1}\right|$ is always equal to $\pi/4$, snow that $x^2 + y^2 - 2y = 1$.

8. If amp $\left(\dfrac{z-i}{z+i}\right) = \dfrac{\pi}{4}$ prove that locus of z on the argand-diagram is a circle whose

   centre has affix $(-1, 0)$ and radius is $\sqrt{2}$.

9. Show that $\dfrac{|z|}{|\bar{z}|} + \dfrac{|\bar{z}|}{|z|} = 2$

10. Show that the triangle whose verticles are the points, $-1, 1$ and $i\sqrt{3}$ in the argand plane is equilateral.

11. If z_1, z_2 and z_3 are the vertices of an isosceles triangle, right angled at the vertices z_2 prove that

$$z_1^2 + 2z_2^2 + z_3^2 = 2z_2(z_1 + z_3).$$

12. Find the locus of the complex number z, if $\arg\left(\dfrac{z - 3i}{z - 3}\right) = \dfrac{\pi}{3}$.

13. Prove that if x, y are real number such that $x^2 + y^2 \neq 0$. Then

$$\left|\frac{x - iy}{x + iy}\right| = 1$$

14. If z_1, z_2 and z_3 are the vertices of an equilateral triangle, prove that

$$z_1^2 + z_2^2 + z_3^2 = z_1 z_2 + z_2 z_3 + z_3 z_1.$$

HINT TO SELECTED PROBLEMS

1. (i) We can write $\sqrt{3} - \sqrt{(-1)} = \sqrt{3} - i = r(\cos\theta + i\sin\theta)$

$$\Rightarrow r\cos\theta = \sqrt{3} \quad \text{and} \quad r\sin\theta = -1 \Rightarrow r = 2, \quad \theta = \frac{11}{6}\pi$$

4. $\left|\dfrac{z - 1}{z - 2}\right| = 2 \Rightarrow \left|\dfrac{x + iy - 1}{x + iy + 1}\right| = 2$

On solving, we get $x^2 + y^2 + \dfrac{10}{3}x + 1 = 0$

Comparing this equation, with $x^2 + y^2 + 2gx + 2fy + c = 0$, we get
$g = 5/3$, $f = 0$, $c = 1$.
Hence, the centre of the circle $= (-g, -f) = (-5/3, 0)$ and radius is $4/3$.

7. On putting $z = x + iy$ in the given equation and after some simplification, we get

$$r\cos\theta = \frac{x^2 + y^2 - 1}{\{(x + 1)^2 + y^2\}}$$

$$r\sin\theta = \frac{2y}{\{(x + 1)^2 - y^2\}}$$

$\Rightarrow \qquad \tan\theta = \dfrac{2y}{x^2 + y^2 - 1}$. According to given question, we have

$$\tan\theta = \tan\frac{\pi}{4} = 1$$

$\Rightarrow \qquad \dfrac{2y}{x^2 + y^2 - 1} = 1 \qquad \text{i.e.} \quad x^2 + y^2 - 24 = 1$

9. L.H.S. $\dfrac{|z|}{|\bar{z}|} + \dfrac{|\bar{z}|}{|z|} = \dfrac{|z|.|z| + |\bar{z}|.|\bar{z}|}{|\bar{z}|.|z|}$

Now, using $z = x + iy$ and $\bar{z} = x - iy$

$\Rightarrow \qquad |z| = \sqrt{x^2 + y^2} = \bar{z}$.

11. Since $BA = BC$.

$$\Rightarrow \quad |z_1 - z_2|^2 = |z_3 - z_2|^2$$

$$\Rightarrow \quad |z_1 - z_2|^2 = |z_3 - z_2|^2$$

$$\Rightarrow \quad (z_1 - z_2)(\bar{z}_1 - \bar{z}_2) = (z_3 - z_2)(\bar{z}_3 - \bar{z}_2).$$

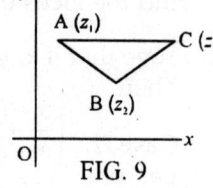

FIG. 9

Also $\angle ABC = \dfrac{\pi}{2}$, Thus $\arg\left(\dfrac{z_1 - z_2}{z_3 - z_2}\right) = \dfrac{\pi}{2}$.

so that $\dfrac{z_1 - z_2}{z_3 - z_2}$ is purely imaginary

$$\frac{z_1 - z_2}{z_3 - z_2} + \frac{\bar{z}_1 - \bar{z}_2}{\bar{z}_3 - \bar{z}_2} = 0 \quad \Rightarrow \quad \frac{z_1 - z_2}{z_3 - z_2} + \frac{z_3 - z_2}{z_1 - z_2} = 0$$

On simplifying, we have

$$z_1^2 + 2z_2^2 + z_3^2 = 2z_2(z_1 + z_2).$$

ANSWERS

(1) (i) $2\left(\cos\dfrac{11\pi}{6} + i\sin\dfrac{11\pi}{6}\right)$,　　(ii) $r = 2\sqrt{2 + \sqrt{3}}$, $\theta = \tan^{-1}(2 - \sqrt{3})$

(iii) $r = 13$, $\theta = \tan^{-1}\left(\dfrac{5}{12}\right)$,　　(iv) $\sqrt{2}\left(\cos\dfrac{7\pi}{6} + i\sin\dfrac{7\pi}{6}\right)$

(v) $2\left(\cos\dfrac{5\pi}{6} + i\sin\dfrac{5\pi}{6}\right)$,　　(vi) 1 or $(\cos 0 + i\sin 0)$

(2) (i) $-5 + 12i$,　　(ii) $-11 - 60i$　　(iii) $-\dfrac{21}{98} + \dfrac{1}{2}i$

(3) (i) $A = \dfrac{8}{65}$, $B = \dfrac{1}{65}$　　(ii) $\dfrac{13}{65}$, $B = \dfrac{-26}{65}$

(12) $x^2 + y^2 - (3 - \sqrt{3})x - (3 - \sqrt{3})y - 3\sqrt{3} = 0.$

8.9 DE'MOIVRE'S THEOREM

If n is any integer, positive or negative, then

$$(\cos\theta + i\sin\theta)^n = \cos n\theta + i\sin n\theta$$

and if n is a fraction, positive or negative, then $\cos n\theta + i\sin n\theta$ is one of the values of $(\cos\theta + i\sin\theta)^n$.

Above statement can also be written as : *If n is any rational number, then $\cos n\theta + i\sin n\theta$ is a value of $(\cos\theta + i\sin\theta)^n$.*

Proof. Case I. *If n be a positive integer.*

$(\cos\alpha + i\sin\alpha)(\cos\beta + i\sin\beta) = \cos\alpha\cos\beta - \sin\alpha\sin\beta + i(\sin\alpha\cos\beta + \cos\alpha\sin\beta)$

$$= \cos(\alpha + \beta) + i\sin(\alpha + \beta).$$

Similarly, multiplying by $(\cos \gamma + i \sin \gamma)$ on both sides, we get

$(\cos \alpha + i \sin \alpha)(\cos \beta + i \sin \beta)(\cos \gamma + i \sin \gamma) = \cos (\alpha + \beta + \gamma) + i \sin (\alpha + \beta + \gamma)$.

Proceeding in this way and continuing upto n factors, we get

$(\cos \alpha + i \sin \alpha)(\cos \beta + i \sin \beta)(\cos \gamma + i \sin \gamma) \ldots$ to n factors

$$\cos (\alpha + \beta + \gamma + \ldots \text{ to } n \text{ terms}) + i \sin (\alpha + \beta + \gamma + \ldots \, n \text{ terms})$$

Put $\alpha = \beta = \gamma = \ldots = \theta.$

Then $(\cos \theta + i \sin \theta)^n = \cos n\theta + i \sin n\theta$

Case II. *If n is negative integer.*

Let n be equal to $-m$, where m is a +ve integer.

$$(\cos \theta + i \sin \theta)^n = (\cos \theta + i \sin \theta)^{-m}$$

$$= \frac{1}{(\cos \theta + i \sin \theta)^m} = \frac{1}{\cos m\theta + i \sin m\theta}$$

$$= \frac{\cos m\theta - i \sin m\theta}{(\cos m\theta + i \sin m\theta)(\cos m\theta - i \sin m\theta)}$$

$$= \frac{\cos m\theta - i \sin m\theta}{\cos^2 m\theta - i^2 \sin^2 m\theta} = \cos m\theta - i \sin m\theta$$

$$= \cos (-m\theta) + i \sin (-m\theta) = \cos n\theta + i \sin n\theta$$

Case III. Suppose n is a fraction and equal to p/q where p is an integer positive or negative and q is a positive integer.

Now $\left(\cos \dfrac{\theta}{q} + i \sin \dfrac{\theta}{q}\right)^q = \cos q.\dfrac{\theta}{q} + i \sin q.\dfrac{\theta}{q} = (\cos \theta + i \sin \theta)$

$\therefore$ $\cos \dfrac{\theta}{q} + i \sin \dfrac{\theta}{q}$ is one of the qth roots of $(\cos \theta + i \sin \theta)$

or $\cos \dfrac{\theta}{q} + i \sin \dfrac{\theta}{q}$ is one of the values of $(\cos \theta + i \sin \theta)^{1/q}$.

Now raising each of these quantities of pth power, we get

$\left(\cos \dfrac{\theta}{q} + i \sin \dfrac{\theta}{q}\right)^p$ is one of the values of $(\cos \theta + i \sin \theta)^{p/q}$.

or $\left(\cos \dfrac{p\theta}{q} + i \sin \dfrac{p\theta}{q}\right)$ is one of the values of $(\cos \theta + i \sin \theta)^{p/q}$.

or $(\cos n\theta + i \sin n\theta)$ is one the values of $(\cos \theta + i \sin \theta)^n$ as per $p/q = n$.

Cor. If we write $-\theta$ for θ, we see at once that $\cos n\theta - i \sin n\theta$ is a value of $(\cos \theta - i \sin \theta)^n$ where n is any rational number.

Remarks

❖ $(\sin \theta + i \cos \theta)^n \neq \sin n\theta + i \cos n\theta$

❖ and $(\cos \theta + i \sin \phi)^n \neq \cos n\theta + i \sin n\phi$.

SOLVED EXAMPLES

Based on the following Results

- If n is any integer, positive or negative then $(\cos\theta + i\sin\theta)^n = \cos n\theta + i\sin n\theta$

- If n is any rational number then $\cos n\theta + i\sin n\theta$ is one of the values of $(\cos\theta + i\sin\theta)^n$

Example 1. *Prove that*

(i) $\left(\dfrac{1+\cos\phi+i\sin\phi}{1+\cos\phi-i\sin\phi}\right)^n = \cos n\phi + i\sin n\phi.$

(ii) $\left(\dfrac{1+\sin\theta+i\cos\theta}{1+\sin\theta-i\cos\theta}\right)^n = \cos\left(\dfrac{n\pi}{2}-n\theta\right)+i\sin\left(\dfrac{n\pi}{2}-n\theta\right).$

Solution. (i) L.H.S. $= \left\{\dfrac{2\cos^2\dfrac{\phi}{2}+2i\sin\dfrac{\phi}{2}\cos\dfrac{\phi}{2}}{2\cos^2\dfrac{\phi}{2}-2i\sin\dfrac{\phi}{2}\cos\dfrac{\phi}{2}}\right\}^n$

$= \left\{\dfrac{\cos\dfrac{\phi}{2}+i\sin\dfrac{\phi}{2}}{\cos\dfrac{\phi}{2}-i\sin\dfrac{\phi}{2}}\right\}^n = \left\{\left(\cos\dfrac{\phi}{2}+i\sin\dfrac{\phi}{2}\right)^2\right\}^n = \cos n\phi + i\sin n\phi$

(ii) Put $\dfrac{\pi}{2}-\theta$ for ϕ in (i).

Example 2. *If* $2\cos\theta = x+\dfrac{1}{x}$, $2\cos\phi = y+\dfrac{1}{y}$, *prove that one of the values of*

(i) $x^m y^n + \dfrac{1}{x^m y^n}$ *is* $2\cos(m\theta + n\phi)$. *and* (ii) $\dfrac{x^m}{y^n}+\dfrac{y^n}{x^m}$ *is* $2\cos(m\theta - n\phi)$

Solution. Since $2\cos\theta = x+\dfrac{1}{x}$, therefore

$x^2 - 2x\cos\theta = -1,$

$x^2 - 2x\cos\theta + \cos^2\theta = -1+\cos^2\theta = -\sin^2\theta,$

$(x^2 - \cos\theta)^2 = -\sin^2\theta$

$x - \cos\theta = i\sin\theta,$

$x = \cos\theta + i\sin\theta,\ \dfrac{1}{x} = \cos\theta - i\sin\theta.$

Similarly, $y = \cos\phi + i\sin\phi,\ \dfrac{1}{y} = \cos\phi - i\sin\phi.$

$\therefore$ (i) $\quad x^m y^n = (\cos\theta + i\sin\theta)^m (\cos\phi + i\sin\phi)^n$

$= (\cos m\theta + i\sin m\theta)(\cos n\phi + i\sin n\phi)$

$= \cos(m\theta + n\phi) + i\sin(m\theta + n\phi).$

$$\frac{1}{x^m} \times \frac{1}{y^n} = x^{-m} \times y^{-n} = \cos(m\theta + n\phi) - i\sin(m\theta + n\phi).$$

$$\therefore \quad x^m y^n + \frac{1}{x^m y^n} = 2\cos(m\theta + n\phi)$$

and (ii) $\dfrac{x^m}{y^n} = \dfrac{(\cos\theta + i\sin\theta)^m}{(\cos\phi + i\sin\phi)^n} = \dfrac{\cos m\theta + i\sin m\theta}{\cos n\phi + i\sin n\phi} = \cos(m\theta - n\phi) + i\sin(m\theta - n\phi).$

Similarly, $\qquad \dfrac{x^n}{y^m} = \cos(m\theta - n\phi) - i\sin(m\theta - n\phi)$

Adding, we get $\quad \dfrac{x^m}{y^n} + \dfrac{y^n}{x^m} = 2\cos(m\theta - n\phi).$

Example 3. *If* $x = \cos\theta + i\sin\theta$ *and* $\sqrt{1-c^2} = nc - 1$, *prove that*

$$1 + c\cos\theta = \frac{c}{2n}(1 + nx)\left(1 + \frac{n}{x}\right).$$

Solution. R.H.S. $= \dfrac{c}{2n}[1 + n(\cos\theta + i\sin\theta)][1 + n(\cos\theta - i\sin\theta)]$

$$= \frac{c}{2n}[1 + n^2(\cos^2\theta + \sin^2\theta) + 2n\cos\theta]$$

$$= \frac{c}{2n}[1 + n^2 + 2n\cos\theta] = \frac{c(1+n^2)}{2n} \, c\cos\theta$$

$$= 1 + c\,\cos\theta \text{ as } 1 - c^2 = (nc - 1)^2.$$

Example 4. *If* $(a_1 + b_1 i)(a_2 + b_2 i) \dots (a_n + b_n i) = A + Bi$,

prove that $\quad (a_1^2 + b_1^2)(a_2^2 + b_2^2) \dots (a_n^2 + b_n^2) = A^2 + B^2$

and $\qquad \tan^{-1}\dfrac{b_1}{a_1} + \tan^{-1}\dfrac{b_2}{a_2} + \dots \tan^{-1}\dfrac{b_n}{a_n} = \tan^{-1}\dfrac{B}{A}$

Solution. Let $\quad \left.\begin{matrix} a_k = r_k\cos\theta_k \\ b_k = r_k\sin\theta_k \end{matrix}\right\}, \quad k = 1, 2, 3, \dots, n$

Then from $(a_1 + b_1 i)(a_2 + b_2 i) \dots (a_n + b_n i) = A + iB$

we have $\quad r_1 r_2 \dots r_n (\cos\theta_1 + i\sin\theta_1)(\cos\theta_2 + i\sin\theta_2) \dots (\cos\theta_n + i\sin\theta_n) = A + iB$

or $\quad r_1 r_2 \dots r_n [\cos(\theta_1 + \theta_2 + - - - + \theta_n) + i\sin(\theta_1 + \theta_2 + \dots \theta_n)] = A + iB.$

Equating real and imaginary parts, we get

$$A = r_1 r_2 \dots r_n \cos(\theta_1 + \theta_2 + - - - + \theta_n)$$
$$B = r_1 r_2 \dots r_n \sin(\theta_1 + \theta_2 + - - - - + \theta_n).$$

$$\therefore \qquad \frac{B}{A} = \tan(\theta_1 + \theta_2 + \dots + \theta_n) \text{ or } \tan^{-1}\frac{B}{A} = \theta_1 + \theta_2 + \dots + \theta_n$$

$$\tan^{-1}\frac{b_1}{a_1} + \tan^{-1}\frac{b_2}{a_2} + \dots \tan^{-1}\frac{b_n}{a_n}$$

and $\qquad A^2 + B^2 = r_1^2 . r_2^2 \dots r_n^2 = (a_1^2 + b_1^2)(a_2^2 + b_2^2) \dots (a_n^2 + b_n^2)$

Example 5. *If n be a positive integer, prove that*

$$(1+i)^n + (1-i)^n = 2^{(n/2)+1}\cos\frac{n\pi}{4}$$

Solution. Suppose
$$1 + i = r(\cos\theta + i\sin\theta)$$
$\therefore \qquad 1 = r\cos\theta = r\sin\theta,$

$$r = \sqrt{2} \text{ and } \theta = \tan^{-1}1 = \frac{\pi}{4}.$$

$\therefore$ L.H.S $\quad = r^n(\cos\theta + i\sin\theta)^n + r^n(\cos\theta - i\sin\theta)^n$

$\qquad\qquad = r^n(\cos n\theta + i\sin n\theta) + r^n(\cos n\theta - i\sin n\theta)$

$\qquad\qquad = r^n . 2\cos n\theta = 2^{n/2} \times 2\cos\frac{n\pi}{4} = 2^{(n/2)+1}\cos\frac{n\pi}{4}.$

Example 6. *If* $\cos\alpha + \cos\beta + \cos\gamma = \sin\alpha + \sin\beta + \sin\gamma = 0,$

prove that $\quad \cos 3\alpha + \cos 3\beta + \cos 3\gamma = 3\cos(\alpha + \beta + \gamma)$

and $\qquad \sin 3\alpha + \sin 3\beta + \sin 3\gamma = 3\sin(\alpha + \beta + \gamma).$

Solution. We know that if $x + y + z = 0,$
$$x^3 + y^3 + z^3 = 3xyz.$$

Let $\qquad x = \text{cis } \alpha, y = \text{cis } \beta, z = \text{cis } \gamma \qquad\qquad (\text{cis } \alpha = \cos\alpha + i\sin\alpha)$

Hence $\quad x + y + z = (\cos\alpha + \cos\beta + \cos\gamma) + i(\sin\alpha + \sin\beta + \sin\gamma)$

$\qquad\qquad\qquad = 0, \text{ by given conditions}$

$\therefore \qquad x^3 + y^3 + z^3 = 3xyz \text{ gives}$

$\qquad\qquad (\text{cis } \alpha)^3 + (\text{cis } \beta)^3 + (\text{cis } \gamma)^3 = 3(\text{cis } \alpha)(\text{cis } \beta)(\text{cis } \gamma)$

or $\qquad\qquad \text{cis }(3\alpha) + \text{cis }(3\beta) + \text{cis }(3\gamma) = 3\text{ cis }(\alpha + \beta + \gamma)$

or $\qquad\qquad (\cos 3\alpha + \cos 3\beta + \cos 3\gamma) + i(\sin 3\alpha + \sin 3\beta + \sin 3\gamma)$

$\qquad\qquad\qquad\qquad = 3[\cos(\alpha + \beta + \gamma) + i\sin(\alpha + \beta + \gamma)]$

Equating real and imaginary parts, we get the required result.

Example 7. *If* $\cos\alpha + \cos\beta + \cos\gamma = \sin\alpha + \sin\beta + \sin\gamma = 0,$

prove that $\quad \cos^2\alpha + \cos^2\beta + \cos^2\gamma = \sin^2\alpha + \sin^2\beta + \sin^2\gamma$

and $\qquad \cos 2\alpha + \cos 2\beta + \cos 2\gamma = \sin 2\alpha + \sin 2\beta + \sin 2\gamma = 0.$

Solution. We know that if
$$x + y + z = 0 \qquad\qquad\qquad\qquad \dots(1)$$

and $\qquad\qquad\qquad \frac{1}{x} + \frac{1}{y} + \frac{1}{z} = 0 \qquad\qquad\qquad\qquad \dots(2)$

then $$x^2 + y^2 + z^2 = 0 \qquad \ldots(3)$$

Let $x = $ cis α, $y = $ cis β, $z = $ cis γ

then (1) and (2) are satisfied due to given conditions. Hence the relation (3) is also satisfied. Now we have

$$x^2 + y^2 + z^2 = 0$$

or $$(\text{cis } \alpha)^2 + (\text{cis } \beta)^2 + (\text{cis } \gamma)^2 = 0$$

or $$\text{cis } (2\alpha) + \text{cis } (2\beta) + \text{cis } (2\gamma) = 0$$

or $$(\cos 2\alpha + \cos 2\beta + \cos 2\gamma) + i(\sin 2\alpha + \sin 2\beta + \sin 2\gamma) = 0$$

$\therefore$ $\cos 2\alpha + \cos 2\beta + \cos 2\gamma = 0$, $\sin 2\alpha + \sin 2\beta + \sin 2\gamma = 0$.

Further,

$$\cos^2 \alpha - \sin^2 \alpha + \cos^2 \beta - \sin^2 \beta + \cos^2 \gamma - \sin^2 \gamma = 0$$
$$\cos^2 \alpha + \cos^2 \beta + \cos^2 \gamma = \sin^2 \alpha + \sin^2 \beta + \sin^2 \gamma.$$

Example 8. *If* $x_r = $ cis $\dfrac{\pi}{2^r}$, *prove that*

$$x_1 x_2 x_3 \ldots \text{ upto infinite terms } = \cos \pi = -1.$$

Solution. We know $x_1 x_2 x_3 \ldots$ upto infinite terms

$$= \left[\text{cis}\frac{\pi}{2}\right]\left[\text{cis}\frac{\pi}{2^2}\right]\left[\text{cis}\frac{\pi}{2^3}\right] \ldots \text{ upto infinite terms}$$

$$= \text{cis}\left[\frac{\pi}{2} + \frac{\pi}{2^2} + \frac{\pi}{2^3} + \ldots \infty\right] = \cos\left[\frac{\pi/2}{1 - \dfrac{1}{2}}\right] = \cos \pi$$

$$= \cos \pi = -1 \qquad\qquad [\because \text{ since } \sin \pi = 0]$$

Example 9. *If* α, β *are roots of the equation* $x^2 - 2x + 4 = 0$, *prove that*

$$\alpha^n + \beta^n = 2^{n+1} \cos \frac{n\pi}{3}.$$

Solution. $x^2 - 2x + 4 = 0$ or $x = 1 \pm i\sqrt{3}$.

Let $\alpha = 1 + i\sqrt{3}$ and $\beta = 1 - i\sqrt{3}$.

then we have to prove that

$$\alpha^n + \beta^n = (1 + i\sqrt{3})^n + (1 - i\sqrt{3})^n = 2^{n+1} \cos \frac{n\pi}{3}.$$

Let $1 = r \cos \theta$, $\sqrt{3} = r \sin \theta$

$\therefore$ $r = 2$, $\theta = \dfrac{\pi}{3}$.

$\therefore$ L.H.S. of (1) $= r^n (\cos \theta + i \sin \theta)^n + r^n (\cos \theta - i \sin \theta)^n$

$$= 2r^n \cos n\theta = 2^{n+1} \cos \frac{n\pi}{3}.$$

Example 10. *If* $\cos(\beta - \gamma) + \cos(\gamma - \alpha) + \cos(\alpha - \beta) = -\dfrac{3}{2}$ *show that*

(i) $\cos\alpha + \cos\beta + \cos\gamma = \sin\alpha + \sin\beta + \sin\gamma = 0,$

and (ii) $\cos n\alpha + \cos n\beta + \cos n\gamma = 3\cos\dfrac{1}{3}n(\alpha + \beta + \gamma)$, *if n is a multiple of 3;*

$$= 0, \ \textit{if } n \textit{ is not a multiple of 3.}$$

Solution. (*i*) We know from given relation

$$3 + 2\cos(\beta - \gamma) + 2\cos(\gamma - \alpha) + 2\cos(\alpha - \beta) = 0$$

or $(\cos^2\alpha + \sin^2\alpha) + (\cos^2\beta + \sin^2\beta) + (\cos^2\gamma + \sin^2\gamma) + 2\cos\beta\cos\gamma$
$\qquad + 2\sin\beta\sin\gamma + 2\cos\gamma\cos\alpha + 2\sin\gamma\sin\alpha + 2\cos\alpha\cos\beta + 2\sin\alpha\sin\beta = 0$

or $\{(\cos^2\alpha + \cos^2\beta + \cos^2\gamma + 2\cos\beta\cos\gamma + ... + ...) + (\sin^2\alpha + \sin^2\beta + \sin^2\gamma$
$\qquad + 2\sin\beta\sin\gamma + ... + ...)\} = 0$

or $(\cos\alpha + \cos\beta + \cos\gamma)^2 + (\sin\alpha + \sin\beta + \sin\gamma)^2 = 0.$

or $\cos\alpha + \cos\beta + \cos\gamma + \sin\alpha + \sin\beta + \sin\gamma = 0.$

[∵ sum of two positive quantities can be zero if they are separately zero]

(*ii*) This part now can be solved as solved Ex. 6 and Ex. 7 with the help of following identity.

If $x + y + z = 0$ and $\dfrac{1}{x} + \dfrac{1}{y} + \dfrac{1}{z} = 0$, then $x^n + y^n + z^n = 3(xyz)^{n/3}$ (if n is multiple of 3)

$$x^n + y^n + z^n = 3(xyz)^{n/3} = 0 \quad \text{(if } n \text{ is not a multiple of 3)}$$

Example 11. *Prove that, if n is a positive integer and*

$$(1 + x)^n = c_0 + c_1 x + c_2 x^2 + ... + c_n x^n,$$

then $c_0 + c_4 + c_8 + ... = 2^{n-2} + 2^{(n/2)-1}\cos\dfrac{n\pi}{4}$

Solution. Putting $x = 1, -1$ and i in the given relation, we get

$$c_0 + c_1 + c_2 + ... = 2^n, \qquad\qquad\qquad\qquad\qquad ...(1)$$

$$c_0 - c_1 + c_2 - c_3 + ... = 0, \qquad\qquad\qquad\qquad\qquad ...(2)$$

$$c_0 + ic_1 - c_2 - c_3 i + ... (1 + i)^n = 2^{n/2}\left(\frac{n\pi}{4} + i\sin\frac{n\pi}{4}\right), \qquad ...(3)$$

Adding (1) and (2) and dividing by 2, we get

$$c_0 + c_2 + c_4 + ... = 2^{n-1}. \qquad\qquad\qquad\qquad\qquad ...(4)$$

Equating real parts in (3), we get

$$c_0 - c_2 + c_4 - c_6 + c_8 - ... = 2^{n/2}\cos\frac{n\pi}{4} \qquad\qquad ...(5)$$

Adding (4) and (5) and dividing by 2, we have the required result.

EXERCISE 8.2

1. Simplify the following

   (a) $\dfrac{(\cos 2\theta - i \sin 2\theta)^7 .(\cos 3\theta + i \sin 3\theta)^{-5}}{(\cos 4\theta + i \sin 4\theta)^{12} .(\cos 5\theta + i \sin 5\theta)^6}$

   (b) $[(\cos \theta + \cos \phi) + i (\sin \theta + \sin \phi)]^n + [(\cos \theta + \cos \phi) - i (\sin \theta + \sin \phi)]^n$

   (c) $[1 + \cos \theta + i \sin \theta]^n + [1 + \cos \theta - i \sin \theta]^n$

   (d) $[(\cos \theta - \cos \phi) + i (\sin \theta - \sin \phi)]^n + [(\cos \theta - \cos \phi) - i (\sin \theta - \sin \phi)]^n$

   (e) $\dfrac{(\cos 3\theta + i \sin 3\theta)^5 .(\cos \theta - i \sin \theta)^3}{(\cos 5\theta + i \sin 5\theta)^7 .(\cos 2\theta - i \sin 2\theta)^5}$

   (f) $\dfrac{(\cos \theta + i \sin \theta)^8 .(\cos 3\theta - i \sin 3\theta)^2}{(\cos 2\theta + i \sin 2\theta)^5 .(\cos 4\theta - i \sin 4\theta)^7}$

2. If x, y, z and u stand for $(\cos \alpha + i \sin \alpha)$, $(\cos \beta + i \sin \beta)$, $(\cos \gamma + i \sin \gamma)$ and $(\cos \delta + i \sin \delta)$ respectively. Find the value of

   (a) $(x + y)(z + u)$ (b) $\dfrac{1}{(x - y)(z - u)}$ (c) $xy + zu$.

3. Prove that $(a + ib)^{m/n} + (a - ib)^{m/n} = 2(a^2 + b^2)^{m/2n} . \cos\left(\dfrac{m}{n} \tan^{-1} \dfrac{b}{a}\right)$

4. If $p = \cos \theta + i \sin \theta$ and $q = \cos \phi + i \sin \phi$, show that

   $$\dfrac{p - q}{p + q} = i \tan\left(\dfrac{\theta - \phi}{2}\right)$$

5. If a denotes $\cos 2\alpha + i \sin 2\alpha$ with similar expression for b, c, d prove that

   (a) $\sqrt{(abcd)} + \dfrac{1}{\sqrt{(abcd)}} = 2 \cos (\alpha + \beta + \gamma + \delta)$

   (b) $\sqrt{\left(\dfrac{ab}{cd}\right)} + \sqrt{\left(\dfrac{cd}{ab}\right)} = 2 \cos (\alpha + \beta - \gamma - \delta)$

   (c) $\sqrt{a^p b^q c^r d^s} + \dfrac{1}{\sqrt{a^p b^q c^r d^s}} = 2 \cos (p\alpha + q\beta + r\gamma - s\delta)$

   (d) $\dfrac{(a + b)(b + c)(c + a)}{abc}$ is real and equal to $8 \cos (\alpha - \beta) \cos (\beta - \gamma) \cos (\gamma - \alpha)$.

6. If $2 \cos \theta = x + \dfrac{1}{x}$, $2 \cos \phi = y + \dfrac{1}{y} - - -$ prove that

(a)　$2 \cos (\theta + \phi + \psi \ldots) = xyz \ldots + \dfrac{1}{xyz} \ldots$

(b)　$2 \cos (p\theta + q\phi + r\psi \ldots) = x^p y^q z^r \ldots + \dfrac{1}{x^p y^q z^r}$

7.　If $x + \dfrac{1}{x} = 2 \cos \theta$, show that

(a).　$x^7 + \dfrac{1}{x^7} = 2 \cos 7\theta$　　　　　　　(b)　$x^n + \dfrac{1}{x^n} = 2 \cos n\theta$

(c)　$x^n - \dfrac{1}{x^n} = 2 i \sin n\theta.$

8.　If α and β are the roots of the equation　$x^2 - 2x \cos \theta + 1 = 0$ form an equation whose roots are α^n and β^n.

9.　If $\cos \alpha + \cos \beta + \cos \gamma = \sin \alpha + \sin \beta + \sin \gamma = 0$, prove that

$$\cos^2 \alpha + \cos^2 \beta + \cos^2 \gamma = \sin^2 \alpha + \sin^2 \beta + \sin^2 \gamma = \frac{3}{2}$$

10.　If $\cos \alpha + 2 \cos \beta + 3 \cos \gamma = 0 = \sin \alpha + 2 \sin \beta + 3 \sin \gamma$, prove that

$$\cos 3\alpha + 8 \cos 3\beta + 27 \cos 3\gamma = 18 \cos (\alpha + \beta + \gamma)$$

and　$\sin 3\alpha + 8 \sin 3\beta + 27 \sin 3\gamma = 18 \sin (\alpha + \beta + \gamma)$

11.　If $(1 + x)^n = p_0 + p_1 x + p_2 x^2 + \ldots$ show that

$$p_0 - p_2 + p_4 \ldots = 2^{n/2} \cos \frac{n\pi}{4}$$

and　　　$p_1 - p_3 + p_5 \ldots = 2^{n/2} \sin \frac{n\pi}{4}$

12.　If $z_n = \cos \dfrac{\pi}{2^n} + i \sin \dfrac{\pi}{2^n}$, prove that $\lim\limits_{n \to \infty} (z_1 . z_2 . z_3 . \ldots z_n) = -1$

HINT TO SELECTED PROBLEMS

3.　Let　　　$a + ib = r(\cos \theta + i \sin \theta)$
$\Rightarrow$　　　　$a - ib = r(\cos \theta - i \sin \theta)$

$\Rightarrow$　　　$r \cos \theta = a$, $r \sin \theta = b$. Also $r = (a^2 + b^2)^{1/2}$ and $\theta = \tan^{-1} \dfrac{b}{a}$

Therefore $(a + ib)^{m/n} + (a - ib)^{m/n} = r^{m/n}(\cos \theta + i \sin \theta)^{m/n} + r^{m/n}(\cos \theta - i \sin \theta)^{m/n}$.
Now simplifying this equation.

4.　Consider　$\dfrac{p - q}{p + q} = \dfrac{(\cos \theta + i \sin \theta) - (\cos \phi + i \sin \phi)}{(\cos \theta + i \sin \theta) + (\cos \phi + i \sin \phi)}$

$$= \frac{[(\cos \theta - \cos \phi) + i(\sin \theta - \sin \phi)]}{[(\cos \theta + \cos \phi) + i(\sin \theta + \sin \phi)]}$$

$$= \frac{2\sin\frac{(\theta+\phi)}{2}.\sin\frac{(\phi-\theta)}{2} + 2i\cos\frac{(\theta+\phi)}{2}.\sin\frac{(\theta-\phi)}{2}}{2\cos\frac{(\theta+\phi)}{2}.\cos\frac{(\theta-\phi)}{2} + i\sin\frac{(\theta+\phi)}{2}.\cos\frac{(\theta-\phi)}{2}}$$

Now simplifying this equation to get the required result.

7. (i) As per given

$$x + \frac{1}{x} = 2\cos\theta \quad \Rightarrow x^2 - 2x\cos\theta + 1 = 0$$

$$\Rightarrow x = \frac{2\cos\theta \pm \sqrt{4\cos^2\theta - 4}}{2} = \cos\theta \pm \sqrt{(-\sin^2\theta)} = \cos\theta \pm i\sin\theta.$$

Taking positive sign only, we get

$$x = \cos\theta + i\sin\theta \quad \Rightarrow \frac{1}{x} = \cos\theta - i\sin\theta$$

$$\Rightarrow \quad x^7 + \frac{1}{x^7} = (\cos\theta + i\sin\theta)^7 + (\cos\theta - i\sin\theta)^7$$

$$= \cos 7\theta + i\sin 7\theta + \cos 7\theta - i\sin 7\theta = 2\cos 7\theta.$$

8. Proceed same as (7) (i) we get

$$x = \cos\theta \pm i\sin\theta$$

Let

$$\alpha = \cos\theta + i\sin\theta, \ \beta = \cos\theta - i\sin\theta.$$

Then required equation is

$$x^n - (\text{sum of the roots}) \, x + \text{product of the roots} = 0$$

i.e.

$$x^n - (\alpha^n + \beta^n)x + \alpha^n\beta^n = 0$$

ANSWERS

(1) (a) $\cos 107\theta - i\sin 107\theta$ 　　　(b) $2^{n+1}\cos\frac{1}{2}(\theta-\phi)\cos\frac{n}{2}(\theta+\phi),$

(c) $2^{n+1}\cos^n\left(\frac{\theta}{2}\right)\cos\left(\frac{n\theta}{2}\right),$ 　(d) $2^{n+1}\sin\left(\frac{\theta-\phi}{2}\right)\left[\cos n\left(\frac{\pi+\theta+\phi}{2}\right)\right]$

(e) $\cos 13\theta - i\sin 13\theta,$ 　　　(f) $\cos 20\theta + i\sin 20\theta.$

(2) (a) $4\cos\left(\frac{\alpha-\beta}{2}\right)\cos\left(\frac{\gamma-\delta}{2}\right)\left[\cos\left(\frac{\alpha+\beta+\gamma+\delta}{2}\right) + i\sin\left(\frac{\alpha+\beta+\gamma+\delta}{2}\right)\right]$

(b) $\frac{1}{4}\cos ec\left(\frac{\alpha-\beta}{2}\right)\cos ec\left(\frac{\gamma-\delta}{2}\right)\left[\cos\left(\frac{\alpha+\beta+\gamma+\delta}{2}\right) + i\sin\left(\frac{\alpha+\beta+\gamma+\delta}{2}\right)\right]$

(c) $2\cos\left(\frac{\alpha+\beta-\gamma-\delta}{2}\right)\left[\cos\left(\frac{\alpha+\beta+\gamma+\delta}{2}\right) + i\sin\left(\frac{\alpha+\beta+\gamma+\delta}{2}\right)\right]$

(8) $x^2 - 2x\cos n\theta + 1 = 0$

8.10 TO SHOW THAT $(\cos θ + i \sin θ)^{p/q}$, WHERE p AND q ARE INTEGERS PRIME TO EACH OTHER, q BEING POSITIVE, HAS q DISTINCT VALUE AND NO MORE

We know that, $\cos\dfrac{pθ}{q}+i\sin\dfrac{pθ}{q}$ is one of the q values of $(\cos θ + i \sin θ)^{n/q}$.

Now we will find other values

$$(\cos θ + i \sin θ)^{p/q} = [\cos (2nπ + θ) + i \sin (2nπ + θ)]^{p/q}$$

where $\sin θ$ and $\cos θ$ remain unaltered when $θ$ is increasing by any multiple of $2π$.

Now one of the values of $(\cos θ + i \sin θ)^{p/q}$ is

$$\cos \frac{p}{q}(2nπ + θ) + i \sin \frac{p}{q}(2nπ + θ).$$

By putting $n = 0, 1, 2, ...$ we find different quantities.

$$\cos\frac{pθ}{q}+i\sin\frac{pθ}{q},$$

$$\cos \frac{p}{q}(2nπ + θ) + i \sin \frac{p}{q}(2nπ + θ)$$

$$...\qquad ...\qquad ...\qquad ...\qquad ...$$

where each one of them is equal to one of the values of $(\cos θ + i \sin θ)^{p/q}$.

This can be easily verified by raising these quantities to the power p/q. In each case, we get

$$\cos θ + i \sin θ$$

But for $n = q, q + 1, q + 2, ...$ the result will be same as for $n = 0, 1, 2, ...$

Hence by giving to n the successive values $0, 1, 2, ... (q - 1)$ in the expression

$$\cos p\left(\frac{2nπ+θ}{q}\right)+i\sin p\left(\frac{2nπ+θ}{q}\right)$$

we obtain q different values for $(\cos θ + i \sin θ)^{p/q}$.

Remarks

❖ $(\text{cis } θ)^{p/q}$ has two meanings:
either $(\text{cis } θ)^{p/q} = \{(\text{cis}θ)^{p/q}\}^{1/q}$ or $(\text{cis } θ)^{p/q} = \{(\text{cis}θ)^{1/q}\}^p$.

❖ If p is not prime to q, let $\dfrac{p'}{q'}$ be their ratio in its lowest terms $e.g.$, $\dfrac{4}{6}=\dfrac{2}{3}$ then

$(\cos θ + i \sin θ)^{p/q}$ will give only 'q'' values and no more, and that these q' values are the values of $(\cos θ + i \sin θ)^{p'/q'}$.

❖ Since $\cos\left\{\dfrac{p}{q}(\theta + 2n\pi)\right\} + i\sin\left\{\dfrac{p}{q}(\theta + 2n\pi)\right\}$

$$= \left\{\cos\left(\dfrac{p\theta}{q}\right) + i\sin\left(\dfrac{p\theta}{q}\right)\right\}\left\{\cos\dfrac{2n\pi p}{q} + i\sin\dfrac{2n\pi p}{q}\right\}$$

$$= \left\{\cos\left(\dfrac{p\theta}{q}\right) + i\sin\left(\dfrac{p\theta}{q}\right)\right\}\left\{\cos\dfrac{2p\pi}{q} + i\sin\dfrac{2p\pi}{q}\right\}^{n}$$

Therefore, the q roots of $(\cos\theta + i\sin\theta)^{p/q}$ may be arranged in the geometrical progression

$$z, \omega z, \omega^2 z, \dots w^{q-1}z,$$

where $\qquad z = \cos\left(\dfrac{p\theta}{q}\right) + i\sin\left(\dfrac{p\theta}{q}\right)$ and $w = \cos\left(\dfrac{2p\pi}{q}\right) + i\sin\left(\dfrac{2p\pi}{q}\right)$

SOLVED EXAMPLES

Example 1. *Find all the values of* $(\cos\alpha + i\sin\alpha)^{2/3}$ *and* $(\cos\alpha + i\sin\alpha)^{4/6}$.

Solution. (i) Defining $(\cos\alpha + i\sin\alpha)^{2/3}$ as $[(\cos\alpha + i\sin\alpha)^2]^{1/3}$, we have, writing cis α for $\cos\alpha + i\sin\alpha$,

$$(\text{cis }\alpha)^{2/3} = (\text{cis }\alpha)^{2/3}\,[\text{cis }(2n\pi) + 2\alpha]^{1/3} = \text{cis }\dfrac{2n\pi + 2\alpha}{3}, \text{ where } n = 0, 1, 2.$$

Thus the three values are

$$\text{cis}\left(\dfrac{2\alpha}{3}\right), \text{cis}\left(\dfrac{2\pi + 2\alpha}{3}\right), \text{cis}\left(\dfrac{4\pi + 2\alpha}{3}\right)$$

(ii) Defining $(\text{cis }\alpha)^{2/3}$ as $[(\text{cis }\alpha)^{1/3}]^2$, we have

$$(\text{cis }\alpha)^{2/3} = [(\text{cis }\alpha)^{1/3}]^2 = [\{\text{cis }(\alpha + 2n\pi)\}^{1/3}]^2$$

$$= \left[\text{cis}\left(\dfrac{2n\pi + 2\alpha}{3}\right)\right]^2 = \text{cis }\dfrac{2}{3}(2n\pi + 2\alpha), \text{ where } n = 0, 1, 2.$$

Thus the three values of $(\text{cis }\alpha)^{2/3}$ are

$$\text{cis}\left(\dfrac{2\alpha}{3}\right), \text{cis}\left(\dfrac{4\pi + 2\alpha}{3}\right), \text{cis}\left(\dfrac{8\pi + 2\alpha}{3}\right)$$

i.e. $\qquad \text{cis}\left(\dfrac{2\alpha}{3}\right), \text{cis}\left(\dfrac{4\pi + 2\alpha}{3}\right), \text{cis}\left(\dfrac{2\pi + 2\alpha}{3}\right)$

which are the same as already obtained in (i). Hence both definitions give the same set of values.

(iii) If $\dfrac{4}{6}$ be reduced to its lowest terms, the values of $(\text{cis } \alpha)^{4/6}$ are the same as $(\text{cis } \alpha)^{2/3}$.

But if we take $\dfrac{4}{6}$ as it is, then the values of $(\text{cis } \alpha)^{4/6}$ are according to second definition :

$$\text{cis}\left(\frac{4\alpha}{6}\right),\ \text{cis}\left(\frac{8\pi+4\alpha}{6}\right)\text{ and cis}\left(\frac{16\pi+4\alpha}{6}\right)$$

$$\text{cis}\left(\frac{24\pi+4\alpha}{6}\right),\ \text{cis}\left(\frac{32\pi+4\alpha}{6}\right),\ \text{cis}\left(\frac{40\pi+4\alpha}{6}\right)$$

i.e. $\text{cis}\left(\dfrac{2\alpha}{3}\right),\ \text{cis}\left(\dfrac{-\pi-2\alpha}{3}\right),\ \text{cis}\left(\dfrac{2\pi+2\alpha}{3}\right)$

$$\text{cis}\left(\frac{2\alpha}{3}\right),\ \text{cis}\left(\frac{4\pi+2\alpha}{3}\right),\ \text{cis}\left(\frac{2\pi+2\alpha}{3}\right)$$

We notice that the values are same as that of $(\text{cis } \alpha)^{2/3}$, but they are repeated. The reason for the repetition is obvious. This shows that p/q should be taken in their lowest terms, i.e. p and q should be prime to each other.

Example 2. *Find all the values of* $(1 + i)^{2/3}$

Solution. Let $1 = r \cos \theta,\ 1 = r \sin \theta,$

so that $r = \sqrt{2}$ and $\theta = \dfrac{\pi}{4}$.

Now $\begin{aligned}(1 + i)^{2/3} &= (r \cos \theta + i \sin \theta)^{2/3}\\ &= r^{2/3}(\cos \theta + i \sin \theta)^{2/3}\\ &= 2^{1/3}[\cos (2n\pi + \theta) + i \sin (2n\pi + \theta)]^{2/3}.\end{aligned}$

$\therefore$ $2^{1/3}\left[\cos\left(2n\pi + \dfrac{\pi}{4}\right) + i \sin\left(2n\pi + \dfrac{\pi}{4}\right)\right]^{2/3}$

$$= 2^{1/3}\left[\cos\frac{2}{3}\left(2n\pi + \frac{\pi}{4}\right) + i \sin\frac{2}{3}\left(2n\pi + \frac{\pi}{4}\right)\right]$$

Giving to n the values 0, 1, and 2 the required values are

$$2^{1/3}\left[\cos\frac{\pi}{6} + i \sin\frac{\pi}{6}\right],\ 2^{1/3}\left[\cos\frac{3\pi}{2} + i \sin\frac{3\pi}{2}\right]$$

and $2^{1/3}\left[\cos\dfrac{17\pi}{6} + i \sin\dfrac{17\pi}{6}\right],\ 2^{1/3}\dfrac{\sqrt{3}+i}{2},\ -2^{1/3}i$ and $2^{1/3}\dfrac{-\sqrt{3}+i}{2}$

Example 3. *By using De Moivre's theorem, solve*

$$x^4 - x^3 + x^2 - x + 1 = 0$$

Solution. We know that

$$x^5 + 1 = (x + 1)(x^4 - x^3 + x^2 - x + 1)$$

or $\qquad x^4 - x^3 + x^2 - x + 1 = \dfrac{x^5 - 1}{x + 1} = 0$

Now $\qquad x^5 + 1 = 0$

$\qquad x^5 = -1 = \cos \pi + i \sin \pi = \cos(2r\pi + \pi) + i \sin(2r\pi + \pi).$

$\therefore \qquad x = [\cos(2r+1)\pi + i\sin(2r+1)\pi]^{1/5} = \cos\dfrac{(2r+1)\pi}{5} + i\sin\dfrac{(2r+1)\pi}{5}$

Giving r the values 0, 1, 2, 3, 4,

$$x = \left(\cos\frac{\pi}{5} + i\sin\frac{\pi}{5}\right), \left(\cos\frac{3\pi}{5} + i\sin\frac{3\pi}{5}\right), (\cos\pi + i\sin\pi),$$

$$\left(\cos\frac{7\pi}{5} + i\sin\frac{7\pi}{5}\right), \left(\cos\frac{9\pi}{5} + i\sin\frac{9\pi}{5}\right)$$

$$= \left(\cos\frac{\pi}{5} \pm i\sin\frac{\pi}{5}\right), \left(\cos\frac{3\pi}{5} \pm i\sin\frac{3\pi}{5}\right) \text{ and } -1.$$

$\because \qquad x = -1$ does not satisfy the given equation, therefore the roots are

$$\cos\frac{\pi}{5} \pm i\sin\frac{\pi}{5}, \cos\frac{3\pi}{5} \pm i\sin\frac{3\pi}{5}.$$

Example 4. *Find the nth roots of unity and show that they form a series in G.P. Also prove that the sum of pth power always vanishes unless p is a multiple of n, p being on integer, and that then the sum is n.*

Solution. We have $(1)^{1/n} = [\cos 2r\pi + i\sin 2r\pi]^{1/n} = \left[\cos\dfrac{2r\pi}{n} + i\sin\dfrac{2r\pi}{n}\right],$

where $r = 0, 1, 2, \ldots n - 1.$

As we know that these roots can be written in the form

$$1, \omega, \omega^2, \omega^3, \ldots \omega^{n-1},$$

where $\qquad \omega = \cos\dfrac{2\pi}{n} + i\sin\dfrac{2\pi}{n}$

(i) If p is not a multiple of n, the sum of the pth powers of these roots

$$= 1^p + \omega^p + \omega^{2p} + \ldots + \omega^{(n-1)p}.$$

$$= \frac{1 - (\omega^p)^n}{1 - \omega^p} \qquad \{\text{if } \omega^p \neq 1, \text{ i.e. } p \text{ is not a multiple of } n\}$$

$$= \frac{1 - \omega^{pn}}{1 - \omega^p} = \frac{1 - (\omega^n)^p}{1 - \omega^p}$$

$$= 0 \qquad (\text{since } \omega^n = 1, \omega \text{ being } n\text{th roots of unity})$$

(ii) If p is a multiple of n say $p = mn$ where m is an integer, then

$$1^p + \omega^p + \omega^{2p} + \ldots + \omega^{(n-1)p}$$

$$= 1 + (\omega^n)^m + (\omega^n)^{2m} + \ldots + (\omega^n)^{(n-1)m}$$

$$= 1 + 1 + 1 + \ldots \text{ to } n \text{ terms} \qquad [\because \omega^n = 1]$$

$$= n$$

This prove the result.

Example 5. *Solve $x^7 = 1$ by the help of De Moivre's theorem, and find the sum of the cubes of the roots.*

Solution. $$x^7 = 1 = (\cos 2n\pi + i \sin 2n\pi)$$

$$\therefore \qquad x = (\cos 2n\pi + i \sin 2n\pi)^{1/7} = \cos\frac{2n\pi}{7} + i \sin\frac{2n\pi}{7}$$

where $\qquad n = 0, 1, 2, 3, 4, 5, 6.$

Hence roots are

$$1, \text{cis}\frac{2\pi}{7}, \text{cis}\frac{4\pi}{7}, \text{cis}\frac{6\pi}{7}, \text{cis}\frac{8\pi}{7}, \text{cis}\frac{10\pi}{7}, \text{cis}\frac{12\pi}{7}$$

Let $\qquad \text{cis}\dfrac{2\pi}{7} = \alpha.$

Sum of cubes of the roots

$$= 1 + \alpha^3 + (\alpha^2)^3 + (\alpha^3)^3 + (\alpha^4)^3 + (\alpha^5)^3 + (\alpha^6)^3$$
$$= 1 + \alpha^3 + \alpha^6 + \alpha^9 + \alpha^{12} + \alpha^{15} + \alpha^{18}$$
$$= 1 + \alpha^3 + \alpha^6 + \alpha^2 + \alpha^5 + \alpha + \alpha^4 \qquad [\because \alpha^7 = 1, \alpha \text{ being 7th roots of unity}$$
$$= 1 + \alpha + \alpha^2 + \alpha^3 + \alpha^4 + \alpha^5 + \alpha^6 = 0$$

since $\qquad \alpha^7 - 1 = 0, \alpha \neq 1$

or $\qquad (\alpha - 1)(\alpha^6 + \alpha^5 + \alpha^4 + \alpha^3 + \alpha^2 + \alpha + 1) = 0$

or $\qquad \alpha^6 + \alpha^5 + \alpha^4 + \alpha^3 + \alpha^2 + \alpha + 1 = 0$

Example 6. *If ω denotes any imaginary with nth root of unity, then show that*
$$1 + \omega + \omega^2 + \dots + \omega^{n-1} = 0$$
or find the sum of n, nth roots of unity.

Solution. Since ω is imaginary nth roots of unity,
$$\omega^n = 1 \text{ or } 1 - \omega^n = 0$$

or $\qquad (1 - \omega)(1 + \omega + \omega^2 + \dots \omega^{n-1}) = 0$

But $\omega \neq 1$(since ω is imaginary), we have
$$1 + \omega + \omega^2 + \dots + \omega^{n-1} = 0$$

Example 7. *Solve $(2x - 1)^5 = (x - 2)^5$*

Solution. Let $y = \dfrac{2x-1}{x-2}$, then $x = \dfrac{2y-1}{y-2}$ ...(1

The given equation becomes $y^5 = 1$, hence values of y can be obtained by putting

$n = 0, 1, 2, 3, 4$ in $\cos\dfrac{2n\pi}{5} + i \sin\dfrac{2n\pi}{5}.$

Substituting these values of y in (1), we get all the required five values of x.

Example 8. *Find the values of x, such that*

$$\frac{(x+a)^n - (x+\beta)^n}{\alpha - \beta} = \frac{\sin n\theta}{\sin^n \theta}$$

where α *and* β *are the roots of the equation* $t^2 - 2t + 2 = 0$

Solution. Solving $t^2 - 2t + 2 = 0$, we have

$$t = \frac{2 \pm \sqrt{4-8}}{2} = 1 \pm i.$$

Let $\qquad \alpha = 1 + i$ and $\beta = 1 - i$.

$\therefore \qquad x + \alpha = x + 1 + i$ and $x + \beta = x + 1 - i$.

Taking $\qquad x + 1 = r \cos\phi$ and $1 = r \sin\phi$ so that $\cot\phi = x + 1.$ $\qquad$ ...(1)

Hence $\dfrac{(x+\alpha)^n - (x+\beta)^n}{\alpha - \beta} = \dfrac{\sin n\theta}{\sin^n \theta}$, gives

$$\frac{r^n \{(\cos\phi + i\sin\phi)^n - (\cos\phi - i\sin\phi)^n}{(1-i) - (1-i)}$$

or $\qquad \dfrac{r^n \{(\cos n\phi + i\sin n\phi) - (\cos n\phi - i\sin n\phi)}{2i} = \dfrac{\sin n\theta}{\sin^n \theta}$

or $\qquad \left(\dfrac{1}{\sin\phi}\right)^n \cdot \dfrac{2i\sin n\phi}{2i} = \dfrac{\sin n\theta}{\sin^n \theta}$ or $\dfrac{\sin n\phi}{\sin^n \phi} = \dfrac{\sin n\theta}{\sin^n \theta}$

Thus $\qquad \phi = \theta$

$\therefore \qquad \cot\theta = x + 1$ $\qquad\qquad$ [by (1)]

or $\qquad x = \cot\theta - 1$

Example 9. *If* ω *is an imaginary cube root of unity, prove that*

$$\frac{1}{1+2\omega} - \frac{1}{1+\omega} + \frac{1}{2+\omega} = 0$$

Solution. We have $\dfrac{1}{1+2\omega} - \dfrac{1}{1+\omega} + \dfrac{1}{2+\omega} = \dfrac{1}{-\omega^2 + \omega} - \dfrac{1}{1+\omega} + \dfrac{1}{1-\omega^2}$

$$= \frac{1 + \omega - \omega + \omega^2 + \omega}{\omega(1 - \omega)(1+\omega)} = 0$$

Example 10. *If* α *and* β *be the imaginary cube roots of unity, prove that*

$$\alpha e^{\alpha x} + \beta e^{\beta x} = -e^{-x/2}\left[\sqrt{3}\sin\frac{\sqrt{(3)}}{2}x + \cos\frac{\sqrt{3}}{2}x\right]$$

Solution. We have

$$\alpha = \frac{-1 + \sqrt{3}i}{2}, \quad \beta = \frac{-1 - \sqrt{3}i}{2}$$

Now $\qquad \alpha e^{\alpha x} + \beta e^{\beta x} = \dfrac{-1+\sqrt{3}i}{2}.e^{\left(\frac{-1+\sqrt{3}i}{2}\right)x} + \left(\dfrac{-1-\sqrt{3}i}{2}\right)e^{\left(\frac{-1-\sqrt{3}i}{2}\right)x}$

$$= e^{-x/2}\left[\left(\frac{-1+\sqrt{3}i}{2}\right)\left(\cos\frac{\sqrt{3}}{2}x+i\sin\frac{\sqrt{3}}{2}x\right)\right]$$

$$+\left(\frac{-1+\sqrt{3}i}{2}\right)\left[\left(\cos\frac{\sqrt{3}}{2}x+i\sin\frac{\sqrt{3}}{2}x\right)\right]$$

$$= -e^{-x/2}\left[\sqrt{3}\sin\frac{\sqrt{(3)}}{2}x+\cos\frac{\sqrt{3}}{2}x\right]$$

EXERCISE 8.3

1. Find all the values of

   (a) $(-1)^{1/3}$ (b) $(1-i\sqrt{3})^{1/5}$ (c) $(8i)^{1/3}$

   (d) $(\sqrt{3}+i)^{1/3}$ (e) $\left(\cos\frac{\pi}{3}+i\sin\frac{\pi}{3}\right)^{1/4}$

   (f) $\left(\cos\frac{\pi}{3}+i\sin\frac{2\pi}{3}\right)^{1/4}$, in a form free from trigonometry expression.

   (g) $(i)^{1/4}$ (h) $(1+i)^{1/3}$ (i) $(1+i)^{1/5}$ (j) $(-1)^{1/6}$

2. Find the continued product of $\left(\cos\frac{\pi}{3}+i\sin\frac{\pi}{3}\right)^{3/4}$

3. Solve the equation $x^{12}-1=0$ and find which of its roots satisfy the equation, $x^4+x^2+1=0$

4. Solve the equations

   (i) $x^7+x^4+x^3+1=0$ (ii) $x^7+1=0$

5. Find the general value of θ which satisfies

   $$(\cos\theta+i\sin\theta)(\cos 2\theta+i\sin 2\theta)\dots(\cos n\theta+i\sin n\theta)=1.$$

6. Prove that the roots of the equation $x^{10}+11x^5-1=0$ are

   $$\frac{\pm\sqrt{5}-1}{2}\left[\cos\frac{2r\pi}{5}+i\sin\frac{2r\pi}{5}\right]$$

7. Prove that n, nth roots of unity form a series in G.P.

8. (a) Find the seven, 7th roots of unity and prove that the sum of their nth powers always vanishes unless n be a multiple of 7, n being an integer, and that then the sum is 7.

   (b) Find the five, 5th roots of unity and prove that the sum of their nth powers always vanishes unless n be a multiple of 5, n being an integer, and that then the sum is 5.

9. Prove that $(a + ib)^{1/n} + (a - ib)^{1/n}$ has n real values and find those of

$$(1 + \sqrt{3}i)^{1/3} + (1 - \sqrt{3}i)^{1/3}$$

10. Prove by the use of De-Moivre's theorem that the roots of the equation $(x - 1)^n = x^n$ (n being a positive integer) are

$$\frac{1}{2}\left\{1 + i \cot\frac{r\pi}{n}\right\} \quad \text{where } n = 0, 1, 2, \dots n - 1.$$

11. Express $P = \dfrac{(\sqrt{3} - 1) + i(\sqrt{3} + 1)}{2\sqrt{2}}$ in the form $r(\cos\theta + i\sin\theta)$ and derive all the six values of $p^{1/6}$.

12. Apply De-Moivre's theorem to solve the following equation

$$x^6 + x^5 + x^4 + x^3 + x^2 + x + 1 = 0$$

13. Find the cube roots of unity.

14. Solve $x^9 - x^5 + x^4 - 1 = 0$ by the De-Moivre's theorem.

15. Find the cube roots of $(1 - \cos\phi - i\sin\phi)$ where ϕ is real.

HINT TO SELECTED PROBLEMS

3. We have $x^{12} - 1 = 0 \implies (x^6 + 1)(x^6 - 1) = 0$

   i.e. $x^6 = -1; \; x^6 = 1$

   Solving to $x^6 = -1$ we have

   $$x = (-1)^{1/6} = [\cos\pi + i\sin\pi]^{1/6} = [\cos(2n\pi + \pi) + i\sin(2n\pi + \pi)]^{1/6}$$

   $$= \cos\frac{(2n + 1)\pi}{6} + i\sin\frac{(2n + 1)\pi}{6}$$

   Now putting $n = 0, 1, 2, 3, 4, 5$ and get the six roots of $x^6 = -1$.

   Similarly we can solve $x^6 = 1$

4. (i) $x^7 + x^4 + x^3 + 1 = 0 \implies (x^4 + 1)(x^3 + 1) = 0$

   i.e. $x = (-1)^{1/4}, \; x = (-1)^{1/3}$. Now proceed same as above.

6. $\dfrac{-11 + 5\sqrt{5}}{2} = \dfrac{-176 \pm 80\sqrt{5}}{32} = \left[\dfrac{\pm\sqrt{5} - 1}{2}\right]^5$

10. $(x - 1)^n = x^n \implies \left(\dfrac{x - 1}{x}\right)^n = 1 \implies \dfrac{x - 1}{x} = 1^{1/n} = [\cos 2r\pi + i\sin 2r\pi]^{1/n}$

    $$= \cos\frac{2r\pi}{n} + i\sin\frac{2r\pi}{n}$$

$$\Rightarrow \quad \cos\frac{2r\pi}{n}+i\sin\frac{2r\pi}{n}$$

$$\Rightarrow \quad 1-\frac{1}{x}=\cos\frac{2r\pi}{n}+i\sin\frac{2r\pi}{n}$$

$$\Rightarrow \quad \frac{1}{x}=1-\cos\frac{2r\pi}{n}-i\sin\frac{2r\pi}{n}$$

14. The given equation can be written as

$$(x^5+1)(x^4-1)=0 \quad \Rightarrow \quad x^5+1=0,\ x^4-1=0.$$

$$\Rightarrow x=(-1)^{1/5},\ x=(1)^{1/4}.$$

ANSWERS

(1) (a) $\dfrac{1+\sqrt{(-3)}}{2},\ -1$ and $\dfrac{1-\sqrt{(-3)}}{2}$,

(b) $2^{1/5}\left[\cos\dfrac{r\pi}{15}-i\sin\dfrac{r\pi}{15}\right]$ where $r=1,7,13$

(c) $-2i$ and $\pm\sqrt{3}+i$,

(d) $2^{1/3}\left[\cos\dfrac{r\pi}{18}+i\sin\dfrac{r\pi}{18}\right]$ where $r=1,13,15$

(e) $\left[\left(\cos\dfrac{\pi}{12}+i\sin\dfrac{\pi}{12}\right),\cos\left(\dfrac{7\pi}{12}\right)+i\sin\left(\dfrac{7\pi}{12}\right),\cos\left(\dfrac{13\pi}{12}\right)+i\sin\dfrac{13\pi}{12}\right.$

$$\left.\cos\left(\dfrac{19\pi}{12}\right)+i\sin\left(\dfrac{19}{12}\right)\pi\right]$$

(f) $\dfrac{\pm(1-i\sqrt{3})}{2},\ \pm(\sqrt{3}+i)/2$, **(g)** $\operatorname{cis}\dfrac{\pi}{8},\operatorname{cis}\dfrac{5\pi}{8},\operatorname{cis}\dfrac{9\pi}{8},\operatorname{cis}\dfrac{13\pi}{8}$,

(h) $2^{1/6}\operatorname{cis}\left(\dfrac{\pi}{12}\right),2^{1/6}\operatorname{cis}\left(\dfrac{9\pi}{12}\right),2^{1/6}\operatorname{cis}\left(\dfrac{17\pi}{12}\right);$

or $2^{1/6}\left[\cos\left(\dfrac{r\pi}{12}\right)+i\sin\left(\dfrac{r\pi}{12}\right)\right]$ where $r=1,9$ and 17

(i) $2^{1/10}[\cos\pi/20+i\sin\pi/20],\ 2^{1/10}[\cos 9\pi/20+i\sin 9\pi/20]$

$2^{1/10}[\cos 17\pi/20+i\sin 17\pi/20],\ 2^{1/10}[\cos 5\pi/4+i\sin 5\pi/4],$

$2^{1/10}\cos\dfrac{33\pi}{20}+i\sin\dfrac{33\pi}{20}$

(j) $\cos\dfrac{\pi}{6}(2n\pi)+i\sin\dfrac{\pi}{6}(2n+1)$, where $r = 0, 1, 2, 3, 4,$ and $5.$

(2) 1

(3) $x = \pm\,1,\,\pm\,i,\,\pm\left(\cos\dfrac{\pi}{6}\pm i\sin\dfrac{\pi}{6}\right),\,\pm\left(\cos\dfrac{\pi}{3}\pm i\sin\dfrac{\pi}{3}\right)$ are the required roots.

(4) (i) $x = \dfrac{1\pm i}{2},\dfrac{1\pm i}{\sqrt{2}},\dfrac{1\pm i\sqrt{3}}{2},-1$

   (ii) -1 and $\left(\cos\dfrac{r\pi}{7}\pm i\sin\dfrac{r\pi}{7}\right)$, where $r = 1, 3$ or $5.$

(5) $\theta = \dfrac{4m\pi}{n(n+1)}$

(9) $2^{4/3}\cos\dfrac{m\pi}{9}$, $m = 1, 7, 13.$

(11) $P = \cos\dfrac{5\pi}{72}+i\sin\dfrac{5\pi}{72},\cos\dfrac{24r+5\pi}{72}+i\sin\dfrac{24r+5\pi}{72},r = 0,1,2,3,4,5$

(12) $1,\,\text{cis}\dfrac{2\pi}{7},\,\text{cis}\dfrac{4\pi}{7},\,\text{cis}\dfrac{6\pi}{7},\,\text{cis}\dfrac{8\pi}{7},\,\text{cis}\dfrac{10\pi}{7},\,\text{cis}\dfrac{12\pi}{7}$

(13) $1,\,\dfrac{-1+i\sqrt{3}}{2},\dfrac{-1-i\sqrt{3}}{2}$

(14) $1,\pm\,1,\pm\,i,\cos\pi/5\pm i\sin\pi/5,\cos\dfrac{3\pi}{5}\pm i\sin\dfrac{3\pi}{5}.$

(15) $(2\sin\phi/2)^{1/3}\left[\cos\left(\dfrac{4n\pi+\pi-\phi}{6}\right)-i\sin\left(\dfrac{4n\pi+\pi-\phi}{6}\right)\right]$, where $n = 0, 1, 2.$

CHAPTER REVIEW : A COMPETITIVE APPROACH

Selected terms and Results

TERMS

- **Complex numbers :** Let x and y be two real numbers. Then the set of ordered pairs (x, y) is called the system of complex numbers.

  i.e. a complex number z can be expressed as

  $z = x + iy \ : \ x, y \in R$

- **Modulus of a complex number :** Let $z = x + iy$ be the given complex number then modulus of z, denoted by $|z|$ is defined by.

  $$|z| = +\sqrt{x^2 + y^2}$$

- **Argument or Amplitude :** Let $z = x + iy$ be the given complex number. Then the value of $\theta = \tan^{-1}\dfrac{y}{x}$ is called the argument or amplitude of z.

- **Conjugate complex number :** Let $z = x + iy$ be the given complex number then the complex number $x - iy$ is called the conjugate of z.

RESULTS

- The form of $r[\cos\theta + i\sin\theta]$ is called modulus amplitude form or polar form or trigonometrical form.
- $|z_1.z_2| = |z_1|.\,|z_2|$
- $\arg(z_1.z_2) = \arg(z_1) + \arg(z_2)$
- Modulus of a complex number is always unique.
- Argument of a complex number is not unique.
- The modulus of the sum of two complex numbers can never exceed the sum of their moduli.

- The modulus of the difference of two complex numbers can never be less than the difference of their moduli.
- The modulus of the sum of two complex numbers is always greater than or equal to the difference of their moduli.
- **De-Moivre's theorem :** If n is any integer then

  $(\cos\theta + i\sin\theta)^n = \cos n\theta + i\sin n\theta$

  and if n is a fraction, positive or negative then $\cos n\theta + i\sin n\theta$ is one of the value of $(\cos\theta + i\sin\theta)^n$.

Review Questions and Project Work

1. Show that every complex number can be represented by a point in the argand plane.

2. Show that a complex number can be represented by a vector.

3. Show that $|z_1 + z_2|^2 + |z_1 - z_2|^2 = 2|z_1|^2 + 2|z_2|^2$

4. If $|z_1| = |z_2|$ and amp z_1 + amp $z_2 = 0$, show that z_1 and z_2 are conjugate numbers.

5. If a, b are complex numbers, find numbers $z_1 z_2$ so that the points $z_1 z_2$ and a, b be opposite corner of a square.

6. Show that $\left|\dfrac{z-1}{z+1}\right| = $ constant and amplitude $\left(\dfrac{z-1}{z+1}\right) = $ constant are orthogonal circles.

7. Show that $\arg(z) + \arg(\bar{z}) = 2n\pi$, $n \in Z$.

8. Show that the origin and the points representing the roots of the equation

$z^2 + p^2 + q = 0$ form an equilateral triangle if $p^2 = 3q$.

Objective Type Questions

Fill in the blanks:

1. A plane whose points are represented by complex number is known as .... plane.

2. Argand's plane also known as .... plane.

3. The modulus of the product of complex number is equal to the .... of moduli of these complex numbers.

4. Argument of the product of complex numbers is equal to the .... of their arguments.

5. Arg. (z_1/z_2) = ...

6. If z be the complex number and $\bar{z}$ be the complex conjugate of z, then $z + \bar{z}$ is always ....

7. The modulus of the sum of two complex numbers can never exceed the .... of their moduli.

8. The modulus of the sum of two complex numbers is always .... than or equal to the difference of their moduli.

9. If Amp. $\left(\dfrac{z-i}{z+i}\right) = \dfrac{p}{4}$, then the locus of z on the argand plane is a ....

10. The principal arguments of $\dfrac{1+i}{1-i}$ is ....

11. $(\cos \theta + i \sin \theta)^n$ = ...

12. If $p = \cos \theta + i \sin \theta$ and $q = \cos \phi + i \sin \phi$, then $\dfrac{p-q}{p+q}$ =

13. The n, nth roots of the unity form a .... series.

14. If $p = \cos \theta + i \sin \theta$, then real part of (p^n) = ...

15. The value of $(\cos \theta + i \sin \theta)^{-n}$ = ...

16. If α, β are the roots of the equation $x^2 - 2x + 4 = 0$ then the value of $\alpha^2 + \beta^2$ is ....

17. If $z = -\dfrac{1}{2} + \dfrac{i\sqrt{3}}{2}$, then $z^n + z^{2n} = 2$. If n equals ....

18. The one value of $(i)^{1/4}$ is ....

True/False: *Write 'T' for true and 'F' for false statement.*

1. Euler was the first, who used the symbol i. (T/F)

2. The value of $i^2 = -1$. (T/F)

3. The Argand plane is also known as complex plane or Gaussian plane. (T/F)

4. Modulus of the product of complex numbers is equal to the product of the moduli of these complex numbers. (T/F)

5. Arguments of the product of complex numbers is equal to the product of their arguments of these complex numbers. (T/F)

6. The modulus of the sum of the two complex numbers exceeds the sum of their moduli. (T/F)

7. The modulus of the difference of two complex numbers can never be less than the difference of their moduli. (T/F)

8. If z_1 and z_2 are two complex numbers, then value of $\left|\dfrac{z_1}{z_2}\right| = |z_1 - z_2|$. (T/F)

9. If the sum of product of the complex numbers is real then two complex numbers are conjugate of each other.
   (T/F)

10. If z_1 and z_2 are two complex numbers then real of $(z_1 + z_2)$ is equal to real of (z_1/z_2).
    (T/F)

11. The De-Moivre's theorem is not true for $n \in Z^-$.
    (T/F)

12. $(\sin \theta + i \cos \theta)^n = \sin n\theta + i \cos n\theta$.
    (T/F)

13. $(\cos \theta + i \sin \phi)^n = \cos n\theta + i \sin n\phi$.
    (T/F)

14. The n, nth roots of unity form a harmonic series.
    (T/F)

15. If $z = -\dfrac{1}{2} + \dfrac{i\sqrt{3}}{2}$, then the value of z^3 is 1.
    (T/F)

16. If p is not prime to q, let $\dfrac{p}{q}$ be their ratio in the lowest term, then $(\cos \theta + i \sin \theta)^{p/q}$ will give only q values and no more.
    (T/F)

17. The q values obtained in 16 are $(\cos \theta + i \sin \theta)^{p/q}$.
    (T/F)

18. If $x + \dfrac{1}{x} = 2 \cos \theta$ then $x^n + \dfrac{1}{x^n} = 2 \cos n\theta$.
    (T/F)

19. If $z_n = \cos \dfrac{p}{2^n} + i \sin \dfrac{p}{2^n}$ then $\lim\limits_{x \to \infty} (z_1 . z_2 \ldots z_n) = 1$.
    (T/F)

20. $(\cos \theta + i \sin \theta)^{p/q}$, where p and q are integers, prime to each other $(q > 0)$ has q distinct values.
    (T/F)

Multiple Choice Questions : *Choose the most appropriate one :*

1. If $z = x + iy$ be a complex number, its polar form is obtained by putting :

(a) $x = r \sin \theta, y = r \cos \theta$

(b) $x = \dfrac{\sin ?}{r}, y = \dfrac{\cos ?}{r}$

(c) $x = \dfrac{r}{\sin ?}, y = \dfrac{r}{\cos ?}$

(d) $x = r \cos \theta, y = r \sin \theta$

(e) none of these

2. If $z = x + iy$ then $|z|$ is given by :

(a) $\sqrt{x^2 - y^2}$

(b) $\sqrt{y^2 - x^2}$

(c) $\sqrt{x^2 + y^2}$

(d) $\sqrt{x + y}$

(e) none of these

3. If $z = x + iy$ then arg (z) is given by :
   (a) $\theta = \tan^{-1}(y/x)$
   (b) $\theta = \tan^{-1}(x/y)$
   (c) $\tan^{-1}(x - y)$
   (d) $\tan^{-1}(y - x)$
   (e) none of these

4. Polar form of the complex number $(-1 + i\sqrt{3})$ is :
   (a) $2(\cos \pi/3 + i \sin \pi/3)$

   (b) $\sqrt{2}\left(\cos \dfrac{2p}{3} + i \sin \dfrac{2p}{3}\right)$

   (c) $2\left(\cos \dfrac{2p}{3} + i \sin \dfrac{2p}{3}\right)$

   (d) $\sqrt{2}\left(\cos \dfrac{2p}{3} - i \sin \dfrac{2p}{3}\right)$

   (e) none of these

5. Modulus of $\left|\dfrac{1+i}{1-i}\right|$ is :

(a) -1 (b) 0

(c) 1 (d) $\dfrac{1}{2}$

(e) none of these

6. Principal argument of $\dfrac{1+i}{1-i}$ is :

   (a) $\pi/3$ (b) $\pi/4$

   (c) $\pi/5$ (d) $\pi/2$

   (e) none of these

7. If z_1 and z_2 are any two complex numbers, then value of $\left|\dfrac{z_1}{z_2}\right|$ is :

   (a) $||z_1||.|z_1|$ (b) $\dfrac{|z_2|}{|z_1|}$

   (c) $\dfrac{|z_1|}{|z_2|}$ (d) $|z_1| + |z_1|$

   (e) none of these

8. If z_1 and z_2 are any two complex numbers then $\arg\left(\dfrac{z_1}{z_2}\right)$ is :

   (a) $\dfrac{\arg(z_1)}{\arg(z_2)}$

   (b) $\arg(z_1) - \arg(z_2)$

   (c) $\arg(z_1) + \arg(z_2)$

   (d) $\arg(z_2) - \arg(z_1)$

   (e) none of these

9. Radius of the circle $\left|\dfrac{z-i}{z+i}\right| = 5$ is :

   (a) $\dfrac{5}{12}$ (b) $\dfrac{12}{5}$

   (c) $\dfrac{5}{11}$ (d) $\dfrac{11}{5}$

   (e) none of these

10. Centre of the circle $\left|\dfrac{z-i}{z+i}\right| = 5$ is :

   (a) $\left(\dfrac{13}{12}, 0\right)$ (b) $\left(-\dfrac{13}{12}, 0\right)$

   (c) $\left(0, \dfrac{13}{12}\right)$ (d) $\left(0, -\dfrac{13}{12}\right)$

   (e) none of these

11. If z_1 and z_2 are any two complex numbers then $\operatorname{Re}(z_1 + z_2)$ is :

   (a) $\operatorname{Re}(z_1) - \operatorname{Re}(z_2)$

   (b) $\operatorname{Re}(z_1) + \operatorname{Re}(z_2)$

   (c) $\operatorname{Re}(z_2) - \operatorname{Re}(z_1)$

   (d) $\operatorname{Re}\left(\dfrac{z_1}{z_2}\right)$

   (e) none of these

12. If z_1 and z_2 are any two complex numbers then $\operatorname{Im}(z_1 + z_2)$ is :

   (a) $\operatorname{Im}(z_1) + \operatorname{Im}(z_2)$

   (b) $\operatorname{Im}(z_1) - \operatorname{Im}(z_2)$

   (c) $-\operatorname{Im}(z_2) - \operatorname{Im}(z_1)$

   (d) $\operatorname{Im}\left(\dfrac{z_1}{z_2}\right)$

   (e) none of these

13. If z_1 and z_2 are any two complex numbers then $\operatorname{Re}(z_1 - z_2)$ is :

   (a) $\operatorname{Re}(z_1) - \operatorname{Re}(z_2)$

   (b) $\operatorname{Re}(z_1) + \operatorname{Re}(z_2)$

   (c) $\operatorname{Re}(z_1) + \operatorname{Re}(-z_2)$

   (d) none of these

14. If z_1 and z_2 are any two complex numbers then $I(z_1 - z_2)$ is :

   (a) $I(z_1) + I(z_2)$

   (b) $I(z_1) - I(z_2)$

   (c) $I(z_2) - I(z_1)$

   (d) none of these

15. If the sum and product of the complex numbers is real then the two complex numbers have the relation :

   (a) no relation

   (b) they are conjugate of each other

   (c) nothing can be said

   (d) none of these

16. If z_1 and z_2 are any two complex numbers, then :

   (a) $|z_1 + z_2| = |z_1| + |z_2|$

   (b) $|z_1 + z_2| = |z_1| - |z_2|$

   (c) $|z_1 + z_2| \geq |z_1| + |z_2|$

   (d) $|z_1 - z_2| \geq |z_1| - |z_2|$

   (e) none of these

17. If z_1 and z_2 are any two complex numbers, then :

(a) $|z_1+z_2|^2+|z_1-z_2|^2=2[|z_1|-|z_2|]$

(b) $|z_1+z_2|^2+|z_1-z_2|^2=2[|z_1|^2-|z_2|^2]$

(c) $|z_1+z_2|^2+|z_1-z_2|^2=2[|z_2|^2-|z_1|^2]$

(d) nothing can be said

(e) none of these

18. Value of $(\sin\theta+i\sin\theta)^n$ is :

(a) $\sin n\theta+i\sin n\theta$

(b) $\cos\theta+i\sin n\theta$

(c) $\cos n\theta-i\sin n\theta$

(d) $\sin n\theta-i\cos n\theta$

(e). none of these

19. Value of $\left(\dfrac{\cos?+i\sin?}{\cos?-i\sin?}\right)^4$ is :

(a) $\cos 8\theta+i\sin 8\theta$

(b) $\sin 8\theta+i\cos 8\theta$

(c) $\sin 8\theta-i\cos 8\theta$

(d) $\cos 8\theta-i\sin 8\theta$

(e) none of these

20. If $a=\cos 2\alpha+i\sin 2\alpha$ with similar expressions for b,c and d, then value

of $\sqrt{abc}+\dfrac{1}{\sqrt{abc}}$ is :

(a) $2\cos(\alpha+\beta-\gamma)$

(b) $2i\sin(\alpha+\beta+\gamma)$

(c) $2\cos(\alpha+\beta+\gamma)$

(d) nothing can be said

(e) none of these

21. For the same values of a,b,c,d as in

Q. 20, the value $\sqrt{\left(\dfrac{ab}{cd}\right)}$ is :

(a) $\cos(\alpha+\beta+\gamma+\delta)$

(b) $\sin(\alpha+\beta+\gamma+\delta)$

(c) $\cos(\alpha+\beta-\gamma+\delta)$

(d) $\cos(\alpha+\beta-\gamma-\delta)$

(e) none of these

22. If $x+\dfrac{1}{x}=2\cos\theta$ and $y+\dfrac{1}{y}=2\cos\phi$

then the value of xy is :

(a) $\cos(\theta+\phi)+i\sin(\theta+\phi)$

(b) $\cos(\theta+\phi)-i\sin(\theta-\phi)$

(c) $\cos(\theta-\phi)+i\sin(\theta-\phi)$

(d) $\sin(\theta+\phi)+i\cos(\theta+\phi)$

(e) none of these

23. For the values of above question the

value of $xy+\dfrac{1}{xy}$ is :

(a) $2\cos(\theta+\phi)$

(b) $2\sin(\theta+\phi)$

(c) $2\cos(\theta-\phi)$

(d) $2\sin(\theta-\phi)$

(e) none of these

24. For the values of Q. 22 the value of

$xy-\dfrac{1}{xy}$ is :

(a) $2\cos(\theta-\phi)$

(b) $2i\sin(\theta-\phi)$

(c) $x\cos(\theta+\phi)$

(d) $2i\sin(\theta+\phi)$

25. Value of $(1+i)^n-(1-i)^n$ is :

(a) $2^{(n/2)+1}\,i\cos\dfrac{np}{4}$

(b) $2^{(n/2)+1}\,i\sin\dfrac{np}{4}$

(c) $2^{(n/2)+1}\,\cos\dfrac{np}{4}$

(d) $2^{(n/2)+1}\,\sin\dfrac{np}{4}$

(e) none of these

26. Value of $(\cos\alpha+i\sin\alpha)^{-n}$ is:

(a) $\cos n\alpha+i\sin n\alpha$

(b) $\sin n\alpha+i\cos n\alpha$

(c) $\cos n\alpha-i\sin n\alpha$

(d) $\sin n\alpha-i\cos n\alpha$

(e) none of these

27. If $\left(1+i\dfrac{x}{a}\right)\left(1+i\dfrac{x}{b}\right)\left(1+i\dfrac{x}{c}\right) = A + iB,$

then value of $A^2 + B^2$, is :

(a) $\left(1+\dfrac{x}{a}\right)\left(1+\dfrac{x}{b}\right)\left(1+\dfrac{x}{c}\right)$

(b) $\left(1-\dfrac{x}{a}\right)\left(1-\dfrac{x}{b}\right)\left(1-\dfrac{x}{c}\right)$

(c) $\left(1+\dfrac{x^2}{a^2}\right)\left(1+\dfrac{x^2}{b^2}\right)$

(d) $\left(1+\dfrac{x^2}{a^2}\right)\left(1+\dfrac{x^2}{b^2}\right)\left(1+\dfrac{x^2}{c^2}\right)$

(e) none of these

28. If $\left(1+i\dfrac{x}{a}\right)\left(1+i\dfrac{x}{b}\right)\left(1+i\dfrac{x}{c}\right) = A + iB,$

then value of $\tan^{-1}\left(\dfrac{B}{A}\right)$ is :

(a) $\tan^{-1}\left(\dfrac{a}{x}\right) + \tan^{-1}\left(\dfrac{b}{x}\right) + \tan^{-1}\left(\dfrac{c}{x}\right)$

(b) $\tan^{-1}\left(\dfrac{x}{a}\right) - \tan^{-1}\left(\dfrac{x}{b}\right) - \tan^{-1}\left(\dfrac{x}{c}\right)$

(c) $\tan^{-1}\left(\dfrac{x}{a}\right) + \tan^{-1}\left(\dfrac{x}{b}\right) + \tan^{-1}\left(\dfrac{x}{c}\right)$

(d) none of these

29. If $\sin\alpha + \sin\beta + \sin\gamma = \cos\alpha + \cos\beta + \cos\gamma = 0$, then the value of $\cos 3\alpha + \cos 3\beta + \cos 3\gamma$ is :
(a) $\cos(3\alpha + 3\beta + 3\gamma)$
(b) $\sin(3\alpha + 3\beta + 3\gamma)$
(c) $3\cos(\alpha + \beta + \gamma)$
(d) $3\sin(\alpha + \beta + \gamma)$
(e) none of these

30. If $\sin\alpha + \sin\beta + \sin\gamma = \cos\alpha + \cos\gamma = 0$, then the value of $\sin 3\alpha + \sin 3\beta + \sin 3\gamma$ is :
(a) $\cos(3\alpha + 3\beta + 3\gamma)$
(b) $\sin(3\alpha + 3\beta + 3\gamma)$
(c) $3\sin(\alpha + \beta + \gamma)$
(d) $3\cos(\alpha + \beta + \gamma)$
(e) none of these

31. If α, β are the roots of the equation $x^2 - 2x\cos\theta + 1 = 0$, then the value of $\alpha^n + \beta^n$ is :

(a) $2^{n+1}\cos\dfrac{p}{3}$

(b) $2^n\cos\dfrac{np}{3}$

(c) $2\cos\dfrac{2np}{3}$

(d) $2^{n+1}\cos\dfrac{np}{3}$

(e) none of these

32. If α and β are the roots of the equation $x^2 - 2x + 4 = 0$, then the value of $\alpha^3 + \beta^3$ is :
(a) 16   (b) -16
(c) 14   (d) -14
(e) none of these

33. If $z = -\dfrac{1}{2} + i\dfrac{\sqrt{3}}{2}$, then the value of z^3 is :
(a) -1 (b) 0
(c) 1   (d) 2
(e) none of these

34. If α, β are the roots of the equation $x^2 - 2x\cos\theta + 1 = 0$, then the equation whose roots are α^n, β^n is :
(a) $x^2 - 2\cos n\theta \cdot x + 1 = 0$
(b) $x^2 + 2\cos n\theta \cdot x + 1 = 0$
(c) $x^2 - 2\sin n\theta \cdot x + 1 = 0$
(d) $x^2 + 2\sin n\theta \cdot x + 1$
(e) none of these

ANSWERS

Fill in the blanks

(1) Argand	(2) Gaussian	(3) Product	(4) Sum
(5) $\text{Arg } z_1 - \text{Arg } z_2$	(6) Real	(7) Sum	(8) Greater

(9) Circle   (10) $\pi/2$ (11) $\cos n\theta + i \sin n\theta$ (12) $i \tan \dfrac{\theta - \phi}{2}$

(13) Geometric   (14) $\cos n\theta$ (15) $\cos n\theta - i \sin n\theta$.

(16) -16 (17) $n = 3m, m \in N$ (18) $\text{cis } \dfrac{\pi}{8}$

True/False

(1) T	(2) F	(3) T	(4) T	(5) F
(6) F	(7) T	(8) F	(9) T	(10) T
(11) F	(12) F	(13) F	(14) F	(15) T
(16) T	(17) T	(18) T	(19) F	(20) T

Multiple Choice Questions

(1) d	(2) c	(3) a	(4) c	(5) c
(6) d	(7) c	(8) b	(9) a	(10) d
(11) b	(12) a	(13) a	(14) b	(15) b
(16) d	(17) b	(18) e	(19) a	(20) c
(21) d	(22) a	(23) a	(24) d	(25) b
(26) c	(27) d	(28) d	(29) c	(30) c
(31) b	(32) b	(33) c	(34) a	

CHAPTER 9

Exponential and Trigonometric Functions of a Complex Variable

❖ Exponential Series

❖ Circular Function

❖ Hyperbolic Function

❖ Exponential Function

❖ Euler's Exponential Value

❖ Period of Hyperbolic Function

9.1 CONVERGENCE

Let the series of complex numbers $z_1 + z_2 + \ldots z_n + \ldots$ where $z_n = x_n + iy_n$, x_n, y_n being real numbers, be denoted by $\sum_{1}^{\infty} z_n$ or simple by Σz_n.

The series Σz_n is said to be convergent, if the series Σx_n and Σy_n are both convergent.

If $\quad \lim_{n \to \infty} (x_1 + x_2 + \ldots + x_n) = R \quad$ and $\quad \lim_{n \to \infty} (y_1 + y_2 + \ldots + y_n) = Q$

then $P + iQ$ is called the sum upto infinity of the series Σz_n.

9.1.1 Absolute Convergence. If series, $|z_1| + |z_2| + \ldots + |z_n| + \ldots$

of positive real numbers, where $|z_n| = \sqrt{x_n^2 + y_n^2}$, be convergent, we say that the series Σz_n of complex numbers $z_1 + z_2 + \ldots z_n + \ldots$ where $z_n \equiv x_n + iy$, is absolutely convergent.

Remark

❖ If a series is absolutely convergent, it is necessarily convergent.

9.2 EXPONENTIAL SERIES OF COMPLEX NUMBERS

We know that the exponential series for all real values of x is given by

$$e^x = 1 + x + \frac{x^2}{2!} + \frac{x^3}{3!} + \ldots \text{ad.inf} \qquad \ldots(1)$$

But where x is complex, the expression e^x has no meaning at present. The series (1) is absolutely convergent for all finite values of x.

Now consider the series

$$E(z) = 1 + z + \frac{z^2}{2!} + \frac{z^3}{3!} + \dots \text{ ad.inf} \qquad \dots(2)$$

where $z = x + iy = r(\cos\theta + i\sin\theta)$ and therefore $|z| = r > 0$.

Let the series of the moduli be

$$1 + |z| + \frac{|z^2|}{2!} + \frac{|z^3|}{3!} + \dots + \frac{|z^n|}{n!} + \dots = 1 + r + \frac{r^2}{2!} + \frac{r^3}{3!} + \dots + \frac{r^n}{n!} + \dots$$

This is a series of positive numbers and convergent and hence the series (2) is absolutely convergent for all finite values of z.

In particular, if $z = x + i0$ which corresponding to the real number x, $E(z)$ assumes the value

$$1 + \frac{x}{1!} + \frac{x^2}{2!} + \dots + \frac{x^n}{n!} + \dots$$

which corresponds to exp (x) or e^x,

where $\qquad e = \left(1 + \frac{1}{1!} + \frac{1}{2!} + \frac{1}{3!} + \dots + \frac{1}{n!} + \dots\right)$

and e^x means $\left(1 + \frac{1}{1!} + \frac{1}{2!} + \frac{1}{3!} + \dots + \frac{1}{n!} + \dots\right)^x$, for all real values of x.

Hence the series (2) is usually written as exp (z) or e^z in close analogy to the exponential series for real number, and also because the fact that

$$E(z) = 1 + z + \frac{z^2}{2!} + \dots + \frac{z^n}{n!} + \dots$$

where $z \equiv x + iy$ when $z \equiv x + i0$, corresponds to exp. (x) or e^x.

It should be clearly understood that the series

$$1 + z + \frac{z^2}{2!} + \dots + \frac{z^n}{n!} + \dots$$

is written as exp. (z) or e^z by definition only, and that it does not mean, unless it is so proved that the exp (z) or e^z stands for

$$\left(1 + \frac{1}{1!} + \frac{1}{2!} + \dots + \frac{1}{n!} + \dots\right)^z, \text{ If } z \text{ is complex.}$$

Hence by definition, if $z = x + iy$,

$$\exp(z) = 1 + z + \frac{z^2}{2!} + \dots + \frac{z^n}{n!} + \dots$$

9.3 PROPERTIES OF EXPONENTIAL FUNCTION OF COMPLEX NUMBERS

THEOREM 1

If z_1 and z_2 are any two complex numbers, then

$$e^{z_1} \cdot e^{z_2} = e^{z_1 + z_2} \quad \text{or} \quad \exp(z_1) \cdot \exp(z_2) = \exp(z_1 + z_2).$$

Proof. By definition we have

$$\exp(z_1) \cdot \exp(z_2) = \left[1 + z_1 + \frac{z_1^2}{2!} + \dots + \frac{z_1^n}{n!} + \dots\right] \times \left[1 + z_2 + \frac{z_2^2}{2!} + \dots + \frac{z_2^n}{n!} + \dots\right]$$

$$= \left[1 + (z_1 + z_2) + \frac{1}{2!}(z_1^2 + 2z_1 z_2 + z_2^2) + \dots\right.$$

$$\left. + \frac{1}{n!}\left(z_1^n + n z_1^{n-1} z_2 + \frac{n(n-1)}{2!} z_1^{n-2} z_2^2 + \dots + z_2^n\right) + \dots\right]$$

$$= \left[1 + (z_1 + z_2) + \frac{(z_1 + z_2)^2}{2!} + \frac{(z_1 + z_2)^3}{3!} + \dots + \frac{(z_1 + z_2)^n}{n!} + \dots\right] = \exp(z_1 + z_2).$$

This series on the R.H.S is absolutely convergent if $\exp(z_1) \exp(z_2)$ are absolutely convergent.

THEOREM 2

If z is a complex number, then $(e^z)^m = e^{mz}$, m being a positive integer.

Proof. If m is positive integer, we have, by repeated application of theorem (1).

$$\exp(z_1) \exp(z_2) \dots \exp(z_m) = \exp(z_1 + z_2 + \dots z_n).$$

If $z_1 = z_2 = \dots z_m = z$, we have

$$(\exp z)^m = \exp(mz).$$

THEOREM 3

$E(z) \neq 0$, for any value of z.

Proof. By the addition theorem, we have

$$E(z) \cdot E(-z) = E\{z + (-z)\} = E(0) = 1$$

since $E(z)$ is well defined for all values of z, therefore, it follows that $E(z) \neq 0$, for any value of z.

Remark

❖ $\{E(z)\}^{-1} = E(-z)$.

9.4 CIRCULAR FUNCTION OF COMPLEX QUANTITIES

For real values of x,

$$\sin x = x - \frac{x^3}{3!} + \frac{x^5}{5!} + \dots + \frac{(-1)^n x^{2n+1}}{(2n+1)!} + \dots$$

and

$$\cos x = 1 - \frac{x^2}{2!} + \frac{x^4}{4!} - \dots + \frac{(-1)^n x^{2n}}{(2n)!} + \dots$$

These definitions are extended for the complex quantity $z = x + iy$, where x and y are real.

$$\sin z = z - \frac{z^3}{3!} + \frac{z^5}{5!} - \dots + \frac{(-1)^n z^{2n-1}}{(2n+1)!} + \dots$$

and $$\cos z = 1 - \frac{z^2}{2!} + \frac{z^4}{4!} - \dots + \frac{(-1)^n z^{2n}}{(2n)!}$$

Similarly, $\tan z = \dfrac{\sin z}{\cos z}$, $\cot z = \dfrac{\cos z}{\sin z}$, $\sec z = \dfrac{1}{\cos z}$ and $\cos ec\, z = \dfrac{1}{\sin z}$

From these definitions we can deduce the fundamental properties of two functions.

(a) $\cos z + i \sin z = e^{iz}$; $\cos z - i \sin z = e^{-iz}$

$\therefore$ $\cos^2 z + i \sin^2 z = e^{iz} e^{-iz} = e^{-0} = 1$.

(b) $\cos z = \dfrac{1}{2}[e^{iz} + e^{-iz}]$; $\sin z = \dfrac{1}{2i}[e^{iz} - e^{-iz}]$.

(c) To prove that

$$\sin[z_1 + z_2] = \sin z_1 \cos z_2 + \cos z_1 \sin z_2.$$

$$\text{R.H.S} = \frac{e^{iz_1} - e^{-iz_1}}{2i} \times \frac{e^{iz_2} + e^{-iz_2}}{2} + \frac{e^{iz_1} + e^{-iz_1}}{2} \cdot \frac{e^{iz_2} - e^{-iz_2}}{2i}$$

$$= \frac{1}{4i}[2e^{i(z_1+z_2)} + e^{i(z_1-z_2)} - e^{-i(z_1-z_2)} - 2e^{-i(z_1+z_2)} - e^{i(z_1-z_2)} + e^{-i(z_1-z_2)}]$$

$$= \frac{1}{2i}[e^{i(z_1+z_2)} - e^{-i(z_1+z_2)}] = \sin(z_1 + z_2)$$

(d)   To prove that $\sin 3z = 3 \sin z - 4 \sin^3 z$.

$$\text{R.H.S.} = 3\left(\frac{e^{iz} - e^{-iz}}{2i}\right) - 4\left(\frac{e^{iz} - e^{-iz}}{2i}\right)^3$$

$$= \frac{1}{2i}(3e^{iz} - 3e^{-iz}) - 4 \cdot \left(\frac{e^{3iz} - 3e^{2iz} \cdot e^{-iz} + 3e^{iz}e^{-2iz} - e^{-3iz}}{8i^3}\right)$$

$$= \frac{1}{2i}(3e^{iz} - 3e^{-iz} + e^{3iz} - 3e^{iz} + 3e^{-iz} - e^{-3iz})$$

$$= \frac{1}{2i}(e^{3iz} - e^{-3iz}) = \sin 3z.$$

Similarly we can derive other results. Generally, these results show the trigonometrical formulae.

9.5 EULER'S EXPONENTIAL VALUE

To prove that $e^{i\theta} = \cos\theta + i \sin\theta$, for any real θ.

Proof. We have $e^z = 1 + z + \dfrac{z^2}{2!} + \dfrac{z^3}{3!} + \dots$

Put $z = i\theta$, where θ is real

$$e^{i\theta} = 1 + i\theta + \frac{i^2\theta^2}{2!} + \frac{i^3\theta^3}{3!} + \dots$$

$$= 1 - \frac{\theta^2}{2!} + \frac{\theta^4}{4!} - \frac{\theta^6}{6!} + \dots + i\left[\theta - \frac{\theta^3}{3!} + \frac{\theta^5}{5!} - \dots\right] \qquad (\because i^2 = -1)$$

$$= \cos\theta + i\sin\theta \qquad \qquad \dots(1)$$

$$\therefore \qquad e^{i\theta} = \cos\theta + i\sin\theta \qquad \qquad \dots(2)$$

Similarly $e^{-i\theta} = \cos\theta - i\sin\theta$

By adding of (1) or (2), we get

$$\cos\theta = \frac{e^{i\theta} + e^{-i\theta}}{2}$$

By subtracting (2) from (1), we get

$$\sin\theta = \frac{e^{i\theta} - e^{-i\theta}}{2i}$$

There results are known as **Euler's exponential values.**

Further more,

$$\tan\theta = \frac{\sin\theta}{\cos\theta} = \frac{(e^{i\theta} - e^{-i\theta})}{i(e^{i\theta} + e^{-i\theta})}$$

$$\cot\theta = \frac{\cos\theta}{\sin\theta} = \frac{i(e^{i\theta} + e^{-i\theta})}{(e^{i\theta} - e^{-i\theta})}$$

Remark

❖ Since $e^{x + iy} = e^x \cdot [\cos y + i\sin y]$, this methods helps in breaking an exponential function into real and imaginary parts.

9.6 PERIODS OF COMPLEX CIRCULAR FUNCTIONS

$$\cos(z + 2n\pi) = \cos z \cos 2n\pi - \sin z \sin 2n\pi = \cos z \qquad \text{[If } n \text{ is an integer]}$$
$$\sin(z + 2n\pi) = \sin z \cos 2n\pi + \cos z \sin 2n\pi = \sin z \qquad \text{[}n \text{ being an integer]}$$

and $\qquad \tan(z + 2n\pi) = \dfrac{\sin(z + n\pi)}{\cos(z + n\pi)} = \dfrac{\pm\sin z}{\pm\cos z} = \tan z$

[according as n is even or odd integer]

Hence, the period of $\cos z$ is 2π, that of $\sin z$ is 2π, whereas the period of $\tan z$ is π.

9.7 PERIOD OF e^z

If $z = x + iy$, then

$$e^{z + 2n\pi i} = e^x \cdot e^{i(y + 2\pi)} = e^x[\cos(y + 2\pi) + i\sin(y + 2\pi)]$$

$$= e^x \cdot [\cos y + i\sin y] = e^x \cdot e^{iy} = e^{x + iy} = e^z$$

Therefore $e^{z + 2n\pi i} = e^z$ for all z.

Hence, the period of exp (z) is $2\pi i$.

9.8   SOME TRIGONOMETRICAL IDENTITIES FOR COMPLEX VARIABLE

To prove that for all values of x, y real or complex.

1. $\cos^2 x + \sin^2 x = 1$
2. $\sin(-x) = -\sin x$
3. $\cos(-x) = \cos x$
4. $\sin 2x = 2 \sin x \cos x$
5. $\cos 2x = \cos^2 x - \sin^2 x = 2\cos^2 x - 1 = 1 - 2\sin^2 x$
6. $\sin 3x = 3 \sin x - 4 \sin^3 x$
7. $\cos 3x = 4 \cos^3 x - 3 \cos x$
8. $\sin x + \sin y = 2\sin\dfrac{x+y}{2}\cos\dfrac{x-y}{2}$
9. $\sin x - \sin y = 2\cos\dfrac{x+y}{2}\sin\dfrac{x-y}{2}$
10. $\cos x + \cos y = 2\cos\dfrac{x+y}{2}\cos\dfrac{y-x}{2}$
11. $\cos x - \cos y = 2\sin\dfrac{x+y}{2}\sin\dfrac{x-y}{2}$
12. $\sin(x \pm y) = \sin x \cos y \pm \cos x \sin y$
13. $\cos(x \pm y) = \cos x \cos y \mp \sin x \sin y$.

Proof. By Euler's exponential values of $\cos x$ and $\sin x$, we have

$$\sin x = \frac{e^{ix} - e^{-ix}}{2i} \quad \text{and} \quad \cos x = \frac{e^{ix} + e^{-ix}}{2}$$

1.
$$\cos^2 x + \sin^2 x = \left[\frac{e^{ix} + e^{-ix}}{2}\right]^2 + \left[\frac{e^{ix} - e^{-ix}}{2i}\right]^2$$

$$= \frac{1}{4}(e^{2ix} + e^{-2ix} + 2) - \frac{1}{4}(e^{2ix} + e^{-2ix} - 2)$$

$$= \frac{1}{2} + \frac{1}{2} = 1$$

2.
$$\sin(-x) = \frac{e^{-(ix)} - e^{-(-ix)}}{2i} = -\left[\frac{e^{ix} - e^{-ix}}{2i}\right] = -\sin x$$

3.
$$\cos(-x) = \frac{e^{i(-x)} + e^{-i(-x)}}{2} = \left[\frac{e^{ix} + e^{-ix}}{2}\right] = \cos x.$$

4.
$$\sin 2x = \left(\frac{e^{2ix} - e^{-2ix}}{2i}\right)$$

$$= \frac{1}{2i}[(e^{ix} + e^{-ix})(e^{ix} - e^{-ix})]$$

$$= 2\left[\left(\frac{e^{ix} + e^{-ix}}{2}\right)\left(\frac{e^{ix} - e^{-ix}}{2i}\right)\right]$$

$$= 2\cos x \sin x.$$

5. $\qquad \cos^2 x - \sin^2 x = \left[\dfrac{e^{ix} + e^{-ix}}{2}\right]^2 - \left[\dfrac{e^{ix} - e^{-ix}}{2i}\right]^2$

$$= \frac{1}{4}(e^{2ix} + e^{-2ix} + 2) + \frac{1}{4}(e^{2ix} + e^{-2ix} - 2)$$

$$= \frac{e^{2ix} + e^{-2ix}}{2}$$

$$= \cos 2x.$$

Also $\qquad 2\cos^2 x - 1 = 2\left[\dfrac{e^{ix} + e^{-ix}}{2}\right]^2 - 1$

$$= \left[\frac{e^{2ix} + e^{-2ix} + 2}{4}\right] - 1$$

$$= \frac{e^{2ix} + e^{-2ix}}{2} = \cos 2x$$

and $\qquad 1 - 2\sin^2 x = 1 - 2\left[\dfrac{e^{ix} - e^{-ix}}{2i}\right]^2$

$$= 1 + 2\left[\frac{e^{2ix} + e^{-2ix} - 2}{4}\right]$$

$$= \left[\frac{e^{2ix} + e^{-2ix}}{2}\right] = \cos 2x.$$

6. $\qquad 3\sin x - 4\sin^3 x = 3\left[\dfrac{e^{ix} - e^{-ix}}{2i}\right] - 4\left[\dfrac{e^{ix} - e^{-ix}}{2i}\right]^3$

$$= \frac{e^{ix} - e^{-ix}}{2i}\left[3 - 4\left(\frac{e^{ix} - e^{-ix}}{2i}\right)^2\right]$$

$$= \frac{e^{ix} - e^{-ix}}{2i}[3 + (e^{2ix} + e^{-2ix} - 2)]$$

$$= \frac{e^{ix} - e^{-ix}}{2i}[e^{2ix} + e^{-2ix} - 1)]$$

$$= \frac{1}{2i}[e^{3ix} + e^{-ix} + e^{ix} - e^{ix} - e^{-3ix} - e^{-ix}]$$

$$= \frac{e^{3ix} - e^{-3ix}}{2i} = \sin 3x.$$

7. $4\cos^3 x - 3\cos x = \cos x\,(4\cos^2 x - 3)$

$$= \frac{e^{ix}+e^{-ix}}{2}\left[4\left(\frac{e^{ix}+e^{-ix}}{2}\right)^2 - 3\right]$$

$$= \frac{e^{ix}+e^{-ix}}{2}\,[e^{2ix}+e^{-2ix}-1]$$

$$= \frac{1}{2}[e^{3ix}+e^{-ix}-e^{ix}+e^{ix}+e^{-3ix}-e^{-ix}]$$

$$= \frac{e^{3ix}+e^{-3ix}}{2} = \cos 3x.$$

8. $2\sin\dfrac{x+y}{2}\cos\dfrac{x-y}{2} = 2\left[\dfrac{e^{\frac{i(x+y)}{2}}-e^{\frac{-i(x+y)}{2}}}{2i}\right]\left[\dfrac{e^{\frac{i(x-y)}{2}}-e^{\frac{-i(x-y)}{2}}}{2}\right]$

$$= \frac{1}{2i}[e^{\frac{i(x+y+x-y)}{2}}+e^{\frac{i(x+y-x+y)}{2}}-e^{\frac{i(-x-y+x-y)}{2}}-e^{\frac{i(-x-y-x+y)}{2}}]$$

$$= \frac{1}{2i}[e^{ix}+e^{iy}+e^{-iy}-e^{-ix}]$$

$$= \frac{e^{ix}-e^{-ix}}{2i}+\frac{e^{iy}-e^{-iy}}{2i}$$

$$= \sin x + \sin y.$$

9. $2\cos\dfrac{x+y}{2}\cos\dfrac{x-y}{2} = 2\left[\dfrac{e^{\frac{i(x+y)}{2}}+e^{\frac{-i(x+y)}{2}}}{2}\right]\left[\dfrac{e^{\frac{i(x-y)}{2}}-e^{\frac{-i(x-y)}{2}}}{2i}\right]$

$$= \frac{1}{2i}[e^{\frac{i(x+y+x-y)}{2}}-e^{\frac{i(x+y-x+y)}{2}}+e^{\frac{i(-x-y+x-y)}{2}}-e^{\frac{i(x-y-x+y)}{2}}]$$

$$= \frac{1}{2i}[e^{ix}-e^{iy}+e^{-iy}-e^{-ix}]$$

$$= \frac{e^{ix}-e^{-ix}}{2i}-\frac{e^{iy}-e^{-iy}}{2i}$$

$$= \sin x - \sin y$$

10. $2\cos\dfrac{x+y}{2}\cos\dfrac{y-x}{2} = 2\left[\dfrac{e^{\frac{i(x+y)}{2}}+e^{\frac{-i(x+y)}{2}}}{2}\right]\left[\dfrac{e^{\frac{i(y-x)}{2}}+e^{\frac{-i(y-x)}{2}}}{2}\right]$

$$= \frac{1}{2}[e^{\frac{i(x+y+y-x)}{2}}+e^{\frac{i(x+y-y-x)}{2}}+e^{\frac{i(-x-y+y-x)}{2}}-e^{\frac{i(-x-y-y-x)}{2}}]$$

$$= \frac{1}{2}[e^{ix}+e^{iy}+e^{-iy}+e^{-ix}]$$

$$= \frac{e^{ix} + e^{-ix}}{2} + \frac{e^{iy} + e^{-iy}}{2}$$

$$= \cos x + \cos y.$$

11. $2\sin\dfrac{x+y}{2}\sin\dfrac{y-x}{2} = 2\left[\dfrac{e^{\frac{i(x+y)}{2}} - e^{\frac{-i(x+y)}{2}}}{2i}\right]\left[\dfrac{e^{\frac{i(x-y)}{2}} - e^{\frac{-i(x-y)}{2}}}{2i}\right]$

$$= \frac{1}{2}\left[e^{\frac{i(x+y+x-y)}{2}} - e^{\frac{i(x+y-x-y)}{2}} - e^{\frac{i(-x-y+x-y)}{2}} - e^{\frac{i(-x-y-x+y)}{2}}\right]$$

$$= \frac{1}{2}[e^{iy} - e^{ix} + e^{-ix} + e^{-iy}] = \cos x - \cos y.$$

$$= \frac{e^{ix} + e^{-ix}}{2} - \frac{e^{iy} + e^{-iy}}{2}$$

12. $\sin x \cos y + \cos x \sin y = \left(\dfrac{e^{ix} - e^{-ix}}{2i}\right)\left(\dfrac{e^{iy} + e^{-iy}}{2}\right) + \left(\dfrac{e^{ix} + e^{-ix}}{2}\right)\left(\dfrac{e^{iy} - e^{-iy}}{2i}\right)$

$$= \frac{1}{4i}\left[e^{i(x+y)} + e^{i(x-y)} - e^{-i(x-y)} - e^{-i(x+y)}\right.$$

$$\left. + e^{i(x+y)} - e^{-i(x-y)} + e^{-i(x-y)} - e^{-i(x+y)}\right]$$

$$= \frac{1}{4}[2e^{i(x+y)} - 2e^{-i(x+y)}]$$

$$= \frac{e^{i(x+y)} - e^{-i(x+y)}}{2i} = \sin(x+y)$$

Similarly we can prove $\sin(x - y) = \sin x \cos y - \cos x \sin y$

13. $\cos x \cos y - \sin x \sin y = \left(\dfrac{e^{ix} + e^{-ix}}{2}\right)\left(\dfrac{e^{iy} + e^{-iy}}{2}\right) + \left(\dfrac{e^{ix} - e^{-ix}}{2i}\right)\left(\dfrac{e^{iy} - e^{-iy}}{2i}\right)$

$$= \frac{1}{4i}[e^{i(x+y)} + e^{i(x-y)} + e^{-i(x-y)} + e^{-i(x+y)} + e^{i(x+y)} + e^{i(x+y)} - e^{-i(x-y)} + e^{-i(x-y)}]$$

$$= \frac{1}{4i}[2e^{i(x+y)} + 2e^{-i(x+y)}]$$

$$= \frac{e^{i(x+y)} + e^{-i(x+y)}}{2i} = \cos(x+y)$$

Similarly, we can prove $\cos(x + y) = \cos x \cos y + \sin x \sin y$.

SOLVED EXAMPLES

Example 1. *Show that* $\exp\left(\pm i\dfrac{\pi}{2}\right) = \pm i$.

Solution. Since $\exp(\pm i\theta) = \cos\theta \pm i\sin\theta$, we have

$$\exp\left(\pm i\frac{\pi}{2}\right) = \cos\frac{\pi}{2} \pm i\sin\frac{\pi}{2} = \pm i.$$

Example 2. *Prove* $\sin(\alpha + n\theta) - e^{-i\alpha} \sin n\theta = e^{-in\theta}\sin\alpha$.

Solution.   L.H.S. $= \sin(\alpha + n\theta) - (\cos\alpha + i\sin\alpha)\sin n\theta$

$= \sin\alpha\cos n\theta + \cos\alpha\sin n\theta - \cos\alpha\sin n\theta - i\sin\alpha\sin n\theta$

$= \sin\alpha(\cos n\theta - i\sin n\theta) = \sin\alpha\, e^{-in\theta} = \text{R.H.S.}$

Example 3. *Prove that*

$$[\sin(\alpha - \theta) + e^{-i\alpha i}\sin\theta]^n = \sin^{n-1}\alpha\,\{\sin(\alpha - n\theta) + e^{-\alpha i}\sin n\theta\}$$

Solution.   L.H.S. $= \{\sin(\alpha - \theta) + e^{-\alpha i}\sin\theta\}^n$

$= \{\sin\alpha\cos\theta - \cos\alpha\sin\theta + (\cos\alpha - i\sin\alpha)\sin\theta\}^n$

$= \{\sin\alpha\cos\theta - i\sin\alpha\sin\theta\}^n = \sin^n\alpha\,\{\cos\theta - i\sin\theta\}^n$

$= \sin^n\alpha\,\{\cos n\theta - i\sin n\theta\}$ [by De Moivre's theorem]

Again R.H.S. $= \sin^{n-1}\alpha\,\{\sin(\alpha - n\theta) + e^{-i\alpha}\sin n\theta\}$

$= \sin^{n-1}\alpha\,\{\sin\alpha\cos n\theta - \cos\alpha\sin n\theta + (\cos\alpha - i\sin\alpha)\sin n\theta\}$

$= \sin^{n-1}\alpha\,\{\sin\alpha\cos n\theta - i\sin\alpha\sin n\theta\}$

$= \sin^n\alpha\,\{\cos n\theta - i\sin n\theta\} = \text{L.H.S.}$

∴   R.H.S. = L.H.S.

Example 4. *If* $\cos\theta + i\sin\theta = x$, $\sqrt{i - c^2} = nc - 1$, *prove that*

$$1 + c\cos\theta = \frac{c}{2n}(1 + nx)\left(1 + \frac{n}{x}\right).$$

Solution. We have $x = \cos\theta + i\sin\theta$,

∴ $x^{-1} = \cos\theta - i\sin\theta$

∴   R.H.S. $= \dfrac{c}{2n}(1 + nx)\left(1 + \dfrac{n}{x}\right)$

$= \dfrac{c}{2n}[\{1 + n(\cos\theta + i\sin\theta)\}\,\{1 + n(\cos\theta - i\sin\theta)\}$

$= \dfrac{c}{2n}[\{1 + ne^{i\theta}\}\,\{1 + ne^{-i\theta}\}] = \dfrac{c}{2n}[1 + ne^{-i\theta} + ne^{i\theta} + n^2]$

$= \dfrac{2}{2n}[1 + n(e^{i\theta} + e^{-i\theta}) + n^2] = \dfrac{c}{2n}[1 + 2n\cos\theta + n^2]$

$= \dfrac{c}{2n}(1 + n^2) + c\cos\theta$ ...(1)

But $\sqrt{1 - c^2} = nc - 1$ or $1 - c^2 = (nc - 1)^2$

Hence from (1), we have

⇒ $c^2(1 + n^2) = 2nc$

$\dfrac{c}{2n} \times 2n + c\cos\theta = 1 + c\cos\theta.$

EXERCISE 9.1

1. Prove that $[\sin (\alpha + \theta) - e^{\alpha i} \sin \theta]^n = \sin^n\alpha \, e^{-n\theta i}$.

2. Prove that $\sin (\alpha + n\theta) - e^{i\alpha} \sin n\theta = e^{-in\theta} \sin \alpha$.

3. If $\tan^{-1}(e^{ix}) - \tan^{-1}(e^{-ix}) = \tan^{-1} i$, find x.

HINT TO SELECTED PROBLEMS

1. L.H.S $= \sin \alpha \cos \theta + \cos \alpha \sin \theta = (\cos \alpha + i \sin \alpha) \sin \theta$
$= \sin^n\alpha \, (\cos \theta - i \sin \theta)^n = \sin^n \alpha \, e^{-in\theta}$.

3. $\tan^{-1} \dfrac{e^{ix} - e^{-ix}}{1 + e^{ix} \cdot e^{-ix}} = \tan^{-1} i$ or $i \sin x = i \implies \sin x = \dfrac{\pi}{2} \implies x = n\pi + (-1)^n \dfrac{\pi}{2}$.

ANSWER

(3) $x = 2n\pi + \dfrac{\pi}{2}$ where n is an integer.

9.9 HYPERBOLIC FUNCTIONS

We have proved that for all values of the argument y (real or complex),

$$\cos y = 1 - \frac{y^2}{2!} + \frac{y^4}{4!} - \frac{y^6}{6!} + \dots \qquad \dots(A)$$

$$\sin y = y - \frac{y^3}{3!} + \frac{y^5}{5!} - \frac{y^7}{7!} + \dots \qquad \dots(B)$$

We notice that in each of these series, the terms are alternatively positive and negative. If we place the positive sign before all the terms, we get two functions of y defined by indefinite series, which are related to the circular functions $\cos y$ and $\sin y$ by interesting properties. These functions are known as hyperbolic cosine and hyperbolic sine of y and are indicated for shortness by $\cosh y$ and $\sinh y$ respectively. Thus.

$$\cosh y = 1 + \frac{y^2}{2!} + \frac{y^4}{4!} + \dots \qquad \dots(1)$$

$$\sinh y = y + \frac{y^3}{3!} + \frac{y^5}{5!} + \frac{y^7}{7!} + \dots \qquad \dots(2)$$

$\therefore \qquad \cosh y + \sinh y = 1 + y + \dfrac{y^2}{2!} + \dfrac{y^3}{3!} + \dots = e^y \qquad \dots(3)$

and $\quad \cosh y - \sinh y = e^{-y} \qquad \dots(4)$

$\therefore \qquad \cosh y = \dfrac{1}{2}[e^y + e^{-y}] \quad$ and $\quad \sinh y = \dfrac{1}{2}[e^y - e^{-y}]$

Now we give formal definition of these functions.

Definition. *The quantity* $\dfrac{e^y - e^{-y}}{2}$, *whether y be real or complex, is called the* **hyperbolic** *sine of y and is written as sinh y.*

Similarly $\dfrac{e^y - e^{-y}}{2}$ is known as **hyperbolic cosine** of y and is written as cosh y.

The hyperbolic tangent, secant, cosecant, and catangent can be obtained with the help of hyperbolic sine and cosine.

$$\tanh y = \frac{\sinh y}{\cosh y} = \frac{e^y - e^{-y}}{e^y + e^{-y}},$$

$$\operatorname{cosech} y = \frac{1}{\sinh y} = \frac{2}{e^y - e^{-y}},$$

$$\operatorname{sech} y = \frac{1}{\cosh y} = \frac{2}{e^y + e^{-y}},$$

$$\coth y = \frac{\cosh y}{\sinh y} = \frac{e^y + e^{-y}}{e^y - e^{-y}}$$

9.10 RELATION BETWEEN HYPERBOLIC AND CIRCULAR FUNCTIONS

Hyperbolic functions can be expressed in terms of corresponding circular functions.

We know $\qquad \sin x = \dfrac{e^{ix} - e^{-ix}}{2i}$, put $x = iy$

$$\sin iy = \frac{e^{i^2 y} - e^{-i^2 y}}{2i} = \frac{i[e^{-y} - e^{y}]}{2i^2}$$

$$\sin iy = \frac{i[e^{y} - e^{-y}]}{2} = i \sinh y.$$

Similarly, $\qquad \cos iy = \dfrac{e^{i^2 y} + e^{-i^2 y}}{2} = \dfrac{e^{y} + e^{-y}}{2} = \cosh y$

and $\qquad \tan iy = \dfrac{\sin iy}{\cos iy} = \dfrac{i \sinh y}{\cosh y} = i \tanh y.$

From (3) and (4), we have

$$(\cosh y + \sinh y)^n = e^{ny} = \cosh ny + \sinh ny \qquad \dots(5)$$

and $\qquad (\cosh y - \sinh y)^n = e^{-ny} = \cosh ny + \sinh ny. \qquad \dots(6)$

These results are analogous to De Moivre's Theorem.

9.11 SOME IMPORTANT RESULTS OF HYPERBOLIC FUNCTIONS

For any real x and y we have
(i) $\sinh 0 = 0$, and $\cosh 0 = 1$, $\tanh 0 = 0$
(ii) $\cosh^2 x - \sinh^2 x = 1$
(iii) $1 - \tanh^2 x = \operatorname{sech}^2 x$

(*iv*) $\coth^2 x - 1 = \text{coshech}^2 x$

(*v*) $\sinh 2x = 2 \sinh x \cosh x = \dfrac{2 \tanh x}{1 - \tanh^2 x}$

(*vi*) $\cosh 2x = \cosh^2 x + \sinh^2 x = 1 + 2\sinh^2 x = 2\cosh^2 x - 1 = \dfrac{1 + \tanh^2 x}{1 - \tanh^2 x}$

(*vii*) $\tan h\, 2x = \dfrac{2 \tan hx}{1 + \tan h^2 x}$

(*viii*) $\sinh 3x = 3 \sin h\, x + 4 \sinh^3 x$

(*ix*) $\cosh 3x = 4 \cosh^3 x - 3 \cosh x$

(*x*) $\tanh 3x = \dfrac{3 \tanh x + \tan h^3 x}{1 + 3 \tanh^2 x}$

(*xi*) $\sin h\,(x + y) = \sinh x \cosh y + \cosh x \cdot \sinh y$

(*xii*) $\cosh (x + y) = \cosh x \cosh y + \sinh x \cdot \sinh y$

(*xiii*) $e^x = \cosh x + \sinh,\ e^{-x} = \cosh x - \sinh x$

Proof. We know that $\sinh x = \dfrac{e^x - e^{-x}}{2},\quad \cosh x = \dfrac{e^x + e^{-x}}{2},\quad \tan x = \dfrac{e^x - e^{-x}}{e^x + e^{-x}}$

(*i*) $\sinh 0 = \dfrac{e^0 - e^{-0}}{2} = \dfrac{1 - 1}{2} = 0$

$\cosh 0 = \dfrac{e^0 + e^{-0}}{2} = \dfrac{1 + 1}{2} = 1$

$\tanh 0 = \dfrac{\sinh 0}{\cosh 0} = \dfrac{0}{1} = 0.$

(*ii*) $\cosh^2 x - \sinh^2 x = \left(\dfrac{e^x + e^{-x}}{2}\right)^2 - \left(\dfrac{e^x - e^{-x}}{2}\right)^2$

$= \dfrac{1}{4}[e^{2x} + e^{-2x} + 2 - e^{2x} - e^{-2x} + 2] = 1$

(*iii*) $1 - \tanh^2 x = 1 - \left(\dfrac{e^x - e^{-x}}{e^x + e^{-x}}\right)^2$

$= \dfrac{(e^x - e^{-x})^2 - (e^x - e^{-x})^2}{(e^x + e^{-x})^2}$

$= \dfrac{(e^{2x} - e^{-2x} + 2) - (e^{2x} + e^{-2x} - 2)}{(e^x + e^{-x})^2}$

$= \dfrac{4}{(e^x + e^{-x})^2} = \dfrac{2}{\left(\dfrac{e^x + e^{-x}}{2}\right)^2} = \dfrac{1}{\cosh^2 x} = \text{sech}^2 x$

(iv) $\coth^2 x - 1 = \left(\dfrac{e^x + e^{-x}}{e^x - e^{-x}}\right)^2 - 1$

$= \dfrac{(e^x + e^{-x})^2 - (e^x - e^{-x})^2}{(e^x - e^{-x})^2}$

$= \dfrac{4}{(e^x - e^{-x})^2} = \dfrac{1}{\left(\dfrac{e^x - e^{-x}}{2}\right)^2} = \dfrac{1}{\sinh^2 x} = \operatorname{cosech}^2 x.$

(v) $2 \sinh x \cosh x = 2\left(\dfrac{e^x - e^{-x}}{2}\right)\left(\dfrac{e^x + e^{-x}}{2}\right)$

$= \dfrac{1}{2}[e^{2x} + 1 - 1 - e^{-2x}]$

$= \dfrac{e^{2x} - e^{-2x}}{2} = \sinh 2x$

Now, $\dfrac{2 \tanh x}{1 - \tanh^2 x} = \dfrac{2\left(\dfrac{e^x - e^{-x}}{e^x + e^{-x}}\right)}{1 - \left(\dfrac{e^x - e^{-x}}{e^x + e^{-x}}\right)^2}$

$= \dfrac{2(e^x - e^{-x})(e^x + e^{-x})}{(e^x + e^{-x})^2 - (e^x - e^{-x})^2}$

$= \dfrac{2(e^{2x} - e^{-2x})}{4} = \dfrac{(e^{2x} - e^{-2x})}{2} = \sinh 2x$

(vi) $\cosh^2 x + \sinh^2 x = \left(\dfrac{e^x + e^{-x}}{2}\right)^2 + \left(\dfrac{e^x - e^{-x}}{2}\right)^2$

$= \dfrac{1}{4}[e^{2x} + e^{-2x} + 2 + e^{2x} + e^{-2x} - 2]$

$= \dfrac{e^{2x} + e^{-2x}}{2} = \cosh 2x$

Now, $1 + 2 \sinh^2 x = 1 + 2\left(\dfrac{e^x - e^{-x}}{2}\right)^2$

$= \dfrac{1}{2}[2 + (e^x - e^{-x})]^2$

$= \dfrac{1}{2}[2 + e^{2x} + e^{-2x} - 2]$

$= \dfrac{e^{2x} + e^{-2x}}{2} = \cosh 2x,$

$$2\cosh^2 x - 1 \;=\; 2\left(\frac{e^x + e^{-x}}{2}\right)^2 - 1$$

$$= \frac{1}{2}[(e^x + e^{-x})^2 - 2]$$

$$= \frac{1}{2}[e^{2x} + e^{-2x} + 2 - 2]$$

$$= \frac{e^{2x} + e^{-2x}}{2} = \cosh 2x$$

and

$$\frac{1 + \tanh^2 x}{1 - \tanh^2 x} \;=\; \frac{1 + \left(\dfrac{e^x - e^{-x}}{e^x + e^{-x}}\right)^2}{1 - \left(\dfrac{e^x - e^{-x}}{e^x + e^{-x}}\right)^2}$$

$$= \frac{(e^x + e^{-x})^2 + (e^x - e^{-x})^2}{(e^x + e^{-x})^2 - (e^x - e^{-x})^2}$$

$$= \frac{2(e^{2x} + e^{-2x})}{4}$$

$$= \frac{(e^{2x} + e^{-2x})}{2} = \cosh 2x$$

(vii)
$$\frac{2\tanh x}{1 + \tanh^2 x} \;=\; \frac{2\left(\dfrac{e^x - e^{-x}}{e^x + e^{-x}}\right)}{1 + \left(\dfrac{e^x - e^{-x}}{e^x + e^{-x}}\right)^2}$$

$$= \frac{2(e^x - e^{-x})(e^x + e^{-x})}{(e^x + e^{-x})^2 + (e^x - e^{-x})^2}$$

$$= \frac{2(e^{2x} - e^{-2x})}{2(e^{2x} + e^{-2x})}$$

$$= \frac{e^{2x} - e^{-2x}}{e^{2x} + e^{-2x}} = \tanh 2x.$$

(viii) $3\sinh x + 4\sinh^3 x \;=\; 3\left(\dfrac{e^x - e^{-x}}{2}\right) + 4\left(\dfrac{e^x - e^{-x}}{2}\right)^3$

$$= \left(\frac{e^x - e^{-x}}{2}\right)\left[3 + 4\left(\frac{e^x - e^{-x}}{2}\right)^2\right]$$

$$= \left(\frac{e^x - e^{-x}}{2}\right)[3 + e^{2x} + e^{-2x} - 2]$$

$$= \left(\frac{e^x - e^{-x}}{2}\right)[1 + e^{2x} + e^{-2x}]$$

$$= \frac{1}{2}[e^{3x} - e^{-3x}] = \sinh 3x.$$

(ix) $\quad 4\cosh^3 x - 3\cosh x \quad = 4\left(\frac{e^x + e^{-x}}{2}\right)^3 - 3\left(\frac{e^x + e^{-x}}{2}\right)$

$$= \left(\frac{e^x + e^{-x}}{2}\right)[(e^x + e^{-x})^2 - 3]$$

$$= \left(\frac{e^x + e^{-x}}{2}\right)[e^{2x} + e^{-2x} + 2 - 3]$$

$$= \left(\frac{e^x + e^{-x}}{2}\right)[e^{2x} + e^{-2x} - 1]$$

$$= \frac{e^{3x} + e^{-3x}}{2} = \cosh 3x$$

(x) $\quad \dfrac{3\tanh x + \tanh^3 x}{1 + 3\tanh^2 x} = \dfrac{3\left(\dfrac{e^x - e^{-x}}{e^x + e^{-x}}\right) + \left(\dfrac{e^x - e^{-x}}{e^x + e^{-x}}\right)^3}{1 + 3\left(\dfrac{e^x - e^{-x}}{e^x + e^{-x}}\right)^2}$

$$= \frac{3(e^x - e^{-x})(e^x + e^{-x})^2 + (e^x - e^{-x})^3}{(e^x + e^{-x})[(e^x + e^{-x})^2 + 3(e^x + e^{-x})^2]}$$

$$= \frac{(e^x - e^{-x})[3(e^{2x} + e^{-2x} + 2) + e^{2x} + e^{-2x} - 2]}{(e^x + e^{-x})[e^{2x} + e^{-2x} + 2 + 3(e^{2x} + e^{-2x} - 2)]}$$

$$= \frac{(e^x - e^{-x})(4e^{2x} + 4e^{-2x} + 4)}{(e^x + e^{-x})(4e^{2x} + 4e^{-2x} - 4)}$$

$$= \frac{(e^x - e^{-x})(e^{2x} + e^{-2x} + 1)}{(e^x + e^{-x})(e^{2x} + e^{-2x} - 1)}$$

$$= \frac{e^{3x} - e^{-3x}}{e^{3x} + e^{-3x}} = \tanh 3x$$

(xi) $\sinh x \cosh y + \cosh x \sinh y = \left(\dfrac{e^x - e^{-x}}{2}\right)\left(\dfrac{e^y + e^{-y}}{2}\right) + \left(\dfrac{e^x + e^{-x}}{2}\right)\left(\dfrac{e^y - e^{-y}}{2}\right)$

$$= \frac{1}{4}[e^{x+y} + e^{x-y} - e^{-(x-y)} - e^{-(x+y)} + e^{x+y} - e^{x-y} + e^{-(x-y)} - e^{-(x+y)}]$$

$$= \frac{1}{4}[2e^{(x+y)} - 2e^{-(x+y)}]$$

$$= \frac{e^{(x+y)} - e^{-(x+y)}}{2} = \sinh(x+y).$$

(xii) $\cosh x \cosh y + \sinh x \sinh y = \left(\dfrac{e^x + e^{-x}}{2}\right)\left(\dfrac{e^y + e^{-y}}{2}\right) + \left(\dfrac{e^x - e^{-x}}{2}\right)\left(\dfrac{e^y - e^{-y}}{2}\right)$

$$= \frac{1}{4}[e^{x+y} + e^{x-y} + e^{-(x-y)} + e^{-(x+y)} + e^{x+y} - e^{x-y} - e^{-(x-y)} + e^{-(x+y)}]$$

$$= \frac{1}{4}[2e^{(x+y)} + 2e^{-(x+y)}]$$

$$= \frac{e^{(x+y)} + e^{-(x+y)}}{2} = \cosh(x+y).$$

(xiii) $\dfrac{e^x - e^{-x}}{2} = \sinh x,\quad \dfrac{e^x + e^{-x}}{2} = \cosh x.$

Adding and substracting respectively, we get

$$e^x = \sinh x + \cosh x,\ e^{-x} = \cosh x - \sinh x$$

9.12 EXPANSIONS OF sinh x AND cosh x

We know that

$$\sinh x = \frac{e^x - e^{-x}}{2}$$

Then $\sinh x = \dfrac{1}{2}\left[\left(1 + x + \dfrac{x^2}{2!} + \dfrac{x^3}{3!} + \dfrac{x^4}{4!} + \ldots\right) - \left(1 - x + \dfrac{x^2}{2!} - \dfrac{x^3}{3!} + \dfrac{x^4}{4!} - \ldots\right)\right]$

$$= \frac{1}{2}\left[2x + 2\frac{x^3}{3!} + 2\frac{x^5}{5!} + \ldots\right]$$

∴ $\sinh x = x + \dfrac{x^3}{3!} + \dfrac{x^5}{5!} + \ldots$ to infinity

Also, $\cosh x = \dfrac{1}{2}(e^x + e^{-x})$

$$= \frac{1}{2}\left[\left(1 + x + \frac{x^2}{2!} + \frac{x^3}{3!} + \frac{x^4}{4!} + \ldots\right) + \left(1 - x + \frac{x^2}{2!} - \frac{x^3}{3!} + \frac{x^4}{4!} - \ldots\right)\right]$$

$$= \frac{1}{2}\left[2 + 2\frac{x^2}{2!} + 2\frac{x^4}{4!} + \ldots\right]$$

∴ $\qquad \cosh x = 1 + \dfrac{x^2}{2!} + \dfrac{x^4}{4!} + \ldots$ to infinity.

9.13 PERIOD OF HYPERBOLIC FUNCTIONS

We know that $\qquad \cos i\theta = \cosh \theta$

Therefore $\quad \cosh (x + iy) = \cos[i \, (x + iy)] = \cos (xi - y) = \cos [-2\pi + ix - y]$

$\qquad\qquad\qquad = \cos [(2\pi i + x + iy) \, i] = \cosh [(2\pi i + x + iy)]$

Hence the hyperbolic cosine is periodic, its period being imaginary and equal to $2\pi i$.

Similarly it can be shown for $\sinh (x + iy)$ that its period is $2\pi i$ and of $\tanh (x + iy)$ is πi.

It is to be noted here that hyperbolic functions differ from the circular functions in having imaginary periods.

Aliter. These results can be obtained in a another way also.

Since $\qquad\qquad e^{2n\pi i} = \cos 2\, n\pi + i \sin 2\, n\pi = 1.$

∴ $\qquad\qquad e^{z + 2n\pi i} = e^z$ and $e^{-z - 2n\pi i} = e^{-z}.$

∴ $\quad e^{z + 2n\pi i} + e^{-z - 2n\pi i} = e^z + e^{-z}$ and $e^{z + 2n\pi i} - e^{-z - 2n\pi i} = e^z - e^{-z}.$

∴ $\qquad \cosh (z + 2n\pi i) = \cosh (z)$ and $\sinh (z + 2n\pi i) = \sinh z$

Next $\qquad\qquad e^{n\pi i} = \cos n\pi + i \sin n\pi = (-1)^n,$

$\qquad\qquad e^{-n\pi i} = \cos n\pi - i \sin n\pi = (-1)^n.$

∴ $\qquad \tanh (z + 2n\pi i) = \dfrac{e^{z + n\pi i} - e^{-z - n\pi i}}{e^{z + n\pi i} + e^{-z - n\pi i}} = \dfrac{e^z - e^{-z}}{e^z + e^{-z}} = \tanh z.$

Hence $\cosh z$ and $\sinh z$ have an imaginary period of $2\pi i$ and $\tanh z$ has an imaginary period of πi.

9.14 GEOMETRICAL ANALOGY BETWEEN CIRCULAR AND HYPERBOLIC FUNCTIONS

If $x = a \cosh u$, and $y = a \sinh u$,

then the point P whose co-ordinates are (x, y) satisfy the rectangular hyperbola

$$x^2 - y^2 = a^2$$

Area APN $\qquad = \displaystyle\int_0^x y \, dx = \int_0^x a \sinh u \, d \, (a \cosh u)$

$\qquad\qquad\qquad = \displaystyle\int_0^x a^2 \sinh^2 u \, du$

$\qquad\qquad\qquad = \dfrac{1}{2} a^2 \displaystyle\int_0^u (\cosh 2u - 1) du$

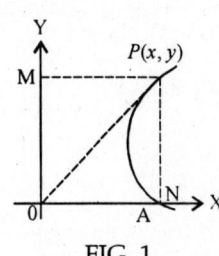

FIG. 1

$$= \frac{1}{2}a^2\left[\frac{1}{2}\sinh 2u - u\right]$$

Area of triangle

$$OPN = \frac{1}{2}ON \times NP = \frac{1}{2} a \cosh u \times a \sinh u = \frac{1}{4}a^2 \sinh 2u.$$

Hence area of the sector

$$OAP = \Delta OPN - \text{area } APN = \frac{1}{2}a^2u.$$

Denoting the area of sector OAP by S, we have

$$\frac{1}{2}a^2u = S \text{ or } u = \frac{2S}{a^2}$$

Hence co-ordinates of any point on hyperbola $x^2 - y^2 = a^2$ may be represented by

$$x = a \cosh \frac{2S}{a^2}, y = a \sinh \frac{2S}{a^2}$$

We notice that this is similar to the expressions for the co-ordinates of a point on the circle $x^2 + y^2 = a^2$,

where $\quad x = a \cos \frac{2S}{a^2}, y = a \sin \frac{2S}{a^2}$

S being the area of the sector $OAP = \frac{1}{2}a^2\theta$, so that $\frac{2S}{a^2} = \theta = $ circular measure of the angle AOP.

It is thus apparent that the hyperbolic functions are connected in the same way with the rectangular hyperbola as the circular functions are with the circle.

SOLVED EXAMPLES

Example 1. *If* $\tan y = \tan \alpha \tanh \beta$, $\tan z = \cot \alpha \tanh \beta$, *prove that*

$$\tan (y + z) = \sinh 2\beta \operatorname{cosec} 2\alpha.$$

Solution.   We have

$$\tan (y + z) = \frac{\tan y + \tan z}{1 - \tan y \tan z}$$

$$= \frac{\tan \alpha \tanh \beta + \cot \alpha \tanh \beta}{1 - \tan \alpha \tanh \beta \times \cot \alpha \tanh \beta} = \frac{\tan \beta\left[\dfrac{\sin \alpha}{\cos \alpha} + \dfrac{\cos \alpha}{\sin \alpha}\right]}{1 - \tanh^2 \beta}$$

$$= \frac{\dfrac{\sinh \beta}{\cosh \beta}}{1 - \dfrac{\sinh^2 \beta}{\cosh^2 \beta}} \times \frac{1}{\sin \alpha \cos \alpha} = \frac{\sinh \beta \cosh \beta}{\sin \alpha \cos \alpha} \times \frac{2}{2} = \frac{\sinh 2\beta}{\sin 2\alpha}$$

$$= \sinh 2\beta \operatorname{cosec} 2\alpha.$$

Example 2. *If* $\cosh x = \sec \theta$, *prove that* $\tanh^2 \dfrac{x}{2} = \tan^2 \dfrac{\theta}{2}$

Solution. We know that

$$\cosh x = \frac{1 + \tanh^2 x/2}{1 - \tanh^2 x/2} = \sec \theta = \frac{1}{\cos \theta}$$

Apply compondendo and dividendo, $\left[\text{i.e. if } \dfrac{a}{b} = \dfrac{c}{d} \text{ then } \dfrac{a-b}{a+b} = \dfrac{c-d}{c+d} \right]$ we have

$$\frac{2 \tanh^2 x/2}{2} = \frac{1 - \cos \theta}{1 + \cos \theta} = \tan^2 \frac{\theta}{2}$$

$$\therefore \qquad \tanh^2 \frac{x}{2} = \tan^2 \frac{\theta}{2}$$

Example 3. *If* θ *is acute and* $x = \log \tan \left(\dfrac{\pi}{4} + \dfrac{\theta}{2} \right)$. *Show that* $\cos \theta \cosh x = 1$

Solution. We have $\quad x = \log \tan \left(\dfrac{\pi}{4} + \dfrac{\theta}{2} \right)$

$$\therefore \qquad e^x = \tan \left(\frac{\pi}{4} + \frac{\theta}{2} \right) = \frac{1 - \tan \theta/2}{1 + \tan \theta/2}$$

Now $\qquad \cosh x = \dfrac{1}{2}[e^x + e^{-x}] = \dfrac{1}{2} \left[\dfrac{1 + \tan \theta/2}{1 - \tan \theta/2} + \dfrac{1 - \tan \theta/2}{1 + \tan \theta/2} \right]$

$$= \frac{1}{2} \left[\frac{(1 + \tan \theta/2)^2 + (1 - \tan \theta/2)^2}{1 - \tan^2 \theta/2} \right] = \frac{1 - \tan^2 \theta/2}{1 + \tan^2 \theta/2} = \sec \theta.$$

$$\therefore \qquad \cosh x \cos \theta = 1.$$

Example 4. *If* $u = \log \tan \left(\dfrac{\pi}{4} + \dfrac{\theta}{2} \right)$, *prove that* $\tanh \dfrac{u}{2} = \tan \dfrac{\theta}{2}$.

Solution. We have $u = \log \tan \left(\dfrac{\pi}{4} + \dfrac{\theta}{2} \right)$

or $\qquad e^u = \tan \left(\dfrac{\pi}{4} + \dfrac{\theta}{2} \right) \quad$ or $\quad \dfrac{e^{u/2}}{e^{-u/2}} = \dfrac{1 + \tan \theta/2}{1 - \tan \theta/2}$

By componendo and dividendo, we have

$$\frac{e^{u/2} - e^{-u/2}}{e^{u/2} + e^{-u/2}} = \frac{2 \tan \theta/2}{2} = \tan \frac{\theta}{2} \quad \text{or} \quad \tanh \frac{1}{2}u = \tan \frac{\theta}{2}.$$

Example 5. *If* $u = \log \tan \left(\dfrac{\pi}{4} + \dfrac{\theta}{2} \right)$, *prove that*

(i) $\sinh u = \tan \theta$ \qquad (ii) $\tanh u = \sin \theta$

(iii) if $u = \theta + a_3\theta^3 + a_5\theta^5 + \ldots$ *show that* $\theta = u - a_3u^3 + a_5u^5 - \ldots$

Solution. (*i*) We have $\qquad u = \log \tan\left(\dfrac{\pi}{4} + \dfrac{\theta}{2}\right)$

or $\qquad\qquad e^u = \tan\left(\dfrac{\pi}{4} + \dfrac{\theta}{2}\right) = \dfrac{1 + \tan\theta/2}{1 - \tan\theta/2}$ \hfill ...(1)

$\therefore \qquad\qquad e^{-u} = \dfrac{1 - \tan\theta/2}{1 + \tan\theta/2}$ \hfill ...(2)

Hence, $\qquad e^u - e^{-u} = \dfrac{1 + \tan\theta/2}{1 - \tan\theta/2} - \dfrac{1 - \tan\theta/2}{1 + \tan\theta/2}$

$$= \frac{(1 + \tan\theta/2)^2 - (1 - \tan\theta/2)^2}{1 - \tan^2\theta/2} = \frac{4\tan\theta/2}{1 - \tan\theta/2} = 2\tan\theta.$$

$\therefore \qquad\qquad \sinh u = \dfrac{1}{2}[e^u - e^{-u}] = \tan\theta.$

(*ii*) From (1) and (2), We have

$$e^u + e^{-u} = \frac{1 + \tan\theta/2}{1 - \tan\theta/2} + \frac{1 - \tan\theta/2}{1 + \tan\theta/2}$$

$$= \frac{(1 + \tan\theta/2)^2 + (1 - \tan\theta/2)^2}{1 - \tan^2\theta/2}$$

$$= \frac{2(1 + \tan^2\theta/2)}{1 - \tan^2\theta/2} = \frac{2}{\cos\theta}$$

$\therefore \qquad\qquad \cosh u = \sec\theta.$

Hence $\tanh u = \cdot\dfrac{e^u - e^{-u}}{e^u + e^{-u}} = \dfrac{\tan\theta}{\sec\theta} = \sin\theta.$

(*iii*) We have

$$\tan\theta = \sinh u \quad \text{or} \quad i\tan\theta = \sin ui \Rightarrow \tanh\theta = \sin ui$$

$\therefore \qquad\qquad \dfrac{e^{i\theta} - e^{-i\theta}}{e^{i\theta} + e^{-i\theta}} = \dfrac{\sin ui}{1}$

By componendo and dividendo, we have

$$\frac{2e^{i\theta}}{2e^{-i\theta}} = \frac{1 + \sin ui}{1 - \sin ui} = \frac{1 - \cos(\pi/2 + iu)}{1 + \cos(\pi/2 + iu)}$$

or $\qquad\qquad e^{2i\theta} = \dfrac{2\sin^2\left(\dfrac{\pi}{4} + \dfrac{iu}{2}\right)}{2\cos^2\left(\dfrac{\pi}{4} + \dfrac{ui}{2}\right)} = \tan^2\left(\dfrac{\pi}{4} + \dfrac{ui}{2}\right)$

$\therefore \qquad\qquad e^{i\theta} = \tan\left(\dfrac{\pi}{4} + \dfrac{ui}{2}\right)$

$$\theta = \frac{1}{i}\log\tan\left(\frac{\pi}{4} + \frac{ui}{2}\right) = \frac{1}{i}[ui + a_3(ui)^3 + a_5(ui)^5 + \ldots] \text{ by definition}$$

$$= u - a_3 u^3 + a_5 u^5 - \dots$$

Example 6. *Show that if n is a positive integer*

$$\cosh n\theta = \cosh^n \theta + {}^nC_2 \cosh^{n-2}\theta \sinh^2 \theta + \dots$$

and $$\sinh n\theta = n \cosh^{n-1}\theta \sinh \theta + {}^nC_3 \cosh^{n-3}\theta \sinh^3 \theta + \dots$$

Solution. We have $\cosh \theta + \sinh \theta = e^\theta$

Then $\cosh \theta - \sinh \theta = e^{-\theta}$ and $\cosh n\theta + \sinh n\theta = e^{n\theta}$

$\cosh n\theta - \sinh n\theta = e^{-n\theta}$

$\therefore$ $\cosh n\theta = \dfrac{1}{2}[e^{n\theta} + e^{-n\theta}] = \dfrac{1}{2}[\cosh \theta + \sinh \theta]^n + \dfrac{1}{2}[\cosh \theta - \sinh \theta]^n$.

$$= \dfrac{1}{2}[\cosh^n \theta + {}^nC_1 \cosh^{n-1}\theta \sinh \theta + {}^nC_2 \cosh^{n-2}\theta \sinh^2 \theta + \dots]$$

$$+ \dfrac{1}{2}[\cosh^n \theta - {}^nC_1 \cosh^{n-1}\theta \sinh \theta + {}^nC_2 \cosh^{n-2}\theta \sinh^2 \theta - \dots]$$

$$= \cosh^n \theta + {}^nC_2 \cosh^{n-2}\theta \sinh^2 \theta + \dots$$

Similarly, $\sinh n\theta = \dfrac{1}{2}[e^{n\theta} - e^{-n\theta}] = \dfrac{1}{2}[\cosh \theta + \sinh \theta]^n - \dfrac{1}{2}[\cosh \theta - \sinh \theta]^n$

$$= n \cosh^{n-1}\theta \sinh \theta + {}^nC_3{}^{n-3}\theta \sinh^3 \theta + \dots$$

Example 7. *Prove that* $\tan\left(\dfrac{u+iv}{2}\right) = \dfrac{\sin u + i \sinh v}{\cos u + \cosh v}$

Solution.   L.H.S. $= \tan\left(\dfrac{u+iv}{2}\right)$

$$= \dfrac{\sin\left[\dfrac{u+iv}{2}\right]}{\cos\left[\dfrac{u+iv}{2}\right]} = \dfrac{2\sin\left(\dfrac{u+iv}{2}\right)\cos\left(\dfrac{u-iv}{2}\right)}{2\cos\left(\dfrac{u+iv}{2}\right)\cos\left(\dfrac{u-iv}{2}\right)}$$

$$= \dfrac{\sin u + \sin iv}{\cos u + \cosh iv} = \dfrac{\sin u + i \sinh v}{\cos u + \cosh v} = \text{R.H.S.}$$

EXERCISE 9.2

Verify the following :

1. (a) $\sinh (x-y) = \sinh x \cosh y - \cosh x \sinh y,$

   (b) $\cosh (x-y) = \cosh x \cosh y - \sinh x \cosh y,$

2. $\tanh (x+y) = \dfrac{\tanh x + \tanh y}{1 + \tanh x \tan y}$

3. (a) $\sinh x - \sinh y = 2\cosh\dfrac{x+y}{2}\sinh\dfrac{x-y}{2}$

   (b) $\cosh x + \cosh y = 2\cosh\dfrac{x+y}{2}\cosh\dfrac{x-y}{2}$

Prove the following :

4. $\sinh(x + y)\cosh(x - y) = \dfrac{1}{2}[\sinh 2x + \sinh 2y]$

5. $\cosh 2x + \cosh 5x + \cosh 8x + \cosh 11x = 4\cosh\dfrac{13x}{2}\cosh\dfrac{3x}{2}$

6. $\sinh x + n\sinh 2x + \dfrac{n(n-1)}{1.2}\sinh 3x + ... + \text{to } (n-1)\text{term}$

$= 2^n \cosh^n\dfrac{x}{2}\sinh\left(\dfrac{n}{2}+1\right)x.$

7. $\sinh\beta\sin\alpha + i\cosh\beta\cos\alpha = i\cos(\alpha + i\beta).$

8. $\sin 2\alpha + i\sinh 2\beta = 2\sin(\alpha + i\beta)\cos(\alpha - i\beta)$

9. $\cos(\alpha + i\beta) + i\sin(\alpha + i\beta) = e^{-\beta}(\cos\alpha + i\sin\alpha).$

10. $\dfrac{1+\tanh x}{1-\tanh x} = \cosh 2x + \sinh 2x.$

11. $\cos(\alpha - i\beta) + i\sin(\alpha - i\beta) = e^{\beta}(\cos\alpha - i\sin\alpha).$

12. If $\cosh\alpha = \sec\theta$, show that $\alpha = \log_e \tan(\pi/4 + \theta/2).$

13. Prove the following

(a) $\dfrac{1}{2}[\sinh x + \sin x] = x + \dfrac{x^5}{5!} + \dfrac{x^9}{9!} + ... \text{ ad inf.}$

(b) $\dfrac{1}{2}[\cosh x + \cos x] = 1 + \dfrac{x^4}{4!} + \dfrac{x^8}{8!} + ... \text{ inf.}$

14. If $\tan\theta = \tanh x\cot y$ and $\tan\phi = \tanh x\tan y$, prove that

$\dfrac{\sin 2\theta}{\sin 2\phi} = \dfrac{\cosh 2x + \cos 2y}{\cosh 2x - \cos 2y}$

15. If n is a positive integer, show that

$2^{n-1}\cosh{}^n\theta = \cosh n\theta +^n C_1\cosh(n-2)\theta +^n C_2\cosh(n-4)\theta + ...$

16. If $\cosh u = \sec\theta$, prove that $\sinh u = \tan\theta.$

HINT TO SELECTED PROBLEMS

10. L.H.S. $= \dfrac{1+\tanh x}{1-\tanh x} = \dfrac{1+\sinh x/\cosh x}{1-\sinh x/\cosh x} = \dfrac{\cosh x + \sinh x}{\cosh x - \sinh x}$

$= \dfrac{e^x}{e^{-x}} = e^{2x} = \cosh 2x + \sinh 2x.$

14. $\dfrac{\sin 2\theta}{\sin 2\phi} = \dfrac{2\tanh\theta}{(1+\tan^2\theta)}\cdot\dfrac{(1+\tan^2\phi)}{2\tan\phi}$

15. $(2\cosh\theta)^n = (e^\theta + e^{-\theta})^n$

$= e^{n\theta} +^n C_1 e^{(n-1)\theta}e^{-\theta} +^n C_2 e^{(n-2)\theta}e^{-2\theta} + ... +^n C_{n-1}e^\theta.e^{-(n-2)\theta} + e^{-2\theta}]$

$= (e^{n\theta} + e^{-n\theta}) +^n C_1(e^{(-2)\theta} + e^{-(-2)\theta}) + ...$

$= \cosh n\theta + {}^n C_1\cosh(n-2)\theta + ...$

CHAPTER REVIEW : A COMPETITIVE APPROACH

Selected terms and Results

TERMS

■ **Euler's Exponential values :**

$$e^{i\theta} = \cos\theta + i\sin\theta.$$

$$\cos\theta = \frac{e^{i?} + e^{-i?}}{2}, \sin\theta = \frac{e^{i?} - e^{-i?}}{2i}$$

■ **Hyperbolic functions :** The quantity

$$\frac{e^{y} - e^{-y}}{2}$$ where y is real or complex is

called the hyperbolic sine of y and is written as sinh y.

Similarly, $\dfrac{e^{y} - e^{-y}}{2}$ is called the hyperbolic cosine of y and is written as cosh y.

RESULTS

■ Let $z_n = x_n + iy_n$. Then the series Σz_n is said to be convergent, if the series Σx_n and Σy_n are both convergent.

■ If a series $|z_1| + |z_n| + ... + |z_n|$ of positive real numbers where $|z_n| = \sqrt{x_n^2 + y_n^2}$ be convergent then the series Σz_n of complex numbers $z_1 + z_2 + ... + z_n$ where $z_n = x_n + y_n$ is absolutely convergent.

■ If a series is absolutely convergent, it is necessarily convergent

$$e_1^{z_1} . e_2^{z_2} = e_1^{z_1 + z_2}$$

■ If z is a complex number, then $(e^z)^m = e^{mz}$, m being a positive integer.

■ The period of cos z and sin z is 2π where as the period of tan z is π.

■ The hyperbolic functions differ from the circular functions is having imaginary periods.

■ The hyperbolic functions are connected in the same way with the rectangular hyperbola as the circular functions are with the circle.

■ cosh z and sinh z have an imaginary period of $2\pi i$ and tanh z has an imaginary period of πi.

Review Questions and Project Work

1. Show that

$$\sinh(x + y)\cosh(x - y) = \frac{1}{2}$$
$$(\sinh 2x + \sinh 2y)$$

2. If cosh α = sec θ then prove that

$$\tanh^2\frac{1}{2}a = \tanh^2\frac{1}{2}\theta$$

3. If tan $(\alpha + i\beta) = x + iy$, prove that
   (i) $x \cot 2\alpha + y \coth 2\beta = 1$
   (ii) $x^2 + y^2 - 2y \coth 2\beta + 1 = 0$

4. If tan $(\theta + i\phi) = \cos\alpha + i\sin\alpha$, prove that

$$\phi = \frac{1}{2}\log\tan\left(\frac{p}{4} + \frac{a}{2}\right).$$

5. Show that $\tan\dfrac{u + iv}{2} = \dfrac{\sin u + i\sinh v}{\cos u + \cosh v}$

6. If α, β are the imaginary cube roots of unity, prove that
   $$\alpha e^{px} + \beta e^{qx}$$
   $$= e^{-x/2}\left[\sqrt{3}\sin\frac{\sqrt{3}}{2}x + \cos\frac{\sqrt{3}}{2}x\right]$$

7. Show that
   $$\frac{\sin(a + i\beta)}{\sin(a - i\beta)} + \frac{\sin(a - i\beta)}{\sin(a + i\beta)} =$$
   $$2.\frac{2 - (e^{2\beta} + e^{-2\beta})\cos 2a}{4\sin^2 a + (e^{\beta} + e^{-\beta})^2}$$

8. If $\tan(\theta + i\phi) = \tan\alpha + i\sec\alpha$, prove that

$$e^{2\phi} = \pm\cot\frac{a}{2} \quad \text{and} \quad 2\theta = np + \frac{1}{2}p + a$$

Objective Type Questions

Fill in the blanks:

1. The series $\Sigma z_n = \Sigma x_n + i\Sigma y_n$ is convergent if Σx_n and Σy_n both are ....

2. Every absolutely convergent series is ....

3. If the series $|\Sigma z_n|$ is convergent, then it is said to be ....

4. $e^{z_1}.e^{z_2} = $ ....

5. $\{E(z)\}^{-1} = E($............$)$

6. The period for $\cos z$ and $\sin z$ is ....

7. The period for $\tan z$ is ....

8. The period for e^z is ....

9. $\cos z$ is an even function and $\sin z$ is .... function.

10. $\cosh^2 y - \sinh^2 y = $ ....

True/False: *Write 'T' for true and 'F' for false statement.*

1. The hyperbolic function differ from the circular function in having imaginary periods. (T/F)

2. $\tanh z$ has an imaginary period of $2\pi i$. (T/F)

3. $\cosh z$ and $\sinh z$ have an imaginary period of $2\pi i$. (T/F)

4. If z is a complex number then $E(z).E(-z)$ is one. (T/F)

5. The value of $\sin x$ is $\dfrac{e^x - e^{-x}}{2}$. (T/F)

Multiple Choice Questions : *Choose the most appropriate one :*

1. If z_1 and z_2 are two complex numbers then the value of $e^{z_1}.e^{z_2}$ is :

(a) $e^{z_1 - z_2}$

(b) $e^{z_1.z_2}$

(c) $e^{z_2 - z_1}$

(d) $e^{z_1 + z_2}$

(e) none of these

2. If z is a complex number then the value of $E(z).E(-z)$ is :

(a) e^{-2z}

(b) e^{2z}

(c) 1

(d) -1

(e) none of these

3. If $e^{i\theta} = \cos\theta + i\sin\theta$, then the value of $\cos\theta$ is :

(a) $\dfrac{e^{i?} - e^{-i?}}{2}$

(b) $\dfrac{e^{i?} + e^{-i?}}{2}$

(c) $\dfrac{e^{-i?} - e^{i?}}{2}$

(d) $\dfrac{e^{i?} + e^{-i?}}{-2i}$

(e) none of these

4. If $e^{i\theta} = \cos\theta + i\sin\theta$, then the value of $\sin\theta$ is :

(a) $\dfrac{e^{i?} + e^{-i?}}{2i}$

(b) $\dfrac{e^{i?} - e^{-i?}}{2i}$

(c) $\dfrac{e^{-i?} - e^{i?}}{2i}$

(d) $\dfrac{e^{i?} + e^{-i?}}{-2i}$

(e) none of these

5. Value of $\cosh x$ is :

(a) $\dfrac{e^x - e^{-x}}{2}$

(b) $\dfrac{e^{-x} - e^x}{2}$

(c) $\dfrac{e^x + e^{-x}}{2}$

(d) $\dfrac{e^x + e^{-x}}{x}$

(e) none of these

6. Value of $\sinh x$ is :

(a) $\dfrac{e^x - e^{-x}}{2}$

(b) $\dfrac{e^{-x} - e^{-x}}{2}$

(c) $\dfrac{e^x + e^{-x}}{2}$

(d) $\dfrac{e^x + e^{-x}}{x}$

(e) none of these

7. If $\cosh \alpha = \sec \theta$, then the value of α is :

    (a) $\log_e \tan\left(\dfrac{\pi}{2}+\dfrac{\theta}{2}\right)$

    (b) $\log_e \tan\left(\dfrac{\pi}{4}+\theta\right)$

    (c) $\log_e \tan\left(\pi+\dfrac{\theta}{2}\right)$

    (d) $\log_e \tan\left(\dfrac{\pi}{4}+\dfrac{\theta}{2}\right)$

    (e) none of these

8. $\cos(A+iB)$ is:

    (a) $\cos A \cos B + i \sin A \sin B$

    (b) $\cos A \sin B + i \sin A \cos B$

    (c) $\sin A \cos B + i \cos A \sin B$

    (d) $\cos A \cosh B - i \sin A \sinh B$

ANSWERS

Fill in the blanks

(1) Convergent

(2) Convergent

(3) Absolutely convergent

(4) $e^{z_1+z_2}$

(5) $-z$

(6) 2π

(7) π

(8) $2\pi i$

(9) odd

(10) 1

True/False

(1) T

(2) F

(3) T

(4) T

(5) F

Multiple choice questions

(1) d

(2) c

(3) b

(4) b

(5) c

(6) a

(7) d

(8) d

CHAPTER 10

Logarithm of Complex Quantities

> ❖ Logarithm of Positive Numbers
> ❖ Logarithm of Negative Numbers
> ❖ Logarithm of any Base
> ❖ General Exponential Function

10.1 INTRODUCTION

We know that if x and y are real quantities and $e^x = y$, then x is said to be the logarithm of y to the base e and is written as

$$x = \log_e y$$

Similarly if $e^{x+iy} = u + iv$, then $x + iy$ is called the logarithm (Napierian) of $u + iv$ to the base e and is written as

$$\log_e(u + iv) = x + iy \qquad \qquad ...(1)$$

Since $\qquad \qquad e^{2n\pi i} = 1 \qquad \qquad$ (where n is an integer or zero)

we have $\qquad e^{x+iy+2n\pi i} = e^{x+iy},$

giving that $\quad \text{Log}_e(u + iv) = 2n\pi i + x + iy \qquad \qquad ...(2)$

$$= x + i(2n\pi + y)$$

This shows that the logarithm of a complex quantity has an infinite number of values and hence is many-valued function. These values are called general values of $\log_e (u + iv)$.

This is known as the general value of the logarithm, the principal value of the logarithm is obtained by putting $n = 0$ in (2).

In order to distinguish between the general value of the logarithm as given by (2) and the principal value as given by (1), we get by putting $n = 0$ in (2), **general value is written as 'Log'** and the **principal values of 'log'.**

Since n can take any integral values, there are an infinite number of logarithms of $x + iy$ and they differ from each other by $2\pi i$.

Remark

❖ The base of a logarithm will be e, unless or otherwise stated.

10.2 LOGARITHM OF A POSITIVE REAL NUMBER

Let x be a positive real number. Then

$$x = x + 0 \cdot i = r(\cos \theta + i \sin \theta)$$

$\Rightarrow \qquad r \cos \theta = x, \quad r \sin \theta = 0$

$\therefore \qquad r = x \text{ and } \theta = 0.$

Now $\qquad \text{Log } x = \text{Log } r (\cos \theta + i \sin \theta)$

$$= 2 n\pi i + \log r (\cos \theta + i \sin \theta)$$

$$= 2 n\pi i + \log r + i \theta$$

$$= 2 n\pi i + \log x + i \cdot 0. \qquad\qquad [\therefore r = x, \theta = 0]$$

$\therefore \qquad \text{Log } x = 2 n\pi i + \log x$

10.3 LOGARITHM OF A NEGATIVE REAL NUMBER

Let x be a positive real number. Then

$$-x = -x + i \cdot 0 = r(\cos \theta + i \sin \theta)$$

$\Rightarrow \qquad r \cos \theta = -x \text{ and } r \sin \theta = 0$

$\Rightarrow \qquad r = x \text{ and } \theta = \theta \text{ (not 0)}$

Now $\qquad \text{Log } (-x) = 2 n\pi i + \log (-x)$

$$= 2 n\pi i + \log r (\cos \theta + i \sin \theta)$$

$$= 2 n\pi i + \log r + i\theta$$

$$= 2 n\pi i + \log x + i\pi$$

$\qquad \text{Log } (-x) = (2n + 1)\pi i + \log x$

10.4 LOGARITHM OF $x + iy$ IN THE FORM OF $A + iB$

Let $\qquad x + iy = r(\cos \theta + i \sin \theta)$

$\Rightarrow \qquad x = r \cos \theta, \quad y = r \sin \theta$

$\Rightarrow \qquad r = \sqrt{x^2 + y^2} \quad \text{and} \quad \theta = \tan^{-1}\left(\dfrac{y}{x}\right)$

Now $\qquad \text{Log } (x + iy) = 2 n\pi i + \log (x + iy)$

$$= 2 n\pi i + \log r (\cos \theta + i \sin \theta)$$

$$= 2 n\pi i + \log r + i\theta$$

$$= 2 n\pi i + \log \sqrt{x^2 + y^2} + i \tan^{-1}\left(\frac{y}{x}\right)$$

$\therefore \qquad \text{Log } (x + iy) = 2 n\pi i + \dfrac{1}{2} \log(x^2 + y^2) + i \tan^{-1}\left(\dfrac{y}{x}\right) \qquad \dots(1)$

This equation gives the general logarithm of $x + iy$.

For the principal value put $n = 0$ in (1), we get

$$\log (x + iy) = \frac{1}{2} \log(x^2 + y^2) + i \tan^{-1}\left(\frac{y}{x}\right) \qquad \text{...(2)}$$

Remark

❖ In (2) if we put $-y$ for y, we get

$$\log (x - iy) = \frac{1}{2} \log(x^2 + y^2) - i \tan^{-1}\left(\frac{y}{x}\right)$$

10.5 SOME IMPORTANT RESULTS

(i) $\log (z_1 z_2) = \log z_1 + \log z_2$

(ii) $\log \dfrac{z_1}{z_2} = \log z_1 - \log z_2.$

Proof. Let $z_1 = r_1 e^{i\theta_1}, \; z_2 = r_2 e^{i\theta_2}.$

Now
$$\log z_1 + \log z_2 = [\log r_1 + i(2m_1\pi + \theta_1)] + [\log r_2 + i(2m_2\pi + \theta_2)]$$
$$= (\log r_1 + \log r_2) + i (\theta_1 + \theta_2 + 2n\pi) \qquad \text{...(1)}$$

where m_1 and m_2 are integers, and $n = m_1 + m_2$

Also
$$\log z_1 z_2 = \log r_1 r_2 \; e^{i(\theta_1 + \theta_2)} = \log r_1 r_2 + i (\theta_1 + \theta_2 + 2m\pi) \qquad \text{...(2)}$$

Since n and m can take up any integral values, it is clear that every value of $\log (z_1 z_2)$ is equal to some value of $\log z_1 + \log z_2$ and that every value of latter is equal to some value of the former.

∴
$$\log z_1 z_2 = \log z_1 + \log z_2 \qquad \text{...(A)}$$

Similarly it can be proved that

$$\log \frac{z_1}{z_2} = \log z_1 - \log z_2 \qquad \text{...(B)}$$

Remark

❖ It is important to note that,

$$\log z_1 z_2 = \log z_1 + \log z_2$$

and
$$\log \frac{z_1}{z_2} = \log z_1 - \log z_2.$$

the principal values of the two sides of these equations need not necessarily be equal, for the simple reason that amp. $(z_1) \pm$ amp. (z_2) need not necessrily lie

between $-\pi$ and $+\pi$, whereas amp. $(z_1 z_2)$ and amp. $\left(\dfrac{z_1}{z_2}\right)$ must lie between

$-\pi$ and $+\pi$.

SOLVED EXAMPLES

Based on the following Results

- $\text{Log}_e\,(u + iv) = 2n\pi i + x + iy = x + i(2n\pi + y)$
- $\text{Log}\,x = 2n\pi i + \log x$
- $\text{Log}\,(-x) = (2n + 1)\pi i + \log x$

- $\text{Log}\,(x \pm iy) = \dfrac{1}{2}\log\,(x^2 + y^2) \pm i\,\tan^{-1}\dfrac{y}{x}$

Example 1. *Find the value of*

$$\text{Log}\,(-1 + \sqrt{3}\,i) + \text{Log}\,(-\sqrt{3} + i)$$

and $\qquad \text{Log}\,[(-1 + \sqrt{3}\,i)(-\sqrt{3} + i)]$

and compare their principal values.

Solution. Let $-1 + \sqrt{3}\,i = r_1\,(\cos\theta_1 + i\sin\theta_1)$ and $-\sqrt{3} + i = r_2\,(\cos\theta_2 + i\sin\theta_2)$.

$\therefore \qquad\qquad r_1 = 2,\ \theta_1 = \dfrac{2}{3}\pi,$ and $\quad \therefore \quad r_2 = 2,\ \theta_2 = \dfrac{5\pi}{6}$

$\therefore \qquad \text{Log}\,(-1 + \sqrt{3}\,i) + \text{Log}\,(-\sqrt{3} + i) = \text{Log}\,2 + \text{Log}\,2 + i\left(\dfrac{2\pi}{3} + \dfrac{5\pi}{6} + 2n\pi\right)$

$$= \text{Log}\,4 + i\left(\dfrac{3\pi}{2} + 2n\pi\right).$$

$\therefore$ The principal value is $\text{Log}\,4 + \dfrac{3\pi i}{2}$. $\hspace{4cm}$ ...(1)

Also, $\qquad \text{Log}\,[(-1 + \sqrt{3}\,i)(-\sqrt{3} + i)] = \text{Log}\,(-4i)$

$$= \text{Log}\,4\left[\cos\left(-\dfrac{\pi}{2}\right) + i\sin\left(-\dfrac{\pi}{2}\right)\right]$$

$$= \text{Log}\,4 + \left(2m\pi - \dfrac{\pi}{2}\right)i$$

$\therefore$ In this case, the principal value is,

$$\text{Log}\,4 - \dfrac{\pi}{2}i \hspace{4cm} ...(2)$$

which is different from (1).

Thus, although

$$\text{Log}\,[(-1 + \sqrt{3}\,i) + \text{Log}\,(-\sqrt{3} + i)] = \text{Log}\,(-1 + \sqrt{3}\,i)(-\sqrt{3} + i),$$

yet the principal values of the two are not equal.

Example 2. *Find the value of* $\text{Log}\,(-3)$.

Solution. Let $\qquad -3 = r\,(\cos\theta + i\sin\theta)$

$$\Rightarrow \qquad r\cos\theta = -3 \text{ and } r\sin\theta = 0$$

$$\Rightarrow \qquad r = 3 \text{ and } \theta = \pi$$

$$\therefore \qquad \text{Log}(-3) = 2n\pi i + \log(-3)$$

$$= 2n\pi i + \log r\,(\cos\theta + i\sin\theta)$$

$$= 2n\pi i + \log re^{i\theta} \qquad\qquad (\therefore\ e^{i\theta} = \cos\theta + i\sin\theta)$$

$$= 2n\pi i + \log r + i\theta$$

$$= 2n\pi i + \log 3 + i\pi$$

Hence, $\qquad \text{Log}(-3) = (2n+1)\pi i + \log 3.$

Example 3. *Prove that* $\text{Log}(1+i) = \dfrac{1}{2}\log 2 + i\left(2n\pi + \dfrac{\pi}{4}\right).$

Solution. $\text{Log}(1+i) = 2n\pi i + \log(1+i)$

$$= 2n\pi i + \frac{1}{2}\log(1^2 + 1^2) + i\tan^{-1}\left(\frac{1}{1}\right)$$

$$= 2n\pi i + \frac{1}{2}\log 2 + i\tan^{-1}(1)$$

$$= 2n\pi i + \frac{1}{2}\log 2 + \frac{i\pi}{4}$$

$$= \frac{1}{2}\log 2 + i\left(2n\pi + \frac{\pi}{4}\right)$$

Example 4. *Show that* $\log(1+e^{i\theta}) = \log\left(2\cos\dfrac{\theta}{2}\right) + \dfrac{1}{2}i\theta$ *where* $-\pi < \theta < \pi.$

Solution. We have $\qquad 1 + e^{i\theta} = 1 + \cos\theta + i\sin\theta$

$$= 2\cos^2\frac{\theta}{2} + 2i\sin\frac{\theta}{2}\cos\frac{\theta}{2}$$

$$= 2\cos\frac{\theta}{2}\left(\cos\frac{\theta}{2} + i\sin\frac{\theta}{2}\right)$$

$$= 2\cos\frac{\theta}{2}e^{i\theta/2}$$

Then $\qquad \log(1 + e^{i\theta}) = \log\left(2\cos\frac{\theta}{2}\right)e^{i\theta/2}$

$$= \log\left(2\cos\frac{\theta}{2}\right) + \frac{i\theta}{2}$$

Example 5. *Show that* $i \log \left(\dfrac{x-i}{x+i} \right) = \pi - 2 \tan^{-1} x = 2 \cot^{-1} x$

Solution. We have

$$i \log \left(\frac{x-i}{x+i} \right) = i \left[\log \frac{i(-1-xi)}{i(1-xi)} \right]$$

$$= i \log \left(\frac{-1-xi}{1-xi} \right)$$

$$= i \left[\log (-1-xi) - \log (1-xi) \right]$$

$$= i \left[\log (-1)(1+xi) - \log (1-xi) \right]$$

$$= i \left[\log (-1) + \log (1+xi) - \log (1-xi) \right]$$

$$= i \left[-i\pi + \frac{1}{2} \log (1+x^2) + i \tan^{-1} x - \frac{1}{2} \log (1+x^2) - i \tan^{-1}(-x) \right]$$

$$= i \left[-i\pi + i \tan^{-1} x - i \tan^{-1}(-x) \right]$$

$$= i \left[-i\pi + i \tan^{-1} x + i \tan^{-1} x \right] \qquad\qquad [\because \ \tan^{-1}(-x) = -\tan^{-1} x]$$

$$= i \left[-i\pi + 2i \tan^{-1} x \right]$$

$$= \pi - 2 \tan^{-1} x = 2 \cot^{-1} x \qquad\qquad \left[\because \ \cot^{-1} x = \frac{\pi}{2} - \tan^{-1} x \right]$$

Example 6. *Prove that* $\sin (\log i^i) = -1$.

Solution. We have $\quad \log i^i = i \log i$

$$= i \left[\tan^{-1} \infty \right]$$

$$= i \left[\frac{i\pi}{2} \right] = -\pi/2$$

$\therefore \qquad\qquad \sin (\log i^i) = \sin (-\pi/2) = -\sin \pi/2 = -1.$

Example 7. *Show that one of the values of* $\log \left[\dfrac{(1+i)(1+i\sqrt{3})}{\sqrt{3}+i} \right]$ *is* $\dfrac{1}{2} \log 2 + \dfrac{5\pi i}{12}$.

Solution. We have $\quad \log \left[\dfrac{(1+i)(1+i\sqrt{3})}{\sqrt{3}+i} \right]$

$$= \log \left[\frac{(1+i)(1+i\sqrt{3})(\sqrt{3}-i)}{3+1} \right]$$

$$= \log \frac{(1+i)(\sqrt{3}-i+3i+\sqrt{3})}{4}$$

$$= \log \frac{(1+i)(2\sqrt{3}+2i)}{4}$$

$$= \log \frac{(1+i)(\sqrt{3}+i)}{2}$$

$$= \log(1+i) + \log(\sqrt{3}+i) - \log 2$$

$$= \frac{1}{2}\log 2 + i\tan^{-1}(1) + \frac{1}{2}\log 4 + i\tan^{-1}\left(\frac{1}{\sqrt{3}}\right) - \log 2$$

$$= \frac{1}{2}\log 2 + \frac{i\pi}{4} + \log 2 + \frac{i\pi}{6} - \log 2$$

$$= \frac{1}{2}\log 2 + i\frac{5\pi}{12}$$

Example 8. *If* $\tan \log(x+iy) = a + ib$, *where* $a^2 + b^2 \neq 1$, *prove that*

$$\tan\{\log(x^2 + y^2)\} = \frac{2a}{1 - a^2 - b^2}$$

Solution. We have $\tan\{\log(x+iy)\} = a + ib$ ...(1)

Then $\tan\{\log(x-iy)\} = a - ib$ ...(2)

Adding (1) and (2), we get

$$\tan\{\log(x+iy)\} + \tan\{\log(x-iy)\} = 2a \qquad \qquad ...(3)$$

Multiplying (1) and (2), we get

$$\tan\{\log(x+iy)\}\tan\{\log(x-iy)\} = a^2 + b^2$$

Now $\dfrac{2a}{1-a^2-b^2} = \dfrac{\tan\{\log(x+iy) + \tan\{\log(x-iy)\}}{1 - \tan\{\log(x+iy)\}\tan\{\log(x-iy)\}}$

$$= \tan[\log(x+iy) + \log(x-iy)]$$
$$= \tan[\log(x+iy)(x-iy)]$$
$$= \tan[\log(x^2+y^2)].$$

Example 9. *If* $\log \log(x+iy) = p + iq$, *prove that* $y = x\tan[\tan q \log \sqrt{x^2+y^2}\,]$.

Solution. We have $\log \log(x+iy) = p + iq$ ...(1)

then, $\log \log(x-iy) = p - iq$ ...(2)

Substracting (2) from (1), we get

$$-2iq = \log \log(x-iy) - \log \log(x+iy)$$

$\Rightarrow$ $-2iq = \log\left[\dfrac{\log(x-iy)}{\log(x+iy)}\right]$

$\Rightarrow$ $e^{-2iq} = \dfrac{\log(x-iy)}{\log(x+iy)}$

$\Rightarrow$ $-\dfrac{e^{-iq}}{e^{iq}} = \dfrac{\log(x-iy)}{\log(x+iy)}$

$$\Rightarrow \qquad \frac{e^{-iq} - e^{iq}}{e^{-iq} + e^{iq}} = \frac{\log(x - iy) - \log(x + iy)}{\log(x - iy) + \log(x + iy)}$$

[By componendo and dividendo theorem]

$$\Rightarrow \qquad -i\tan q = \frac{\frac{1}{2}\log(x^2 + y^2) - i\tan^{-1}\frac{y}{x} - \frac{1}{2}\log(x^2 + y^2) - i\tan^{-1}\left(\frac{y}{x}\right)}{\log(x^2 + y^2)}$$

$$\Rightarrow \qquad -i\tan q = \frac{-2i\tan^{-1}\left(\dfrac{y}{x}\right)}{\log(x^2 + y^2)}$$

$$\Rightarrow \qquad \tan^{-1}\frac{y}{x} = \tan q.\frac{1}{2}\log(x^2 + y^2)$$

$$\Rightarrow \qquad \tan^{-1}\frac{y}{x} = \tan q.\log\sqrt{(x^2 + y^2)}$$

Hence $\qquad y = x\tan[\tan q.\log\sqrt{(x^2 + y^2)}]$.

Example 10. *Find the general value of* $\text{Log}_e\,(1 + \cos\theta + i\sin\theta)$.

Solution. We have

$$\text{Log}_e\,(1 + \cos\theta + i\sin\theta) = \log_e\left(2\cos^2\frac{\theta}{2} + 2i\sin\frac{\theta}{2}\cos\frac{\theta}{2}\right) + 2n\pi i$$

$$= \log_e 2\cos\frac{\theta}{2}\left(\cos\frac{\theta}{2} + i\sin\frac{\theta}{2}\right) + 2n\pi i$$

$$= \log_e 2\cos\frac{\theta}{2} + \log_e e^{i\theta/2} = \log_e 2\cos\frac{\theta}{2} + i\left(\frac{\theta}{2} + 2n\pi\right).$$

when n is any integer positive or negative or zero.

Example 11. *Prove that* $\log_e \tan\left(\dfrac{\pi}{4} + \dfrac{x}{2}i\right) = i\tan^{-1}(\sinh x)$

Solution. L.H.S. $= \log_e \tan\left(\dfrac{\pi}{4} + \dfrac{x}{2}i\right) = \log_e \dfrac{\sin\left(\dfrac{\pi}{4} + \dfrac{x}{2}i\right)}{\cos\left(\dfrac{\pi}{4} + \dfrac{x}{2}i\right)}$

$$= \log_e\left\{\frac{2\sin\left(\dfrac{\pi}{4} + \dfrac{x}{2}i\right)\cos\left(\dfrac{\pi}{4} - \dfrac{x}{2}i\right)}{2\cos\left(\dfrac{\pi}{4} + \dfrac{x}{2}i\right)\cos\left(\dfrac{\pi}{4} - \dfrac{x}{2}i\right)}\right\}$$

$$= \log_e \frac{\sin\dfrac{\pi}{2} + \sin xi}{\cos\dfrac{\pi}{2} + \cos xi} = \log_e \frac{1 + i\sinh x}{\cosh x} \qquad (\text{as } \sin ix = i\sinh x)$$

$$= \frac{1}{2}\log_e \frac{1 + \sinh^2 x}{\cosh^2 x} + i\tan^{-1}(\sinh x)$$

$$= \frac{1}{2}\log_e \frac{\cosh^2 x}{\cosh^2 x} + i\tan^{-1}(\sinh x)$$

$$= \frac{1}{2}\log_e 1 + i\tan^{-1}(\sinh x) = i\tan^{-1}(\sinh x).$$

Example 12. *If $u = \log\tan\left(\dfrac{\pi}{4} + \dfrac{\theta}{2}\right)$, prove that*

$$\theta = -i\log\tan\left(\frac{\pi}{4} + \frac{iu}{2}\right).$$

Hence or otherwise, prove that if also $u = \theta + a_3\theta^3 + a_5\theta^5 + \dots$

then $\qquad\qquad\qquad \theta = u - a_3 u^3 + a_5 u^5 - \dots$

Solution. We have $\qquad\qquad \log\tan\left(\dfrac{\pi}{4} + \dfrac{\theta}{2}\right) = u$

or $\qquad\qquad\qquad\qquad \tan\left(\dfrac{\pi}{4} + \dfrac{\theta}{2}\right) = e^u$

$\therefore \qquad\qquad\qquad \dfrac{1 + \tan\dfrac{\theta}{2}}{1 - \tan\dfrac{\theta}{2}} = \dfrac{e^{u/2}}{e^{-u/2}}$

Applying dividendo and componendo, we have

$$\tan\frac{\theta}{2} = \frac{e^{u/2} - e^{-u/2}}{e^{u/2} + e^{-u/2}} = \tanh\frac{u}{2}$$

or $\qquad\qquad \tan\dfrac{\theta}{2} = -i\tanh\dfrac{ui}{2}$

$\therefore \qquad \dfrac{e^{i\theta/2} - e^{-i\theta/2}}{i(e^{i\theta/2} + e^{-i\theta/2})} = -i\tanh\dfrac{ui}{2}$

or $\qquad \dfrac{e^{i\theta/2} - e^{-i\theta/2}}{(e^{i\theta/2} + e^{-i\theta/2})} = \dfrac{\tan ui/2}{1}$

Applying again componendo and dividendo, we get

$$\frac{e^{i\theta/2}}{e^{-i\theta/2}} = \frac{1+\tan ui/2}{1-\tan ui/2}$$

or $\qquad\qquad e^{i\theta} = \tan\left(\frac{\pi}{4}+\frac{ui}{2}\right)$

$\therefore\qquad\qquad \theta = -i\log\tan\left(\frac{\pi}{4}+\frac{ui}{2}\right)$

Next, $\qquad\qquad \theta = -i[ui + a_3(ui)^3 + a_5(ui)^5 + ...]$

$\therefore\qquad\qquad \theta = u - a_3 u^3 + a_5 u^5 - ...$

Example 13. *Show that* $\tan\left(i\log\dfrac{a-ib}{a+ib}\right) = \dfrac{2ab}{a^2-b^2}$

Solution. Let $\qquad a = r\cos\theta, b = r\sin\theta.$

$\therefore\qquad \dfrac{a-ib}{a+ib} = \dfrac{r(\cos\theta - i\sin\theta)}{r(\cos\theta + i\sin\theta)} = \dfrac{e^{-i\theta}}{e^{i\theta}} = e^{-2i\theta}$

$\therefore\qquad \tan\left[i\log\dfrac{a-ib}{a+ib}\right] = \tan[i(-2i\theta)]$

$$= \tan 2\theta = \frac{2\tan\theta}{1-\tan^2\theta} = \frac{2\dfrac{b}{a}}{1-\dfrac{b^2}{a^2}} = \frac{2ab}{a^2-b^2}$$

Example 14. *If* $\log_e \sin(x+iy) = \alpha + i\beta$*, show that*

(i) $\alpha = \dfrac{1}{2}\log\dfrac{1}{2}(\cosh 2y - \cos 2x)$ $\qquad$ (ii) $2\cos 2x = e^{2y} + e^{-2y} - 4e^{2\alpha},$

(iii) $\beta = \tan^{-1}(\cos x \tanh y),$ $\qquad\qquad$ (iv) $\cos(x-\beta) = e^{2y}\cos(x+\beta)$

Solution. (i) We have $\qquad \log_e \sin(x+iy) = \alpha + i\beta$

$\therefore\qquad\qquad \sin(x+iy) = e^{\alpha+i\beta}$

or $\qquad \sin x \cosh y + i\cos x \sinh y = e^\alpha(\cos\beta + i\sin\beta).$

Equating real and imaginary parts, we get

$\therefore\qquad\qquad\qquad \sin x \cosh y = e^\alpha \cos\beta,$ $\qquad\qquad$ ...(1)

$\qquad\qquad\qquad\qquad \cos x \sinh y = e^\alpha \sin\beta$ $\qquad\qquad$ ...(2)

Squaring and adding (1) and (2), we get

$\qquad\qquad \sin^2 x \cosh^2 y + \cos^2 x \sinh^2 y = e^{2\alpha}(\cos^2\beta + \sin^2\beta)$

or $\qquad \dfrac{1}{4}(1-\cos 2x)(1+\cosh 2y) + \dfrac{1}{4}(1+\cos 2x)(\cosh 2y - 1) = e^{2\alpha}$

or $$\frac{1}{2}\left[\cosh 2y - \cos 2x\right] = e^{2\alpha} \qquad \text{...(A)}$$

$$\therefore \qquad \alpha = \frac{1}{2}\log\frac{1}{2}\left[\cosh 2y - \cos 2x\right]$$

(*ii*) From result (A),

$$\frac{1}{2}\left[\frac{e^{2y} + e^{-2y}}{2}\right] - \frac{1}{2}\cos 2x = e^{2\alpha}$$

or $$2\cos 2x = e^{2y} + e^{-2y} - 4e^{2\alpha}.$$

(*iii*) From (1) or (2), we get on dividing

$$\tan \beta = \frac{\cos x \sinh y}{\sin x \cosh y} \qquad \text{...(B)}$$

or $$\tan \beta = \cot x \tanh y.$$

$$\therefore \qquad \beta = \tan^{-1}(\cot x \tanh y)$$

(*iv*) From result (B)

$$\frac{\cosh y}{\sinh y} = \frac{\cos x \cos \beta}{\sin x \sin \beta}$$

By componendo and dividendo, we have

$$\frac{\cosh y + \sinh y}{\cosh y - \sinh y} = \frac{\cos x \cos \beta + \sin x \sin \beta}{\cos x \cos \beta - \sin x \sin \beta}$$

or $$\frac{e^y}{e^{-y}} = \frac{\cos(x - \beta)}{\cos(x + \beta)}$$

$$\therefore \qquad e^{2y}\cos(x + \beta) = \cos(x - \beta)$$

Example 15. *If $a + ib = e^{x + iy}$ prove that* $\dfrac{y}{x} = \dfrac{2\tan^{-1}b/a}{\log(a^2 + b^2)}$

Solution. We have $\qquad a + ib = e^{x + iy}$

Then $\qquad\qquad x + iy = \log(a + ib)$

$$= \frac{1}{2}\log(a^2 + b^2) + i\tan^{-1}\frac{b}{a}$$

Separating real and imaginary parts, we have

$$x = \frac{1}{2}\log(a^2 + b^2)$$

$$y = \tan^{-1}\frac{b}{a}$$

Therefore, $$\frac{y}{x} = \frac{2\tan^{-1}b/a}{\log(a^2 + b^2)}$$

EXERCISE 10.1

Prove the following :

1. (a) $\text{Log } i = \dfrac{1}{2}(4n+1)\pi i$

   (b) $\text{Log } 3i = \log 3 + \left(2n\pi + \dfrac{1}{2}\pi\right)i$

   (c) $\text{Log}\left(\dfrac{a+ib}{a-ib}\right) = 2i\tan^{-1}\left(\dfrac{b}{a}\right)$

2. (a) $\text{Log }(-i) = \dfrac{1}{2}(4n-1)\pi i.$ (b) $\text{Log } \sqrt{i} = \dfrac{1}{4}(4n+1)\pi i.$

3. $\log(1+i\tan\theta) = \log_e \sec\theta + i\theta.$

4. Show that $\log_e \dfrac{1}{1-e^{i\alpha}} = \log_e\left[\dfrac{1}{2}\operatorname{cosec}\dfrac{\alpha}{2}\right] + i\left(\dfrac{\pi}{2} - \dfrac{\alpha}{2}\right).$

5. Show that $\log\log(x+iy) = \dfrac{1}{2}\log(\alpha^2+\beta^2) + \tan^{-1}\dfrac{\beta}{\alpha}$

   where $2\alpha = \log_e(x^2+y^2)$ and $\beta = \tan^{-1}\dfrac{y}{x}$

6. (a) If $(a_1 + ib_1)(a_2 + ib_2) \dots (a_n + ib_n) = A + iB$, prove that

   $$\tan^{-1}\dfrac{b_1}{a_1} + \tan^{-1}\dfrac{b_2}{a_2} + \dots + \tan^{-1}\dfrac{b_n}{a_n} = \tan^{-1}\dfrac{B}{A}$$

   and $(a_1^2 + b_1^2).(a_2^2 + b_2^2) \dots (a_n^2 + b_n^2) = A^2 + B^2$

   (b) If $(1+i)(1+2i)(1+3i) \dots (1+ni) = A + iB$
   Show that $2 . 5 . 10 \dots (1+n^2) = A^2 + B^2.$

7. Prove that $\log\dfrac{\cos(x-iy)}{\cos(x+iy)} = 2i\tan^{-1}(\tan x \tanh y)$

8. Prove that the value of $\log\log\sin(x+iy)$ is $\dfrac{1}{2}\log(u^2+v^2) + i\tan^{-1}\dfrac{v}{u}$,

   where $u = \dfrac{1}{2}\log\dfrac{\cosh 2y - \cos 2x}{2}$ and $v = \tan^{-1}(\cot x \tanh y)$

9. If $\log_e \log_e \log_e(\alpha + i\beta) = p + iq$, prove that

   (i) $\exp.(e^p.\cos q)\cos(e^p\sin q) = \dfrac{1}{2}\log(\alpha^2 + \beta^2)$

   (ii) $\exp.(e^p\cos q)\sin(e^p\sin q) = \tan^{-1}\dfrac{\beta}{\alpha}$

10. Show that $\log \cos (x + iy) = \frac{1}{2}\log\left\{\frac{1}{2}(\cosh 2y + \cos 2x)\right\} - i \tan^{-1}(\tan x \tanh y)$.

11. Prove that $\log\left[\dfrac{\sin(x+iy)}{\sin(x-iy)}\right] = 2i \tan^{-1}(\cot x \tanh y)$.

12. Prove that $\log (1 + \cos 2\theta + i \sin 2\theta) = \log (2 \cos \theta) + i\theta$, if $-\pi < \theta < \pi$.

13. Prove that $\log (i\beta) = \log |\beta| \dfrac{\pi}{2}$, positive or negative sign being taken as β is positive or negative.

14. Prove that $\log\dfrac{(a-b)+i(a+b)}{(a+b)+i(a-b)} = i\left[2n\pi + \tan^{-1}\dfrac{2ab}{a^2-b^2}\right]$

15. Find the general value of $\log \sqrt{i}$.

HINT TO SELECTED PROBLEMS

3. L.H.S. $= \log (1 + i \tan \theta) = \log 1 + i\,\dfrac{\sin\theta}{\cos\theta} = \log\left(\dfrac{\cos\theta + i\sin\theta}{\cos\theta}\right)$

$\qquad = \log\dfrac{e^{i\theta}}{\cos\theta} = \log(\sec\theta.e^{i\theta}) = \log\sec\theta + \log e^{i\theta} = \log\sec\theta + i\theta.$

4. L.H.S. $= \log\dfrac{1}{1 - \cos\alpha - i\sin\alpha} = \log\dfrac{1}{2\sin^2\dfrac{\alpha}{2} - 2i\sin\dfrac{\alpha}{2}\cos\dfrac{\alpha}{2}}$

$\qquad = \log\left[\dfrac{1}{2}\csc\dfrac{\alpha}{2}\right] - \log\left[\sin\dfrac{\alpha}{2} - i\cos\dfrac{\alpha}{2}\right]$

$\qquad = \log\left(\dfrac{1}{2}\csc\dfrac{\alpha}{2}\right) - \log\left[\cos\left(\dfrac{\pi}{2} - \dfrac{\alpha}{2}\right) - i\sin\left(\dfrac{\pi}{2} - \dfrac{\alpha}{2}\right)\right].$

5. Using log $\log (x + iy) = \log\left[\dfrac{1}{2}\log(x^2 + y^2) + i\left(\tan^{-1}\dfrac{y}{x}\right)\right].$

7. $\log\dfrac{\cos x - iy}{\cos x + iy} = \log[\cos x \cosh y + i \sin x \sinh y] - \log[\cos x \cosh y - i \sin x \sinh y].$

10. $\log \cos (x + iy) = \log (\cos x \cosh y - i \sin x \sinh y)$

$\qquad = \dfrac{1}{2}\log(\cos^2 x \cosh x + \sin^2 x \sinh^2 y) - i\tan^{-1}\left(\dfrac{\sin x \sinh y}{\cos x \cosh y}\right)$

$\qquad = \dfrac{1}{2}\log\left[\dfrac{1}{2}(1 + \cos 2x)\cosh^2 y + \dfrac{1}{2}(1 - \cos 2x)\sinh^2 y - i\tan^{-1}(\tan x \tanh y)\right]$

10.6 GENERAL EXPONENTIAL FUNCTION

The general exponential function is defined as :

$$a^z = e^{z \operatorname{Log} a} \qquad \qquad ...(1)$$

where a and z are any two complete numbers.

The function a^z is many valued function as $\log a$

(i) *General value of a^z*

From (1), we have

$$a^z = e^{z \operatorname{Log} a}$$
$$= \exp [z \operatorname{Log} a]$$
$$a^z = \exp [z (\log a + 2 n\pi i)] \qquad \qquad ...(2)$$
$$[\because \operatorname{Log} a = \log a + 2n\pi i]$$

(ii) *Principal value of a^z*

putting $n = 0$ in (2), we get

$$a^z = \exp [z \log a]$$

which is the principal value of a^z.

10.7 REAL AND IMAGINARY PART OF $(\alpha + i\beta)^{p + iq}$

Let $\qquad (\alpha + i\beta)^{p + iq} = A + iB$

Now $\qquad (\alpha + i\beta)^{p + iq} = \exp \{(p + iq) \operatorname{Log}_e(\alpha + i\beta)\}$

$$= \exp\left[(p+iq)\left\{ \frac{1}{2}\log(\alpha^2 +\beta^2)+i\left(\tan^{-1}\frac{\beta}{\alpha}+2n\pi \right)\right\}\right]$$

$$= \exp\left[\frac{1}{2}p\log(\alpha^2 +\beta^2)-q\left(\tan^{-1}\frac{\beta}{\alpha}+2n\pi \right)\right.$$

$$\left. +i\left\{\frac{1}{2}q\log_e(\alpha^2 +\beta^2)+p\left(\tan^{-1}\frac{\beta}{\alpha}+2n\pi \right)\right\}\right]$$

Hence $\qquad (\alpha + i\beta)^{p + iq} = e^x (\cos y + i \sin y)$

where $\qquad\qquad x = \frac{1}{2}p\log_e(\alpha^2 +\beta^2)-q\left(\tan^{-1}\frac{\beta}{\alpha}+2n\pi \right)$

and $\qquad\qquad y = \frac{1}{2}q\log_e(\alpha^2 +\beta^2)+p\left(\tan^{-1}\frac{\beta}{\alpha}+2n\pi \right)$

The real part $\qquad A = e^x \cos y$ and the imaginary part $B = e^x \sin y$

Remark

❖ If the only principal value is required, we put $n = 0$ in the above result.

10.8 LOGARITHMS OF ANY BASE

If z, w and σ be any three complex numbers, and if

$$\sigma^w = z,$$ ...(1)

then we define that w is a logarithm of z to the base σ, and we write

$$\text{Log}_\sigma z = w$$ ...(2)

But we have already defined σ^w as $e^{w\,\text{Log}_e\,\sigma}$

$\therefore$ $e^{w\,\text{Log}_e\,\sigma} = z$ or $w\text{Log}_e\sigma = \text{Log}_e z$

or $$w = \frac{\text{Log}_e z}{\text{Log}_e \sigma}$$ ...(3)

From (2) and (3), we have

$$\text{Log}_\sigma z = \frac{\text{Log}_e z}{\text{Log}_e \sigma}$$ ...(A)

With the help of formula (A), we can write logarithm of any base to base 'e'. The principal value of $\text{Log}_\sigma z$ is defined by

$$\text{Log}_\sigma z = \frac{\text{Log}_e z}{\text{Log}_e \sigma}$$ ...(B)

SOLVED EXAMPLES

Example 1. *Prove that* $\text{Log}_i i = \dfrac{4m+1}{4n+1}$, *where m and n are integers.*

Solution. We know that

$$\text{Log}_a b = \frac{\text{Log}\,b}{\text{Log}\,a}$$

Then $$\text{Log}_i i = \frac{\text{Log}\,i}{\text{Log}\,i}$$

$$= \frac{\log i + 2m\pi i}{\log i + 2n\pi i}, \quad m, n \in Z$$

$$= \frac{i\pi/2 + 2m\pi i}{i\pi/2 + 2n\pi i} = \frac{4m+1}{4n+1}$$

Example 2. *Find the general and principal value of* $(i)^i$.

Solution. We know that

$$a^z = e^{z\text{Log}\,a}$$

So, $$(i)^i = e^{i\text{Log}\,i}$$

$$= e^{i[\log i + 2n\pi i]} \qquad \left[\because \log i = \frac{i\pi}{2}\right]$$

$$= e^{i[i\pi/2 + 2n\pi i]}$$

$$= e^{-(\pi/2 + 2n\pi)}$$

$$\therefore \qquad (i)^i = e^{-(4n + 1)\pi/2}. \qquad \qquad \dots(1)$$

For principal value, put $n = 0$ in (1), we get

$$i^i = e^{-\pi/2}$$

Also putting $n = 0, 1, 2, 3, \dots$ in (1), the various values of i^i are $e^{-\pi/2}, e^{-5\pi/2}, e^{-9\pi/2}, e^{-13\pi/2},$ $\dots$, which form a geometric progression with common ratio $e^{-\pi/2}$.

Example 3. *If $i^{\alpha + i\beta} = e^x(\cos y + i \sin y)$, then prove that*

$$x = -\frac{1}{2}(4n + 1)\pi\beta \qquad and \quad y = \frac{1}{2}(4n + 1)\pi\alpha$$

Solution. We know that

$$a^z = e^{z \, \text{Log } a}$$

So, $\qquad i^{\alpha + i\beta} = e^{(\alpha + i\beta) \, \text{Log } i}$

$$= e^{(\alpha + i\beta) \, [\log i + 2n\pi i]}$$

$$= e^{(\alpha + i\beta) \, [i\pi/2 + 2n\pi i]}$$

$$= e^{(\alpha + i\beta) \frac{i(4n+1)\pi}{2}}$$

$$= e^{i(4n+1)\pi\alpha/2} e^{-\left(\frac{(4n+1)\pi}{2}\right)\pi\beta}$$

$$\therefore \qquad i^{\alpha + i\beta} = e^{-\frac{1}{2}(4n+1)\pi\beta} \left[\cos\left\{\frac{1}{2}(4n + 1)\pi\alpha\right\} + i\sin\left\{\frac{1}{2}(4n + 1)\pi\alpha\right\} \right]$$

But $\qquad i^{\alpha + i\beta} = e^x (\cos y + i \sin y)$

$$\Rightarrow \quad e^x (\cos y + i \sin y) = e^{-\frac{1}{2}(4n+1)\pi\beta} \left[\cos\left\{\frac{1}{2}(4n + 1)\pi\alpha\right\} + i\sin\left\{\frac{1}{2}(4n + 1)\pi\alpha\right\} \right]$$

$$\Rightarrow \qquad x = -\frac{1}{2}(4n + 1)\pi\beta \quad and \quad y = \frac{1}{2}(4n + 1)\pi\alpha$$

Example 4. *If $a^{\alpha + i\beta} = (x + iy)^{p + iq}$, principal values only being considered, prove that*

(i) $\alpha = \frac{1}{2}p\log_a(x^2 + y^2) - q\tan^{-1}\left(\frac{y}{x}\right)\log_a e$

(ii) $\dfrac{2(\alpha p + \beta q)}{p^2 + q^2} = \log_a(x^2 + y^2)$

Solution. We know that

$$a^z = e^{z \, \text{Log } a}$$

For principal value, we have

$$a^z = e^{z \, \log a}$$

Since, $\qquad a^{\alpha + i\beta} = (x + iy)^{p + iq}$

$\Rightarrow \qquad e^{(\alpha + i\beta)\log a} = e^{(p + iq)\log(x + iy)}$

$\therefore \qquad (\alpha + i\beta)\log a = (p + iq)\log(x + iy)$ ...(1)

(i) From (1), we have

$$\alpha \log a + i\beta \log a = (p + iq)\left[\frac{1}{2}\log(x^2 + y^2) + i\tan^{-1}\frac{y}{x}\right]$$

Equating real parts on both sides, we get

$$\alpha \log a = \frac{1}{2}p\log(x^2 + y^2) - q\tan^{-1}\left(\frac{y}{x}\right)$$

$\Rightarrow \qquad \alpha = \dfrac{1}{2}\dfrac{p\log(x^2 + y^2)}{\log a} - \dfrac{q\tan^{-1}\dfrac{y}{x}}{\log a}$

$$= \frac{1}{2}p\log_a(x^2 + y^2) - q\tan^{-1}\left(\frac{y}{x}\right)\log_a e$$

(ii) From (1), we have

$$\frac{\alpha + i\beta}{p + iq} = \frac{\log(x + iy)}{\log a} = \log_a(x + iy)$$

or $\qquad \dfrac{\alpha + i\beta}{p + iq} = \dfrac{1}{2}\log_a(x^2 + y^2) + i\tan^{-1}\left(\dfrac{y}{x}\right)\log_a e$

or $\qquad \dfrac{(\alpha + i\beta)(p - iq)}{p^2 + q^2} = \dfrac{1}{2}\log_a(x^2 + y^2) + i\tan^{-1}\left(\dfrac{y}{x}\right)\log_a e$

Equating real parts on both sides, we get

$$\frac{\alpha p + \beta q}{p^2 + q^2} = \frac{1}{2}\log_a(x^2 + y^2)$$

$\therefore \qquad \dfrac{2(\alpha p + \beta q)}{p^2 + q^2} = \log_a(x^2 + y^2)$

Example 5. *If* $\sin(\log i^i) = a + ib$, *find a and b. Hence find* $\cos(\log i^i)$.

Solution. We have $\qquad \log i^i = i\log i$

$$= i[i\tan^{-1}\infty]$$

$$= i[i\pi/2] = -\pi/2$$

$\therefore \qquad \sin(\log i^i) = \sin(-\pi/2) = -1.$

But $\qquad \sin(\log i^i) = a + ib$

$\Rightarrow \qquad a = -1, \ b = 0.$

Also,　　　$\cos(\log i^i) = \sqrt{1 - \sin^2(\log i^i)}$

$$= \sqrt{1 - (-1)^2} = \sqrt{1-1} = 0.$$

Example 6. *If* $i^{i^{i\ldots ad.\ inf}} = A + iB$, *principal values only being considered, prove that*

(i)　$\tan\dfrac{1}{2}\pi A = \dfrac{B}{A}$　　　　　　　　　　(ii)　$A^2 + B^2 = e^{-\pi B}$.

Solution. We have

$$i^{i^{i\ldots ad.\ inf}} = A + iB$$

$\Rightarrow$　　　　　　　　$i^{(A + iB)} = A + iB$

$\Rightarrow$　　　　　$e^{(A + iB)\log i} = A + iB$　　　　　　　　[∵ principal value being taken]

$\Rightarrow$　　　　　$e^{(A+iB)\left(\dfrac{i\pi}{2}\right)} = A + iB$

$\Rightarrow$　　　　　$e^{-\frac{\pi B}{2}} e^{\frac{i\pi A}{2}} = A + iB$

$\Rightarrow$　　　$e^{-\frac{\pi B}{2}}\left(\cos\dfrac{\pi A}{2} + i\sin\dfrac{\pi A}{2}\right) = A + iB$

Separating the real and imaginary parts, we get

$$e^{-\frac{\pi B}{2}}\cos\dfrac{\pi A}{2} = A \qquad\qquad\qquad \ldots(1)$$

and　　　　　$e^{-\frac{\pi B}{2}}\sin\dfrac{\pi A}{2} = B. \qquad\qquad\qquad \ldots(2)$

(i)　Dividing (2) by (1), we get

$$\frac{\sin\dfrac{\pi A}{2}}{\cos\dfrac{\pi A}{2}} = \frac{B}{A}$$

∴　　　　　　　　$\tan\dfrac{1}{2}\pi A = \dfrac{B}{A}$

(ii)　Squaring (1) and (2) and adding, we get

$$e^{-\frac{\pi B}{2}}\left(\cos^2\dfrac{\pi A}{2} + \sin^2\dfrac{\pi A}{2}\right) = A^2 + B^2$$

∴　　　　　　　　$A^2 + B^2 = e^{-\pi B}$

Example 7. *If* $i^{x + iy} = x + iy$, *prove that* $x^2 + y^2 = e^{-(4n + 1)\pi y}$

Solution. We have

$$i^{(x + iy)} = x + iy$$

$$\Rightarrow \quad e^{(x + iy)\text{Log } i} = x + iy$$

$$\Rightarrow \quad e^{(x + iy)(2n\pi i + \log i)} = x + iy$$

$$\Rightarrow \quad e^{(x+iy)\left(2n\pi i + \frac{i\pi}{2}\right)} = x + iy$$

$$\Rightarrow \quad e^{-(4n+1)\pi y/2}e^{(4n+1)i\pi x/2} = x + iy$$

$$\Rightarrow \quad e^{-(4n+1)\pi y/2}\left[\cos\left\{(4n+1)\frac{\pi x}{2}\right\} + i\sin\left\{(4n+1)\frac{\pi x}{2}\right\}\right] = x + iy$$

Separating the real and imaginary parts, we get

$$\Rightarrow \quad e^{-(4n+1)\pi y/2}\cos\left\{(4n+1)\frac{\pi x}{2}\right\} = x$$

and

$$e^{-(4n+1)\pi y/2}\sin\left\{(4n+1)\frac{\pi x}{2}\right\} - y$$

Squaring and adding, we get

$$x^2 + y^2 = e^{-(4n+1)\pi y}.$$

Example 8. *If* $(a + ib)^p = m^{x + iy}$, *then prove that* $\dfrac{y}{x} = \dfrac{2\tan^{-1}(b / a)}{\log(a^2 + b^2)}$ *where only principal values are considered.*

Solution. We have $\quad (a + ib)^p = m^{x + iy}$

Taking logarithm both sides, we get

$$p \log (a + ib) = (x + iy) \log m$$

$$\Rightarrow \quad p\left[\frac{1}{2}\log(a^2 + b^2) + i\tan^{-1}\left(\frac{b}{a}\right)\right] = x \log m + iy \log m.$$

Separating real and imaginary parts, we get

$$x \log m = \frac{p}{2}\log(a^2 + b^2) \qquad \qquad ...(1)$$

and

$$y \log m = p\tan^{-1}\left(\frac{b}{a}\right) \qquad \qquad ...(2)$$

Dividing (2) by (1), we get

$$\frac{y}{x} = \frac{2\tan^{-1}(b / a)}{\log(a^2 + b^2)}$$

Example 9. *Prove that if* $(1 + i \tan \alpha)^{1 + i \tan \beta}$ *can have real values, and one of them is* $(\sec \alpha)^{\sec^2 \beta}$.

Solution. $(1 + i \tan \alpha)^{1 + i \tan \beta}$

$$= e^{(1 + i \tan \alpha) \log (1 + i \tan \alpha)} \qquad \text{[if principal values are considered]}$$

$$= e^{(1 + i \tan \alpha) (\log \sec \alpha + i\alpha)}$$

$$= e^{(\log \sec \alpha - \alpha \tan \beta)} . e^{i(\alpha + \tan \beta \log \sec \alpha)}$$

$$\therefore \ (1 + i \tan \alpha)^{1 + i \tan \beta} = e^{(\log \sec \alpha - \alpha \tan \beta)} [\cos (\alpha + \tan \beta \log \sec \alpha)]$$
$$+ i \sin (\alpha + \tan \beta \log \sec \alpha)] \qquad \qquad ...(1)$$

Now $(1 + i \tan \alpha)^{1 + i \tan \beta}$ will have real values if

$$\sin (\alpha + \tan \beta \log \sec \alpha) = 0$$

$$\Rightarrow \qquad\qquad \alpha + \tan \beta \log \sec \alpha = 0$$

$$\Rightarrow \qquad\qquad\qquad \alpha = - \tan \beta \log \sec \alpha.$$

Putting this value of α in R.H.S. of (1), we get

$$(1 + i \tan \alpha)^{1 + i \tan \beta} = e^{(\log \sec \alpha + \tan^2 \beta \log \sec \alpha)}$$

$$= e^{\sec^2 \beta \log \sec \alpha}$$

$$= e^{\log (\sec \alpha)^{\sec^2 \beta}}$$

$$= (\sec \alpha)^{\sec^2 \beta}$$

Hence one of real values is $(\sec \alpha)^{\sec^2 \beta}$.

Example 10. *Prove that the real part of the principal value of* $(i)^{\log(1 + i)}$ *is*

$$e^{-\pi^2 / 8} \cos\left(\frac{1}{4}\pi \log 2\right)$$

Solution. Let $\qquad (i)^{\log(1 + i)} = A + iB$

$$\Rightarrow \qquad e^{\log(1 + i) \log i} = A + iB \qquad\qquad\qquad \text{[Principal values are considered]}$$

$$\Rightarrow \qquad\qquad e^{\log(1+i)\left(\frac{i\pi}{2}\right)} = A + iB$$

$$\Rightarrow \qquad\qquad e^{\frac{i\pi}{2}\left[\frac{1}{2}\log 2 + \frac{i\pi}{4}\right]} = A + iB$$

$$\Rightarrow \qquad\qquad e^{-\pi^2 / 8} e^{\frac{i\pi}{4}\log 2} = A + iB$$

$$\Rightarrow \qquad e^{-\pi^2 / 8}\left[\cos\left(\frac{\pi}{4}\log 2\right) + i\sin\left(\frac{\pi}{4}\log 2\right)\right] = A + iB$$

Equating real parts, we get

$$A = e^{-\pi^2 / 8} \cos\left(\frac{1}{4}\pi \log 2\right)$$

Example 11. *Find the general value of $(\alpha + i\beta)^{(p + iq)}$ and show that the sum of the moduli of the values of $(1 + i)^{1 + i}$ which are less than unity is $\dfrac{1}{\sqrt{2}}e^{3\pi/4}$ cosech π.*

Solution. Let

$$(\alpha + i\beta)^{(p + iq)} = e^x (\cos y + i \sin y) \qquad \text{...(1)}$$

Then

$$e^{(p + iq)\, \text{Log}\, (\alpha + i\beta)} = e^x (\cos y + i \sin y)$$

$$\Rightarrow \qquad e^{(p + iq)\, [\log (\alpha + i\beta) + 2n\pi i]} = e^x (\cos y + i \sin y)$$

$$\Rightarrow \qquad e^{(p+iq)\left[\frac{1}{2}\log(\alpha^2 +\beta^2)+i\tan^{-1}\frac{\beta}{\alpha}+2n\pi i\right]} = e^x (\cos y + i \sin y)$$

$$\Rightarrow \quad e^{\frac{p}{2}\log(\alpha^2 +\beta^2)-q\left(2n\pi+\tan^{-1}\frac{\beta}{\alpha}\right)}\, e^{i\left[\frac{q}{2}\log(\alpha^2 +\beta^2)+ 2n\pi p + p\tan^{-1}\frac{\beta}{\alpha}\right]}$$

$$= e^x (\cos y + i \sin y)$$

Separating real and imaginary parts, we have

$$x = \frac{1}{2}p\log(\alpha^2 +\beta^2)-q\left(2n\pi + \tan^{-1}\frac{\beta}{\alpha}\right) \qquad \text{...(2)}$$

and

$$y = \frac{1}{2}q\log(\alpha^2 +\beta^2)+p\left(2n\pi + \tan^{-1}\frac{\beta}{\alpha}\right) \qquad \text{...(3)}$$

Hence the general value of $(\alpha + i\beta)^{p + iq}$ is $e^x (\cos y + i \sin y)$ where x and y are given by (2) and (3).

Now putting $\alpha = 1$, $\beta = 1$ and $p = 1$, $q = 1$ in (2) and (3), we get

$$x = \frac{1}{2}\log 2 -\left(2n\pi + \frac{\pi}{4}\right) \qquad \text{...(4)}$$

and

$$y = \frac{1}{2}\log 2 +\left(2n\pi + \frac{\pi}{4}\right) \qquad \text{...(5)}$$

Also, from (1), we have

$$(1 + i)^{1 + i} = e^x (\cos y + i \sin y) = e^{x + iy}$$

∴ The moduli of $(1 + i)^{1 + i}$ are $e^x = e^{\frac{1}{2}\log 2 -\left(2n\pi + \frac{\pi}{4}\right)}$

Putting $n = 0, 1, 2, 3, \dots..$ the moduli are

$\sqrt{2}e^{-\pi/4}, \sqrt{2}e^{-9\pi/4}, \sqrt{2}e^{-17\pi/4}, \sqrt{2}e^{-25\pi/4}, \dots$ which are less than unity as $\sqrt{2}e^{-(2n\pi+\pi/4)} <$ 1 for $n = 0, 1, 2, 3, \dots$

Thus the sum of all moduli

$$= \sqrt{2}e^{-\pi/4} + \sqrt{2}e^{-9\pi/4} + \sqrt{2}e^{-17\pi/4} + \dots$$

$$= \sqrt{2}[e^{-\pi/4} + e^{-9\pi/4} + e^{-17\pi/4} + \dots]$$

$$= \sqrt{2}\left[\frac{e^{-\pi/4}}{1-e^{-2\pi}}\right] \qquad\qquad = \sqrt{2}\left(\frac{e^{3\pi/4}}{e^{\pi}-e^{-\pi}}\right)$$

$$= \sqrt{2}e^{3\pi/4}\frac{1}{2}\operatorname{cosech}\pi \qquad = \sqrt{2}e^{3\pi/4}\operatorname{cosech}\pi$$

Example 12. *If* $\dfrac{(1+i)^{p+iq}}{(1-i)^{p-iq}} = x + iy$, *prove that one of the values of* $\tan^{-1}\left(\dfrac{y}{x}\right)$ *is* $\dfrac{1}{2}p\pi + q\log 2$.

Solution. We have

$$\frac{(1+i)^{p+iq}}{(1-i)^{p-iq}} = x + iy$$

Taking logarithm of both sides, we get (considering only principal values)

$$(p + iq)\log(1 + i) - (p - iq)\log(1 - i) = \log(x + iy)$$

$$\Rightarrow \quad (p+iq)\left[\frac{1}{2}\log 2 + \frac{i\pi}{4}\right] - (p-iq)\left[\frac{1}{2}\log 2 - \frac{i\pi}{4}\right] = \frac{1}{2}\log(x^2 + y^2) + i\tan^{-1}\left(\frac{y}{x}\right)$$

Equating imaginary parts, we get

$$\tan^{-1}\left(\frac{y}{x}\right) = \frac{1}{2}q\log 2 + \frac{p\pi}{4} + \frac{1}{2}q\log 2 + \frac{p\pi}{4} = q\log 2 + \frac{1}{2}p\pi$$

Example 13. *Find the general value of* $\text{Log}_4(-2)$.

Solution. We know that

$$\text{Log}_a b = \frac{\text{Log}\,b}{\text{Log}\,a}$$

Then

$$\text{Log}_4(-2) = \frac{\text{Log}(-2)}{\text{Log}\,4}$$

$$= \frac{\log(-2) + 2m\pi i}{\log 4 + 2n\pi i}$$

$$= \frac{\log(2e^{i\pi}) + 2m\pi i}{\log 2^2 + 2n\pi i}$$

$$= \frac{\log 2 + i\pi + 2m\pi i}{2\log 2 + 2n\pi i}$$

$$= \frac{\log 2 + (2m + 1)\pi i}{2\log 2 + 2n\pi i}$$

$$= \frac{[\log 2 + (2m + 1)\pi i][\log 2 - n\pi i]}{2(\log 2)^2 + 2n^2\pi^2}$$

$$\therefore \text{Log}_4(-2) = \left[\frac{(\log 2)^2 + (2m + 1)n\pi^2}{2(\log 2)^2 + 2n^2\pi^2}\right] + i\left[\frac{(2m + 1 - n)\pi\log 2}{2(\log 2)^2 + 2n^2\pi^2}\right]$$

Example 14. *Find the general value of* $\text{Log}_2(5i)$.

Solution. $\quad \text{Log}_2(5i) = \dfrac{\text{Log}(5i)}{\text{Log}\,2}$

$$= \dfrac{\log 5i + 2n\pi i}{\log 2 + 2m\pi i}$$

$$= \dfrac{\log 5 + \log i + 2n\pi i}{\log 2 + 2m\pi i}$$

$$= \dfrac{\log 5 + \dfrac{i\pi}{2} + 2n\pi i}{\log 2 + 2m\pi i}$$

$$= \dfrac{\log 5 + \left(2n + \dfrac{1}{2}\right)\pi i}{\log 2 + 2m\pi i}$$

$$= \dfrac{\left[\log 5 + \left(2n + \dfrac{1}{2}\right)\pi i\right][\log 2 - 2m\pi i]}{(\log 2)^2 + 4m^2\pi^2}$$

$$\therefore \ \text{Log}_2(5i) = \left[\dfrac{(\log 5)(\log 2) + (4n+1)m\pi^2}{(\log 2)^2 + 4m^2\pi^2}\right] + i\left[\dfrac{\left(2n + \dfrac{1}{2}\right)\pi\log 2 - 2m\pi\log 5}{(\log 2)^2 + 4m^2\pi^2}\right]$$

Example 15. *Prove that* $(x + ix\tan y)^{\log(x\sec y) - iy}$ *is real when only principal values are considered.*

Solution. We know that

$$a^z = e^{z\text{Log}\,a}$$

For principal value, we have

$$a^z = e^{z\text{Log}\,a} = \exp(z\log a)$$

Then, $(x + ix\tan y)^{\log(x\sec y) - iy}$

$$= \exp\left[\{(\log(x\sec y) - iy\}\log(x + ix\tan y)\right]$$

$$= \exp\left[\{(\log(x\sec y) - iy\}\log(x\sec y) + iy\right]$$

$$= \exp\left[\{(\log(x\sec y)\}^2 + y^2\right], \text{ which is real.}$$

EXERCISE 10.2

1. If $(i^i)^i = \cos\theta - i\sin\theta$, prove that $\theta = \dfrac{1}{2}\pi(4n+1)$

2. If $i^{i^i} = \cos\theta + i\sin\theta$, prove that $\theta = \left(2m + \dfrac{1}{2}\right)\pi\exp\left[-\left(2n + \dfrac{1}{2}\right)\pi\right]$

3. Separate $(1 - i)^i$ into real and imaginary parts.

4. Prove that $i^a = \cos\left[\left(2m + \dfrac{1}{2}\right)\pi a\right] + i\sin\left[\left(2m + \dfrac{1}{2}\right)\pi a\right]$

5. Show that the principal value of $\dfrac{(a + ib)^{p+iq}}{(a - ib)^{p-iq}}$ is $\cos 2(p\alpha + q \log r) + i \sin 2\,(p\alpha + q \log r)$

   where $r = \sqrt{a^2 + b^2}$ and $\alpha = \tan^{-1}\left(\dfrac{b}{a}\right)$.

6. Prove that the principal value of $(a + ib)^{c + id}$ is purely real and imaginary according as

   $\dfrac{1}{2}d \log (a^2 + b^2) + c \tan^{-1}\left(\dfrac{b}{a}\right)$ is an even or odd multiple of $\dfrac{1}{2}\pi$.

   In case it is purely real that $(a + ib)^{c \pm id} = (a^2 + b^2)^{(c^2 + d^2)/2c}$

7. Show that the ratio of the principal values of $(1 + i)^{1 - i}$ and $(1 - i)^{1 + i}$ is $\sin (\log 2) + i \cos (\log 2)$.

8. Prove that the general value of $(1 + i \tan \alpha)^{-i}$ is
   $e^{(\alpha + 2m\pi)} [\cos (\log \cos \alpha) + i \sin (\log \cos \alpha)]$.

9. If $[\cos (\theta - i\phi)]^{x + iy} = A + iB$ and principal values are taken into consideration, then

   $\tan^{-1}\left(\dfrac{B}{A}\right) = \dfrac{1}{2}y \log (\cosh^2 \phi - \sin^2\theta) + x \tan^{-1} (\tan \theta \tanh \phi)$.

10. If $\left[\dfrac{a + x + iy}{a - x - iy}\right]^{\lambda + \mu i} = X + iY$, prove that one of the values of $\tan^{-1}\left(\dfrac{Y}{X}\right)$ is

    $\lambda \tan^{-1}\left[\dfrac{2ay}{a^2 - x^2 - y^2}\right] + \dfrac{1}{2}\mu \log\left[\dfrac{(a + x)^2 + y^2}{(a - x)^2 + y^2}\right].$

11. Prove that : $\log\left[\dfrac{a + ib - x}{a + ib + x}\right] + \dfrac{1}{2}\log\left[\dfrac{(a - x)^2 + b^2}{(a + x)^2 + b^2}\right] + i\left[\tan^{-1}\left(\dfrac{b}{a - x}\right) - \tan^{-1}\left(\dfrac{b}{a + x}\right)\right].$

12. If $x^{x^{x^{\cdots \text{ad.inf.}}}} = a (\cos \theta + i \sin \theta)$, show that the general value of x is $r (\cos \phi + i \sin \phi)$, where $a \log r = (2n\pi + \theta) \sin \theta + \cos \theta \log a$ and $a\phi = (2n\pi + \theta) \cos \theta - \sin \theta \log a$.

HINT TO SELECTED PROBLEMS

1. $(i^i)^i = \cos \theta + i \sin \theta$.

   Taking logarithm both sides, we get

   $$i \log i^i = \log (\cos \theta + i \sin \theta)$$

   $\Rightarrow \qquad i \log (e^{-(4n + 1)\pi/2}) = i\theta$ $\qquad\qquad\qquad\qquad$ $[\because\ i^i = e^{-(4n + 1)\pi/2}]$

   $\Rightarrow \qquad \theta = -(4n + 1)\pi/2$

5. $(a + ib)^{p + iq}$ $= \exp[(p + iq)\log(a + ib)]$

$$= \exp\left[(p + iq)\left\{\frac{1}{2}\log(a^2 + b^2) + i\tan^{-1}\left(\frac{b}{a}\right)\right\}\right]$$

$$= \exp\left[\frac{1}{2}p\log(a^2 + b^2) - q\tan^{-1}\left(\frac{b}{a}\right)\right]$$

$$\exp\left[i\left\{\frac{1}{2}q\log(a^2 + b^2) + p\tan^{-1}\left(\frac{b}{a}\right)\right\}\right]$$

Similarly $(a + ib)^{p + iq} = \exp\left[\frac{1}{2}p\log(a^2 + b^2) - q\tan^{-1}\left(\frac{b}{a}\right)\right]$

$$\exp\left[i\left\{-p\tan^{-1}\left(\frac{b}{a}\right) - \frac{1}{2}q\log(a^2 + b^2)\right\}\right]$$

Now $\dfrac{(a + ib)^{p+iq}}{(a - ib)^{p-iq}} = \exp\left[i\left\{2p\tan^{-1}\left(\frac{b}{a}\right) + q\log(a^2 + b^2)\right\}\right]$

$$= \exp[i(2p\alpha + 2q\log r)], \quad r = \sqrt{a^2 + b^2} \quad \alpha = \tan^{-1}\left(\frac{b}{a}\right)$$

8. $(1 + i\tan\alpha)^{-i}$ $= \exp[-i\operatorname{Log}(1 + i\tan\alpha)]$
$= \exp[-i\{\log(1 + i\tan\alpha) + 2m\pi i]$
$= \exp[-i(\log\sec\alpha + i\alpha + 2m\pi i]$
$= \exp(\alpha + 2m\pi) . \exp[-i\log\sec\alpha]$
$= \exp(\alpha + 2m\pi)\exp[i\log\cos\alpha]$
$= e^{(\alpha + 2m\pi)}[\cos(\log\cos\alpha) + i\sin(\log\cos\alpha)]$

11. $\log\left[\dfrac{a + ib - x}{a + ib + x}\right] = \log\left[\dfrac{(a - x) + ib}{(a + x) + ib}\right]$

$$= \log[(a - x) + ib] - \log[(a + x) + ib]$$

$$= \frac{1}{2}\log[(a - x)^2 + b^2] + i\tan^{-1}\left(\frac{b}{a - x}\right) - \frac{1}{2}\log[(a + x)^2 + b^2] - i\tan^{-1}\left(\frac{b}{a + x}\right)$$

$$= \frac{1}{2}\log\left[\frac{(a - x)^2 + b^2}{(a + x)^2 + b^2}\right] + i\left[\tan^{-1}\left(\frac{b}{a - x}\right) - \tan^{-1}\left(\frac{b}{a + x}\right)\right]$$

ANSWERS

(3) Real part $= e^{\pi/4}.\cos\left(\frac{1}{2}\log 2\right)$, Imaginary part $= e^{\pi/4}\sin\left(\frac{1}{2}\log 2\right)$.

CHAPTER REVIEW : A COMPETITIVE APPROACH

Selected terms and Results

TERMS

- **General exponential function :** It is defined by

$$a^z = e^{z \mathrm{Log}\, a}$$

  where a and z are any two complex numbers.

- **General values of a^z :**

$$a^z = \exp[z(\log a + 2n\pi i)]$$
$$= \exp[z\, \mathrm{Log}\, a]$$

- **Principal value of a^z :**

$$a^z = \exp(z \log a)$$

- **Logarithm to any base :** If z, w and σ be any three complex numbers and if $e^w = z$ then we define that w is a logarithm of z to the base σ and we write $\mathrm{Log}_\sigma z = w$.

RESULTS

- The logarithm of a complex quantity has an infinite number of values and hence is many valued function.

- The general value of logarithm is written as 'Log' and the principal values as 'log'.

- $\log(x + iy) = \dfrac{1}{2}\log(x^2 + y^2) + i\tan^{-1}\dfrac{y}{x}$

- $\log(x - iy) = \dfrac{1}{2}\log(x^2 + y^2) - i\tan^{-1}\dfrac{y}{x}$

- The function a^z is a many valued function.

- The principal value of $\mathrm{Log}_\sigma z$ is defined by

$$\log_\sigma z = \dfrac{\log_e z}{\log_e s}$$

Review Questions and Project Work

1. Prove that $\log\left(\dfrac{1}{1 - e^{ia}}\right) =$

$$\log\left(\dfrac{1}{2}\operatorname{cosec}\dfrac{a}{2}\right) + i\left(\dfrac{p}{2} - \dfrac{a}{2}\right).$$

2. Prove that $\log\dfrac{(a - b) + i(a + b)}{(a + b) + i(a - b)}$

$$= i\left\{2np + \tan^{-1}\dfrac{2ab}{a^2 - b^2}\right\}$$

3. Prove that $\log\tan\left(\dfrac{p}{4} + \dfrac{a}{2}i\right)$

$$= i\tan^{-1}(\sinh a).$$

4. Prove that $i\log\dfrac{x - i}{x + i} = p - 2\tan^{-1}x$

5. Prove that $\tan\left(i\log\dfrac{a - ib}{a + ib}\right) = \dfrac{2ab}{a^2 - b^2}$

6. Prove that $i^{(1-i)} = ie^{2np + \frac{p}{2}}$.

7. If $p^{a+i\beta} = (x + iy)^{m+in}$ and only the principal values are considered, prove that

$$(a + ib)^{(c+id)} = (a^2 + b^2)^{\left(\frac{c^2 + d^2}{2c}\right)}$$

8. Show that the principal value of

$$\dfrac{(a + ib)^{p+iq}}{(a - ib)^{p-iq}} \text{ is } \cos 2(p\alpha + q\log r)$$

$$+ r\sin 2(p\alpha + q\log r).$$

Objective Type Questions

Fill in the blanks:

1. The principal values of a logarithm of i should lie between ....

2. The principal value of a logarithm of a negative quantity is logarithm of the positive quantity added with......

3. $\log (1 + i \tan \theta) = \log_e \sec \theta + ...$

4. The value of $\log \tan \left(\dfrac{p}{4} + \dfrac{ia}{2} \right) =$

   $i \tan^{-1} ....$

5. Every value of $p \log z$ is equal to same value of ...., when p is any rational number.

True/False: *Write 'T' for true and 'F' for false statement.*

1. The logarithm function is one-one function.   (T/F)

2. The logarithm function is many-one function.   (T/F)

3. The principal value of $\log z_1 z_2$ is always equal to the $\log z_1 + \log z_2$.   (T/F)

4. The principal value of $\log \dfrac{z_1}{z_2}$ is not necessarily equal to the principal value of $\log z_1 - \log z_2$.   (T/F)

5. $\log (1 + i \tan \theta) = \log_e \sec \theta + i\theta$.   (T/F)

Multiple Choice Questions : *Choose the most appropriate one :*

1. If z_1 and z_2 are complex numbers then $\log (z_1 . z_2)$ is :
   (a) $\log z_1 - \log z_2$
   (b) $\log z_2 - \log z_1$
   (c) $\log z_1 + \log z_2$
   (d) $\log (z_1 - z_2)$
   (e) none of these

2. If z_1 and z_2 are complex numbers then

   $\log \left(\dfrac{z_1}{z_2} \right)$ is :

   (a) $\log z_1 - \log z_2$
   (b) $\log z_2 - \log z_1$

   (c) $\log z_1 + \log z_2$
   (d) $\log (z_1 + z_2)$
   (e) none of these

3. General value of
   $\log_e (1 + \cos \theta + i \sin \theta)$ is :
   (a) $\log 2 \cos \theta/2 - i(\theta/2 + 2n\pi)$
   (b) $\log 2 \sin \theta/2 - i(\theta/2 + 2n\pi)$
   (c) $\log 2 \cos \theta/2 - i(\theta + 2n\pi)$
   (d) $\log 2 \cos \theta/2 + i(\theta/2 + 2n\pi)$
   (e) none of these

4. Value of $\tan \left[i \log \left(\dfrac{a - ib}{a + ib} \right) \right]$ is :

   (a) $\dfrac{2ab}{b^2 - a^2}$ (b) $\dfrac{2ab}{a^2 + b^2}$

   (c) $\dfrac{-2ab}{a^2 + b^2}$ (d) $\dfrac{2ab}{a^2 - b^2}$

   (e) none of these

5. Value of $\text{Log} (1 + i)$ is :

   (a) $\dfrac{1}{2}\log 2 + i(2np + p/4)$

   (b) $\dfrac{1}{2}\log 2 - i(2np + p/4)$

   (c) $\dfrac{1}{3}\log 3 + i(2np + p/4)$

   (d) $\dfrac{1}{3}\log 3 - i(2np + p/4)$

   (e) none of these

6. Value of $\log \tan (\pi/4 + ix/2)$ is :
   (a) $\tan^{-1} (\sinh x)$
   (b) $\tan^{-1} (\cosh x)$
   (c) $i \tan^{-1} (\cosh x)$
   (d) $i \tan^{-1} (\sinh x)$
   (e) none of these

7. Principal value of $\log (- i)$ is :

   (a) $-\dfrac{p}{2}i$ (b) $\dfrac{p}{2}i$

   (c) $-\dfrac{2}{p}i$ (d) $\dfrac{2}{p}i$

   (d) none of these

8. Principal value of log (– 1) is :
   (a) π   (b) πi   (c) $\dfrac{p}{2}i$ (d) $\dfrac{2i}{p}$

   (c) $\dfrac{1}{i}p$ (d) $\dfrac{2}{p}i$ (e) none of these

   (e) none of these   10. The general value of log (– i) is :

9. Principal value of log (i) is :
                                                                     (a) $\dfrac{(2n+1)}{2}pi$ (b) $\left(\dfrac{2n+1}{2}\right)pi$

   (a) πi   (b) $\dfrac{i}{p}$ (c) $\dfrac{(3n-1)}{2}pi$ (d) $\left(\dfrac{4n-1}{2}\right)pi$

ANSWERS

Fill in the blanks

(1) – π to π   (2) πi   (3) $i\theta$
(4) sin hα   (5) log z^p

True/False

(1) F   (2) T   (3) F
(4) T   (5) T

Multiple choice questions

(1) c   (2) a   (3) d
(4) d   (5) a   (6) d
(7) a   (8) b   (9) c
(10) d

CHAPTER 11

Inverse Circular and Hyperbolic Functions of a Complex Number

> ❖ Inverse circular functions
> ❖ Relation between inverse circular functions and inverse hyperbolic functions
>
> ❖ Inverse hyperbolic function

11.1 INVERSE CIRCULAR FUNCTIONS

1. Inverse Sine of $(x + iy)$

If $\sin(u + iv) = x + iy$, then $u + iv$ is called an **inverse sine** of $(x + iy)$, which is denoted by $\sin^{-1}(x + iy)$ and can be written as

$$u + iv = \sin^{-1}(x + iy)$$

Also, we know that if $\sin(u + iv) = x + iy$, then

$$x + iy = \sin\{n\pi + (-1)^n (u + iv)\}, n \in Z.$$

Thus, the general value of inverse sine of $(x + iy)$ is

$$n\pi + (-1)^n (u + iv) = n\pi + (-1)^n \sin^{-1}(x + iy)$$

which is denoted by $\sin^{-1}(x + iy)$ and is written as

$$\mathrm{Sin}^{-1}(x + iy) = n\pi + (-1)^n \sin^{-1}(x + iy), \ n \in Z.$$

Remarks

❖ For general value of inverse sine, the first letter 'S' being written capital.

❖ For principal value of inverse sine, the first letter 's' being written small and the real part lies between $-\dfrac{\pi}{2}$ and $\dfrac{\pi}{2}$.

2. Inverse Cosine of $(x + iy)$

If $\cos(u + iv) = x + iy$, then $u + iv$ is called an **inverse cosine** of $x + iy$, which is denoted by $\cos^{-1}(x + iy)$ and can be written as

$$u + iv = \cos^{-1}(x + iy)$$

Also, if $\cos(u + iv) = x + iy$, then

$$\cos\{2n\pi \pm (u + iv)\} = x + iy, n \in Z.$$

Thus, the general value of inverse cosine of $(x + iy)$ is

$$2n\pi \pm (u + iv) = 2n\pi \pm \cos^{-1}(x + iy)$$

which is denoted by $\cos^{-1}(x + iy)$ and is written as

$$\text{Cos}^{-1}(x + iy) = 2n\pi \pm \cos^{-1}(x + iy), \; n \in Z.$$

Remarks

❖ For general value of inverse cosine the first letter 'C' being written capital and for principal value, the first letter 'c' being written small.

❖ Principal value is that value of $2n\pi + \cos^{-1}(x + iy)$ in which the real part lies between 0 and π.

3. Inverse Tangent of $(x + iy)$

If $\tan(u + iv) = x + iy$, then $u + iv$ is called an inverse tangent of $(x + iy)$, which is denoted by $\tan^{-1}(x + iy)$,

i.e. $u + iv = \tan^{-1}(x + iy)$

Also, if $\tan(u + iv) = x + iy$, then

$$\tan[n\pi + (u + iv)\}] = x + iy, \; n \in Z.$$

Thus, the general value of inverse tangent of $(x + iy)$ is

$$n\pi + (u + iv) = n\pi + \tan^{-1}(x + iy)$$

which is denoted by $\tan^{-1}(x + iy)$ and is written as

$$\text{Tan}^{-1}(x + iy) = n\pi + \tan^{-1}(x + iy), \; n \in Z.$$

Remarks

❖ For general value of inverse tangent, the first letter 'T' being written capital and for principal value, the first letter 't' being written small.

❖ Principal value is that value of $n\pi + \tan^{-1}(x + iy)$ in which the real part lies between $-\dfrac{\pi}{2}$ and $\dfrac{\pi}{2}$.

Similarly, other inverse circular function are

$$\text{Cot}^{-1}(x + iy) = n\pi + \cot^{-1}(x + iy)$$
$$\text{Sec}^{-1}(x + iy) = 2n\pi \pm \sec^{-1}(x + iy)$$
$$\text{Cosec}^{-1}(x + iy) = n\pi + (-1)^n \, \text{cosec}^{-1}(x + iy).$$

Remarks

❖ In case of cot and cosec, the principal value is that value for which the real part lies between $-\pi/2$ and $\pi/2$.

❖ In case of sec, the principal value is that value for which the real part lies between 0 and π.

11.2 INVERSE HYPERBOLIC FUNCTION

Let z and w be any two complex numbers. If $\sinh w = z$ then w is called the **inverse hyperbolic sine of z,** which is denoted by $\sinh^{-1} z$ and is written as

$$w = \sinh^{-1}z$$

Similarly, other inverse hyperbolic functions viz. $\cosh^{-1} z$, $\tanh^{-1} z$, $\coth^{-1} z$, $\operatorname{sech}^{-1} z$, and $\operatorname{cosech}^{-1} z$ can be defined.

(i) *To prove* $\sinh^{-1} z = \log[z + \sqrt{z^2 + 1}]$

Proof. Let $\qquad \sinh^{-1} z = w$ $\qquad\qquad$ ...(1)

then, $\qquad\qquad \sinh w = z$

and $\qquad\qquad \cosh w = \sqrt{\sinh^2 w + 1} = \sqrt{z^2 + 1}$

We know that $\qquad e^w = \sinh w + \cosh w = z + \sqrt{z^2 + 1}$

$\Rightarrow \qquad\qquad w = \log[z + \sqrt{z^2 + 1}]$

$\therefore \qquad\qquad \mathbf{sinh^{-1} z = log[z + \sqrt{z^2 + 1}]}$ $\qquad\qquad$ [using (1)]

(ii) *To prove* $\cosh^{-1} z = \log[z + \sqrt{z^2 - 1}]$

Proof. Let $\qquad \cosh^{-1} z = w$ $\qquad\qquad$ ...(1)

then, $\qquad\qquad \cosh w = z$ and $\sinh w = \sqrt{\cosh^2 w - 1} = \sqrt{z^2 - 1}$

We know that

$$e^w = \sinh w + \cosh w$$

$\Rightarrow \qquad\qquad e^w = \sqrt{z^2 - 1} + z$

$\Rightarrow \qquad\qquad w = \log[z + \sqrt{z^2 - 1}]$

$\therefore \qquad\qquad \mathbf{cosh^{-1} z = log[z + \sqrt{z^2 - 1}]}$ $\qquad\qquad$ [using (1)]

(iii) *To prove* $\tanh^{-1} z = \dfrac{1}{2}\log\left[\dfrac{1+z}{1-z}\right]$

Proof. Let $\qquad \tanh^{-1} z = w$ $\qquad\qquad$ ...(1)

then, $\qquad\qquad \tanh w = z$

$\Rightarrow \qquad\qquad \dfrac{e^w - e^{-w}}{e^w + e^{-w}} = \dfrac{z}{1}$

Applying componendo and dividendo theorem, we get

$$\frac{(e^w + e^{-w}) + (e^w - e^{-w})}{(e^w + e^{-w}) - (e^w - e^{-w})} = \frac{1+z}{1-z}$$

$$\Rightarrow \qquad \frac{2e^w}{2e^{-w}} = \frac{1+z}{1-z}$$

$$\Rightarrow \qquad e^{2w} = \frac{1+z}{1-z}$$

$$\Rightarrow \qquad 2w = \log\left[\frac{1+z}{1-z}\right]$$

$$\Rightarrow \qquad w = \frac{1}{2}\log\left[\frac{1+z}{1-z}\right]$$

$$\therefore \qquad \mathbf{tanh^{-1}\,z} = \frac{1}{2}\log\left[\frac{1+z}{1-z}\right] \qquad\qquad \text{[using (1)]}$$

(iv) *To prove* $\coth^{-1} z = \dfrac{1}{2}\log\left[\dfrac{z+1}{z-1}\right]$

Proof. Let $\qquad \coth^{-1} z = w$ $\qquad\qquad\qquad\qquad\qquad$ ...(1)

then, $\qquad\qquad \coth w = z$

$$\Rightarrow \qquad \frac{e^w + e^{-w}}{e^w - e^{-w}} = \frac{z}{1}$$

Applying componendo and dividendo, we get

$$\frac{(e^w + e^{-w}) + (e^w - e^{-w})}{(e^w + e^{-w}) - (e^w - e^{-w})} = \frac{z+1}{z-1}$$

$$\Rightarrow \qquad \frac{2e^w}{2e^{-w}} = \frac{z+1}{z-1}$$

$$\Rightarrow \qquad e^{2w} = \frac{z+1}{z-1}$$

$$\Rightarrow \qquad 2w = \log\left[\frac{z+1}{z-1}\right]$$

$$\Rightarrow \qquad w = \frac{1}{2}\log\left[\frac{z+1}{z-1}\right]$$

$$\therefore \qquad \mathbf{coth^{-1}\,z} = \frac{1}{2}\log\left[\frac{z+1}{z-1}\right] \qquad\qquad \text{[using (1)]}$$

11.3 RELATION BETWEEN INVERSE HYPERBOLIC FUNCTION AND INVERSE CIRCULAR FUNCTIONS

(i) *To prove* $\sinh^{-1} x = -i \sin^{-1} (ix)$

Proof. Let $\quad \sinh^{-1}x = y$ $\qquad\qquad$ ...(1)

Then $\qquad\qquad \sinh y = x$

$\Rightarrow \qquad\qquad ix = i \sinh y = \sin (iy)$

$\Rightarrow \qquad\qquad iy = \sin^{-1} (ix)$

$\Rightarrow \qquad\qquad y = \dfrac{1}{i} \sin^{-1} (ix)$

$\Rightarrow \qquad\qquad y = -i \sin^{-1} (ix)$

$\therefore \qquad\qquad \sinh^{-1} x = -i \sin^{-1} (ix)$ $\qquad\qquad$ [using (1)]

(ii) *To prove* $\cosh^{-1} x = -i \cos^{-1} x$

Proof. Let $\quad \cosh^{-1}x = y$ $\qquad\qquad$ ...(1)

Then $\qquad\qquad \cosh y = x$

$\Rightarrow \qquad\qquad \cos (iy) = x$

$\Rightarrow \qquad\qquad iy = \cos^{-1} x$

$\Rightarrow \qquad\qquad y = -i \cos^{-1} x$

$\therefore \qquad\qquad \cosh^{-1} x = -i \cos^{-1} x$ $\qquad\qquad$ [using (1)]

(iii) *To prove* $\tanh^{-1} x = -i \tanh^{-1} (ix)$

Proof. Let $\quad \tanh^{-1}x = y$ $\qquad\qquad$ ...(1)

Then $\qquad\qquad \tanh y = x$

$\Rightarrow \qquad\qquad ix = i \tanh y$

$\Rightarrow \qquad\qquad ix = \tan (iy)$

$\Rightarrow \qquad\qquad iy = \tan^{-1} (ix)$

$\Rightarrow \qquad\qquad y = -i \tan^{-1} (ix)$

$\therefore \qquad\qquad \tanh^{-1} x = -i \tan^{-1} (ix)$ $\qquad\qquad$ [using (1)]

SOLVED EXAMPLES

Example 1. *Express* $\tan^{-1} (\alpha + i\beta)$ *in the form of* $A + iB$.

Solution. Let $\qquad\qquad \tan^{-1} (\alpha + i\beta) = A + iB$ $\qquad\qquad$ ...(1)

Then $\qquad\qquad \tan^{-1} (\alpha - i\beta) = A - iB$ $\qquad\qquad$ ...(2)

Adding (1) and (2), we get

$$2A = \tan^{-1} (\alpha + i\beta) + \tan^{-1} (\alpha - i\beta)$$

$$= \tan^{-1} \left[\frac{(\alpha - i\beta + \alpha - i\beta)}{1 - (\alpha + i\beta)(\alpha - i\beta)} \right]$$

$$\therefore \qquad A = \frac{1}{2}\tan^{-1}\left(\frac{2\alpha}{1-\alpha^2-\beta^2}\right)$$

Subtracting (2) from (1), we get

$$2iB = \tan^{-1}(\alpha+i\beta) - \tan^{-1}(\alpha-i\beta)$$

$$= \tan^{-1}\left[\frac{(\alpha+i\beta)-(\alpha-i\beta)}{1+(\alpha+i\beta)(\alpha-i\beta)}\right]$$

$$2iB = \tan^{-1}\left(\frac{2i\beta}{1+\alpha^2+\beta^2}\right)$$

$$\Rightarrow \qquad \tan(2iB) = \frac{2i\beta}{1+\alpha^2+\beta^2}$$

$$\Rightarrow \qquad i\tanh 2B = \frac{2i\beta}{1+\alpha^2+\beta^2}$$

$$\Rightarrow \qquad \tanh 2B = \frac{2\beta}{1+\alpha^2+\beta^2}$$

$$\therefore \qquad B = \frac{1}{2}\tanh^{-1}\left(\frac{2i\beta}{1+\alpha^2+\beta^2}\right)$$

Hence, $\tan^{-1}(\alpha+i\beta) = \frac{1}{2}\tan^{-1}\left(\frac{2\alpha}{1-\alpha^2-\beta^2}\right) + i.\frac{1}{2}\tanh^{-1}\left(\frac{2\beta}{1+\alpha^2+\beta^2}\right)$

Remark

❖ The general value of $\tan^{-1}(\alpha+i\beta)$ *i.e.,* $\mathrm{Tan}^{-1}(\alpha+i\beta)$ is given by

$$\mathrm{Tan}^{-1}(\alpha+i\beta) = n\pi + \frac{1}{2}\tan^{-1}\left(\frac{2\alpha}{1-\alpha^2-\beta^2}\right) + \frac{1}{2}i\tanh^{-1}\left(\frac{2\beta}{1+\alpha^2+\beta^2}\right)$$

Example 2. *Express* $\cos^{-1}(x+iy)$ *in the form of* $A+iB$.

Solution. Let $\quad \cos^{-1}(x+iy) = A+iB$ $\qquad\qquad\qquad\qquad$ ...(1)

Then $\qquad\qquad\qquad x+iy = \cos(A+iB)$

$\Rightarrow \qquad\qquad\qquad x+iy = \cos A\cosh B - i\sin A\sinh B.$

Separating real and imaginary parts, we get

$$x = \cos A\cosh B \qquad\qquad\qquad\qquad ...(2)$$

$$y = -\sin A\sinh B \qquad\qquad\qquad\qquad ...(3)$$

Now first, eliminating B Between (2) and (3), we have

From (2), $\qquad\qquad\qquad \cosh B = \frac{x}{\cos A}$

From (3),
$$\sinh B = -\frac{y}{\sin A}$$

$\Rightarrow$ $\cosh^2 B - \sinh^2 B = 1$

$\Rightarrow$ $\dfrac{x^2}{\cos^2 A} - \dfrac{y^2}{\sin^2 A} = 1$

$\Rightarrow$ $x^2 \sin^2 A - y^2 \cos^2 A = \cos^2 A \sin^2 A$

$\Rightarrow$ $x^2 \sin^2 A - y^2 (1 - \sin^2 A) = (1 - \sin^2 A)\sin^2 A$

$\Rightarrow$ $\sin^4 A + (x^2 + y^2 - 1)\sin^2 A - y^2 = 0$

It is quadratic in $\sin^2 A$, so we have

$$\sin^2 A = \frac{-(x^2 + y^2 - 1) \pm \sqrt{(x^2 + y^2 - 1)^2 + 4y^2}}{2}$$

or $\sin^2 A = \dfrac{-(x^2 + y^2 - 1) + \sqrt{(x^2 + y^2 - 1)^2 + 4y^2}}{2}$ $[\because \sin^2 A \geq 0]$

or $\sin A = \pm \left[\dfrac{\sqrt{(x^2 + y^2 - 1)^2 + 4y^2} - (x^2 + y^2 - 1)}{2}\right]^{1/2}$

$\therefore$ $A = \pm \sin^{-1}\left[\dfrac{\sqrt{(x^2 + y^2 - 1)^2 + 4y^2} - (x^2 + y^2 - 1)}{2}\right]^{1/2}$

Next, we eliminating A between (2) and (3), we have

From (2), $\cos A = \dfrac{x}{\cosh B}$

From (3), $\sin A = \dfrac{-y}{\sinh B}$

$\Rightarrow$ $\sin^2 A + \cos^2 A = 1$

$\Rightarrow$ $\dfrac{x^2}{\cosh^2 B} + \dfrac{y^2}{\sinh^2 B} = 1$

$\Rightarrow$ $x^2 \sinh^2 B + y^2 \cosh^2 B = \cosh^2 B \sinh^2 B$

$\Rightarrow x^2 \sinh^2 B + y^2 (1 + \sinh^2 B) = (1 + \sinh^2 B)\sinh^2 B$

$\Rightarrow$ $\sinh^4 B + (1 - x^2 - y^2)\sinh^2 B - y^2 = 0$

It is quadratic in $\sinh^2 B$, so we have

$$\sin h^2 B = \frac{-(1 - x^2 - y^2) \pm \sqrt{(1 - x^2 - y^2)^2 + 4y^2}}{2}$$

or $\sin h^2 B = \dfrac{-(1 - x^2 - y^2) + \sqrt{(1 - x^2 - y^2)^2 + 4y^2}}{2}$ $[\because \sinh^2 B \geq 0]$

$$\therefore \quad \sinh B = \pm \left[\frac{\sqrt{(1-x^2-y^2)^2+4y^2}-(1-x^2-y^2)}{2} \right]^{1/2}$$

$$\therefore \quad B = \pm\sinh^{-1} \left[\frac{\sqrt{(1-x^2-y^2)^2+4y^2}-(1-x^2-y^2)}{2} \right]^{1/2}$$

Putting the value of A and B in (1), we get the required form.

Rĕmark

❖ The general value of $\cos^{-1}(x+iy)$ is given by

$$\text{Cos}^{-1}(x+iy) = 2n\pi \pm \cos^{-1}(x+iy) = 2n\pi \pm (A+iB)$$

Example 3. *Express* $\cosh^{-1}(x+iy)$ *in the form of $A+iB$.*

Solution. Let $\qquad\qquad \cosh^{-1}(x+iy) = A+iB$ ...(1)

Then $\qquad\qquad x+iy = \cosh^{-1}(A+iB)$

$\Rightarrow \qquad\qquad x+iy = \cos(i(A+iB))$

$\Rightarrow \qquad\qquad x+iy = \cos(-B+iA)$

$\Rightarrow \qquad\qquad x+iy = \cos B \cosh A - i \sin(-B) \sinh A$

$\Rightarrow \qquad\qquad x+iy = \cos B \cosh A + i \sin B \sinh A.$

Separating real and imaginary parts, we get

$$x = \cos B \cosh A \qquad\qquad\qquad\qquad ...(2)$$

$$y = \sin B \sinh A \qquad\qquad\qquad\qquad ...(3)$$

Now, we first eliminate B between (2) and (3), we have

From (2) $\qquad \cos B = \dfrac{x}{\cosh A}$

From (3) $\qquad \sin B = \dfrac{y}{\sinh A}$

$$\therefore \qquad \frac{x^2}{\cosh^2 A} + \frac{y^2}{\sinh^2 A} = \cos^2 B + \sin^2 B = 1$$

$\Rightarrow \qquad x^2 \sinh^2 A + y^2 \cosh^2 A = \cosh^2 A \sinh^2 A$

As the reference of example 2, we get

$$A = \pm\sinh^{-1} \left[\frac{\sqrt{(1-x^2-y^2)^2+4y^2}-(1-x^2-y^2)}{2} \right]^{1/2}$$

Next, we elimate A between (2) and (3), we have

From (2) $\qquad \cosh A = \dfrac{x}{\cos B}$

From (3) $\qquad \sinh A = \dfrac{y}{\sin B}$

$\therefore \qquad \dfrac{x^2}{\cos^2 B} - \dfrac{y^2}{\sin^2 B} = 1$

$\Rightarrow \qquad x^2 \sin^2 B - y^2 \cos^2 B = \cos^2 B \sin^2 B$

As the reference of example 2, we get

$$B = \pm \sin^{-1} \left[\frac{\sqrt{(x^2 - y^2 - 1)^2 + 4y^2} - (x^2 - y^2 - 1)}{2} \right]^{1/2}$$

Hence, putting the value of A and B in (1), we get the required form.

Example 4. *Show that* $\sinh^{-1}x = \tanh^{-1}\left(\dfrac{x}{\sqrt{1+x^2}} \right)$

Solution. Let $\qquad \sinh^{-1}x = y$

Then $\qquad\qquad x = \sinh y$

So, $\qquad \dfrac{x}{\sqrt{1+x^2}} = \dfrac{\sinh y}{\sqrt{1+\sinh^2 y}} = \dfrac{\sinh y}{\cosh y} = \tanh y$

$\Rightarrow \qquad\qquad y = \tanh^{-1}\left(\dfrac{x}{\sqrt{1+x^2}} \right)$

$\therefore \qquad\qquad \sinh^{-1} x = \tanh^{-1}\left(\dfrac{x}{\sqrt{1+x^2}} \right) \qquad\qquad$ [using (1)]

Example 5. *Show that* $\operatorname{Sin}^{-1}(ix) = n\pi + i(-1)^n \log [x + \sqrt{1+x^2}\,]$.

Solution. Let general value of $\sin^{-1}(ix)$ is

$\qquad\qquad \operatorname{Sin}^{-1}(ix) = n\pi + (-1)^n \sin^{-1}(ix) \qquad\qquad$ ...(1)

Let $\qquad\qquad \sin^{-1}(ix) = y$

$\Rightarrow \qquad\qquad ix = \sin y$

$\Rightarrow \qquad \cos y = \sqrt{1 - \sin^2 y} = \sqrt{1 - (-ix)^2} = \sqrt{1 + x^2}$

Now, $\qquad\qquad e^{iy} = \cos y + i \sin y = \sqrt{1+x^2} - x$

$\Rightarrow \qquad\qquad iy = \log [\sqrt{1+x^2} - x]$

$$\Rightarrow \qquad y = -i \log [\sqrt{1+x^2} - x]$$

$$\Rightarrow \qquad y = i \log [\sqrt{1+x^2} - x]^{-1}$$

$$\Rightarrow \qquad y = i \log [\sqrt{1+x^2} + x]$$

$$\therefore \qquad \sin^{-1}(ix) = i \log [\sqrt{1+x^2} + x]$$

Putting the value of $\sin^{-1}(ix)$ in (1), we get

$$\text{Sin}^{-1}(ix) = n\pi + i(-1)^n \log [x + \sqrt{1+x^2}]$$

Example 6. *Show that* $\tan^{-1}\left(i.\dfrac{x-a}{x+a}\right) = -\dfrac{1}{2}i\log\dfrac{a}{x}$

Solution. Let $\qquad \tan^{-1}\left(i.\dfrac{x-a}{x+a}\right) = y$ ...(1)

Then $\qquad \dfrac{i(x-a)}{(x+a)} = \tan y$

$$\dfrac{x-a}{x+a} = \dfrac{1}{i}\tan y = -i \tan y$$

$$\Rightarrow \qquad -\dfrac{x-a}{x+a} = \dfrac{e^{iy} - e^{-iy}}{e^{iy} + e^{-iy}}$$

Applying componendo and dividendo, we get

$$\dfrac{(a-x)+(x+a)}{(a-x)-(x+a)} = \dfrac{(e^{iy} - e^{-iy})+(e^{iy} + e^{-iy})}{(e^{iy} - e^{-iy})-(e^{iy} + e^{-iy})}$$

$$\Rightarrow \qquad \dfrac{2a}{-2x} = \dfrac{2e^{iy}}{-2e^{-iy}}$$

$$\Rightarrow \qquad e^{2iy} = \dfrac{a}{x}$$

$$\Rightarrow \qquad 2yi = \log\left(\dfrac{a}{x}\right)$$

$$\Rightarrow \qquad iy = \dfrac{1}{2}\log\left(\dfrac{a}{x}\right)$$

$$\Rightarrow \qquad y = -\dfrac{1}{2}\log\left(\dfrac{a}{x}\right)$$

Putting the value of y in (1), we get

$$\tan^{-1}\left(i.\dfrac{(x-a)}{(x+a)}\right) = -\dfrac{1}{2}i\log\left(\dfrac{a}{x}\right)$$

Example 7. *Show that* $\sinh^{-1}(\cot x) = \log(\cot x + \csc x)$.

Solution. Let $\quad \sinh^{-1}(\cot x) = y$ $\qquad\qquad$ ...(1)

Then $\qquad\qquad\qquad \cot x = \sinh y$

and $\qquad\qquad\qquad \csc x = \sqrt{1+\cot^2 x} = \sqrt{1+\sinh^2 y} = \cosh y.$

We know that

$$e^y = \sinh y + \cosh y$$
$$\Rightarrow \qquad\qquad e^y = \cot x + \csc x$$
$$\Rightarrow \qquad\qquad y = \log(\cot x + \csc x)$$
$$\therefore \qquad \sinh^{-1}(\cot x) = \log(\cot x + \csc x) \qquad\qquad \text{[using (1)]}$$

Example 8. *Prove that*

$$\sin^{-1}(\cos\theta + i\sin\theta) = \cos^{-1}(\sqrt{\sin\theta}) + i\log[\sqrt{\sin\theta} + \sqrt{1+\sin\theta}]$$

Solution. Let $\qquad \sin^{-1}(\cos\theta + i\sin\theta) = x + iy$ $\qquad\qquad$ ...(1)

Then $\quad \cos\theta + i\sin\theta = \sin(x+iy)$

$\Rightarrow \qquad \cos\theta + i\sin\theta = \sin x \cosh y + i\cos x \sinh y$

Separating the real and imaginary parts, we get

$$\cos\theta = \sin x \cosh y \qquad\qquad \text{...(2)}$$
$$\sin\theta = \cos x \sinh y \qquad\qquad \text{...(3)}$$

Squaring (2) and (3) and adding, we get

$$1 = \cos^2\theta + \sin^2\theta = \sin^2 x \cosh^2 y + \cos^2 x \sinh^2 y$$
$$\Rightarrow \qquad 1 = \sin^2 x(1+\sinh^2 y) + \cos^2 x \sinh^2 y$$
$$\Rightarrow \qquad 1 = \sin^2 x + \sin^2 x \sinh^2 y + \cos^2 x \sinh^2 y$$
$$\Rightarrow 1 - \sin^2 x = (\sin^2 x + \cos^2 x)\sinh^2 y$$
$$\Rightarrow \qquad\qquad \cos^2 x = \sinh^2 y$$
$$\Rightarrow \qquad\qquad \sinh y = \cos x \qquad\qquad \text{...(4)}$$

From (3) and (4), we get

$$\cos^2 x = \sin\theta \qquad\qquad [\because \theta \text{ is a acute and positive}]$$
$$\Rightarrow \qquad\qquad \cos x = \pm\sqrt{\sin\theta} \qquad\qquad \text{...(5)}$$

Since, x being real part of $\sin^{-1}(\cos\theta + i\sin\theta)$, lies between $-\pi/2$ and $\pi/2$.

$\therefore \qquad\qquad x = \cos^{-1}(\sqrt{\sin\theta})$

From (2) and (5), we get

$$\sinh y = \sqrt{\sin\theta}$$
$$\Rightarrow \qquad\qquad \frac{e^y - e^{-y}}{2} = \sqrt{\sin\theta}$$
$$\Rightarrow \qquad\qquad e^y - e^{-y} = 2\sqrt{\sin\theta}$$

$\Rightarrow e^{2y} - 2\sqrt{\sin\theta}.e^y - 1 = 0$

It is quadratic in e^y, so we have

$$e^y = \frac{2\sqrt{\sin\theta} \pm \sqrt{4\sin\theta + 4}}{2}$$

$$\Rightarrow \qquad e^y = \sqrt{\sin\theta} \pm \sqrt{1 + \sin\theta}$$

$$\Rightarrow \qquad e^y = \sqrt{\sin\theta} + \sqrt{1 + \sin\theta} \qquad\qquad [\because e^y > 0]$$

$$\Rightarrow \qquad y = \log[\sqrt{\sin\theta} + \sqrt{1 + \sin\theta}]$$

Putting the values of $\dot{x}$ and y in (1), we get

$$\sin^{-1}(\cos\theta + i\sin\theta) = \cos^{-1}(\sqrt{\sin\theta}) + i\log[\sqrt{\sin\theta} + \sqrt{1 + \sin\theta}].$$

Example 9. *Show that* $\tan^{-1}\left(\dfrac{3 - 2i}{3 + 2i}\right) = \dfrac{\pi}{4} - \dfrac{i}{2}\log 5$

Solution. Let $\quad \tan^{-1}\left(\dfrac{3 - 2i}{3 + 2i}\right) = x + iy$ ...(1)

Then $\qquad\qquad \tan^{-1}\left(\dfrac{3 + 2i}{3 - 2i}\right) = x - iy$ ...(2)

Adding (1) and (2), we get

$$2x = \tan^{-1}\left(\frac{3 - 2i}{3 + 2i}\right) + \tan^{-1}\left(\frac{3 + 2i}{3 - 2i}\right)$$

$$= \tan^{-1}\left[\frac{\dfrac{3 - 2i}{3 + 2i} + \dfrac{3 + 2i}{3 - 2i}}{1 - \left(\dfrac{3 - 2i}{3 + 2i}\right)\left(\dfrac{3 + 2i}{3 - 2i}\right)}\right]$$

$$= \tan^{-1}\infty = \pi/2$$

$$\therefore \qquad\qquad x = \pi/4.$$

Subtracting (2) from (1), we get

$$2iy = \tan^{-1}\left(\frac{3 - 2i}{3 + 2i}\right) - \tan^{-1}\left(\frac{3 + 2i}{3 - 2i}\right)$$

$$\Rightarrow \qquad 2iy = \tan^{-1}\left[\frac{\left(\dfrac{3 - 2i}{3 + 2i}\right) - \left(\dfrac{3 + 2i}{3 - 2i}\right)}{1 + \left(\dfrac{3 - 2i}{3 + 2i}\right)\left(\dfrac{3 + 2i}{3 - 2i}\right)}\right]$$

$$\Rightarrow \qquad \tan(2\,iy) = \frac{(3 - 2i)^2 - (3 + 2i)^2}{2(3 + 2i)(3 - 2i)}$$

$\Rightarrow$ $\qquad\qquad i\tanh 2y = \dfrac{-24i}{26} = -\dfrac{12i}{13}$

$\Rightarrow$ $\qquad\qquad \tanh 2y = -\dfrac{12}{13}$

$\Rightarrow$ $\qquad\qquad \dfrac{e^{2y}-e^{-2y}}{e^{2y}+e^{-2y}} = -\dfrac{12}{13}$

Applying componendo and dividendo, we get

$$\dfrac{2e^{2y}}{-2e^{-2y}} = \dfrac{1}{-25}$$

$\Rightarrow$ $\qquad\qquad e^{4y} = \dfrac{1}{25}$

$\Rightarrow$ $\qquad\qquad e^{2y} = \dfrac{1}{5}$

$\Rightarrow$ $\qquad\qquad 2y = \log\left(\dfrac{1}{5}\right)$

$\therefore$ $\qquad\qquad y = \dfrac{1}{2}\log\left(\dfrac{1}{5}\right) = -\dfrac{1}{2}\log 5$

Putting the value of x and y in (1), we get

$$\tan^{-1}\left(\dfrac{3-2i}{3+2i}\right) = \dfrac{\pi}{4} - \dfrac{i}{2}\log 5$$

Example 10. *Prove that $Tan^{-1}(\cos\theta + i\sin\theta) = n\pi + \dfrac{\pi}{4} + \dfrac{1}{2}i\log\tan\left(\dfrac{\pi}{4}+\dfrac{\theta}{2}\right)$*

Solution. We know that

$$Tan^{-1}(z) = n\pi + \tan^{-1}z$$

$\therefore$ $\qquad Tan^{-1}(\cos\theta + i\sin\theta) = n\pi + \tan^{-1}(\cos\theta - i\sin\theta)$ $\qquad$ ...(1)

Let $\qquad \tan^{-1}(\cos\theta + i\sin\theta) = A + iB$ $\qquad$ ...(2)

Then $\qquad \tan^{-1}(\cos\theta - i\sin\theta) = A - iB$ $\qquad$ ...(3)

Adding (2) and (3) we get

$$2A = \tan^{-1}(\cos\theta + i\sin\theta) + \tan^{-1}(\cos\theta - i\sin\theta)$$

$$= \tan^{-1}\left[\dfrac{2\cos\theta}{1-(\cos^2\theta + \sin^2\theta)}\right]$$

$$= \tan^{-1}\left(\dfrac{2\cos\theta}{0}\right) = \tan^{-1}\infty = \pi/2$$

$\therefore$ $\qquad A = \pi/4$.

Subtracting (3) from (2), we get

$$2iB = \tan^{-1}(\cos\theta + i\sin\theta) - \tan^{-1}(\cos\theta - i\sin\theta)$$

$$= \tan^{-1}\left(\frac{2i\sin\theta}{1+\cos^2\theta+\sin^2\theta}\right) = \tan^{-1}\left(\frac{2i\sin\theta}{2}\right)$$

$\Rightarrow$ $\qquad 2\,iB = \tan^{-1}(i\sin\theta)$

$\Rightarrow$ $\qquad \tan(2\,iB) = i\sin\theta$

$\Rightarrow$ $\qquad i\tanh 2\,B = i\sin\theta$

$\Rightarrow$ $\qquad \tanh 2\,B = \sin\theta$

$\Rightarrow$ $\qquad \dfrac{e^{2B}-e^{-2B}}{e^{2B}+e^{-2B}} = \dfrac{\sin\theta}{1}$

Applying componendo and dividendo, we get

$$\frac{2e^{2B}}{-2e^{-2B}} = \frac{1+\sin\theta}{\sin\theta-1}$$

$\Rightarrow$ $\qquad e^{4B} = \dfrac{1+\sin\theta}{1-\sin\theta}$

$\Rightarrow$ $\qquad e^{2B} = \sqrt{\dfrac{1+\sin\theta}{1-\sin\theta}}$

$\Rightarrow$ $\qquad e^{2B} = \dfrac{1+\sin\theta}{\cos\theta}$

$\Rightarrow$ $\qquad e^{2B} = \dfrac{1-\cos\left(\dfrac{\pi}{2}+\theta\right)}{\sin\left(\dfrac{\pi}{2}+\theta\right)} = \tan\left(\dfrac{\pi}{4}+\dfrac{\theta}{2}\right)$

$\Rightarrow$ $\qquad 2B = \log\tan\left(\dfrac{\pi}{4}+\dfrac{\theta}{2}\right)$

$\therefore$ $\qquad B = \dfrac{1}{2}\log\tan\left(\dfrac{\pi}{4}+\dfrac{\theta}{2}\right)$

Putting the value of A and B in (1), we get

$$\text{Tan}^{-1}(\cos\theta+i\sin\theta) = n\pi+\frac{\pi}{4}+\frac{i}{2}\log\tan\left(\frac{\pi}{4}+\frac{\theta}{2}\right)$$

Example 11. *Prove that* $\tan^{-1}(\sinh\theta) = -i\log\tan\left(\dfrac{\pi}{4}+\dfrac{i\theta}{2}\right)$

Solution. Let $\qquad \tan^{-1}(\sinh\theta) = \phi$ $\hfill ...(1)$

Then $\qquad\qquad \sinh\theta = \tan\phi$

$\Rightarrow$ $\qquad -i\sin(i\,\theta) = \dfrac{e^{i\phi}-e^{-i\phi}}{i(e^{i\phi}+e^{-i\phi})}$

$$\Rightarrow \qquad \frac{\sin(i\,\theta)}{1} = \frac{e^{i\phi} - e^{-i\phi}}{e^{i\phi} + e^{-i\phi}}$$

Applying componendo and dividendo, we get

$$\frac{1 + \sin(i\theta)}{1 - \sin(i\theta)} = \frac{e^{i\phi}}{e^{-i\phi}} = e^{2i\phi}$$

$$\Rightarrow \qquad \frac{1 - \cos\left(\dfrac{\pi}{2} + i\theta\right)}{1 + \cos\left(\dfrac{\pi}{2} + i\theta\right)} = e^{2i\phi}$$

$$\Rightarrow \qquad \tan^2\left(\frac{\pi}{4} + \frac{i\theta}{2}\right) = e^{2i\phi} \quad \Rightarrow \quad \tan\left(\frac{\pi}{4} + \frac{i\theta}{2}\right) = e^{i\phi}$$

$$\Rightarrow \qquad i\phi = \log\tan\left(\frac{\pi}{4} + \frac{i\theta}{2}\right)$$

$$\Rightarrow \qquad \phi = -i\log\tan\left(\frac{\pi}{4} + \frac{i\theta}{2}\right)$$

$$\therefore \qquad \tan^{-1}(\sinh\theta) = -i\log\tan\left(\frac{\pi}{4} + \frac{i\theta}{2}\right). \qquad\qquad \text{[using (1)]}$$

Example 12. *If* $\cosh^{-1}(x + iy) + \cosh^{-1}(x - iy) = \cosh^{-1} a$, *show that*
$$2(a - 1)x^2 + 2(a + 1)y^2 = a^2 - 1$$

Solution. Let
$$\cosh^{-1}(x + iy) = A + iB \qquad\qquad\qquad ...(1)$$
$$\Rightarrow \qquad (x + iy) = \cosh(A + iB)$$
$$\Rightarrow \qquad (x + iy) = \cos(i(A + iB))$$
$$\Rightarrow \qquad (x + iy) = \cos(-B + iA)$$
$$\Rightarrow \qquad (x + iy) = \cos B \cosh A + \sin B \sinh A$$

Separating real and imaginary parts, we get
$$x = \cos B \cosh A \qquad\qquad\qquad ...(2)$$
and
$$y = \sin B \sinh A \qquad\qquad\qquad ...(3)$$

Eliminating B between (2) and (3), we have

$$\frac{x^2}{\cosh^2 A} + \frac{y^2}{\sinh^2 A} = 1$$

$$\Rightarrow \qquad x^2 \sinh^2 A + y^2 \cosh^2 A = \cosh^2 A \sinh^2 A \qquad\qquad ...(4)$$

Also,
$$\cosh^{-1}(x - iy) = A - iB \qquad\qquad\qquad ...(5)$$

Adding (1) and (5), we get
$$2A = \cosh^{-1}(x + iy) + \cosh^{-1}(x - iy) = \cosh^{-1} a \ (\text{given})$$

$$\Rightarrow \qquad a = \cosh 2A$$

$$\Rightarrow \qquad a + 1 = 1 + \cosh 2A = 2\cosh^2 A$$

$$\therefore \quad \cosh^2 A = \frac{1}{2}(a + 1)$$

and $\quad a - 1 = \cosh 2A - 1 = 2 \sinh^2 A$

$$\therefore \quad \sinh^2 A = \frac{1}{2}(a - 1)$$

Putting the values of $\cosh^2 A$ and $\sinh^2 A$ in (4), we get

$$\frac{1}{2}(a - 1)x^2 + \frac{1}{2}(a + 1)y^2 = \frac{1}{2}(a + 1)\frac{1}{2}(a - 1)$$

or $\qquad\qquad 2(a - 1)x^2 + 2(a + 1)y^2 = a^2 - 1$

Example 13. *If* $\sin^{-1}(\theta + i\phi) = \alpha + i\beta$, *then prove that* $\sin^2 \alpha$ *and* $\cosh^2 \beta$ *are the roots of the equation* $x^2 - x(1 + \theta^2 + \phi^2) + \theta^2 = 0$.

Solution. We have

$$\sin^{-1}(\theta + i\phi) = \alpha + i\beta$$

$\Rightarrow \qquad\qquad \theta + i\phi = \sin(\alpha + i\beta)$

$\Rightarrow \qquad\qquad \theta + i\phi = \sin \alpha \cosh \beta + i \cos \alpha \sinh \beta$

Separating real and imaginary parts, we get

$$\theta = \sin \alpha \cosh \beta$$

$$\phi = \cos \alpha \sinh \beta$$

Then $\qquad 1 + \theta^2 + \phi^2 = 1 + \sin^2 \alpha \cosh^2 \beta + \cos^2 \alpha \sinh^2 \beta$

$$= 1 + \sin^2 \alpha \cosh^2 \beta + (1 - \sin^2 \alpha)(\cosh^2 \beta - 1)$$

$$= \sin^2 \alpha \cosh^2 \beta$$

and $\qquad\qquad \theta^2 = \sin^2 \alpha \cosh^2 \beta$

A quadratic equation whose roots are $\sin^2 \alpha$ and $\cosh^2 \beta$, is given by

$$x^2 - x(\sin^2 \alpha + \cosh^2 \beta) + \sin^2 \alpha \cosh^2 \beta = 0$$

$\Rightarrow \qquad\qquad x^2 - x(1 + \theta^2 + \phi^2) + \theta^2 = 0$

EXERCISE 11.1

1. Express $\sin^{-1}(x + iy)$ in the form of $A + iB$.

2. Show that $\operatorname{Sin}^{-1}(i) = 2n\pi - i \log\left(\sqrt{2} - 1\right)$.

3. Explain the meaning of $\sin^{-1} x$, when x is real and greater than 1, and show that
$$\operatorname{Sin}^{-1}(2) = n\pi + (-1)^n \left[\pm \frac{\pi}{2} + i \log(\sqrt{3} \pm 2) \right]$$

4. If $\cos^{-1}(\alpha + i\beta) = \theta + i\phi$, prove that
   (i) $\alpha^2 \operatorname{sech}^2 \phi + \beta^2 \operatorname{cosech}^2 \phi = 1$
   (ii) $\alpha^2 \sec^2 \theta - \beta^2 \operatorname{cosec}^2 \theta = 1$

5. If $\sin^{-1}(x + iy) = \tan^{-1}(u + iv)$, show that $[(x - 1)^2 + y^2][(x + 1)^2 + y^2] = \dfrac{(x^2 + y^2)^2}{(u^2 + v^2)^2}$

6. Prove that one of the values of $\sin^{-1}(\cos\theta + i\sin\theta)$ is
$$\cos^{-1}(\sqrt{\sin\theta}) + i\log[\sqrt{\sin\theta} + \sqrt{1 + \sin\theta}]$$

7. Show that : $\cos^{-1}(\cos\theta + i\sin\theta) = \sin^{-1}(\sqrt{\sin\theta}) + i\log[\sqrt{1 + \sin\theta} - \sqrt{\sin\theta}]$

8. If $\cos^{-1}(\alpha + i\beta) = \theta + i\phi$, prove that $\alpha\tan\theta + \beta\coth\phi = 0$.

9. Prove that : $\sin^{-1}(\operatorname{cosec}\theta) = n\pi + (-1)^n\left[\dfrac{\pi}{2} + i\log\cot\dfrac{\theta}{2}\right]$

10. Prove that $\tan^{-1}(\cos\theta + i\sin\theta) = n\pi + \dfrac{\pi}{4} - \dfrac{1}{2}i\log\tan\left(\dfrac{\pi}{4} - \dfrac{\theta}{2}\right)$

11. If $x > y$, then show that $\tan^{-1}\left(\dfrac{x + iy}{x - iy}\right) = \dfrac{\pi}{4} + \dfrac{i}{2}\log\left(\dfrac{x + y}{x - y}\right)$

12. If $\tan(\alpha + i\beta) = \tan^{-1}(x - iy)$, where $x^2 + y^2 \neq 1$, prove that
$$\tan[\log(\alpha^2 + \beta^2)] = \dfrac{2x}{1 - x^2 - y^2}.$$

13. Express $\tanh^{-1}(x + iy)$ in the form of $A + iB$.

14. If $\cosh x = \sec\theta$, then prove that $x = \log(\sec\theta \pm \tan\theta)$

15. Prove that $\tanh^{-1}x = \sinh^{-1}\left(\dfrac{x}{\sqrt{1 - x^2}}\right)$

16. Prove that $\cosh^{-2}(2/x) = \sinh^{-1}\left(\dfrac{x}{\sqrt{4 - x^2}}\right)$

17. Prove that $\tan^{-1}(\cot\theta\tanh\phi) = -\dfrac{i}{2}\log\left[\dfrac{\sin(\theta + i\phi)}{\sin(\theta - i\phi)}\right]$

18. Prove that $\operatorname{Tan}^{-1}\left(\dfrac{\tan 2\theta + \tanh 2\phi}{\tan 2\theta - \tanh 2\phi}\right) + \operatorname{Tan}^{-1}\left(\dfrac{\tan\theta - \tanh\phi}{\tan\theta + \tanh\phi}\right) = \operatorname{Tan}^{-1}(\cot\theta\coth\phi)$.

HINT TO SELECTED PROBLEMS

1. $\sin^{-1}(x + iy) = \dfrac{\pi}{2} - \cos^{-1}(x + iy)$
   and for $\cos^{-1}(x + iy)$, see example 2.

3. $\operatorname{Sin}^{-1}(2) = n\pi + (-1)^n\sin^{-1}(2)$
   Let $\qquad\qquad \sin^{-1}(2) = x + iy$
   $\Rightarrow \qquad\qquad \sin(x + iy) = 2$
   $\Rightarrow \quad \sin x\cosh y + i\cos x\sinh y = 2$
   $\Rightarrow \qquad \sin x\cosh y = 2$ and $\cos x\sinh y = 0$

Now $\cos x \sinh y = 0 \Rightarrow \cos x = 0$ and $\sinh y \neq 0$ as $\sin^{-1}(2)$ is not real.

$$\therefore \qquad\qquad \cos x = 0 \Rightarrow x = \pm\pi/2$$

Then $\cosh y = \pm 2 \Rightarrow y = \cosh^{-1}(\pm 2) = \log(\sqrt{3} \pm 2)$

$\therefore$ Hence we obtain the required result.

8. $\cos^{-1}(\alpha + i\beta) = \theta + i\phi$

$\Rightarrow \qquad \alpha + i\beta = \cos(\theta + i\phi) = \cos\theta\cosh\phi - i\sin\theta\sinh\phi$

$\Rightarrow \qquad \alpha = \cos\theta\cosh\phi$ and $\beta = -\sin\theta\sinh\phi$

$\therefore \qquad \alpha\tan\theta + \beta\coth\phi = 0.$

17. $\tan^{-1}(\cot\theta\tanh\phi) = x$

$\Rightarrow \qquad\qquad\qquad \cot\theta\tanh\phi = \tan x$

$\Rightarrow \qquad\qquad \dfrac{\cos\theta\sinh\phi}{\sin\theta\cosh\phi} = \dfrac{e^{ix} - e^{-ix}}{i(e^{ix} + e^{-ix})}$

$\Rightarrow \qquad\qquad \dfrac{e^{ix} + e^{-ix}}{e^{ix} - e^{-ix}} = \dfrac{\sin\theta\cosh\phi}{i\cos\theta\sinh\phi}$

Applying componendo and dividendo, we get

$$\dfrac{2e^{ix}}{2e^{-ix}} = \dfrac{\sin\theta\cosh\phi + i\cos\theta\sinh\phi}{\sin\theta\cosh\phi - i\cos\theta\sinh\phi} = \dfrac{\sin(\theta + i\phi)}{\sin(\theta - i\phi)}$$

$$= -\dfrac{1}{2}i\log\left[\dfrac{\sin(\theta + i\phi)}{\sin(\theta - i\phi)}\right]$$

$$\therefore \; \tan^{-1}(\cot\theta\tanh\phi) = -\dfrac{1}{2}i\log\left[\dfrac{\sin(\theta + i\phi)}{\sin(\theta - i\phi)}\right]$$

ANSWERS

(1) $\sin^{-1}(x + iy) = \dfrac{\pi}{2} \pm \sin^{-1}\left[\dfrac{\sqrt{(x^2 + y^2 - 1)^2 + 4y^2} - (x^2 + y^2 - 1)}{2}\right]^{1/2}$

$\pm i\sinh^{-1}\left[\dfrac{\sqrt{(1 - x^2 - y^2)^2 + 4y^2} - (1 - x^2 - y^2)}{2}\right]^{1/2}$

(3) When x is real and greater than 1, then $\sin^{-1}x$ will be a complex number.

(13) $\tanh^{-1}(x + iy) = \dfrac{1}{2}\tanh^{-1}\left(\dfrac{2x}{1 + x^2 + y^2}\right) + \dfrac{i}{2}\tan^{-1}\left(\dfrac{2y}{1 - x^2 - y^2}\right)$

CHAPTER REVIEW : A COMPETITIVE APPROACH

Selected terms and Results

TERMS

- **Inverse Sine of $(x + iy)$:** If $\sin(u + iv) = x + iy$ then $u + iv$ is called an inverse sine of $x + iy$.

- **Inverse cosine of $(x + iy)$:** If $\cos(u + iv) = x + iy$ then $u + iv$ is called an inverse cosine of $x + iy$.

- **Inverse hyperbolic function :** Let z and w be any two complex numbers. If $\sinh w = z$ then w is called the inverse hyperbolic sine of z.

RESULTS

- For general value of inverse sine, the first letter S being written capital.

- For principal value of inverse sine, the first letter 's' being written small and the real part lies between $-\dfrac{p}{2}$ and $\dfrac{p}{2}$.

- $\text{Sin}^{-1}(x + iy) = n\pi + (-1)^n \sin^{-1}(x + iy) : n \in \mathbb{Z}$.

- $\text{Cos}^{-1}(x + iy) = 2n\pi \pm \cos^{-1}(x + iy) : n \in \mathbb{Z}$.

- $\sinh^{-1} z = \log[z + \sqrt{z^2 + 1}]$

- $\cosh^{-1} z = \log[z + \sqrt{z^2 - 1}]$

- $\tanh^{-1} z = \dfrac{1}{2}\log\left[\dfrac{1+z}{1-z}\right]$.

- $\coth^{-1} z = \dfrac{1}{2}\log\left(\dfrac{z+1}{z-1}\right)$.

- $\sinh^{-1} x = -i \sin^{-1}(ix)$
- $\cosh^{-1} x = -i \cos^{-1} x$
- $\tanh^{-1} x = -i \tan^{-1}(ix)$

Review Questions and Project Work

1. Prove that $\sinh^{-1}(\cot x) = \log(\cot x + \operatorname{cosec} x)$

2. Prove that $\sin^{-1}(\cot \theta + i \sin \theta) =$

$$\cos^{-1}\sqrt{\sin?} + i\log(\sqrt{\sin?+1}) - \sqrt{\sin?}$$

3. If $\cos^{-1}(u + iv) = \alpha + i\beta$, prove that $\cos^2\alpha$ and $\cosh^2\beta$ are the roots of the equation $x^2 - (1 + u^2 + v^2)x + u^2 = 0$

4. If $\cosh^{-1}(x+iy) + \cosh^{-1}(x-iy) = \cosh^{-1} a$, show that
$$2(a-1)x^2 + 2(a+1)y^2 = a^2 - 1$$

5. Prove that $\sinh^{-1}(\cot x) = \log(\cot x + \operatorname{cosec} x)$

Objective Type Questions

Fill in the blanks :

1. $\sinh^{-1} z = \log[\ldots\ldots\ldots\ldots]$
2. $\cosh^{-1} z = \log[\ldots\ldots\ldots\ldots]$
3. $\sinh^{-1} x = \ldots\ldots\ldots \sin^{-1}(ix)$
4. $\cosh^{-1} x = \ldots\ldots\ldots \cos^{-1} x$
5. $\tanh^{-1} x = \ldots\ldots\ldots \tan^{-1} x$

True/False: *Write 'T' for true and 'F' for false statement.*

1. $\tanh^{-1} x = i \tan^{-1}(ix)$ (T/F)

2. $\tanh^{-1} z = \dfrac{1}{2}\log\left[\dfrac{1+z}{1-z}\right]$ (T/F)

3. $\coth^{-1} z = \dfrac{1}{2}\log\left[\dfrac{z+1}{z-1}\right]$ (T/F)

4. $\sinh^{-1} z = \log[z + \sqrt{z^2 + 1}]$ (T/F)

Multiple Choice Questions : *Choose the most appropriate one :*

1. In case of cot and cosec, the principal value is that value for which the real part lies between :

   (a) $-\dfrac{\pi}{2}$ and $\dfrac{\pi}{2}$

   (b) $-\pi$ and π

   (c) 0 and π

   (d) none of these

2. In case of sec, the principal value is that value for which the real part lies between :

   (a) $-\dfrac{\pi}{2}$ and $\dfrac{\pi}{2}$

   (b) $-\pi$ and π

   (c) 0 and π

   (d) none of these

3. The value of $\coth^{-1} z =$

   (a) $\log\left[\dfrac{z+1}{z-1}\right]$

   (b) $\dfrac{1}{2}\log\left[\dfrac{z+1}{z-1}\right]$

   (c) $\dfrac{1}{2}\log\left[\dfrac{z-1}{z+1}\right]$

   (d) none of these

4. The value of $\tanh^{-1} x =$

   (a) $i\tan^{-1}(ix)$

   (b) $-i\tan^{-1}(ix)$

   (c) $-i\tan^{-1}(x)$

   (d) none of these

5. The value of $\sinh^{-1}(\cot x) =$

   (a) $\log(\cot x + \operatorname{cosec} x)$

   (b) $\log \cot x + \operatorname{cosec} x$

   (c) $\log \cot x$

   (d) none of these

ANSWERS

Fill in the blanks

(1) $z+\sqrt{z^2+1}$

(2) $z+\sqrt{z^2-1}$

(3) $-i$

(4) $-i$

(5) $-i$

True/False

(1) F

(2) T

(3) T

(4) T

Multiple choice questions

(1) a

(2) c

(3) b

(4) b

(5) a

CHAPTER 12

Polynomial with Real Coefficients

❖ Transformation of equations	❖ Roots of Coefficients
❖ Descartes' Rule of Signs	❖ Multiple Roots
❖ Rolle's Theorem	❖ Limit of the Roots

12.1 INTRODUCTION

A function $f(x)$ of the form

$$f(x) = a_0x^n + a_1x^{n-1} + a_2x^{n-2} + \ldots + a_{n-1}x + a_n, \; a_0 \neq 0$$

of degree n is said to be a rational integral function of x if all the coefficients $a_0, a_1, a_2,$ $\ldots a_{n-1}, a_n$ are supposed to be rational.

Definition. *The equation $f(x) = 0$ is called the general form of rational integral equation of nth degree.*

Definition. *Any value of x for which the value of $f(x)$ comes out to be zero, then this value of x is called a **root** of the equation $f(x) = 0$.*

Since when $f(x)$ is divided by the factor $(x - a)$ then $f(a)$ is obtained as a remainder. If this remainder $f(a)$ becomes zero, then a is a root of the function $f(x) = 0$. Therefore, we can say that if 'a' is a root of the equation $f(x) = 0$, then $f(a) = 0$

12.2 NUMBER OF ROOTS OF ANY EQUATION

THEOREM-1

Every equation of degree n has n roots and no more.

Proof. Let the equation of degree n be

$$f(x) = a_0x^n + a_1x^{n-1} + a_2x^{n-2} + \ldots + a_{n-1}x + a_n = 0 \qquad \ldots(1)$$

provided $a_0 \neq 0$.

The equation $f(x) = 0$ has the roots, real as well as imaginary. Therefore, if α_1 is any root of the equation (1), then $f(x)$ can be written as

$$f(x) = (x - \alpha_1)(a_0x^{n-1} + \ldots)$$

or $\qquad f(x) = (x - \alpha_1)\phi_1(x) \qquad \ldots(2)$

where $\phi_1(x)$ is a function of x of degree $n - 1$, such that $\phi_1(\alpha_1) \neq 0$. Further let α_2 be a root of $\phi_1(x) = 0$, then $\phi_1(x)$ can be written

$$\phi_1(x) = (x - \alpha_2)\phi_2(x)$$

$\therefore \qquad f(x) = (x - \alpha_1)(x - \alpha_2)\phi_2(x). \hspace{3cm} \text{...(3)}$

Continuing this process upto n times, we obtain

$$f(x) = a_0(x - \alpha_1)(x - \alpha_2) \dots (x - \alpha_n). \hspace{2cm} \text{...(4)}$$

From equation (4) it is clear that when x take the values from α_1 to α_n, $f(x)$ comes out be zero.

Hence the equation $f(x)$ has n roots. Moreover if x takes any value different from α_1, α_2, ... α_n, $f(x)$ can not be zero so that $f(x) = 0$ has exactly n roots.

12.3   RELATION BETWEEN THE ROOTS AND COEFFICIENTS

Let the general equation of degree n be given by

$$a_0x^n + a_1x^{n-1} + a_2x^{n-2} + \dots + a_{n-1}x + a_n = 0 \hspace{2cm} \text{...(1)}$$

where $a_0, a_1, a_2, \dots a_n$ are the coefficients and $a_0 \neq 0$ and let $\alpha_1, \alpha_2, \alpha_3 \dots \alpha_n$ be the roots of the equation (1). Then the equation (1) can be identically written as

$$a_0x^n + a_1x^{n-1} + a_2x^{n-2} + \dots + a_{n-1}x + a_n = a_0(x - \alpha_1)(x - \alpha_2) \dots (x - \alpha_n)$$

or $\qquad a_0x^n + a_1x^{n-1} + a_2x^{n-2} + \dots + a_{n-1}x + a_n$

$$= a_0[x^n - (\Sigma\alpha_1)x^{n-1} + (\Sigma\alpha_1\alpha_2)x^{n-2} + \dots + (-1)^n \alpha_1\alpha_2 \dots \alpha_n]$$

where $\Sigma\alpha_1 = \alpha_1 + \alpha_2 + \dots + \alpha_n$

$\qquad \Sigma\alpha_1\alpha_2 = \alpha_1\alpha_2 + \alpha_1\alpha_3 + \dots$ etc.

Now equating the coefficients of like power of x of both sides we get

$$\left. \begin{array}{l} \Sigma\alpha_1 = -\dfrac{a_1}{a_0} \\[3mm] \Sigma\alpha_1\alpha_2 = \dfrac{a_2}{a_0} \\[3mm] \Sigma\alpha_1\alpha_2\alpha_3 = -\dfrac{a_3}{a_0} \\[3mm] \vdots \\[3mm] \alpha_1\alpha_2\alpha_2 \dots \alpha_n = (-1)^n \dfrac{a_n}{a_0} \end{array} \right\} \hspace{2cm} \text{...(2)}$$

Hence the equation (2) gives the required relation between the roots and the coefficients of equation.

Remark

❖   If the equation is not complete *i.e.*, some of the terms are missing, then we should first make this equation complete by adding the missing terms with zero coefficients.

12.4 IMPORTANT RESULTS

1. In an equation with real coefficients, imaginary roots occur in pair, that is if $\alpha + i\beta$ is one of the root of the equation $f(x) = 0$, then $\alpha - i\beta$ will also be a root of that equation.

2. If the equation $f(x) = 0$ has a pair of complex (imaginary) roots $\alpha \pm i\beta$, then $(x - \alpha)^2 + \beta^2$ will be a factor of $f(x)$.

3. If $\alpha + \sqrt{\beta}$ is a root of the equation $f(x) = 0$, then $\alpha - \sqrt{\beta}$ will also be a root of $f(x) = 0$

4. Every equation of odd degree with real coefficients has at least one real root with the sign opposite to that of its last term.

5. Every equation of even degree with negative last term has at least two real roots with contrary sign.

6. If the equation $f(x) = 0$ and $g(x) = 0$ have common roots and there common roots are the roots of $h(x) = 0$, then $h(x)$ will be H.C.F. (G.C.D.) of $f(x)$ and $g(x)$.

7. If the equation $f(x) = 0$ has two roots equal, then the equation $f(x) = 0$ and $f'(x) = 0$ must have a common root.

12.5 HORNER'S SYNTHETIC DIVISION

In order to find the quotient and the remainder when a polynomial

$$f(x) = a_0 x^n + a_1 x^{n-1} + a_2 x^{n-2} + \dots + a_{n-1} x + a_n, (a_0 \neq 0) \qquad \dots(1)$$

of degree n is divided by a linear factor $(x - \alpha)$, we use a method given by **Horner,** called *synthetic division*. This method is being discussed as follows :

$$
\begin{array}{c|cccccc}
\alpha & a_0 & a_1 & a_2\dots & a_{n-1} & a_n \\
 & & \alpha a_0 & \alpha b_1\dots & \alpha b_{n-2} & \alpha b_{n-1} \\
\hline
 & a_0 & b_1 & b_2 & b_{n-1} \mid & R
\end{array}
$$

Step (1) If the equation (1) is not complete, then first make it complete by adding missing terms with zero coefficient.

Step (2) In the first horizontal line (row) we should write the coefficients $a_0, a_1, a_2, \dots a_{n-1}, a_n$ of the polynomial $f(x)$.

Step (3) Since we have to divide to polynomial $f(x)$ by $x - \alpha$, so we should write α to the left of the vertical line as shown above.

Step (4) In the third horizontal line (row) we should write a_0 and the first term of the second horizontal line (row) is obtained by multiplying a_0 to α and then add this term with a_1 we obtain b_1 which is the second term of the third row. Next, we multiply b_1 and α and obtained the second term of the second row now adding this αb_1 with a_2 we obtain third terms of the third row. Continue the process in the same way we obtain the last term in the third row which is in fact the remainder R while the second last term in the same is b_{n-1}.

Remark

❖ If the remainder R comes out be zero, then α will be a root of the equation $f(x) = 0$.

12.6 TRANSFORMATION OF EQUATION

Sometimes there arises some difficulties to find the roots of a given equation. In that case a process of transformation of a given equation into another equation plays an important role for finding the roots of given equation.

In this section we shall discuss some important transformation.

(i) *To transform an equation into another equation whose roots are the roots of the given equation with different sign.*

Let the given equation be

$$f(x) = a_0x^n + a_1x^{n-1} + a_2x^{n-2} + \ldots + a_{n-1}x + a_n = 0 \qquad \ldots(1)$$

and let $\alpha_1, \alpha_2, \alpha_3, \ldots, \alpha_n$ be the roots of the equation (1).

Now put $x = -y$ in (1), we get

$$f(-y) = a_0(-y)^n + a_1(-y)^{n-1} + a_2(-y)^{n-2} + \ldots + a_{n-1}(-y) + a_n = 0$$

or $\quad f(-y) = (-1)^n[a_0y^n - a_1y^{n-1} + a_2y^{n-2} - \ldots + (-1)^{n-1}a_{n-1}y + (-1)^na_n = 0 \qquad \ldots(2)$

This is the transformed equation.

Now we shall have to show that the equation (2) has the roots $-\alpha_1, -\alpha_2, -\alpha_3, \ldots, -\alpha_n$.

Since $\alpha_1, \alpha_2, \ldots \alpha_n$ are the roots of equation (1), then (1) can also be written as

$$a_0x^n + a_1x^{n-1} + a_2x^{n-2} + \ldots + a_{n-1}x + a_n = a_0(x - \alpha_1)(x - \alpha_2) \ldots (x - \alpha_n).$$

Now putting $x = -y$ in both sides, we get

$$a_0(-y)^n + a_1(-y)^{n-1} + a_2(-y)^{n-2} + \ldots + a_{n-1}(-y) + a_n$$
$$= a_0(-y - \alpha_1)(-y - \alpha_2) \ldots (-y - \alpha_n)$$

or $\quad (-1)^n[a_0y^n - a_1y^{n-1} + a_2y^{n-2} - \ldots + (-1)^{n-1}a_{n-1}y + (-1)^na_0]$
$$= a_0(-1)^n(y + \alpha_1)(y + \alpha_2) \ldots (y + \alpha_n).$$

Using (2) $\qquad\qquad f(-y) = a_0(-1)^n(y + \alpha_1)(y + \alpha_2) \ldots (y + \alpha_n)$

Thus the roots of the equation $f(-y) = 0$ are given by

$$(y + \alpha_1)(y + \alpha_2) \ldots (y + \alpha_n) = 0 \quad \text{or} \quad y = -\alpha_1, -\alpha_2, \ldots -\alpha_n$$

Hence the roots of the transformed equation (2) are the roots of the given equation with different sign.

(ii) *To transform an equation into another equation whose roots are equal to the roots of the given equation multiplied by a given constant number m.*

Let the given equation be

$$f(x) = a_0x^n + a_1x^{n-1} + a_2x^{n-2} + \ldots + a_{n-1}x + a_n = 0 \qquad \ldots(1)$$

and let $\alpha_1, \alpha_2, \ldots, \alpha_n$ be its roots, then (1) can be written as

$$a_0x^n + a_1x^{n-1} + a_2x^{n-2} + \ldots + a_{n-1}x + a_n = a_0(x - \alpha_1)(x - \alpha_2) \ldots (x - \alpha_n) \ldots(2)$$

Putting $y = mx$ or $x = \dfrac{y}{m}$ in (1), we get

$$f\left(\frac{y}{m}\right) = a_0\left(\frac{y}{m}\right)^n + a_1\left(\frac{y}{m}\right)^{n-1} + a_2\left(\frac{y}{m}\right)^{n-2} + \ldots + a_{n-1}\left(\frac{y}{m}\right) + a_n$$

or $\quad f\left(\dfrac{y}{m}\right) = \dfrac{1}{m^n}[a_0 y^n + m a_1 y^{n-1} + m^2 a_2 y^{n-2} + \ldots + m^{n-1} y a_{n-1} + m^n a_n] = 0$

or $\quad a_0 y^n + m a_1 y^{n-1} + m^2 a_2 y^{n-2} + \ldots + m^{n-1} a_{n-1} y + m^n a_n = 0 \qquad \ldots(3)$

This is the transformed equation. Now we shall show that the transformed equation

has the roots $m\alpha_1, m\alpha_2, \ldots m\alpha_n$. For this let us put $x = \dfrac{y}{m}$ in (2), we get

$$a_0\left(\frac{y}{m}\right)^n + a_1\left(\frac{y}{m}\right)^{n-1} + a_2\left(\frac{y}{m}\right)^{n-2} + \ldots + a_{n-1}\left(\frac{y}{m}\right) + a_n$$

$$= a_0\left(\frac{y}{m} - \alpha_1\right)\left(\frac{y}{m} - \alpha_2\right)\ldots\left(\frac{y}{m} - \alpha_n\right)$$

or $\quad \dfrac{1}{m^n}[a_0 y^n + m a_1 y^{n-1} + m^2 a_2 y^{n-2} + \ldots + m^{n-1} a_{n-1} y + m^n a_n]$

$$= a_0 \frac{1}{m^n}(y - m\alpha_1)(y - m\alpha_2)\ldots(y - m\alpha_n)$$

or $\quad a_0 y^n + m a_1 y^{n-1} + m^2 a_2 y^{n-2} + \ldots + m^{n-1} a_{n-1} y + m^n a_n$

$$= a_0(y - m\alpha_1)(y - m\alpha_2)\ldots(y - m\alpha_n).$$

This shows that the transformed equation (3) has the roots $m\alpha_1, m\alpha_2, \ldots m\alpha_n$.

(iii) *To transform an equation into another equation whose roots are the reciprocals of the roots of the given equation.*

Let the given equation be

$$f(x) = a_0 x^n + a_1 x^{n-1} + a_2 x^{n-2} + \ldots + a_{n-1} x + a_n = 0 \qquad \ldots(1)$$

and let $\alpha_1, \alpha_2, \ldots \alpha_n$ be its roots, then we have

$$a_0 x^n + a_1 x^{n-1} + a_2 x^{n-2} + \ldots + a_{n-1} x + a_n = a_0(x - \alpha_1)(x - \alpha_2)\ldots(x - \alpha_n)\ldots(2)$$

Putting $x = \dfrac{1}{y}$ in (1), we get

$$f\left(\frac{1}{y}\right) = a_0\left(\frac{1}{y}\right)^n + a_1\left(\frac{1}{y}\right)^{n-1} + a_2\left(\frac{1}{y}\right)^{n-2} + \ldots + a_{n-1}\left(\frac{1}{y}\right) + a_n = 0$$

or $\quad f\left(\dfrac{1}{y}\right) = \dfrac{1}{y^n}[a_0 + a_1 y + a_2 y^2 + \ldots + a_{n-1} y^{n-1} + a_n y^n] = 0$

or $\quad a_n y^n + a_{n-1} y^{n-1} + \ldots + a_1 y + a_0 = 0 \qquad \ldots(3)$

This is the transformed equation. Now we shall show that this equation (3) has the roots

$$\frac{1}{\alpha_1}, \frac{1}{\alpha_2}, \ldots \frac{1}{\alpha_n}$$

Let us put $x = \dfrac{1}{y}$ in (2), we get

$$a_0 \left(\frac{1}{y}\right)^n + a_1 \left(\frac{1}{y}\right)^{n-1} + a_2 \left(\frac{1}{y}\right)^{n-2} + \ldots + a_{n-1}\left(\frac{1}{y}\right) + a_n$$

$$= a_0 \left(\frac{1}{y} - \alpha_1\right)\left(\frac{1}{y} - \alpha_2\right)\ldots\left(\frac{1}{y} - \alpha_n\right)$$

or

$$\frac{1}{y^n}[a_0 + a_1 y + a_2 y^2 + \ldots + a_{n-1}y^{n-1} + a_n y^n]$$

$$= \frac{a_0}{y^n}(1 - \alpha_1 y)(1 - \alpha_2 y)\ldots(1 - \alpha_n y)$$

or

$$a_0 + a_1 y + a_2 y^2 + \ldots + a_{n-1}y^{n-1} + a_n y^n$$

$$= a_0 (1 - \alpha_1 y)(1 - \alpha_2 y)\ldots(1 - \alpha_n y)$$

This shows that the equation (3) has the roots $\dfrac{1}{\alpha_1}, \dfrac{1}{\alpha_2}, \ldots \dfrac{1}{\alpha_n}$.

(iv) **Reciprocal equation.** *An equation which remains unchanged when x is replaced by* $\dfrac{1}{x}$, *is called a reciprocal equation.*

Let the given equation be

$$f(x) \equiv a_0 x^n + a_1 x^{n-1} + a_2 x^{n-2} + \ldots + a_{n-1}x + a_n = 0 \qquad \ldots(1)$$

Replace x by $\dfrac{1}{x}$, we obtain

$$f\left(\frac{1}{x}\right) \equiv a_0 + a_1 x + a_2 x^2 + \ldots + a_{n-1}x^{n-1} + a_n x^n = 0 \qquad \ldots(2)$$

The equation (2) is an equation whose roots are the reciprocal of the roots of the equation (1). If both equations are same, then by comparing the coefficients of like powers of x we obtain

$$\frac{a_0}{a_n} = \frac{a_1}{a_{n-1}} = \frac{a_2}{a_{n-2}} \ldots \frac{a_{n-1}}{a_1} = \frac{a_n}{a_0}$$

From first and last fraction, we get

$$\frac{a_0}{a_n} = \frac{a_n}{a_0} \quad \text{or} \quad a_n^2 = a_0^2 \quad \text{or} \quad a_n = \pm a_0$$

Therefore from this result we have $a_n = a_0$, $a_n = -a_0$ and thus there are two classes of the reciprocal equations.

(a) If $a_n = a_0$, then

$$\frac{a_1}{a_{n-1}} = \frac{a_2}{a_{n-2}} = \ldots 1 \quad \text{or} \quad a_1 = a_{n-1}, \ a_2 = a_{n-2} \ldots$$

This is, the coefficients of the terms in the equation equidistant from the beginning and the end are equal and the equation is therefore called the **first class**.

(b) If $a_n = -a_0$, then

$$\frac{a_1}{a_{n-1}} = \frac{a_2}{a_{n-2}} = \ldots = -1 \quad \text{or} \quad a_1 = -a_{n-1}, \ a_2 = -a_{n-2} \ldots$$

That is, the coefficients of the terms in the equation equidistant from the beginning and the end are equal in magnitude and opposite in sign. Therefore the reciprocal is called **second class**. In this case if the degree of the equation is **2m** (even) then, $a_m = -a_m$ or $a_m = 0$. Thus we can say that if the equation of second class and of even degree, then the middle term of the equation will be absent.

(v) *Standard form of the reciprocal equation.* Let $f(x) = 0$ be a reciprocal equation and if $f(x) = 0$ is of first class and of an odd degree, then one of the roots of this equation $f(x) = 0$ must be its own reciprocal so it has a root -1 and thus $f(x)$ is divisible by the factor $x + 1$. If $\phi(x)$ is the quotient, then $\phi(x) = 0$ will be a reciprocal equation of first class and of an even degree.

On the other hand if the equation $f(x) = 0$ is of second class and of an odd degree, then it will have the root $+1$, and therefore $f(x)$ is divisible by the factor $x - 1$. If $\phi(x)$ is the quotient, then

$$f(x) = (x - 1)\phi(x)$$

Thus $\phi(x) = 0$ is a reciprocal equation of first class and of even degree.

And if the equation $f(x) = 0$ is of the second class and of an even degree, then it will have two roots -1 and $+1$. Therefore, $f(x)$ is divisible by $(x + 1)$ and $(x - 1)$ or divisible by $(x^2 - 1)$. If $\phi(x)$ is the quotient, then

$$f(x) = (x^2 - 1)\ \phi(x)$$

from this equation it is obvious that $\phi(x) = 0$ will be a reciprocal equation of first class and of even degree. Hence from above discussion we can say that every reciprocal equation can be reduced to a reciprocal of first class and of even degree which is known as the *standard form.*

(vi) *Every reciprocal equation of the standard form can be reduced to an equation of degree half of the degree of the original equation.*

Let the reciprocal equation of the standard form be given by

$$a_0 x^{2m} + a_1 x^{2m-1} + a_2 x^{2m-2} + \ldots + a_m x^m + \ldots + a_2 x^2 + a_1 x + a_0 = 0 \qquad \ldots(1)$$

Divide this equation by x^m, we get

$$a_0 x^m + a_1 x^{m-1} + a_2 x^{m-2} + \ldots + a_m + \ldots + a_2 \cdot \frac{1}{x^{m-2}} + \frac{a_1}{x^{m-1}} + \frac{a_0}{x^m} = 0$$

$$a_0\left(x^m + \frac{1}{x^m}\right) + a_1\left(x^{m-1} + \frac{1}{x^{m-1}}\right) + a_2\left(x^{m-2} + \frac{1}{x^{m-2}}\right) + \ldots + a_m = 0 \qquad \ldots(2)$$

Since, we know that

$$x^{k+1} + \frac{1}{x^{k+1}} = \left(x^k + \frac{1}{x^k}\right)\left(x + \frac{1}{x}\right) - \left(x^{k-1} + \frac{1}{x^{k-1}}\right)$$

putting $x + \frac{1}{x} = y$ for $k = 1, 2, 3, ...$

for $k = 1$, $x^2 + \frac{1}{x^2} = \left(x + \frac{1}{x}\right)\left(x + \frac{1}{x}\right) - (1+1) = y^2 - 2$

for $k = 2$, $x^3 + \frac{1}{x^3} = \left(x^2 + \frac{1}{x^2}\right)\left(x + \frac{1}{x}\right) - \left(x + \frac{1}{x}\right) = (y^2 - 2)y - y = y^3 - 3y$

for $k = 3$, $x^4 + \frac{1}{x^4} = \left(x^3 + \frac{1}{x^3}\right)\left(x + \frac{1}{x}\right) - \left(x^2 + \frac{1}{x^2}\right) = (y^3 - 3y)y - (y^2 - 2) = y^4 - 4y^2 + 2$

an so on, we obtain $\left(x^m + \frac{1}{x^m}\right)$ is a polynomial of degree 'm'. Hence the equation (2) is obtained an equation of degree m which is half of the degree of the equation (1).

(vii) *To transform an equation into another equation whose roots are any powers of the roots of the given equation.*

Let the given equation be

$$f(x) \equiv a_0 x^n + a_1 x^{n-1} + a_2 x^{n-2} + ... + a_{n-1} x + a_n = 0 \qquad ...(1)$$

and let $\alpha_1, \alpha_2, ... \alpha_n$ be it's roots, then we have

$$f(x) \equiv a_0 x^n + a_1 x^{n-1} + a_2 x^{n-2} + ... + a_{n-1} x + a_n = a_0(x - \alpha_1)(x - \alpha_2) ... (x - \alpha_n).$$
$$...(2)$$

The equation (2) can be modified as follows

$$f(x) \equiv a_0(x - \alpha_1)\left(\frac{x^m - \alpha_1^m}{x^m - \alpha_1^m}\right)(x - \alpha_2)\left(\frac{x^m - \alpha_2^m}{x^m - \alpha_2^m}\right)...(x - \alpha_n)\left(\frac{x^m - \alpha_n^m}{x^m - \alpha_n}\right)$$

or $\qquad a_0(x^m - \alpha_1^m)(x^m - \alpha_2^m)...(x^m - \alpha_n^m) = f(x)\left(\frac{x^m - \alpha_1^m}{x - \alpha_1}\right)\left(\frac{x^m - \alpha_2^m}{x - \alpha_2}\right)...\left(\frac{x^m - \alpha_n^m}{x - \alpha_n}\right)$

or $\qquad a_0(x^m - \alpha_1^m)(x^m - \alpha_2^m)...(x^m - \alpha_n^m) = f(x)\,[x^{m+1} + \alpha_1 x^{m-2} + ... + \alpha_1^{m-1}]$

$$[x^{m-1} + \alpha_2 x^{m-2} + ... + \alpha_2^{m-1}]...[x^{m-1} + \alpha_n x^{m-2} + ... + \alpha_n^{m-1}] \quad ...(3)$$

Let us assume

$$\phi(x^m) = a_0(x^m - \alpha_1^m)(x^m - \alpha_2^m)...(x^m - \alpha_n^m) \qquad ...(4)$$

It is obvious that for $x = \alpha_1, \alpha_2, ... \alpha_n$ equation (3) gives the identity so that the equation (4) is the transformed equation whose roots are $\alpha_1^m, \alpha_2^m, ... \alpha_n^m$. Hence if we put $x^m = y$ in (4), we obtain (4) as follow :

$$\phi(y) = a_0(y - \alpha_1^m)(y - \alpha_2^m)...(y - \alpha_n^m)$$

(vii) *To transform an equation into another equation whose roots exceed the roots of the given equation by a constant h.*

Let the given equation be

$$f(x) \equiv a_0 x^n + a_1 x^{n-1} + a_2 x^{n-2} + \dots + a_{n-1} x + a_n = 0 \qquad \dots(1)$$

and let $\alpha_1, \alpha_2, \dots \alpha_n$ be it's roots. We have

$$f(x) \equiv a_0 x^n + a_1 x^{n-1} + a_2 x^{n-2} + \dots + a_{n-1} x + a_n$$
$$= a_0 (x - \alpha_1)(x - \alpha_2) \dots (x - \alpha_n). \qquad \dots(2)$$

putting $y = x + h$, i.e. $x = y - h$ in (1); we get

$$f(y - h) = a_0 (y - h)^n + a_1 (y - h)^{n-1} + a_2 (y - h)^{n-2} + \dots + a_{n-1}(y - h) + a_n = 0$$

The equation can be written in decending powers of y as follows :

$$A_0 y^n + A_1 y^{n-1} + A_2 y^{n-2} + \dots + A_{n-1} y + A_n = 0 \qquad \dots(3)$$

where $A_0, A_1, A_2, \dots A_n$ are coefficients and constants and whose values depend upon $a_0, a_1, a_2, \dots a_n$.

Now put $y = x + h$ in (3), we get

$$f(x) = A_0 (x + h)^n + A_1 (x + h)^{n-1} + \dots + A_{n-1}(x + h) + A_n = 0$$
$$f(x) = (x + h)[A_0 (x + h)^{n-1} + A_1 (x + h)^{n-2} + \dots + A_{n-1}] + A_n.$$

This equation gives that if $f(x)$ is divided $x + h$, then A_n is obtained as remainder and the quotient is

$$A_0 (x + h)^{n-1} + A_1 (x + h)^{n-2} + \dots + A_{n-2}(x + h) + A_{n-1}$$

similarly if this quotient is divided by $(x + h)$, then we obtain A_{n-1} as remainder.

Continuing this process until we get all the constants $A_n, A_{n-1}, \dots A_2, A_1$ and we also obtain $A_0 = a_0$.

Hence the transformed equation is

$$f(y - h) \equiv A_0 y^n + A_1 y^{n-1} + A_2 y^{n-2} + \dots + A_{n-1} y + A_n = 0$$

Now we have to show that $\alpha_1 + h, \alpha_2 + h, \dots \alpha_n + h$ are the roots of this transformed equation.

$\therefore$ put $x = y - h$ in (2), we get

$$f(y - h) \equiv a_0 (y - h - \alpha_1)(y - h - \alpha_2) \dots (y - h - \alpha_n)$$
$$\equiv a_0 (y - (\alpha_1 + h))(y - (\alpha_2 + h)) \dots (y - (\alpha_n + h)).$$

Hence, $\alpha_1 + h, \alpha_2 + h, \dots \alpha_n + h$ are the roots of the transformed equation.

12.7 REMOVAL OF TERMS OF AN EQUATION

Let the given equation be

$$f(x) \equiv a_0 x^n + a_1 x^{n-1} + a_2 x^{n-2} + \dots + a_{n-1} x + a_n = 0 \qquad \dots(1)$$

If we put $x = y + h$, we get

$$a_0 (y + h)^n + a_1 (y + h)^{n-1} + a_2 (y + h)^{n-2} + \dots + a_{n-1}(y + h) + a_n = 0$$

This equation can be written in the decending powers of y as follows :

$$a_0 y^n + (na_0 h + a_1) y^{n-1} + \left\{ \frac{n(n-1)}{2!} a_0 h^2 + (n-1)a_1 h y^{n-1} + a_2 \right\} + \dots = 0$$

Now, we want to remove second term, then we shall equal to zero the coefficient of y^{n-1}, we get

$$na_0h + a_1 = 0 \quad \text{or} \quad h = -\frac{a_1}{na_0}.$$

Hence we decreased all the roots of the given equation by a constant $-\dfrac{a_1}{na_0}$, the second term of the given equation can be removed.

Similarly if we want to remove third term, we put

$$\frac{n(n-1)}{2!}a_0h^2 + (n-1)a_1h + a_2 = 0$$

Solve this equation we get two values of h and similarly we can remove any term of the given equation.

(i) *To remove the second term of the equation*
$$a_0x^3 + 3a_1x^2 + 3a_2x + a_3 = 0$$
and form the equation with integral coefficients having leading coefficient unity.
Since the equation is
$$f(x) \equiv a_0x^3 + 3a_1x^2 + 3a_2x + a_3 = 0 \qquad \text{...(1)}$$
Let $\alpha_1, \alpha_2, \alpha_3$ be its roots
put $x = y + h$ in (1), we get
$$a_0(y+h)^3 + 3a_1(y+h)^2 + 3a_2(y+h) + a_3 = 0$$
or $\quad a_0(y^3 + h^3 + 3y^2h + 3yh^2) + 3a_1(y^2 + h^2 + 2yh) + 3a_2(y+h) + a_3 = 0$
or $\quad a_0y^3 + (3ha_0 + 3a_1)y^2 + (3h^2a_0 + 6a_1h + 3a_2)y + (a_0h^3 + 3a_1h^2 + 3a_2h + a_3) = 0 \quad \text{...(2)}$
Now we want to remove second term, then put

$$3ha_0 + 3a_1 = 0 \quad \text{or} \quad h = -\frac{a_1}{a_0}$$

Substitute the value of h in (2), we get

$$a_0y^3 + \left(\frac{3a_1^2}{a_0} - \frac{6a_1^2}{a_0} + 3a_2\right)y + \left(-\frac{a_1^3}{a_0^2} + \frac{3a_1^3}{a_0^2} - \frac{3a_1a_2}{a_0} + a_3\right) = 0$$

or $\qquad a_0y^3 + \dfrac{3(a_0a_2 - a_1^2)}{a_0}y + \dfrac{(a_0^2a_3 - 3a_0a_1a_2 + 2a_1^3)}{a_0^2} = 0$

or $\qquad a_0y^3 + \dfrac{3H}{a_0}y + \dfrac{G}{a_0^2} = 0 \qquad\qquad\qquad \text{...(3)}$

where $\quad H = a_0a_2 - a_1^2, \ G = a_0^2a_3 - 3a_0a_1a_2 + 2a_1^3$

Thus the equation (3) is a transformed equation. Further, make all the coefficients of (3) integers, so that (3) can be written as

$$a_0^3 y^3 + 3Ha_0 y + G = 0$$

let us put $z = a_0 y$.

$$\therefore \qquad z^3 + 3Hz + G = 0 \qquad\qquad\qquad\qquad \text{...(4)}$$

This is the transformed equation with integral coefficients and having leading coefficient unity. Now the roots of (4) are obtained by the transformation

$$z = a_0 y = a_0 (x - h) = a_0 \left(x + \frac{a_1}{a_0} \right) \qquad\qquad \left[\because h = -\frac{a_1}{a_0} \right]$$

$$z = a_0 x + a_1$$

Since $\alpha_1, \alpha_2, \alpha_3$ are the roots of equation (1), then the roots of (4) are $a_0 \alpha_1 + a_1$, $a_0 \alpha_2 + a_1$, $a_2 \alpha_3 + a_1$.

Further since we know that

$$\alpha_1 + \alpha_2 + \alpha_3 = -\frac{3a_1}{a_0} \quad \text{or} \quad \frac{a_1}{a_0} = -\frac{\alpha_1 + \alpha_2 + \alpha_3}{3}$$

then $\qquad a_0 \alpha_1 + a_1 = a_0 \left(\alpha_1 + \frac{a_1}{a_0} \right) = a_0 \left(\alpha_1 - \frac{\alpha_1 + \alpha_2 + \alpha_3}{3} \right) = \frac{a_0}{3}(2\alpha_1 - \alpha_2 - \alpha_3)$

Similarly, $\qquad G_0 \alpha_2 + a_1 = \frac{a_0}{3}(2\alpha_2 - \alpha_1 - \alpha_3)$

$$a_0 \alpha_3 + a_1 = \frac{a_0}{3}(2\alpha_3 - \alpha_1 - \alpha_2)$$

Hence the roots of (4) can also be taken as

$$\frac{a_0}{3}(2\alpha_1 - \alpha_2 - \alpha_3), \; \frac{a_0}{3}(2\alpha_2 - \alpha_1 - \alpha_3), \; \frac{a_0}{3}(2\alpha_3 - \alpha_1 - \alpha_2)$$

Now if we put $z = a_0 x + a_1$ in (4), we get

$$(a_0 x + a_1)^3 + 3H(a_0 x + a_1) + G \equiv a_0^2 [a_0 x^3 + 3a_1 x^2 + 3a_2 x + a_3]$$

(ii) *To remove the second term in the equation*

$$a_0 x^4 + 4a_1 x^3 + 6a_2 x^2 + 4a_3 x + a_4 = 0$$

with binomial coefficients and to form the equation with integral coefficients having leading coefficients unity.

Since the equation is

$$f(x) \equiv a_0 x^4 + 4a_1 x^3 + 6a_2 x^2 + 4a_3 x + a_4 = 0 \qquad\qquad \text{...(1)}$$

and let $\alpha_1, \alpha_2, \alpha_3, \alpha_4$ be it's roots

Put $x = y - h$ in (1), we obtain

$$f(y - h) \equiv a_0 (y - h)^4 + 4a_1 (y - h)^3 + 6a_2 (y - h)^2 + 4a_3 (y - h) + a_4 = 0$$

or $\quad f(y - h) \equiv a_0 y^4 + 4 (a_0 h + a_1) y^3 + 6(a_0 h^2 + 2a_1 h + a_2)y^2 + 4(a_0 h^3 + 3a_1 h^2 + 3a_2 h + a_3)$
$$+ (a_0 h^4 + 4a_1 h^3 + 6a_2 h^2 + 4a_3 h + a_4) = 0 \qquad\qquad \text{...(2)}$$

Now we want to remove second term by putting

$$4(a_0h + a_1) = 0 \quad \text{or} \quad h = -\frac{a_1}{a_0}$$

Substitute the value of h in (2), we obtain

$$a_0 y^4 + \frac{6H}{a_0} y^2 + \frac{4G}{a_0^2} y + \frac{(a_0^2 I - 3H^2)}{a_0^3} = 0 \qquad \text{...(3)}$$

where $H = a_0 a_1 - a_1^2, \quad G = a_0^2 a_3 - 3a_0 a_1 a_2 + 2a_1^3$

and $I = a_0 a_4 - 4a_1 a_3 + 3a_2^2$

Equation (3) can also be written as

$$a_0^4 y^4 + 6H a_0^2 y^2 + 4G a_0 y + (a_0^2 I - 3H^2) = 0$$

let us put $z = a_0 y$, we get

$$\therefore \qquad z^4 + 6Hz^2 + 4Gz + (a_0^2 I - 3H^2) = 0 \qquad \text{...(4)}$$

This is the transformed equation whose leading coefficients being unity and all other coefficients are integers. Since we have

$$z = a_0 y = a_0 (x - h) = a_0 \left(x + \frac{a_1}{a_0} \right) \qquad \left[\because h = -\frac{a_1}{a_0} \right]$$

$$\therefore \qquad z = a_0 x + a_1.$$

Thus the roots of the equation (4) are obtained by the transformation $z = a_0 x + a_1$.

Since $\alpha_1, \alpha_2, \alpha_3, \alpha_4$ are the roots of (1), then $a_0\alpha_1 + a_1, a_0\alpha_2 + a_2, a_0\alpha_3 + a_3$ and $a_0\alpha_4 + a_4$ are the roots of (4). Further since we know that

$$\alpha_1 + \alpha_2 + \alpha_3 + \alpha_4 = -\frac{3a_1}{a_0}$$

or

$$\frac{a_1}{a_0} = -\frac{\alpha_1 + \alpha_2 + \alpha_3 + \alpha_4}{4}$$

$$\therefore \qquad a_0\alpha_1 + a_1 = a_0 \left(\alpha_1 + \frac{a_1}{a_0} \right) = a_0 \left(\alpha_1 - \frac{\alpha_1 + \alpha_2 + \alpha_3 + \alpha_4}{4} \right)$$

$$= \frac{a_0}{4} (3\alpha_1 - \alpha_2 - \alpha_3 - \alpha_4)$$

Similarly, $a_0\alpha_2 + a_1 = \dfrac{a_0}{4} (3\alpha_2 - \alpha_1 - \alpha_3 - \alpha_4)$

$$a_0\alpha_3 + a_1 = \frac{a_0}{4} (3\alpha_3 - \alpha_1 - \alpha_2 - \alpha_4)$$

and $a_0\alpha_4 + a_1 = \dfrac{a_0}{4} (3\alpha_4 - \alpha_1 - \alpha_2 - \alpha_3)$

Hence the roots of equation (4) can also be taken as $\dfrac{a_0}{4}(3\alpha_1 - \alpha_2 - \alpha_3 - \alpha_4)$,

$\dfrac{a_0}{4}(3\alpha_2 - \alpha_1 - \alpha_3 - \alpha_4)$, $\dfrac{a_0}{4}(3\alpha_3 - \alpha_1 - \alpha_2 - \alpha_4)$, and $\dfrac{a_0}{4}(3\alpha_4 - \alpha_1 - \alpha_2 - \alpha_3)$.

12.8 AN IMPORTANT RELATION

In order to discuss the biquadratic equation, a function of its coefficients plays a key role. This function is taken as

$$J = a_0 a_2 a_4 + 2a_1 a_2 a_3 - a_0 a_3^2 - a_1^2 a_4 - a_2^3$$

which can also be written in the form of a determinant as follows :

$$J = \begin{vmatrix} a_0 & a_1 & a_2 \\ a_1 & a_2 & a_3 \\ a_2 & a_3 & a_4 \end{vmatrix}$$

Further, we have an important relation between H, G, I and J as follows :

$$G^2 + 4H^3 = a_0^2(HI - a_0 J)$$

Verification :

$$
\begin{aligned}
\text{L.H.S.} = G^2 + 4H^3 &= (a_0^2 a_3 - 3a_0 a_1 a_2 + 2a_1^3)^2 + 4(a_0 a_2 - a_1^2)^3 \\
&= (a_0^4 a_3^2 + 9a_0^2 a_1^2 a_2^2 + 4a_1^6 - 6a_0^3 a_1 a_2 a_3 + 4a_0^2 a_1^3 a_3 - 12a_0 a_1^4 a_2) \\
&\quad + 4\,(a_0^3 a_2^3 - a_1^6 - 3a_0^2 a_2^2 a_1^2 + 3a_0 a_2 a_1^4) \\
&= a_0^4 a_3^2 - 3a_0^2 a_1^2 a_2^2 - 6a_0^3 a_1 a_2 a_3 + 4a_0^2 a_1^3 a_3 + 4a_0^3 a_2^3 \\
&= a_0^2(a_0^2 a_3^2 - 3a_1^2 a_2^2 - 6a_0 a_1 a_2 a_3 + 4a_1^3 a_3 + 4a_0 a_2^3)
\end{aligned}
$$

$$
\begin{aligned}
\text{R.H.S.} = a_0^2(HI - a_0 J) &= a_0^2[(a_0 a_2 - a_1^2)(a_0 a_4 - 4a_1 a_3 + 3a_2^2) \\
&\qquad - a_0(a_0 a_2 a_4 + 2a_1 a_2 a_3 - a_0 a_3^2 - a_1^2 a_4 - a_2^3) \\
&= a_0^2[a_0^2 a_2 a_4 - 4a_0 a_1 a_2 a_3 + 3a_0 a_2^3 - a_0^2 \quad + 4a_1^3 a_3 - 3a_1^2 a_2^2 \\
&\qquad - a_0^2 a_2 a_4 - 2a_0 a_1 a_2 a_3 + a_0^2 a_3^2 + a_0 a_1^2 a_4 + a_0 a_2^3] \\
&= a_0^2(a_0^2 a_3^2 - 3a_1^2 a_2^2 - 6a_0 a_1 a_2 a_3 + 4a_1^3 a_3 + 4a_0 a_2^3)
\end{aligned}
$$

Hence L.H.S. = R.H.S.

12.9 GENERAL METHOD OF TRANSFORMATION

Let the given equation be

$$f(x) = 0 \qquad\qquad\qquad ...(1)$$

and suppose y is a root of transformed equation such that x and y are related by some relation

$$\phi(x, y) = 0 \qquad \qquad ...(2)$$

Eliminating x between (1) and (2), we get the transformed equation.

(i) *To form the equation whose roots are $(\alpha - \beta)^2$, $(\beta - \gamma)^2$, $(\gamma - \alpha)^2$, where α, β, γ, are the roots of the given equation $a_0 x^3 + 3a_1 x^2 + 3a_2 x + a_3 = 0$ and to discuss the nature of the roots of the given equation.*

Since the given equation is

$$a_0 x^3 + 3a_1 x^2 + 3a_2 x + a_3 = 0 \qquad \qquad ...(1)$$

First remove the second term of (1) by diminishing it's roots by h, we obtain

$$y^3 + \frac{3H}{a_0^2} y + \frac{G}{a_0^3} = 0 \quad \text{or} \quad y^3 + 2y + r = 0 \qquad \qquad ...(2)$$

where

$$q = \frac{3H}{a_0^2}, \quad r = \frac{G}{a_0^3}$$

and also

$$h = -\frac{a_1}{a_0}, \quad H = a_0 a_2 - a_1^2, \quad G = a_0^2 a_3 - 3a_0 a_1 a_2 + 2a_1^3$$

The roots of the transformed equation (2) are $\alpha - h$, $\beta - h$, $\gamma - h$ respectively.

For simplicity let us take $\alpha_1 = \alpha - h$, $\beta_1 = \beta - h$, and $\gamma_1 = \gamma - h$.

Now $\quad (\gamma - \beta)^2 = (\alpha - h - \beta + h)^2 = [(\alpha - h) - (\beta - h)]^2 = (\alpha_1 - \beta_1)^2$.

Similarly $\quad (\beta - \gamma)^2 = (\beta_1 - \gamma_1)^2$ and $(\gamma - \alpha)^2 = (\gamma_1 - \alpha_1)^2$

Hence the equation of the squared differences of (1) is same as that of the equation (2). Therefore if α_1, β_1, γ_1 are the roots of (2), then we have to find the equation whose roots are $(\alpha_1 - \beta_1)^2$, $(\beta_1 - \gamma_1)^2$, $(\gamma_1 - \alpha_1)^2$. Let z be one of the roots of required equation

$$\therefore \qquad z = (\alpha_1 - \beta_1)^2 = \alpha_1^2 + \beta_1^2 - 2\alpha_1 \beta_1 = (\alpha_1 + \beta_1)^2 - 4\alpha_1 \beta_1$$

$$= (-\gamma_1)^2 - \frac{4\alpha_1 \beta_1 \gamma_1}{\gamma_1} \qquad \qquad [\because \alpha_1 + \beta_1 + \gamma_1 = 0]$$

$$z = \gamma_1^2 + \frac{4r}{\gamma_1} \qquad \qquad [\because \alpha_1 \beta_1 \gamma_1 = -r]$$

or $\qquad \gamma_1^3 + z\gamma_1 - 4r = 0$

Since γ_1 is the root of equation (2) so put $\gamma_1 = y$.

$$\therefore \qquad y^3 + zy - 4r = 0 \qquad \qquad ...(3)$$

Eliminating y between (2) and (3), we get

$$z^3 + 6qz^2 + 9q^2 z + 27r^2 + 4q^3 = 0 \qquad \qquad ...(4)$$

putting $q = \dfrac{3H}{a_0^2}, r = \dfrac{G}{a_0^3}$ we get

$$z^3 + \frac{18H}{a_0^2} z^2 + \frac{81H^2}{a_0^4} z + \frac{27}{a_0^6}(G^2 + 4H^3) = 0 \qquad \qquad ...(5)$$

Hence $(\alpha - \beta)^2$, $(\beta - \gamma)^2$, $(\gamma - \alpha)^2$ are the roots of (5)

$$\therefore \qquad (\alpha - \beta)^2(\beta - \gamma)^2(\gamma - \alpha)^2 = -\frac{27}{a_0^6}(G^2 + 4H^3) \qquad \qquad ...(6)$$

12.9.1 Nature of Roots of the Given Equation.

Since the degree of the given equation is odd so it has at least one of the roots α, β, γ, say α real, and we know that complex roots lie in pair. If β, γ are also real, then $(\alpha - \beta)^2(\beta - \gamma)^2(\gamma - \alpha)^2$ must be positive. If β, γ are supposed to be imaginary and if $\beta = a + ib$, then $\gamma = a - ib$, where a, b are real.

$$\therefore \qquad (\alpha - \beta)^2(\beta - \alpha)^2(\gamma - \alpha)^2 = (\alpha - a - ib)^2 (2ib)^2(a - ib - \alpha)$$
$$= -4b^2 [(\alpha - a)^2 + b^2] < 0.$$

Thus $(\alpha - \beta)^2(\beta - \gamma)^2(\gamma - \alpha)^2$ is negative.

Therefore we can say that if $(\alpha - \beta)^2(\beta - \gamma)^2(\gamma - \alpha)^2$ is positive then the roots of the given equation will be real and if $(\alpha - \beta)^2(\beta - \gamma)^2(\gamma - \alpha)^2$ is negative, then two roots of the given equation will be imaginary. But we have

$$(\alpha - \beta)^2(\beta - \gamma)^2(\gamma - \alpha)^2 = -\frac{27}{a_0^6}(G^2 + 4H^3)$$

so that we can discuss the nature of the roots of the given equation as follows :

1. If $G^2 + 4H^3 > 0$, then two roots of the given equation will be imaginary.

2. If $G^2 + 4H^3 < 0$, then all the roots of the given equation will be real.

3. If $G^2 + 4H^3 = 0$, then two roots of the given equation will be equal.

4. If $G = 0$, $H = 0$, then all the three roots of the given equation will be equal.
On the other hand we can discuss this case as follows:

Since we know that $G^2 + 4H^3 = a_0^2(HI - a_0J)$

$\therefore$ If $G = 0$, $H = 0$, then $a_0^2(HI - a_0J) = 0$

or $\qquad a_0^2\Delta = 0 \qquad$ or $\qquad \Delta = 0$

where Δ is the *discriminant* of the given equation. Hence we can say that if the discriminant of the cubic is zero, then all the roots of the equation will be equal.

SOLVED EXAMPLES

Example 1. *Change the signs of the roots of the equation*
$$x^7 + 5x^5 - x^3 + x^2 + 7x + 3 = 0$$

Solution. First making the equation complete by adding missing terms with zero coefficients, we get

$$f(x) \equiv x^7 + 0.x^6 + 5x^5 + 0.x^4 - x^3 + x^2 + 7x + 3 = 0 \qquad \qquad ...(1)$$

Put $x = -y$, in (1), we get

$$(-y)^7 + 0.(-y)^6 + 5(-y)^5 + 0.(-y)^4 - (-y)^3 + (-y)^2 + 7(-y) + 3 = 0$$

or $\qquad\qquad\qquad\qquad -y^7 + 0.y^6 - 5y^5 + 0.y^4 + y^3 + y^2 - 7y + 3 = 0y^7 + 5y^5 -$
$$y^3 - y^2 + 7x - 3 = 0$$

This is the required equation whose roots are same to the roots of the given equation with contrary signs.

Example 2. *Transform the equation $72x^3 - 54x^2 + 45x - 7 = 0$ into another equation with integral coefficients and having the leading coefficient unity.*

Solution. The given equation can be written as

$$x^3 - \frac{54}{72}x^2 + \frac{45}{72}x - \frac{7}{72} = 0$$

or $\qquad\qquad x^3 - \frac{3}{4}x^2 + \frac{5}{8}x - \frac{7}{72} = 0 \qquad\qquad\qquad ...(1)$

Put $y = xm$ or $x = \dfrac{y}{m}$ in (1), we get

$$\left(\frac{y}{m}\right)^3 - \frac{3}{4}\left(\frac{y}{m}\right)^2 + \frac{5}{8}\left(\frac{y}{m}\right) - \frac{7}{72} = 0$$

or $\qquad\qquad y^3 - \frac{3}{4}my^2 + \frac{5}{8}m^2y - \frac{7}{72}m^3 = 0 \qquad\qquad ...(2)$

Now to remove fractional coefficients let us put $m = 12$ in (2), we get

$$y^3 - \frac{3}{4}(12)y^2 + \frac{5}{8}(12)^2 y - \frac{7}{72}(12)^3 = 0$$

or $\qquad\qquad y^3 - 9y^2 + 90y - 168 = 0$

This is the required equation.

Example 3. *Form the equation whose roots are the reciprocals of the roots of the equation*
$$x^4 - 3x^3 + 7x^2 + 5x - 2 = 0$$

Solution. The given equation is
$$x^4 - 3x^3 + 7x^2 + 5x - 2 = 0 \qquad\qquad\qquad ...(1)$$

Putting $x = \dfrac{1}{y}$ in (1), we get

$$\left(\frac{1}{y}\right)^4 - 3\left(\frac{1}{y}\right)^3 + 7\left(\frac{1}{y}\right)^2 + 5\left(\frac{1}{y}\right) - 2 = 0$$

or $\qquad\qquad 1 - 3y + 7y^2 + 5y^3 - 2y^4 = 0$

or $\qquad\qquad 2y^4 - 5y^3 - 7y^2 + 3y - 1 = 0$

This is the required equation whose roots are the reciprocal of the roots of (1).

Example 4. *Remove the fractional coefficients from the equation*

$$2x^3 - \frac{3}{2}x^2 - \frac{1}{8}x + \frac{3}{16} = 0$$

Solution. The given equation is

$$2x^3 - \frac{3}{2}x^2 - \frac{1}{8}x + \frac{3}{16} = 0 \qquad \qquad ...(1)$$

Putting $x = \dfrac{y}{m}$ in (1) we get

$$2\left(\frac{y}{m}\right)^3 - \frac{3}{2}\left(\frac{y}{m}\right)^2 - \frac{1}{8}\left(\frac{y}{m}\right) + \frac{3}{16} = 0$$

or $\qquad\qquad 2y^3 - \dfrac{3}{2}my^2 - \dfrac{1}{8}m^2y + \dfrac{3}{16}m^3 = 0$

Let us put $m = 4$, we get

$$2y^3 - \frac{3}{2}(4)y^2 - \frac{1}{8}(4)^2y + \frac{3}{16}(4)^3 = 0$$

or $\qquad\qquad 2y^3 - 6y^2 - 2y + 12 = 0 \qquad$ or $\qquad y^3 - 3y^2 - y + 6 = 0$

This is the required equation.

Example 5. *Solve the following reciprocal equation*

$$x^4 - 10x^3 + 26x^2 - 10\,x + 1 = 0$$

Solution. The given equation can be written as

$$x^4 + 1 - 10(\,x^3 + x) + 26x^2 = 0$$

Divide by x^2, we get

$$\left(x^2 + \frac{1}{x^2}\right) - 10\left(x + \frac{1}{x}\right) + 26 = 0 \qquad\qquad ...(1)$$

Let us put $x + \dfrac{1}{x} = y$ and $x^2 + \dfrac{1}{x^2} = y^2 - 2$ in (1), we get

$$y^2 - 2 - 10y + 26 = 0 \quad \text{or} \quad y^2 - 10y + 24 = 0$$

or $\qquad\qquad y^2 - 6y - 4y + 24 = 0 \quad \text{or} \quad (y - 6)(y - 4) = 0$

or $\qquad\qquad\qquad\qquad y = 4,\,6.$

Since $\qquad x + \dfrac{1}{x} = y$ if $y = 4$, then $x + \dfrac{1}{x} = 4$

or $\qquad x^2 - 4x + 1 = 0$ or $x = \dfrac{4 \pm \sqrt{16 - 4}}{2} = \dfrac{4 \pm 2\sqrt{3}}{2} = 2 \pm \sqrt{3}$

If $\qquad y = 6$, then $x + \dfrac{1}{x} = 6$

or $\qquad x^2 - 6x + 1 = 0$ or $x = \dfrac{6 \pm \sqrt{36 - 4}}{2} = \dfrac{6 \pm 4\sqrt{2}}{2} = 3 \pm 2\sqrt{2}$

Hence, the roots of the given equation are $2 \pm \sqrt{3},\, 3 \pm 2\sqrt{2}$.

Example 6. *Find the equation whose roots are the cubes of the roots of the equation*

$$x^4 - x^3 + 2x^2 + 3x + 1 = 0.$$

Solution. Since the equation is

$$x^4 - x^3 + 2x^2 + 3x + 1 = 0 \qquad \qquad ...(1)$$

This equation can be written as

$$(1 - x^3) + x(x^3 + 3) + 2x^2 = 0 \qquad \qquad ...(2)$$

Let $\qquad P = (1 - x^3), \ Q = x(x^3 + 3), \ R = 2x^2.$

Then we have

$$P + Q + R = 0 \quad \text{or} \quad P + Q = -R.$$

Cubing of both sides, we get

$$(P + Q)^3 = -R^3 \quad \text{or} \quad P^3 + Q^3 + 3PQ(P + Q) = -R^3$$

or $\qquad P^3 + Q^3 + 3PQ(-R) = -R^3 \qquad \qquad [\because P + Q = -R]$

or $\qquad P^3 + Q^3 + R^3 - 3PQR = 0 \qquad \qquad ...(3)$

Now substitute the values of P, Q and R in (3), we get

$$(1 - x^3)^3 + x^3(x^3 + 3)^3 + (2x^2)^3 - 3(1 - x^3)x(x^3 + 3)(2x^2) = 0$$

or $\qquad (1 - x^3)^3 + x^3(x^3 + 3)^3 + 8(x^3)^2 - 6(1 - x^3)(x^3 + 3)x^3 = 0$

Let us put $x^3 = y$, we get

$$(1 - y)^3 + y(y + 3)^3 + 8y^2 - 6(1 - y)(y + 3)y = 0$$

$$1 - y^3 - 3y + 3y^2 + y(y^3 + 27 + 9y^2 + 27y) + 8y^2 - 6y(y + 3 - y^2 - 3y) = 0$$

$$1 - y^3 - 3y + 3y^2 + y^4 + 27y + 9y^3 + 27y^2 + 8y^2 - 6y^2 - 18y + 6y^3 + 18y^2 = 0$$

or $\qquad y^4 + 14y^3 + 50y^2 + 6y + 1 = 0$

This is the required equation whose roots are the cube of the roots of the given equation.

Example 7. *Find the equation whose roots are the roots of the equation*

$$x^5 - 4x^4 + 3x^2 - 4x + 6 = 0$$

diminished by 3.

Solution. First complete the given equation,

$$f(x) \equiv x^5 - 4x^4 + 0x^3 + 3x^2 - 4x + 6 = 0 \qquad \qquad ...(1)$$

Suppose the required equation is

$$A_0 y^5 + A_1 y^4 + A_2 y^3 + A_3 y^2 + A_4 y + A_5 = 0 \qquad \qquad ...(2)$$

where $A_0, A_1, A_2, ... A_5$ are the constants which can be determined as follows :

Using *synthetic division method* :

3	1	-4	0	3	-4	6
		3	-3	-9	-18	-66
3	1	-1	-3	-6	-22	$-60 = A_5$
		3	6	9	9	
3	1	2	3	3	$-13 = A_4$	
		3	15	54		
3	1	5	18	$57 = A_3$		
		3	24			
3	1	8	$42 = A_2$			
		3				
	1	$11 = A_1$				
	$1 = A_0$					

∴ $A_0 = 1, A_1 = 11, A_2 = 42, A_3 = 57, A_4 = -13, A_5 = -60.$

Thus the required equation is

$$y^5 + 11y^4 + 42y^3 + 57y^2 - 13y - 60 = 0$$

Example 8. *If α, β, γ are the roots of the cubic $x^3 - px^2 + qx - r = 0$, form the equation whose roots are*

$$\beta\gamma + \frac{1}{\alpha}, \gamma\alpha + \frac{1}{\beta}, \alpha\beta + \frac{1}{\gamma}$$

Solution. Since the given equation is

$$x^3 - px^2 + qx - r = 0 \qquad \qquad ...(1)$$

and α, β, γ are its roots, then

$$\alpha + \beta + \gamma = p, \quad \alpha\beta + \beta\gamma + \alpha\gamma = q, \quad \alpha\beta\gamma = r.$$

Let y be a root of the required equation. Then

$$y = \beta\gamma + \frac{1}{\alpha} = \frac{\alpha\beta\gamma + 1}{\alpha}$$

$$y = \frac{r+1}{\alpha} \quad \Rightarrow \quad y = \frac{r+1}{x} \qquad \qquad [\because x = \alpha]$$

∴ $$x = \frac{r+1}{y}$$

Substitute this value of x in (1), we get

$$\left(\frac{r+1}{y}\right)^3 - p\left(\frac{r+1}{y}\right)^2 + q\left(\frac{r+1}{y}\right) - r = 0$$

or $$\frac{(r+1)^3}{y^3} - \frac{p(r+1)^2}{y^2} + \frac{q(r+1)}{y} - r = 0$$

or $\quad (r + 1)^3 - p(r + 1)^2 y + q(r + 1)y^2 - ry^3 = 0$

or $\quad ry^3 - q(r + 1)y^2 + p(r + 1)^2 y - (r + 1)^3 = 0$

This is the required equation.

Example 9. *Remove the second term of the equation*

$$x^4 + 4x^3 + 2x^2 - 4x - 2 = 0$$

Solution. Suppose the roots of the given equation are diminished by h so put $y = x - h$ or $x = y + h$ in the given equation, we get

$$(y + h)^4 + 4(y + h)^3 + 2(y + h)^2 - 4(y + h) - 2 = 0$$

$$(y^4 + 4hy^3 + 6h^2 y^2 + 4h^3 y + h^4) + 4(y^3 + 3hy^2 + 3h^2 y + h^3)$$
$$+ 2(y^2 + 2yh + h^2) - 4y - 4h - 2 = 0$$

or $\quad y^4 + (4h + 4)y^3 + (6h^2 + 12h + 2)y^2 + (4h^3 + 12h^2 + 4h - 4)y$

$$+ (h^4 + 4h^3 + 2h^2 - 4h - 2) = 0 \qquad ...(1)$$

In order to remove the second term let us put

$$4h + 4 = 0 \quad \text{or } h = -1$$

substitute this value of h in (1), we get

$$y^4 - 4y^2 + 1 = 0.$$

Example 10. *If α, β, γ are the roots of the equation $x^3 + qx + r = 0$. Find the equation whose roots are*

$$(\alpha - \beta)^2, (\beta - \gamma)^2, (\gamma - \alpha)^2.$$

Solution. Since the given equation is

$$x^3 + qx + r = 0 \qquad ...(1)$$

and α, β, γ are its roots, then, we have

$$\alpha + \beta + \gamma = 0$$
$$\alpha\beta + \beta\gamma + \alpha\gamma = q$$
$$\alpha\beta\gamma = -r$$

let y be a root of the required equation, then

$$y = (\alpha - \beta)^2 = \alpha^2 + \beta^2 - 2\alpha\beta = \alpha^2 + \beta^2 + \gamma^2 - \gamma^2 - 2\alpha\beta$$

$$= (\alpha + \beta + \gamma)^2 - 2(\alpha\beta + \beta\gamma + \alpha\gamma) - \gamma^2 - \frac{2\alpha\beta\gamma}{\gamma}$$

$$y = 0 - 2q - \gamma^2 + \frac{2r}{\gamma} \quad \text{or} \quad y + 2q = -\gamma^2 + \frac{2r}{\gamma}$$

or $\quad \gamma^3 + (y + 2q)\gamma - 2r = 0$

since γ is a root of (1) so we have $x = \gamma$, then

$$x^3 + (y + 2q)x - 2r = 0 \qquad ...(2)$$

Substract (2) from (1) we get

$$-(y + q)x + 3r = 0 \quad \text{or} \quad x = \frac{3r}{y + q}$$

Now substitute this value of x in (1), we get

$$\left(\frac{3r}{y+q}\right)^3 + q\left(\frac{3r}{y+q}\right) + r = 0$$

or $\qquad (3r)^3 + 3qr(y+q)^2 + r(y+q)^3 = 0$

or $\qquad (y+q)^3 + 3q(y+q)^2 + 27r^2 = 0$

$\qquad y^3 + q^3 + 3y^2q + 3yq^2 + 3qy^2 + 3q^3 + 6q^2y + 27r^2 = 0$

or $\qquad y^3 + 6qy^2 + 9q^2y + (4q^3 + 27r^2) = 0$

This is the required equation.

EXERCISE 12.1

1. Change the sign of the roots of the equation $x^5 - 4x^3 + 3x^2 + 8x - 9 = 0$

2. Transform the equation $x^3 - 4x^2 + \frac{1}{4}x - \frac{1}{9} = 0$ into another equation with integral coefficients and having leading coefficient unity.

3. Transform the equation $3x^4 - 5x^3 + x^2 - x + 1 = 0$ into another equation with integral coefficients having leading coefficient unity.

4. Find the equation whose roots are twice the reciprocals of the roots of
$$x^4 + 3x^3 - 6x^2 + 2x - 4 = 0$$

5. Remove the fractional coefficients from the equation
$$x^3 - \frac{5}{2}x^2 - \frac{7}{18}x + \frac{1}{108} = 0$$

6. Remove the fractional coefficients from the equation
$$x^4 - \frac{5}{6}x^3 - \frac{13}{12}x^2 + \frac{1}{300} = 0$$

7. Solve the following reciprocal equations :

   (i) $\cdot$ $6x^6 - 25x^5 + 31x^4 - 31x^2 + 25x - 6 = 0$

   (ii) $x^5 - 5x^4 + 9x^3 - 9x^2 + 5x - 1 = 0$

8. Reduce the equation $4x^4 - 85x^3 + 357x^2 - 340x + 64 = 0$ into a reciprocal equation.

9. Find the equation whose roots are the squares of the roots of the equation
$$x^4 + x^3 + 2x^2 + x + 1 = 0$$

10. Find the equation whose roots are the cubes of the roots of the following equations:

    (i) $x^3 + ax^2 + bx + ab = 0$ $\qquad$ (ii) $x^3 + 3x^2 + 2 = 0$

11. Remove the second term form the following equations :

    (i) $x^3 - 6x^2 + 10x - 3 = 0$ $\qquad$ (ii) $x^4 + 8x^3 + x - 5 = 0$

    (iii) $x^5 + 5x^4 + 3x^3 + x^2 + x - 1 = 0$ $\qquad$ (iv) $x^4 + 20x^3 + 143x^2 + 430x + 462 = 0$

    (v) $x^6 - 12x^5 + 3x^2 - 17x + 300 = 0$

12. Find the equation each of whose roots is greater than unity then a root of the equation $x^3 - 5x^2 + 2x - 3 = 0$.

13. Transform the equation $x^3 - \dfrac{x}{4} - \dfrac{3}{4} = 0$ into an equation whose roots increased by $\dfrac{3}{2}$ the corresponding roots of the given equation.

14. Find the equation whose roots are the roots of $3x^3 - 2x^2 + x - 9 = 0$ each diminished by 5.

15. If α, β, γ are the roots of the equation $x^3 + xq + r = 0$, form the equation whose roots are

    (i) $\alpha(\beta + \gamma), \beta(\gamma + \alpha), \gamma(\alpha + \beta)$

    (ii) $\left(\dfrac{\beta}{\gamma} + \dfrac{\gamma}{\beta}\right), \left(\dfrac{\gamma}{\alpha} + \dfrac{\alpha}{\gamma}\right), \left(\dfrac{\alpha}{\beta} + \dfrac{\beta}{\alpha}\right)$

    (iii) $\left(\alpha - \dfrac{1}{2}\right), \left(\beta - \dfrac{1}{2}\right), \left(\gamma - \dfrac{1}{2}\right)$

16. If α, β, γ are the roots of the equation $x^3 + px^2 + qx + r = 0$, form the equation whose roots are

$$\alpha - \dfrac{1}{\beta\gamma}, \ \beta - \dfrac{1}{\gamma\alpha}, \ \gamma - \dfrac{1}{\alpha\beta}$$

17. Show that the same transformation removes both second and fourth terms of the equation

$$x^4 + 16x^3 + 83x^2 + 152x + 84 = 0$$

18. Find the condition that the second and third terms of the equation $a_0x^3 + 3a_1x^2 + 3a_2x + a_3 = 0$ are removed by the same transformation.

19. If α, β, γ are the roots of the equation $x^3 - 6x^2 + 11x - 6 = 0$, form the equation whose roots are

$$\beta^2 + \gamma^2, \ \gamma^2 + \alpha^2, \ \alpha^2 + \beta^2.$$

20. If α, β, γ are the roots of the equation $2x^3 + x^2 + x + 1 = 0$, form the equation whose roots are

$$\dfrac{1}{\beta^2} + \dfrac{1}{\gamma^2} - \dfrac{1}{\alpha^2}; \ \dfrac{1}{\gamma^2} + \dfrac{1}{\alpha^2} - \dfrac{1}{\beta^2}; \ \dfrac{1}{\alpha^2} + \dfrac{1}{\beta^2} - \dfrac{1}{\gamma^2}$$

21. If α, β, γ are the roots of the equation $x^3 + px^2 + qx + r = 0$, form the equation whose roots are

    (i) $\alpha + \dfrac{1}{\beta\gamma}, \ \beta + \dfrac{1}{\gamma\alpha}, \ \gamma + \dfrac{1}{\alpha\beta}$

    (ii) $\dfrac{\alpha}{\beta + \gamma}, \ \dfrac{\beta}{\gamma + \alpha}, \ \dfrac{\gamma}{\alpha + \beta}$

22. If α, β, γ are the roots of the equation $x^3 + px^2 + qx + r = 0$, form the equation whose roots are

$$\alpha^2 + 2\beta\gamma, \ \beta^2 + 2\alpha\gamma, \ \gamma^2 + 2\alpha\beta.$$

23. If α, β, γ are the roots of the equation $x^3 - px^2 + qx - r = 0$, form the equation whose roots are $\beta + \gamma - \alpha$, $\gamma + \alpha - \beta$, $\alpha + \beta - \gamma$,

also find the value of $(\beta + \gamma - \alpha)(\gamma + \alpha - \beta)(\alpha + \beta - \gamma)$

24. If the roots of $x^3 + 3px^2 + 2qx + r = 0$ are in harmonic progression, show that $2q^3 = r(3\,pq - r)$

HINT TO SELECTED PROBLEMS

1. Put $x = -y$ in the given equation.

2. Put $y = mx$, i.e. $x = \dfrac{y}{m}$ in the given equation.

7. (i) The given equation can be written as

$$6(x^6 - 1) - 25x(x^4 - 1) + 31x^2(x^2 - 1) = 0$$

9. The given equation can be written as

$$(x^4 + 2x^2 + 1) = -x(x^2 + 1)$$

On squaring both sides, we get

$$x^8 + 3x^6 + 4x^4 + 3x^2 + 1 = 0$$

Now put $x^2 = y$.

17. Suppose the roots of the given equation, are diminished by h, put $y = x - h$, or $x = y + h$ in the given equation.

21. If the roots of the given equation are α, β, γ . Then

$$\alpha + \beta + \gamma = -p$$
$$\alpha\beta + \beta\gamma + \gamma\alpha = q$$
$$\alpha\beta\gamma = -r$$

If y be the root of the required equation, the $y = \alpha + \dfrac{1}{\beta\gamma}$

ANSWERS

(1) $y^5 - 4y^3 - 3y^2 + 8y + 9 = 0$ (2) $y^3 - 24y^2 + 9y - 24 = 0$

(3) $y^4 - 5y^3 + 3y^2 - 9y + 27 = 0$ (4) $y^4 - y^3 + 6y^2 - 6y - 4 = 0$

(5) $y^3 - 15y^2 - 14y^2 + 2 = 0$ (6) $y^4 - 25y^3 - 975y^2 + 2700 = 0$

(7) (i) $\pm 1, 2, \dfrac{1}{2}, \dfrac{5 \pm i\sqrt{11}}{6}$ (ii) $1, \dfrac{1}{2}(1 \pm i\sqrt{3}), \dfrac{1}{2}(3 \pm \sqrt{5})$

(8) $16y^4 - 170y^3 + 357y^2 - 170y + 16 = 0$ (9) $y^4 + 3y^3 + 4y^2 + 3y + 1 = 0$

(10) (i) $y^3 + a^3y^2 + b^3y + a^3b^3 = 0$ (ii) $y^3 + 33y^2 + 12y + 8 = 0$

(11) (i) $y^3 - 2y + 1 = 0$ (ii) $y^4 - 24y^2 + 65y - 55 = 0$

     (iii) $y^5 - 7y^3 + 12y^2 - 7y + 2 = 0$ (iv) $y^4 - 7y^2 + 12 = 0$

     (v) $y^6 - 60y^4 - 320y^3 - 717y^2 - 773y - 42 = 0$

(12) $y^3 - 8y^2 + 19y - 15 = 0$ **(13)** $y^3 - \dfrac{9}{2}y^2 + \dfrac{13}{2}y - \dfrac{15}{4} = 0$

(14) $3y^3 + 43y^2 + 206y + 321 = 0$

(15)   (i) $y^3 - 2qy^2 + q^2 y + r^2 = 0$

(ii) $r^3 y^3 + 3r^2 y^2 + (3r^2 + q^3)y + (r^2 + 2q^3) = 0$

(iii) $8y^3 + 12y^2 + (6 + 8q)y + (8r + 4q + 1) = 0$

(16) $r^3 y^3 + pr(1 + r)y^2 + q(1 + r)^2 y + (1 + r)^3 = 0$

(18) $a_0 a_2 - a_1^2 = 0$

(19) $y^3 - 28y^2 + 245y - 650 = 0$ **(20)** $z^3 + z^2 - 13z + 19 = 0$

(21)   (i) $r^2 y^3 + pr(1 - r)y^2 + q(1 - r)^2 y + (1 - r)^3 = 0$

(ii) $(pq - r)y^3 + (2pq - p^3 - 3r)y^2 + (pq - 3r)y - r = 0$

(22) $y^3 - p^2 y^2 + q(2p^2 - 3q)y - (4p^3 r - 18pqr + 2q^3 + 27r^2) = 0$

(23) $y^3 - py^2 + (4q - p^2)y + (8r - 4pq + p^3) = 0;\ 4pq - p^3 - 8r.$

12.10   DESCARTES' RULE OF SIGNS

We know that an equation $f(x) = 0$ cannot have more positive roots than the number of changes of signs from positive to negative or from negative to positive in terms of its first number and an equation $f(x) = 0$ cannot have more negative roots than the number of changes of sign $f(-x) = 0$.

We shall simply verify the above statement.

Let the signs of a polynomial be

$$+ \ + \ - \ + \ - \ + \ - \ -.$$

The given polynomial has five changes of signs. Now we shall multiply the given polynomial by binomial $x - h$ corresponding to the +ive root h. The signs of this binomial are $+ \ -$. We are concerned only with the signs and hence we multiply as below :

```
        + + - + - -              5 changes of sign

        + -
        ─────────────────
        + + - + - + - -

          - - + - - + +
        ─────────────────
        + ± - + - + - ± +
```

The resulting polynomial has two ambiguous signs and we can write in four different ways as follows :

$$+ \ + \ - \ + \ - \ + \ - \ + \ + \qquad \text{A 6 changes of sign}$$
$$+ \ + \ - \ + \ - \ + \ - \ - \ + \qquad \text{A 6 changes of sign}$$
$$+ \ - \ - \ + \ - \ + \ - \ - \ + \qquad \text{A 6 changes of sign}$$
$$+ \ - \ - \ + \ - \ + \ - \ + \ + \qquad \text{A 6 changes of sign}$$

Thus we see that in all the four possible ways the resulting polynomial has six changes of signs, i.e. one more than the number of changes of signs in the original polynomial. Hence we conclude that corresponding to the introduction of a positive root the resulting polynomial has one more change of sign. Now if $\phi(x)$ be the product of factors corresponding to –ive and complex roots and α, β, γ ... be the +ive roots, then if $\phi(x)$ be multiplied by $(x - \alpha)$, $(x - \beta)$, $(x - \gamma)$ ... in succession, then each multiplication will introduce one more change of sign. Hence the number of positive roots cannot exceed the number of changes of signs in $f(x) = 0$.

12.10.1. Negative Roots

We know that negative roots of $f(x) = 0$, are positive roots of $f(-x) = 0$ and as such the number of negative roots of $f(x) = 0$ cannot exceed the number of changes of signs of $f(-x) = 0$.

12.10.2 Complex Roots

If $f(x) = 0$ be an equation of nth degree and if it be *complete*, then the number of changes of signs in $f(x)$, i.e. +ive roots and numbers of sign in $f(-x)$ i.e. –ive roots is equal to n the degree of the equation and as such we cannot draw any definite conclusion regarding the existence of imaginary roots. In case the equation be *incomplete*, then the number of changes of signs in $f(x)$, i.e. +ive roots and the number of changes of signs in $f(-x)$, i.e. –ive roots is less than the degree n of the equation. If a and b be the number of changes of signs in $f(x)$ and $f(-x)$ respectively, i.e. greatest number of +ive roots is a and that of –ive roots is b, then $n - (a + b)$ is the least number of imaginary roots. For example, consider the equation

$$f(x) = x^7 - 3x^5 + 4x^4 + 2x^3 - 11 = 0$$

The above equation has three changes of sign and as such if cannot have more than three +ive roots. Again $f(-x) = 0$, i.e. $x^7 - 3x^5 + 4x^4 + 2x^3 + 11 = 0$ has only two changes of sign and as such $f(x) = 0$ can not have more than two –ive roots. Thus, the max. number of real roots is 3 + 2, i.e. 5 and the degree of the equation being 7 we conclude that the equation must have at least two imaginary roots.

12.11 CHANGE OF SIGN

Let two real numbers a and b be substituted for x in the polynomial $f(x)$ and $f(a)$ and $f(b)$ are found to be of opposite signs, then at least one or an odd number of real roots of the equation $f(x) = 0$ lie between a and b. In case $f(a)$ and $f(b)$ of the same sign, then either no real root or an even number of roots of $f(x) = 0$ lie between a and b.

Case I. $f(a)$ and $f(b)$ of opposite signs.

Let $y = f(x)$ be a continuous function of x, it should assume all values between $f(a)$ and $f(b)$.

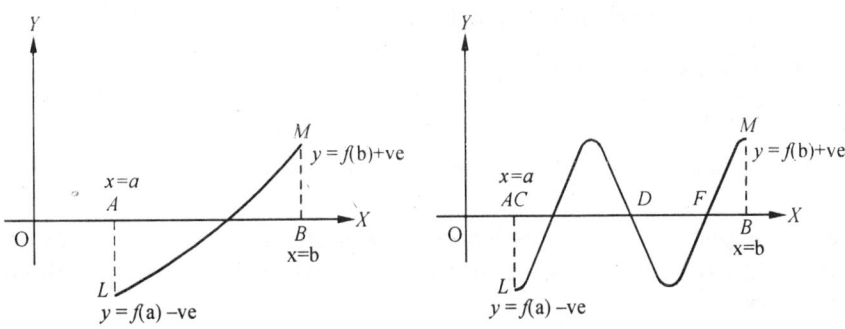

FIG. 1   FIG. 2

Now $f(a)$, $f(b)$ are of opposite signs, i.e. values of y corresponding to the values of x, i.e. a and b are of opposite signs. From one side of x-axis to the other side of x-axis the curve $y = f(x)$ must cross the axis of x at least once as in Fig. 1 at C or an odd number of times as in Fig. 2 at C, D and E at all such points where the curve crosses the axis of x, $y = 0$, i.e. $f(x) = 0$ which means that $f(x)$ vanishes either at one or an odd number of times for values of x between a and b. Hence at least one or an odd number of roots of $f(x) = 0$ lie between a and b.

Case II. $f(a)$ and $f(b)$ of same sign.

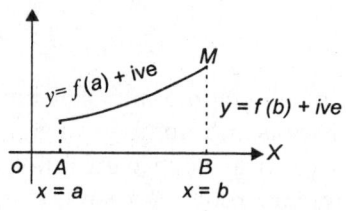

FIG. 3

Here $f(a)$ and $f(b)$ are of the same sign, the values of y corresponding to the values of x, i.e. a and b are of the same sign which means that in passing from a point on one side of x-axis to the other point on the same side either the curve $y = f(x)$ will not cross the x-axis as in Fig. 3 where $y = 0$, i.e. $f(x) = 0$ or, it must cross an even number of times of Fig. 4. Hence we conclude that either $f(x)$ does not vanish for values of x between a and b. or, if it vanishes, it must vanish for even number of times.

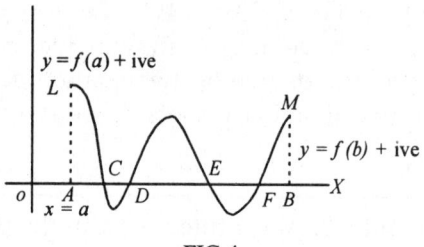

FIG. 4

THEOREM-1

Every equation of an odd degree has at least one real root whose sign is opposite to that of its last term, the coefficient of the first term being +ive.

Proof. Let the equation be

$$x^n + a_1 x^{n-1} + a_2 x^{n-1} + \dots + a_n = 0 \ (n \text{ odd})$$

$$f(-\infty) = -\text{ive} \ (n \text{ odd}), \ f(0) = a_n, \ f(\infty) = +\text{ive}.$$

Case I. a_n is positive. In this case $f(\infty)$ and $f(0)$ are of opposite signs and hence at least open real root must lie between $-\infty$ and 0, i.e. it must be negative opposite to the sign of a_n which is +ive.

Case II. a_n **is negative.** Here $f(0) = a_n = -$ ive and $f(\infty) = +$ive, i.e. they are of opposite signs and hence at least one real root must lie between 0 and ∞. This root is clearly, +ive, i.e. of sign opposite to that of a_n which is –ive.

THEOREM-2

Every equation of an even degree whose last term is negative and the coefficient of the first term positive, has at least two real roots, one positive and one negative.

Proof. Let the equation be

$$x^n + a_1 x^{n-1} + a_2 x^{n-2} + \ldots + a_n = 0 \ (n \text{ even})$$

$$f(-\infty) = + \text{ positive}; f(0) = a_n, \text{ i.e. } -\text{negative}; f(\infty) = +\text{ive}.$$

Hence $f(-\infty)$ and $f(0)$ are of opposite signs; therefore at least one real root must lie between $-\infty$ and 0 and it is –negative. Again $f(0)$ and $f(\infty)$ are of opposite signs and hence at least one real root must lie between 0 and ∞ and it is +ive. Thus the equation must have two real roots, one + ive and the other –ive.

THEOREM-3

If an equation has only one change of sign, it must have one positive root and no more.

Proof. Taking the leading coefficient to be positive, the equation must have a set of positive terms followed by a set of –ive terms (i.e. last term –ive) since there is only one change of sign.

$\therefore \ f(\infty) = $ positive; and $f(0) = $ last term –ive, i.e. $f(0)$ and $f(\infty)$ are of opposite signs and as such there must lie at least one or an odd number of roots of the equation $f(x) = 0$ between 0 and ∞. But as there is only one change of sign in $f(x)$, the number of positive roots cannot be greater than one. Hence the equation **must have only** one positive root.

THEOREM-4

If all the terms of an equation are positive and the equation, involves no odd powers of x, then all its root are complex.

Proof. Clearly both $f(x)$ and $f(-x)$ will have no change of signs and hence by Descarte's rule there will not be any positive or negative roots. Therefore all the roots must be complex.

Remark

❖ If all the terms of an equation are positive and all involve odd powers of x, then 0 is the only real root.

SOLVED EXAMPLES

Based on the following Results

- An equation $f(x) = 0$ can not have more positive roots then the number of changes of signs from +ve to −ve or from −ve to +ve in terms of it first member.

- An equation $f(x) = 0$ cannot have more negative roots than the number of changes of sign in $f(-x) = 0$

- Every equation of an odd degree has at least one real root whose sign is opposite to that if its, last term the coefficient of the first term being positive.

- Every equation of an even degree whose last term is negative and the coefficient of the first term positive has at least two real roots, one positive and one negative.

- If all the terms of an equation are positive and the equation involves no odd powers of x then all its roots are complex.

- If all the terms of an equation are positive and all involve odd powers of x then 0 is the only root.

Example 1. *Apply Descarte's rule of signs to discuss the nature of the roots of the equation*
$$x^4 + 15x^2 + 7x - 11 = 0$$

Solution. Let $f(x) = x^4 + 15x^2 + 7x - 11 = 0$. It has only one change of sign and hence it must have one +ive root.

Now, $f(x) = x^4 + 15x^2 + 7x - 11 = 0$.

As above it must have one +ive root *i.e.*, $f(x) = 0$ must have one −ive root. Thus, the equation has two real roots, one +ive and one −ive and hence the other two roots must be imaginary.

Example 2. *Show that the equation* $f(x) = x^5 + x^3 - 8x - 5 = 0$ *cannot have more than three real roots and prove that it must have three real roots.*

Solution. Let $f(x) = 0$ has only one change of sign and hence the equation must have one positive root.

Now, $f(-x) = -x^5 - x^3 + 8x - 5 = 0$

has two changes of sign and as such it can have at the most two +ive roots or the maximum number of −ive roots of $f(x) = 0$ is two.

Again $f(0) = $ −ive, $f(-1) = $ +ive, $f(-\infty) = $ −ive.

Since $f(0)$ and $f(-\infty)$ are of the same sign, so there lies either none or an even number of roots between 0 and $-\infty$ and now $f(0)$ and $f(-1)$ are of opposite signs and also $f(-1)$ and $f(-\infty)$ too are of opposite signs and hence one −ive root lies between 0 and -1 and the other between -1 and $-\infty$. Thus, we conclude that the equation must have three real roots and hence two complex.

Example 3. *Show that the equation* $f(x) = x^{12} - x^4 - x^3 - x^2 + 1 = 0$ *has at least six complex roots.*

Solution. Here $f(x) = 0$ has four changes and hence it can have at the most four +ive roots

Now, $\quad f(-x) = x^{12} - x^4 + x^3 - x^2 + 1 = 0$

has two changes and it can at the most have two +ive roots or $f(x) = 0$ can at the most have two –ive roots. Hence, the maximum number of real roots of $f(x) = 0$ can be six; but there being twelve roots, the minimum number of complex roots must be six.

Example 4. *Find the minimum number of imaginary roots which equation* $f(x) = 2x^7 - x^4 + 4x^3 - 5 = 0$ *must possess.*

Solution. Here $f(x) = 0$ has three changes of signs and as such it can at the most have three +ive roots.

Now, $\quad f(-x) = -2x^7 - x^4 - 4x^3 - 5 = 0$

or $\quad 2x^7 + x^4 + 4x^3 + 5 = 0$

Clearly, $f(-x) = 0$ has no changes of signs and as such it will have no +ive root which means that $f(x) = 0$ has no –ive root. Hence, the given equation can at the most have three real, *i.e.*, +ive roots and it being of 7th degree and hence the minimum number of imaginary roots is $7 - 3$, *i.e.*, 4.

Example 5. *Find the least possible number of imaginary roots of the equation*

$$x^9 - x^5 + x^4 + x^2 + 1 = 0$$

Solution. Clearly, $f(x) = 0$ has two changes of signs and hence 2 is the maximum number of +ive roots.

$$f(-x) = -x^9 - x^5 + x^4 + x^2 + 1 = 0$$

or $\quad x^9 + x^5 - x^4 - x^2 - 1 = 0$

$f(-x) = 0$ has only one change of sign and hence it has only one +ive root or $f(x) = 0$ has only one –ive root. Thus, the max. number of real roots is $2 + 1 = 3$ and the equation being of 9th degree will have at least $9 - 3 = 6$ imaginary roots.

Example 6. *Prove that the equation* $x^5 - x + 16 = 0$ *has two pairs of complex roots.*

Solution. Let $f(x) = x^5 - x + 16 = 0$

$f(x)$ has got two changes of signs and as such it cannot have more than two +ive roots. Again $f(0) = $ +ive and $f(\infty) = $ +ive, since both $f(0)$ and $f(\infty)$ are of the same sign. Also we observe that for all values of x between 0 and ∞, $f(x)$ remain +ive always, i.e. the graph of the curve $y = f(x)$ never crosses the x-axis which in other words means that y or $f(x)$ never becomes negative or y or $f(x)$ is always +ive for all values of x. Hence the equation $f(x) = 0$ has no +ive roots.

If $f(a)$ and $f(b)$ are of the same sign, then either no root or in general even number of roots of $f(x) = 0$ lie between a and b. In the later case the curve crosses the x-axis even times; but here we have shown $f(x)$ always remains +ive and hence we have established that the given equation has no +ive root even though $f(x) = 0$ has two changes of signs.

Again $\quad f(x) = -x^5 + x + 16 = 0$ or $x^5 - x - 16 = 0$

$f(-x) = 0$ has only one change of sign and as such it must have one +ive root or $f(x) = 0$ must have one –ive root.

Thus, in all the given equation has only one real root which is –ive and it being of fifth degree we conclude that the remaining four roots must be imaginary. Again since imaginary roots occur in conjugate pairs we say that the given equation has two pairs of complex roots.

Example 7. *Show that the equation $f(x) = x^3 - qx + r = 0$ where q and r are essentially positive has one negative root and that the other two roots are either imaginary or both +ive.*

Solution. Clearly $f(x) = 0$ has got only two changes of signs and hence the number of +ive roots cannot exceed two.

Now, $f(-x) = -x^3 + qx + r = 0$ or $x^3 - qx - r = 0$

and it has only one change of sign; hence, $f(x) = 0$ must have one –ive root. Otherwise the equation must have one real root whose sign is opposite to that of the last terms and hence it should be –ive.

Hence we conclude that one root of $f(x) = 0$ is essentially –ive and therefore the remaining two are either both +ive or imaginary.

Example 8. *Correct the mistakes if any in the following :*

(i) *If n be the degree of $f(x)$ and μ and μ' the number of changes of signs in $f(x)$ and $f(-x)$ respectively then if $\mu + \mu' < n$ the equation $f(x) = 0$ has exactly $n - (\mu + \mu')$ imaginary roots.*

Solution. Replace the word 'exactly' by 'at least'.

If two numbers a and b be substituted for x in polynomial $f(x)$ give results with the same sign no real root lies between them.

Replace 'no real root' by 'either no real root' or 'an even number of real roots lie between a and b'.

(iii) *$f(x)$ cannot have a greater number of negative roots then there are changes of sign in the terms of polynomial of $f(x)$.*

Replace 'polynomial $f(x)$' by 'polynomial $f(-x)$'.

Example 9. *Prove that if n be even the equation $x^{n-1} = 0$ has two and only two real roots, one –ive and one +ive and the rest complex and when n is odd the real roots is unity and rest are complex.*

Solution. If n even : Then $f(x) = 0$ and $f(-x) = 0$ both have only one change of sign each and hence the equation must have one +ive and one –ive root and no more. Thus, only two real and rest complex roots.

If n odd : $f(x)$ has one change of sign and hence $f(x) = 0$ has one and only one positive root. Again $f(-x) = x^{n-1} = 0$ has no change of sign and hence $f(x) = 0$ has no negative root. Thus, an odd degree equation has only one positive root which is clearly seen to be unity and the rest are therefore complex which will be even in number.

Example 10. *Find the situation of the roots of the equation*

$$f(x) = x^3 + x^2 - 2x - 1 = 0$$

Solution. Clearly, $f(x)$ has only one change of sign and hence it must have one +ive root. Also $f(0) = -1, f(1) = -1, f(2) = +7$. Since $f(1)$ and $f(2)$ are of opposite signs therefore the +ive root lies between 1 and 2.

$$F(x) = f(-x) = -x^3 + x^2 + 2x - 1 = 0$$
or $F(x) = 0 \implies x^3 - x^2 - 2x + 1 = 0.$

$F(x)$ has got two changes of sign; so at the most it can have two +ive roots.

$$F(0) = 1, \quad F(1) = -1, \quad F(2) = +1.$$

$F(0)$ and $F(1)$ are of opposite signs and again $F(1)$ and $F(2)$ are of opposite signs

∴ $F(x) = 0$ has two +ive root between $(0, 1)$ and $(1, 2)$. But $F(x) = f(-x)$.

$f(x) = 0$ has two –ive root one between $(0, -1)$ the other between $(-1, -2)$.

EXERCISE 12.2

1. Show that the equation $x^7 - 3x^4 + 12x^2 + 5x - 4 = 0$ has at least two imaginary roots.

2. Show that the equation $2x^7 + 3x^4 + 3x + k = 0$ has at least four imaginary roots for all values of k (constant).

3. Prove if $q > r > 0$ the cubic $x^3 + 9x + r = 0$ has one negative and two imaginary roots.

4. Prove that the equation $x^9 - x^5 + x^4 + x^2 + 1$ has one real root which is negative and eight imaginary roots.

5. Find an equation whose roots are the squares of the roots $x^3 - x^2 + 8x - 6 = 0$ and hence deduce that the equation must have a pair of imaginary roots.

6. Find the equation whose roots are the squares of the roots of the equation $x^3 + 4x^2 + 9x + 10 = 0$ and hence find the nature of the roots of the given equation.

ANSWERS

(5) $y^3 + 15y^2 + 52y - 36 = 0$ **(6)** One real, two complex

---

12.12   MULTIPLE ROOTS

If an equation $f(x) = 0$ has exactly m roots equal to α, then $f(x)$ and its first $(m - 1)$ derivatives all vanish for $x = \alpha$ but the mth and all the following derivatives do not vanish.

and if $f(x)$ and its first $(m - 1)$ derivatives vanish for $x = \alpha$, then $f(x) = 0$ has m roots equal to α.

Proof. Let $f(x) = (x - \alpha)^m \phi(x)$ ...(1)

where $\phi(x)$ does not vanish for $x = \alpha$, for if it vanishes, then it will contain the factor $x - \alpha$ and then $f(x)$ will have more than m equal roots.

Differentiating both sides of (1), we get

$$f'(x) = m(x - \alpha)^{m-1} \phi(x) + (x - \alpha)^m \phi'(x)$$
$$= (x - \alpha)^{m-1} [m\phi(x) + (x - \alpha)\phi'(x)]$$
$$= (x - \alpha)^{m-1} \psi_1(x),$$ ...(2)

where $\psi_1(x)$ does not vanish for $x = \alpha$, because $\psi_1(\alpha) = m\phi(\alpha)$ and it can be zero only if $\phi(\alpha) = 0$ which is contrary to the supposition. From (2), we observe that a root which occurs exactly m times in $f(x) = 0$ occurs exactly $(m - 1)$ times in $f'(x) = 0$. Hence we have the following :

A multiple root of order m of the equation $f(x) = 0$ is a multiple root of order $(m - 1)$ of the first derived equation $f'(x) = 0$ and hence $f(x)$ and $f'(x)$ have the common factor $(x - \alpha)^{m-1}$.

Again if we differentiate (2), we get

$$f''(x) = (x - \alpha)^{m-2} [(m - 1)\psi_1(x) + (x - \alpha)\psi_1'(x)]$$

$$= (x - \alpha)^{m-2}\psi_2(x)$$

where $\psi_2(x)$ does not vanish for $x = \alpha$, i.e. $f''(x) = 0$ has exactly $(m - 2)$ equal roots α. Similarly we can show $f'''(x)$ will have the factor $(x - \alpha)^{m-3}$ and so on till $f^{m-1}(x)$ will have the following :

Any root which occurs m times in f(x) = 0 occurs in degree of multiplicity diminishing by it in the first derived function.

Conversely Let $f(x) = f\{\alpha + (x - \alpha)\}$.

Expanding the R.H.S. by Taylor's theorem, we get

$$f(x) = f(\alpha) + (x - \alpha)f'(\alpha) + \frac{(x - \alpha)^2}{2!}f''(\alpha) + \dots$$

$$\frac{(x - \alpha)^{m-1}}{(m - 1)!}f^{m-1}(\alpha) + \frac{(x - \alpha)^m}{m!}f^m(\alpha) + \dots + \frac{(x - \alpha)^n}{n!}f^m(\alpha)$$

Now $f(\alpha) = f'(\alpha) = f''(\alpha) = \dots = f^{m-1}(\alpha) = 0$ (given)

$$\therefore \qquad f(x) = \frac{(x - \alpha)^m}{m!}f^m(\alpha) + \dots + \frac{(x - \alpha)^n}{n!}f^m(\alpha)$$

Above shows that $(x - \alpha)^m$ is a factor of $f(x) = 0$ which therefore has m roots equal to α.

12.12.1 Determination of Multiple Roots

We have discussed above that multiple root of the order m is multiple root of order $(m - 1)$ of the equation $f'(x) = 0$. In order to find such roots we should find the H.C.F. of $f(x)$ and $f'(x)$. This H.C.F. will give the multiple root of $f(x)$ each repeated $(m - 1)$ times. Thus if $(x - 2)$ is the H.C.F. of $f(x)$ and $f'(x)$, then $f(x)$ contains $(x - 2)^2$, as a factor or 2 is a double root of $f(x) = 0$. If $(x - 2)^2 (x - 1)^5$ is the H.C.F. of $f(x)$ and $f'(x)$, then $f(x) = 0$ has three roots equal to 2 and six roots equal to 1.

Remark

❖ In case $f(x)$ and $f'(x)$ have no common factor, then clearly, $f(x)$ has no equal roots.

12.12.2 Condition for Two or Three Equal Roots.

If α be a double root of $f(x) = 0$, then $f(\alpha) = 0$ and $f'(\alpha) = 0$ and the required condition is obtained by eliminating α between $f(\alpha) = 0$ and $f'(\alpha) = 0$. Similarly if α is a triple root the required condition can be obtained by eliminating α between $f(\alpha) = 0$, $f'(\alpha) = 0$, and $f''(\alpha) = 0$.

SOLVED EXAMPLES

Based on the following Results

- If an equation $f(x) = 0$ has exactly m roots equal to α then $f(x)$ and its first $(m - 1)$ derivatives all vanish for $x = \alpha$ but the mth and all the following derivatives do not vanish.

- If $f(x)$ and its $(m - 1)$ derivatives vanish for $x = \alpha$ then $f(x) = 0$ has m roots equal to α.

Example 1. *Solve the equation $x^5 - 15x^3 + 10x^2 + 60x - 72 = 0$ be testing for equal roots.*

Solution. Let $f(x) = x^5 - 15x^3 + 10x^2 + 60x - 72 = 0$

$$f'(x) = x^4 - 9x^2 + 4x + 12 = 0$$

Let us find the H.C.F. of $f(x) = 0$ and $f'(x) = 0$

x	$x^5 - 15x^3 + 10x^2 + 60x - 72$	$x^4 - 9x^2 + 4x + 12$	x
	$x^5 - 9x^3 + 4x^2 + 12x$	$x^4 - x^3 - 8x^2 + 12x$	
	$-6x^3 + 6x^2 + 48x - 72$	$x^3 - x^2 - 8x + 12$	1
or	$x^3 - x^2 - 8x + 12$	$x^3 - x^2 - 8x + 12$	

Thus we find that H.C.F. of $f(x) = 0$ and $f'(x) = 0$ is

$$x^3 - x^2 - 8x + 12 = 0 \text{ or } (x - 2)(x^2 + x - 6) = 0$$

or $(x - 2)^2(x + 3) = 0$

giving $n = 2, 2, - 3$. Now we know that HCF gives a multiple root of $f(x) = 0$, $(m - 1)$ times. Hence 2 is triple root and $- 3$ is a double root of $f(x) = 0$

Example 2. *Solve the equation $f(x) = 4x^3 + 20x^2 - 3x + 6 = 0$ given that two of its roots are equal.*

Solution. The H.C.F. of $f(x)$ and $f'(x)$ is found to be $2x - 1 = 0$ giving $x = \dfrac{1}{2}$.

$\therefore$ $\dfrac{1}{2}$ is double root of $f(x) = 0$

Let the third root by γ; then product of all the roots is

$$\frac{1}{2} \cdot \frac{1}{2} \gamma = -\frac{6}{4} \quad \therefore \gamma = -6$$

$\therefore$ Roots are $\dfrac{1}{2}, \dfrac{1}{2}, - 6.$

Example 3. *Solve the equation $x^4 - 6x^3 + 12x^2 - 10x + 3 = 0$ which has equal roots.*

Solution. The H.C.F. is $(x - 1)^2 = 0$ giving 1 as a triple root of $f(x) = 0$

Also $\alpha\beta\gamma\delta = 3$ or $1.1.1.\delta = 3$; $\therefore$ $\delta = 3$

$\therefore$ Roots are 1, 1, 1, 3.

Example 4. *Factorize the following* $2x^5 - x^4 + 5x^2 - 4x + 3$.

Solution. The H.C.F. of $f(x)$ and $f'(x)$ is found to be $x^2 - x + 1$. Hence $(x^2 - x + 1)^2$ is a factor of $f(x)$.

Let the other factor which will be linear is say $x + k$.

$\therefore$ $f(x) = 2(x^2 - x + 1)^2 (x + k)$ as the coefficient of x^5 is 2.

Comparing constant term on either side, we get

$$3 = 2k; \quad \therefore \quad k = \frac{3}{2}.$$

$\therefore$ $f(x) = (x^2 - x + 1)^2(2x + 3)$

Example 5. *Show that the equation* $x^m - qx^{n-m} + r = 0$ *has two equal roots if*

$$\left\{\frac{q}{n}(n - m)\right\}^n = \left\{\frac{r}{m}(n - m)\right\}^m$$

Solution. Let $f(x) = x^m - qx^{n-m} + r = 0$ ...(1)

and $f'(x) = nx^{n-1} - q(n - m)x^{n-m-1} = 0$ ...(2)

The required condition is obtained by eliminating x between (1) and (2).

From (2), $n = \dfrac{q(n - m)}{x^m}$ or $x^m = \dfrac{q}{n}(n - m)$ ...(3)

and from (1) with the help of (3), we get

$$x^n\left[1 - \frac{n}{n - m}\right] + r = 0$$

or $x^n = \dfrac{r}{m}(n - m)$ ...(4)

or $(x^m)^n = (x^n)^m$ etc. from (3) and (4).

Example 6. *If the equation* $x^4 - 4p^3x + 1$ *has a pair of equal roots, find the value of* p *and solve the equation completely.*

Solution. Let $f(x) = x^4 - 4p^3x + 0 \Rightarrow f'(x) = 4(x^3 - p^3) = 0$ giving $x = p$.

Eliminating x between $f(x)$ and $f'(x) = 0$, we get

$$p^4 - 4p^4 + 1 = 0$$

or $3p^4 = 1$ or $p = 3^{-1/4}$ ...(1)

Now let us find the H.C.F. of $f(x)$ and $f'(x)$.

$$x^3 - p^3 \overline{)x^4 - 4p^3x + 1}(x$$
$$\underline{x^4 - p^3x}$$
$$-3p^3x + 1$$

or $-3p^4x + p$ or $-x + p;$

$\therefore$ $3p^4 = 1$, by (1)

$$-x + p \overline{)x^3 - p^3} (-x^2 - px - p^2$$
$$\underline{x^3 - p^3}$$
$$\times \cdot$$

or

Thus the H.C.F. is $(x - p)$ and hence two equal roots are p and q, their sum being $2p$ and product p^2. Let the other roots be α and β.

Now, sum of all the roots $= 0$.

$\therefore \qquad \alpha + \beta + p + p = 0$ or $\alpha + \beta = -2p$

and $\alpha \beta pp = $ product of all the roots $= 1$.

$\therefore \qquad \alpha \beta = \dfrac{1}{p^2} = 3p^2$ from (1), $\because 3p^4 = 1$

$\therefore \qquad \alpha$ and β are the roots of the quadratic $t^2 + 2pt + 3p^2 = 0$

or $\qquad t = \dfrac{-2p \pm \sqrt{(4p^2 - 12p^2)}}{2} = p \pm ip\sqrt{2}$

Hence the four roots are $p, p, -p \pm ip\sqrt{2}$ where $p = 3^{-(1/4)}$.

Example 7. *If the equation $x^n + p_1 x^{n-1} + p_2 x^{n-2} + \ldots + p_n = 0$ has two roots and equal to α, prove that α is also a root of the equation,*

$$p_1 x^{n-1} + 2p_2 x^{n-2} + 3p_3 x^{n-3} + \ldots + np_n = 0$$

Solution. Let

$f(x) = x^n + p_1 x^{n-1} + p_2 x^{n-2} + \ldots + p_{n-2} x^2 + p_{n-1} x + p_n = 0$

$\Rightarrow \quad f'(x) = -nx^{n-1} + (n-1)p_1 x^{n-2} + (n-2)p_2 x^{n-3} + \ldots + 2xp_{n-2} + p_{n-1} = 0$

and $\quad f''(x) = n(n-1)x^{n-2} + (n-1)(n-2)p_1 x^{n-3} + (n-2)(n-3)p_2 x^{n-4}$
$\qquad\qquad\qquad + \ldots + 2.p_{n-2} = 0$

If α is a double root of $f(x) = 0$ then it is also a root of $f'(x) = 0$. Hence α will also be a root of $nf(x) - xf'(x) = 0$ as both $f(x)$ and $f'(x)$ vanish for $x = \alpha$. Putting the values of $f(x)$ and $f'(x)$ and simplifying, we get the required equation as given.

Example 8. *In case the given equation has three roots equal to α then α is a root of the equation*

$$n^2 x^{n-1} + (n-1)^2 p_1 x^{n-2} + (n-p)^2 p_2 x^{n-3} + \ldots + p_{n-1} = 0$$

Solution. In case α be a triple root of $f(x) = 0$, then α is a double root of $f'(x) = 0$ and is a single root of $f''(x) = 0$. Hence, α is a root of $f'(x) + xf''(x) = 0$ as both $f'(x)$ and $f''(x)$ vanish for $x = \alpha$.

or $\quad \{nx^{n-1} + (n-1)p_1 x^{n-2} + \ldots + 2xp_{n-2} + p_{n-1}\}$
$\qquad\qquad + x\{n(n-1)x^{n-2} + (n-1)(n-2)p_1 x^{n-3} + 2p_{n-2}\} = 0$

or $\quad \{n + n^2 - n\}x^{n-1} + (n-1)(1 + n - 2)p_1 x^{n-2}$
$\qquad\qquad + (n-2)\{1 + n - 3)p_2 x^{n-3} + (2 + 2)xp_{n-2} + p_{n-1} = 0$

or $\quad n^2 x^{n-1} + (n-1)^2 p_1 x^{n-2} + (n-2)^2 p_2 x^{n-2}$
$\qquad\qquad + \ldots + 2^2 xp_{n-2} + p_{n-1} = 0$

Example 9. *Prove that the equation $x^5 + 5px^3 + 5p^2 x + q = 0$ has a pair of equal roots when $p^2 + 4q^5 = 0$ and that if it has one pair of equal roots, it must have a second pair.*

Solution. Let us find the H.C.F. of $f(x) = 0$ and $f'(x) = 0$ in order to find the repeated roots

$x^4 + 3px^2 + p^2$	$x^5 + 5px^3 + 5p^2x + q$	x
$2p$	$x^5 + 3px^3 + p^2x$	
$2px^4 + 6p^2x^2 + 2p^3$	$2px^3 + 4p^2 + q$	
$2px^4 + 4p^2x^2 + qx$	p	
$2p^2x^2 - qx + 2p^3$	$2p^2x^3 + 4p^3x + pq$	x
x	$2p^2x^3 - qx^2 + 2p^3x$	
	$qx^2 + 2p^3x + pq$	
	$2p^2$	
	$2p^2qx^2 + 4p^5x + 2p^3q$	q
	$2p^2qx^2 - q^2x + 2p^3q$	
	$x(4p^5 + q^2)$	

In case the given equation has a pair of equal roots then the H.C.F. should be linear. Therefore $x(4p^5 + q) = 0$ is the H.C.F. giving that $x = 0$ is the H.C.F. If $x = 0$ is the H.C.F. then x^2 should be a factor of $f(x)$ which is not. Hence when $x(4p^5 + q^2) = 0$. $x \neq 0$ but $4p^5 + q^2 = 0$.

Also when $4p^5 + q^2 = 0$ then the H.C.F. will be $2p^2x^2 - qx + 2p^3$ so that $(2p^2x^2 - qx + 2p^3)^2$ is a factor giving that $f(x) = 0$ has two pairs of equal roots.

Example 10. *Prove that if the coefficients of a given equation are all integers, an integral root is an exact divisor of the absolute term.*

Solution. Let the equation be

$$a_0x^n + a_1x^{n-1} + a_2x^{n-2} + \dots + a_{n-1}x + a_n = 0$$

where all the a's are integers. If α be an integral root of above equation, then

$$a_0\alpha^n + a_1\alpha^{n-1} + a_2\alpha^{n-2} + \dots + a_{n-1}\alpha + a_n = 0$$

or $\alpha(a_0\alpha^{n-1} + a_1\alpha^{n-2} + \dots + a_{n-1}) = -a_n$

or $\dfrac{a_n}{\alpha} = -(a_0\alpha^{n-1} + a_1\alpha^{n-2} + \dots + a_{n-1}) =$ an integer.

Hence $\dfrac{a_n}{\alpha} =$ an integer. Therefore the integral root α divides a_n exactly.

Example 11. *Find the values of a, for which the equation $ax^3 - 9x^2 + 12x - 5 = 0$ has two equal roots and solve the equation completely in one case.*

Solution. Let $f(x) = ax^3 - 9x^2 + 12x - 5 = 0$ ...(1)

$\Rightarrow$ $f'(x) = 3(ax^2 - 6x + 4) = 0$ or $ax^2 - 6x + 4 = 0$ ...(2)

Multiplying (2) by x and subtracting from (1), we get

$$3x^2 - 8x + 5 = 0$$

Solving (2) and (3) by the method of cross-multiplication, we get

$$\frac{x^2}{-30+32} = \frac{x}{12-5a} = \frac{1}{18-8a}$$

Eliminating x, we get

$$(12-5a)^2 = 2(18-8a) \text{ or } 25a^2 - 104a + 108 = 0$$

or $(a-2)(25a-54) = 0; \therefore a = 2 \text{ or } \dfrac{54}{25}$

Putting $a = 2$, we get

$$f(x) = 2x^3 - 9x^2 + 12x + 5 = 0,$$
$$\Rightarrow \quad f'(x) = x^2 - 3x + 2 = 0 \text{ or } (x-1)(x-2) = 0$$

whose roots are 1 and 2. Either can be a double root of the given equation $f(x) = 0$. Since (2) does not divide the absolute term of the given equation, as such it cannot be its root. Hence (1) is a double root of the given equation corresponding to the value 2 of a.

Example 12. *If the equation $x^4 + ax^3 + bx^2 + cx + d = 0$ had three equal roots show that each of them is $\dfrac{(6c-ab)}{(3a^2-8b)}$.*

Solution. In the process of finding H.C.F. of $f(x)$ and $f'(x)$, the 1st quotient is $x + a$ and quadratic remainder is

$$F(x) = (8b - 3a^2)x^2 + 2(6c - ab)x + (16d - ac).$$

If the equation is to have three equal roots then the H.C.F. should be perfect square of a linear expression in x. Hence $F(x)$ should be the H.C.F. and must have equal roots the condition for which is

$$4(6c - ab)^2 = 4(8b - 3a^2)(16d - ac)$$

· Multiplying $F(x)$ by $8b - 3a^2$, it becomes with the help of the above condition,

$$[(8b - 3a^2)x + (6c - ab)]^2$$

and hence the equal root is given by

$$F(x) = 0 \quad \text{or} \quad x = \frac{6c - ab}{3a^2 - 8b}$$

Example 13. *Find the condition that the cubic $a_0x^3 + 3a_1x^2 + 3a_2x + a_3 = 0$. may have two roots equal and find its value. Also, find the condition that the cubic may have all the three roots equal.*

Solution. The above cubic can be reduced to the form

$$f(z) = z^3 + 3Hz + G = 0 \quad \text{where } z = a_0x + a_1$$
$$f'(z) = 3(z^2 + H) = 0 \quad \text{or} \quad z^2 + H = 0$$

The H.C.F. of $f(z)$ and $f'(z)$ is found below :

$z^2 + H$	$z^3 + 3Hz + G$	z
$2H$	$z^3 + Hz$	

	$2Hz^2 + 2H^2$	$2Hz + G$
z	$2Hz^2 + Gz$	1st quotient

$- Gz + 2H^2$
$- 2H$

$2HGz - 4H^3$
$2HGz + G^2$

| $- (G^2 + 4H^3)$ | Remainder |

From above the remainder is $- (G^2 + 4H^3)$ and the first quotient is $2Hz + G$. If the equation has two equal roots then the H.C.F. should be of first degree and hence the remainder should be zero. Therefore $G^2 + 4H^3$ is the required condition and the equal root is given by equating the H.C.F.

Now, $2Hz + G = 0$

$$\therefore \quad z = -\frac{G}{2H} = a_0x + a_1; \quad \therefore \quad x = -\frac{1}{a_0}\left[a_1 + \frac{G}{2H}\right]$$

In case all the three roots are equal then the H.C.F. should be of 2nd degree and that too a perfect square and hence the first degree remainder, i.e. $2Hz + G = 0$ should be zero. Now the H.C.F. in this case will be $z^2 + H$ and it will be a perfect square only if $H = 0$ and putting $H = 0$ in $2Hz + G = 0$ we get $G = 0$. Hence the required conditions in this case are
$$G = 0, H = 0$$

Example 14. *Find the conditions that a biquadratic*
$$f(x) = a_0x^4 + 4a_1x^3 + 6a_2x^2 + 4a_3x + a_4 = 0$$
may have three roots equal and show that in this case it can be expressed in the form
$$a_0^3 f(x) = [a_0x + a_1 + \sqrt{(-H)}]^3 [a_0x + a_1 - 3\sqrt{(-H)}].$$

Solution. The given equation can be reduced to the form
$$F(z) = z^4 + 6Hz^2 + 4Gz + a_0^2 - 3H^2 = 0 \quad \text{where } z = a_0x + a_1.$$
$$F'(z) = z^3 + 3Hz^2 + G = 0$$

$z^3 + 3Hz + G = 0$	$z^3 + 6Hz^2 + 4Gz$ z
$3H$	$+(a_0^2 I - 3H^2)$
z $\quad 3Hz^3 + 9H^2z + 3GH$	$z^4 + 3Hz^2 + Gz$
$3Hz^3 + 3Gz^2 + (a_0^2 I - 3H^2)z$.	$3Hz^2 + 3Gz +$
	$(a_0^2 I - 3H^2) = \phi(z)$
$-3Gz^2 + (12H^3 - a_0^2 I)z + 3GH$	
$-H$	
G $\quad 3GHz^2 - (12H^3 - a_0^2 IH)z + 3GH^2$	
$3GHz^3 + 3G^2z + G(a_0^2 I - 3H^2)$	
$z\{a_0^2 HI - 12H^3 - 3G^2\} - a_0^2 IG$	

If $F(z)$ is to have three roots equal then the H.C.F. should be of second degree and at the same time a perfect square. Also the linear remainder should be zero.

i.e. $$9G^2 - 4.3H(a_0^2 - 3H^2) = 0$$

for perfect square $$b^2 - 4ac = 0 \qquad \qquad ...(1)$$

and $$z\{a_0^2 HI - 3(G^2 + 4H^3)\} - a_0^2 GI = 0$$

so that $$z\{a_0^2 HI - 4a_0^2 HI\} - a_0^2 IG = 0, \text{ by (1)} \qquad \qquad ...(2)$$

or $$a_0^2 I(3Hz + G) = 0$$

Now $3Hz + G$ cannot be zero for in that case $F_1(z)$ becomes z^3, i.e. a perfect cube so that all the four roots of $F(z) = 0$ are equal.

Hence we have $a_0^2 I = 0$, i.e. $I = 0$ and hence from (1) $G^2 + 4H^3 = 0$.

Also we know that

$$G^2 + 4H^3 = a_0^2 HI - a_0^3 J \quad \text{or} \quad 0 = 0 - a_0^3 J; \qquad \therefore \quad J = 0$$

$\therefore$ $I = 0$ and $J = 0$ are the required conditions and the condition $G^2 + 4H^3 = 0$ is included in them. With the help of this condition, we get the H.C.F.

$\therefore$ $$\phi(z) = 3Hz^2 + 3Gz - 3H^2 = 0 \qquad [\because I = 0]$$

or $$H^2 z^2 + GHz - H^3 = 0, \qquad \text{Put} \quad G^2 = -4H^3$$

or $$H^2 z^2 + GHz + \frac{G^2}{4} = 0 \qquad \text{or} \qquad \left(Hz + \frac{G}{2}\right)^2 = 0$$

Hence each of the triple root is

$$z = \frac{-G}{2H} = \frac{\sqrt{(-4H^3)}}{\sqrt{(4H^2)}} = -\sqrt{(-H)}$$

If the fourth root be α, then sum of the roots is

$$\alpha - 3\sqrt{(-H)} = 0, \qquad \therefore \qquad \alpha = 3\sqrt{(-H)}$$

Hence $f(z) \equiv [z + \sqrt{(-H)}]^3 [z - 3\sqrt{(-H)}]$

or $$a_0^3 f(x) = [a_0 x + a_1 + \sqrt{(-H)}]^3 [a_0 x + a_1 - 3\sqrt{(-H)}]$$

Example 15. *Find the conditions which must be satisfied when the biquadratic of the last example be a perfect square and prove that in that case* $a_0^3 f(x) = \{(a_0 x + a_1)^2 + 3H\}^2$

Solution. In this case H.C.F. $\phi(z)$ of last example should be quadratic (but not perfect square).

$\therefore$ $$3Hz^2 + 3Gz + (a_0^2 I - 3H^2) = 0 \quad \text{should be the H.C.F}$$

or $$z^2 + \frac{G}{H} z + \frac{a_0^2 I - 3H^2}{3H} = 0$$

and hence $z^4 + 6Hz^2 + 4Gz + a_0^2 I - 3H^2 \equiv \left[z^2 + \dfrac{G}{Hz} + \dfrac{a_0^2 I - 3H^2}{3H} \right]^2$

Comparing the coefficient of like powers of z, we get

$$\frac{2G}{H} = 0; \quad \therefore \ G = 0$$

and $\qquad \dfrac{G^2}{H^2} + \dfrac{2(a_0^2 I - 3H^2)}{3H} = 6H. \ \ \text{Put} \ G = 0$

$\therefore \qquad\qquad\qquad a_0^2 I - 3H^2 = 9H^2 \quad \text{or} \quad a_0^2 I = 12H^2$

Under these conditions the H.C.F. is reduced to $(z^2 + 3H)$ and hence

$$F(z) = (z^2 + 3H)^2. \ \text{Put} \ z = a_0 x + a_1$$

or $\qquad\qquad a_0^3 f(x) = \{(a_0 x + a_1)^2 + 3H\}^2$

Example 16. *Show that* $a\alpha + b = \dfrac{-GI}{2HI - 3aJ}$ *or* $-\sqrt{(-H)}$ *according as* α *is a double or triple root of the biquadratic of the last example.*

Solution. The 2nd part is done in Ex. 14. For the first part, find the H.C.F. of $F(z)$ and $F'(z)$ and if the biquadratic has two equal roots, the H.C.F. should be linear.

As done earlier the H.C.F. of $F(z)$ and $F'(z)$ is $z[a_0^2 HI - 3(G^2 + 4H^3)] - a_0^2 IG = 0$ giving the value of z and hence of x.

Putting $\qquad G^2 + 4H^3 = a_0^2 HI - a_0^3 J$, we get

$$z(a_0^2 HI - 3a_0^2 HI + 3a_0^2 J) - a_0^2 IG = 0$$

$\Rightarrow \qquad a_0\alpha + \alpha_1 = z = \dfrac{GI}{3a_0 J - 2HI} = -\dfrac{GI}{2HI - 3a_0 J}$

Example 17. *If* $\alpha, \beta, \gamma, \ldots$ *be the roots of* $f(x) = 0$ *and* $\alpha', \beta', \gamma', \ldots$ *of* $f'(x) = 0$ *prove that*

$$f'(\alpha)f'(\beta)f'(\gamma)f'(\delta) = n^n f(\alpha')f(\beta')f(\gamma')f(\delta') \cdots$$

and that each is equal to the absolute term in the equation whose roots are the squares of the difference of the roots of $f(x) = 0$.

Solution. Let $f(x) = 0$ be an equation of nth degree so that

$\qquad f(x) = (x - \alpha)(x - \beta)(x - \gamma) \ldots n$ factors $\qquad\qquad$ ...(1)

$\qquad x^n + p_1 x^{n-1} + p_2 x^{n-2} + \ldots \qquad\qquad$ (say)

then $\qquad f'(x) = nx^{n-1} + (n-1)p_1 x^{n-2} + \ldots$

But we are given that the roots of

$\qquad f'(x) = 0 \ \text{ are } \alpha', \beta', \gamma', \ldots$

$\therefore \qquad f'(x) = k \ (x - \alpha')(x - \beta')(x - \gamma') \ldots (n-1)$ factors.

Comparing the coefficients of x^{n-1} in the two expressions for $f'(x)$, we find that $k = n$

$\therefore \qquad f'(x) = n \ (x - \alpha')(x - \beta')(x - \gamma') \ldots (n-1)$ factors $\qquad\qquad$ ...(2)

Putting in (2) the n quantities α, β, γ, ... successively for x, we get

$$f'(\alpha) = n \, (\alpha - \alpha')(\alpha - \beta')(\alpha - \gamma') \, ... \, (n-1) \text{ factors}$$

$$f'(\beta) = n \, (\beta - \alpha')(\beta - \beta')(\beta - \gamma') \, ... \, (n-1) \text{ factors}$$

... ...

... ...

n such equations as α, β, γ, ... are n in number. Multiplying the above vertically,

$$f'(\alpha)f'(\beta)f'(\gamma) ... \, n \text{ factors} = n \begin{vmatrix} (\alpha - \alpha')(\beta - \alpha')(\gamma - \alpha') \, ... \, n \text{ factors} \\ (\alpha - \beta')(\beta - \beta')(\gamma - \beta') \, ... \, n \text{ factors} \\ (\alpha - \gamma')(\beta - \gamma')(\gamma - \gamma') \, ... \, n \text{ factors} \\ ... \quad ... \quad ... \quad ... \quad ... \\ ... \quad ... \quad ... \quad ... \quad ... \\ (n-1) \text{ such factors } \alpha'\beta'\gamma' \text{ are} \\ (n-1) \text{ in number.} \end{vmatrix}$$

Again putting the $(n-1)$ quantities α', β', γ', ... respectively for x in (1), we get

$$f(\alpha') = (-1)^n \, (\alpha - \alpha')(\beta - \alpha')(\gamma - \alpha') \, ... \, n \text{ factors}$$

$$f(\beta') = (-1)^n \, (\alpha - \beta')(\beta - \beta')(\gamma - \beta') \, ... \, n \text{ factors}$$

$$f(\gamma') = (-1)^n \, (\alpha - \gamma')(\beta - \gamma')(\gamma - \gamma') \, ... \, n \text{ factors}$$

... ...

$(n-1)$ such equations.

Hence from (3) and (4), we get

$$f(\alpha') \, f(\beta')f(\gamma') \, ... \, (n-1) \text{ factors} = \frac{(-1)^{n(n-1)}}{n^n} \{f(\alpha') \, f(\beta')f(\gamma') \, ... \, n \text{ factors}\}$$

Now $n(n-1)$ being the product of two consecutive numbers must be even and as such $(-1)^{n(n-1)} = 1$.

$\therefore \qquad n^n f(\alpha') \, f(\beta')f(\gamma') \, ... = f'(\alpha)f'(\beta)f'(\gamma) \, ...$

Now we have to prove that $f'(\alpha)f'(\beta)f'(\gamma) \, ... =$ absolute term in the equation whose roots are the squares of the differences of the roots of the equation $f(x) = 0$.

$$f(x) = (x - \alpha)(x - \beta)(x - \gamma) \, ...$$

$$\log f(x) = \log (x - \alpha) + \log (x - \beta) + \log (x - \gamma) \, ...$$

Differentiating we get

$$f'(x) = \frac{f(x)}{(x-\alpha)} + \frac{f(x)}{(x-\beta)} + \frac{f(x)}{(x-\gamma)} + ...$$

$$= (x - \beta)(x - \gamma) \, ... + (x - \alpha)(x - \gamma) \, ... + (x - \alpha)(x - \beta) \, ... + \, ...$$

Putting $x = \alpha$, β, γ, we get

$$f'(\alpha) = (\alpha - \beta)(\alpha - \gamma)(\alpha - \delta) \, ...$$

$$f'(\beta) = (\beta - \alpha)(\beta - \gamma)(\beta - \delta) \, ...$$

$$f'(\gamma) = (\gamma - \alpha)(\gamma - \beta)(\gamma - \delta) \, ...$$

∴ $f'(\alpha)f'(\beta)f'(\gamma)$ ...

$= (1)^{n(n-1)/2}(\alpha - \beta)^2(\beta - \gamma)^2(\alpha - \gamma)^2$ ...

= Product of all the roots of the equation of squared difference

= absolute term (numerically) of the equation of square difference.

Example 18. *If* α, β, γ, δ *be the roots of* $f(x) = 0$, *prove that* $f'(\alpha) + f'(\beta) + f'(\gamma) + f'(\delta)$ *can be expressed as the product of three factors.*

Solution. Proceed same as in previous example we get

$f'(\alpha) + f'(\beta) + f'(\gamma) + f'(\delta)$

$= (\alpha - \beta)(\alpha - \gamma)(\alpha - \delta) + (\beta - \alpha)(\beta - \gamma)(\beta - \delta)$
$\qquad + (\gamma - \alpha)(\gamma - \beta)(\gamma - \delta) + (\delta - \alpha)(\delta - \beta)(\delta - \gamma)$

$= (\alpha - \beta)\{(\alpha^2 - \gamma\alpha - \delta\alpha + \gamma\delta) - (\beta^2 - \gamma\beta - \delta\beta + \gamma\delta)\}$
$\qquad + (\gamma - \delta)\{(\gamma^2 - \alpha\gamma - \beta\gamma + \alpha\beta) - (\delta^2 - \alpha\delta - \beta\delta + \alpha\beta)\}$

$= (\alpha - \beta)\{\alpha^2 - \beta^2 - (\alpha - \beta)(\gamma + \delta)\} + (\gamma - \delta)\{(\gamma^2 - \delta^2) - (\gamma - \delta)(\alpha + \beta)\}$

$= (\alpha - \beta)^2(\alpha + \beta - \gamma - \delta) - (\gamma - \delta)^2 - (\alpha + \beta - \gamma - \delta)$

$= (\alpha + \beta - \gamma - \delta)\{(\alpha - \beta)^2 - (\gamma - \delta)^2\}$

$= (\alpha + \beta - \gamma - \delta)(\alpha - \beta - \gamma + \delta)(\alpha - \beta + \gamma - \delta)$

$= (\alpha + \beta - \gamma - \delta)(\alpha + \gamma - \beta - \delta)(\alpha + \delta - \beta - \gamma)$

EXERCISE 12.3

1. Factorize : $x^4 + 12x^3 + 32x^2 - 24x + 4$

2. Solve the equation :

   (i) $x^4 - 8x^3 + 24x^2 - 32x + 16 = 0$

   (ii) $x^6 - 6x^4 - 4x^3 + 9x^2 + 12x + 4 = 0$

3. Find the multiple root of :

   (i) $x^4 + 3x^3 - 7x^2 - 15x + 18 = 0$

   (ii) $x^4 + 7x^3 + 17x^2 + 17x + 6 = 0$

   (iii) $x^5 - x^4 + 2x^3 - 2x^2 + x - 1 = 0$

4. Show that the equation $x^n - nax + (n - 1)b = 0$ will have a pair of equal roots of $a^n = b^{n-1}$.

5. Show that the equation

   (i) $x^n - 1 = 0$ (ii) $1 + \dfrac{x}{1!} + \dfrac{x^2}{2!} + ... + \dfrac{x^n}{n!} = 0$

   can not have equal roots.

6. If $x^5 + 5px^3 + 5qx^2 + r = 0$ has two equal roots prove that either of them is a root of the quadratic
   $$3q\dot{p}x^2 - 6p^2x - 4pq + r = 0$$

ANSWERS

(1) $(x^2 + 6x - 2)^2$

(2) (i) All roots are equal to 2
 (ii) $2, 2, -1, -1, -1, -1)$

(3) (i) $-3, -3, 1, 2$
 (ii) $-1, -1, -2, -3)$
 (iii) $\pm i, \pm i, 1$

12.13 MAXIMUM AND MINIMUM VALUES OF $f(x)$

We know from the definition of Max. and Min. that as x varies in the interval $(x - h, x + h)$, where h is small, then $f(x)$ is greater than both $f(x - h)$ and $f(x + h)$ if $f(x)$ be Max., i.e. $f(x - h) - f(x)$ and $f(x + h) - f(x)$ are of the same sign, i.e. –ive. Similarly if $f(x)$ is Min., then $f(x)$ is less than both $f(x - h)$ and $f(x + h)$, i.e. $f(x - h) - f(x)$ and $f(x + h) - f(x)$ are of the same sign, i.e. +ive.

Now by Taylor's Theorem,

$$f(x + h) - f(x) = hf'(x) + \frac{h^2}{2!}f''(x) + \frac{h^3}{3!}f'''(x) + \frac{h^4}{4!}f^{iv}(x) + \dots$$

$$f(x - h) - f(x) = -hf'(x) + \frac{h^2}{2!}f''(x) - \frac{h^3}{3!}f'''(x) + \frac{h^4}{4!}f^{iv}(x) - \dots$$

Now when h is made sufficiently small the sign of the right hand side of each equation in (1) and therefore of the left hand side is ultimately dependent upon that of $hf'(x)$, that being the term of lowest degree in h. But both $f(x + h) - f(x)$ and $f(x - h) - f(x)$ are of opposite signs whereas they should have the same sign in $f(x)$ is either Max. or Min. It is therefore necessary that $f'(x)$ should vanish so that the lowest term of the right hand side of equation in (1) should depend upon an even power of h.

Hence $f'(x) = 0$ is the essential condition for the occurrence of Max. and Min. values of $f(x)$.

Let the roots of $f'(x) = 0$ be a, b, c etc.

Consider one of the roots of the above equation say b; then putting $x = b$ and $f'(x) = 0$ at $x = b$ in (1), we get

$$f(b + h) - f(b) = \frac{h^2}{2!}f''(b) + \frac{h^3}{3!}f'''(b) + \frac{h^4}{4!}f^{iv}(b) + \dots$$

$$f(b - h) - f(x) = \frac{h^2}{2!}f''(b) - \frac{h^3}{3!}f'''(b) + \frac{h^4}{4!}f^{iv}(b) - \dots$$

Now if $f''(b)$ is –ive, then clearly $f(b + h)$ and $f(b - h)$ are both less than $f(b)$ so that $f(x)$ is maximum. at $x = b$. In case $f''(b)$ is positive, then both $f(b + h)$ and $f(b - h)$ are greater than $f(b)$, showing that $f(x)$ is minimum. at $x = b$.

SOLVED EXAMPLES

Example 1. *Show that the Max. and Min. values of the biquadratic* $ax^4 + 4bx^3 + 6cx^2 + 4dx + e$ *are the roots of the equation* $a^2k^3 - 3(a^2I - 9H^2)k^2 + 3(aI^2 - 18HJ)k - \Delta = 0$, *where* Δ *is discriminant of the quadratic.*

Solution. We know that the discriminant Δ of the quadratic is $I^3 - 27J^2$, where $I = ae - 4bd + 3c^2$ and $J = ace + 2bcd - ad^2 - b^2e - c^3$.

Let $f(x) = ax^4 + 4bx^3 + 6cx^2 + 4dx + e$.

Now if the curve $y = f(x)$ be moved parallel to the axis of x through a distance k, where k is the max. or min. value of $f(x)$, then x-axis will become a tangent, *i.e.*, the two values of x will coincide. Now by moving the curve parallel to x-axis through a distance k, y becomes $y - k$. Then the equation $f(x) - k = 0$ will have two roots equal, the condition for which is that its discriminant should be zero.

$$f(x) - k \equiv ax^4 + 4bx^3 + 6cx^2 + 4dx + (e - k) = 0$$

Its discriminate is $I'^3 - 27J'^2$, where in the usual values of I and J, we have to put e equal to $e - k$ in order to get I' and J'.

$$\therefore \qquad I' = a(e - k) - 4bd + 3c^2 = I - ak.$$
$$J' = ac(e - k) + 2bcd - ad^2 - b^2(e - k) - c^3$$
$$= J - (ack - b^2k) = J - kH. \qquad\qquad \because H = ac - b^2$$
$$\therefore \qquad I'^3 - 27J'^2 = 0 \text{ gives}$$
$$I^3 - 3I^2ak + 3Ia^2k^3 - a^3k^3 - 27J^2 + 54JkH - 27k^2H^2 = 0$$

Cancelling minus sign and putting $I^3 - 27J^2 = \Delta$, we get

$$a^3k^3 - 3(a^2I - 9H^2)k^2 + 3(aI^2 - 18HJ)k - \Delta = 0.$$

The above equation gives the value of k.

12.14. ROLLE'S THEOREM

Between two consecutive real roots of the equation $f(x) = 0$ *there lies at least one real root of the equation* $f'(x) = 0$.

Proof. If a and b are consecutive roots of $f(x) = 0$ and as x varies from a to b, $f(x)$ varies continuously from $f(a)$ to $f(b)$. It will therefore vary either by increasing first and then decreasing or by decreasing first and then increasing. It must therefore pass through at least one (in general odd) max. or min. values during its variation from $f(a)$ to $f(b)$ and let α be the value of x between a and b where $f(x)$ is either max. or min. and from the definition of max. or min. $f'(x) = 0$ at $x = \alpha$. Thus a number α lying between a and b is a root of equation $f'(x) = 0$.

DEDUCTION. *To prove that between any two consecutive roots of the equation* $f'(x) = 0$ *there lies at the most one real root of the equation* $f(x) = 0$ *or they may not lie any.*

Proof. Let x_1 and x_2 be the two consecutive roots of the equation $f'(x) = 0$ lying between the intervals (α, β) and (β, γ) where α, β, γ are three consecutive roots of the equation $f(x) = 0$ \qquad\qquad [Rolle's Theorem]

Hence $\alpha < x_1 < \beta < x_2 < \gamma$. This relation shows that β, a root of $f(x) = 0$, lies between x_1 and x_2 the consecutive roots of $f'(x) = 0$

Uniqueness. If possible, it is suppose that there are two roots β and β' of the equation $f(x) = 0$ which lie between x_1 and x_2 the consecutive roots of $f'(x) = 0$, so that $x_1 < \beta < \beta' < x_2$; then β and β' being consecutive roots of $f(x) = 0$ must have between them a root of the equation $f'(x) = 0$ which is contrary to the supposition that x_1 and x_2 are consecutive roots of $f(x) = 0$. Hence there cannot lie more than one root of $f(x) = 0$ between two consecutive root of the equation $f'(x) = 0$. In this case $f(x_1)$ and $f(x_2)$ should be of opposite signs and if they are of the same sign, then no root of $f(x) = 0$ will lie between x_1 and x_2.

SOLVED EXAMPLES

Example 1. *The equation $x^4 - 15x^3 + 75x^2 - 145x + 84 = 0$ has its roots $1, 2, 3, 4, 7$. Locate the roots of the equation $4x^3 - 45x^2 + 150x - 145 = 0$.*

Solution. We have $\qquad f(x) = x^4 - 15x^3 + 75x^2 - 145x + 84 = 0$, then

$\Rightarrow \qquad\qquad f'(x) = 4x^3 - 45x^2 + 150x - 145 = 0.$

Now the roots of $f(x) = 0$ are $1, 3, 4, 7$ arranged in order and by Rolle's theorem we know that between the consecutive roots of the equation $f(x) = 0$ there lies at least one real root of the equation $f'(x) = 0$. Hence there will be a root each of $f'(x) = 0$ between 1, $3; 3, 4$ and $4, 7$.

12.15 LIMITS OF THE ROOTS OF AN EQUATION

1. **Superior or upper limit of positive roots :** A number which is greater than all the positive roots of a given equation is called upper or superior limit of the positive roots of that equation.

2. **Inferior or lower limit of positive roots :** A number which is less than all the +ive roots of a given equation is called the inferior or lower limit of the positive roots of that equation.

3. **Superior or upper limit of negative roots :** A number which is greater than all the negative roots of a given equation is called an upper or superior limit of the negative roots of that equation. In other words, superior limit of negative roots of a given equation is that negative number below which (numerically) lie all the –ive roots.

4. **Inferior or lower limit of negative roots :** A number which is less than all the negative roots of a given equation is called the inferior or lower limit of negative roots of that equation. In other words, inferior limit of the –ive roots of a given equation is that negative number above which (numerically) lie all the –ive roots.

From the above definitions we can say that the superior limit of the real roots of a given equation is that number which is greater than all the real roots of that equation and the inferior limit of the real roots in that number which is less than all the real roots of that equation. Thus superior limit of positive roots is the superior limit of the real roots and the inferior limit of the negative roots is the inferior limit of the real roots of a given equation.

THEOREM-1

If the polynomial $a_0x^n + a_1x^{n-1} + a_2x^{n-2} + \ldots + a_{n-1}x + a_n$

the value of $\dfrac{a_k}{a_0} + 1$ *or any greater value be substituted for x where* a_k *is that one of the coefficient* $a_1, a_2, \ldots a_n$ *whose numerical value is greatest, irrespective of sign, the term containing the highest power of x will exceed the sum of all the terms which follow.*

Proof. We have a_0x^n is the term containing the highest power of x and it will exceed the sum of all the terms that follows it if

$$a_0x^n > a_1x^{n-1} + \ldots + a_2x^{n-2} + a_{n-1}x + a_n.$$

The above inequality is satisfied for any value of x which makes

$$a_0x^n > a_k(x^{n-1} + x^{n-2} + \ldots + x + 1)$$

where a_k is the greatest among the coefficients $a_1, a_2, \ldots a_n$ without any regard to sign

or $\qquad a_0x^n > a_k\left(\dfrac{x^n - 1}{x - 1}\right) \qquad$ (Sum of G.P.)

or $\qquad x^n > \dfrac{a_k}{a_0}\left(\dfrac{x^n - 1}{x - 1}\right)$

Now $x^n > x^n - 1$ and hence above will be satisfied if

$$\dfrac{a_k}{a_0(x-1)} \le 1 \quad \text{or} \quad a_k \le a_0(x-1)$$

or $\qquad x - 1 \ge \dfrac{a_k}{a_0} \quad \text{or} \quad x \ge \dfrac{a_k}{a_0} + 1$

THEOREM-2

If in the polynomials $a_0x^n + a_1x^{n-1} + \ldots + a_{n-1}x + a_n$ *the value* $\dfrac{a_n}{a_n + a_k}$ *or any smaller value be substituted for w where* a_k *is the greatest coefficient exclusive of* a_n, *the term* a_n *will numerically be greater than sum of all the others.*

Proof. Let us put $x = \dfrac{1}{y}$ in the given polynomial so that it becomes

$$\dfrac{1}{y^n}[a_ny^n + a_{n-1}\cdot y^{n-1} + a_{n-2}y^{n-2} + \ldots + a_1y + a_0]$$

Hence by Theorem 1, we have

$$a_ny^n < a_{n-1}y^{n-1} + a_{n-2}y^{n-2} + \ldots + a_1y + a_0$$

for all values of $y \ge \left(\dfrac{a_k}{a_n} + 1\right)$ where a_k is numerically the greatest of all the coefficients

$a_0, a_1, a_2, \ldots a_{n-1}$.

Dividing by y^n, we get from (1),

$$a_n > \frac{a_{n-1}}{y} + \frac{a_{n-2}}{y^2} + \ldots + \frac{a_0}{y^n} \qquad \ldots(2)$$

or

$$> a_k \left(\frac{a}{y} + \frac{1}{y^2} + \ldots + \frac{1}{y^n} \right)$$

or

$$> a_k \frac{1}{y} \left\{ \frac{1 - \left(\frac{1}{y^n} \right)}{1 - \left(\frac{1}{y} \right)} \right\} \qquad \text{or} \qquad > \frac{a_k}{y^n} \cdot \left(\frac{y^n - 1}{y - 1} \right)$$

or

$$y^n > \frac{a_k}{a_n(y-1)} (y - 1)$$

Now y^n is greater than y^{n-1} and hence above will be satisfied if

$$\frac{a_k}{a_n(y-1)} \le 1 \qquad \text{or} \qquad a_k \le a_n(y-1)$$

or

$$y - 1 \ge \frac{a_k}{a_n} \qquad \text{or} \qquad y \ge \frac{a_k}{a_n} + 1 \qquad \text{or} \qquad y \ge \frac{a_k + a_n}{a_n}$$

or

$$\frac{1}{x} \ge \frac{a_k + a_n}{a_n} \qquad \text{or} \qquad x \le \frac{a_n}{a_k + a_n}$$

Hence the value of x equal to or less than $\dfrac{a_n}{a_k + a_n}$ will make

$$a_n > a_{n-1}x + a_{n-2}x^2 + \ldots + a_0x^n \quad \text{[from (2)]}$$

12.16 METHOD OF FINDING THE UPPER LIMIT OF THE ROOTS

THEOREM-1

In any equation

$$x^n + p_1 x^{n-1} + p_2 x^{n-2} + \ldots + p_{n-1}x + p_n = 0$$

if the first negative terms occur in $(r + 1)$th term, i.e. $- p_r x^{n-r}$ and if the greatest negative coefficient be k^{th}, say $- p_k$, then $k^{1/r} + 1$ or $(p_k)^{1/r} + 1$ is a superior limit of positive roots.

Proof. Any value of x which makes

$$x^n > p_k(x^{n-r} + x^{n-r-1} + \ldots + x + 1) \qquad \ldots(1)$$

where p_k is the greatest negative coefficient and $p_r x^{n-r}$ is the first negative term, will make $f(x)$ positive.

Summing a G.P., we have from (1), taking $x > 1$,

$$x^n > p_k \frac{x^{n-r+1} - 1}{x - 1} \qquad \text{or} \qquad x^n(x - 1) > p_k(x^{n-r-1})$$

or

$$x^n(x - 1) > p_k \frac{x^n}{x^{r-1}}$$

or $\qquad x^{r-1}(x-1) > p_k$ $\qquad\qquad\qquad\qquad\qquad\qquad\qquad$ ...(2)

·Now $\qquad x^{r-1} > (x-1)^{r-1}, x > 1.$

Hence the inequality is satisfied if we write

$$(x-1)^{r-1}(x-1) \geq p_k \quad \text{or} \quad (x-1)^r \geq p_k$$

or $\qquad x - 1 \geq (p_k)^{1/r} \quad \text{or} \quad x \geq 1 + (p_k)^{1/r}$

Remark

❖ In case $f(x) = a_0 x^n + a_1 x^{n-1} + ... + a_n = 0$, we can divide by a_0 to make the coefficient of highest term unity; then p_k will be $\dfrac{a_k}{a_0}$; so the superior limit

will be $1 + \left(\dfrac{a_k}{a_0}\right)^{1/r}$

SOLVED EXAMPLES

Example 1. *Find the superior limit of the positive roots of the equation*

$$x^4 - 5x^3 + 40x^2 - 8x + 23 = 0$$

Solution. Here the first negative term is 2nd, i.e. $r + 1 = 2$.

$$r = 1$$

Also the greatest –ive coefficient $p_k = 8$.

∴ Superior limit of +ive roots is

$$1 + (p_k)^{1/r} = 1 + (8)^1 = 0$$

Example 2. *Find a superior limit of the positive roots of the equation*

$$x^8 + 20x^7 + 4x^6 + 11x^5 - 120x^4 + 13x - 25 = 0$$

Solution. Here $p_k = 120$, $(r + 1)$ the term = first –ive term = 4th term.

∴ $r = 3$

Hence the superior limit is $\geq (p_k)^{1/r} + 1 \geq (120)^{1/3} + 1$ *i.e.,* 6.

THEOREM-2

If in any equation each negative coefficient be taken positive, and divided by sum of all the +ive coefficients which precede it, the greatest quotient thus formed increased by unity is a superior limit of positive roots.

Proof. Let $f(x) = a_0 x^n + a_1 x^{n-1} + a_2 x^{n-2} + ... + a_r x^{n-r} + ... + a_n = 0$ $\qquad\qquad$ ...(1)

Let us regard the rth coefficient as negative and in general the negative coefficient be $-a_r$.

Now we know that $\dfrac{x^m - 1}{x - 1} = (x^{m-1} + x^{m-2} + ... + x + 1)$

so that $x^m = (x - 1)(x^{m-1} + x^{m-2} + ... + x + 1) + 1$

or we can write

$$a_m x^m = a_m(x - 1)(x^{m-1} + x^{m-2} + ... + x + 1) + a_m \qquad\qquad ...(2)$$

We shall now express every positive term of (1) by the help of (2).

i.e.
$$a_0 x^n = a_0(x-1)(x^{n-1} + x^{n-2} + \dots + x + 1) + a_0$$
$$= a_0(x-1)x^{n-1} + a_0(x-1)x^{n-2} + \dots$$
$$\dots + a_0(x-1)x^{n-r} + \dots + a_0$$

Therefore,
$$f(x) = a_0(x-1)x^{n-1} = a_0(x-1)x^{n-2} + a_2(x-1)x^{n-3}$$
$$+ \dots + a_0(x-1)x^{n-r} + a_0$$
$$+ a_1(x-1)x^{n-2} + a_1(x-1)x^{n-3} + \dots + a_1(x-1)x^{n-r} + a_1$$
$$+ a_2(x-1)x^{n-3} + \dots + a_2(x-1)x^{n-r} + \dots + a_2$$
$$- a_3(x-1)x^{n-3} + \dots - a_r x^{n-r} + \dots + a_n.$$

We have not changed the negative coefficients and the fourth coefficient being taken as negative and in general a_r is negative coefficient.

Adding above, we get
$$f(x) = a_0(x-1)x^{n-r} + (a_0 + a_1)(x-1)x^{n-2} + [(a_0 + a_1 + a_2)(x-1) - a_3]x^{n-3}$$
$$+ [(a_0 + a_1 + a_2 + \dots + a_{r-1})(x-1) - a_r]x^{n-r} + \dots (a_0 + a_1 + a_2 + \dots + a_n)$$

Now any value of $x > 1$ is sufficient to make +ive every term in which no –ive coefficient a_3, a_r etc. occur. The term containing these –ive coefficients will be +ive if

$$(a_0 + a_1 + a_2)(x-1) - a_3 > 0$$

and
$$(a_0 + a_1 + a_2 + \dots a_{r-1})(x-1) - a_r > 0$$

or
$$x > \frac{a_3}{a_0 + a_1 + a_2} + 1 \qquad \dots(A)$$

$$\dots \qquad \dots \qquad \dots \qquad \dots$$

or
$$x > \frac{a_r}{a_0 + a_1 + a_2 + \dots + a_{r-1}} + 1$$

We should choose the greatest of the values of x given by relations in (A) so that every term of $f(x)$ is +ive. This greatest value of x found from (A) will be the superior limit.

SOLVED EXAMPLES

Example 1. *Find a superior limit of the positive roots of*
$$x^7 + 4x^6 - 3x^5 + 5x^4 + 9x^3 - 11x^2 + 6x - 8 = 0$$

Solution. Superior limit of positive roots
$$= \frac{\text{The Ist negative coefficient taken positive}}{\text{Sum of all the positive coefficients that precede it}}$$

The greatest number obtained by above rule is the superior limit of positive roots.

$$\frac{3}{1+4} + 1, \quad \frac{9}{1+4+5} + 1, \quad \frac{11}{1+4+5} + 1, \quad \frac{8}{1+4+5+6} + 1$$

The greatest expression is $\dfrac{11}{1+4+5} + 1$, i.e. > 2.

Hence 3 is the superior limit of +ive roots.

Example 2. *Find the superior limit of the positive roots of*
$$x^8 + 20x^7 + 4x^6 - 11x^5 - 120x^4 + 13x - 25 = 0$$

Solution. Proceed same as in example-1, we get

$$\frac{11}{1+20+4}+1, \quad \frac{120}{1+20+4}+1, \quad \frac{25}{1+20+4+13}+1,$$

The greatest expression is $\dfrac{120}{25}+1>5$. Hence the superior limit is 6.

THEOREM-3

The upper limit of +ive roots of $f(x) = 0$ in which the leading coefficients is unity is equal to the numerically greatest coefficient increased by one.

Proof. Let $f(x) = x^n + p_1 x^{n-1} + p_2 x^{n-2} + ... + p_{n-1} x + p_n = 0$

Suppose $x > 1$; then $f(x) > 0$ if $x^n > p(x^{n-1} + x^{n-2} + ... + 1)$

where p is the greatest negative coefficient

$$x^n > p\frac{x^n - 1}{x-1} \qquad \qquad \text{[Sum of G.P.]}$$

or if $\qquad \qquad x^n > \dfrac{x^n}{x-1} \qquad \qquad$ or if $(x-1) > p$

or $\qquad \qquad x > p+1$

THEOREM-4 (NEWTON'S METHOD)

If h is a number which makes $f(x)$ and all its derivatives positive then h is an upper limit of the positive roots of $f(x) = 0$, it is also the upper limit of real roots of $f(x) = 0$.

Proof. Diminish the roots of the equation by h, so that

$$y = x - h \quad \text{or } x = y + h$$

$\therefore \qquad \qquad f(y + h) = 0$

or $\qquad f(h) + yf_1(h) + \dfrac{y^2}{2!} f_2(h) + ... + \dfrac{y^n}{n!} f_n(h) = 0 \qquad \qquad$ ..(1)

Now if h is a number such that $f(h), f_1(h), f_2(h), ...$ are all positive then $f(y + h) > 0$ and as such the equation $f(y + h) = 0$ cannot have a positive root. In other words, it means that the equation $f(x) = 0$ has no root greater than h, therefore h is the superior limit of positive roots.

WORKING PROCEDURE

Write down all the derivatives. Take the smallest integer which makes the last but one derivative positive and proceed with the values of x to find that the other derivatives written before are also positive. If any of the derivatives become negative for a particular value of x being tried, increase the integer by unity successively till it makes all the functions positive.

SOLVED EXAMPLES

Example 1. *Find by Newton's method the superior limit of positive roots of the equation*
$x^4 - 2x^3 - 3x^2 - 15x - 3 = 0$.

Solution. We have
$$f(x) = x^4 - 2x^3 - 3x^2 - 15x - 3,$$
$$f_1(x) = 4x^3 - 6x^2 - 6x - 15,$$
$$f_2(x) = 12x^2 - 12x - 6,$$
$$f_3(x) = 24x - 12,$$
$$f_4(x) = 24x,$$

Clearly $x = 1$ makes $f_3(x)$ positive but it makes $f_2(x)$ negative. Now increase the value of x by 1, i.e. try $x = 2$, it makes $f_2(x)$ positive. Now check $f_1(x)$ for $x = 2$ which is negative for $x = 2$ increase again by one the value of x, i.e. $x = 3$ now makes $f_1(x)$ positive. Try $x = 3$ on $f(x)$ and it becomes negative. Increase the value of x by 1, i.e. $x = 4$ and it makes $f(x)$ positive. Thus 4 is the value of x which makes $f(x)$ and all its derivatives positive. Hence 4 is the superior limit of positive roots of $f(x) = 0$.

Example 2. *Find a superior limit of positive roots of the equation*
$$x^5 + 4x^4 - 2x^3 + 10x^2 - 2x - 962 = 0$$

Solution. We have
$$f(x) = x^5 + 4x^4 - 2x^3 + 10x^2 - 2x - 962,$$
$$f_1(x) = 5x^4 + 16x^3 - 6x^2 + 20x - 2,$$
$$f_2(x) = 20x^3 + 48x^2 - 12x + 20,$$
$$f_3(x) = 60x^2 + 96x - 12,$$
$$f_4(x) = 120x + 96,$$
$$f_5(x) = 120,$$

$f_4(x)$ is positive for $x = 1$; $f_3(x)$, $f_2(x)$, $f_1(x)$ are all positive for $x = 1$ but $f(x)$ is negative for $x = 1$. Increase the value of x by one and try $x = 2$ on $f(x)$, $f(x)$ is again negative for $x = 2$. Again try $x = 3$; $f(x)$ is negative for $x = 3$. Now try $x = 4$ on $f(x)$ and it becomes positive. Thus $x = 4$ makes $f(x)$ and its derivatives positive. Hence 4 is the superior limit of positive roots.

12.16.1 Method of Grouping

This method consists in grouping the terms of the given equation and if h be the least number which makes the sum of terms in each group positive, then h is the superior limit of positive roots. This method will give a closer limit than the preceding methods which will given quite big number for superior limit. Group negative term with positive term.

SOLVED EXAMPLES

Example 1. *Find a superior limit of positive roots of*
$$x^5 + 3x^4 + x^3 - 8x^2 - 51x + 18 = 0$$
by method of grouping.

Solution. We have $f(x) = x^2(x^3 - 8) + 3x(x^3 - 17) + x^3 + 18 = 0$, $x = 3$ makes the sum of terms in each group positive.

Hence 3 is the superior limit of positive roots.

Example 2. *Find a superior limit of the positive roots of* $x^3 - 10x^2 - 11x - 100 = 0$

Solution. Here the number of negative terms is greater than positive terms. In such cases multiply the given equation by suitable number and distribute positive terms among the various negative terms.

We have $3f(x) = 3x^3 - 30x^2 - 33x - 300$.

Write $3x^3$ as $x^3 + x^3 + x^3$

$\therefore$ $3f(x) = x^2(x - 30) + x(x^2 - 33) + (x^3 - 300)$.

Clearly any number greater than or equal to 30 makes the sum of terms in each group positive. Hence 30 is a superior limit.

If we multiply $f(x)$ by 6,

$$6f(x) = 6x^3 - 60x^2 - 66x - 600$$
$$= 4x^2(x - 15) + x(x^2 - 66) + (x^3 - 600)$$

Here we find that any value of x greater than or equal to 15 makes the sum of terms in each group positive. Hence 15 is still a closer superior limit of positive roots of $f(x) = 0$.

Example 3. *Find a superior limit of positive roots of the equation*
$$5x^5 - 7x^4 - 10x^3 - 23x^2 - 90x - 317 = 0$$

Solution. Distribute $5x^5$ with each negative term,
$$f(x) = x^4(x - 7) + x^3(x^2 - 10) + x^2(x^3 - 23) + x(x^4 - 90) + (x^5 - 317)$$

Clearly $x \geq 7$ makes the sum of terms in each group positive. Hence 7 is the superior limit of positive roots.

Example 4. *Find a superior limit of positive roots of the equation*
$$x^4 - 4x^3 + 33x^2 - 2x + 18 = 0$$

Solution. We have $f(x) = x^4 - 4x^3 + 5x^2 + 28x^2 + 2x + 18$

We have splitted $33x^2$ into $5x^2 + 28x^2$.

$$f(x) = x^2(x^2 - 4x + 5) + 28x\left(x - \frac{1}{14}\right) + 18.$$

The first group is a quadratic $x^2 - 4x + 5$ whose roots are imaginary, i.e. $b^2 - 4ac$ is $16 - 20$, i.e. negative and hence it is positive for all real values of x. The other group $x - \dfrac{1}{14}$ is positive for $x = 1$. Hence 1 is the superior limit of positive roots.

Example 5. *Find a superior limit of positive roots of the equation*
$$x^4 - x^3 - 2x^2 - 4x - 24 = 0.$$

Solution. We have $4f(x) = 4x^4 - 4x^3 - 8x^2 - 16x - 96$

Write $4x^4$ as $x^4 + x^4 + x^4 + x^4$

$x^3(x - 4) + x^2(x^2 - 8) + x(x^3 - 16) + (x^4 - 96)$

Any value of $x \geq 4$ makes the sum of all the terms in each group positive. Hence 4 is the superior limit of positive roots.

12.17 LOWER LIMIT DEDUCED FROM UPPER LIMIT

1. If k is upper limit of the positive roots of the equation $f(-x) = 0$, then $-k$ is the lower limit to the real roots of $f(x) = 0$, because negative roots of $f(x) = 0$ are positive roots of $f(-x) = 0$

2. If k is the upper limit of +ive roots of $f\left(\dfrac{1}{x}\right) = 0$, then $\dfrac{1}{k}$ is the lower limit of the positive roots of $f(x) = 0$

3. Limit of negative roots reduced from the limit of positive roots.

Remark

❖ We know that negative roots of $f(x) = 0$ are positive roots of $f(-x) = 0$. Then if h and k are the upper and lower limits of the positive roots of $f(-x) = 0$, then $-h$, $-k$, are the lower and upper limits respectively of the negative roots of $f(x) = 0$.

SOLVED EXAMPLES

Example 1. *Find a superior and inferior limit of positive and negative roots of*
$$x^5 + 4x^4 + 2x^3 - 18x^2 - 36x - 72 = 0$$

Solution. Superior limit of +ive roots is the greatest of the integers given by
$$\frac{18}{1+4+2}+1, \frac{36}{1+4+2}+1, \frac{72}{1+4+2}+1,$$

i.e. $11\dfrac{2}{7}$ or 12 is the superior limit of positive roots.

Lower or inferior limit of positive roots.

Putting $x = \dfrac{1}{y}$ in the given equation, we get
$$72y^2 + 36y^4 + 18y^3 - 2y^2 - 4y - 1 = 0$$

Superior limit of y = reciprocal of the inferior limit of x.

By Theorem-2 superior limit of y is the greatest of the integers given by
$$\frac{2}{72+36+18}+1, \frac{4}{72+36+18}+1, \frac{1}{72+36+18}+1$$

$$\frac{4}{126}+1 \qquad \text{or} \qquad \frac{65}{63}$$

∴ lower limit of positive roots of $f(x)$ is $\dfrac{65}{63}$ or 1.

Limits of negative roots.
Put $x = -y$ in $f(x)$ and we get
$$y^5 - 4y^4 + 2y^3 + 18y^2 - 36y + 72 = 0 \qquad \qquad \text{...(1)}$$

Clearly superior limit of y is $\dfrac{4}{1}+1 = 5$

Hence the inferior limit of negative roots of $f(x)$ is -5.

Putting $y = \dfrac{1}{z}$ in (1), we get

$$72z^5 - 36z^4 + 18z^3 + 2z^2 - 4z + 1 = 0$$

Superior limit of z = reciprocal of inferior limit of y.

By theorem-2, superior limit of z is $\dfrac{36}{72} + 1 = \dfrac{3}{2}$.

Hence $\dfrac{2}{3}$ is the lower limit of positive roots of y in (1),

∴　　　　by definition $-\dfrac{2}{3}$ is the superior limit of negative roots of $f(x)$.

Hence the negative root lies between $\left(-\dfrac{2}{3}, -5 \right)$

In general all the real root lie between

$$-5 \text{ and } 11\dfrac{2}{7} \quad \text{or} \quad -5 \text{ or } 12.$$

EXERCISE 12.4

1. Show that the maximum and minimum values of the cubic $ax^3 + 3bx^2 + 3cx + d = 0$ are the roots of the equation $a^2b^2 - 2Gp + \Delta = 0$ where Δ is the discriminant.

2. Find a superior limit of the roots of the equation
   (i)　　$x^4 - 2x^3 + 3x^2 - 5x + 1 = 0$
   (ii)　　$x^4 + 4x^3 - 11x^2 - 9x - 50 = 0$

3. Find the superior limit of the positive roots of
   (i)　　$x^4 - 2x^3 + 3x^2 - 5x + 1 = 0$
   (ii)　　$x^4 + 4x^3 - 11x^2 - 9x - 50 = 0$

4. Find the superior limit of positive roots of the following equations by Newton's method
   (i)　　$x^4 - 4x^3 - 3x + 23 = 0$
   (ii)　　$x^4 - 2x^3 - 3x^2 - 15x - 3 = 0$
   (iii)　　$x^5 + 6x^4 - 10x^3 - 112x^2 - 207x - 110 = 0$

5. Find the limits between which the real roots of the equation
   $$x^5 + 5x^4 + x^3 - 16x^2 - 20x - 16 = 0 \text{ lie.}$$

ANSWERS

(2)　(i)　6　　　(ii)　8　　　　　　　　　(3)　(i)　3　　　(ii)　11

(4)　(i)　4　　　(ii)　4　　　(iii)　5

(5)　positive roots lies between $\dfrac{52}{57}$ and $3\dfrac{6}{7}$.

　　　negative root lies between $-\dfrac{4}{9}$ and -6.

12.18 STURM'S METHOD OF FINDING THE EXACT NUMBER OF REAL ROOTS OF AN EQUATION

By Descarte's rule of signs we cannot get the exact number of real roots of a given equation $f(x) = 0$. Sturm's Theorem gives us the exact number of real roots of a given equation.

12.19 STURM'S FUNCTIONS

Let $f(x)$ be any function of x of degree n and $f'(x)$ be its first derivative. Divide $f(x)$ by $f'(x)$ and let the remainder with sign changed be denoted by $f_2(x)$.

Again divide $f'(x)$ by $f_2(x)$ and let the remainder with sign changed be denoted by $f_3(x)$.

Continue the process till you get the last remainder whose sign is also to be changed.

The above process is the same as that of finding the H.C.F. of $f(x)$ and $f'(x)$ with the modification that the sign of each remainder is to be changed before it becomes the divisor. Also we know that in the process of finding the H.C.F. we can multiply and divide any remainder by any constant before using it as a divisor; but here in the above process we can only multiply or divide the remainder only by a positive constant before using it as a divisor or by a polynomial of x which is always positive for real values of x say of the type $x^2 + 1$ or $x^4 + x^2 + 1$ etc.

The series of functions

$$f(x), f'(x), f_2(x), f_3(x), \ldots f_r(x)$$

consisting of the given function, its derivative and remainder with their sign changed in process of finding the H.C.F. of $f(x)$ and $f'(x)$ are called Sturm's functions.

The functions $f_2(x), f_3(x), \ldots f_r(x)$ are called auxiliary functions.

(i) $f(x) = 0$ having equal roots.

In case $f(x) = 0$ has equal roots, then we know that $f(x)$ and $f'(x)$ will have some H.C.F. and hence the last of the Sturm's remainder will be a function of x.

(ii) $f(x) = 0$ having no equal roots.

In this case evidently the last of the Sturm's remainder will be numerical for if it is some function of x, then it would mean that $f(x)$ has got equal roots. In this case there will be $(n + 1)$ Sturm's function.

i.e. $\qquad f(x), f'(x), f_2(x), f_3(x), \ldots f_n(x).$

12.20 STURM'S THEOREM

Case-I All roots unequal

If $f(x)$ is a polynomial and a and b be any two real numbers, then the number of distinct real roots of the equation $f(x) = 0$ lying between a and b is exactly equal to the difference between the number of changes of sign when x is put equal to a and the number when x is put equal to b in the $(n + 1)$. Sturm's functions $f(x), f'(x), f_2(x), f_3(x), \ldots f_n(x)$ consisting of the given function, its derivative and the $(n - 1)$ remainders with their sign changed in the process of finding the H.C.F. of $f(x)$ and $f'(x)$.

From the definition of Sturm's functions we can establish the following relations between them :

Dividend = Q × Divisor + Remainder :

$$\left.\begin{aligned}
f(x) &= q_1 f'(x) - f_2(x), \\
f'(x) &= q_2 f_2(x) - f_3(x), \\
f_2(x) &= q_3 f_3(x) - f_4(x), \\
\cdots \quad \cdots \quad \cdots \quad \cdots \quad \cdots \\
\cdots \quad \cdots \quad \cdots \quad \cdots \quad \cdots \\
\cdots \quad \cdots \quad \cdots \quad \cdots \quad \cdots \\
f_{r-1}(x) &= q_r f_r(x) - f_{r+1}(x) \\
\cdots \quad \cdots \quad \cdots \quad \cdots \quad \cdots \\
f_{n-2}(x) &= q_{n-1} f_{n-1}(x) - f_n(x)
\end{aligned}\right\} \qquad \ldots(A)$$

From the relation (A) we have the following observations :

1. As $f(x) = 0$ has no equal roots, $f(x)$ and $f'(x)$ have no common factors and consequently $f_n(x)$ the last Sturm's function is numerical having a definite sign + or −.

2. No two consecutive auxiliary functions vanish for the same value of x. If possible, suppose that when $x = \alpha$, both $f_2(x)$ and $f_3(x)$ vanish which shows that $f'(x)$ contains the factor $x - \alpha$ and from first of the relations (A), we find that $(x - \alpha)$ is also a factor of $f(x)$. Thus $(x - \alpha)$ is the H.C.F. of $f(x)$ and $f'(x)$ showing the existence of equal roots which is contrary to the hypothesis.

3. If any of the auxiliary functions vanishes, then the two adjacent functions, *i.e.*, one which precedes it and the one which follows it must have opposite signs. For example suppose that $f_3(x) = 0$, when $x = \alpha$; then from relation (A), $f_2(x) = - f_4(x)$ for $x = \alpha$, showing that $f_2(x)$ and $f_4(x)$ have opposite signs for $x = \alpha$.

4. The same reasoning applies if any of the q's vanishes. Since all the functions are polynomials in x, no one can changes sign as x increases continuously from a to b, when x assumes a value which causes that function to vanish.

5. Now we shall show that when x in passing from a to b takes a value α such that $f_r(\alpha) = 0$, then no change of sign is gained or lost. Now since $f_r(\alpha) = 0$, we have from (3), $f_{r-1}(\alpha)$ and $f_{r+1}(\alpha)$ must be of opposite signs. Now as $f_r(x)$ passes through zero, it changes sign either from + to − or from − to +. But $f_{r-1}(x), f_{r+1}(x)$ are continuous at $x = \alpha$; so that each of them has an invariable sign near $x = \alpha$. Thus the three functions $f_{r-1}(x)$, $f_r(x)$ and $f_{r+1}(x)$ will have only one change of sign just before $x = \alpha$ and just after $x = \alpha$, *i.e.*, their signs can be either + +, −, − − + +, + − − , − + +.

Thus we find that whatever sign we may place between two unlike signs, we have only one change of sign. Hence no change of sign in either lost or gained among Sturm's functions.

In case the value of x, i.e. α be such that it causes more than one of the functions to vanish, then they cannot be consecutive, for in that case, by (2), the equation $f(x) = 0$ will have equal roots.

6. Now we shall show that when x is passing from a to b takes a value α which causes $f(x)$ to vanish, i.e. α be a root of $f(x) = 0$, then a change of sign is lost. Now by Taylor's Theorem,

$$f(\alpha - h) = 0 - hf'(\alpha) + \frac{h^2}{2!} f''(\alpha)... \quad \because f(\alpha) = 0$$

$$f(\alpha + h) = 0 + hf'(\alpha) + \frac{h^2}{2!} f''(\alpha)... \quad \because f(\alpha) = 0$$

Now let h be sufficiently small so that the sign of the L.H.S. is made to depend on the first term of R.H.S.

Therefore if $f'(\alpha)$ be positive, then $f(\alpha - h)$ is –ive and $f(\alpha + h)$ is +ive i.e. in this case the signs of $f(x)$ and $f'(x)$ will $- +$ just before $x = \alpha$ and $+ +$ just after $x = \alpha$. Thus one change of sign is lost. Again if $f'(x)$ be –ive, then $f(\alpha - h)$ is +ive and $f(\alpha + h)$ is –ive, i.e. in this case the signs of $f(x)$ and $f'(x)$ will be $+ -$ just before $x = \alpha$ and $- -$ just after $x = \alpha$. Here also one change of sign is lost. Thus we conclude that when x passes through a root α of the equation $f(x) = 0$, one change of sign is lost whether $f'(\alpha)$ be +ive or –ive. Hence the number of variations lost as x goes from a real value a to a real value b is exactly equal to the number of real root; of the equation $f(x) = 0$ between a and b.

Case-II Equal roots. If $f(x) = 0$ be an equation having equal roots and the Sturm's functions be found as $f, f', f_2, ... f_r$, the last of these being the H.C.F. of $f(x)$ and $f'(x)$, then the difference between the number of changes of signs when a and b are substituted in the Sturm's functions is equal to the number of real roots of the equation $f(x) = 0$ which lie between a and b each multiple root being counted once only.

Let $f(x) = (x - \alpha)^p (x - \beta)^q (x - \gamma)(x - \delta) ...$, then clearly $f(x)$ and $f'(x)$ will have an H.C.F. $(x - \alpha)^{p-1}(x - \beta)^{q-1}$ which may be denoted by H.

Thus H.C.F. will be a factor of all the Sturm's functions $f, f', f_2, f_3, ... f_r$. Again let $\psi(x)$ stand for

$$(x - \alpha)\ (x - \beta)\ (x - \gamma)\ (x - \delta)$$

then clearly $f(x) = H.\ \psi(x)$, $f'(x) = H.\ \psi'(x)$, $f_2(x) = H.\ \psi_2(x)...$ For any value of x, the number of changes of sign in the sequence of f's is the same as that in sequence of ψ's.

Now $\psi(x) = 0$ has all its roots unequal and all its roots are the same as those of $f(x) = 0$ only with the change that the multiple roots of $f(x) = 0$ occur in $\psi(x) = 0$ only once.

Now applying the reasoning of 1st case on $\psi(x) = 0$ we can say that difference between the number of changes of signs when x is put equal to a and b in the sequence of ψ's (hence of f's) represents exactly the number of real roots of the equation $\psi(x) = 0$ lying between a and b or the number of real roots of $f(x) = 0$ that lie between a and b but each multiple roots being counted only once.

TIME SAVING TRICKS:

Trick 1. When $f(x) = 0$ has no repeated root, i.e. the last or Sturm's function be numerical, and we are concerned only with its sign. In order to get its sign we put $f_{n-1}(x) = 0$ and find the value of x; then we know that for this value of x, $f_{n-2}(x)$ and $f_n(x)$ must have the opposite sign. Thus if the value of x obtained from $f_{n-1}(x) = 0$ makes $f_{n-2}(x)$ +ive, then $f_n(x)$ is –ive and if it makes $f_{n-2}(x)$ –ive, then $f_n(x)$ is +ive. This device saves us the labour of actually calculating the value of $f_n(x)$.

Trick 2. If any of the Sturmain function say $f_2(x)$ has all roots imaginary, we may stop further calculation and we should see $f, f', f_2, f_3, \ldots f_r$ functions only for Sturm's theorem, because in this theorem the last of the function should be of invariable sign for all real values of x and we know from the properties of equation that if $f(x) = 0$ has all its roots imaginary, then $f(x)$ is always +ive for all real values of x. The quadratic $ax^2 + bx + c = 0$ has its roots imaginary if $b^2 - 4ac < 0$.

Similarly the calculation of Sturmain functions will stop at the stage when any of them becomes a perfect squares for it too will have an in variable sign for all real values of x.

Trick 3. In case any of the Sturm's functions $f_r(x)$ vanishes for $x = \alpha$; then for counting the number of changes of sign in the sequence

$$f(\alpha), f'(\alpha), f_2(\alpha), f_3(\alpha), \ldots,$$

we may regard the sign of $f_r(\alpha)$ either +ive or –ive because the function that precedes it and the one which follows it have opposite signs.

WORKING PROCEDURE

Step 1. Find the Sturm's functions as explained.

Step 2. The last Sturm's function will be numerical in case $f(x) = 0$ has no equal roots. Otherwise it will be some function of x.

Step 3. The calculation of Sturm's functions should stop at the stage when any of them is either a perfect square or has all its roots imaginary.

Step 4. Make a table as explained below:

(i) In the first column write down

$$x, f, f', f_2, f_3, f_4, \ldots$$

In the first row write down x and the various values that you may give to x.

(ii) In the various other columns write down the signs of the Sturm's functions corresponding to the value of x written at the top that column.

(iii) At the bottom of each column write down the number of changes of signs in that particular column.

(iv) If we are to find only number of real roots, we put $x = \infty$, and $-\infty$ and find out the difference of the changes of signs corresponding to the two values.

(v) If we want to find out the +ive and –ive roots, we put $x = \infty$, 0 and $-\infty$ and proceeding as above, we get the number of +ive roots which will lie between 0 and ∞ –ive roots which lie between 0 and $-\infty$.

(vi) If we want to find interval in which the roots lie, then for +ive roots, we put $x = 1, 2, 3, \ldots$ and for –ive roots, we put $x = -1, -2, -3, \ldots$

(vii) If corresponding to the two values of x say 2 and 3 the changes of sign in Sturm's function be same, i.e., their differences be zero, then no root will lie between 2 and 3.

(viii) We know that if $f(a)$ and $f(b)$ are of opposite signs, then **at least one** or in general odd number of roots of the equation $f(x) = 0$ lie between a and b. In case they be of the same sign, then either no root or an even number of roots of $f(x) = 0$ lie between a and b.

In case there be only one +ive roots, then in order to find its location we need not put $x = 1, 2, 3 \ldots$ in all the Sturm's functions. We shall put only in $f(x)$ and in case of any two consecutive values of x the values of $f(x)$ are of opposite signs; then by the above theorem the root will lie between those two consecutive numbers. This will save us the labour of putting the consecutive numbers in all the series of Sturm's functions.

SOLVED EXAMPLES

Example 1. *Find the number and position of the real roots of the equation $x^6 - 2x^2 + 3x - 4 = 0$.*

Solution : We have

$$f(x) = x^6 - 2x^2 + 3x - 4$$
$$f'(x) = 6x^5 - 4x + 3.$$

Multiplying $f(x)$ by 6 and then dividing by $f'(x)$

$$6x^5 - 4x + 3 \overline{)6x^6 - 12x^2 + 18x - 24} (x$$

$6x^5 - 4x + 3) 6x^6 - 12x^2 + 18x - 24(x$

$$\dfrac{6x^5 - 4x^2 - 3x}{8x^2 + 15x - 24}$$

Changing the sign of this remainder, we get

$$f_2(x) = 8x^2 - 15x + 24.$$

Now since $b^2 - 4ac$, i.e. $(15)^2 - 4.8.24$ is –ive, we conclude that the roots of the Sturm's functions $f_2(x) = 0$ are imaginary and hence we stop further calculations.

x	$-\infty$	0	∞	1	2	-1	-2
$f(x)$	$+$	$-$	$+$	$-$	$+$	$-$	$+$
$f'(x)$	$-$	$+$	$+$	$+$	$+$	$+$	$-$
$f_2(x)$	$+$	$+$	$+$	$+$	$+$	$+$	$+$
No. of changes of sign	2	1	0	1	0	1	2

12.20.1 Nature of the Roots :

1. There are only two real roots (from column 2 and 4) as the difference of changes of signs when x is put $-\infty$ and ∞ in sturm's functions $2 - 0 = 2$.

2. One of them is +ive (from column 3 and 4) which lies between 1 and 2 (from columns 5 and 6).

3. One of them is –ive (from columns 2 and 3) which lies between -1 and -2 (from columns 7 and 8).

4. The equation being of sixth degree has only two real roots; therefore remaining four roots are imaginary.

Remark

❖ We have found above that the above equations has only one +ive root and only one –ive root. Also we know that if $f(a)$ and $f(b)$ be of opposite signs then at least one root and in general odd number of roots must lie between a and b.

Now $f(0) = -$ and $f(1) = -$.

From above we could easily conclude that +ive root does not lie between 0 and 1 as $f(0)$ and $f(1)$ are of the same sign. This would save us from calculating the signs of all the Sturm's functions corresponding to $x = 1$.

Again $f(1) = $ –ive, $f(2) = $ +ive. Since $f(1)$ and $f(2)$ are of opposite signs, hence the only +ive root of $f(x) = 0$ must lie between 1 and 2. This has saved us the labour of calculating the signs of all Sturm's functions corresponding to $x = 2$.

Again $f(0) = -$ and $f(-1) = -$.

Therefore arguing as above –ive root does not lie between 0 and -1.

Again $f(-1) = $ –ive and $f(-2) = $ +ive.

$\therefore$ –ive root lies between $-1, -2$

Hence the above chart could be easily put as

x	$-\infty$	0	∞	1	2	-1	-2
$f(x)$	$+$	$-$	$+$	$-$	$+$	$-$	$+$
$f'(x)$	$-$	$+$	$+$				
$f_2(x)$	$+$	$+$	$+$				
No. of changes of sign	2	1	0				

SOLVED EXAMPLES

Example 1. *Find the number and position of the real roots of the equation $x^3 - 7x + 7 = 0$.*
Solution : We have $f(x) = x^3 - 7x + 7, f'(x) = 3x^2 - 7$.

Let us find the H.C.F. of $f(x)$ and $f'(x)$ and change the sign of remainders to get Sturm's functions.

	$f(x)$	$f'(x)$	
	$x^3 - 7x + 7$	$3x^2 - 7$	
	3	2	
	$3x^3 - 21x + 21$	$6x^2 - 14$	
x	$3x^3 - 7x$	$6x^2 - 9x$	$3x$
	$-14x + 21$	$9x - 14$	
	$or - 2x + 3$	2	
	$\therefore f_2(x) = 2x - 3$	$18x - 28$	9
	after changing sign	$18x - 27$	
		-1	
		$\therefore f_2(x)$	
		after changing sign	

∴ Sturm's functions are

$$x^3 - 7x + 7, \qquad 3x^2 - 7, 2x - 3, 1$$

x	$-\infty$	0	∞	1	2	-1	-2	-3	-4
$f(x) = x^3 - 7x + 7$	$-$	$+$	$+$	$+$	$+$	$+$	$+$	$+$	$-$
$f'(x) = 3x^2 - 7$	$+$	$-$	$+$	$-$	$+$				
$f_2(x) = 2x - 3$	$-$	$-$	$+$	$-$	$+$				
$f_3(x) = 1$	$+$	$+$	$+$	$+$	$+$				
No. of changes of sign	3	2	0	2	0				

Above table shows that all the three roots are real out of which one is –ive and two are +ive which clearly lie between 1 and 2. Also the –ive root lies between – 3 and – 4. We have not tabulated the signs of all Sturm's functions for the only –ive root because $f(0)$, $f(-1)$, $f(2)$, $f(-3)$ are all of same signs and hence there arises no question of a negative root lying between these numbers. Again $f(-3) = $ +ive and $f(-4) = $ –ive and hence a root lies between – 3 and – 4.

Example 2. *Find the number of distinct real roots of the equation $x^3 - 3x + 1 = 0$ and locate them.*

Solution : We have $f(x) = x^3 - 3x + 1$,

$f'(x) = 3(x^2 - 1)$ or $(x^2 - 1)$ on dividing by 3.

$f_2(x) = 2x - 1$ [after changing sign of the remainder].

$f_3(x) = 3$ [after changing the sign].

x	$-\infty$	0	∞	1	2	-1	-2
$f(x)$	$-$	$+$	$+$	$+$	$+$	$+$	$-$
$f'(x)$	$+$	$-$	$+$	$+$ or $-$	$+$	$+$ or $-$	$+$
$f_2(x)$	$-$	$-$	$+$	$+$	$+$	$-$	$-$
$f_3(x)$	$+$	$+$	$+$	$+$	$+$	$+$	$+$
No. of changes of sign	3	2	0	1	0	2	3

As in example 1 the above table shows that the given equation has all its roots real and distinct out of which one is –ive and the other two +ive. The –ive root lies in the interval $(- 2, - 1)$ and one of the +ive roots lies in the interval $(1, 2)$ and the other in the interval $(0, 1)$.

Since $f(0) = $ +ive and $f(- 1) = $ +ive, i.e. they are of the same sign, hence the root does not lie between 0 and – 1 and as such we need not have completed wholly the column corresponding to $x = - 1$.

Again $f(- 1) = $ +ive and $f(- 2) = $ –ive. Hence the only –ive root lies between $(- 1, - 2)$ and we need not complete the column corresponding to $x = - 2$.

Example 3. *Apply Sturm's theorem to the analysis of the equation*

$$x^4 - 2x^3 - 3x^2 + 10x - 4 = 0.$$

Solution : We have $f(x) = x^4 - 2x^3 - 3x^2 + 10x - 4$,

$f'(x) = 2x^3 - 3x^2 - 3x + 5$, after cancelling 2.

$f_2(x) = 9x^2 - 27x + 11$, after changing sign.

$f_3(x) = -8x - 3$, after changing sign.

Now $f_4(x)$ will be numerical and we are concerned only with its sign.

Putting $f_3(x) = 0$, we get $x = -\dfrac{3}{8}$. This value of x makes $f_2(x) = 9 \cdot \dfrac{3}{64} + \dfrac{81}{8} + 11$, i.e. +ive

and we know that when $f_3(x) = 0$, then $f_2(x)$ and $f_4(x)$ are of opposite signs. Since $f_2(x)$ is +ive therefore $f_4(x)$ must be –ive.

Remark

❖ If we actually proceed to calculate the value of $f_4(x)$, it will be -1433 (after changing the sign).

x	$-\infty$	0	∞	1	-1	-2	3
$f(x)$	$+$	$-$	$+$	$+$	$-$	$-$	$+$
$f'(x)$	$-$	$+$	$+$				
$f_2(x)$	$+$	$-$	$+$				
$f_3(x)$	$+$	$-$	$-$				
$f_4(x)$	$-$	$-$	$-$				
No. of changes of sign	3	2	1				

From above table we observe that there are only $3 - 1 = 2$ real roots. One of them is +ive and the other –ive. Again $f(0) = -$ ive and $f(1) = +$ive, i.e. they are of opposite signs; so we conclude that the positive root lies between 0 and 1. Hence we need not complete the column corresponding to $x = 1$.

Again $f(0) = $ –ive and $f(-1) = $ –ive, i.e. they are of the same sign. Hence either no root or an even no. of roots of $f(x) = 0$, lie between 0 and -1; but as there is only one –ive root, as such it can not lie between 0 and -1; therefore we have not completed the column corresponding to $x = -1$. Similarly for column corresponding to $x = -2$.

Again $f(-2) = $ –ive and $f(-3) = +$ive.

As they are of opposite signs, therefore the only –ive root lies between -2 and -3, and we need not complete the column corresponding to $x = -3$.

Example 4. *Discuss the nature and position of the roots of the equation $x^4 - 12x^2 + 12x - 3 = 0$ by means of Sturmain functions.*

Solution : We have $f(x) = x^4 - 12x^2 + 12x - 3$,

$f'(x) = 4(x^3 - 6x + 3)$ or $x^3 - 6x + 3$, on dividing by 4.

$$
\begin{array}{r}
x^3 - 6x + 3 \overline{)x^4 - 12x^2 + 12x - 3}(x \\
\underline{x^4 - 6x^2 + 3x} \\
-6x^2 + 9x - 3
\end{array}
$$

changing the sign and dividing by 3.

$$f_2(x) = \overset{}{2x^2 - 3x + 1} \overline{)x^3 - 6x + 3}\, (x + 3$$
$$ 2$$

$$\underline{2x^3 - 12x + 6}$$

$$2x^3 - 3x^2 + 6$$
$$\underline{3x^2 - 13x + 6}$$
$$2$$

$$\underline{6x^2 - 26x + 12}$$
$$6x^2 - 9x + 3$$
$$\underline{- 17x + 9}$$

changing the sign

$$f_3(x) = 17x - 9.$$

Now $f_4(x)$ will be numerical and we are concerned only with its sign, $f_3(0) = 0$ gives

$x = \dfrac{9}{17}$, i.e. slightly $> \dfrac{1}{2}$ say $\dfrac{1}{2} + k$ where k is small. This value of x when substituted in $f_2(x)$

makes it

$$2 \left(\frac{1}{2} + k\right)^2 - 3\left(\frac{1}{2} + k\right) + 1 = 2k^2 - k$$

i.e. –ive for small values of k and hence $f_4(x)$ is +ive and it may be verified by actual calculation that $f_4(x) = 8$.

x	$-\infty$	0	∞	1	2	3	-1	-2	-3	-4
$f(x)$	+	–	+	–	–	+	–	–	–	+
$f'(x)$	–	+	+	–	–	+	+	+	–	–
$f_2(x)$	+	+	+	+ or –	+	+	+	+	+	+
$f_3(x)$	–	–	+	+	+	+	–	–	–	–
$f_4(x)$	+	+	+	+	+	+	+	+	+	+
No. of changes of sign	4	3	0	1	1	0	3	3	3	4

The above table shows that the given equation has all its real roots; one of them is -ive lying in the interval $(-4, -3)$. The other there are +ive lying in the interval $(2, 3)$ and two in the interval $(0, 1)$.

Iere we need not commute fully the columns corresponding to $x = -1, -2, -3, -4,$ for in all these cases $f(0), f(-1), f(-2), f(-3)$ are all –ive i.e. of the same sign and there being only one –ive root, it cannot lie between any of the two consecutive numbers. Again $f(-3) = $ –ive and $f(-4) = $ +ive, i.e. opposite signs, therefore the only –ive root lies between -3 and -4.

Example 5. *Find the number and position of roots of*
$$f(x) = x^4 - 3x^3 - 2x^2 + 7x + 3 = 0.$$

Solution : We have $f'(x) = 4x^3 - 9x^2 - 4x + 7,$
$$f_2(x) = 43x^2 - 72x - 69,$$
$$f_3(x) = 83x - 191,$$

$f_4(x)$ = +ive as shown

	$f(x)$	$f'(x)$	
	$x^4 - 3x^3 - 2x^2 + 7x + 3$	$4x^4 - 9x^2 - 4x + 7$	
	4	43	
x	$4x^4 - 12x^3 - 8x^2 + 28x + 12$	$172x^3 - 387x^2 - 172x + 301$	4
	$4x^4 - 9x^3 - 4x^2 + 7x$	$172x^3 - 288x^2 - 276x$	
	$-3x^3 - 4x^2 + 21x + 12$	$-99x^2 + 104x + 301$	
	4	43	
-3	$-12x^3 - 16x^2 + 84x + 48$	$-4257x^2 + 4472x + 12943$	-99
	$-12x^3 + 27x^2 + 12x - 21$	$-4257x^2 + 7128x + 6831$	
	$\quad - 43x^2 + 72x + 69$	$32) - 2656x + 6112$	
	$\quad 43x^2 - 72x - 69$	$\overline{-83x + 191}$	
$f_2(x)$	sign changed	$83x - 191$	$f_3(x)$
		sign changed	

Since $f_4(x)$ will be numerical and we are concerned with its sign, only $f_3(x) = 0$, gives

$x = \dfrac{191}{88}$ which is slightly less than $\dfrac{7}{3}$.

Let $x = \dfrac{7}{3} - h$ where h is small

Putting for x in $f_2(x)$, we get

$$f_2(x) = 43\left(\frac{7}{3} - h\right)^2 - 72\left(\frac{7}{3} - h\right) - 69$$

$$= 43h^2 - 128\frac{2}{3}h - 2\frac{8}{9}.$$

Since h is small the sign of $f_2(x)$ is the same as that of the last term which is –ive and hence $f_4(x)$ = +ive.

x	$-\infty$	0	∞	1	2	3	-1	-2
$f(x)$	+	+	+	+	+	+	–	+
$f'(x)$	–	+	+	–	–	+	–	–
$f_2(x)$	+	–	+	–	–	+	+	+
$f_3(x)$	–	–	+	–	–	+	–	–
$f_4(x)$	+	+	+	+	+	+	+	+
No. of changes of sign	4	2	0	2	2	0	3	4

Above table shows that $f(x) = 0$ has got all its roots real out of which two are +ive and two –ive. Both the +ive roots lie between 2 and 3, one –ive root lies between $(-1, 0)$ and the other between -2 and -1.

Example 6. *Find the number and position of the roots of $x^4 + 3x^3 + 7x^2 + 10x + 1$.*

Solution : We have $f(x) = x^4 + 3x^3 + 7x^2 + 10x + 1$

$$f'(x) = 4x^3 + 9x^2 + 14x + 10,$$
$$f_2(x) = -29x^2 - 78x + 14,$$
$$f_3(x) = -1086x - 481, \text{ as shown below.}$$

	$f(x)$	$f'(x)$	
	$x^4 + 3x^3 + 7x^2 + 10x + 1$	$4x^3 + 9x^2 + 14x + 10$	
	4	29	
x	$4x^4 + 12x^3 + 28x^2 + 40x + 4$	$116x^3 + 261x^2 + 406x + 290$	
	$4x^4 + 9x^3 + 14x^2 + 10x$	$116x^3 + 312x^2 - 6x$	$-4x$
	$3x^3 + 14x^2 + 30x + 4$	$-51x^2 + 462x + 290$	
	4	29	
3	$12x^3 + 56x^2 + 120x + 16$	$-1479x^2 + 13398x + 8410$	51
	$12x^3 + 27x^2 + 42x + 30$	$-1479x^2 - 3978x + 714$	
	$29x^2 + 78x - 14$	$16)17376x + 7696$	
$\therefore f_2(x)$	$= -29x^2 - 78x + 14$	$1086x + 481$	
	after changing sign	$-1086x - 481$	$f_3(x)$
		after changing sign	

Now it is clear that $f_4(x)$ is numerical and we are concerned with its sign.

$f_3(x) = 0$ gives $x = -\dfrac{481}{1086}$ which is slightly less than $-\dfrac{1}{2}$. Let $x = -\dfrac{1}{2} - h$ where h is small.

This value of x makes

$$f_2(x) = -29\left(-\frac{1}{2} - h\right)^2 - 78\left(-\frac{1}{2} - h\right) + 14$$

$$= -29h^2 + 49h + 6\frac{3}{4}$$

Since h is small clearly $f_2(x) = +$ive.

x	$-\infty$	-0	$+\infty$	-1	-2
$f(x)$	$+$	$+$	$+$	$-$	$+$
$f'(x)$	$-$	$+$	$+$	$+$	$-$
$f_2(x)$	$-$	$+$	$-$	$+$	$+$
$f_3(x)$	$+$	$-$	$-$	$+$	$+$
$f_4(x)$	$-$	$-$	$-$	$-$	$-$
No. of changes of sign	3	1	1	2	3

Above table shows that there are only $3 - 1 = 2$ real roots which are clearly –ive. One of them lies between $(0, -1)$ and other between $(-1, -2)$.

Example 7. *Apply Sturm's theorem to find the location of the real roots of the equation*
$$x^4 - 2x^3 + 5x^2 - 4x - 8 = 0.$$

Solution : Clearly, $f'(x) = 2x^3 - 3x^2 + 5x - 2$, $f_2(x) = -7x^2 + 7x + 34$

$f_3(x) = -2x + 1$, $f_4(x) = $ –ive, as below.

	$f(x)$	$f'(x)$	
	$x^4 - 2x^3 + 5x^2 - 4x - 8$	$4x^3 - 6x^2 + 10x - 4$	
	2	$2x^3 - 3x^2 + 5x - 2$	
x	$2x^4 - 4x^3 + 10x^2 - 8x - 16$	7	
	$2x^4 - 3x^3 + 5x^2 - 2x$	$14x^3 - 21x^2 + 35x - 14$	$2x + 1$
	$-x^3 + 5x^2 - 6x - 16$	$-14x^3 - 14x^2 - 68x$	
	2	$-7x^2 + 103x - 14$	
-1	$-2x^3 + 10x^2 - 12x - 32$	$-7x^2 + 7x + 34$	
	$-2x^3 + 3x^2 - 5x + 2$	$48)96x - 48$	
$f_2(x)$	$7x^2 - 7x - 34$	$2x - 1$	$f_3(x)$
	$-7x^2 + 7x + 34$	$-2x + 1$	
	sign changed	sign changed	

$f_3(x) = 0$ gives $x = \dfrac{1}{2}$ and $f_2(x) = -7\left(\dfrac{1}{4}\right) + 7\left(\dfrac{1}{2}\right) + 34 = $ +ive.

Since $f_2(x)$ is +ive; $\therefore$ $f_4(x) = -$ ive.

x	$-\infty$	0	∞	1	2	-1
$f(x)$	$+$	$-$	$+$	$-$	$+$	$+$
$f'(x)$	$-$	$-$	$+$			
$f_2(x)$	$-$	$+$	$-$			
$f_3(x)$	$+$	$+$	$-$			
$f_4(x)$	$-$	$-$	$-$			
No. of changes of sign	3	2	1			

We find that there are only two real roots one –ive and one +ive. Positive root lies between $(1, 2)$ and –ive between $(0, -1)$.

Here you need not compute fully the columns corresponding to $x = 1, 2, -1$.

$\therefore$ $f(0) = -, f(1) = -, f(2) = -, f(-1) = +$

Since $f(0)$ and $f(1)$ are of the same sign therefore either no root or even number of roots lie between $0, 1$; but since there is only one +ive root, it cannot lie between $(0, 1)$ and as such we need not calculate column corresponding to $x = 1$, $f(1)$ and $f(2)$ being of opposite the signs, the +ive root lies between 1 and 2. Also $f(0)$ and $f(-1)$ being of opposite signs, the –ive root lies between $(0, -1)$.

Example 8. *Find the nature of the roots of the equation* $x^3 + x^2 + x - 100 = 0$.

Solution : We have $f'(x) = 3x_2 + 2x + 1, f_2(x) = -4x + 901$.

$$f_2(x) = 0 \text{ gives } x = \frac{901}{4} = 200 \text{ app. and this value makes}$$

$f'(x)$ +ive, $\therefore$ $f_3(x)$ which will be numerical is –ive.

x	$-\infty$	0	∞	1	2	3	4	5
$f(x)$	–	–	+	–	–	–	–	+
$f'(x)$	+	+	+					
$f_2(x)$	+	+	–					
$f_3(x)$	–	–	–					
No. of changes of sign	2	2	1					

Above table shows that there is only one real root which is positive and that it lies between 4 and 5 as $f(4)$ and $f(5)$ are of opposite signs.

Example 9. *Find the nature of the roots of the equation*

$$x^4 - 5x^3 + 9x^2 - 7x + 2 = 0.$$

Solution : We have $f(x) = x^4 - 5x^3 + 9x^2 - 7x + 2$

$$f'(x) = 4x^3 - 15x^2 + 18x - 7,$$
$$f_2(x) = x^2 - 2x + 1, \text{ as shown below.}$$

	$f(x)$	$f'(x)$	$4x - 7$
	$x^4 - 5x^3 + 9x^2 - 7x + 2$	$4x^3 - 15x^2 + 18x - 7$	
	4	$4x^3 - 8x^2 + 4x$	
x	$4x^4 - 20x^3 + 36x^2 - 28x + 8$	$-7x^2 + 14x - 7$	
	$4x^4 - 15x^3 + 18x^2 - 7x$	$-7x^2 + 14x - 7$	
	$-5x^3 + 18x^2 - 21x + 8$	$\times$	
	4		
-5	$-20x^3 + 72x^2 - 84x + 32$		
	$-20x^3 + 75x^2 - 90x + 35$		
$f_2(x)$	$-3x^2 + 6x - 3$		
	$3x^2 - 6x + 3$		
	(sign changed)		
	or $x^2 - 2x + 1$		

Now $f_2(x)$ divides $f'(x)$ without any remainder and as such we conclude that $x^2 - 2x + 1$, i.e. $(x - 1)^2$ is the H.C.F. of $f(x)$ and $f'(x)$. Hence $(x - 1)^3$ must be a factor of $f(x)$ showing that $f(x) = 0$ has three roots each equal to 1.

x	$-\infty$	0	∞	2
$f(x)$	$+$	$+$	$+$	$0 \therefore 2$ is a root.
$f'(x)$	$-$	$-$	$+$	
$f_2(x)$	$+$	$+$	$+$	
No. of changes of sign	2	2	0	

From the above table, we find that there are only two distinct real roots which are positive. Since $f(2) = 0$, $\therefore$ 2 is one of the roots and the other root is 1 which is a triple root. Thus the equation has all its roots real, three of them being each equal to 1 and the fourth equal to 2.

Example 10. *By Sturm's method prove that the equation*

$$x^4 - 6x^3 + 13x^2 - 12x + 4, \text{ has two pairs of equal roots.}$$

Solution : We have $\quad f(x) = x^4 - 6x^3 + 13x^2 - 12x + 4,$

$$f'(x) = 2x^3 - 9x^2 + 13x - 6 \text{ (cancelling 2)},$$

$$f_2(x) = x^2 - 3x + 2, \text{ as shown below.}$$

	$f(x)$	$f'(x)$	
	$x^4 - 6x^3 + 13x^2 - 12x + 14$	$2x^3 - 9x^2 + 13x - 6$	$2x - 3$
	4	$2x^3 - 6x^2 + 4x$	
x	$2x^4 - 12x^3 + 26x^2 - 24x + 8$	$-3x^2 + 9x - 6$	
	$2x^4 - 9x^3 + 13x^2 - 6x$	$-3x^2 + 9x - 6$	
	$-3x^2 + 13x^2 - 18x + 8$	$\times$	
	2		
-3	$-3x^3 + 26x^2 - 36x + 16$		
	$-6x^3 + 27x^2 - 39x + 18$		
$f_2(x)$	$-x^2 + 3x - 2$		
	$x^2 - 3x + 2$		
	sign changed		

From above we conclude that $x^2 - 3x + 2$ or $(x - 2) \cdot (x - 1)$ is the H.C.F. of $f(x)$ and $f'(x)$. Hence the given equation has all the four roots real each root accruing twice, i.e. 2, 2, 1, 1. But Sturm's function will show only distinct type of roots as shown below :

x	$-\infty$	0	∞	1	2	
$f(x)$	$+$	$+$	$+$	0	0	1 and 2 are
$f'(x)$	$-$	$-$	$+$			the roots of
$f_2(x)$	$+$	$+$	$+$			$f(x)$
No. of changes of sign	2	2	0			

Example 11. *Using Sturm's theorem prove that the equation* $2x^6 - 18x^5 + 60x^4 - 120x^3 - 30x^2 + 18x - 5 = 0$ *has four imaginary roots and that its +ive roots lies in the interval* (5, 6).

Solution : Here proceeding as usual, we find that

$$f_2(x) = 5x^4 + 220x^2 + 1$$

and this expression is always +ive for all real values of x and we can divide by it taking $f_2(x) = 1$, Now proceed as usual.

x	$-\infty$	0	$-\infty$	1	2	3	4	5	6	-1
$f(x)$	+	$-$	+	$-$	$-$	$-$	$-$	$-$	+	+
$f'(x)$	$-$	+	+	+					+	$-$
$f_2(x)$	+	+	+	+					+	+
No. of changes of sign	2	1	0	1					0	2

Above table shows that there are only two real roots, one +ive and one –ive, and the examining four imaginary. Since $f(0)$, $f(1)$, $f(2)$, $f(3)$, $f(4)$, $f(5)$ are all –ive and as such root cannot lie between any two consecutive numbers from 0 to 5 and we need not compute these column fully. Again $f(5)$ = –ive and $f(6)$ = +ive; the root lies between 5 and 6. Similarly $f(0)$ = –ive and $f(-1)$ = +ive; the only –ive root lies between 0 and – 1. Also it may be noted that difference in changes of signs corresponding to $x = 0$ and $x = -1$ is 2 = 1 + 1, showing thereby that one –ive root lies between 0 and – 1.

We may also not compute fully the column for $x = 6$ and $x = -1$ in the above table.

Example 12. *By Sturm's method find the number and location of the real roots of the equation* $x^5 - 10x^3 + 6x + 1 = 0$.

Solution : We have
$$f(x) = x^5 - 10x^3 + 6x + 1,$$
$$f'(x) = 5x^4 - 30x^2 + 6,$$
$$f_2(x) = 20x^3 - 24x - 5,$$
$$f_3(x) = 96x^2 - 5x - 24,$$
$$f_4(x) = 43651x - 10920,$$
$$f_5(x) = \text{+ive as shown below.}$$

	$f(x)$	$f'(x)$	
	$x^5 - 10x^3 + 6x + 1$	$5x^4 - 30x^2 + 6$	
	5	4	
x	$5x^5 - 50x^3 + 30x + 5$	$20x^4 - 120x^2 + 24$	x
	$5x^5 - 30x^3 + 6x$	$20x^4 - 24x^2 - 5x$	
$f_2(x)$	$-20x^2 + 24x + 5$	$-96x^2 + 5x + 24$	$f_3(x)$
	$20x^3 - 24x - 5$	$96x^2 - 5x - 24$	
	sign changed	sign changed	
	24		
	$480x^3 - 576x - 120$		
$5x$	$480x^3 - 25x^2 - 120x$		
	$25x^2 - 456x - 120$		
	96		
25	$2400x^3 - 42706x - 11520$		
	$2400x^3 - 125x - 600$		
	$-43651x - 10920$		
$f_4(x)$	$43651x + 10920$		
	sign changed		

Clearly $f_5(x)$ will be numerical and we are concerned only with its sign. From $f_4(x) = 0$

we get $x = -\dfrac{10920}{43651}$, i.e. slightly greater than $-\dfrac{1}{4}$. Let it be $-\dfrac{1}{2} + h$ where h is small.

$$\therefore \qquad f_3\left(-\frac{1}{4}+h\right) = 96\left(-\frac{1}{4}+h\right)^2 - 5\left(-\frac{1}{4}+h\right) - 24$$

$$= 96h^2 - 53h - \frac{67}{4}.$$

For small value of h, $f_3\left(-\dfrac{1}{4}+h\right)$ is clearly –ive and hence $f_4(x)$ will be +ive.

x	$-\infty$	0	∞	1	2	3	4	–1	–2	–3	–4
$f(x)$	–	+	+	–	–	–	+	+	+	+	–
$f'(x)$	+	+	+	–	–	+	+	–	–	–	+
$f_2(x)$	–	–	+	–	+	+	+	–	–	–	–
$f_3(x)$	+	–	+	+	+	+	+	+	+	+	+
$f_4(x)$	–	+	+	+	+	+	+	–	–	–	–
$f_5(x)$	+	+	+	+	+	+	+	+	+	+	+
No. of changes of sign	5	2	0	1	1	1	0	4	4	4	5

From the above table it is clear that the roots are all real out of which 3 are –ive two +ive. One +ive root lies between 0 and 1. We need not compute fully the columns for $x = 1, 2,$ and 3.

$\because$ $f(1)$, $f(2)$ and $f(3)$ are all –ive and as such the remaining +ive root cannot lie between any of these numbers. Again since $f(4)$ comes out to be +ive whereas $f(3)$ is –ive and hence the only remaining +ive root lies between 3 and 4 and the columns under $x = 4$ also need not be fully tabulated. Again from the table is clear that two of the –ive roots lie between 0 and – 1 and we have to search for one –ive root more. Since $f(- 1)$, $f(- 2)$, $f(- 3)$ are all +ve as such the remaining –negative root cannot lie between any of them. Again since $f(- 3) = $ +ive and $f(- 4)$ is –ive therefore the remaining –ive root lies between – 3 and – 4. Hence we need not compute fully the columns for $x = - 2, - 3, - 4$.

Example 13. *Calculate Sturm's function for the quartic*

$$z^4 + 6Hz^2 + 4Gz + a^2I - 3H^2 = 0.$$

Solution : We have
$$f(z) = z^4 + 6Hz^2 + 4Gz + a^2I - 3H^2$$
$$f'(z) = z^3 + 3Hz + G, \text{ on dividing by (4)}$$

$$z^3 + 3Hz + G\overline{)z^4 + 6Hz^2 + 4Gz + a^2I - 3H^2}(z$$

$$\underline{z^4 + 3Hz^2 + Gz}$$
$$3Hz^2 + 3Gz + a^2I - 3H^2$$

$$\therefore \quad f_2(z) = -3Hz^2 - 3Gz - (a^2I - 3H^2))\overline{z^3 + 3Hz + G}(-Hz + G$$

Multiple by a +ive quantity $3H^2$

$$\overline{3H^3z^3 + 9H^3z + 3H^2G}$$

$$3H^3z^2 + 3GHz^2 + (a^2HI - 3H^3)z$$

$$\overline{-3GHz^2 - (a^2IH - 12H^3)z + 3H^2G}$$

$$-3GHz^2 - 3G^2z - (a^2I - 3H^2)G$$

$$\overline{z\{3(G^2 + 4H^3) - a^2IH\} + a^2IG}$$

$= za^2 (2HI - 3aJ) + a^2 IG,$ $\quad \because \quad$ $G^2 + 4H^3 = a^3 (HI - aJ)$

cancelling a^2 and changing the sign, we get

$$3H (2HI - 3aJ)z + 3G (HI - 3aJ)$$

$f_3(z) = -z \{(2HI - 3aJ)\} - IG \overline{) - 3Hz^2 - 3Gz - (a^2I - 3H^2) (}$

Multiplying by the +ive quantity $(2HI - 3aJ)^2$

$$\overline{-3H(2HI - 3aJ)^2 z^2 - 3G(2HI - 3aJ)^2 z - (a^2I - 3H^2)(2HI - 3aJ)^2}$$

$$-3H(2HI - 3aJ)^2 z^2 - 3HIG(2HI - 3aJ)z$$

$$\overline{-3Gz(2HI - 3aJ)[HI - 3aJ] - (a^2I - 3H^2)(2HI - 3aJ)^2}$$

$$-3Gz(2HI - 3aJ)[HI - 3aJ] - 3G^2I(HI - 3aJ)$$

$$\overline{-3G^2I(HI - 3aJ) - (a^2I - 3H^2)(2HI - 3aJ)^2}$$

$= 3G^2 HI^2 - 9aG^2IJ - [(a^2I - 3H^2)(4H^2 I^2 - 12HIJa + 9a^2 J^2]$

$3G^2 (HI^2 - 9aG^2 IJ) - 4a^2 H^2 I^3 + 12H^4 + I^2 12a^3 HI^2 J$

$$- 36H^3 IJa - a^4IJ^2 + 27a^2 H^2 J^2$$

$= 3HI^2 (G^2 + 4H^3) - 9aIJ (G^2 + 4H^3) - 4a^2 I^3 H^2 - 9a^4IJ^2$

$$+ 12a^3 HI^2 J + 27H^2 a^2 J^2$$

Putting $G^2 + 4H^3 = a^2 HI - a^3J$,

$= (3HI^2 - 9aIJ (a^2HI - a^3 J) - 4a^2 I^3 H^2 - 9^4aIJ^2 + 12a^3 HI^2 J + 27H^2 a^2 J^2$

$= 3a^2 H^2 I^3 - (9a^3HI^2J - 3a^3HI^2J) + \overline{9a^4IJ^2}$

$$- 4a^2I^3 H^2 - 9a^4IJ^2 + \underline{12a^3HI^2J} + 27H^2 a^2 J^2$$

$= - a^2I^3H^2 + 27H^2 a^2 J^2.$

Cancelling the +ive factor a^2H^2 and changing the sign, we get $f_4'(z) \equiv I^3 - 27 J^2$ which is called the discriminant of the biquadratic.

12.21 CONDITION FOR ALL THE ROOTS REAL AND DISTINCT

Case 1. In case the equation has all its roots distinct, then the sequence of Sturm's function must in general consist of $(n + 1)$ functions, i.e. the given function, its derivatives and $(n - 1)$ Sturm's remainders.

Case 2. In case the equation has all its roots real, then the difference of the number of changes of signs when x is put ∞ in the sequence over the number when x is put $-\infty$ should be n. In other words it means that when x is put $+\infty$ no change of signs occurs, i.e. the number of changes of signs be zero and when $x = -\infty$ in the $(n+1)$ functions, it should be alternately +ive and −ive so that there be n changes of signs and the difference of these changes of signs be $n - 0 = n$, the number of real and distinct roots.

The above conditions are satisfied if we say that the leading coefficients of all the the Sturm's functions be +ive (We always take the leading coefficients of a given equation to be +ive).

SOLVED EXAMPLES

Example 1. *Use Sturm's theorem to show that the equation $z^3 + Hz + G = 0$ has all the three roots real and distinct if and only if $G^2 + 4H^3 < 0$.*

Solution : The leading coefficients of all the Sturm's functions is +ive we get the condition as

$$-H + \text{ive}, \ -(G^2 + 4H^3) \ +\text{ive}$$

i.e. H − ive and $G^2 + 4H^3$ −ive.

But $G^2 + 4H^3$ can be −ive only if H is −ive and hence the former condition is implied in the latter. Hence $G^2 + 4H^3$ is −ive, i.e. < 0 if all the roots of the given cubic are real and distinct.

12.22   NATURE OF THE ROOTS OF BIQUADRATIC

By the help of Sturm's theorem, discuss the nature of the roots of the equation $ax^4 + 4bx^3 + 6cx^2 + 4dx + e = 0$

The above equation can be reduced to the form

$$f(z) = z^4 + 6Hz^2 + 4Gz + a^2I - 3H^2 = 0$$

where $z = ax + b$ and the nature of the roots of the given equation and of the transformed equation is same the Sturm's functions are

$$f(z) = z^4 + 6Hz^2 + 4Gz + a^2I - 3H^2$$
$$f'(z) = z^3 + 3Hz + G,$$
$$f_2(z) = -3Hz^2 - 3Gz - (a^2I - 3H^2),$$
$$f_3(z) = -z(2HI - 3aJ) - IG,$$
$$f_4(z) = I^3 - 27J^2 = \Delta, \text{ the discriminate.}$$

1. **All roots real and distinct.** The leading coefficients of Sturm's functions should be +ive.

$\therefore$ H negative, $2HI - 3aJ$ negative, $I^3 - 27J^2$ positive.

2. **All roots imaginary.**

$I^3 - 27J^2$ positive, and either H positive or $2HI - 3aJ$ positive.

In this case corresponding to $z = \infty$, the number of changes of sign in the above sequence will be

$$z = \infty + + - (+ \text{ or } -) + 2 \text{ changes}$$
$$z = -\infty + - - (+ \text{ or } -) + 2 \text{ changes}$$

The difference corresponding to the values ∞ and $-\infty$ of z is zero and hence the equation has no real roots. Thus all its roots are imaginary.

3. Two roots real and two imaginary.

$I^3 - 27J^2$ –ive : In this case we shall find that the difference of the changes of signs corresponding to $z = \infty$ and $-\infty$ in the above sequence is always 2. We may give H and $2HI - 3aJ$ any sign we like. Thus the equation has two real and two imaginary roots.

4. Two equal roots.

In this case clearly $I^3 - 27\,J^2 = 0$ and then $f_3(z) = 0$ will give the H.C.F. of $f(z)$ and $f'(z)$, and thus proving the existence of two equal roots.

5. Three equal roots or two pairs of equal roots.

In this case the H.C.F. should be of 2nd degree and also a perfect square or composed of two unequal factors according as three roots are equal to two pairs of equal roots exist. Hence $f_2(z)$ should vanish identically which will happen when either:

(1) $I = 0$ and $J = 0$ or

(2) $G = 0$ and $2HI - 3aJ = 0$.

when $I = 0$ and $J = 0$, we get $G^2 + 4H^3 = a^2\,(HI - aJ) = 0$ and $f_2(z) = 0$ becomes

$$3Hz^2 + 3Gz - 3H^2 = 0$$

and its discriminant is $9\,(G^2 + 4H^3)$ which is zero and hence its roots are equal, $i.e.$, it is a perfect square. Since the H.C.F. is a perfect square, $f(z) = 0$, has three equal roots.

When $G = 0$ and $2HI - 3aJ = 0$, then $f_2(z)$ is the H.C.F. but is not a perfect square and hence the equation will have a pair of equal roots.

Remark

❖ We have $G^2 + 4h^3 = a^2\,HI - a^3 J$. Putting $G = 0$ and $J = \dfrac{2HI}{3a}$, we get

$$4H^3 = a^2\,HI - a^3\,\dfrac{2HI}{3a} \text{ or } a^2 I = 12H^2 \text{ and this is same as the condition we had found.}$$

6. **All roots equal.** If $I = 0$, $H = 0$ and $G = 0$, then $f_3(z)$ vanishes identically and the H.C.F. comes out to be $f'(z)$, i.e. z^5 showing that all the roots are equal.

EXERCISE 12.5

1. Find the number and position of real roots of the following equations :

    (i) $x^4 + 15x^2 + 7x - 11 = 0$ (ii) $x^4 - 6x^3 + 5x^2 + 14x - 1 = 0$

    (iii) $x^4 - 4x^3 + 7x^2 - 6x - 4 = 0$ (iv) $x^4 - 8x^3 + 25x^3 - 36x + 8 = 0$

    (v) $x^3 + x^2 - 2x - 1 = 0$ (vi) $x^4 - 7x^2 + 18x - 8 = 0$

2. Apply sturm's theorem to prove that the equation $x^3 - 2x - 5 = 0$ has only one positive real root lying between 2 and 3.

3. Prove that the equation $x^5 - x + 16 = 0$ has two pair of complex roots.

4. Prove by Sturm's method that the equation $x^4 - 6x^3 + 13x^2 + 4 = 0$ has two pair of equal roots.

5. Varify by means of Sturm's remainders, the condition which must be fulfilled when the biquadratic $9x^4 + 4bx^3 + 6cx^3 + 4dx + e = 0$ is a perfect square and prove in that case $a^3 f(x) = \{(ax + b)^2 + 3H\}^2$

6. Prove that when the biquadratic of the last example has a triple factor it may be expressed in the form

$$a^3 f(x) = \{ax + b + \sqrt{(-H)}\}^3 \{ax + b - 3\sqrt{(-H)}\}$$

7. Find Sturm's function for the cubic $z^3 + 3Hz + G = 0$

ANSWERS

(1) (i)   One negative in $(-1, -2)$ and one positive in $(0, 1)$

(ii) One negaive in $(-1, -2)$, three +ive out of which one is in $(0, 1)$ and two in $(2, 4)$

(iii) One –ive in $(0, -1)$ and one +ive in $(2, 3)$ and two imaginary

(iv) Two imaginary, two real +ive, one in $(0, 1)$ and other two in $(3, 4)$

(v)   All real, one +ve in $(1, 2)$, two negative in $(0, 1)$ and other in $(-1, -2)$.

(vi) Two imaginary, two real one +ve in $(0, 1)$ one –ve in $(3, 4)$

(4) 1, 1, 2, 2

(7)

$$f(z) = z^3 + 3Hz + G$$
$$f_1(z) = z^2 + H, \quad f_2(z) = -2Hz - G$$
$$f_3(z) = -(G^2 + 4H^3)$$

CHAPTER REVIEW : A COMPETITIVE APPROACH

Selected terms and Results

TERMS

- **Rational integral function :** A function $f(x) = a_0 x^n + a_1 x^{n-1} + \ldots + a_{n-1}x + a_n$, $a_0 \neq 0$ of degree n is called rational integral function of x. if all the coefficients $a_0, a_1, a_2, \ldots a_{n-1}, a_n$ are supposed to be rational.

- **Roots of the equation :** Any value of x for which the value of $f(x)$ comes out to be zero, then this value of x is called a root of the equation $f(x) = 0$.

- **Reciprocal equation :** An equation which remains unchanged when x is replaced by $\dfrac{1}{x}$ is called a reciprocal equation.

RESULTS

- Every equation of degree n has n roots and no more.

- In an equation with real coefficients, imaginary roots occur in pair, that is if $\alpha + i\beta$ is one of the root of the equation $f(x) = 0$, then $\alpha - i\beta$ will also be a root of that equation.

- If the equation $f(x) = 0$ has a pair of complex (imaginary) roots $\alpha \pm i\beta$, then $(x - \alpha)^2 + \beta^2$ will be a factor of $f(x)$.

- If $\alpha + \sqrt{\beta}$ is a root of the equation $f(x) = 0$, then $\alpha - \sqrt{\beta}$ will also be a root of $f(x) = 0$.

- Every equation of odd degree with real coefficients has at least one real root with the sign opposite to that of its last term.

- Every equation of even degree with negative last term has at least two real roots with contrary sign.

- If the equation $f(x) = 0$ and $g(x) = 0$ have common roots and there common roots are the roots of $h(x) = 0$, then $h(x)$ will be H.C.F. (G.C.D.) of $f(x)$ and $g(x)$.

- If the equation $f(x) = 0$ has two roots equal, then the equation $f(x) = 0$ and $f'(x) = 0$ must have a common root.

- Every reciprocal equation of the standard form can be reduced to an equation of degree half of the degree of the original equation.

- An equation $f(x) = 0$ cannot have more positive roots than the number of changes of signs from positive to negative.

- An equation $f(x) = 0$ can not have more negative roots than the number of changes of sign in $f(-x) = 0$.

- If two real numbers a and b be substituted for x in the polynomial $f(x)$ and $f(a)$ and $f(b)$ are of opposite signs then at least one or odd number of real roots of the equation $f(x) = 0$ lie between a and b.

- Every equation of an odd degree has at least one real root whose sign is positive to that of its last term, the coefficient of the first term being positive.

- Every equation of an even degree whose last term is negative and the coefficient of the first term is positive, has at least two real roots, one is positive and one is negative.

- If an equation has only one change of sign it must have one positive root and no more.

- If all the terms of an equation are positive and the equation involves no odd powers of x, then all roots are complex.

- If all the terms of an equation are positive and involve odd powers of x then 0 is the only root.

- If two numbers a and b be substituted for x in polynomial $f(x)$ give results with the same sign no real root lies between them.

- If an equation $f(x) = 0$ has exactly m roots equal to α, then $f(x)$ and its first $(m-1)$ derivative all vanish for $x = \alpha$ but the mth and other higher order derivative do not vanish.

- If $f(x)$ and its first $(m-1)$ derivatives vanish for $x = \alpha$, then $f(x) = 0$ has m roots equal to α.

- A multiple root of order m of the equation $f(x) = 0$ is a multiple root of order $(m-1)$ of the first derived equation $f'(x) = 0$ and hence $f(x)$ and $f'(x)$ have the common factor $(x - \alpha)^{m-1}$.

- If α be a double root of $f(x) = 0$ and $f(\alpha) = 0$ and $f'(x) = 0$. Then condition for two or more equal roots is obtained by eliminating α between $f(x) = 0$ and $f'(x) = 0$

- A number which is greater than all the positive roots of a given equation is called upper or superior limit of the positive root of the equation.

- A number which is greater than all the negative roots of a given equation is called upper or superior limit of the negative roots of that equation.

- A number which is less than all the positive roots of a given equation is called the inferior or lower limit of the positive roots of that equation.

- A number which is less than all the negative roots of that equation is called the inferior or lower limit of negative roots of that equation.

- The series of functions $f(x), f'(x), f_2(x)$, $f_3(x) \dots f_r(x)$ consisting of the given function, its derivative and remainders with their sign changed in the process of finding H.C.F. of $f(x)$ and $f'(x)$ are called Sturm's function.

- The function $f_2(x), f_3(x) \dots f_r(x)$ are called auxiliary function.

- If $f(x) = 0$ having equal roots then Sturm's remainder will be a function of x.

- If $f(a)$ and $f(b)$ are of opposite signs, then at least one or in general odd number of roots of the equation $f(x) = 0$ lie between a and b. In case they be of the same sign, then either no root or an even number of roots of $f(x) = 0$ lie between a and b.

- When $f(x) = 0$ has no repeated roots i.e. the last or Sturm's function be numerical.

- In case the equation has all its root distinct, then the sequence of Sturm's function must in general consist of $(n + 1)$ functions.

Review Questions and Project Work

1. Show that the equation $x^4 - ax^3 - bx - c = 0$ has one positive root, one negative root and two complex roots for all positive values of a, b and c.

2. Prove that $4x^3 - 13x^2 - 13x - 275 = 0$ has one positive root and show that it lies between 6 and 7.

3. Prove that the condition for the equation $x^5 - 10a^3x^2 + b^4x + c^5 = 0$ to have three equal roots is $ab^4 - 9a^5 + c^5$.

4. Use Sturm method, to show that equation $x^4 - 12x + 7 = 0$ has a root method 2 and 3.

Objective Type Questions

Fill in the blanks :

1. If $\alpha_1, \alpha_2, \dots \alpha_n$ are the roots of equation $f(x) = 0$ where $f(x) \equiv a_0 x^n + a_1 x^{n-1} + \dots a_n$, then the product of the roots is ....

2. If $1, \alpha_1, \alpha_2, \dots \alpha_{n-1}$ are the roots of $x^n - 1 = 0$, then the value of $(1 - \alpha_1)$. $(1 - \alpha_2) \dots (1 - \alpha_{n-1})$ is ....

3. If α, β are the roots of $ax^2 + bx + c = 0$, then the equation whose roots are $\dfrac{1}{\alpha}, \dfrac{1}{\beta}$ is ....

4. If α, β, γ are the roots of equation $x^3 - 5x - 3 = 0$, then the equation whose roots are $-\alpha, -\beta, -\gamma$ is ....

5. If α, β, γ are the roots of equation $x^3 - 5x^2 + 6x - 7 = 0$, the equation whose roots are $3\alpha, 3\beta, 3\gamma$ is ....

6. If α, β, γ are the roots of equation $x^3 + qx + r = 0$, then the equation whose roots are $\dfrac{1}{\alpha}, \dfrac{1}{\beta}, \dfrac{1}{\gamma}$ is ....

7. To remove the second term of the equation $x^4 - 5x^3 + 7x^2 - 17x + 11 = 0$ we diminish its all roots by diminishing $h = $ ....

8. If the three roots of the equation $x^5 - 5x^4 + 9x^3 - 9x^2 + 5x - 1 = 0$ are 1, $\dfrac{1}{2}(1 - i\sqrt{3}), \dfrac{1}{2}(3 + \sqrt{5})$, then its other roots are ....

9. The equation $x^3 + qx + r = 0$ may have all its roots real if $27r^2 + 4q^3 < $ ....

10. If α, β, γ be the roots of $x^3 + 2x^2 - 3x - 1 = 0$, then all the value of $\alpha^{-3} + \beta^{-3} + \gamma^{-3}$ is ....

11. Every equation of an odd degree has at least .... real root.

12. Every equation of even degree having its last term negative has at least .... real roots.

13. If α, β, γ be the roots of the equation $x^3 + px + r = 0$, then the equation whose roots are $\alpha^2, \beta^2, \gamma^2$ is ....

14. If the two roots of the equation $x^3 + px + r = 0$ are 2 and -1, then the value of p and r are ....

15. The value of H and G for the cubic $x^3 + 3x^2 + 4x - 10 = 0$ are ....

16. The equation whose roots are the cube of the roots of the equation $x^3 + 3x^2 + 2 = 0$ is ....

17. Let $\alpha, \beta, \gamma, \delta$ be the roots of $ax^4 + bx^3 + cx^2 + dx + c = 0$. Then $\Sigma\alpha = $ ....

18. If $f(x) = x^4 - 3x^2 - 6x - 2 = (x^2 + 2x + 2)$ $(x^2 - 2x - 1)$ then roots of $f(x) = 0$ are ....

True/False: *Write 'T' for true and 'F' for false statement.*

1. Every equation of odd degree has at least two real roots.   (T/F)

2. Every equation of even degree with last term negative has at least two real roots.   (T/F)

3. To remove the second term of the equation $a_0 x^n + a_1 x^{n-1} + \dots + a_n = 0$, we diminish its all roots by $h = \dfrac{-a_1}{na_0}$.   (T/F)

4. If α and β are the roots of $x^2 + bx + c = 0$, then the equation $x^2 + (b - 2)x + c - b + 1 = 0$ has the roots of $\alpha + 1$, $\beta + 1$.   (T/F)

5. If α, β, γ are the roots of $x^3 + 3x + 3 = 0$, then the equation whose roots are $\dfrac{1}{\alpha}, \dfrac{1}{\beta}, \dfrac{1}{\gamma}$ is $3x^3 + 3x + 1 = 0$.   (T/F)

6. If α, β, γ are the roots of the equation $x^3 + qr + r = 0$, then the equation

whose roots are $\alpha + \beta$, $\beta + \gamma$, $\gamma + \alpha$, is $x^3 + qr - r = 0$

7. If equation $z^3 + 3Hz + g = 0$ has two roots equal if $G^2 + 4H^3 \neq 0$

Multiple Choice Questions : *Choose the most appropriate one :*

1. If $1, \alpha_1, \alpha_2, \ldots \alpha_{n-1}$ are the roots of $x^n - 1 = 0$, then the value of $(1 - \alpha_1)$ $(1 - \alpha_2) \ldots (1 - \alpha_{n-1})$ is :
   (a) $n - 1$
   (b) n
   (c) $n + 1$
   (d) n^2.

2. If α, β, γ, are the roots of the equation $x^3 + qx + r = 0$, then the equation whose roots are $\dfrac{1}{\alpha}, \dfrac{1}{\beta}, \dfrac{1}{\gamma}$:
   (a) $rx^3 + qx^2 + 1 = 0$
   (b) $rx^3 - qx^2 + 1 = 0$
   (c) $rx^3 + qx^2 - 1 = 0$
   (d) $rx^3 - qx^2 + r = 0$

3. If α, β are the roots of $x^2 - x + 1 = 0$, then the equation whose roots are α^2, β^2 is :
   (a) $x^4 + x^2 + 1 = 0$
   (b) $x^2 + x + 1 = 0$
   (c) $x^4 - x^2 + 1 = 0$
   (d) $x^2 + x - 1 = 0$

4. To remove second term of the equation $x^4 + 8x^3 + x - 5 = 0$ we diminish its all roots by :
   (a) 2 (b) 3
   (c) -2 (d) -3

5. If α, β, γ are the roots of the equation $x^3 + px + r = 0$ then $\alpha + \beta + \gamma$ is :
   (a) p (b) $-p$
   (c) 0 (d) 1

6. If α, β, γ be the roots of the equation $x^3 + 2x^2 - 3x - 1 = 0$ then $\alpha^{-3} + \beta^{-3} + \gamma^{-3}$ is :
   (a) 42 (b) 41
   (c) -42 (d) -41

ANSWERS

Fill in the blanks :

(1) $(-1)^n \dfrac{a_n}{a_0}$

(2) n

(3) $cx^2 + bx + a = 0$

(4) $x^3 - 5x + 3 = 0$

(5) $x^3 - 15x^2 + 54x - 189 = 0$ (6) $rx^3 + qx^2 + 1 = 0$

(7) $\dfrac{5}{4}$

(8) $\dfrac{1}{2}(1 + i\sqrt{3}), \dfrac{1}{2}(3 - \sqrt{5})$

(9) 0

(10) -42

(11) one

(12) two

(13) $y^3 + 2py^2 + p^2 y - r^2 = 0$ (14) $p = -3, r = -2$

(15) $H = \dfrac{1}{3}, G = -12$

(16) $y^3 + 33y^2 + 12y + 8 = 0$ (17) $-b/a$

(18) $1 \pm \sqrt{2}i$

True/False

(1) F   (2) T   (3) T   (4) T   (5) T   (6) T   (7) F

Multiple choice questions

(1) b (2) a (3) b (4) c (5) c (6) c

CHAPTER 13

Algebraic Solution of Cubic Equations

- ❖ Cardan's Method
- ❖ Hessian of the Cubic Equation
- ❖ Application of Cardan's Method
- ❖ Sum or Differences of Two Cubes

13.1 CARDAN'S METHOD TO FIND THE ROOTS OF A CUBIC EQUATION

Let the general cubic equation be

$$a_0x^3 + 3a_1x^2 + 3a_2x + a_3 = 0 \qquad \ldots(1)$$

First reduce this equation (1) into an equation having no second degree term, i.e. $3a_1x^2$. The equation (1) is reduced to the following equation.

$$Z^3 + 3HZ + G = 0 \qquad \ldots(2)$$

. where $H = a_0a_2 - a_1^2$, $G = a_0^2a_3 - 3a_0a_1a_2 + 2a_1^3$ and $Z = a_0x + a_1$.

Let us assume $z = u + v$. $\qquad \ldots(3)$

Cubing both the sides of (3), we get

$$z^3 = (u + v)^3 = u^3 + v^3 + 3uv(u + v) = u^3 + v^3 + 3uv(z)$$
$$z^3 = u^3 + v^3 + 3uvz \quad \text{or} \quad z^3 - 3uvz - (u^3 + v^3) = 0 \qquad \ldots(4)$$

Comparing (2) and (4), we get

$$uv = -H, \ u^3 + v^3 = -G \quad \text{or} \quad u^3v^3 = (-H)^3, \ u^3 + v^3 = -G$$

hence u^3, v^3 are the roots of the quadratic equation given by

$$t^2 + Gt - H^3 = 0 \qquad \ldots(5)$$

Solving (5), we get

$$t = \frac{-G \pm \sqrt{G^2 + 4H^3}}{2}$$

$$\therefore \quad u^3 = \frac{-G + \sqrt{G^2 + 4H^3}}{2} \qquad \ldots(6)$$

and $\quad v^3 = \dfrac{-G - \sqrt{G^2 + 4H^3}}{2} \qquad \ldots(7)$

From (3), we get

$$z = \left\{-\frac{G}{2}+\frac{1}{2}\sqrt{G^2+4H^3}\right\}^{1/3} + \left\{-\frac{G}{2}-\frac{1}{2}\sqrt{G^2+4H^3}\right\} \qquad \ldots(8)$$

From (6) and (7) it is obvious that each u and v will have three cube roots and hence from (8) z will have nine values. But the degree of the equation (2) in z is three so it must have three roots, i.e. *three values of z*. Since we have that $uv = -H$, therefore the cube roots are taken in pairs so that $uv = -H$. Hence we shall take the pair of cube roots as

$$u, v;\ u\omega,\ v\omega^2;\ u\omega^2,\ v\omega$$

where ω and ω^2 are the imaginary cube roots of unity. Therefore the roots of the equation (2) are

$$u + v,\ u\omega + v\omega^2,\ u\omega^2 + v\omega$$

and hence we can find the roots of the equation (1) by the relation $z = a_0x + a_1$ corresponding to $u + v$, $u\omega + v\omega^2$ and $u\omega^2 + v\omega$.

13.2 APPLICATIONS OF CARDAN'S METHOD

From equation (6) and (7), we have

$$u^3 = \frac{-G+\sqrt{G^2+4H^3}}{2} \quad \text{and } v^3 = \frac{-G-\sqrt{G^2+4H^3}}{2}$$

Case-I. If $G^2 + 4H^3 > 0$, i.e. *the cubic equation (1) has a pair of imaginary roots*, then R.H.S. of above two equations are real and hence by the some method we can extract the cube root of real quantities and consequently we get the values of z from equation (8).

Case-II. If $G^2 + 4H^3 < 0$, i.e. if the roots of the cubic equation (1) are all real, then from above equations u^3 and v^3 *are imaginary and there is no method in general for extracting the cube root of imaginary number so that Cardan's method fails to give the roots of the given cubic* (1). However, in this case we use De Moivre's theorem to find the cubic root of imaginary number.

Let us assume

$$-G = P \text{ and } G^2 + 4H^3 = -Q, \text{ then from equation (8), we get}$$

$$z = \left(\frac{P}{2}+\frac{i}{2}Q\right)^{1/3} + \left(\frac{P}{2}-\frac{i}{2}Q\right)^{1/3} \quad \text{where } i = \sqrt{-1}$$

Now put $\dfrac{P}{2} = r\cos\theta$, $\dfrac{Q}{2} = r\sin\theta$

$$\therefore \qquad r^2 = \frac{P^2+Q^2}{4} = \frac{G^2-G^2-4H^3}{4} = -H^3$$

and $\quad \tan\theta = \dfrac{Q/2}{P/2} = \dfrac{Q}{P}$

$$\tan\theta = \frac{-\sqrt{-(G^2+4H^3)}}{G}$$

$\therefore \qquad z = (r\cos\theta + ir\sin\theta)^{1/3} + (r\cos\theta - ir\sin\theta)^{1/3}$

$$= r^{1/3}\left[\cos\left(\frac{2n\pi+\theta}{3}\right) + i\sin\left(\frac{2n\pi+\theta}{3}\right) + \cos\left(\frac{2n\pi+\theta}{3}\right) - i\sin\left(\frac{2n\pi+\theta}{3}\right)\right]$$

$$= 2r^{1/3}\cos\left(\frac{2n\pi+\theta}{3}\right), \quad \text{where } n = 0, 1, 2.$$

Hence the roots of the equation (2) are

$$z_1 = 2r^{1/3}\cos\left(\frac{\theta}{3}\right) = 2(-H)^{1/2}\cos\left(\frac{\theta}{3}\right)$$

$$z_2 = 2r^{1/3}\cos\left(\frac{2\pi+\theta}{3}\right) = 2(-H)^{1/2}\cos\left(\frac{2\pi+\theta}{3}\right)$$

$$z_3 = 2r^{1/3}\cos\left(\frac{2\pi-\theta}{3}\right) = 2(-H)^{1/2}\cos\left(\frac{2\pi-\theta}{3}\right)$$

where $\qquad \theta = \tan^{-1}\left[\dfrac{-\sqrt{-(G^2+4H^3)}}{G}\right]$

Consequently by the relation $z = a_0 x + a$, we can find all the roots of the given cubic.

13.3 METHOD BY EXPRESSING THE EQUATION AS SUM OR DIFFERENCE OF TWO CUBES

Let the given cubic equation be

$$a_0 x^3 + 3a_1 x^2 + 3a_2 x + a_3 = 0 \qquad \qquad ...(1)$$

Let us suppose the equation (1) can be expressed as follows :

$$a_0 x^3 + 3a_1 x^2 + 3a_2 x + a_3 \equiv A(x-a)^3 + B(x-b)^3 \qquad ...(2)$$

or

$$a_0 x^3 + 3a_1 x^2 + 3a_2 x + a_3 \equiv (A+B)x^3 - 3(Aa+Bb)x^2$$
$$+ 3(Aa^2+Bb^2)x - (Aa^3+Bb^3).$$

Now equating the coefficients of like powers of x, we get

$$A + B = a_0 \qquad \qquad ...(3)$$
$$Aa + Bb = -a_1 \qquad \qquad ...(4)$$
$$Aa^2 + Bb^2 = a_2 \qquad \qquad ...(5)$$
$$Aa^3 + Bb^3 = -a_3 \qquad \qquad ...(6)$$

Multiplying (3), (4), (5) by a and subtracting respectively (4), (5) and (6), we get

$$B(a-b) = a_0 a + a_1 \qquad \qquad ...(7)$$
$$-Bb(a-b) = a_1 a + a_2 \qquad \qquad ...(8)$$
$$Bb^2(a-b) = a_2 a + a_3 \qquad \qquad ...(9)$$

From (7), (8) and (9), we get

$$(a_1 a + a_2)^2 = (a_0 a + a_1)(a_2 a + a_3) \qquad \qquad ...(10)$$

Similarly multiply (3), (4), (5) by b and subtracting respectively (4), (5) and (6), we get

$$(a_1b + a_2)^2 = (a_0b + a_1)(a_2b + a_3) \qquad \qquad ...(11)$$

From equation (10) and (11) it is concluded that a and b are the roots of the equation

$$(a_1x + a_2)^2 = (a_0x + a_1)(a_2x + a_3) \qquad \qquad ...(12)$$

or $\qquad (a_0a_2 - a_1^2)x^2 + (a_0a_3 - a_1a_2)x + (a_1a_3 - a_2^2) = 0 \qquad ...(13)$

This equation (13) is called the **Hessian of the cubic.**

From (13) we obtain the values of a and b and then we can find A and B from (3) and (4). Substitute these values of a, b, A and B in (2) we get the given equation as the sum or difference of two cubes as

$$A(x - a)^3 + B(x - b)^3 = 0$$

or $\qquad \left(\dfrac{x-a}{x-b}\right)^3 = \left(-\dfrac{B}{A}\right) \quad$ or $\quad \left(\dfrac{x-a}{x-b}\right) = \left(-\dfrac{B}{A}\right)^{1/3} = -\left(\dfrac{B}{A}\right)^{1/3}k \qquad ...(14)$

where $k = 1, \omega, \omega^2$.

Therefore if a and b (provided $a \neq b$) are the roots of the equation,

$$(a_1x + a_2)^2 = (a_0x + a_1)(a_2x + a_3)$$

Then the given equation (1) can be reduced to the form given by

$$A(x - a)^3 + B(x - b)^3 = 0$$

Now from (14) one values of x is obtained and then divide the given equation by so obtained factor we get a quadratic as quotient and the remaining roots of the given equation can be obtained from this quadratic quotient.

Further, if the obtained values of a and b are imaginary and so A and B, then we can not find the cube roots of complex number. Hence by above method we can find the roots of the given cubic if the roots of the quadratic equation (13) are real and distinct.

That is, *if the following condition holds*

$$(a_0a_3 - a_1a_2)^2 - 4(a_1a_3 - a_2^2)(a_0a_2 - a_1^2) > 0$$

then the equation (13) will have both real and distinct roots.

13.4   HESSIAN OF THE CUBIC EQUATION

The quadratic equation (13) of § 13.3 is given by

$$(a_0a_2 - a_1^2)x^2 + (a_0a_3 - a_1a_2)x + (a_1a_3 - a_2^2) = 0$$

is called the *Hessian of the cubic equation*

$$a_0x^3 + 3a_1x^2 + 3a_2x + a_3 = 0$$

This Hessian of cubic can also be obtained as follows.

First making the given cubic homogenous by introducing a new variable t as

$$f(x, t) \equiv a_0x^3 + 3a_1x^2t + 3a_2xt^2 + a_3t^3 = 0 \qquad ...(1)$$

Now differentiating (1) partially w.r.t. x and t and respectively, we get

$$\frac{\partial f}{\partial x} = 3a_0x^2 + 6a_1xt + 3a_2t^2 \qquad ...(2)$$

$$\frac{\partial f}{\partial t} = 3a_1 x^2 + 6a_2 xt + 3a_3 t^2 \qquad ...(3)$$

Again differentiating (2) w.r.t. x and (3) w.r.t, t we get

$$\frac{\partial^2 f}{\partial x^2} = 6a_0 x + 6a_1 t \qquad ...(4)$$

$$\frac{\partial^2 f}{\partial t^2} = 6a_2 x + 6a_3 t \qquad ...(5)$$

and differentiating (3) w.r.t. x, we get

$$\frac{\partial^2 f}{\partial x\, \partial t} = 6a_1 x + 6a_2 t \qquad ...(6)$$

Now putting $t = 1$ and observed that the expression

$$\left(\frac{\partial^2 f}{\partial x\, \partial t} \right)^2 = \left(\frac{\partial^2 f}{\partial x^2} \right) \cdot \left(\frac{\partial^2 f}{\partial t^2} \right) \qquad ...(7)$$

gives the Hessian of the cubic. Hence the Hessian of the cubic can also be obtained by (7).

Remark

❖ Hessian of the cubic can also be expressed in the form of a determinant as follows :

$$\begin{vmatrix} 1 & -x & x^2 \\ a_0 & a_1 & a_2 \\ a_1 & a_2 & a_3 \end{vmatrix} = 0$$

SOLVED EXAMPLES

Example 1. *Solve the equation* $x^3 - 15x - 126 = 0$ *by Cardan's method.*

Solution. Since the given equation is

$$x^3 - 15x - 126 = 0 \qquad ...(1)$$

and let the solution of (1) be $\qquad x = u + v \qquad ...(2)$

Cubing (2), we get

$$x^3 = (u + v)^3 = u^3 + v^3 + 3uv(u + v)$$

or $\qquad x^3 = u^3 + v^3 + 3uv(x) \qquad\qquad [\because x = u + v]$

or $\qquad x^3 - 3uvx - (u^3 + v^3) = 0 \qquad ...(3)$

The equations (1) and (3) are same so comparing the coefficient of like terms, we get

$$3uv = 15 \quad \text{or } uv = \frac{15}{3} \quad \text{or} \quad u^3 v^3 = 125$$

and $\qquad u^3 + v^3 = 126$

hence u^3, v^3 are the roots of the quadratic

$$t^2 - 126t + 125 = 0$$

$\therefore$ $\qquad$ $(t - 125)(t - 1) = 0, \ t = 125, \ t = 1$

$\therefore$ $\qquad$ $u^3 = 125, \ v^3 = 1 \ $ or $ \ u = 5, \ v = 1$

Thus the roots of (1) are given by

$$u + v, \ u\omega + v\omega^2, \ u\omega^2 + v\omega$$

where $\qquad$ $\omega = -\dfrac{1}{2} + \dfrac{i\sqrt{3}}{2}$

$\therefore$ $\qquad$ $u + v = 5 + 1 = 6$

$$u\omega + v\omega^2 \ = \ 5\omega + \omega^2 = 4\omega + \omega + \omega^2$$

$$= \ 4\omega - 1 \qquad\qquad (\because 1 + \omega + \omega^2 = 0)$$

$$= \ 4\left(-\dfrac{1}{2} + \dfrac{i\sqrt{3}}{2}\right) - 1 = -3 + i2\sqrt{3}$$

and $\qquad$ $u\omega^2 + v\omega \ = \ 5\omega^2 + \omega = 4\omega^2 + \omega^2 + \omega = 4\omega^2 - 1$

$$= \ 4\left(-\dfrac{1}{2} + \dfrac{i\sqrt{3}}{2}\right)^2 - 1 = -3 - i2\sqrt{3}$$

Hence, roots are $6, \ -3 + 2i\sqrt{3}, -3 - 2i\sqrt{3}$.

Example 2. *Solve the equation* $x^3 - 15x^2 - 33x + 847 = 0$ *by Cardan's method.*

Solution. Since the given equation is

$$x^3 - 15x^2 - 33x + 847 = 0 \qquad\qquad ...(1)$$

First we remove the second term, i.e. $- 15x^2$ by diminishing each of its roots by the constant

$$h = -\dfrac{a_1}{na_0} = -\dfrac{-15}{3 \times 1} = 5$$

Now using synthetic division method

5	1	-15	-33	847
		5	-50	-418
	1	-10	-83	432
		$+5$	-25	
	1	-5	-108	
		5		
	1	0		
	1			

Thus the transformed equation is (without second degree term)

$$z^3 - 108\,z + 432 = 0 \qquad\qquad ...(2)$$

where $\qquad$ $z = x - 5$

Let the solution of (2) be

$$z = u + v \qquad\qquad ...(3)$$

Cubing (3) of both sides, we get

$$z^3 - 3uvz - (u^3 + v^3) = 0 \qquad \text{...(4)}$$

The equation (2) and (4) are same so we have

$$uv = 36, \ u^3 + v^3 = -432 \ \text{ or } \ u^3 v^3 = (36)^3.$$

∴ u^3, v^3 are the roots of the equation $t^2 + 432\,t + (36)^3 = 0$

∴
$$t = \frac{-432 \pm \sqrt{(432)^2 - 4(36)^3}}{2} = -\frac{432}{2} = -216$$

∴
$$u^3 = -216, \ v^3 = -216$$

∴
$$u = (-216)^{1/3} = -6, \ v = (-216)^{1/3} = -6$$

∴   The roots of (2) are

$$u + v, \ u\omega + v\omega^2, \ u\omega^2 + v\omega \quad \text{i.e. } z_1 = u + v = -6 - 6 = -12$$
$$z_2 = -6\omega - 6\omega^2 = -6(\omega + \omega^2) = -6(-1) = 6$$
$$z_3 = -6\omega^2 - 6\omega = -6(\omega^2 + \omega) = -6(-1) = 6$$

Therefore the roots of given equation (1) are

$$x_1 = z_1 + 5 = -12 + 5 = -7$$
$$x_2 = z_2 + 5 = 6 + 5 = 11$$
$$x_3 = z_3 + 5 = 6 + 5 = 11$$

Hence the roots of the given cubic equation, are – 7, 11, 11.

Example 3. *Show that the roots of the equation $x^3 - 3x + 1 = 0$ are*

$$2\cos\frac{2\pi}{9}, \ 2\cos\frac{8\pi}{9}, \ \cos\frac{14\pi}{9}$$

Solution. Since the given equation is

$$x^3 - 3x + 1 = 0 \qquad \text{...(1)}$$

Let
$$x = u + v \qquad \text{...(2)}$$

Cubing (2) of both sides, we get

$$x^3 - 3uvx - (u^3 + v^3) = 0 \qquad \text{...(3)}$$

Since (1) and (3) are same so we have

$$uv = 1, \ u^3 + v^3 = -1$$

or
$$u^3 v^3 = 1, \ u^3 + v^3 = -1$$

∴ u^3, v^3 are the roots of the following equation $t^2 + t + 1 = 0$

∴
$$t = \frac{-1 \pm \sqrt{1 - 4}}{2}$$

$$t = \frac{-1 \pm i\sqrt{3}}{2}$$

∴
$$u^3 = -\frac{1}{2} + \frac{i}{2}\sqrt{3}, \ v^3 = -\frac{1}{2} - \frac{i}{2}\sqrt{3}$$

From (2), we get

$$x = \left(-\frac{1}{2} + \frac{i}{2}\sqrt{3}\right)^{1/3} + \left(-\frac{1}{2} - \frac{i}{2}\sqrt{3}\right)^{1/3} \qquad \text{...(4)}$$

Change the complex number of R.H.S. of (4) into polar form by putting

$$-\frac{1}{2} = r\cos\theta, \quad \frac{\sqrt{3}}{2} = r\sin\theta$$

$\therefore \qquad r^2 = 1 \Rightarrow r = 1$

and $\qquad \tan\theta = -\sqrt{3} \Rightarrow \theta = \dfrac{2\pi}{3}$

$\therefore \qquad x = (r\cos\theta + ir\sin\theta)^{1/3} + (r\cos\theta - ir\sin\theta)^{1/3}$

$$= r^{1/3}\left[\cos\frac{2n\pi+\theta}{3} + i\sin\frac{2n\pi+\theta}{3} + \cos\frac{2n\pi+\theta}{3} - i\sin\frac{2n\pi+\theta}{3}\right]$$

$$= 2r^{1/3}\cos\frac{2n\pi+\theta}{3}, \quad n = 0, 1, 2$$

Therefore, $\qquad x_1 = 2(1)^{1/3}\cos\dfrac{\theta}{3} = 2\cos\dfrac{2\pi}{9}$

$$x_2 = 2\cos\frac{2\pi+\theta}{3} = 2\cos\left(\frac{2\pi}{3} + \frac{2\pi}{9}\right) = 2\cos\frac{8\pi}{9}$$

and $\qquad x_3 = 2\cos\dfrac{4\pi+\theta}{3} = 2\cos\left(\dfrac{4\pi}{3} + \dfrac{2\pi}{9}\right) = 2\cos\dfrac{14\pi}{9}$

Example 4. *Solve the equation* $28x^3 - 9x^2 + 1 = 0$ *by Cardan's method.*

Solution. Since the given equation is

$$28x^3 - 9x^2 + 1 = 0 \qquad\qquad\qquad ...(1)$$

First remove the second terms, i.e. $-9x^2$ by putting $x = \dfrac{1}{z}$, we get

$$\frac{28}{z^3} - \frac{9}{z^2} + 1 = 0 \quad \text{or} \quad 28 - 9z + z^3 = 0$$

or $\qquad z^3 - 9z + 28 = 0 \qquad\qquad\qquad ...(2)$

where $\qquad z = \dfrac{1}{x}$

Let the solution of (2) be

$$z = u + v \qquad\qquad\qquad ...(3)$$

Cubing (3), we get

$$z^3 - 3uvz - (u^3 + v^3) = 0 \qquad\qquad\qquad ...(4)$$

The equation (2) and (4) are same so, we have

$$uv = 3, u^3 + v^3 = -28$$

or $\qquad u^3v^3 = 27, \ u^3 + v^3 = -28$

$\therefore \quad u^3, v^3$ are the root of the equation

$$t^2 + 28t + 27 = 0 \quad \text{or} \quad (t+27)(t+1) = 0$$

$$t = -27, \ t = -1$$

$\therefore \qquad u^3 = (-27), v^3 = -1$

or $\qquad u = (-27)^{1/3} = -3, \ v = (-1)^{1/3} = -1$

Thus the roots of (2) are

$$z_1 = u + v = -3 - 1 = -4$$

$$z_2 = u\omega + v\omega^2 = -3\omega - \omega^2$$
$$= -2\omega - (\omega + \omega^2) = -2\omega + 1 \qquad [\because 1 + \omega + \omega^2 = 0]$$
$$= -2\left(-\frac{1}{2} + \frac{i}{2}\sqrt{3}\right) + 1 = 2 - i\sqrt{3}$$

$$z_3 = u\omega^2 + v\omega = -3\omega^2 - \omega = 2 + i\sqrt{3}$$

Hence the roots of the given cubic equation are

$$x_1 = \frac{1}{z_1} = \frac{1}{-4} = -\frac{1}{4}.$$

$$x_2 = \frac{1}{z_2} = \frac{1}{2 - i\sqrt{3}} = \frac{2 + i\sqrt{3}}{4 + 3} = \frac{2}{7} + \frac{i}{7}\sqrt{3}$$

$$x_3 = \frac{1}{z_3} = \frac{1}{2 + i\sqrt{3}} = \frac{2 - i\sqrt{3}}{4 + 3} = \frac{2}{7} - \frac{i}{7}\sqrt{3}$$

Example 5. *Solve the cubic $x^3 - 3a^2x - 2a^3 \cos 3A = 0$ by Cardan's method.*

Solution. Since the given cubic

$$x^3 - 3a^2x - 2a^3 \cos 3A = 0 \qquad \qquad ...(1)$$

Let the solution of (1) be

$$x = u + v \qquad \qquad ...(2)$$

Cubing of both sides of (2), we get

$$x^3 - 3uvx - (u^3 + v^3) = 0 \qquad \qquad ...(3)$$

The equations (1) and (3) are same so we have

$$uv = a^2, \quad u^3 + v^3 = 2a^3 \cos 3A$$

or $$u^3v^3 = a^6, \quad u^3 + v^3 = 2a^3 \cos 3A.$$

$\therefore$ u^3, v^3 are the roots of the following equation

$$t^3 - 2a^3 \cos 3A + a^6 = 0 \qquad \qquad ...(4)$$

$\therefore$
$$t = \frac{2a^3 \cos 3A \pm \sqrt{(4a^6 \cos^2 3A - 4a^6)}}{2} = \frac{2a^3 \cos 3A \pm 2a^3 i \sin 3A}{2}$$

$$t = a^3 (\cos 3A \pm i \sin 3A)$$

$\therefore$ $u^3 = a^3 (\cos 3A + i \sin 3A), \quad v^3 = a^3 (\cos 3A - i \sin 3A)$

or $u = a (\cos 3A + i \sin 3A)^{1/3}, \quad v = a (\cos 3A - i \sin 3A)^{1/3}$

Substitute these values of u and v in (2), we get

$$x = a (\cos 3A + i \sin 3A)^{1/3} + a (\cos 3A - i \sin 3A)^{1/3}$$

$$= a\left[\cos\left(\frac{2n\pi + 3A}{3}\right) + i\sin\left(\frac{2n\pi + 3A}{3}\right)\right.$$

$$\left. + \cos\left(\frac{2n\pi + 3A}{3}\right) - i\sin\left(\frac{2n\pi + 3A}{3}\right)\right]$$

$$= 2a\cos\left(\frac{2n\pi + 3A}{3}\right), \quad \text{where } n = 0, 1, 2$$

$\therefore$ $x_1 = 2a \cos A$ when $n = 0$

$$x_2 = 2a\cos\left(\frac{2\pi + 3A}{3}\right), \text{ when } n = 1$$

$$= 2a\cos\left(\frac{2\pi}{3} + 1\right)$$

and $$x_3 = 2a\cos\left(\frac{4\pi + 3A}{3}\right), \text{ when } n = 2$$

$$= 2a\cos\left(\frac{4\pi}{3} + A\right) = 2a\cos\left\{2\pi - \left(\frac{2\pi}{3} - A\right)\right\} = 2a\cos\left(\frac{2\pi}{3} - A\right)$$

Hence the solution of the given cubic is given by

$$2a\cos A, \; 2a\cos\left(\frac{2\pi}{3} \pm A\right)$$

Example 6. *Solve the equation $9x^3 - 30x^2 + 36x - 16 = 0$ by expressing it as the sum or difference of two cubes.*

Solution. Since given cubic is

$$9x^3 - 30x^2 + 36x - 16 = 0 \qquad\qquad\qquad ...(1)$$

Compare this equation with the equation

$$a_0 x^3 + 3a_1 x^2 + 3a_2 x + a_3 = 0$$

we get $$a_0 = 9, a_1 = -10, \; a_2 = 12, a_3 = -16$$

Let the given cubic (1) can be expressed as

$$9x^3 - 30x^2 + 36x - 16 \equiv A(x - a)^3 + B(x - b)^3 = 0 \qquad\qquad ...(2)$$

Then a, b are the roots of the following equation

$$(a_1 x + a_2)^2 = (a_0 x + a_1)(a_2 x + a_3)$$

or $$(a_0 a_2 - a_1^2)x^2 + (a_0 a_3 - a_1 a_2)x + (a_1 a_3 - a_2^2) = 0 \qquad\qquad ...(3)$$

Substitute the values of a_0, a_1, a_2, a_3 in (3), we get

$$(108 - 100)x^2 + (-144 + 120)x + (160 - 144) = 0$$

or $$8x^2 - 24x + 16 = 0 \quad \text{or} \quad x^2 - 3x + 2 = 0$$

or $$(x - 1)(x - 2) = 0 \quad \text{or} \quad x = 1, 2$$

∴ $$a = 1, b = 2$$

Now substitute these values of a and b in (2) and equating the coefficients of like powers of x of both sides we get

$$9x^3 - 30x^2 + 36x - 16 \equiv A(x - 1)^3 + B(x - 2)^3 = 0 \qquad\qquad ...(4)$$

Taking the coefficients of x^3 and x^2, we get

$$A + B = 9$$

$$A + 2B = 10$$

Solving these two equation, we get

$$A = 8, \; B = 1$$

Substitute the values of A and B in (4), we get the given cubic as

$$8(x - 1) + (x - 2)^3 = 0 \quad \text{or} \quad \left(\frac{x - 1}{x - 2}\right)^3 = -\frac{1}{8}$$

or $$\frac{x - 1}{x - 2} = \left(-\frac{1}{8}\right)^{1/3}$$

or
$$\frac{x-1}{x-2} = \left(-\frac{1}{2}\right)^{1/3}(1)^{1/3} = \left(-\frac{1}{2}\right)k$$

where $k = 1, \omega, \omega^2$ and $\omega = \left(\frac{-1}{2} + \frac{i}{2}\sqrt{3}\right)$

or $\qquad 2(x-1) = -k(x-2) \qquad$ or $\quad 2x - 2 = -kx + 2k$

or $\qquad x(2+k) = 2k + 2 \qquad$ or $\quad x = \dfrac{2k+2}{2+k} = \dfrac{2(k+1)}{(2+k)}$

when $\qquad k = 1, \ x_1 = \dfrac{2(1+1)}{2+1} = \dfrac{4}{3}$

when $k = \omega$ $\qquad x_2 = \dfrac{2(\omega+1)}{(2+\omega)} = \dfrac{2(-\omega^2)}{(1-\omega^2)}$ $\qquad [\because 1 + \omega + \omega^2 = 0 \text{ and } \omega^3 = 1]$

$$= \frac{2(-\omega^3)}{(\omega - \omega^3)} = -\frac{2}{\omega - 1} = \frac{2}{1-\omega} = \frac{2}{1 - \left(-\dfrac{1}{2} + \dfrac{i}{2}\sqrt{3}\right)} = \frac{2}{\dfrac{3}{2} - \dfrac{i}{2}\sqrt{3}}$$

$$= \frac{4}{3 - i\sqrt{3}} = \frac{4(3 + i\sqrt{3})}{9 + 3} = \frac{3 + i\sqrt{3}}{3} \qquad \text{when } k = \omega^2,$$

$$x_3 = \frac{2(\omega^2 + 1)}{(2+\omega^2)} = \frac{2(\omega^3 + \omega)}{(2\omega + \omega^3)} = \frac{2(1+\omega)}{2\omega + 1} \qquad [\because \omega^3 = 1]$$

$$= \frac{2\left(1 - \dfrac{1}{2} + \dfrac{i}{2}\sqrt{3}\right)}{2\left(-\dfrac{1}{2} + \dfrac{i}{2}\sqrt{3}\right) + 1} \qquad \left[\because \omega = \left(-\frac{1}{2} + \frac{i}{2}\sqrt{3}\right)\right]$$

$$= \frac{1 + i\sqrt{3}}{i\sqrt{3}} = \frac{i\sqrt{3} - 3}{-3} = \frac{3 - i\sqrt{3}}{3}$$

Hence the roots of the given cubic are $\dfrac{4}{3}, \ \dfrac{3 \pm i\sqrt{3}}{3}$

EXERCISE 13.1

Solve the following cubic equations by Cardan's method

1. $x^3 + 6x^2 + 9x + 4 = 0$
2. $x^3 + 6x^2 - 12x + 32 = 0$
3. $x^3 - 21x - 344 = 0$
4. $x^3 - 12x^2 - 6x - 10 = 0$
5. $27x^3 + 54x^2 + 198x - 73 = 0$
6. $x^3 - 18x - 35 = 0$
7. $9x^3 + 6x^2 - 1 = 0$
8. $x^3 - 15x^2 - 357x + 5491 = 0$
9. $x^3 + 3x^2 - 27x + 104 = 0$
10. $x^3 - 6x - 9 = 0$
11. $2x^3 + 3x^2 + 3x + 1 = 0$
12. $8a^3x^3 - 6ax + 2\sin 3A = 0$
13. $64x^3 - 144x^2 + 108x - 27 = 0$
14. $x^3 + 3ax^2 + 3(a^2 - bc)x + a^3 + b^3 + c^3 - 3abc = 0$

Solve the following equations by expressing them as the sum or difference of two cubes:

15. $x^3 - 3x^2 + 33x - 1 = 0$
16. $2x^3 + 3x^2 - 21x + 19 = 0$
17. $x^3 + 3x^2 - 27x + 104 = 0$
18. $152x^3 - 60x^2 - 606x - 485 = 0$

HINT TO SELECTED PROBLEMS

1. $h = -\dfrac{a_0}{na_0} = -\dfrac{6}{3 \times 1} = -2$

The transformed equation is $z^3 - 3z + 2 = 0$

where $z = x + 2$

Assume that the solution of transformed equation is $z = u + v$.

11. $h = -\dfrac{1}{2}$ Transformed equation is $2z^3 + \dfrac{3}{2}z = 0$

13. $h = \dfrac{3}{4}$ Transformed equation is $z^3 = 0$

14. $h = -a$ Transformed equation is $z^3 - 3bcz + b^3 + c^3 = 0$

18. The given equation can be expressed as

$$152x^3 - 60x^2 - 606x - 485 = A(x - a)^3 + B(x - b)^3$$

Then a, b are the roots of the following equation

$$(a_0a_1 - a_1^2)x^2 + (a_0a_3 - a_1a_2)x + (a_1x_3 - a_2^2) = 0$$

ANSWERS

(1) $-4, -1, -1.$ **(2)** $-8, (1 \pm i\sqrt{3})$ **(3)** $8, (-4 \pm i3\sqrt{3})$

(4) $4 + 3(2)^{1/3} + 3(4)^{1/3}, 4 + 3\omega(2)^{1/3} + 3\omega^2(4)^{1/3}, 4 + 3\omega^2(2)^{1/3}$

$$+ 3\omega(4)^{1/3} \text{ where } \omega = \left(-\frac{1}{2} \pm \frac{i}{2}\sqrt{3}\right)$$

(5) $\dfrac{1}{3}\left(-\dfrac{7}{6} \pm \dfrac{3\sqrt{3}}{2}i\right)$ **(6)** $5, \left(-\dfrac{5}{2} \pm \dfrac{i\sqrt{3}}{2}\right)$ **(7)** $\dfrac{1}{3}, \dfrac{-3 \pm i\sqrt{3}}{6}$

(8) $-19, 17, 17$ **(9)** $-8, \dfrac{1}{2}, (5 \pm i3\sqrt{3})$ **(10)** $3, \left(-\dfrac{3}{2} \pm \dfrac{i\sqrt{3}}{2}\right)$

(11) $-\dfrac{1}{2}\left(-\dfrac{1}{2} \pm \dfrac{i\sqrt{3}}{2}\right)$ **(12)** $\dfrac{1}{a}\sin A, \dfrac{1}{a}\sin\left(\dfrac{\pi}{3} - A\right), -\dfrac{1}{a}\sin\left(\dfrac{\pi}{3} + A\right)$

(13) $\dfrac{3}{4}, \dfrac{3}{4}, \dfrac{3}{4}$ **(14)** $-(a + b + c), -(a + b\omega + c\omega^2), -(a + b\omega^2 + c\omega)$

(15) $\dfrac{7 - 7k(2)^{1/3}(5)^{2/3} + 7k^2(2)^{2/3}(5)^{1/3}}{7}, k = 1, \omega, \omega^2.$

(16) $\dfrac{-1 - 3^{1/3}(5)^{2/3}\alpha - (3)^{2/3}(5)^{1/3}\alpha}{2}, \alpha = 1, \omega, \omega^2$

(17) $-8, \dfrac{1}{2}(5 \pm i3\sqrt{3})$ **(18)** $\dfrac{5}{2}, \dfrac{-40 \pm i7\sqrt{7}}{38}$

CHAPTER REVIEW : A COMPETITIVE APPROACH

Selected terms and Results

TERMS

- **Hessian of the cubic :** The equation $(a_0a_2 - a_1^2)x^2 + (a_0a_3 - a_1a_2)x + (a_1a_3 - a_2^2) = 0$ is called the Hessian of the cubic.

RESULTS

- If the condition $(a_0a_3 - a_1a_2)^2 - 4(a_1a_3 - a_2^2)(a_0a_2 - a_1^2) > 0$, then the given equation will have both real and distinct roots.
- Hessian of the cubic can also be obtained by the equation given by

$$\left(\frac{\partial^2 f}{\partial x\, \partial t}\right)^2 = \left(\frac{\partial^2 f}{\partial x^2}\right) \cdot \left(\frac{\partial^2 f}{\partial t^2}\right)$$

- Hessian of the cubic can also be expressed in the form of determinant as follows:

$$\begin{vmatrix} 1 & -x & x^2 \\ a_0 & a_1 & a_2 \\ a_1 & a_2 & a_3 \end{vmatrix} = 0$$

Review Questions and Project Work

1. Show that the roots of the cubic $x^3 - 3a^2x - 2a^2\cos 3A = 0$ are given by $3a \cos A$, $2a \cos (120 + A)$

2. Prove that the roots of the equation $x^3 - 3x + 1 = 0$ are

$$2\cos\frac{2p}{9}, 2\cos\frac{8p}{9}, 2\cos\frac{14p}{9} .$$

3. Show that the roots of the equation $64x^3 - 144x^2 + 108x - 27 = 0$ are

$$\frac{3}{4}, \frac{3}{4}, \frac{3}{4}.$$

4. If $ax^3 + 3bx^2 + 3cx + d + k(x - r)^3$ be a perfect cube, prove that $(ac - b^2)r^2 + (ad - bc)r + (bd - c^2) = 0$

5. Prove that if a cubic has three roots equal, its Hessian quadratic vanishes identically and conversly.

6. If the cubic $ax^3 + 3bx^2 + 3cx + d = l(x + \theta)^3 + m(x + \phi)^3$, show that δ and ϕ are the roots of the equation.

$$\begin{vmatrix} 1 & x & x^2 \\ a & b & c \\ a & c & d \end{vmatrix} = 0$$

7. Show that the condition that the cubic $ax^3 + 3bx^2 + 3cx + d = 0$ may be capable of being written under the form $l(x - \alpha_1)^3 + m(x - \beta_1)^3 + n(x - \gamma_1)^3$ where $\alpha_1, \beta_1, \gamma_1$, are the roots of cubic $a_1x^3 + 3b_1x^2 + 3c_1x + d_1 = 0$ is $(ad_1 - a_1d) = 3(bc_1 - b_1c)$

8. Show that the relation between q and r in order that equation $x^3 + qx + r = 0$ may be put in the form $x^4 = (x^2 + ax + b)^2$ is given by $8r^2 + q^3$.

Objective Type Questions

Fill in the blanks :

1. To solve the cubic equation $a_0x^3 + 3a_1x^2 + 3a_2x + a_3 = 0$ by Cardan's method, we first remove the second term by diminishing its roots by $h =$ ....

2. The cubic equation $a_0x^3 + 3a_1x^2 + 3a_2x + a_3 = 0$ reduces to $z^3 + 3Hz + G = 0$ by $z = a_0x + a_1$, then G equals ....

3. If $z = u + v$ is a solution of the cubic equation $z^3 + 3Hz + G$, then $u^3 + v^3 =$ .... and $u^3v^3 =$ ....

4. If $x = u + v$ is a solution of $x^3 - 15x - 126 = 0$, then u^3 and v^3 are the roots of the quadratic equation ....

5. The roots of the cubic $z^3 + 3Hz + G = 0$ are all real if $G^2 +$ .... ≤ 0.

6. If $G^2 + 4H^3 > 0$, then $z^3 + 3Hz + G = 0$ has two .... roots.

7. If $G^2 + 4H^3 = 0$, then $z^3 + 3Hz + G = 0$ has two .... roots.

8. If the two roots of $28x^3 - 9x^2 + 1 = 0$ are $\frac{1}{7}(2 \pm i\sqrt{3})$, then its third root is ....,

9. If $x^3 - 3x^2 + 33x - 1 \equiv A(x - u)^3 - B(x - v)^3 = 0$, then u and v are the roots of the equation $x^2 + x + \dots \equiv 0$

10. If the cubic equation $a_0x^3 + 3a_1x^2 + 3a_2x + a_3 = 0$, has two roots equal to α, then $\alpha = \dots$

True/False: *Write 'T' for true and 'F' for false statement.*

1. The equation $z^3 + 3Hz + G = 0$ has two equal roots if $G^2 + 4H^3 = 0$. (T/F)
2. A cubic equation with real coefficient has at least one real root.   (T/F)
3. To solve the cubic equation $a_0x^3 + 3a_1x^2 + 3a_2x + a_3 = 0$ by Cardan's method we first reduce the given cubic into $z^3 + 3Hz^2 + Gz = 0$ (T/F)
4. The equation $x^3 + 3Hx + G = 0$ all its roots real if $G^2 + 4H^3 > 0$.   (T/F)
5. If $a_0x^3 + 3a_1x^2 + 3a_2x + a_3 \equiv A(x - u)^3 + B(x - v)^3$, then
$$(a_0a_2 - a_1^2)x^2 + (a_0a_3 - a_1a_2)x + (a_1a_3 - a_2^2) = 0$$ is known as Hessian of the given cubic.   (T/F)
6. If $z = u + v$ is a solution of $z^3 + 3Hz + G = 0$, then $z = u\omega + \omega^2$ is also the solution of given equation where ω is cube root

of unity.   (T/F)

7. The equation $x^3 + 3bx + c = 0$ has two imaginary roots of $c^2 + 4b^2 > 0$ (T/F)

Multiple Choice Questions : *Choose the most appropriate one :*

1. If $z = u + v$ is a solution of $z^3 + 3Hz + G = 0$, then $u^3 + v^3$ equals :
   (a) G (b) $-G$ (c) H (d) $-H$

2. If $z = u + v, z = u\omega + v\omega^2$ are the roots of $z^3 + 3Hz + G = 0$ then its third root is :
   (a) $u\omega^2 + v\omega$ (b) $u\omega^2 - v\omega$
   (c) $u\omega - v\omega^2$ (d) $u - v$

3. If $z = u + v$ is a solution $z^3 - 12z - 65 = 0$, then u and v are the roots of the quadratic :
   (a) $t^2 + 65t - 64 = 0$
   (b) $t^2 - 65t + 64 = 0$
   (c) $t^2 - 64t + 65 = 0$
   (d) $t^2 + 64t + 65 = 0$

4. To remove the second term of the cubic $27x^3 + 54x^2 + 198x - 73 = 0$ we diminish its roots by diminishing $h = \dots$
   (a) $\frac{2}{3}$ (b) $-\frac{3}{2}$ (c) $-\frac{2}{3}$ (d) $\frac{3}{2}$

5. If $x^3 + 3x^2 - 27x + 104 \equiv A(x - u)^3 + B(x - v)^3$, then the Hessian of the given cubic is :
   (a) $10x^2 - 113x - 23 = 0$
   (b) $10x^2 + 113x - 23 = 0$
   (c) $10x^2 + 113x + 23 = 0$
   (d) $10x^2 - 113x + 23 = 0$

ANSWERS

Fill in the blanks

(1) $\dfrac{-a_1}{na_0}$ (2) $a_0^2a_3 - 3a_0a_1a_2 + 2a_1^3$ (3) $-G, -H^3$ (4) $t^2 - 126t + 125 = 0$ (5) $4H^3$

(6) Imaginary   (7) equal   (8) $-\dfrac{1}{4}$ (9) -12 (10) $a = \dfrac{a_1a_2 - a_0a_3}{2(a_0a_2 - a_1^2)}$

True/False

(1) T   (2) T   (3) F   (4) F   (5) T   (6) T   (7) T

Multiple Choice Questions

(1) b (2) a (3) b (4) c (5) a

CHAPTER 14

Solution of Biquadratic Equations

> ❖ Euler's Cubic
>
> ❖ Descartes' Method
>
> ❖ Biquadratic Equation
>
> ❖ Ferrari's Method

14.1 REDUCTION OF BIQUADRATIC EQUATION INTO EULER'S CUBIC AND REDUCING CUBIC

(i) **Reduction into Euler's Cubic.** Let the biquadratic equation be
$$a_0 x^4 + 4a_1 x^3 + 6a_2 x^2 + 4a_3 x + a_4 = 0 \qquad \text{...(1)}$$
First we remove the second term, i.e. $4a_1 x^3$ by diminishing each of its roots by a

constant $h = -\dfrac{a_1}{4a_0}$, we get

$$z^4 + 6Hz^2 - 4Gz + (a_0^2 I - 3H^2) = 0 \qquad \text{...(2)}$$

where $\quad z = a_0 x + a_1, \ H = a_0 a_2 - a_1^2, \ G = a_0^2 a_3 - 3a_0 a_1 a_2 + 2a_1^3$

and $\quad I = a_0 a_4 - 4a_1 a_3 + 3a_2^2.$

Let the solution of (2) be $z = \sqrt{a} + \sqrt{b} + \sqrt{c}$. $\qquad \text{...(3)}$

Squaring of both sides of (3), we get

$$z^2 = a + b + c + 2(\sqrt{a}\sqrt{b} + \sqrt{b}\sqrt{c} + \sqrt{c}\sqrt{a})$$

or $\qquad z^2 - (a + b + c) = 2(\sqrt{a}\sqrt{b} + \sqrt{b}\sqrt{c} + \sqrt{c}\sqrt{a})$

Again squaring of both sides, we get

$$[z^2 - (a + b + c)^2] = 4[ab + bc + ca + 2\sqrt{a}\sqrt{b}\sqrt{c}(\sqrt{a} + \sqrt{b} + \sqrt{c})]$$

or $\quad z^4 + (a + b + c)^2 - 2(a + b + c)z^2 = 4(ab + bc + ca) + 8\sqrt{a}\sqrt{b}\sqrt{c}(z)$

or $\quad z^4 - 2(a + b + c)z^2 - 8z\sqrt{a}\sqrt{b}\sqrt{c} + (a + b + c)^2 - 4(ab + bc + ca)^2 = 0. \quad \text{...(4)}$

The equation (2) and (4) are same, so comparing the coefficients of like powers of z, we get

$$(a + b + c) = -3H, \ \sqrt{a}\sqrt{b}\sqrt{c} = -\frac{G}{2}$$

and $\qquad (a + b + c)^2 - 4(ab + bc + ca) = a_0^2 I - 3H^2$

or $\qquad \Sigma a = -3H, \ abc = \dfrac{G^2}{4} s$

and $\quad\quad (\Sigma a)^2 - 4(\Sigma ab) = a_0^2 I - 3H^2 \quad$ or $\quad \Sigma ab = 3H^2 - \dfrac{a_0^2 I}{4}$

Therefore a, b, c are the roots of the equation
$$t^2 - (\Sigma a)t^2 + (\Sigma ab)t - abc = 0$$

or $\quad\quad t^3 + 3Ht^2 + \left(3H^2 - \dfrac{a_0^2 I}{4}\right)t - \dfrac{G^2}{4} = 0 \quad\quad\quad$ ...(5)

This equation is called **Euler's cubic of** biquadratic equation **(1)**.

(ii) **Reducing cubic.** Since we have a relation
$$G^2 + 4H^3 = a_0^2(HI - a_0 J)$$

where $\quad\quad\quad\quad J = a_0 a_2 a_4 + 2a_1 a_2 a_3 - a_0 a_3^2 - a_1^2 a_4 - a_2^3$

$\therefore \quad\quad \dfrac{G^2}{4} + H^3 = \dfrac{a_0^2 HI}{4} - \dfrac{a_0^3 J}{4} \quad$ or $\quad -\dfrac{G^2}{4} = H^3 - \dfrac{a_0^2 HI}{4} + \dfrac{a_0^3 J}{4}$

Substitute the value of $-\dfrac{G^2}{4}$ in (5), we get

$$t^3 + 3Ht^2 + \left(3H^2 - \dfrac{a_0^2 I}{4}\right)t + H^3 - \dfrac{a_0^2 HI}{4} + \dfrac{a_0^3 J}{4} = 0$$

or $\quad\quad t^3 + H^3 + 3H^2 t + 3Ht^2 - \dfrac{a_0^2 I}{4}t - \dfrac{a_0^2 HI}{4} + \dfrac{a_0^3 J}{4} = 0$

or $\quad\quad\quad\quad (t + H)^3 - \dfrac{a_0^2 I}{4}(t + H) + \dfrac{a_0^3 J}{4} = 0$

Let us put $t + H = a_0^2 \theta$ we get

$$a_0^6 \theta^3 - \dfrac{a_0^2 I}{4}(a_0^2 \theta) + \dfrac{a_0^3 J}{4} = 0 \quad \text{or} \quad 4a_0^3 \theta^3 - Ia_0 \theta + J \quad\quad\quad ...(6)$$

This cubic equation in θ is called the reducing cubic of the biquadratic equation (1).

14.2 RELATION BETWEEN THE ROOTS OF BIQUADRATIC AND EULER'S CUBIC

Since the equation of the biquadratic and Euler's cubic are respectively given by
$$a_0 x^4 + 4a_1 x^3 + 6a_2 x^2 + 4a_3 x + a_4 = 0 \quad\quad\quad ...(1)$$

and $\quad\quad t^3 + 3Ht^2 + \left(3H^2 - \dfrac{a_0^2 I}{4}\right)t - \dfrac{G^2}{4} = 0 \quad\quad\quad ...(2)$

Let $\alpha, \beta, \gamma, \delta$ be the roots of the biquadratic and a, b, c the roots of the Euler's Cubic. Since we have taken $\quad\quad\quad\quad z = \sqrt{a} + \sqrt{b} + \sqrt{c} \quad\quad\quad\quad\quad ...(3)$

and $\quad\quad\quad\quad\quad\quad\quad\quad z = a_0 x + a_1 \quad\quad\quad\quad\quad\quad ...(4)$

It has been observed from equation (2) that z will have eight values because $\sqrt{a}$, $\sqrt{b}$ and $\sqrt{c}$ will have double signs. But we have
$$\sqrt{a}\sqrt{b}\sqrt{c} = -\dfrac{G}{2}$$

$\therefore$ $$\sqrt{c} = \frac{G}{2\sqrt{a}\sqrt{b}}$$

$\therefore$ from (2), we get $\qquad z = \sqrt{a} + \sqrt{b} - \dfrac{G}{2\sqrt{a}\sqrt{b}}$ ...(5)

Now from this equation it is observed that z will have four values. The signs of $\sqrt{a}$, $\sqrt{b}$ and $\sqrt{c}$ are taken in such a way that equation (4) should be satisfied. Therefore if G is negative, then $-\dfrac{G}{2}$ will be positive. Thus we should take the signs of $\sqrt{a}$, $\sqrt{b}$ and $\sqrt{c}$ so as to make the quantity $\sqrt{a}\sqrt{b}\sqrt{c}$ positive. Let z_1, z_2, z_3 and z_4 be the four values of z as the equation (5) indicates.

$$\left.\begin{aligned}
z_1 &= a_0\alpha + a_1 = \sqrt{a} - \sqrt{b} - \sqrt{c} \\
z_2 &= a_0\beta + a_1 = -\sqrt{a} + \sqrt{b} - \sqrt{c} \\
z_3 &= a_0\gamma + a_1 = -\sqrt{a} - \sqrt{b} + \sqrt{c} \\
z_4 &= a_0\delta + a_1 = \sqrt{a} + \sqrt{b} + \sqrt{c}
\end{aligned}\right\} \dots(A)$$

(Using (2) and (3) and $\sqrt{a}$ is taken positive)

($\sqrt{b}$ taken positive)

($\sqrt{c}$ is taken positive)

(all are taken positive)

Adding first two equations and adding last two equations of above system of equations respectively, we get

$$a_0(\alpha + \beta) + 2a_1 = -2\sqrt{c}$$

and $\qquad a_0(\gamma + \delta) + 2a_1 = 2\sqrt{c}$

Now substracting these equations, we get

$$a_0(\alpha + \beta - \gamma - \delta) = -4\sqrt{c}$$

Squaring of both sides, we get

$$c = \frac{a_0^2}{16}(\alpha + \beta - \gamma - \delta)^2$$

Similarly we get $\qquad a = \dfrac{a_0^2}{16}(\beta + \gamma - \alpha - \delta)^2$

and $\qquad b = \dfrac{a_0^2}{16}(\gamma + \alpha - \beta - \delta)^2$

Hence we obtained the required relations as follows

$$a = \frac{a_0^2}{16}(\beta + \gamma - \alpha - \delta)^2, \; b = \frac{a_0^2}{16}(\gamma + \alpha - \beta - \delta)^2, \; c = \frac{a_0^2}{16}(\alpha + \beta - \gamma - \delta)^2$$

If G is a positive, then $-\dfrac{G}{2}$ will be negative. Therefore we will take the signs of $\sqrt{a}$, $\sqrt{b}$, $\sqrt{c}$ either all negative or two positive and one negative. Thus we shall obtain

$$\left.\begin{aligned}
z_1 &= a_0\alpha + a_1 = -\sqrt{a} + \sqrt{b} + \sqrt{c} \\
z_2 &= a_0\beta + a_1 = \sqrt{a} - \sqrt{b} + \sqrt{c} \\
z_3 &= a_0\gamma + a_1 = \sqrt{a} + \sqrt{b} - \sqrt{c} \\
z_4 &= a_0\delta + a_1 = -\sqrt{a} - \sqrt{b} - \sqrt{c}
\end{aligned}\right\} \dots(B)$$

and

Solving these equations we obtain the relations as the same as obtained above.

14.3   RELATION BETWEEN THE ROOTS OF BIQUADRATIC AND THE REDUCING CUBIC

The equations of biquadratic and reducing cubic are respectively given by

$$a_0 x^4 + 4a_1 x^3 + 6a_2 x^2 + 4a_3 x + a_4 = 0 \qquad \text{...(1)}$$

and

$$4a_0^3 \theta^3 - I a_0 \theta + J = 0 \qquad \text{...(2)}$$

Let $\alpha, \beta, \gamma, \delta$ be the roots of (1) and $\theta_1, \theta_2, \theta_3$ be the roots of (2) in the system of equations (A) of § 14.2, we have

$$a_0(\alpha - \beta) = 2(\sqrt{a} - \sqrt{b}) \quad \text{and} \quad a_0(\gamma - \beta) = -2(\sqrt{a} + \sqrt{b})$$

$$\therefore \qquad a_0^2 (\alpha - \beta)(\gamma - \delta) = 4(a - b)$$

But we have $t + H = a_0^2 \theta$

$$\therefore \qquad a + H = a_0^2 \theta_1$$
$$b + H = a_0^2 \theta_2$$
$$c + H = a_0^2 \theta_3$$

$$\therefore \qquad (a - b) = a_0^2 (\theta_1 - \theta_2)$$

$$\left. \begin{aligned} 4(a - b) &= 4a_0^2(\theta_1 - \theta_2) = -a_0^2(\alpha - \beta)(\gamma - \delta) \\ 4(b - c) &= 4a_0^2(\theta_2 - \theta_3) = -a_0^2(\beta - \gamma)(\alpha - \delta) \\ 4(c - a) &= 4a_0^2(\theta_3 - \theta_1) = -a_0^2(\gamma - \alpha)(\beta - \delta) \end{aligned} \right\} \qquad \text{...(A)}$$

Similarly

Substracting first equation of (A) from third equation of (A), we get

$$4 a_0^2 (\theta_3 - \theta_1 - \theta_1 + \theta_2) = a_0^2 [(\alpha - \beta)(\gamma - \delta) - (\gamma - \alpha)(\beta - \delta)]$$

$$4 a_0^2 (\theta_2 + \theta_3 - 2\theta_1) = a_0^2 [(\alpha - \beta)(\gamma - \delta) - (\gamma - \alpha)(\beta - \delta)]$$

But we have $\theta_1 + \theta_2 + \theta_3 = 0$, $\theta_2 + \theta_3 = -\theta_1$

$$\therefore \qquad 4 a_0^2 (-3\theta_1) = a_0^2 [(\alpha - \beta)(\gamma - \delta) - (\gamma - \alpha)(\beta - \delta)]$$

or

$$12\theta_1 = (\gamma - \alpha)(\beta - \delta) - (\alpha - \beta)(\gamma - \delta)$$

Similarly

$$12\theta_2 = (\alpha - \beta)(\gamma - \delta) - (\beta - \gamma)(\alpha - \delta)$$
$$12\theta_3 = (\beta - \gamma)(\alpha - \delta) - (\gamma - \alpha)(\beta - \delta)$$

Hence the required relations are

$$\theta_1 = \frac{1}{12} [(\gamma - \alpha)(\beta - \delta) - (\alpha - \beta)(\gamma - \delta)]$$

$$\theta_2 = \frac{1}{12} [(\alpha - \beta)(\gamma - \delta) - (\beta - \gamma)(\alpha - \delta)]$$

$$\theta_3 = \frac{1}{12} [(\beta - \gamma)(\alpha - \delta) - (\gamma - \alpha)(\beta - \delta)]$$

14.4   DESCARTES' METHOD FOR FINDING THE ROOTS OF A BIQUADRATIC

Let the equation of a biquadratic be

$$a_0 x^4 + 4a_1 x^3 + 6a_2 x^2 + 4a_3 x + a_4 = 0 \qquad \text{...(1)}$$

First we remove the second term, i.e. $4a_1x^3$ from (1), diminishing each of root of (1) by a constant $h = -\dfrac{a_1}{na_0}$, we get

$$z^4 + 6Hz^2 + 4Gz + a_0^2 I - 3H^2 = 0 \qquad \dots(2)$$

where $H = a_0a_2 - a_1^2$, $G = a_0^2a_3 - 3a_0a_1a_2 + 2a_1^3$, $I = a_0a_4 - 4a_1a_3 + 3a_2^2$ and $z = a_0x + a_1$.

Let us assume

$$z^4 + 6Hz^2 + 4Gz + a_0^2 I - 3H^2 \equiv (z^2 + kz + l)(z^2 - kz + m)$$

Now equating the coefficients of like powers of z, we get

$$l + m - k^2 = 6H, \quad k(m - 1) = 4G, \quad lm = a_0^2 I - 3H^2.$$

Solving first two of these equations for l and m, we get

$$\left.\begin{aligned} 2l &= k^2 + 6H - \dfrac{4G}{k} \\[2mm] \text{and} \quad 2m &= k^2 + 6H + \dfrac{4G}{k} \end{aligned}\right\} \qquad \dots(A)$$

Substitute these values of l, m in the following equation $lm = a_0^2 I - 3H^2$, we get

$$\left(k^2 + 6H - \dfrac{4G}{k}\right)\left(k^2 + 6H + \dfrac{4G}{k}\right)$$

$$= 4(a_0^2 I - 3H^2)$$

$$(k^3 + 6Hk - 4G)(k^3 + 6Hk + 4G) = 4(a_0^2 I - 3H^2)k^2$$

$$k^6 + 12Hk^4 + 4k^2(12H^2 - a_0^2 I) - 16G^2 = 0. \qquad \dots(3)$$

This is a cubic equation in k^2 so it will always have one positive real value of k^2 when k^2 is known, then the values of l and m are obtained from the equation (A). Thus the biquadratic (2) is obtained as the product of quadratics $(z^2 + kz + l)$ and $(z^2 - kz + m)$.

Now solving these two quadratics

$$z^2 + kz + l = 0 \quad \text{and} \quad z^2 - kz + m = 0$$

and finally from the transformation $z = a_0x + a_1$ we obtain the solution of the given biquadratic (1) corresponding to the roots of the equations

$$z^2 + kz + l = 0 \quad \text{and} \quad z^2 - kz + m = 0$$

14.5 FERRARI'S METHOD FOR FINDING THE ROOTS OF A BIQUADRATIC EQUATION

Let the equation of a biquadratic be

$$x^4 + 2a_1x^3 + a_2x^2 + 2a_3x + a_4 = 0 \qquad \dots(1)$$

Now adding $(ax + b)^2$ to each side of (1), we get

$$x^4 + 2a_1x^3 + a_2x^2 + 2a_3x + a_4 + (ax + b)^2 = (ax + b)^2$$

or $\quad x^4 + 2a_1x^3 + (a_2 + a^2)x^2 + 2(a_3 + ab)x + (a_4 + b^2) = (ax + b)^2 \qquad \dots(2)$

In order to determine a and b make the left side of above equation *a perfect square.* Suppose the perfect square of left side of (2) is $(x^2 + a_1x + k)^2$, then

$$x^4 + 2a_1x^3 + (a_2 + a^2)x^2 + 2(a_3 + ab)x + (a_4 + b^2) \equiv (x^2 + a_1x + k)^2 \qquad \dots(3)$$

Comparing the coefficients of like powers of x of (3), we get

$$a_1^2 + 2k = a_2 + a^2, \quad a_1k = a_3 + ab, \quad k^2 = a_4 + b^2.$$

Eliminating a and b between these equations, we get

$$(2k + a_1^2 - a_2)(k^2 - a_4) = (a_1 k - a_3)^2$$

or $$2k^3 - a_2 k^2 + 2(a_1 a_3 - a_4)k - a_1^2 a_4 + a_2 a_4 - a_3^2 = 0 \qquad \text{...(4)}$$

This is a cubic equation in k so it must have one real values of k. This real value is obtained by trial method. Once we obtained the value of k we thus obtain a and b and then put these values in (3) and using (2), we get

$$(x^2 + a_1 x + k^2) = (ax + b)^2$$

or $$x^2 + a_1 x + k = \pm(ax + b)$$

Thus the given biquadratic is obtained as the product of two quadratics

$$\left. \begin{array}{l} x^2 + (a_1 - a)x + (k - b) = 0 \\ x^2 + (a_1 + a)x + (k + b) = 0 \end{array} \right\} \qquad \text{...(5)}$$

and

On solving these quadratics we finally obtained the solution of the given quadratic.

SOLVED EXAMPLES

Example 1. *Solve the equation* $x^4 - 3x^2 - 42x - 40 = 0$ *by Descartes' method.*

Solution. Since the given equation is

$$x^4 - 3x^2 - 42x - 40 = 0 \qquad \text{...(1)}$$

Let us assume $$x^4 - 3x^2 - 42x - 40 \equiv (x^2 + kx + l)(x^2 - kx + m) = 0 \qquad \text{...(2)}$$

Equating the coefficients of like powers of x, we get

$$l + m - k^2 = -3 \qquad \text{or} \qquad l + m = -3 + k^2 \qquad \text{...(3)}$$

and $$k(m - l) = -42 \quad , \quad \text{or} \qquad m - l = -\frac{42}{k} \qquad \text{...(4)}$$

and $$lm = -40 \qquad \text{...(5)}$$

Solving (3) and (4), we get

$$2m = -3 + k^2 - \frac{42}{k}$$

and $$2l = -3 + k^2 + \frac{42}{k}$$

Substitute the values of l and m in (5) we get

$$\left(-3 + k^2 - \frac{42}{k}\right)\left(-3 + k^2 + \frac{42}{k}\right) = 4(-40)$$

or $$(k^3 - 3k - 42)(k^3 - 3k + 42) = -160k^2$$

or $$(k^3 - 3k)^2 - (42)^2 = -160k^2 \qquad \text{or} \quad k^6 - 6k^4 + 169k^2 - 1764 = 0$$

Let $k^2 = t$, then we get

$$t^3 - 6t^2 + 169t - 1764 = 0$$

By trial method it is obvious that $t = 9$ satisfies above equation.

Hence $$k^2 = 9 \text{ or } k = \pm 3$$

Taking $k = 3$, in (3) and (4), we get

$$l + m = 6 \quad \text{and} \quad m - l = -14$$

Solving these equation for l and m, we get $l = 10$, $m = -4$ therefore from (2) we obtain the given biquadratic as the product of two quadratics

$$(x^2 + 3x + 10)(x^2 - 3x - 4) = 0$$

Solving these quadratics respectively we get the required solutions

$$x = 4, -1, \frac{-2 \pm i\sqrt{31}}{2}$$

Example 2. *Solve the equation* $x^4 + 8x^3 + 9x^2 - 8x - 10 = 0$ *by Descartes' method.*

Solution. The given equation is

$$x^4 + 8x^3 + 9x^2 - 8x - 10 = 0 \qquad \qquad ...(1)$$

First we remove the second term, i.e. by diminishing each of its roots by a constant

$$h = -\frac{a_1}{na_0} = -\frac{8}{4.1} = -2$$

Using synthetic division method:

```
-2 |  1      8       9     -8    -10
   |        -2     -12      6      4
   |--------------------------------
      1 .    6      -3     -2    | -6
            -2      -8     22
      --------------------------
      1      4     -11   | 20
            -2      -4
      ------------------
      1      2    | -15
            -2
      ----------
      1    | 0
      ---
      1
```

Thus the transformed equation is

$$z^4 - 15z^2 + 20z - 6 = 0 \qquad \qquad ...(2)$$

where $\qquad \qquad z = x + 2$

Let us assume $\qquad z^4 - 15z^2 + 20z - 6 \equiv (z^2 + kz + l)(z^2 - kz + m) = 0 \qquad ...(3)$

Comparing the coefficients of like powers of z, we get

$$l + m - k^2 = -15 \qquad \text{or} \qquad l + m = -15 + k^2 \qquad \qquad ...(4)$$

$$k(m - l) = 20 \qquad \text{or} \qquad m - l = \frac{20}{k} \qquad \qquad ...(5)$$

and $\qquad \qquad lm = -6$

Solving (4) and (5), we get

$$2l = k^2 - 15 - \frac{20}{k}$$

$$2m = k^2 - 15 + \frac{20}{k}$$

Substitute these values of l and m in (6), we get

$$\left(k^2 - 15 - \frac{20}{k}\right)\left(k^2 - 15 + \frac{20}{k}\right) = -24$$

or $\qquad (k^3 - 15k - 20)(k^3 - 15k + 20) = -24k^2$

or $\qquad (k^3 - 15k)^2 - 400 = -24k^2 \quad$ or $\quad k^6 - 30k^4 + 249k^2 - 400 = 0$

Let $k^2 = t$, then $\qquad t^3 - 30t^2 + 249t - 400 = 0 \qquad \qquad ...(7)$

From (7) it is obvious that $t = 16$ satisfies the equation (7)

∴ $k^2 = 16$ or $k = \pm 4$

Taking $k = 4$, in (4) and (5), we get

$$l + m = 1$$
$$m - l = 5$$

On solving these equations, we get

$$l = -2, \ m = 3.$$

Substitute the values of l, m and k in (3), we get

$$z^4 - 15z^2 + 20z - 6 \equiv (z^2 + 4z - 2)(z^2 - 4z + 3) = 0$$

∴ $(z^2 + 4z - 2)(z^2 - 4z + 3) = 0$ and $z = 1, 3, -2 \pm \sqrt{6}$

But $z = x + 2$

$$x = z - 2$$

Hence the solutions of the given biquadratic are

$$x = -1, 1, -4 \pm \sqrt{6}.$$

Example 3. *Solve the equation* $x^4 - 2x^3 - 5x^2 + 10x - 3 = 0$ *by Ferrari's method.*

Solution. Since the equation is

$$x^4 - 2x^3 - 5x^2 + 10x - 3 = 0 \qquad \qquad ...(1)$$

Adding $(ax + b)^2$ of both sides we get

$$x^4 - 2x^3 - 5x^2 + 10x - 3 + (ax + b)^2 = (ax + b)^2$$

or $x^4 - 2x^3 + (a^2 - 5)x^2 + 2(ab + 5)x + b^2 - 3 = (ax + b)^2 \qquad ...(2)$

Let us assume that L.H.S. of (2) must be a perfect square, let us suppose $(x^2 - a_1x + k^2)$ is a perfect square of L.H.S. of (2)

∴ $x^4 - 2x^3 + (a^2 - 5)x^2 + 2(ab + 5)x + b^2 - 3 \equiv (x^2 - x + k)^2 \qquad ...(3)$

$$(\because a_1 = -1)$$

Equating the coefficients of like powers of x, we get

$$a^2 = 2k + 6, \ ab = -k - 5, \ b^2 = k^2 + 3$$

Now eliminating a and b between these three equations, we get

$$(2k + 6)(k^2 + 3) = (k + 5)^2 \quad \text{or} \quad 2k^3 + 5k^2 - 4k - 7 = 0$$

It is a cubic in k so it must have one real root, then by trial method, we get

$$k = -1$$

and hence $a^2 = 4, \ b^2 = 4, \ ab = -4$ or $a = 2, \ b = -2$

Substitute the values of k, a, and b in (3) and (4), we get

$$(x^2 - x - 1)^2 = (2x - 2)^2 \quad \text{or} \quad x^2 - x - 1 = \pm (2x - 2)$$

or $x^2 - 3x + 1 = 0$ and $x^2 + x - 3 = 0$

Solving these quadratics, we get $x = \dfrac{3 \pm \sqrt{5}}{2}, \ \dfrac{-1 \pm \sqrt{13}}{2}$

These are the solutions of the given biquadratic equation.

Example 4. *Solve the equation* $x^4 + 2x^3 - 7x^2 - 8x + 12 = 0$ *by Ferrari's method.*

Solution. Since the given biquadratic is

$$x^4 + 2x^3 - 7x^2 - 8x + 12 = 0 \qquad \qquad ...(1)$$

Adding $(ax + b)^2$ of both sides of (1), we get

$$x^4 + 2x^3 - 7x^2 - 8x + 12 + (ax + b)^2 = (ax + b)^2$$

or $\qquad x^4 + 2x^3 + (a^2 - 7)x^2 + (2ab - 8)x + b^2 + 12 = (ax + b)^2$ ...(2)

In order to determine a and b make the L.H.S. of (2) a perfect square. Let the perfect square be $(x^2 + a_1 x + k)^2$.

$\therefore \qquad x^4 + 2x^3 + (a^2 - 7)x^2 + (2ab - 8)x + b^2 + 12 \equiv (x^2 + a_1 x + k)^2$

or $\qquad x^4 + 2x^3 + (a^2 - 7)x^2 + (2ab - 8)x + b^2 + 12 \equiv (x^2 + x + k)^2$ ...(3)

$(\because a_1 = 1 \text{ from (1)})$

Equating the coefficients of like powers of x, we get

$$a^2 - 7 = 2k + 1, \ 2ab - 8 = 2k, \ b^2 + 12 = k^2$$

Eliminating a and b between above three equations, we get

$$(k + 4)^2 = (2k + 8)(k^2 - 12)$$

or $\qquad k^3 + 16 + 8k = 2k^2 + 8k^2 - 24k - 96 \quad$ or $\quad 2k^3 + 7k^2 - 32k - 112 = 0$

This is a cubic in k so it must have one real root. By trial method, $k = -7/2$ satisfies above cubic.

Then $\qquad\qquad a^2 = 1, b^2 = \dfrac{1}{4}$

$\therefore \qquad\qquad a = 1, \ b = \dfrac{1}{2}$

Now substitute the values of k, a and b in (3) and using (2), we get

$$\left(x^2 + x - \frac{7}{2}\right)^2 = \left(x + \frac{1}{2}\right)^2 \quad \text{or} \quad x^2 + x - \frac{7}{2} = \pm \left(x + \frac{1}{2}\right)$$

or $\qquad x^2 - 4 = 0 \quad$ and $\quad x^2 + 2x - 3 = 0$

Solving these quadratics, we get

$$x = -2, \ 2, \text{ and } x = 1, -3$$

Hence the solution of given biquadratic are

$$x = -3, -2, 1, 2$$

Example 5. *Show that the equation $x^4 + px^3 + qx^2 + rx + s = 0$ may be solved as a quadratic if $r^2 = p^2 s$.*

Solution. Since the given equation is

$$x^4 + px^3 + qx^2 + rx + s = 0 \qquad\qquad ...(1)$$

Let us assume $\qquad x^4 + px^3 + qx^2 + rx + s \equiv \left(x^2 + \frac{p}{2}x + l\right)^2 = 0$

Comparing the coefficients of like powers of x, we get

$$2l\left(q^2 - \frac{p^2}{4}\right), \ pl = r, \ l^2 = s.$$

Eliminating l between last two equations, we get

$$pl = r \implies p^2 l^2 = r^2 \quad \text{or} \quad p^2 s = r^2$$

$\because \qquad\qquad l^2 = s.$

EXERCISE 14.1

Solve the following biquadratic equation by Descartes' method :

1. $x^4 - 6x^3 - 9x^2 + 66x - 22 = 0$
2. $x^4 - 8x^2 - 24x + 7 = 0$
3. $x^4 - 10x^2 - 20x - 16 = 0$
4. $x^4 + 2x^3 - 7x^2 - 8x + 12 = 0$

5. $x^4 - 8x^3 - 12x^2 + 60x + 63 = 0$

6. $x^4 - 3x^2 - 6x - 2 = 0$

7. $x^4 - 5x^2 - 6x - 5 = 0$

8. $x^4 - 12x - 5 = 0$

9. $4x^4 - 20x^3 + 33x^2 - 20x + 4 = 0$.

10. $x^4 + 8x^3 + 9x^2 - 8x - 10 = 0$

11. $x^4 - 2x^2 + 8x - 3 = 0$

Solve the following biquadratic equation by Ferrari's method :

12. $x^4 - 8x^3 - 12x^2 + 60x + 63 = 0$

13. $x^4 + 12x - 5 = 0$

14. $x^4 - 2x^3 - 5x^2 + 10x - 3 = 0$

15. $x^4 - 3x^2 - 42x - 40 = 0$

16. $x^4 + 9x^3 + 12x^2 - 80x - 192 = 0$

17. $x^4 - 2x^3 - 12x^2 + 10x + 3 = 0$

18. $x^4 - 10x^3 + 44x^2 - 104x + 96 = 0$

19. $x^4 + 4x^3 + 12x^2 - 8x + 95 = 0$

20. $x^4 + 3x^3 + x^2 - 2 = 0$

21. $2x^4 + 6x^3 - 3x^2 + 2 = 0$

HINT TO SELECTED PROBLEMS

1. $h = \dfrac{3}{2}$ The transfomed equation is

$$y^4 - \frac{90}{4}y^2 + \frac{96}{8}y + \frac{665}{16} = 0$$
$$\Rightarrow \quad (2y)^4 - 90(2y)^2 + 96(2y) + 665 = 0$$

Put $2y = z$.

5. $h = 2$

The transformed equation is

$$z^4 - 36z^2 - 52z + 87 = 0, \quad \text{where } z = x - 2$$

Let us write $\quad z^4 - 36z^2 + 52z + 89 = (z^2 + kz + l)(z^2 - kz + m) = 0$

11. Assume that $\quad x^4 - 2x^2 + 8x - 3 = (x^2 + kx + l)(x^2 - kx + m) = 0$

13. Adding $(ax + b)^2$ of both the sides of the given equation.

ANSWERS

(1) $\pm\sqrt{11}, 3 \pm \sqrt{7}$

(2) $-2 \pm i\sqrt{3}, 2 \pm \sqrt{3}$

(3) $4, -2, -1 \pm i$

(4) $\pm 2, -3, 1$

(5) $-1, 3, 3 \pm \sqrt{30}$

(6) $1, \pm\sqrt{2}, -1 \pm i$

(7) $\dfrac{-1 \pm i\sqrt{3}}{2}, \dfrac{1 \pm \sqrt{21}}{2}$

(8) $-1 \pm 2i, 1 \pm \sqrt{2}$

(9) $2, 2, \dfrac{1}{2}, \dfrac{1}{2}$

(10) $\pm 1, -4 \pm \sqrt{6}$

(11) $1 \pm i\sqrt{2}, -1 \pm \sqrt{2}$

(12) $-1, 3, 3 \pm \sqrt{30}$

(13) $-1 \pm \sqrt{2}, -1 \pm 2i$

(14) $\dfrac{3 \pm \sqrt{5}}{2}, \dfrac{-1 \pm \sqrt{13}}{2}$

(15) $4, -1, -\dfrac{1}{2}(3 \pm i\sqrt{31})$

(16) $-4, -4, -4, 3$

(17) $1, -3, 2 \pm \sqrt{5}$

(18) $2, 4, 2 \pm i2\sqrt{2}$

(19) $-3 \pm i\sqrt{10}, 1 \pm 2i$

(20) $-1 \pm \sqrt{3}, \dfrac{-1 \pm i\sqrt{3}}{2}$

(21) $-2 \pm \sqrt{2}, \dfrac{1}{2}(1 \pm i)$

CHAPTER REVIEW : A COMPETITIVE APPROACH

Selected terms and Results

TERMS

- **Reducing cubic :** The equation $4a_0^3?^3 - Ia_0? + J = 0$ is called reducing cubic of biquadratic equation.

- **Euler's cubic :** The equation

$$t^3 + 3Ht^2 + \left(3H^2 - \frac{a_0^2 I}{4}\right)t - \frac{G^2}{4} = 0 \quad \text{is}$$

called Euler's cubic.

RESULTS

- When the roots of biquadratic are all reals, the roots of Euler's cubic are all real and positive.

- When the quadratic has all its roots imaginary, the roots of Euler's cubic are all real, two being negative and one positive.

- When the biquadratic has two roots real and two imaginary then Euler's cubic has two imaginary and one real positive root.

Review Questions and Project Work

1. Prove that if the biquadratic has two distinct pair of equal roots, then

   $a_0^2 I = 12H^2$ and $a^3 J = 8H^3$

2. When a biquadratic has two equal roots, prove that Euler's cubic has two equal roots whose common value is $\dfrac{3aJ - 2IH}{2I}$ and hence show that the remaining two roots of the biquadratic in this case are real, equal

or imaginary according as $3aJ - 2HI$ is positive, zero or negative.

3. If the roots of α, β, γ, δ of the biquadratic $ax^4 + 4bx^3 + 6cx^2 + 4dx + e = 0$ be so related that $\alpha - \delta$, $\beta - \delta$ and $\gamma - \delta$ be in H.P, show that $ace + 2bcd - ad^2 - b^2e - c^3 = 0$

4. If the roots of a biquadratic α, β, γ, δ represent the distances of four points from an origin on a right line, prove that when these points from a harmonic division on the line, the roots of Euler's cubic are in A.P.

Objective Type Questions

Fill in the blanks:

1. To solve the biquadratic equation $a_0x^4 + a_1x^3 + a_2x^2 + a_3x + a_4 = 0$ by Descartes' method, we first remove its second term by diminishing its roots by $h = \ldots\ldots\ldots\ldots$

2. The biquadratic equation $a_0x^4 + 4a_1x^3 + 6a_2x^2 + 4a_3x + a_4 = 0$ reduces to the cubic

$$t^3 + 3Ht^2 + \left(3H^2 - \frac{a_0^2 I}{4}\right)t - \frac{G^2}{4} = 0 \ .$$

   Then this cubic is known as $\ldots\ldots\ldots\ldots$

3. To solve $x^4 - 8x^3 - 12x^2 + 60x + 63 = 0$ by Descartes' method we first remove the second term by diminishing its roots by $h = \ldots\ldots\ldots\ldots$

4. If the two roots of $x^4 + 12x - 5 = 0$ are $-1 + \sqrt{2}$ and $1 - 2i$ then its other roots are $\ldots\ldots\ldots\ldots$

5. If $x^4 - 2x^2 + 8x - 3 \equiv (x^2 + kx + l)(x^2 - kx + m)$, then $l + m - k^2 = \ldots\ldots\ldots\ldots$ and $k(m - l) = \ldots\ldots\ldots\ldots$ and $lm = \ldots\ldots\ldots\ldots$

6. If $x^4 - 2x^3 - 5x^2 + 10x - 3 \equiv (x^2 - x - 1)^4 - (2x - 2)^2 = 0$, then the roots of the $x^4 - 2x^3 - 5x^2 + 10x - 3 = 0$ are ....

7. The biquadratic equation $a_0x^4 + 4a_1x^3 + 6a_2x^2 + 4a_3x + a_4 = 0$ can be reduced to $z^4 + 6Hz^2 + 4Gz + (a_0^2I - 3H^2) = 0$ by $z =$ ....

8. If $f(x) \equiv x^4 - 3x^2 - 6x - 2 = (x^2 - 2x + 2)(x^2 - 2x - 1)$, then the roots of $f(x) = 0$ are ....

True/False: *Write 'T' for true and 'F' for false statement.*

1. If the two roots of equation $x^4 - 3x^2 - 6x - 2 = 0$ are $-1 + i$ and $1 + \sqrt{2}$, then its other roots are $1 + i$ and $1 - \sqrt{2}$. (T/F)

2. The equation $a_0x^4 + 4a_1x^3 + 6a_2x^2 + 4a_3x + a_4 = 0$ reduces to $z^4 + 6Hz^2 + 4Gz + (a_0^2I - 3H^2) = 0$ by $z = a_0x + a_1$. (T/F)

3. Solve the equation $x^4 - 6x^3 - 9x^2 + 66x - 22 = 0$ by Descartes' method we first remove the second term by diminishing its root by $h = \dfrac{3}{2}$. (T/F)

Multiple Choice Questions : *Choose the most appropriate one :*

1. If $x^4 - 2x^2 + 8x - 3 \equiv (x^2 + 2x + l)(x^2 - 2x + m)$, then the values of l and m are :
   (a) $-1, -3$ (b) $-1, 3$
   (c) $1, 3$ (d) $1, -3$

2. If two roots of $x^4 - 3x^2 - 6x - 2 = 0$ are $-1 + i$ and $1 + \sqrt{2}$ then its other two roots are :
   (a) $-1 - i, -1 + \sqrt{2}$
   (b) $-1 - i, -1 - \sqrt{2}$
   (c) $-1 - i, 1 - \sqrt{2}$
   (d) $1 + i, 1 - \sqrt{2}$

3. If $a_0x^4 + 4a_1x^3 + 6a_2x^2 + 4a_3x + a_4 = 0$ reduces to $z^4 + 6Hz^2 + 4Gz + (a_0^2I - 3H^2) = 0$ by $z = a_0x + a_1$, then H equals:
   (a) $a_0a_2 - a_1^2$ (b) $a_0a_1 - a_2^2$
   (c) $a_1a_2 - a_0^2$ (d) $a_1a_0 - a_2$

4. The sum of all the four roots of $a_0x^4 + a_1x^3 + a_2x^2 + a_3x + a_4 = 0$ is :
   (a) a_1/a_0 (b) $-a_1/a_0$
   (c) a_2/a_0 (d) $-a_1/a_2$

ANSWERS

Fill in the blanks :

(1) $h = -\dfrac{a_1}{4a_0}$ (2) Euler's cubic   (3) 2   (4) $-1 - \sqrt{2}, 1 + 2i$

(5) $-2, 8, -3$ (6) $\dfrac{3 \pm \sqrt{5}}{2}, \dfrac{-1 \pm \sqrt{13}}{2}$ (7) $a_0x + a_1$ (8) $-1 \pm i, 1 \pm \sqrt{2}$

True/False

(1) F   (2) T   (3) T

Multiple choice questions

(1) b   (2) c   (3) a   (4) b

CHAPTER 15

INEQUALITIES

❖ Inequality	❖ Greatest Value
❖ Maxima and Minima	❖ Some Important Inequalities

15.1 INTRODUCTION

A statement that two functions are unequal is called an inequality.

Such a statement is more definite if one of the functions is said to be greater than or less than the other. Here, we shall use the following symbols, to represent the words written against them :

$>$ represents greater than.

$<$ represents less than.

a is said to be greater than b if $a - b$ is +ive and a is said to be less than b if $a - b$ is negative.

i.e. $\qquad 5 > 2 \qquad \because \qquad 5 - 2 = 3$, a positive quantity.

$\qquad -3 < -1 \qquad \because \qquad -3 - (-1) = -2$, a negative quantity.

15.2 FUNDAMENTAL PROPERTIES OF INEQUALITIES

(1) If $a > b$ and c is any +ive number, then

    (i) $a + c > b + c$.   (ii) $a - c > b - c$

    (iii) $ac > bc$,   (iv) $\dfrac{a}{c} > \dfrac{b}{c}$

i.e. in an inequality we can always add, subtract, multiply and divide each side by the same +ive quantity, without affecting the sign of the inequality.

(2) If $a - c > b$ then by (1) we can add c to both sides.

$\therefore \qquad\qquad a - c + c > b + c$, i.e. $a > b + c$

which means that a term on one side of an inequality can be transposed on the other side provided its sign is changed.

(3) If $a > b$, then $a - b$ is positive.

Hence $-(a - b)$ will be negative.

$\therefore$ $(-a) - (-b)$ will be negative. $\qquad \therefore \qquad -a < -b$.

which means that if the sign of all the terms of the inequality be changed the sign of the inequality is also changed.

(4) If $a > b$, then $(a - b)$ is positive

$\therefore$ $c(a - b)$ is also positive.

But $- c (a - b)$ is negative.

$\therefore$ $(- ac) - (- bc)$ is negative. $\therefore$ $- ac < - bc$.

which means that if we multiply both sides of an inequality by some negative quantity, the sign of the inequality must also be changed.

(5) If $a_1 > b_1, a_2 > b_2, a_3 > b_3$ and so on, then

$$a_1 + a_2 + a_3 + \ldots a_n > b_1 + b_2 + b_3 + \ldots + b_n$$

and

$$a_1 \cdot a_2 \cdot a_3 \ldots a_n > b_1 \cdot b_2 \cdot b_3 \ldots b_n$$

(6) If $a > b$, then $a^n > b^n$, provided n is positive.

If $a > b$ then $\dfrac{1}{a} < \dfrac{1}{b}$ and $\dfrac{1}{a^n} < \dfrac{1}{b^n}$

(7) If $a > b$, then $e^a > e^b$ and $\log a > \log b$.

15.3 RELATION BETWEEN ARITHMETIC MEAN AND GEOMETRIC MEAN

Arithmetic mean of any two positive quantities is greater than their geometric mean.

We know that the square of every real number is positive.

$$(a - b)^2 \text{ is positive, i.e.} > 0.$$

or $a^2 + b^2 - 2ab > 0$ or $a^2 + b^2 > 2ab$

or $\dfrac{a^2 + b^2}{2} > ab$. Now put $x = a^2$ and $y = b^2$.

$\therefore$ $\dfrac{x + y}{2} > \sqrt{(xy)}$

15.4 SOME IMPORTANT RESULTS

If the sum of two positive quantities is constant, then their product is maximum when the quantities are equal and if the product is constant, then their sum is minimum when the quantities are equal.

Proof. Let a and b be two positive quantities whose product ab is denoted by P and their sum $a + b$ by S.

Now $4ab = (a - b)^2 - (a - b)^2$

i.e. $4P = S^2 - (a - b)^2$ or $S^2 = 4P + (a - b)^2$ ...(1)

(a) Now if S is constant, then from (1) P will be maximum when $(a - b)^2 = 0$, i.e. when $a = b.$, Which means when the quantities become equal.

(b) If P is constant, then from (1) S will be minimum when $(a - b)^2 = 0.$, i.e. when $a = b$ Which means when the quantities become equal.

FOR EXAMPLE

(i) Consider a sum 20. We can make a number of pairs of the form

(16, 4), (12, 8), (15, 5), (10, 10).

Product of these pairs whose sum in each case is equal to 20, i.e. constant is 64, 96, 75 and 100 respectively. We note the maximum product is that of equal factors (10, 10).

Hence if the sum is constant, then product of equal factors is greatest.

(*ii*) Consider a product 100. Then, we can make a number of factors of the form.

$$(20, 5), (25, 4), (10, 10), (50, 2).$$

Sum of these factors whose product in each case is equal to 100, *i.e.*, constant (given) is 25, 29, 20 and 52 respectively we observe that minimum sum is that of equal factors (10, 10).

Hence if the product is constant, then the sum of equal quantities is least.

THEOREM-1

Find the greatest value of a product, the sum of whose factors is constant, and hence show that arithmetic mean of n positive quantities is greater than their geometric mean.

Proof. Let there be n factors $a, b, c, ... k$, such that $a + b + c + d + ... + k = S$ (constant)

Consider the product $abcd ... k$. In this if we replace the two unequal quantities a and b by two equal quantities $\dfrac{a+b}{2}, \dfrac{a+b}{2}$ then the value of product will be increased without affecting the sum of the factors, because the factors $a, b, c, d, ... k$ have also a sum

$$= a + b + c + d + ... + k = S \text{ and the factors } \dfrac{a+b}{2}, \dfrac{a+b}{2}, c, d ... k \text{ have a sum}$$

$$= a + b + c + d + ... + k = S$$

Similarly, we can further increase the value of the product by replacing two unequal factors c and d by $\dfrac{c+d}{2}, \dfrac{c+d}{2}$ without causing any change in the sum of the factors which will be constant.

Hence so long as any two of the factors $a, b, c, d ... k$ are unequal, the value of the product can be increased by replacing them by equal factors, without altering the sum. Therefore the product will be maximum when all the factors are equal. In this case each of the factors $a, b, c, ... k$ will be

$$\dfrac{a+b+c+...+k}{n}, \dfrac{a+b+c+...+k}{n}, ... \dfrac{a+b+c+...+k}{n},$$

i.e.

$$\dfrac{S}{n}, \dfrac{S}{n}, \dfrac{S}{n}, ... \dfrac{S}{n}$$

and their sum $= n\dfrac{S}{n} = S$, the given constant.

Hence the maximum value of the product $(a, b, c, ... k)$ is

$$\left(\dfrac{S}{n} \cdot \dfrac{S}{n} \cdot \dfrac{S}{n} ... \dfrac{S}{n}\right), \quad \text{i.e.} \left(\dfrac{S}{n}\right)^n$$

$\therefore$ The maximum. product $\left(\dfrac{S}{n}\right)^n$ is greater than the product $(abc ... k)$

i.e. $\qquad \left(\dfrac{a+b+c\ldots k}{n}\right)^n > (abc\ldots k)$

or $\qquad \dfrac{a+b+c\ldots k}{n} > (abc\ldots k)^{1/n}$

$\therefore$ Arithmetic mean of n +ive quantities is greater than their Geometric mean, i.e. $A > G$.

...(1)

Remark

❖ If we consider the quantities $\dfrac{1}{a}, \dfrac{1}{b}, \dfrac{1}{c}, \ldots$ then

$$\dfrac{\dfrac{1}{a}+\dfrac{1}{b}+\dfrac{1}{c}\ldots\dfrac{1}{k}}{n} > \left(\dfrac{1}{a}\cdot\dfrac{1}{b}\cdot\dfrac{1}{c}\ldots\dfrac{1}{k}\right)^{1/n}$$

$$\Rightarrow \qquad \dfrac{n}{\dfrac{1}{a}+\dfrac{1}{b}+\dfrac{1}{c}\ldots\dfrac{1}{k}} < (abc\ldots k)^{1/n}$$

or $\qquad (abc\ldots k)^{1/n} < \dfrac{n}{\dfrac{1}{a}+\dfrac{1}{b}+\dfrac{1}{c}\ldots\dfrac{1}{k}}$ ...(2)

Again harmonic mean of n positive quantities is reciprocal of the arithmetic mean of the reciprocals of n quantities.

Hence from (2), we get $G > H$

$\therefore \qquad$ From (1) and (2), we conclude that $A > G > H$.

SOLVED EXAMPLES

Example 1. *If a, b, c are three positive quantities then prove the following :*

$a^2 + b^2 + c^2 > ab + bc + ca$ *and hence prove that* $\dfrac{a^2+b^2+c^2}{ab+bc+ca} > 1$ *and* < 2. *If a, b, c represent the sides of triangle. Also deduce that* $a^3 + b^3 + c^3 > 3abc$.

Solution. We have $\dfrac{a^2+b^2}{2} > (a^2b^2)^{1/2} \qquad$ or $\quad > ab \;\; (\because \text{ A.M.} > \text{G.M.})$

$\therefore \qquad a^2 + b^2 > 2ab$ $\qquad\qquad$ ...(1)

Similarly $\quad b^2 + c^2 > 2bc$ $\qquad\qquad$ ...(2)

and $\qquad\qquad c^2 + a^2 > 2ca$ $\qquad\qquad$ ...(3)

Adding (1), (2) and (3) we get

$\qquad\qquad 2(a^2 + b^2 + c^2) > 2(ab + bc + ca)$

or $\qquad\qquad a^2 + b^2 + c^2 > ab + bc + ca$ $\qquad\qquad$ ...(4)

Hence $\qquad \dfrac{a^2+b^2+c^2}{ab+bc+ca} > 1.$ $\qquad\qquad$ ...(A)

Since, as a, b, c are sides of a triangle then by cosine formula $b^2 + c^2 - a^2 = 2bc \cos A$. But $\cos A < 1$.

$\therefore$ $b^2 + c^2 - a^2 < 2bc$ and similarly $c^2 + a^2 - b^2 < 2ca$.
$$a^2 + b^2 - c^2 > 2ab.$$

Adding we get $(a^2 + b^2 + c^2) < 2(ab + bc + ca)$

$\therefore$
$$\frac{a^2 + b^2 + c^2}{ab + bc + ca} < 2 \qquad \qquad ...(B)$$

Hence from (A) and (B) we get the required result.

Now $a^3 + b^3 + c^3 - 3abc \;\; = (a + b + c)(a^2 + b^2 + c^2 - ab - bc - ca)$

$\qquad\qquad\qquad\qquad\quad = (+\text{ive})\ (+\text{ive})$ by $(4) = +\text{ive}$, i.e. > 0

$\therefore \qquad a^3 + b^3 + c^3 > 3abc.$

Remark

❖ $(a + b + c)^2 > 3(ab + bc + ca)$

Example 2. *Prove that*
$$(ab + bc + ca) < (a + b - c)^2 + (b + c - a)^2 + (c + a - b)^2$$

Solution. On simplification we have to prove that
$$(ab + bc + ca) \; < \; 3(a^2 + b^2 + c^2) - 2(ab + bc + ca)$$
or $\qquad\qquad 3(ab + bc + ca) \; < \; 3(a^2 + b^2 + c^2)$
or $\qquad\qquad a^2 + b^2 + c^2 \; > \; ab + bc + ca$

which is true as shown in example 1.

Example 3. *Prove that* $(x^2y + y^2z + z^2x)(xy^2 + yz^2 + zx^2) > 9x^2y^2z^2$
$$\frac{x^2y + y^2z + z^2x}{3} > (x^3y^3z^3)^{1/3} \ or \ > xyz$$
and $\qquad \dfrac{xy^2 + yz^2 + zx^2}{3} > (x^3y^3z^3)^{1/3} \ or \ > xyz$

Solution. Multiplying the given inequalities, we get
$$\frac{(x^2y + y^2z + z^2x)(xy^2 + yz^2 + zx^2)}{9} > x^2y^2z^2$$
$\therefore \quad (x^2y + y^2z + z^2x)(xy^2 + yz^2 + zx^2) > 9x^2y^2z^2$

15.5   THE GREATEST VALUE

To find the greatest value of a^m, b^n, c^p ... when $a + b + c$ ... is constant and m, n, p ... are positive integers.

Since m, n, p are all constants, the expression $a^m . b^n . c^p$ ... will be maximum, when
$$\left(\frac{a}{m}\right)^m \left(\frac{b}{n}\right)^n \left(\frac{c}{p}\right)^p \text{ ... is maximum}$$

or $\dfrac{a}{m}\cdot\dfrac{a}{m}\ldots m$ factor $\times \dfrac{b}{n}\cdot\dfrac{b}{n}\ldots n$ factors $\times \dfrac{c}{p}\cdot\dfrac{c}{p}\ldots p$ factors is maximum.

Now above is the product of $m + n + p + \ldots$ factors whose sum is $\dfrac{a}{m} + \dfrac{a}{m} + \ldots m$ factors

$+ \ldots \dfrac{b}{n} + \dfrac{b}{n} + \ldots n$ factors $+ \dfrac{c}{p} + \dfrac{c}{p} \ldots p$ factors $+ \ldots$

i.e. $m\left(\dfrac{a}{m}\right) + n\left(\dfrac{b}{n}\right) + p\left(\dfrac{c}{p}\right) + \ldots$

$= a + b + c + \ldots$ constant (given).

and hence the product will be maximum when all the factors are equal.

i.e. $\dfrac{a}{m} = \dfrac{b}{n} = \dfrac{c}{p} = \ldots = \dfrac{a+b+c}{m+n+p+\ldots}$

$\therefore$ $a = m.\dfrac{\Sigma a}{\Sigma m}, \quad b = n.\dfrac{\Sigma a}{\Sigma m}, \quad c = p.\dfrac{\Sigma a}{\Sigma m}$

Maximum value of $a^m. b^n. c^p \ldots$

$= m^m \left(\dfrac{\Sigma a}{\Sigma m}\right)^m .n^n \left(\dfrac{\Sigma a}{\Sigma m}\right)^n .p^p \left(\dfrac{\Sigma a}{\Sigma m}\right)^p \ldots$

$= m^m .n^n .p^p \ldots \left(\dfrac{\Sigma a}{\Sigma m}\right)^{m+n+p}$

$= m^n n^n p^n \ldots \left(\dfrac{a+b+c\ldots}{m+n+p\ldots}\right)^{m+n+p}$

Remark

❖ $\dfrac{p_1 a_1 + p_2 a_2 + \ldots p_n a_n}{p_1 + p_2 + \ldots + p_n} > \left[a_1^{p_1} a_2^{p_2} \ldots a_n^{p_n}\right]^{\frac{1}{p_1 + p_2 + \ldots p_n}}$

SOLVED EXAMPLES

Example 1. *Show that unless $p = q = r$ or $x = 1$,*

$$px^{q-r} + qx^{r-p} + rx^{p-q} > p + q + r$$

Solution. We know that A.M. > G.M.

$$\dfrac{(x^{q-r} + x^{q-r} + \ldots p \text{ times}) + (x^{r-p} + x^{r-p} + \ldots q \text{ times}) + (x^{p-q} + x^{p-q} + \ldots r \text{ times})}{p+q+r}$$

$$> \left\{ (x^{q-r}.x^{q-r} \ldots p \text{ factors}) + (x^{r-p}.x^{r-p} \ldots q \text{ factors}) + (x^{p-q}.x^{p-q} \ldots r \text{ factors}) \right\}^{1/(p+q+r)}$$

or $\dfrac{px^{q-r} + qx^{r-p} + rx^{p-q}}{p+q+r} > \{x^{p(q-r) + q(r-q) + r(p-q)}\}^{1/(p+q+r)}$

or
$$\frac{px^{q-r} + qx^{r-p} + rx^{p-q}}{p+q+r} > [x^0]^{1/(p+q+r)} \text{ or } > 1.$$

or
$$px^{q-r} + qx^{r-p} + rx^{p-q} > p + q + r$$

Example 2. *Prove that*

$$\left(\frac{x^2 + y^2 + z^2}{x+y+z}\right)^{x+y+z} > x^x y^y z^z > \left(\frac{x+y+z}{3}\right)^{x+y+z}$$

Solution. Consider x numbers each equal to x, y numbers each equal to y and z numbers each equal to z. Apply A.M. > G.M. on these $x + y + z$ numbers.

$$\therefore \quad \frac{(x+x+x\ldots x \text{ times}) + (y+y+y+\ldots y \text{ times}) + (z+z+z+\ldots z \text{ times})}{x+y+z}$$

$$> \{(x.x \ldots x \text{ factors})(y.y \ldots \text{ factors})(z.z \ldots \text{ factors})\}^{1/(x+y+z)}$$

But $\quad x + x + x \ldots x$ times $= x.x. = x^2$

$\qquad x.x.x \ldots x$ factors $= x^x$

$$\therefore \quad \frac{x^2 + y^2 + z^2}{x+y+z} > (x^x y^y z^z)^{1/(x+y+z)}$$

or
$$\left(\frac{x^2 + y^2 + z^2}{x+y+z}\right)^{x+y+z} > x^x y^y z^z \tag{A}$$

Now, consider x numbers each equal to $\dfrac{1}{x}$, y numbers each equal to $\dfrac{1}{y}$, z numbers each equal to $\dfrac{1}{z}$. Apply A.M. > G.M. on these $x + y + z$ numbers.

$$\therefore \quad \frac{\left(\dfrac{1}{x}+\dfrac{1}{x}+\ldots x \text{ times}\right)+\left(\dfrac{1}{y}+\dfrac{1}{y}+\ldots y \text{ times}\right)+\left(\dfrac{1}{z}+\dfrac{1}{z}+\ldots z \text{ times}\right)}{x+y+z}$$

$$> \left[\left(\dfrac{1}{x}.\dfrac{1}{x}\ldots x \text{ factors}\right)\left(\dfrac{1}{y}.\dfrac{1}{y}\ldots y \text{ factors}\right)\left(\dfrac{1}{z}.\dfrac{1}{z}\ldots z \text{ factors}\right)\right]^{1/(x+y+z)}$$

or
$$\frac{\dfrac{1}{x}.x + \dfrac{1}{y}.y + \dfrac{1}{z}.z}{x+y+z} > \left[\dfrac{1}{x^x}.\dfrac{1}{y^y}.\dfrac{1}{z^z}\right]^{1/(x+y+z)}$$

or
$$\left(\frac{3}{x+y+z}\right)^{x+y+z} > \frac{1}{x^x y^y z^z}$$

$\Rightarrow$
$$\left(\frac{x+y+z}{3}\right)^{x+y+z} < x^x y^y z^z$$

or $\qquad$ $x^x y^y z^z < \left(\dfrac{x+y+z}{3}\right)^{x+y+z}$ $\qquad$ ...(B)

Remark

❖ We can generalized the above result as follows.

$$\left(\frac{a^2 + b^2 + c^2 + \dots k^2}{a+b+c+\dots k}\right)^{a+b+c+\dots k} > a^a b^b c^c \dots k^k > \left(\frac{a+b+c+\dots +k}{n}\right)^{a+b+c+\dots +k}$$

where $a, b, c \dots k$ are n +ive quantities.

Example 3. *Prove that* $\left\{\dfrac{bc+ca+ab}{a+b+c}\right\}^{a+b+c} > \sqrt{[(bc)^a (ca)^b (ab)^c]}$

or $\qquad$ $\left[\dfrac{bc+ca+ab}{a+b+c}\right]^{2(a+b+c)} > (a^{b+c} b^{c+a} c^{a+b})$

Solution. Here, it is clear that we must consider $2(a + b + c)$ numbers and $(b + c)$ should be equal to a, $(c + a)$ equal to b and $(a + b)$ equal to c.

∵ A.M. > G.M., we have

$$\frac{[(a+a+\dots(b+c)\text{ times}]+[(b+b+\dots(c+a)\text{ times}]+[c+c+\dots(a+b)\text{ times}]}{b+c+c+a+a+b}$$
$$> (a^{b+c} b^{c+a} c^{a+b})^{1/2(a+b+c)}$$

or $\qquad$ $\dfrac{a(b+c)+b(c+a)+c(a+b)}{2(a+b+c)} > (a^{b+c} b^{c+a} c^{a+b})^{1/2(a+b+c)}$

or $\qquad$ $\left(\dfrac{ab+bc+ca}{a+b+c}\right)^{2(a+b+c)} > a^{b+c} b^{c+a} c^{a+b}$

or $\qquad$ $\left(\dfrac{ab+bc+ca}{a+b+c}\right)^{a+b+c} > \sqrt{[(bc)^a (ca)^b (ab)^c]}$

THEOREM-1

If a and b are positive and unequal, then $\dfrac{a^m + b^m}{2} > \left(\dfrac{a+b}{2}\right)^m$ except when m is a positive proper fraction.

Proof. We can write $\qquad$ $a^m = \left(\dfrac{a+b}{2}+\dfrac{a-b}{2}\right)^m = \left(\dfrac{a+b}{2}\right)^m \left[1+\dfrac{a-b}{a+b}\right]^m$

$$b^m = \left(\dfrac{a+b}{2}-\dfrac{a-b}{2}\right)^m = \left(\dfrac{a+b}{2}\right)^m \left[1-\dfrac{a-b}{a+b}\right]^m$$

Now $\quad (1 + t)^m + (1 - t)^m = 2\left[1+\dfrac{m(m-1)}{2!}t^2 + \dfrac{m(m-1)(m-2)(m-3)t^4}{4!}+\dots\right]$

because the alternate terms cancel,

∴ $\qquad$ $a^m + b^m = 2\left[\dfrac{a+b}{2}\right]^m \left[1+\dfrac{m(m-1)}{2!}t^2 + \dfrac{m(m-1)(m-2)(m-3)t^4}{4!}+\dots\right]$

where $t = \dfrac{a-b}{a+b}$

Case I. *When m is positive integer or a negative quantity.*

Suppose $m = 3$, then all the three terms in the above bracket will be positive. All other terms that follow the third term will have factor $m - 4$ and hence will vanish.

Again if $m = $ –ive quantity, then

$m(m - 1)$ will be positive.

Similarly all other terms in the bracket will be positive.

$\therefore \quad a^m + b^m = 2\left(\dfrac{a+b}{2}\right)^m$ multiplied by something positive but greater than 1.

$\therefore \quad \dfrac{a^m + b^m}{2} > \left[\dfrac{a+b}{2}\right]^m$

Case II. *When m is a positive proper fraction,* i.e. *it lies between 0 and 1.*

In this case all the terms in the bracket except the first will be negative because m being positive and less than 1, $(m - 1)$ will be negative.

$\therefore \qquad\qquad m(m - 1)$ is negative.

Similarly $m(m - 1)(m - 2)(m - 3)$ will be negative.

$\therefore \quad$ It follows clearly that $\dfrac{a^m + b^m}{2} < \left(\dfrac{a+b}{2}\right)^m$

Case III. *When m > 1.*

Now the case when m is a +ive integer greater than 1, has been dealt in Case I, so here we shall take up the case when m is a fraction which is greater than 1.

Put $m = \dfrac{1}{n}$, then n must be < 1.

By Case II, when n is < 1.

$$\frac{A^n + B^n}{2} < \left(\frac{A+B}{2}\right)^n$$

Putting $A = a^m$, $B = b^m$, we get

$$\frac{a^{mn} + b^{mn}}{2} < \left(\frac{a^m + b^m}{2}\right)^n$$

or $\qquad \dfrac{a+b}{2} < \left(\dfrac{a^m + b^m}{2}\right)^{1/m}$ $\qquad\qquad \because mn = 1.$

$\therefore \qquad \left(\dfrac{a+b}{2}\right)^m < \left(\dfrac{a^m + b^m}{2}\right)$

Hence $\qquad \left(\dfrac{a^m + b^m}{2}\right) > \left(\dfrac{a+b}{2}\right)^m$

THEOREM-2

If there are m positive quantities a, b, c, ... k, then

$$\frac{a^m + b^m + c^m + ... + k^m}{n} > \left(\frac{a+b+c+...+k}{n}\right)^m$$

except when m lies between 0 and 1, i.e. m is a positive proper fraction.

or

The arithmetic mean of mth powers of n quantities is greater than the arithmetic mean raised to power m except when m lies between 0 and 1.

Proof. Suppose m does not lie between 0 and 1; then consider the expression

$$a^m + b^m + c^m + ... \; k^m \qquad\qquad ...(A)$$

If in above we replace the unequal quantities a and b by two equal quantities $\dfrac{a+b}{2}, \dfrac{a+b}{2}$, causing thereby no change in sum of the numbers which will be

$$\frac{a+b}{2} + \frac{a+b}{2} + c + ... + k$$

i.e. $a + b + c + ... + k$, the same as before, we shall have the new form of expression (A) as

$$\left(\frac{a+b}{2}\right)^m + \left(\frac{a+b}{2}\right)^m + c^m + ... + k^m \qquad\qquad ...(B)$$

Now $\qquad\qquad a^m + b^m > 2\left(\dfrac{a+b}{2}\right)^m$

$\therefore$ expression (A) will be greater than expression (B).

The effect of replacing unequal quantities of (A) by equal quantities is that the expression (A) has been diminished.

Hence so long as any two of the quantities $a, b, c, ... k$ are unequal, the expression (A) can be diminished by replacing them by equal quantities causing thereby no change in the value of $a + b + c ... k$

Therefore the value of $a^m + b^m + c^m + ... + k^m$ will be least when all the quantities $a, b, c, ...$ k are equal.

In this case each of the quantities will be

$$\frac{a+b+c+...+k}{n}, \text{ i.e. } \frac{S}{n}$$

And the least value of (A) will be

$$\left[\frac{S}{n}\right]^m + \left[\frac{S}{n}\right]^m + \left[\frac{S}{n}\right]^m + ... n \text{ terms}$$

$$= n\left[\frac{S}{n}\right]^m \qquad\qquad ...(C)$$

Now since (C) is the least value of (A)

$$\therefore \qquad A > C, \text{ i.e. } a^m + b^m + c^m + ... + k^m > n\left[\frac{S}{n}\right]^m$$

i.e. $$\frac{a^m + b^m + c^m + ... + k^m}{n} > \left[\frac{a+b+c+...+k}{n}\right]^m$$

THEOREM-3

If a and b are positive integers, a > b and if x be a positive quantity then

$$\left(1+\frac{x}{a}\right)^a > \left(1+\frac{x}{b}\right)^b$$

Proof. We have $\left(1+\dfrac{x}{a}\right)^a = 1+x+\dfrac{a(a-1)}{2!}\dfrac{x^2}{a^2}+\dfrac{a(a-1)(a-2)}{3!}\dfrac{x^3}{a^3}+...$

$$= 1+x+\left(1-\frac{1}{a}\right)\frac{x^2}{2!}+\left(1-\frac{1}{a}\right)\left(1-\frac{2}{a}\right)\frac{x^3}{3!}+... \qquad ...(1)$$

Similarly, $\left(1+\dfrac{x}{b}\right)^b = 1+x+\left(1-\dfrac{1}{b}\right)\dfrac{x^2}{2!}+\left(1-\dfrac{1}{b}\right)\left(1-\dfrac{2}{a}\right)\dfrac{x^3}{3!}+... \qquad ...(2)$

Now (1) contains $a + 1$ terms whereas (2) contains $b + 1$ terms. Because $a > b$, $\therefore$ (1) has greater number of terms than (2) and also after second term each term of (1) is greater than the corresponding terms of (2). Hence (1) > (2).

SOLVED EXAMPLES

Example 1. *Prove that* $16(a^3 + b^3 + c^3 + d^3) > (a + b + c + d)^3$

Solution. We have $\dfrac{a^3+b^3+c^3+d^3}{4} > \left[\dfrac{a+b+c+d}{4}\right]^3$

$\therefore \quad 16(a^3 + b^3 + c^3 + d^3) > (a + b + c + d)^3$.

Example 2. *Prove that* $n(n + 1)^3 < 8(1^3 + 2^3 + ... + n^3)$

Solution. We have $\dfrac{1^3+2^3+...n^3}{n} > \left[\dfrac{1+2+3+...+n}{n}\right]^3$

$$\left[\frac{n(n+1)}{2n}\right]^3 \text{ or } > \frac{(n+1)^3}{8}$$

$\therefore \quad 8(1^3 + 2^3 + ... n^3) > n(n + 1)^3$

Example 3. *If n quantities a, b, c, ... k are in A.P. prove that*

$$\frac{1}{a}+\frac{1}{b}+\frac{1}{c}+...\frac{1}{k} > \frac{2a}{a+k}$$

Solution. We have $(a^{-1} + b^{-1} + c^{-1} + ... + k^{-1}) > n.\left[\dfrac{a+b+c...+k}{n}\right]^{-1} \qquad ...(1)$

But $a, b, c, ... k$ are in A.P. and hence its sum is

$$\frac{n}{2}(a+k) \text{ using } S_n = \frac{n}{2}(a+l)$$

Putting in (1), we get

$$\frac{1}{a}+\frac{1}{b}+\frac{1}{c}+...\frac{1}{k} > n.\left[\frac{n}{2}.(a+k)\frac{1}{n}\right]^{-1} \quad \text{or} \quad \frac{2n}{a+k}$$

Example 4. *If α and β are +ive quantities and $\alpha > \beta$, show that*

$$\left(1+\frac{1}{\alpha}\right)^{\alpha} > \left(1+\frac{1}{\beta}\right)^{\beta}$$

Hence show that if $n > 1$, then the value of $\left(1+\frac{1}{n}\right)^{n}$ lies between 2 and 2.178.

Solution. The 1st part follows from theorem 3

Now we know that if $\alpha > \beta$, then $\left(1+\frac{1}{\alpha}\right)^{\alpha} > \left(1+\frac{1}{\beta}\right)^{\beta}$

Put $\alpha = n$ and $\beta = 1$. $\therefore$ $\left(1+\frac{1}{n}\right)^{n} > 2.$ ...(1)

Again let $\alpha \to \infty$ and $\beta = n$.

$$\lim_{\alpha \to \infty}\left(1+\frac{1}{\alpha}\right)^{\alpha} > \left(1+\frac{1}{n}\right)^{n}$$

but $\lim_{\alpha \to \infty}\left(1+\frac{1}{\alpha}\right)^{\alpha} = e = 2.718$

$\therefore$ $2.718 > \left(1+\frac{1}{n}\right)^{n}$ ..(2)

From (1) and (2) it follows that $2 < \left(1+\frac{1}{n}\right)^{n} < 2.718.$

$\therefore$ $\left(1+\frac{1}{n}\right)^{n}$ lies between 2 and 2.718.

Example 5. *Prove that* $\left(\frac{1+x}{1-x}\right)^{1/x} > \left(\frac{1+y}{1-y}\right)^{1/y}$ *where both x and y are positive proper fractions and $x > y$.*

Solution. Let $P = \left(\frac{1+x}{1-x}\right)^{1/x}$ and $Q = \left(\frac{1+y}{1-y}\right)^{1/y}$

$$\log P = \frac{1}{x}[\log(1+x)-\log(1-x)]$$

$$= \frac{1}{x}\left[2\left(x+\frac{x^3}{3}+\frac{x^5}{5}...\right)\right]$$

$$= 2\left(1+\frac{x^2}{3}+\frac{x^4}{5}+\cdots\right)$$

Similarly, $\qquad \log Q = 2\left(1+\frac{y^2}{3}+\frac{y^4}{5}+\cdots\right)$

Now x being $> y$, $\therefore$ $\log P > \log Q$, i.e. $P > Q$

or $\qquad \left(\dfrac{1+x}{1-x}\right)^{1/x} > \left(\dfrac{1+y}{1-y}\right)^{1/y}$

Example 6. *If a, b, c are in descending order of magnitude, show that* $\left(\dfrac{a+c}{a-c}\right)^a < \left(\dfrac{b+c}{b-c}\right)^b$.

Solution. We are given that $a > b > c$; therefore both $\dfrac{c}{a}$ and $\dfrac{c}{b}$ are positive proper fractions

and $\dfrac{c}{a}$ is less than $\dfrac{c}{b}$. Let us denote them by y and x respectively.

$\therefore \quad x > y.$

We have to prove that $\left(\dfrac{a+c}{a-c}\right)^a < \left(\dfrac{b+c}{b-c}\right)^b$

or $\qquad \left\{\dfrac{1+\dfrac{c}{a}}{1-\dfrac{c}{a}}\right\}^a < \left\{\dfrac{1+\dfrac{c}{b}}{1-\dfrac{c}{b}}\right\}^b$ $\qquad$ or $\qquad \left\{\dfrac{1+\dfrac{c}{a}}{1-\dfrac{c}{a}}\right\}^{a/c} < \left\{\dfrac{1+\dfrac{c}{b}}{1-\dfrac{c}{b}}\right\}^{b/c}$

or $\qquad \left(\dfrac{1+y}{1-y}\right)^{1/y} < \left(\dfrac{1+x}{1-x}\right)^{1/x}$ $\qquad$ or $\qquad \left(\dfrac{1+x}{1-x}\right)^{1/x} < \left(\dfrac{1+y}{1-y}\right)^{1/y}$

where x and y are both positive proper fractions and $x > y$.

This we have proved above in example 2.

Example 7. *Prove* $(1 + x)^{1+x} (1 - x)^{1-x} > 1$ *if x is a positive proper fraction and deduce that*

$a^a.b^b > \left(\dfrac{a+b}{2}\right)^{a+b}$

Solution. If $P > 1$ then $\log P$ must be positive.

Let $P = (1 + x)^{1+x} (1 - x)^{1-x}$, then

$$\log P = (1 + x) \log (1 + x) + (1 - x) \log (1 - x)$$
$$= \{\log (1 + x) + \log (1 - x)\} + x\{\log (1 + x) - \log (1 - x)\}$$
$$= 2\left(-\frac{x^2}{2}-\frac{x^4}{4}-\frac{x^6}{6}\cdots\right)+2x\left(x+\frac{x^3}{3}+\frac{x^5}{5}\cdots\right)$$
$$= 2\left\{\left(-\frac{x^2}{2}-\frac{x^4}{4}-\frac{x^6}{6}\cdots\right)+\left(\frac{x^2}{1}+\frac{x^4}{3}+\frac{x^6}{5}\cdots\right)\right\}$$

$$= 2\left\{\left(\frac{x^2}{1} - \frac{x^2}{2}\right) + \left(\frac{x^4}{3} - \frac{x^4}{4}\right) + \left(\frac{x^6}{5} - \frac{x^6}{6}\right) + \ldots\right\}$$

$$= 2\left(\frac{x^2}{1.2} + \frac{x^4}{3.4} + \frac{x^6}{5.6} \ldots\right)$$

which is definitely a positive quantity.

$\Rightarrow$ $\qquad$ log P being +ive quantity and hence $P > 1$.

Deduction. $a^a b^b > \left(\dfrac{a+b}{2}\right)^{a+b}$

Putting $x = \dfrac{z}{u}$ where $u > z$ because $x < 1$, then

$$\left(1 + \frac{z}{u}\right)^{1+(z/u)} \left(1 - \frac{z}{u}\right)^{1-(z/u)} > 1$$

or $\qquad$ $\left(\dfrac{u+z}{u}\right)^{(u+z)/u} \left(\dfrac{u-z}{u}\right)^{(u-z)/u} > 1$

or $\qquad$ $\left(\dfrac{u+z}{u}\right)^{u+z} \left(\dfrac{u-z}{u}\right)^{u-z} > 1^u$ or 1

or $\qquad$ $(u+z)^{u+z} (u-z)^{u-z} > u^{u+z+u-z} = u^{2u}.$ $\qquad$ ...(1)

Put $u + z = a$ and $u - z = b$; then $u = \dfrac{a+b}{2}$ $\qquad$ ...(2)

$\therefore$ $\qquad$ $a^a b^b > \left(\dfrac{a+b}{2}\right)^{a+b}$ $\qquad\qquad$ by (1) and (2)

Example 8. *Prove* $(1 + x)^{1-x} (1 - x)^{1+x} < 1$; *where x is a positive proper fraction and deduce that* $a^b . b^a < \left(\dfrac{a+b}{2}\right)^{a+b}$.

Solution. left for the reader.

Example 9. *Prove independently that* $a^a b^b > \left(\dfrac{a+b}{2}\right)^{a+b} > a^b . b^a$.

Solution. Consider a quantities each equal to $\dfrac{1}{a}$, b quantities each equal to $\dfrac{1}{b}$, then since A.M. > G.M., we have

$$\frac{\left[\dfrac{1}{a} + \dfrac{1}{a} + \dfrac{1}{a} \ldots a \text{ times}\right] + \left[\dfrac{1}{b} + \dfrac{1}{b} + \ldots b \text{ times}\right]}{a+b}$$

$$> \left\{\left(\frac{1}{a} \cdot \frac{1}{a} \cdot \frac{1}{a} \ldots a \text{ factors}\right) \times \left(\frac{1}{b} \cdot \frac{1}{b} \cdot \frac{1}{b} \ldots b \text{ factors}\right)\right\}^{1/(a+b)}$$

or

$$\frac{a.\dfrac{1}{a}+b.\dfrac{1}{b}}{a+b} > \left\{\frac{1}{a^a}\frac{1}{b^b}\right\}^{1/(a+b)}$$

or

$$\left\{\frac{2}{a+b}\right\} > \left\{\frac{1}{a^a.b^b}\right\}^{1/(a+b)}$$

or

$$\left\{\frac{2}{a+b}\right\}^{a+b} > \frac{1}{a^a.b^b}$$

Taking reciprocal and hence changing the sign of inequality, we get

$$\left\{\frac{a+b}{2}\right\}^{a+b} < a^a.\,b^b \quad \text{or} \quad a^a.\,b^b > \left\{\frac{a+b}{2}\right\}^{a+b}$$

Example 10. *If a_1, a_2 are positive numbers (including zero) and p, q, r ... are positive integers, then prove*

$$\frac{a_1^{p+q}+a_2^{p+q}}{2} > \left[\frac{a_1^p+a_2^p}{2}\right]\left[\frac{a_1^q+a_2^q}{2}\right] \qquad \text{...(A)}$$

and

$$\frac{a_1^{p+q+r+\cdots}+a_2^{p+q+r+\cdots}}{2} \geq \left(\frac{a_1^p+a_2^p}{2}\right)\left(\frac{a_1^q+a_2^q}{2}\right)\left(\frac{a_1^r+a_2^r}{2}\right)\cdots$$

Solution. We have $a_1^p - a_2^p$ and $a_1^q - a_2^q$ have the same sign or each of them is equal to zero; therefore

$$(a_1^p - a_2^p)(a_1^q - a_2^q) \geq 0$$

or

$$a_1^{p+q}+a_2^{p+q} \geq a_1^p a_2^q + a_1^q a_2^p$$

∴

$$2(a_1^{p+q}+a_2^{p+q}) \geq a_1^{p+q}+a_2^{p+q}+a_1^p a_2^q + a_1^q a_2^p$$

or

$$\geq (a_1^p + a_2^p)(a_1^q + a_2^q)$$

Dividing by 4, we get

$$\frac{a_1^{p+q}+a_2^{p+q}}{2} \geq \left(\frac{a_1^p + a_2^p}{2}\right)\left(\frac{a_1^q + a_2^q}{2}\right)$$

Also,

$$\frac{a_1^{p+r}+a_2^{p+r}}{2} \geq \left(\frac{a_1^p + a_2^p}{2}\right)\left(\frac{a_1^r + a_2^r}{2}\right)$$

Put $P = p + q$.

$$\frac{a_1^{p+q+r}+a_2^{p+q+r}}{2} \geq \left(\frac{a_1^{p+q}+a_2^{p+q}}{2}\right)\left(\frac{a_1^r + a_2^r}{2}\right) \geq \left\{\frac{a_1^p+a_2^p}{2}\right\}\left\{\frac{a_1^q+a_2^q}{2}\right\}\left\{\frac{a_1^r+a_2^r}{2}\right\}$$

The above formula is true if instead of two quantities a_1, a_2, we have n quantities a_1, a_2, ... a_n.

Example 11. *Prove that $a^7 + b^7 + c^7 > abc (a^4 + b^4 + c^4)$*

Solution. We have $\dfrac{a^7 + b^7 + c^7}{3} > \dfrac{a^3 + b^3 + c^3}{3} \cdot \dfrac{a^4 + b^4 + c^4}{3}$

$$> [(a^3 b^3 c^3)]^{1/3} \dfrac{a^4 + b^4 + c^4}{3} \qquad \because \text{A.M.} > \text{G.M.}$$

$$\therefore \qquad a^7 + b^7 + c^7 > abc\,(a^4 + b^4 + c^4).$$

EXERCISE 15.1

Prove the following :

1. $(ab + xy)(ax + by) > 4abxy$ 2. $(b + c)(c + a)(a + b) > 8abc$

3. $(a + b + c)(bc + ca + ab) > 9abc$ 4. $a^2cd + b^2ad + c^2ab + d^2cb > 4acd.$

5. $\dfrac{a}{b} + \dfrac{b}{c} + \dfrac{c}{d} + \dfrac{d}{a} > 4$ 6. $\left(\dfrac{a}{e} + \dfrac{b}{f} + \dfrac{c}{g}\right)\left(\dfrac{e}{a} + \dfrac{f}{b} + \dfrac{g}{c}\right) > 9$

7. $\left(\dfrac{a^2 - b^2}{a + b}\right)^{a+b} > a^a . b^b$

8. *If n is a positive integer and x, y, z are not all equal prove that*
$$3(x^{n+1} + y^{n+1} + z^{n+1}) > (x^n + y^n + z^n)(x + y + z)$$

HINT TO SELECTED PROBLEMS

1. $ab + xy > 2\sqrt{(abxy)}$ $\therefore$ A.M. > G.M. and $ax + by > 2\sqrt{(abxy)}$
   Multiplying, $(ab + xy)(ax + by) > 4abxy.$
3. Apply A.M. > G.M. on each factor and multiply.
4. A.M. > G.M.
5. Apply A.M. > G.M.

15.6 SOME MORE RESULTS

THEOREM-1

If sum of any two of the quantities x, y, z be greater than the third then
   (a) $(x + y + z)^3 > 27(y + z - x)(z + x - y)(x + y - z)$
and (b) $xyz > 27(y + z - x)(z + x - y)(x + y - z)$

Proof: $x + y > z \implies x + y - z$ is positive.

Similarly all other factors are positive.

Put $y + z - x = a, z + x - y = b, x + y - z = c,$

$$z = \dfrac{a + b}{2}, x = \dfrac{b + c}{2}, y = \dfrac{c + a}{2}$$

$$\therefore \qquad x + y + z = a + b + c$$

Substituting in (a), we have to prove

$(a + b + c)^3 > 27abc$.

But we know that $(a + b + c) > 3(abc)^{1/3}$. Hence on taking cube of both sides, we get the required result.

(b) we have to prove $\dfrac{a+b}{2} \cdot \dfrac{b+c}{2} \cdot \dfrac{c+a}{2} > abc$ which is clearly true if we apply A.M. > G.M. on each term of L.H.S. and multiply.

THEOREM-2

If A be the area and $2s$ be the sum of three sides of a triangle then $A \leq \dfrac{s^2}{3\sqrt{3}}$.

Proof. Let the sides be x, y, z so that $s = \dfrac{x + y + z}{2}$

Also $s - x = \dfrac{y + z - x}{2}$, $s - y = \dfrac{z + x - y}{2}$, $s - z = \dfrac{x + y - z}{2}$ We have to prove that $\dfrac{s^4}{27} \geq A^2$

or $\dfrac{s^4}{27} \geq s(s - x)(s - y)(s - z)$

as $\qquad A^2 = s(s - x)(s - y)(s - z)$, for a triangle.

or $\qquad s^2 \geq 27(s - x)(s - y)(s - z)$

or $\qquad \dfrac{(x + y + z)^3}{8} \geq 27 \dfrac{(y + z - x)}{2} \dfrac{(z + x - y)}{2} \dfrac{(x + y - z)}{2}$

or $\qquad (x + y + z)^3 \geq 27(y + z - x)(z + x - y)(x + y - z)$

where x, y, z the sides of a triangle are such that sum of any two is greater than the third.

THEOREM-3

If the sum of any three of the quantities a, b, c, d be greater than twice the fourth, prove that $abcd > (b + c + d - 2a)(c + d + a - 2b)(d + a + b - 2c)(a + b + c - 2d)$.

Proof. Since $b + c + d > 2a$ given.

$$b + c + d - 2a \text{ is positive.}$$

Similarly all other factors are positive.

Put $b + c + d - 2a = x$, $c + d + a - 2b = y$, $\qquad$ ...(2)

$$d + a + b - 2c = z, \qquad \text{...(3)}$$
$$a + b + c - 2d = w. \qquad \text{...(4)}$$

Adding (2), (3), (4) we get $a = \dfrac{y + z + w}{3}$

Similarly $b = \dfrac{z + w + x}{3}$. $c = \dfrac{w + x + y}{3}$, $d = \dfrac{x + y + z}{3}$

$\therefore$ We are to prove that

$$\left(\dfrac{y + z + w}{3}\right)\left(\dfrac{z + w + x}{3}\right)\left(\dfrac{w + x + y}{3}\right)\left(\dfrac{x + y + z}{3}\right) > xyzw$$

Now applying A.M. > G.M. on each bracket of L.H.S. and multiplying, we get the required result.

EXERCISE 15.2

1. Prove that $6abc < bc\,(b + c) + ca\,(c + a) + ab(a + b)$

   or   prove $b^2c + c^2b + c^2a + a^2c + a^2b + b^2a > 6abc$

2. Prove that $6abc < a^2(1 + b^2) + b^2(1 + c^2) + c^2(1 + a^2)$

3. Prove that $b^2c^2 + c^2a^2 + a^2b^2 > abc\,(a + b + c)$

4. Prove that $\dfrac{bc}{a^3} + \dfrac{ca}{b^3} + \dfrac{ab}{c^3} > \dfrac{1}{a} + \dfrac{1}{b} + \dfrac{1}{c}$

5. Prove that $\dfrac{1}{a} + \dfrac{1}{b} + \dfrac{1}{c} > \dfrac{1}{\sqrt{(bc)}} + \dfrac{1}{\sqrt{(ca)}} + \dfrac{1}{\sqrt{(ab)}}$

6. Prove that if $x + y + z = 1$, then $(1 - x)(1 - y)(1 - z) > 8xyz$.

7. Prove that if $a + b + c = 1$, then

$$\frac{8}{27abc} > \left\{\frac{1}{a} - 1\right\}\left\{\frac{1}{b} - 1\right\}\left\{\frac{1}{c} - 1\right\} > 8$$

8. Prove that $2(a^3 + b^3 + c^3) > bc\,(b + c) + ca\,(c + a) + ab\,(a + b)$

9. Prove that $3(a^3 + b^3 + c^3) > (a + b + c)(a^2 + b^2 + c^2)$

10. Prove that If a, b, c, are positive then

$$\frac{a}{b+c} + \frac{b}{c+a} + \frac{c}{a+b} > \frac{3}{2}$$

11. Prove that $n^n > 1.3.5. \ldots (2n - 1)$

12. Prove that $(n\,!)^2 > n^n$.

13. Prove that $2.4.6. \ldots 2n < (n + 1)^n$.

14. Prove that $(1^r + 2^r + 3^r + \ldots n^r)^n > n^n\,(n\,!)^r$

15. Prove $(n\,!)^3 < \left(\dfrac{n+1}{2}\right)^{2n} n^n,$

16. Prove that $2^n > 1 + n\sqrt{(2^{n-1})}$ or $\dfrac{2^n - 1}{2 - 1} > n.2^{(n-1)/2}$

17. Prove that $(1 + x^3)(1 + y^3)(1 + z^3) > (1 + xyz)^3$

HINT TO SELECTED PROBLEMS

4. $\dfrac{bc}{a^3} + \dfrac{ca}{b^3} > 2\left(\dfrac{bac^2}{a^3 b^3}\right)^{1/2}$ or $> \dfrac{2c}{ab}$

Similarly writing other inequalities and adding, we get

$$\dfrac{bc}{a^3} + \dfrac{ca}{b^3} + \dfrac{ab}{c^3} > \dfrac{c}{ab} + \dfrac{b}{ca} + \dfrac{a}{bc} \qquad \text{...(A)}$$

Again $\dfrac{c}{ab} + \dfrac{b}{ca} > 2\left[\dfrac{bc}{bc.a^2}\right]^{1/2}$ or $2.\dfrac{1}{a}$

Similarly writing other inequalities, and adding, we get

$$\dfrac{c}{ab} + \dfrac{b}{ca} + \dfrac{a}{bc} > \dfrac{1}{a} + \dfrac{1}{b} + \dfrac{1}{c} \qquad \text{...(B)}$$

$\therefore \quad \dfrac{bc}{a^3} + \dfrac{ca}{b^3} + \dfrac{ab}{c^3} > \dfrac{1}{a} + \dfrac{1}{b} + \dfrac{1}{c}$ from (A) and (B)

5. $\dfrac{1}{b} + \dfrac{1}{c} > 2\left[\dfrac{1}{b}.\dfrac{1}{c}\right]^{1/2}$ or $> 2\dfrac{1}{\sqrt{(bc)}}$ and so on.

7. On multiplying, by abc, we have to prove that

$$\dfrac{8}{27} > (1-a)(1-b)(1-c) > 8abc$$

Now $\qquad \dfrac{(1-a) + (1-b) + (1-c)}{3} > \{(1-a)(1-b)(1-c)\}^{1/3}$

or $\qquad \dfrac{3 - (a+b+c)}{3} > \{(1-a)(1-b)(1-c)\}^{1/3}$

or $\qquad \left(\dfrac{2}{3}\right)^3 > (1-a)(1-b)(1-c) \qquad\qquad \because a+b+c = 1$

which proves the Ist part.

8. The latter expression can also be put as

$\qquad a^2(b+c) + b^2(c+a) + c^2(a+b)$

or $\qquad a(b^2 + c^2) + b(c^2 + a^2) + c(a^2 + b^2)$

We know that $a^2 + b^2 > 2ab$

$\Rightarrow \qquad\qquad a^2 + b^2 - ab > ab.$

Multiplying both sides by $(a+b)$, we get

$$(a+b)(a^2 - ab + b^2) > ab(a+b)$$

or $\qquad a^3 + b^3 > ab(a+b)$

Similarly $b^3 + c^3 > bc(b+c)$

$\qquad c^3 + a^3 > ca(c+a)$

Adding, we get

$$2(a^3 + b^3 + c^3) > ab(a+b) + bc(b+c) + ca(c+a)$$

9. We have to prove that

$$3(a^3 + b^3 + c^3) > a^3 + b^3 + c^3 + a(b^2 + c^2) + b(c^2 + a^2) + c(a^2 + b^2)$$

or $\qquad 2(a^3 + b^3 + c^3) > a(b^2 + c^2) + b(c^2 + a^2) + c(a^2 + b^2)$

which we have already proved above.

The question can also be written as

$$\frac{a^3 + b^3 + c^3}{c} > \frac{a+b+c}{3} \cdot \frac{a^2 + b^2 + c^2}{3}$$

10. $\quad \dfrac{(a+b)+(b+c)+(c+a)}{3} > [(a+b)(b+c)(c+a)]^{1/3}$

$$\frac{\dfrac{1}{a+b} + \dfrac{1}{b+c} + \dfrac{1}{c+a}}{3} > \left[\frac{1}{(a+b)} + \frac{1}{(b+c)} + \frac{1}{(c+a)}\right]^{1/3}$$

Multiplying the two inequalities, we get

$$\frac{2}{3}.(a+b+c).\frac{1}{3}\left[\frac{1}{a+b} + \frac{1}{b+c} + \frac{1}{c+a}\right] > 1$$

or $\qquad \left(1 + \dfrac{c}{a+b}\right) + \left(1 + \dfrac{a}{b+c}\right) + \left(1 + \dfrac{1}{c+a}\right) > \dfrac{9}{2}$

or $\qquad \dfrac{a}{b+c} + \dfrac{b}{c+a} + \dfrac{c}{a+b} > \dfrac{9}{2} - 3 \ \text{ or } \ > \dfrac{3}{2}$

11. Apply A.M. of 1, 3, 5 ... $(2n - 1)$, i.e. of n numbers is > G.M. and remember that $1 + 3 + 5 \dots$ upto n terms

$$= \frac{n}{2}\{2a + (n-1)d\} = n^2 \ \text{ as } a = 1 \ \text{ and } d = 2$$

12. We know $n - r > \left(\dfrac{n-r}{r}\right)$ where r is positive integer.

$\therefore \qquad r(n - r) > n - r \qquad\qquad$ or $\quad r(n - r + 1) > n$

Put $r = 1, 2, 3, \dots n.$

$\qquad\qquad\qquad\qquad$ 1. $n = n$ $\qquad\qquad\qquad\qquad\qquad\qquad$ (a case of equality)

$\qquad\qquad\qquad\qquad$ 2. $(n - 1) > n$

$\qquad\qquad\qquad\qquad$ 3. $(n - 2) > n$

$\qquad\qquad\qquad\qquad\qquad$ ....

$\qquad\qquad\qquad\qquad$ $(n - 1).2 > n$

$n.1 = n \quad$ (a case of equality)

Multiplying, we get

$$(n\,!).(n\,!) > n^n \ \text{ or } (n\,!)^2 > n^n$$

14. $\quad \dfrac{1^r + 2^r + 3^r + \dots n^r}{n} > [1^r.\, 2^r.\, 3^r \ \dots n^r)^{1/n}$

or $\qquad (1^r + 2^r + 3^r + \dots n^r)^n > n^n \,(n\,!)^r$

15. Apply A.M. of $1^3.2^3. \ \dots \ n^3$ is greater than their G.M. and remember that

$$1^3 + 2^3 + \dots n^3 = \left\{\frac{n(n+1)}{2}\right\}^2$$

16. We know that $\dfrac{a(r^n - 1)}{r - 1}$ is the sum of a G.P. whose first term is a and common ratio is r.

In the above $a = 1$ and $r = 2$ ∴ We have to prove that
$$1 + 2 + 2^2 + 2^3 + \dots + 2^{n-1} > n.2^{(n-1)/2}$$

Now $\dfrac{1 + 2 + 2^2 + 2^3 + \dots + 2^{n-1}}{n} > (1.\,2.\,2^2.\,2^3\dots 2^{n-1})^{1/n}$

or $> (2^{1+2+3\,\dots\,n-1})^{1/n}$ ∵ A.M. > G.M.

or $> [2^{(n-1)n/2}]^{1/n}$ or $2^{(n-1)/2}$

∴ $1 + 2 + 2^2 + \dots + 2^{n-1} > n.2^{(n-1)/2}$

17. On simplification, we have to prove that
$$1 + (x^3 + y^3 + z^3) + (x^3y^3 + y^3z^3 + z^3x^3) + x^3y^3z^3 > 1 + 3xyz + 3x^2y^2z^2 + x^3y^3z^3.$$

Now apply A.M. > G.M. on each bracket in L.H.S.

ADDITIONAL SOLVED EXAMPLES

Example 1. *Prove that* (a) $\dfrac{bc}{b+c} + \dfrac{ca}{c+a} + \dfrac{ab}{a+b} < \dfrac{1}{2}(a+b+c)$

(b) $\dfrac{2}{b+c} + \dfrac{2}{c+a} + \dfrac{2}{a+b} < \dfrac{1}{a} + \dfrac{1}{b} + \dfrac{1}{c}$

Solution. We know that $(b + c) > 2\sqrt{(bc)}$

Square both sides.

∴ $4bc < (b + c)^2$ or $\dfrac{4bc}{b+c} \ (b + c)$ for (a)

and $\dfrac{4}{b+c} < \dfrac{b+c}{bc}$ or $< \dfrac{1}{b} + \dfrac{1}{c}$ for (b)

Similarly write other similar inequalities and add.

Example 2. *Prove that* $abcd > 81(s - a)(s - b)(s - c)(s - d)$ *where* $3x = a + b + c + d,$

Solution. $a = 3s - b - c - d$ or $= (s - b) + (s - c) + (s - d)$

$> 3 \{(s - b)(s - c)(s - d)\}^{1/3},$ ∵ A.M. > G.M.

Similarly write other inequalities for b, c and d and then multiply.

Example 3. *Prove that* $a_1 a_2 \dots a_n > (n - 1)^n (s - a_1)(s - a_2) \dots (s - a_n)$

where $(n - 1)s = a_1 + a_2 + a_3 + \dots a_n$

Solution. Above is generalised form of previous example

$a_1 = (n - 1)s - (a_2 + a_3 + \dots a_n)$

$= (s - a_2) + (s - a_3) + \dots (s - a_n)$

$> (n - 1)\{(s - a_2) + (s - a_3) + \dots (s - a_n)\}^{1/(n-1)}$

Similarly write n inequalities of the above form and multiply. Remember that the factors $(s - a_1)$ on the R.H.S. will occur $(n - 1)$ times and similarly other factors.

$$\therefore \qquad a_1 a_2 a_3 \ldots a_n > (n-1)^n \{(s-a_1)^{n-1}(s-a_2)^{n-1} \ldots (s-a_n)^{n-1}\}^{(1/n-1)}$$

$$\text{or} \qquad > (n-1)^n (s-a_1)(s-a_2) \ldots (s-a_n)$$

Example 4. *If a, b, c are in H.P. prove that* $\dfrac{a+b}{2a-b} + \dfrac{c+b}{2c-b} > 4$

Solution. As a, b, c are in H.P., $\therefore b = \dfrac{2ac}{a+c}$

$$\frac{a+b}{2a-b} = \frac{a + \dfrac{2ac}{a+c}}{2a - \dfrac{2ac}{a+c}} = \frac{a^2 + 3ac}{2a^2} = \frac{1}{2} + \frac{3}{2} \cdot \frac{c}{a}$$

Similarly $\qquad \dfrac{c+b}{2c-b} = \dfrac{1}{2} + \dfrac{3}{2} \cdot \dfrac{a}{c}$

Hence we have to prove that $\dfrac{1}{2} + \dfrac{3}{2}\left[\dfrac{c}{a} + \dfrac{a}{c}\right] + \dfrac{1}{2} > 4$

or $\qquad \dfrac{3}{2}\left[\dfrac{c}{a} + \dfrac{a}{c}\right] > 3 \quad$ or $\quad \dfrac{c^2 + a^2}{2ac} > 4$

or $\qquad \dfrac{c^2 + a^2}{2} > ac \quad$ which is true. $\qquad\qquad \because$ A.M. > G.M.

Example 5. *If a > b and n is a +ive integer, then prove that*
$$(a^n - b^n) > n(ab)^{(n-1)/2}(a - b)$$

Solution. Since $a > b$, $\therefore$ $(a - b)$ is +ive and hence we can divide the inequality by $(a - b)$ and we have to prove that

$$\frac{a^n - b^n}{a - b} > n(ab)^{(n-1)/2}$$

By actual division,

$$\frac{a^n - b^n}{a - b} = a^{n-1} + ba^{n-2} + b^2 a^{n-3} + \ldots + a^2 b^{n-3} + ab^{n-2} + b^{n-1} \qquad \ldots(A)$$

Apply A.M. of the n numbers of $A >$ their G.M.

$$\therefore \qquad A > n[a^{(n-1)+(n-2)\ldots 3+2+1} b^{(n-1)+(n-2)\ldots 3+2+1}]^{1/n}$$

or $\qquad n\left[ab^{\tfrac{n(n-1)}{2}}\right]^{1/n} \quad$ or $\quad > n(ab)^{\tfrac{n-1}{2}}$

Example 6. *Show that* $2(a^2 + b^2 + c^2)(a + b + c) > a^3 + b^3 + c^3 + 15abc$,
or $2(a^2 + b^2 + c^2)(a + b + c) > a^3 + b^3 + c^3 - 3abc + 18abc$.

Solution. We have
$$a^3 + b^3 + c^3 - 3abc = (a + b + c)(a^2 + b^2 + c^2 - ab - bc - ca)$$

Transporting on the other side and taking $a + b + c$ common, we have to prove that
$$(a + b + c)(2a^2 + 2b^2 + 2c^2 - a^2 - b^2 - c^2 + ab + bc + ca) > 18abc$$

or $\quad (a + b + c)(a^2 + b^2 + c^2 + ab + bc + ca) > 18abc$.

Now apply A.M. > G.M. on each bracket on the left and multiply

$\therefore \quad (a + b + c)(a^2 + b^2 + c^2 + ab + bc + ca) > 3(abc)^{1/3} \cdot 6(a^4b^4c^4)^{1/6}$

or $\quad > 18(abc)^{1/3} \cdot (abc)^{2/3}$ or $> 18abc$.

Example 7. *If A, B, C be the angles of a triangle then prove the following :*

    (i) $\cot A \cot B \cot C \le 1/3\sqrt{3}$

    (ii) $\cot A/2 \cot B/2 \cot C/2 \ge 3\sqrt{3}$

Solution. We know from Trigonometry that

$$\tan (A + B + C) = \frac{S_1 - S_3}{1 - S_2} = \tan \pi = 0 \quad \because S_1 = S_3$$

or $\qquad \tan A + \tan B + \tan C = \tan A \tan B \tan C \qquad$ ...(1)

$\qquad S_1 \ge 3(S_3)^{1/3} \qquad \because \quad$ A.M. $>$ G.M.

or $\qquad S_3 \ge 3(S_3)^{1/3} \quad$ by (1)

Cubing both sides we get

$$S_2^3 \ge 27S_3 \quad \text{or} \quad S_2^3 > 27 \quad \text{or} \quad S_3 \ge 3\sqrt{3}$$

or $\qquad \tan A \tan B \tan C \ge 3\sqrt{3}$

Taking reciprocal and changing the sign of inequality, we get

$$\cot A \cot B \cot C \le 1/3\sqrt{3}$$

(ii) $\tan (A/2 + B/2 + C/2) = \dfrac{S_1 - S_3}{1 - S_2} = \tan 90 = \infty$

$\therefore \quad 1 - S_2 = 0$

$\tan A/2 \tan B/2 + \tan B/2 \tan C/2 + \tan C/2 \tan A/2 = 1$

Dividing by $\tan A/2 \tan B/2 \tan C/2$ we get

$\cot A/2 + \cot B/2 + \cot C/2 = \cot A/2 \cot B/2 \cot C/2 \quad$ or $S_1 = S_3$

Now proceed as above.

Example 8. *If n is a positive integer and $x < 1$, show that $\dfrac{1 - x^{n+1}}{n+1} < \dfrac{1 - x^n}{n}$.*

Solution. We have to prove that $\dfrac{1 - x^{n+1}}{n+1} < \dfrac{1 - x^{n+1}}{n}$

Now $\qquad \dfrac{1 - x^{n+1}}{1 - x^n} = \dfrac{1 - x^n + x^n - x^{n+1}}{1 - x^n} = 1 + \dfrac{x^n(1-x)}{1 - x^n}$

$$= 1 + \frac{x^n}{\dfrac{1 - x^n}{1 - x}} = 1 + \frac{x^n}{1 + x + x^3 + \dots x^{n-1}}$$

$$= 1 + \cfrac{1}{\dfrac{1}{x^n} + \dfrac{1}{x^{n-1}} + \dfrac{1}{x^{n-2}} + \dots + \dfrac{1}{x}} \qquad \text{...(1)}$$

Since $x < 1$, $\qquad \therefore \quad \dfrac{1}{x} > 1$

Similarly $\dfrac{1}{x^2}, \dfrac{1}{x^3} \ldots \dfrac{1}{x^n}$ are all > 1.

$\therefore$
$$\frac{1}{x^n} + \frac{1}{x^{n-1}} + \frac{1}{x^{n-2}} + \ldots \frac{1}{x} > n$$

Taking reciprocal and changing the sign of inequality we get

$$\frac{1}{\dfrac{1}{x^n} + \dfrac{1}{x^{n-1}} + \dfrac{1}{x^{n-2}} + \ldots \dfrac{1}{x}} < \frac{1}{n}$$

or
$$1 + \frac{1}{\dfrac{1}{x^n} + \dfrac{1}{x^{n-1}} + \dfrac{1}{x^{n-2}} + \ldots \dfrac{1}{x}} < 1 + \frac{1}{n} \qquad \text{by (1)}$$

$\therefore$ by (1)
$$\frac{1 - x^{n+1}}{1 - x^n} < \frac{n+1}{n}$$

$$\frac{1 - x^{n+1}}{n+1} < \frac{1 - x^n}{n}$$

Example 9. *Prove that* $\dfrac{\sqrt{(n-1)}}{2n+1} < \dfrac{1}{2} \cdot \dfrac{3}{4} \cdot \dfrac{5}{6} \ldots \dfrac{2n-1}{2n} < \dfrac{1}{\sqrt{(2n+1)}}$

Solution. For second part squaring both sides, then we have to prove that

$$\frac{1^2}{2^2} \frac{3^2}{4^2} \frac{5^2}{6^2} \ldots \frac{(2n-1)}{(2n)^2}(2n+1) < 1$$

or
$$\left(\frac{1.3}{2^2}\right)\left(\frac{3.5}{4^2}\right)\left(\frac{5.7}{6^2}\right) \ldots \left[\frac{(2n-3)(2n-1)}{(2n-2)^2}\right]\left[\frac{(2n-1)(2n+1)}{(2n^2)}\right] < 1$$

which is clearly true as each bracket is less than 1, and hence their product is less than 1.

Again $n + (n+1) > 2\sqrt{[n(n+1)]}$ $\because$ A.M. > G.M.

or
$$\frac{2n-1}{2n} > \sqrt{\left(\frac{n+1}{n}\right)} \quad \text{or} \quad \sqrt{\left(\frac{n+1}{n}\right)} < \frac{2n-1}{2n} \qquad \ldots(A)$$

Put $n = 1, 2, 3, \ldots (n-1), n$ in (A) and multiply the inequalities.

Example 10. *Prove that* $\dfrac{3.7.11. \dots (4n-1)}{5.9.13. \dots (4n+1)} < \sqrt{\left(\dfrac{3}{4n+3}\right)}$

Solution. We have $\dfrac{(4n-1)+(4n+3)}{2} > \sqrt{[(4n-1)(4n+3)]}$ $(\because \text{A.M.} > \text{G.M.})$

or $\qquad \dfrac{4n+1}{4n-1} > \sqrt{\left(\dfrac{4n+3}{4n-1}\right)}; \quad \therefore \quad \dfrac{4n-1}{4n+1} < \sqrt{\left(\dfrac{4n-1}{4n+3}\right)}$

Now put $n = 1, 2, 3, \dots n$ and multiply to get the required result.

Example 11. *Prove that* $(a_1^2 + a_2^2 + \dots a_n^2)(b_1^2 + b_2^2 + \dots b_n^2) \geq (a_1 b_1 + a_2 b_2 + \dots a_n b_n)^2$

Solution. We know that $(a_1 x + b_1)^2$ is positive, i.e. ≥ 0 [a perfect square]

Similarly $\quad (a_2 x + b_2)^2 \geq 0$

$\qquad \dots \dots \dots \dots \dots$

$\qquad (a_n x + b_n)^2 \geq 0$

Adding, we get

$(a_1^2 + a_2^2 + \dots a_n^2)x^2 + 2x(a_1 b_1 + a_2 b_2 + \dots a_n b_n) + (b_1^2 + b_2^2 + \dots b_n^2) \geq 0$ \qquad ...(1)

L.H.S. is of the form $Ax^2 + Bx + C$ which can be written as

$$\frac{1}{A}[A^2 x^2 + ABC + AC]$$

or $\qquad \dfrac{1}{A}\left\{\left(Ax + \dfrac{B}{2}\right)^2 + \dfrac{4AC - B^2}{4}\right\}$

which is positive only when $4AC - B^2$ is positive, i.e. $4AC > B^2$. \qquad ...(2)

$\therefore \qquad$ L.H.S. +ive if $4(a_1^2 + a_2^2 + \dots a_n^2)(b_1^2 + b_2^2 + \dots b_n^2) \geq 4(a_1 b_1 + a_2 b_2 + \dots a_n b_n)^2$ etc.

Example 12. *If a, b, c are +ive and the sum of any two is greater than the third, then prove that*

$$\left(1 + \frac{b-c}{a}\right)^a \left(1 + \frac{c-a}{b}\right)^b \left(1 + \frac{a-b}{c}\right)^c < 1 \qquad \text{...(A)}$$

Solution. We have $1 + \dfrac{b-c}{a} = \dfrac{a+b-c}{a}$

Now, $a + b > c$ given and a is positive.

$\therefore \qquad \dfrac{a+b-c}{a}$ is positive.

Now consider a factors each equal to $1 + \dfrac{b-c}{a}$ and b factors each equal to $1 + \dfrac{c-a}{b}$ and

c factors each equal to $1 + \dfrac{a-b}{c}$; then because A.M. > G.M. we have

$$\frac{\left\{\left(1 + \dfrac{b-c}{a}\right) + \left(1 + \dfrac{b-c}{a}\right) + \dots a \text{ times} \begin{cases} +(\dots b \text{ times}) \\ +(\dots c \text{ times}) \end{cases}\right\}}{a+b+c}$$

$$> \left[\left\{\left(1+\frac{b-c}{a}\right)\ldots a \text{ factors}\right\}\left\{\left(1+\frac{c-a}{b}\right)\ldots b \text{ factors}\right\}\right.$$

$$\left.\times\left\{\left(1+\frac{a-b}{c}\right)\ldots c \text{ factors}\right\}\right]^{1/(a+b+c)}$$

or $\left\{\dfrac{a\left(\dfrac{a+b-c}{a}\right)+b\left(\dfrac{b+c-a}{b}\right)+c\left(\dfrac{c+a-b}{a}\right)}{a+b+c}\right\}^{a+b+c} > \left(1+\dfrac{b-c}{a}\right)^a\left(1+\dfrac{c-a}{b}\right)^b\left(1+\dfrac{a-b}{c}\right)^c$

or $\left(\dfrac{a+b+c}{a+b+c}\right)^{a+b+c} > \left(1+\dfrac{b-c}{a}\right)^a\left(1+\dfrac{c-a}{b}\right)^b\left(1+\dfrac{a-b}{c}\right)^c$

$\therefore \left(1+\dfrac{b-c}{a}\right)^a\left(1+\dfrac{c-a}{b}\right)^b\left(1+\dfrac{a-b}{c}\right)^c < 1$

Example 13. *If there are n positive numbers $a_1, a_2 \ldots a_n$ and if the square root of all their products taken together be found, prove that*

$$\sqrt{(a_1a_2)} + \sqrt{(a_1a_3)} + \ldots < \frac{n-1}{2}(a_1 + a_2 + \ldots a_n)$$

and hence prove that the arithmetic mean of the square roots of the products taken two together is less than the arithmetic mean of given quantities.

Solution.   Out of n letters the combination of 2 at a time will be $^nC_2$, i.e. $\dfrac{n(n-1)}{2}$.

Now $\sqrt{(a_1a_2)} < \dfrac{a_1+a_2}{2}$ $\because$ G.M. < A.M.

$\sqrt{(a_1a_3)} < \dfrac{a_1+a_3}{2}$

Adding we get

$$\sqrt{(a_1a_2)} + \sqrt{(a_1a_3)} + \ldots < \frac{1}{2}(n-1)(a_1+a_2+\ldots a_n)$$

$\therefore$ in all these inequalities in R.H.S. $a_1, a_2, \ldots a_n$ all occur $(n-1)$ times each.

We have proved that $\left(\sqrt{(a_1a_2)} + \sqrt{(a_1a_3)} + \ldots \dfrac{n(n-1)}{2} \text{ terms}\right) < \dfrac{(n-1)}{2}(a_1+a_2+\ldots a_n)$

$\therefore$ $\dfrac{\sqrt{(a_1a_2)} + \sqrt{(a_1a_3)} + \ldots \dfrac{n(n-1)}{2} \text{ terms}}{\dfrac{n(n-1)}{2}} < \dfrac{a_1+a_2+\ldots a_n}{n}$

or A.M. of square root of products taken two together out of n quantities is less than A.M. of the n quantities itself.

Example 14. *Prove that* $(a + b + c + d)(a^3 + b^3 + c^3 + d^3) > (a^2 + b^2 + c^2 + d^2)^2$

Solution. We have to prove that

$(a + b + c + d)(a^3 + b^3 + c^3 + d^3) - (a^2 + b^2 + c^2 + d^2)^2 > 0$

i.e., +ive

Simplifying and cancel a^4, b^4, c^4, d^4 we get

$\quad (a^3b + b^3a - 2a^2b^2) + \ldots > 0$

or $\quad ab(a^2 - 2ab + b^2) + \ldots > 0$

or $\quad ab(a - b)^2 + \ldots > 0$

But $ab(a - b)^2$ is +ive quantity, because a, b, c, d are all positive. Similarly all other five expressions will be +ive which **proves** the theorem.

Example 15. *Prove that* $a^2(a - b)(a - c) + b^2(b - c)(b - a) + c^2(c - a)(c - b) > 0$

and $\quad a(a - b)(a - c) + b(b - c)(b - a) + c(c - a)(c - b) > 0$ *i.e., +ive.*

Solution. Since the letters a, b, c in the above expression occur symmetrically, we may suppose that $a > b > c$.

$\therefore \qquad\qquad c^2(c - a)(c - b)$ is +ive $\hspace{5cm}$ ...(1)

Now, $\qquad a^2(a - b)(a - c) + b^2(b - c)(b - a)\cdot$

$\qquad = (a - b)(a^3 - ca^2 - b^3 + cb^2)$

$\qquad = (a - b)\{(a^3 - b^3) - c(a^2 - b^2)\}$

$\qquad = (a - b)^2\{a^2 + b^2 + ab - ca - cb\}$ $\hspace{3cm}$ ...(2)

Now $(a - b)^2$ being a perfect square is positive ab is also positive.

$\quad a^2 - ac = a(a - c)$ is also positive. $\hspace{2.5cm} \because\ a > b > c$

$\quad b^2 - cb = b(b - c)$ is also positive. $\hspace{2.5cm} \because\ a > b > c.$

Adding, we get (B) is positive.

$\therefore\quad A + B$ is also positive, i.e. given expression is positive, i.e. > 0.

The second expression may be similarly proved to be positive

Example 16. *Prove that* $\dfrac{1}{m} \log(1 + a^m) < \dfrac{1}{n} \log(1 + a^n)$ *if* $m > n$.

Solution. Case 1. $a < 1$; then because $m > n$, $\hspace{1.5cm} \therefore\quad a^m < a^n$

or $\quad 1 + a^m < 1 + a^n$

Taking log of both sides, $\log(1 + a^m) < \log(1 + a^n)$ $\hspace{3cm}$ ...(1)

$\quad m > n;\hspace{4cm} \therefore\qquad \dfrac{1}{m} < \dfrac{1}{n}$ $\hspace{2.5cm}$ ...(2)

Multiplying the respective sides of (1) and (2), we get

$$\frac{1}{m} \log(1 + a^m) < \frac{1}{n} \log(1 + a^n)$$

Case II. $a > 1$; then $\dfrac{1}{a} < 1$

Put $\dfrac{1}{a} = A < 1$; then by case 1.

$$\frac{1}{m} \log(1 + A^m) < \frac{1}{n} \log(1 + A^n)$$

Putting $A = \dfrac{1}{a}$, we get

$$\frac{1}{m} \log\left(1 + \frac{1}{a^m}\right) < \frac{1}{n} \log\left(1 + \frac{1}{a^n}\right)$$

or $\dfrac{1}{m}[\log(1 + a^m) - \log a^m] < \dfrac{1}{n}[\log(1 + a^n) - \log a^n]$

or $\dfrac{1}{m}\log(1 + a^m) - \dfrac{1}{m}m\log a < \dfrac{1}{n}\log(1 + a^n) - \dfrac{1}{n}n\log a$

or $\dfrac{1}{m}\log(1 + a^m) < \dfrac{1}{n}\log(1 + a^n)$.

Example 17. *Show that* $(x^m + y^m)^n < (x^n + y^n)^m$ *if* $m > n$.

Solution. We have to prove $x^{mn}\left[1 + \left(\dfrac{y}{x}\right)^m\right]^n < x^{mn}\left[1 + \left(\dfrac{y}{x}\right)^n\right]^m$

Taking log of both sides, $n\log\left[1 + \left(\dfrac{y}{x}\right)^m\right] < m\log\left[1 + \left(\dfrac{y}{x}\right)^n\right]$

Put $\dfrac{y}{x} = a;$ $\therefore$ $n\log(1 + a^m) < m\log(1 + a^n)$

Dividing by mn, we have to prove that

$$\dfrac{1}{m}\log(1 + a^m) < \dfrac{1}{n}\log(1 + a^n)$$

which we have already proved in previous example.

Example 18. *If* a, b, c *denote the sides of a triangle, show that*

(i) $a^2(p - q)(p - r) + b^2(q - r)(q - p) + c^2(r - p)(r - q) > 0$, p, q, r *being any real numbers.*

(ii) $a^2yz + b^2zx + c^2xy$ *can not be positive if* $x + y + z = 0$.

Solution. (i) Let $p > q > r$; then $(p - q)(p - r)$ is positive and $(r - p)(r - q)$ is also positive because both the factors are negative and $(p - r)(q - p)$ is negative because one factor is positive and the other negative.

Let $a^2(p - q)(p - r) + b^2(q - r)(q - p) + c^2(r - p)(r - q) = A$

The expression (A) has got Ist and the last term +ive, and the middle is – ive.

Now a, b, c being the sides of a triangle.

$\therefore$ $b < a + c;$ or $b^2 < (a + c)^2$ ...(1

Now we know that if we multiply an inequality by a –ive quantity, the sign of inequality is changed.

Multiply both sides of (1) by the –ive quantity $(q - r)(q - p)$ and hence changing the sign we get

$$b^2(q - r)(q - p) > (a + c)^2(q - r)(q - p)$$ ...(2

Adding 1st and 3rd terms of (A), in both sides of (2) we find that

$$A > a^2(p - q)(p - r) + (a + c)^2(q - r)(q - p) + c^2(r - p)(r - q)$$

or $> a^2(p - q)\{(p - r) - (q - r)\} + c^2(r - q)\{(r - p) - (q - p)\} + 2ac(q - r)(q - p)$

or $> a^2(p - q)^2 + c^2(r - q)^2 + 2ac(p - q)(r - q)$

or $> \{a(p - q) + c(r - q)\}^2$ or $>$ *a* positive quantity,

∴ $A > 0$.

(ii) We have to prove that if $x + y + z = 0$ then

$a^2yz + b^2zx + c^2xy$ cannot be positive.

Now $z = -(x + y)$ ∴ we have to prove that $-(a^2y + b^2x)(x + y) + c^2xy$ cannot be positive

or $-[(a^2y^2 + b^2x^2) + (a^2 + b^2)xy] + c^2xy$ can not be positive or changing the sign.

$a^2y^2 + b^2x^2 - 2abxy + (a^2 + 2ab + b^2)xy + c^2xy$ cannot be negative.

or $(ay - bx)^2 + [(a + b)^2 - c^2]xy$ can not be negative.

Now $(ay - bx)^2$ being a perfect square cannot be negative. ...(1)

In a triangle $a + b > c$; ∴ $(a + b)^2 - c^2$ is positive.

∴ $[(a + b)^2 - c^2]xy$ cannot be negative. (2)

Adding (1) and (2), we find that

$$(ay - bx)^2 + [(a + b)^2 - c^2]xy \text{ cannot be negative.}$$

Example 19. *If x is positive, then show that $\log(1 + x) < x$ and $> \dfrac{x}{1 + x}$*

Solution. We have $\log(1 + x)$ will be less than x, if

$$1 + x + e^x \text{ or } < 1 + x + \frac{x^2}{2!} + \frac{x^3}{3!} \text{ which is quite clear.}$$

Again $\log(1 + x)$ will be greater than $\dfrac{x}{1 + x}$.

If $1 + x > e^{x/(1 + x)}$, put $\dfrac{x}{1 + x} = y$; ∴ $x = \dfrac{y}{1 - y}$

We have now to prove that $1 + \dfrac{y}{1 - y} > e^y$

or $\dfrac{1}{1 - y} > e^y$ or $(1 - y)^{-1} > e^y$

or $1 + y + y^2 + y^3 + y^4 + \dots > 1 + \dfrac{y}{1!} + \dfrac{y^2}{2!} + \dfrac{y^3}{3!} + \dfrac{y^4}{4!} + \dots$

which is quite clear, because all the terms in L.H.S. after the second term are greater than the corresponding terms in the R.H.S.

Example 20. *Prove that $(n\,!)^2 < r\,!(2n - r)!$ where $n > r$.*

or we have to prove that $\dfrac{r!(2n - r)}{n!} > n!$

Solution. We have $\dfrac{r!(2n - r)}{n!}$

$$= \frac{1.2.3.\dots r(2n - r)(2n - r - 1)\dots(n + 1)(n!)}{n!}$$

$$= (1.2.3.\dots r)\{(n + 1)(n + 2)\dots(2n - r)\}$$

$$> (1.2.3. ... r) \{(r + 1)(r + 2) ... n\} \qquad \left[\begin{array}{l} \because n > r \ \therefore \ (n+1) > (r+1) \text{ etc.} \\ \text{and } n - r > 0 \ \therefore \ 2n - r > n \end{array} \right]$$

or $\qquad > 1.2.3. ... r(r + 1)(r + 2) ... n$ or $> n! \Rightarrow \dfrac{r!(2n-r)!}{n!} > n!$

or $\qquad r! (2n - 1)! > (n!)^2$ or $(n!)^2 < r! (2n - r)!$

Example 21. *Show that* $1!3!5! ... (2n - 1)! > (n!)^n$.

Solution. We have already proved that

$$r! (2n - r)! > (n!)^2, \text{ where } n > r.$$

Give r successively the values $1, 3, 5, ... (2n - 1)$

$$1! (2n - 1)! > (n!)^2$$
$$3! (2n - 3)! > (n!)^2$$
$$5! (2n - 5)! > (n!)^2$$

$$.............................$$

$$(2n - 1)! 1! > (n!)^2.$$

Multiplying the respective sides, we get

$$\{1!3!5! \ 7! ... (2n - 1)!\}^2 > (n!)^{2n}$$

Taking square root of both sides,

$$1!3!5! ... (2n - 1)! > (n!)^n$$

EXERCISE 15.3

·1. Prove : $8(a^2 + b^2)(a^3 + b^3) > (a + b)^5$

2. Prove : $n(n + 1)^3 < 8(1^3 + 2^3 + ... n^3)$

3. If a, b, c are in H.P. and n does not lie between 0 and 1, then $a^n + c^n > 2b^n$.

4. Show that sum of the mth powers of n even number is $> n(n + 1)^m$ if $m > 1$. ·

5. If $x + y + z = 1$, then show that $\dfrac{1}{x} + \dfrac{1}{y} + \dfrac{1}{z} > 9$

6. Prove that $(a_1 + a_2 ... + a_n) \left(\dfrac{1}{a_1} + \dfrac{1}{a_2} + ... \dfrac{1}{a_n} \right) > n^2$

7. Find the least value of $\dfrac{1}{x} + \dfrac{1}{y} + \dfrac{1}{z}$ for positive values of x, y, z which satisfy the condition

    $x + y + z = c$.

8. If a and b are positive and $a + b = 1$, show that $\left(a + \dfrac{1}{a} \right)^2 + \left(b + \dfrac{1}{b} \right)^2 > \dfrac{25}{2}$

9. If a, b, c be equal positive quantities such that the sum of any two is greater than the third, then

$$\frac{1}{b+c-a} + \frac{1}{c+a-b} + \frac{1}{a+b-c} > \frac{9}{a+b+c}$$

10. Prove : $\dfrac{2}{a+b} + \dfrac{2}{b+c} + \dfrac{2}{c+a} > \dfrac{9}{a+b+c}$

11. Prove : $\dfrac{3}{b+c+d} + \dfrac{3}{c+d+a} + \dfrac{3}{d+a+b} + \dfrac{3}{a+b+c} > \dfrac{16}{a+b+c+d}$

12. Prove : $a^4 + b^4 + c^4 > abc(a + b + c)$

13. Prove : $a^5 + b^5 + c^5 + d^5 > abcd(a + b + c + d)$

14. Prove : $a^5 + b^5 + c^5 > abc(ab + bc + ca)$

15. Prove : $\dfrac{a^8 + b^8 + c^8}{a^3 b^3 c^3} > \dfrac{1}{a} + \dfrac{1}{b} + \dfrac{1}{c}$

   or $a^8 + b^8 + c^8 > a^2 b^2 c^2 (bc + ca + ab)$

16. Prove : $\dfrac{b^2 + c^2}{b+c} + \dfrac{c^2 + a^2}{c+a} + \dfrac{a^2 + b^2}{a+b} > a + b + c$

17. Prove : $\dfrac{b^4 + c^4}{b+c} + \dfrac{c^4 + a^4}{c+a} + \dfrac{a^4 + b^4}{a+b} > 3abc$

18. If $a_1^2 + a_2^2 + \ldots + a_n^2 = A$ then $nA > (a_1 + a_2 + \ldots a_n)^2 > A$.

HINT TO SELECTED PROBLEMS

1. Apply A.M. of mth powers $> $ (A.M.)^m each bracket of L.H.S., i.e.

$$a^2 + b^2 > 2\left(\frac{a+b}{2}\right)^2 \text{ and } a^3 + b^3 > 2\left(\frac{a+b}{2}\right)^3$$

   Now multiply.

2. Apply A.M. of $1^3, 2^3 \ldots n^3$ is $>$ cube of A.M. of $1, 2, 3, \ldots n$ and

$$1 + 2 + 3 + \ldots n = \frac{n}{2}(n+1)$$

3. $\dfrac{a^n + c^n}{2} > \left(\dfrac{a+c}{2}\right)^n$, If n does not lie between 0 and 1.

   But we know that A.M. $>$ G.M. $>$ H.M.

$\therefore$ $\dfrac{a+c}{2}$ the A.M. of a and c is $> b$, i.e. H.M. of a and c because a, b, c are given to be in H.P.

$\therefore$ $\left(\dfrac{a+c}{2}\right)^n > b^n$

$\therefore$ $\dfrac{a^n + c^n}{2} > \left(\dfrac{a+c}{2}\right)^n > b^n$ $\therefore$ $a^n + c^n > 2b^n$

4. $\dfrac{2^m + 4^m \ldots (2n)^m}{n} > \left(\dfrac{2+4+6+\ldots 2n}{n}\right)^m$

   or $2^m + 4^m \ldots (2n)^m > (2n)^m > n\left[\dfrac{2n(n+1)}{2n}\right]^m$

or $> n(n + 1)^m$ $\because$ $\Sigma n = \dfrac{n(n+1)}{2}$

5. $\dfrac{x^{-1}+y^{-1}+z^{-1}}{3} > \left(\dfrac{x+y+z}{3}\right)^{-1}$ or $> \left(\dfrac{1}{3}\right)^{-1}$ or > 3

$\therefore$ $x^{-1}+y^{-1}+z^{-1} > 9$

or $\dfrac{1}{x}+\dfrac{1}{y}+\dfrac{1}{z} > 9$, i.e. least value of L.H.S. is 9.

6. Either apply A.M. $>$ G.M. on each bracket and multiply

or $\dfrac{\Sigma 1/a_1}{n} = \dfrac{\Sigma a_1^{-1}}{n} > \left(\dfrac{\Sigma a_1}{n}\right)^{-1}$ or $> \dfrac{n}{(\Sigma a_1)}$

$\therefore$ $(\Sigma a_1)(\Sigma a_1^{-1}) > n^2$

8. $\dfrac{1}{a}+\dfrac{1}{b} = a^{-1}+b^{-1} > 2\left[\dfrac{a+b}{2}\right]^{-1}$ or $\geqslant 2\left[\dfrac{2}{a+b}\right]$ or > 4 ...(1)

Now $\left(a+\dfrac{1}{a}\right)^2 + \left(b+\dfrac{1}{b}\right)^2 > 2\left[\dfrac{a+\dfrac{1}{a}+b+\dfrac{1}{b}}{2}\right]^2$

or $> 2\left[\dfrac{a+b+\dfrac{1}{a}+\dfrac{1}{b}}{2}\right] > 2\left(\dfrac{1+4}{2}\right)^2$ by(1)

or $> 2\dfrac{25}{4}$ or $\dfrac{25}{2}$

9. Since $b + c > a$, $\therefore$ $b + c - a =$ positive

Similarly all other factors are positive

Put $b + c - a = x$ ...(1)
                $c + a - b = y$ ...(2)
                $a + b - c = z$ ...(3)
                $a + b + c = x + y + z$

Substituting in the given inequality, we have to prove that

$$\dfrac{1}{x}+\dfrac{1}{y}+\dfrac{1}{z} > \dfrac{9}{x+y+z}$$

Now $\dfrac{x^{-1}+y^{-1}+z^{-1}}{3} > \left(\dfrac{x+y+z}{3}\right)^{-1}$ or, $> \dfrac{3}{x+y+z}$

$\therefore$ $\dfrac{1}{x}+\dfrac{1}{y}+\dfrac{1}{z} > \dfrac{9}{x+y+z}$

10. We know that

$$\dfrac{(a+b)^{-1}+(b+c)^{-1}+(c+a)^{-1}}{3} > \left(\dfrac{a+b+b+c+c+a}{3}\right)^{-1}$$

or $\qquad \dfrac{1}{a+b}+\dfrac{1}{b+c}+\dfrac{1}{c+a} > 3\dfrac{3}{2(a+b+c)}$ etc.

12. $\dfrac{a^4+b^4+c^4}{3} > \left(\dfrac{a+b+c}{3}\right)^4$ or $> \left(\dfrac{a+b+c}{3}\right)\left(\dfrac{a+b+c}{3}\right)^2$

$$> \dfrac{a+b+c}{3}\{(abc)^{1/3}\}^3; \qquad \because \text{ A.M.} > \text{G.M.}$$

or $\qquad > \left\{\dfrac{a+b+c}{3}\right\}abc$

$\therefore \qquad a^4+b^4+c^4 > abc(a+b+c)$

14. $\dfrac{a^5+b^5+c^5}{3} > \left(\dfrac{a+b+c}{3}\right)^5$ or $> \left(\dfrac{a+b+c}{3}\right)^3\left(\dfrac{a+b+c}{3}\right)^2$

or $\qquad > [(abc)^{1/3}]^3\left(\dfrac{a^2+b^2+c^2+2ab+2bc+2ca}{9}\right)$

But we know that $a^2+b^2+c^2 > ab+bc+ca$

$\therefore \qquad \dfrac{a^5+b^5+c^5}{3} > abc\dfrac{3(ab+bc+ca)}{3}$ etc.

15. $\dfrac{a^8+b^8+c^8}{3} > \left(\dfrac{a+b+c}{3}\right)^8$ or $> \left(\dfrac{a+b+c}{3}\right)^6\left(\dfrac{a+b+c}{3}\right)^2$

$$> [(abc)^{1/3}]^6\left(\dfrac{a^2+b^2+c^2+2ab+2bc+2ca}{\cdot 9}\right) \qquad \because \text{ A.M.} > \text{G.M.}$$

$\because \qquad a^2+b^2+c^2 > ab+bc+ca$

$\therefore \qquad \dfrac{a^8+b^8+c^8}{3} > a^2b^2c^2\dfrac{(3ab+3bc+3ca)}{9}$

or $\qquad \dfrac{a^8+b^8+c^8}{a^3b^3c^3} < \dfrac{ab+bc+ca}{abc}$ or $> \dfrac{1}{a}+\dfrac{1}{b}+\dfrac{1}{c}$

Aiter. $\qquad \dfrac{a^8+b^8+c^8}{3} > \dfrac{a^9+b^9+c^9}{3}\cdot\dfrac{a^{-1}+b^{-1}+c^{-1}}{3}$

or $\qquad > (a^9b^9c^9)^{1/3}\cdot\dfrac{1}{3}\left(\dfrac{1}{a}+\dfrac{1}{b}+\dfrac{1}{c}\right)$

We have applied A.M. > G.M. on Ist bracket

16. $\dfrac{b^2+c^2}{2} < \left[\dfrac{b+c}{2}\right]^2 \qquad \because \quad$ A.M. of mth powers > $(\text{A.M.})^m$

$\therefore \qquad \dfrac{b^2+c^2}{b+c} > \dfrac{b+c}{2}$

Similarly write other inequalities and add.

17. $b^4+c^4 > 2\left(\dfrac{b+c}{2}\right)^4 \qquad \therefore \qquad \dfrac{b^4+c^4}{b+c} > \left(\dfrac{b+c}{2}\right)^3 > \{\sqrt{(bc)}\}^3$

Similarly write other inequalities and add.

$\therefore$ L.H.S. $>$ $\{\sqrt{(bc)}\}^3 + \{\sqrt{(ca)}\}^3 + \{\sqrt{(ab)}\}^3$

                $>$ $3[\{\sqrt{(bc)}.\sqrt{(ca)}.\sqrt{(ab)}\}^3]^{1/3}$ $\therefore$ A.M. $>$ G.M.

or $>$ $3abc$

18. $\dfrac{a_1^2 + a_2^2 + \dots a_n^2}{n} > \left[\dfrac{a_1 + a_2 + \dots a_n}{n}\right]^2$

or $n(a_1^2 + a_2^2 + \dots a_n^2)^2 > (a_1 + a_2 + \dots a_n)^2$

or $nA > (a_1 + a_2 + \dots a_n)^2$

which is the Ist part of the inequality.

Again $(a_1 + a_2 + \dots a_n)^2 > a_1^2 + a_2^2 + \dots a_n^2 + 2\Sigma a_1 a_2$

$\therefore$ $(a_1 + a_2 + \dots a_n)^2 > a_1^2 + a_2^2 + \dots a_n^2$

because we have neglected all terms of the form $2a_1 a_2$ and etc. included in $2\Sigma a_1 a_2$. which proves the 2nd part of the inequality.

ANSWER

(7) $9/c$

15.7 MAXIMA AND MINIMA

Suppose $x, y, z \dots w$ are n positive quantities and S denotes their sum and P their product; then

1. **(a)** If $x + y + z \dots w = p$, a constant the value of $xyzw \dots$ is maximum when the quantities are equal.

i.e., $x = y = z = \dots w = \dfrac{S}{n}$ and maximum. value becomes $\left[\dfrac{S}{n}\right]^n$

   (b) The greatest value of $a^m b^n c^p \dots$ when $a + b + c + \dots$ is constant.

   $a^m b^n c^p \dots$ will be greatest when

$$\left(\frac{a}{m}\right)^m \cdot \left(\frac{b}{n}\right)^n \cdot \left(\frac{c}{p}\right)^p$$

is greatest. But this is the product of $m + n + p + \dots$ factors whose sum is

$$m.\frac{a}{m} + n.\frac{b}{n} + p.\frac{c}{p} + = a + b + c \dots$$

which is given to be constant. Hence the product will be greater when all the factors are equal

i.e., $\dfrac{a}{m} = \dfrac{b}{n} = \dfrac{c}{p} = \dots$

2. If $xyz \dots w = p$, is constant, then the value of $x + y + z + \dots w$ is least when the or quantities are equal.

i.e., minimum when $x = y = z = ... = w = P^{1/n}$ and the least value of $x = y = z = ... = w$ will be $nP^{1/n}$.

3. (a) If $x + y + z ... + w = S$, then the least value of $x^m + y^m + z^m ... w^m$ when m does not lie between 0 and 1 occurs when $x = y = z ... = S/n$ and the least value becomes $n\left(\dfrac{S}{n}\right)^m$

(b) If $x + y + z + ... + w = S$, then greatest value of $x^m + y^m + z^m + ... w^m$ when m lies between 0 and 1, occurs when $x = y = z = ... = w = \left(\dfrac{S}{n}\right)$ and the greatest value becomes $n\left(\dfrac{S}{n}\right)^m$.

4. (a) If $x^m + y^m + z^m + ... + w^m = P$, a constant, then the greatest value of $x + y + z + ...$ w when m does not lie between 0 and 1 occurs when $x = y = z = ... w = \left(\dfrac{P}{n}\right)^{1/m}$ and the greatest value becomes $n\left(\dfrac{P}{n}\right)^{1/m}$ and the greatest value becomes $n\left(\dfrac{P}{n}\right)^{\frac{1}{m}}$.

(b) If $x^m + y^m + z^m + ... w^m = P$, a constant, then the least value of $x + y + z + ...$ $+ w$ when m lies between 0 and 1 and occurs when $x = y = z = ... = w = \left(\dfrac{P}{n}\right)^{1/m}$ and the least value becomes $n\left(\dfrac{P}{n}\right)^{1/m}$

SOLVED EXAMPLES

Example 1. *Find the maximum value of $(a + x)^3 (a - x)^4$ for any real value of x numerically less than a.*

Solution. Let $A = a + x$ and $B = a - x$

Then $\qquad A + B = 2a.$

Also since $x < a \qquad \therefore$ both A and B are positive.

We have to find maximum value of $A^3 B^4$ which will be maximum when $\left(\dfrac{A}{3}\right)^3 \left(\dfrac{B}{4}\right)^4$ is maximum.

But sum of the above seven factors is

$$3.\frac{A}{3} + 4.\frac{B}{4} = A + B = 2a$$

$\therefore$ maximum value will occurs when they are equal

i.e., $\qquad \dfrac{A}{3} = \dfrac{B}{4} = \dfrac{A+B}{7} = \dfrac{2a}{7} \qquad \therefore \qquad A = \dfrac{6a}{7}$ and $B = \dfrac{8a}{7}$

Then maximum value of $A^3 B^4$ is $\dfrac{6^3 . 8^4}{7^7} a^7$.

Example 2. *Find the greatest value of $a^2b^3c^4$ subject to the condition $a + b + c = 18$.*

Solution. $a^2b^3c^4$ is maximum when $\left(\dfrac{a}{2}\right)^2 \left(\dfrac{b}{3}\right)^3 \left(\dfrac{c}{4}\right)^4$ is maximum.

Since sum of factors is

$$2.\frac{a}{2} + 3.\frac{b}{3} + 4.\frac{c}{4} = a + b + c = 18 \text{ i.e., constant.}$$

∴   Product is maximum when factors are equal

$$\therefore \quad \frac{a}{2} = \frac{b}{3} = \frac{c}{4} = \frac{a+b+c}{2+3+4} = \frac{18}{9} = 2$$

∴ $a = 4, b = 6, c = 8$ ∴ $a^2b^3c^4 = 4^2.6^3.8^4$.

Example 3. *Find the maximum value of $(8 - x)^3 (x + 6)^4$ when it is given that x lies between $- 6$ and 8.*

Solution. Since x lies between $- 6$ and 8, both $8 - x$ and $x + 6$ are positive.

Hence proceeding exactly as in Ex. 1. we find that maximum value of $(8 - x)^3 (x + 6)^4$ is $6^3 8^4$.

Example 4. *Find the maximum value of $(7 - x)^4 (2 + x)^5$ when x lies between 7 and $- 2$.*

Solution. Because x lies between 7 and $- 2$, both $7 - x$ and $2 + x$ are positive; hence proceeding exactly as in Ex. 1 we find that the maximum value of $(7 - x)^4(2 + x)^5$ is $4^4.5^5$.

Example 5. *Find the maximum value of $x^3 (4a - x)^5$ if x is positive and less than $4a$.*

Solution. Since x is positive and less than $4a$, then both x and $4a - x$ are positive. Hence proceeding as above we find that the maximum value of $x^3(4a - x)^5$ becomes

$$\left[\frac{3a}{2}\right]^3 \left[\frac{5a}{2}\right]^5 = 3^3.5^5 \frac{a^8}{2^8}$$

Example 6. *Find the maximum value of $x^{1/2} [1 - x]^{1/3}$ when x is a proper fraction.*

Solution. Because x is a proper fraction.

∴   Both x and $1 - x$ are positive.

Now $x^{1/2} (1 - x)^{1/3}$ is maximum, when $x^3(1 - x)^2$ is maximum (on taking sixth power) and $x^3(1 - x)^2$ is maximum when

$$\left[\frac{x}{3}\right]^3 \left[\frac{1-x}{2}\right]^2 \text{ is maximum} \tag{...(1)}$$

But (1) is the product of $3 + 2 = 5$ factors whose sum is

$$3.\frac{x}{2} + 2.\frac{1-x}{2} = 1 = \text{constant}.$$

and hence the product will be maximum when

$$\frac{x}{3} = \frac{1-x}{2} = \frac{1}{5} \quad \text{or} \quad \text{when } x = \frac{3}{5}$$

Maximum value of $x^{1/2}(1 - x)^{1/3}$ is $\sqrt{\dfrac{3}{5}} \cdot \left(\dfrac{2}{5}\right)^{1/3}$

Example 7. *Find the greatest value of x^2y^3 where $3x + 2y = 1$.*

Solution. Clearly x^2y^3 is greatest when $\left(\dfrac{3x}{2}\right)^2\left(\dfrac{2y}{3}\right)^3$ is maximum.

Now sum of the factors in the latter product is

$$2\frac{3x}{2}+3\frac{2y}{3} = 3x + 2y = 1 \text{ given } i.e., \text{ a constant}$$

Value will be maximum when they are equal.

$$\therefore \quad \frac{3x}{2} = \frac{2y}{3} = \frac{3x+2y}{5} = \frac{1}{5}, \qquad \therefore \quad x = \frac{2}{3.5}, \ y = \frac{3}{2.5}$$

$$\therefore \quad \text{Maximum value of } x^2y^3 \text{ is } \frac{2^2}{3^2.5^2}\cdot\frac{3^2}{2^3.5^3} = \frac{3}{2.5^5}$$

Example 8. *If $a^2x^4 + b^2y^4 = c^4$, show that maximum value of xy is $\dfrac{c^3}{\sqrt{(2ab)}}$.*

Solution. Let us put $x^4 = X$, $y^4 = Y$ and we are given that

$$a^2X + b^2Y = c^6 \qquad \qquad ...(1)$$

we have to find the maximum value of xy i.e., $(XY)^{1/4}$ which will be maximum when XY is maximum.

Now XY will be maximum when $\left(\dfrac{a^2X}{1}\right)\left(\dfrac{b^2Y}{1}\right)$ is maximum.

But since $\dfrac{a^2X}{1}+\dfrac{b^2Y}{1} = c^6$ i.e., constant, by (1) therefore above will be maximum when factors are equal.

i.e., $\qquad \dfrac{a^2X}{1} = \dfrac{b^2Y}{1} = \dfrac{a^2X+b^2Y}{1+1} = \dfrac{c^6}{2}$

$$\therefore \quad X = \frac{c^6}{2a^2}, \ Y = \frac{c^6}{2b^2} \text{ or } x^4 = \frac{c^6}{2a^2}, \ y^4 = \frac{c^6}{2b^4}$$

$$\therefore \quad \text{Maximum value of } XY \text{ i.e., } x^4y^4 \text{ is } \frac{c^6}{2a^2}\cdot\frac{c^6}{2b^2} = \frac{c^{12}}{4a^2b^2}$$

$$\therefore \quad \text{Maximum value of } xy = \frac{c^3}{\sqrt{(2b)}}.$$

Example 9. *Find the maximum value of xyz when $\dfrac{x^2}{a^2}+\dfrac{y^2}{b^2}+\dfrac{z^2}{c^2} = 1$.*

Solution. Let $\dfrac{x^2}{a^2} = X$, $\dfrac{y^2}{b^2} = Y$, $\dfrac{z^2}{c^2} = Z$.

$\therefore$ We are given that $X + Y + Z = 1$ i.e., constant and we are to find maximum value of xyz, i.e., $abc\sqrt{(XYZ)}$.

Now $abc\sqrt{(XYZ)}$ is maximum when XYZ is maximum and XYZ is maximum when $X/$

$1 = Y/1 = Z/1 = \dfrac{1}{3}$ because $X + Y + Z = 1$, constant given

$$X = \frac{x^2}{a^2} = \frac{1}{3}, \qquad \therefore \qquad x = \frac{a}{\sqrt{3}}$$

Similarly $y = \dfrac{b}{\sqrt{3}}, z = \dfrac{c}{\sqrt{3}}$

$\therefore$ Maximum value of xyz is $\dfrac{abc}{3\sqrt{3}}$.

Example 10. *Find the greatest value of* $(a - x)(b - y)(c - z)(ax + by + cz)$, *where a, b, c are known positive quantities and* $a - x, b - x, c - z$ *are also positive.*

Solution. Multiply above and below of the given expression by abc.

Now given expression is greatest when the numerator of

$$\frac{(a^2 - ax)(b^2 - by)(c^2 - cz)(ax + by + cz)}{abc}$$ is greatest, as its denominator is constant.

The numerator being the product of four factors, whose sum is $a^2 + b^2 + c^2$, i.e., constant because a, b, c are known and hence will be maximum when all the factors are equal,

i.e., $$\frac{a^2 - ax}{1} = \frac{b^2 - by}{1} = \frac{c^2 - cz}{1} = \frac{ax + by + cz}{1} = \frac{a^2 + b^2 + c^2}{4}$$

Hence the maximum value of the given expression is

$$\frac{1}{abc}\left[\frac{a^2 + b^2 + c^2}{4}\right]^4$$

Example 11. *Prove that cube is the rectangular parallelepiped of maximum volume for given surface and of minimum surface for given volume.*

Solution. Let the three adjacent edges of a rectangular parallelepiped be x, y, z; then its surface S and volume V are given by

$$S = 2(xy + yz + zx), \text{ and } V = xyz$$

Putting $X = yz, Y = zx, Z = xy$.

$$S = 2(X + Y + Z) \text{ (given) and } V = \sqrt{(XYZ)}$$

Now V is maximum when XYZ is maximum and XYZ is maximum when $X = Y = Z$ provided $X + Y + Z =$ constant which is given to us.

$\therefore$ for maximum value when surface is given

$$X = Y = Z \text{ or } yz = zx = xy$$

or $$\frac{1}{x} = \frac{1}{y} = \frac{1}{z} \qquad \text{ or } \quad x = y = z,$$

i.e., all the three edges are equal.

i.e., the rectangular parallelepiped is cube.

Similarly if $V = \sqrt{(XYZ)}$ (given) then

$$S = 2(X + Y + Z) \text{ is to be minimum}$$

Now V is given : $\therefore$ XYZ is given.

And we have to find the minimum value of S or $2(X + Y + Z)$ or of $X + Y + Z$ which will be minimum when

$$X = Y = Z \text{ or as above } x = y = z.$$

i.e., rectangular parallelepiped is a cube.

Example 12. *Prove that the equilateral triangle has maximum area for given perimeter and minimum perimeter for given area.*

Solution. I. Perimeter constant.

Let a, b, c be the sides, then $2s = a + b + c$

Area $A = \sqrt{[s(s-a)(s-b)(s-c)]}$

Perimeter $P = a + b + c = 2s$.

Put $s - a = x$, $s - b = y$, $s - c = z$, then adding

$$3s - (a + b + c) = x + y + z, \text{ i.e., } s = x + y + z$$
$$A = \sqrt{[(x+y+z)xyz]} = \sqrt{(sxyz)}$$

and $s = (x + y + z) = $ constant.

Now if perimeter is constant, then $x + y + z$ is constant then A will be maximum when A^2 or when $sxyz$ is maximum. But s being constant, we should find the maximum value of xyz which will be so when $x = y = z$ or when $s - a = s - b = s - c$ or $a = b = c$, *i.e.,* Δ is equilateral.

II. Area constant

$$A^2 = (x + y + z)xyz = x^2yz + y^2zx + z^2xy$$
$$= X + Y + Z, \text{ where } X = x^2yz, \text{ etc.}$$
$$P = 2s = 2(x + y + z) = 2\left(\frac{X}{xyz} + \frac{Y}{xyz} + \frac{Z}{xyz}\right)$$
$$= \frac{2A^2}{xyz} = \frac{2A^2}{(XYZ)^{1/4}}, \qquad \because \quad XYZ = x^4y^4z^4$$

Now area is constant, *i.e.,* $X + Y + Z$ is constant and then perimeter will be minimum when $\dfrac{2A^2}{(XYZ)^{1/4}}$ is minimum or $(XYZ)^{1/4}$ is Max. or XYZ is Max. which will be Max only when $X = Y = Z$.

i.e., $\qquad x^2yz = y^2zx = z^2xy \text{ or } x = y = z \qquad \text{or } a = b = c,$

i.e., $\qquad \Delta$ is equilateral.

Example 13. *Find the minimum value of $bcx + cay + abz$ when $xyz = abc$.*

Solution. Put $bcx = X$, $cay = Y$, and $abz = Z$; then we are to find the minimum value of $X + Y + Z$ when

$$xyz = abc \qquad \text{or when} \quad \frac{X}{bc}\frac{Y}{ca}\frac{Z}{ab} = abc$$

i.e., $\qquad XYZ = a^3b^3c^3 = $ constant.

Now, if $XYZ \ldots = $ constant P, then the value of $X + Y + Z$ is least when $X = Y = Z = \ldots p^{1/n}$.

Now $X + Y + Z$ will be least when

$$X = Y = Z = (XYZ)^{1/3} = (a^3b^3c^3)^{1/3} = abc$$

and hence least value is $abc + abc + abc = 3abc$.

Example 14. *Show that least value of $3x + 4y$, for positive value of x and y subject to $x^2 y^2 = 6$ is 10.*

Solution. Let $A = \dfrac{3}{2} x$ and $B = \dfrac{4}{3} y$.

$\therefore \qquad x^2 y^3 = \dfrac{4A^2}{9} \cdot \dfrac{27}{64} B^3 = 6$ or $A^2 B^3 = 32$ ...(1)

Also $3x + 4y = 2A + 3B$ ...(2)

Now $A.A.B.B.B = 32$ and we are to find the least value of $2A + 3B$ i.e., $A + A + B + B + B$ which will be so, when all the quantities are equal i.e., $A = B$.

$\therefore \qquad\qquad A^5 = 32 \qquad$ or $A = 2 = B \qquad\qquad$ from (1)

Hence from (2), least value of $3x + 4y$ i.e., of $2A + 3B = 2.2 + 3.2 = 10$

Example 15. *Find the minimum value of $P = \dfrac{(5+x)(2+x)}{1+x}$*

Solution. Putting $1 + x = y$, then $x = y - 1$.

$\therefore \qquad P = \dfrac{(4+y)(1+y)}{y} = \dfrac{y^2 + 5y + 4}{y}$

$\qquad\qquad = y + \dfrac{4}{y} + 5 = \left(y + \dfrac{4}{y} - 4 \right) + 9$

or $\qquad\qquad \left(\sqrt{y} - \dfrac{2}{\sqrt{y}} \right)^2 + 9$

Clearly, P is minimum when $\sqrt{y} - \dfrac{2}{\sqrt{y}}$ is zero.

i.e., $y = 2; \qquad \therefore \ x = y - 1 = 1$

and minimum value of P then is 9.

EXERCISE 15.3

1. Prove that
   (i) $(a^2 + b^2 + c^2)(x^2 + y^2 + z^2) > (ax + by + cz)^2$
   (ii) If $\Sigma a^2 = 1$ and $\Sigma x^2 = 1$, then $(ax + by + cz) > 1$.
2. If $2x + 5y = 3$, find the maximum value of $x^3 y^4$.
3. If a, b, c are positive rationals and x, y, z are positive quantities, such that $x + y + z$ is
   constant, show that $x^a y^b z^c$ is maximum when $\dfrac{x}{a} = \dfrac{y}{b} = \dfrac{z}{c}$ and its value is
   $a^a b^b c^c \left(\dfrac{x+y+z}{a+b+c} \right)^{a+b+c}$
4. Find the greatest value of abc for positive values of a, b, c subject to the condition $bc + ca + ab = 12$

HINT TO SELECTED PROBLEMS

4. Put $bc = X, ca = Y$ and $ab = Z$ so that $abc = \sqrt{(XYZ)}$ where $X + Y + Z = 12$. For maximum $X = Y = Z = 4$;

CHAPTER REVIEW : A COMPETITIVE APPROACH

Selected terms and Results

TERMS

- **Inequality :** A statement that two functions are unequal is called an inequality.

RESULTS

- If $a > b$ and $c > 0$ then
  $a + c > b + c$; $a - c > b - c$, $ac > bc$ and
  $\dfrac{a}{c} > \dfrac{b}{c}$

- If the sum of two positive quantities is constant, then their product is maximum when the quantities are equal and if the product is constant, then their sum is minimum when the quantities are equal.

- If a and b are positive and unequal then
  $\dfrac{a^m + b^m}{2} > \left(\dfrac{a+b}{2}\right)^m$ except when m is a proper fraction.

- The arithmetic mean of mth powers of n quantities is greater than the arithmetic raised to power m except when n lies between 0 and 1.

Review Questions and Project Work

1. Prove that $\dfrac{a_1}{a_2} + \dfrac{a_2}{a_3} + ... + \dfrac{a_{n-1}}{a_n} + \dfrac{a_n}{a_1} > n$

2. If $s = a + b + c + d$, prove that
   $(s - a)(s - b)(s - c)(s - d) > 81abcd$
   $s - a = b + c + d > 3(bcd)^{1/3}$

3. Prove that
   $(10x^2 + 5y^2 + 13z^2) > 2 (xy + 4yz + 9zx)$

4. Prove that the greatest value of a^2b^3 subject to $a + b = c$, where c is constant is given by $\dfrac{108}{3125} c^5$.

Objective Type Questions

Fill in the blanks:

1. If $a > b$, $c > 0$ then $a + c$ ... $b + c$

2. If the sum of two positive quantities is constant then their product is maximum when the quantities are ....

3. If m is a positive integer, then
   $\dfrac{a^m + b^m}{2}$ .... $\left[\dfrac{a+b}{2}\right]^m$

4. $a^a b^b$ .... $\left(\dfrac{a^2 + b^2}{a+b}\right)^{a+b}$

True/False: *Write 'T' for true and 'F' for false statement.*

1. $a^a b^b > \left(\dfrac{a+b}{2}\right)^{a+b} > a^b.b^a$ (T/F)

2. $a^7 + b^7 + c^7 < abc (a^4 + b^4 + c^4)$ (T/F)

3. $\dfrac{a}{b} + \dfrac{b}{c} + \dfrac{c}{d} + \dfrac{d}{a} > 4$ (T/F)

4. $6abc < bc(b + c) + ca(c + a) + ab(a + b)$ (T/F)

Multiple Choice Questions : *Choose the most appropriate one :*

1. The value of $a^2(1 + b^2) + b^2(1 + c^2) + c^2(1 + a^2)$ is
   - (a) $< 6abc$
   - (b) $> 6abc$
   - (c) $= 6abc$
   - (d) none of these

2. If a, b, c are positive then
   $\dfrac{a}{b+c} + \dfrac{b}{c+a} + \dfrac{c}{a+b} >$
   - (a) 3
   - (b) 4
   - (c) $\dfrac{3}{2}$
   - (d) none of these

3. The value of $n^n \left(\dfrac{n+1}{2}\right)^{2n}$ is greater than :

 (a) $(n!)^3$ (b) $(n!)^4$
 (c) $(n!)^5$ (d) none of these

4. The value of $1 + n\sqrt{2^{n-1}}$
 (a) $< 2^n$ (b) $> 2^n$
 (c) $= 2^n$ (d) none of these

5. If a, b, c are in H.P., then
$$\frac{a+b}{2a-b} + \frac{c+b}{2c-b} >$$
 (a) 2   (b) 3
 (c) 4
 (d) none of these

ANSWERS

Fill in the blanks

(1) >   (2) equal   (3) >   (4) <

True/False

(1) T   (2) F   (3) T   (4) T

Multiple choice questions

(1) b (2) c (3) a (4) a (5) c

CHAPTER 16

Determinants

- ❖ Determinant
- ❖ Minors
- ❖ Cofactors
- ❖ Some Properties

16.1 INTRODUCTION

The present chapter is devoted to a brief discussion of determinants and their elementary properties.

Consider two homogeneous linear equations.

$$a_1 x + b_1 y = 0$$
$$a_2 x + b_2 y = 0$$

Multiplying the first equation by b_2, the second by b_1, subtracting and dividing by x, we obtained

$$a_1 b_2 - a_2 b_1 = 0$$

This result is sometimes written as

$$\begin{vmatrix} a_1 & b_1 \\ a_2 & b_2 \end{vmatrix} = 0$$

and the expression on the left is called the determinant.

A determinant also is an arrangement of numbers in rows and columns but it always has a square form and can be reduced to a single value. Therefore, a determinant is distinct from matrix in the sense that the determinant is always in square shape and it has a numerical value. The arrangement of the numbers of a determinant is enclosed within two vertical parallel lines.

16.2 ORDER OF A DETERMINANT

The determinant of a square matrix of order n is known as determinant of order n.

16.3 DETERMINANT OF ORDER TWO

Let $a_{11}, a_{12}, a_{21}, a_{22}$ be any four numbers (real or complex). Then

$$|A| = \begin{vmatrix} a_{11} & a_{12} \\ a_{21} & a_{22} \end{vmatrix}$$

represent the number $a_{11}a_{22} - a_{21}a_{12}$ and is called a determinant of order two.

For example :
$$|A| = \begin{vmatrix} 5 & 2 \\ 3 & -7 \end{vmatrix} = (5)(-7) - (3)(2)$$

$$= -35 - 6 = -41$$

16.4 DETERMINANT OF ORDER THREE

$$|A| = \begin{vmatrix} a_{11} & a_{12} & a_{13} \\ a_{21} & a_{22} & a_{23} \\ a_{31} & a_{32} & a_{33} \end{vmatrix}$$

is called a determinant of order 3 and its value can be obtained as follows :

$$|A| = a_{11} \begin{vmatrix} a_{22} & a_{23} \\ a_{32} & a_{33} \end{vmatrix} - a_{12} \begin{vmatrix} a_{21} & a_{23} \\ a_{31} & a_{33} \end{vmatrix} + a_{13} \begin{vmatrix} a_{21} & a_{22} \\ a_{31} & a_{32} \end{vmatrix}$$

$$= a_{11}(a_{22}a_{33} - a_{32}a_{23}) - a_{12}(a_{21}a_{33} - a_{31}a_{23}) + a_{13}(a_{21}a_{32} - a_{31}a_{22})$$

For example
$$|A| = \begin{vmatrix} 2 & 3 & 4 \\ -1 & 2 & 3 \\ 4 & -2 & 1 \end{vmatrix}$$

$$= 2 \begin{vmatrix} 2 & 3 \\ -2 & 1 \end{vmatrix} - 3 \begin{vmatrix} -1 & 3 \\ 4 & 1 \end{vmatrix} + 4 \begin{vmatrix} -1 & 2 \\ 4 & -2 \end{vmatrix}$$

$$= 2(2 + 6) - 3(-1 - 12) + 4(2 - 8)$$

$$= 16 + 39 - 24 = 31$$

Remarks :

❖ The value of a determinant is not changed if it is expanded along any row or column.

❖ When no reference of the corresponding matrix is needed, we may denote a determinant by Δ.

❖ It is adviseble to expand a determinant along that row (or column) which contain maximum number of zeroes.

❖ The determinant of a square zero matrix is zero.

SOLVED EXAMPLES

Example 1. *Find the value of* $|A|$ *if A* $= \begin{vmatrix} \cos \alpha & -\sin \alpha \\ \sin \alpha & \cos \alpha \end{vmatrix}$

Solution.
$$|A| = \begin{vmatrix} \cos \alpha & -\sin \alpha \\ \sin \alpha & \cos \alpha \end{vmatrix}$$

$$= \cos^2 \alpha - (-\sin^2 \alpha) = \cos^2 \alpha + \sin^2 \alpha$$

$$= 1$$

Example 2. *Find the value of* $\begin{vmatrix} 1 & \omega \\ \omega & -\omega \end{vmatrix}$

Solution. $\quad |A| = \begin{vmatrix} 1 & \omega \\ \omega & -\omega \end{vmatrix} = -\omega - \omega^2 = -(\omega + \omega^2) = -(-1) = 1.$

Example 3. *Solve for* x : $\begin{vmatrix} x & 3 \\ 5 & 2x \end{vmatrix} = \begin{vmatrix} 5 & -4 \\ 5 & 3 \end{vmatrix}$

Solution. We have $\quad \begin{vmatrix} x & 3 \\ 5 & 2x \end{vmatrix} = \begin{vmatrix} 5 & -4 \\ 5 & 3 \end{vmatrix}$

$\Rightarrow \qquad\qquad 2x^2 - 15 = 15 + 20$

$\Rightarrow \qquad\qquad 2x^2 = 50 \quad \Rightarrow \quad x^2 = 25 \quad \Rightarrow \quad x = \pm 5.$

Example 4. *Find the value of* $\begin{vmatrix} a & h & g \\ h & b & f \\ g & f & c \end{vmatrix}$

Solution. Let $\quad \Delta = \begin{vmatrix} a & h & g \\ h & b & f \\ g & f & c \end{vmatrix}$

We expand Δ along first row, we get

$$\Delta = a(-1)^2 \begin{vmatrix} b & f \\ f & c \end{vmatrix} + h(-1)^3 \begin{vmatrix} h & f \\ g & c \end{vmatrix} + g(-1)^4 \begin{vmatrix} h & b \\ g & f \end{vmatrix}$$

$$= a(bc - f)^2 - h(hc - fg) + g(hf - bg)$$
$$= abc - af^2 - ch^2 + fgh + fgh - bg^2$$
$$= abc + 2fgh - af^2 - bg^2 - ch^2$$

Example 5. *Find the value of* $\begin{vmatrix} 0 & 1 & \sec\theta \\ \tan\theta & -\sec\theta & \tan\theta \\ 1 & 0 & 1 \end{vmatrix}$

Solution. Let $\quad \Delta = \begin{vmatrix} 0 & 1 & \sec\theta \\ \tan\theta & -\sec\theta & \tan\theta \\ 1 & 0 & 1 \end{vmatrix}$

expand along R_1

$$\Delta = 0 + 1(-1)^3 \begin{vmatrix} \tan\theta & \tan\theta \\ 1 & 1 \end{vmatrix} + \sec(-1)^4 \begin{vmatrix} \tan\theta & -\sec\theta \\ 1 & 0 \end{vmatrix}$$

$$= -(\tan\theta - \tan\theta) + \sec\theta\,(0 + \sec\theta) = 0 + \sec\theta = \sec^2\theta.$$

16.5 CO-FACTOR AND MINORS OF AN ELEMENT

If in the expansion of a determinant $|a_{ij}|$, all the terms containing a_{ij} as a factor, are collected and their sum is denoted by $a_{ij}A_{ij}$ then the factor A_{ij} is called the co-factor of the element a_{ij}. Hence, in a determinant of order n

$$|a_{ij}| = a_{i1}A_{i1} + a_{i2}A_{i2} + ... + a_{in}A_{in} = \sum_{j=1}^{n} a_{ij}A_{ij}$$

Now, let M_{ij} be the $(n-1) \times (n-1)$ sub-matrix of $|a_{ij}|_{n \times n}$ obtained by deleting the ith row and jth column. Then $|M_{ij}|$ is called the minor of the element a_{ij} in the determinant $|a_{ij}|$ of order n. Thus we can express the determinant as a linear combination of the minors of the elements of any row or any column.

Remark

❖ $(-1)^{i+j}$ is 1 or -1 according as $i+j$ is even or odd.
  ∴ A_{ij} and M_{ij} coincides if $i+j$ is even and if $i+j$ is odd then we have $A_{ij} = -M_{ij}$.

SOLVED EXAMPLES

Example 1. *Find the minors and cofactors of elements of the determinant* $\begin{vmatrix} 5 & -2 \\ 3 & 7 \end{vmatrix}$

Solution.   Minor of the element a_{11} is $M_{11} = |7| = 7$
                                 Minor of the element a_{12} is $M_{12} = 3$
                                 Minor of the element a_{21} is $M_{21} = -2$
                                 Minor of the element a_{22} is $M_{22} = 5$

Hence, Cofactors $A_{11} = (-1)^{1+1} M_{11} = 7$

$$A_{12} = (-1)^{1+2} M_{12} = -3$$
$$A_{21} = (-1)^{2+1} M_{21} = 2$$
and $$A_{22} = (-1)^{2+2} M_{22} = 5$$

Example 2. *Find all the minors and cofactors of the elements in* $\begin{vmatrix} 4 & 3 & 1 \\ 1 & 3 & 2 \\ 2 & 1 & 5 \end{vmatrix}$

Solution. Here $a_{11} = 4$ $a_{12} = 3$ $a_{13} = 1$
                       $a_{21} = 1$ $a_{22} = 3$ $a_{23} = 2$
                       $a_{31} = -2$ $a_{32} = 1$ $a_{33} = 5$

$$M_{11} = \begin{vmatrix} 3 & 2 \\ 1 & 5 \end{vmatrix} = 15 - 2 = 13$$

$$M_{12} = \begin{vmatrix} 1 & 2 \\ 2 & 5 \end{vmatrix} = 5 - 4 = 1$$

$$M_{13} = \begin{vmatrix} 1 & 3 \\ 2 & 1 \end{vmatrix} = 1 - 6 = -5$$

$$M_{21} = \begin{vmatrix} 3 & 1 \\ 1 & 5 \end{vmatrix} = 15 - 1 = 14$$

$$M_{22} = \begin{vmatrix} 4 & 1 \\ 2 & 5 \end{vmatrix} = 20 - 2 = 18$$

$$M_{23} = \begin{vmatrix} 4 & 3 \\ 2 & 1 \end{vmatrix} = 4 - 6 = -2$$

$$M_{31} = \begin{vmatrix} 3 & 1 \\ 3 & 2 \end{vmatrix} = 6 - 3 = 3$$

$$M_{32} = \begin{vmatrix} 4 & 1 \\ 1 & 2 \end{vmatrix} = 8 - 1 = 7$$

$$M_{33} = \begin{vmatrix} 4 & 3 \\ 1 & 3 \end{vmatrix} = 12 - 3 = 9$$

The co-factors are

$$A_{11} = (-1)^{1+1} M_{11} = 1 \times 13 = 13$$
$$A_{12} = (-1)^{1+2} M_{12} = -1 \times 1 = -1$$
$$A_{13} = (-1)^{1+3} M_{13} = 1 \times (-5) = -5$$
$$A_{21} = (-1)^{2+1} M_{21} = -1 \times 14 = -14$$
$$A_{22} = (-1)^{2+2} M_{22} = 1 \times 18 = 18$$
$$A_{23} = (-1)^{2+3} M_{23} = -1 \times (-2) = 2$$
$$A_{31} = (-1)^{3+1} M_{31} = 1 \times 3 = 3$$
$$A_{32} = (-1)^{3+2} M_{32} = -1 \times 7 = -7$$
$$A_{33} = (-1)^{3+3} M_{33} = 1 \times 9 = 9$$

Example 3. *Find the minor and co-factors of elements of the following determinant* $\begin{vmatrix} 2 & -3 & 5 \\ 6 & 0 & 4 \\ 1 & 5 & -7 \end{vmatrix}$

Solution. We have

$$M_{11} = \begin{vmatrix} 0 & 4 \\ 5 & -7 \end{vmatrix} = 0 - 20 = -20 \text{ and } A_{11} = -20$$

By using

$$A_{ij} = (-1)^{i+j} M_{i_j}$$

$$M_{12} = \begin{vmatrix} 6 & 4 \\ 1 & -7 \end{vmatrix} = -42 - 4 = -46 \text{ and } A_{12} = 46$$

$$M_{13} = \begin{vmatrix} 6 & 0 \\ 1 & 5 \end{vmatrix} = 30 - 0 = 30 \text{ and } A_{13} = 30$$

$$M_{21} = \begin{vmatrix} -3 & 5 \\ 5 & -7 \end{vmatrix} = 21 - 25 = -4, \quad A_{21} = 4$$

$$M_{22} = \begin{vmatrix} 2 & 5 \\ 1 & -7 \end{vmatrix} = -14 - 5 = -19, \quad A_{22} = -19$$

$$M_{23} = \begin{vmatrix} 2 & -3 \\ 1 & 5 \end{vmatrix} = 10 + 3 = 13, \quad A_{23} = -13$$

$$M_{31} = \begin{vmatrix} -3 & 5 \\ 0 & 4 \end{vmatrix} = -12 - 0 = -12, \quad A_{31} = -12$$

$$M_{32} = \begin{vmatrix} 2 & 5 \\ 6 & 4 \end{vmatrix} = 8 - 30 = -22, \quad A_{32} = 22$$

$$M_{33} = \begin{vmatrix} 2 & -3 \\ 6 & 0 \end{vmatrix} = 0 + 18 = 18, \quad A_{33} = 18$$

Example 4. *Write the co-factors of elements of the second row of the following determinant and hence evaluate them* $\begin{vmatrix} 1 & a & bc \\ 1 & b & ca \\ 1 & c & ab \end{vmatrix}$

Solution. Let $\Delta = \begin{vmatrix} 1 & a & bc \\ 1 & b & ca \\ 1 & c & ab \end{vmatrix}$

$$A_{12} = (-1)^{2+1} \begin{vmatrix} a & bc \\ c & ab \end{vmatrix} = -(a^2 b - bc^2)$$

$$A_{22} = (-1)^{2+2} \begin{vmatrix} 1 & bc \\ 1 & ab \end{vmatrix} = ab - bc$$

$$A_{23} = (-1)^{2+3} \begin{vmatrix} 1 & a \\ 1 & c \end{vmatrix} = -(c - a) = a - c$$

$$\Delta = -(a^2 b - bc^2) + b(ab - bc) + ca(a - c)$$
$$= bc^2 - a^2 b + ab^2 - b^2 c + a^2 c - ac^2.$$

16.6   PROPERTIES OF DETERMINANTS

THEOREM-1

The value of a determinant does not change when rows and columns are interchanged.

Proof. Let $|A| = \begin{vmatrix} a_1 & b_1 & c_1 \\ a_2 & b_2 & c_2 \\ a_3 & b_3 & c_3 \end{vmatrix}$ be a determinant of order three.

Expanding $|A|$ along the first row, we get

$$|A| = a_1(b_2c_3 - b_3c_2) - b_1(a_2c_3 - a_3c_2) + c_1(a_2b_3 - a_3b_2)$$
$$= a_1(b_2c_3 - b_3c_2) - a_2(b_1c_3 - b_3c_1) + a_3(b_1c_2 - b_2c_1)$$

(by rearrangement of the terms)

$$= \begin{vmatrix} a_1 & a_2 & a_3 \\ b_1 & b_2 & b_3 \\ c_1 & c_2 & c_3 \end{vmatrix}$$

Hence, the theorem is proved.

Remark

❖ If A be an n-rowed square matrix, then $|A| = |A'|$.

THEOREM-2

If any two rows [or columns] of a determinant are interchanged, the sign of the determinant is changed.

Proof. Let $|A| = \begin{vmatrix} a_1 & b_1 & c_1 \\ a_2 & b_2 & c_2 \\ a_3 & b_3 & c_3 \end{vmatrix}$ be a determinant of order three.

Expanding $|A|$ along the first row, we get

$$|A| = a_1(b_2c_3 - b_3c_2) - b_1(a_2c_3 - a_3c_2) + c_1(a_2b_3 - a_3b_2)$$
$$= -\{a_3(b_2c_1 - b_1c_2) - b_3(a_2c_1 - a_1c_2) + c_3(a_2b_1 - a_1b_2)\}$$

(by rearrangement of the terms)

$$= - \begin{vmatrix} a_1 & b_1 & c_1 \\ a_2 & b_2 & c_2 \\ a_3 & b_3 & c_3 \end{vmatrix} = (-1) |A|$$

THEOREM-3

If two rows or two columns of the determinant are identical, then the value of determinant is vanishes, i.e.,

$$|A| = \begin{vmatrix} a_1 & b_1 & c_1 \\ a_2 & b_2 & c_2 \\ a_1 & b_1 & c_1 \end{vmatrix} = 0$$

Proof. Let $|A|$ be a determinant of order 3 whose first and third row are identical. If we interchange the two identical rows, then obviously there will be no change in the value of $|A|$. But by theorem 2, the value of A is multiplied by -1 if we interchange two rows. Therefore, we get

$$|A| = -|A|$$

or $$2|A| = 0 \Rightarrow |A| = 0$$

THEOREM-4

If all the elements of any row, or any column, of a determinant are multiplied by the same number then the determinant is multiplied by that number.

Proof. Let $|A| = \begin{vmatrix} a_{11} & a_{12} & \cdots & a_{1n} \\ a_{21} & a_{22} & \cdots & a_{2n} \\ \cdots & \cdots & \cdots & \cdots \\ a_{n1} & a_{n2} & \cdots & a_{nn} \end{vmatrix}$ be a determinant of order n

We have $\begin{vmatrix} ma_{11} & a_{12} & \cdots & a_{1n} \\ ma_{21} & a_{22} & \cdots & a_{2n} \\ \cdots & \cdots & \cdots & \cdots \\ ma_{n1} & a_{n2} & \cdots & a_{nn} \end{vmatrix} = ma_{i1}A_{i1} + ma_{i2}A_{i2} + \ldots ma_{in}A_{in}$

$$= m|A|$$

(where $A_{i1}, A_{i2} \ldots A_{in}$ be the cofactors of elements $a_{i1}, a_{i2}, \ldots a_{in}$ of ith row of $|A|$.

THEOREM-5

If in a determinant each element in any row (or column) consists of two terms, then the determinant can be expressed as the sum of two other determinants.

Proof. Let $|A| = \begin{vmatrix} a_1 + \alpha_1 & b_1 & c_1 \\ a_2 + \alpha_2 & b_2 & c_2 \\ a_3 + \alpha_3 & b_3 & c_3 \end{vmatrix}$

Expanding $|A|$ along the first column, we get

$$|A| = (a_1 + \alpha_1)\begin{vmatrix} b_2 & c_2 \\ b_3 & c_3 \end{vmatrix} - (a_2 + \alpha_2)\begin{vmatrix} b_1 & c_1 \\ b_3 & c_3 \end{vmatrix} + (a_3 + \alpha_3)\begin{vmatrix} b_1 & c_1 \\ b_2 & c_2 \end{vmatrix}$$

$$= \left\{ a_1\begin{vmatrix} b_2 & c_2 \\ b_3 & c_3 \end{vmatrix} - a_2\begin{vmatrix} b_1 & c_1 \\ b_3 & c_3 \end{vmatrix} + a_3\begin{vmatrix} b_1 & c_1 \\ b_2 & c_2 \end{vmatrix} \right\}$$

$$+ \left\{ \alpha_1\begin{vmatrix} b_2 & c_2 \\ b_3 & c_3 \end{vmatrix} - \alpha_2\begin{vmatrix} b_1 & c_1 \\ b_3 & c_3 \end{vmatrix} + \alpha_3\begin{vmatrix} b_1 & c_1 \\ b_2 & c_2 \end{vmatrix} \right\}$$

$$= \begin{vmatrix} a_1 & b_1 & c_1 \\ a_2 & b_2 & c_2 \\ a_3 & b_3 & c_3 \end{vmatrix} + \begin{vmatrix} \alpha_1 & b_1 & c_1 \\ \alpha_2 & b_2 & c_2 \\ \alpha_3 & b_3 & c_3 \end{vmatrix}$$

THEOREM-6

In a determinant, If the elements of a row are added m and n times the corresponding elements of the another rows (or columns), the value of the determinant does not change in particular.

$$\begin{vmatrix} a_1 + mb_1 + nc_1 & b_1 & c_1 \\ a_2 + mb_2 + nc_2 & b_2 & c_2 \\ a_3 + mb_3 + nc_3 & b_3 & c_3 \end{vmatrix} = \begin{vmatrix} a_1 & b_1 & c_1 \\ a_2 & b_2 & c_2 \\ a_3 & b_3 & c_3 \end{vmatrix}$$

Proof. We have

$$\begin{vmatrix} a_1 + mb_1 + nc_1 & b_1 & c_1 \\ a_2 + mb_2 + nc_2 & b_2 & c_2 \\ a_3 + mb_3 + nc_3 & b_3 & c_3 \end{vmatrix} = \begin{vmatrix} a_1 & b_1 & c_1 \\ a_2 & b_2 & c_2 \\ a_3 & b_3 & c_3 \end{vmatrix} + \begin{vmatrix} mb_1 & b_1 & c_1 \\ mb_2 & b_2 & c_2 \\ mb_3 & b_3 & c_3 \end{vmatrix} + \begin{vmatrix} nc_1 & b_1 & c_1 \\ nc_2 & b_2 & c_2 \\ nc_3 & b_3 & c_3 \end{vmatrix}$$

$$= \begin{vmatrix} a_1 & b_1 & c_1 \\ a_2 & b_2 & c_2 \\ a_3 & b_3 & c_3 \end{vmatrix} + m\begin{vmatrix} b_1 & b_1 & c_1 \\ b_2 & b_2 & c_2 \\ b_3 & b_3 & c_3 \end{vmatrix} + n\begin{vmatrix} c_1 & b_1 & c_1 \\ c_2 & b_2 & c_2 \\ c_3 & b_3 & c_3 \end{vmatrix} \text{ (By theorem (4)}$$

$$= \begin{vmatrix} a_1 & b_1 & c_1 \\ a_2 & b_2 & c_2 \\ a_3 & b_3 & c_3 \end{vmatrix} + m(0) + n(0) \qquad\qquad \text{(By theorem 3)}$$

$$= \begin{vmatrix} a_1 & b_1 & c_1 \\ a_2 & b_2 & c_2 \\ a_3 & b_3 & c_3 \end{vmatrix}$$

16.6.1 Some more Properties

1. If any row or column of a determinant Δ be passed over n rows or columns, the resulting determinant will be $(-1)^n.\Delta$.

2. If A be n-rowed square matrix and k be any scaler then $|kA| = k^n|A|$

3. If all the elements of any row (or any column) of a determinant are zero, the value of the determinant is zero.

4. In a determinant, the sum of the product of the elements of any row (or any column) with the co-factors of the corresponding elements of any other row (or column) is zero.

SOLVED EXAMPLES

Example 1. *Evaluate the following determinant* $\begin{vmatrix} 3 & -2 \\ 4 & 5 \end{vmatrix}$.

Solution. We have $= |A| = \begin{vmatrix} 3 & -2 \\ 4 & 5 \end{vmatrix}$

$$= 3 \times 5 - 4 \times (-2) = 15 + 8 = 23.$$

Example 2. *Find the value of the determinant* $|A| = \begin{vmatrix} 1 & 2 & 3 \\ 2 & 3 & 1 \\ 3 & 1 & 2 \end{vmatrix}$

Solution. We have $\qquad |A| = \begin{vmatrix} 1 & 2 & 3 \\ 2 & 3 & 1 \\ 3 & 1 & 2 \end{vmatrix}$

On expanding the determinant along the first row, we get

$$= 1\begin{vmatrix} 3 & 1 \\ 1 & 2 \end{vmatrix} - 2\begin{vmatrix} 2 & 1 \\ 3 & 2 \end{vmatrix} + 3\begin{vmatrix} 2 & 3 \\ 3 & 1 \end{vmatrix}$$

$$= 1.\,(6-1) - 2.\,(4-3) + 3.\,(2-9) = -18$$

Example 3. *Evaluate the determinant of* $\begin{vmatrix} 4 & 1 & 4 \\ 0 & 1 & 0 \\ 1 & 2 & 1 \end{vmatrix}$

Solution. We have $\qquad |A| = \begin{vmatrix} 4 & 1 & 4 \\ 0 & 1 & 0 \\ 1 & 2 & 1 \end{vmatrix}$

On expanding the determinants along first column, we get

$$= 4\begin{vmatrix} 1 & 0 \\ 2 & 1 \end{vmatrix} - 0\begin{vmatrix} 1 & 4 \\ 2 & 1 \end{vmatrix} + 1\begin{vmatrix} 1 & 4 \\ 1 & 0 \end{vmatrix}$$

$$= 4(1-0) - 0 + 1\,(0-4)$$

$$= 4 - 4 = 0$$

Example 4. *Show that :* $\begin{vmatrix} 1 & x & y \\ 0 & \cos x & \sin y \\ 0 & \sin x & \cos y \end{vmatrix} = \cos(x+y)$

Solution. $\qquad$ L.H.S. $= \begin{vmatrix} 1 & x & y \\ 0 & \cos x & \sin y \\ 0 & \sin x & \cos y \end{vmatrix}$

On expanding the given determinants along first column, we get

$$= 1\begin{vmatrix} \cos x & \sin y \\ \sin x & \cos y \end{vmatrix} - 0\begin{vmatrix} x & y \\ \sin x & \sin y \end{vmatrix} + 0\begin{vmatrix} x & y \\ \cos x & \cos y \end{vmatrix}$$

$$= \cos x \cos y - \sin x \sin y = \cos(x+y) = \text{R.H.S.}$$

Example 5. *Show that :* $\begin{vmatrix} 1 & 1 & 1 \\ 1 & 1+x & 1 \\ 1 & 1 & 1+y \end{vmatrix} = xy$

Solution. We have L.H.S. $= \begin{vmatrix} 1 & 1 & 1 \\ 1 & 1+x & 1 \\ 1 & 1 & 1+y \end{vmatrix}$

Applying $C_2 - C_1$ and $C_3 - C_1$ in the given determinant, we get

$$= \begin{vmatrix} 1 & 0 & 0 \\ 1 & x & 0 \\ 1 & 0 & y \end{vmatrix}$$

On expanding the determinant along the first row, we get

$$= 1 \begin{vmatrix} x & 0 \\ 0 & y \end{vmatrix} - 0 \begin{vmatrix} 1 & 0 \\ 1 & y \end{vmatrix} - 0 \begin{vmatrix} 1 & x \\ 1 & 0 \end{vmatrix}$$

$$= xy = \text{R.H.S.}$$

Example 6. *Without expanding, show that* $\begin{vmatrix} b-c & c-a & a-b \\ c-a & a-b & b-c \\ a-b & b-c & c-a \end{vmatrix} = 0$

Solution. We have

$$\text{L.H.S} = \begin{vmatrix} b-c & c-a & a-b \\ c-a & a-b & b-c \\ a-b & b-c & c-a \end{vmatrix} = \begin{vmatrix} 0 & c-a & a-b \\ 0 & a-b & b-c \\ 0 & b-c & c-a \end{vmatrix}$$

$$= 0 \qquad\qquad (\text{Operating } C_1 \to C_1 + C_2 + C_3)$$

Example 7. *Without expanding, show that* $\begin{vmatrix} b^2c^2 & bc & b+c \\ c^2a^2 & ca & c+a \\ a^2b^2 & ab & a+b \end{vmatrix} = 0$

Solution. Consider

$$\begin{vmatrix} b^2c^2 & bc & b+c \\ c^2a^2 & ca & c+a \\ a^2b^2 & ab & a+b \end{vmatrix} = \begin{vmatrix} b^2c^2 & bc & b+c \\ c^2a^2 & ca & c+a \\ a^2b^2 & ab & a+b \end{vmatrix}$$

$$(\text{Multiplying } R_1 \text{ by } a, R_2 \text{ by } b \text{ and } R_3 \text{ by } c)$$

$$= \begin{vmatrix} ab^2c^2 & abc & ab+ca \\ bc^2a^2 & abc & bc+ab \\ ca^2b^2 & abc & ca+bc \end{vmatrix}$$

$$= abc.abc \begin{vmatrix} bc & 1 & ab+ca \\ ca & 1 & bc+ab \\ ca & 1 & ca+bc \end{vmatrix} \qquad (\text{Take } abc \text{ out from } C_1 \text{ and } C_2)$$

$$= a^2b^2c^2 \begin{vmatrix} bc & 1 & ab+bc+ca \\ ca & 1 & ab+bc+ac \\ ca & 1 & ab+ca+bc \end{vmatrix} \qquad (\text{Operate } C_3 \to C_3 + C_1)$$

$$= a^2b^2c^2(ab + bc + ca) \begin{vmatrix} bc & 1 & 1 \\ ca & 1 & 1 \\ ca & 1 & 1 \end{vmatrix}$$

$$= a^2b^2c^2(ab + bc + ca) \times 0$$

$$= 0$$

Example 8. *If a, b, c are in A.P., prove that* $\begin{vmatrix} x+1 & x+2 & x+a \\ x+2 & x+3 & x+b \\ x+3 & x+4 & x+c \end{vmatrix} = 0$

Solution. Given a, b, c are in A.P. therefore $a + c = 2b$

$\Rightarrow$ $\qquad\qquad\qquad a + c - 2b = 0$

Operating $R_1 \rightarrow R_1 + R_3 - 2R_2$, we get

$$\begin{vmatrix} x+1 & x+2 & x+a \\ x+2 & x+3 & x+b \\ x+3 & x+4 & x+c \end{vmatrix} = \begin{vmatrix} 0 & 0 & a+c-2b \\ x+2 & x+3 & x+b \\ x+3 & x+4 & x+c \end{vmatrix}$$

$$= \begin{vmatrix} 0 & 0 & 0 \\ x+2 & x+3 & x+b \\ x+3 & x+4 & x+c \end{vmatrix} = 0$$

Example 9. *Prove that*

$$\begin{vmatrix} a & b & c \\ a^2 & b^2 & c^2 \\ a^3 & b^3 & c^3 \end{vmatrix} = abc \begin{vmatrix} 1 & 1 & 1 \\ a & b & c \\ a^2 & b^2 & c^2 \end{vmatrix} = abc(a - b)(b - c)(c - a)$$

Solution. We have $\qquad |A| = \begin{vmatrix} a & b & c \\ a^2 & b^2 & c^2 \\ a^3 & b^3 & c^3 \end{vmatrix} = abc \begin{vmatrix} 1 & 1 & 1 \\ a & b & c \\ a^2 & b^2 & c^2 \end{vmatrix}$

Now again $\qquad\qquad |A| = abc \begin{vmatrix} 1 & 1 & 1 \\ a & b & c \\ a^2 & b^2 & c^2 \end{vmatrix}$

Applying $C_2 - C_1$ and $C_3 - C_1$, we get

$$|A| = abc \begin{vmatrix} 1 & 0 & 0 \\ a & b-a & c-a \\ a^2 & b^2-a^2 & c^2-a^2 \end{vmatrix}$$

On expanding along the first row, we get

$$= abc \begin{vmatrix} b-a & c-a \\ b^2-a^2 & c^2-a^2 \end{vmatrix}$$

$$= abc\ [(b-a)(c^2-a^2)-(b^2-a^2)(c-a)]$$
$$= abc\ [(b-a)(c-a)\ \{(c+a)-(b+a)\}]$$
$$= abc\ (b-a)(c-a)\ (c+a)-b-a)$$
$$= abc\ (a-b)(b-c)\ (c-a)$$

Example 10. *Prove that*

$$\begin{vmatrix} a+b+2c & a & b \\ c & b+c+2a & b \\ c & a & c+a+2b \end{vmatrix} = 2(a+b+c)^3$$

Solution. Let $|A| = \begin{vmatrix} a+b+2c & a & b \\ c & b+c+2a & b \\ c & a & c+a+2b \end{vmatrix}$

Adding C_2 and C_3 in C_1, we get

$$= \begin{vmatrix} 2(a+b+c) & a & b \\ 2(a+b+c) & b+c+2a & b \\ 2(a+b+c) & a & c+a+2b \end{vmatrix}$$

$$= 2(a+b+c)\begin{vmatrix} 1 & a & b \\ 1 & b+c+2a & b \\ 1 & a & c+a+2b \end{vmatrix}$$

Applying $(R_2 - R_1)$ and $(R_3 - R_1)$, we get

$$|A| = 2(a+b+c)\begin{vmatrix} 1 & a & b \\ 0 & b+c+a & 0 \\ 0 & 0 & c+a+b \end{vmatrix}$$

On expanding the determinant along the first column, we get

$$|A| = 2(a+b+c)\begin{vmatrix} b+c+a & 0 \\ 0 & a+b+c \end{vmatrix}$$

$$= 2(a+b+c)(a+b+c)^2$$
$$= 2(a+b+c)^3$$

Example 11. *Prove that* $\begin{vmatrix} 1+a & 1 & 1 \\ 1 & 1+b & 1 \\ 1 & 1 & 1+c \end{vmatrix} = abc\left(1+\dfrac{1}{a}+\dfrac{1}{b}+\dfrac{1}{c}\right)$

Solution.   Operating $C_1 \rightarrow C_1 - C_3$ and $C_2 \rightarrow C_2 - C_3$, we get

$$\begin{vmatrix} 1+a & 1 & 1 \\ 1 & 1+b & 1 \\ 1 & 1 & 1+c \end{vmatrix} = \begin{vmatrix} a & 0 & 1 \\ 0 & b & 1 \\ -c & -c & 1+c \end{vmatrix}$$

$$= a[b.(1+c) - c(-c).1] + 1[0.(-c) - (-c)b]$$

$$= a(b + bc + c) + bc$$

$$= abc + bc + ca + ab$$

$$= abc\left(1 + \frac{1}{a} + \frac{1}{b} + \frac{1}{c}\right)$$

Example 12. *Prove that*

$$\begin{vmatrix} a-b-c & 2a & 2a \\ 2b & b-c-a & 2b \\ 2c & 2c & c-a-b \end{vmatrix} = (a+b+c)^3$$

Solution. Operating $R_1 \to R_1 + R_2 + R_3$, we get

$$\begin{vmatrix} a-b-c & 2a & 2a \\ 2b & b-c-a & 2b \\ 2c & 2c & c-a-b \end{vmatrix} = \begin{vmatrix} a+b+c & a+b+c & a+b+c \\ 2b & b-c-a & 2b \\ 2c & 2c & c-a-b \end{vmatrix}$$

[Take $(a + b + c)$ out from R_1]

$$= (a+b+c)\begin{vmatrix} 1 & 1 & 1 \\ 2b & b-c-a & 2b \\ 2c & 2c & c-a-b \end{vmatrix}$$

(Operate $C_2 \to C_2 - C_1$ and $C_3 \to C_3 - C_1$)

$$= (a+b+c)\begin{vmatrix} 1 & 0 & 0 \\ 2b & b-c-a & 0 \\ 2c & 0 & -a-b-c \end{vmatrix} \qquad \text{(expand by } R_1)$$

$$= (a+b+c).1.(-a-b-c)(-a-b-c)$$

$$= (a+b+c)^3$$

Example 13. *Show that*

$$\begin{vmatrix} a & b & c \\ a-b & b-c & c-a \\ b+c & c+a & a+b \end{vmatrix} = a^3 + b^3 + c^3 - 3abc$$

Solution. Operating $R_2 \to R_2 - R_1$ and $R_3 \to R_3 + R_1$, we get

$$\begin{vmatrix} a & b & c \\ a-b & b-c & c-a \\ b+c & c+a & a+b \end{vmatrix} = \begin{vmatrix} a & b & c \\ -b & -c & -a \\ a+b+c & a+b+c & a+b+c \end{vmatrix}$$

[Take $(a + b + c)$ out from R_3 and (-1) from R_2]

$$= -(a+b+c).\begin{vmatrix} a & b & c \\ b & c & a \\ 1 & 1 & 1 \end{vmatrix} \qquad \text{(expand by } R_3\text{)}$$

$$= -(a+b+c).[1.(ab-c^2) - 1(a^2-bc) + 1.(ca-b^2)]$$
$$= -(a+b+c).(ab+bc+ca-a^2-b^2-c^2)$$
$$= (a+b+c).(a^2+b^2+c^2-ab-bc-ca)$$
$$= a^3 + b^3 + c^3 - 3abc$$

Example 14. *Find the value of x if*

$$\begin{vmatrix} 3+x & 5 & 2 \\ 1 & 7+x & 6 \\ 2 & 5 & 3+x \end{vmatrix} = 0$$

Solution. We have

$$\begin{vmatrix} 3+x & 5 & 2 \\ 1 & 7+x & 6 \\ 2 & 5 & 3+x \end{vmatrix} = 0$$

Applying $(R_1 - R_3)$, we get

$$\begin{vmatrix} 1+x & 0 & -1-x \\ 1 & 7+x & 6 \\ 2 & 5 & 3+x \end{vmatrix} = 0$$

Applying $C_3 \rightarrow C_3 + C_1$, we get

$$\begin{vmatrix} 1+x & 0 & 0 \\ 1 & 7+x & 7 \\ 2 & 5 & 5+x \end{vmatrix} = 0$$

On expanding the determinant along the first row, we get

$$(1+x)\begin{vmatrix} 7+x & 7 \\ 5 & 5+x \end{vmatrix} = 0$$

$$(1+x)[(7+x)(5+x) - 35] = 0$$

or $$(1+x)(x^2 + 12x) = 0$$

or $$x(1+x)(x+12) = 0$$

$\therefore$ $$x = 0, -1, -12$$

Example 15. *Evaluate :*
$$\begin{vmatrix} 3 & 2 & 1 & 4 \\ 15 & 29 & 2 & 14 \\ 16 & 19 & 3 & 17 \\ 23 & 39 & 8 & 38 \end{vmatrix}$$

Solution. Applying $C_1 \rightarrow C_1 - 3C_2$, $C_2 \rightarrow C_2 - 3C_3$, $C_4 \rightarrow C_4 - 4C_3$, we get

$$|A| = \begin{vmatrix} 0 & 0 & 1 & 0 \\ 9 & 25 & 2 & 6 \\ 7 & 13 & 3 & 5 \\ 9 & 23 & 8 & 6 \end{vmatrix}$$

On expanding the determinant along first row, we get

$$= 1\begin{vmatrix} 9 & 25 & 6 \\ 7 & 13 & 5 \\ 9 & 23 & 6 \end{vmatrix}$$

Applying $R_1 \to R_1 - R_3$, we get

$$= 1\begin{vmatrix} 0 & 2 & 0 \\ 7 & 13 & 5 \\ 9 & 23 & 6 \end{vmatrix}$$

On expanding the determinant along the first row, we get

$$= -2\begin{vmatrix} 7 & 5 \\ 9 & 6 \end{vmatrix} = -2(42 - 45) = 6$$

Example 16. *Using properties of determinants, solve the following determinant for x.*

$$\begin{vmatrix} a+x & a-x & a-x \\ a-x & a+x & a-x \\ a-x & a-x & a+x \end{vmatrix} = 0$$

Solution. Given

$$\begin{vmatrix} a+x & a-x & a-x \\ a-x & a+x & a-x \\ a-x & a-x & a+x \end{vmatrix} = 0$$

$$\Rightarrow \qquad \begin{vmatrix} 3a-x & a-x & a-x \\ 3a-x & a+x & a-x \\ 3a-x & a-x & a+x \end{vmatrix} = 0 \qquad \text{(Operate } C_1 \to C_1 + C_2 + C_3\text{)}$$

$$\Rightarrow \qquad (3a-x)\begin{vmatrix} 1 & a-x & a-x \\ 1 & a+x & a-x \\ 1 & a-x & a+x \end{vmatrix} = 0 \qquad \text{(Operate } R_2 \to R_2 - R_1, R_3 \to R_3 - R_1\text{)}$$

$$\Rightarrow \qquad (3a-x)\begin{vmatrix} 1 & a-x & a-x \\ 0 & 2x & 0 \\ 0 & 0 & 2x \end{vmatrix} = 0 \qquad \text{(expand by } C_1\text{)}$$

$$\Rightarrow \qquad (3a - x) \cdot 1 \cdot \begin{vmatrix} 2x & 0 \\ 0 & 2x \end{vmatrix} = 0$$

$$\Rightarrow \qquad (3a - x) \cdot (4x^2 - 0) = 0$$

$$\Rightarrow \qquad 4x^2 (3a - x) = 0$$

$$\Rightarrow \qquad x^2 = 0 \quad \text{or} \quad 3a - x = 0$$

$$\Rightarrow \qquad x = 0, 0, 3a$$

Hence, the values of x are 0, 0, 3a.

Example 17. *Using properties of determinants, prove that*

$$\begin{vmatrix} 1 & 1 & 1 \\ \alpha & \beta & \gamma \\ \beta\gamma & \gamma\alpha & \alpha\beta \end{vmatrix} = (\alpha - \beta)(\beta - \gamma)(\gamma - \alpha)$$

Solution. Operate $C_1 \rightarrow C_2 - C_1$ and $C_3 \rightarrow C_3 - C_1$, we get

$$\begin{vmatrix} 1 & 1 & 1 \\ \alpha & \beta & \gamma \\ \beta\gamma & \gamma\alpha & \alpha\beta \end{vmatrix} = \begin{vmatrix} 1 & 0 & 0 \\ \alpha & \beta - \alpha & \gamma - \alpha \\ \beta\gamma & \gamma(\alpha - \beta) & \beta(\alpha - \gamma) \end{vmatrix}$$

[Take $(\alpha - \beta)$ out from C_2 and $(\gamma - \alpha)$ out from C_3]

$$= (\alpha - \beta)(\gamma - \alpha) \begin{vmatrix} 1 & 0 & 0 \\ \alpha & -1 & 1 \\ \beta\gamma & \gamma & -\beta \end{vmatrix} \qquad \text{(expand by } C_1)$$

$$= (\alpha - \beta)(\gamma - \alpha) \cdot 1 \cdot \begin{vmatrix} -1 & 1 \\ \gamma & -\beta \end{vmatrix}$$

$$= (\alpha - \beta)(\gamma - \alpha)(\beta - \gamma)$$

$$= (\alpha - \beta)(\beta - \gamma)(\gamma - \alpha)$$

Example 18. *Prove that*

$$\begin{vmatrix} a^2 + 1 & ab & ac \\ ab & b^2 + 1 & bc \\ ac & bc & c^2 + 1 \end{vmatrix} = 1 + a^2 + b^2 + c^2$$

Solution. We have $|A| = \begin{vmatrix} a^2 + 1 & ab & ac \\ ab & b^2 + 1 & bc \\ ac & bc & c^2 + 1 \end{vmatrix}$

Now multiply the column 1st, 2nd and 3rd by a, b and c respectively, we get

$$|A| = \frac{1}{abc} \begin{vmatrix} a(a^2+1) & ab^2 & ac^2 \\ a^2b & b(b^2+1) & bc^2 \\ a^2c & b^2c & c(c^2+1) \end{vmatrix}$$

To take a, b, c common from 1st, 2nd and 3rd rows respectively, we get

$$= \frac{abc}{abc} \begin{vmatrix} a^2+1 & b^2 & c^2 \\ a^2 & b^2+1 & c^2 \\ a^2 & b^2 & c^2+1 \end{vmatrix}$$

Now Applying $C_1 \rightarrow C_1 + C_2 + C_3$, we get

$$= \begin{vmatrix} a^2+b^2+c^2+1 & b^2 & c^2 \\ a^2+b^2+c^2+1 & b^2+1 & c^2 \\ a^2+b^2+c^2+1 & b^2 & c^2+1 \end{vmatrix}$$

$$= (a^2+b^2+c^2+1) \begin{vmatrix} 1 & b^2 & c^2 \\ 1 & b^2+1 & c^2 \\ 1 & b^2 & c^2+1 \end{vmatrix}$$

Now apply $R_2 \rightarrow R_2 - R_1$ and $R_3 \rightarrow R_3 - R_1$, we get

$$= (a^2+b^2+c^2+1) \begin{vmatrix} 1 & 0 \\ 0 & 1 \end{vmatrix}$$

$$= (a^2+b^2+c^2+1) \cdot 1 \qquad \qquad \dots(1)$$

$$= (a^2+b^2+c^2+1)$$

Example 19. *Prove that*

$$\begin{vmatrix} b+c & c+a & a+b \\ q+r & r+p & p+q \\ y+z & z+x & x+y \end{vmatrix} = 2 \begin{vmatrix} a & b & c \\ p & q & r \\ x & y & z \end{vmatrix}$$

Solution. We have

$$\text{L.H.S} = \begin{vmatrix} b+c & c+a & a+b \\ q+r & r+p & p+q \\ y+z & z+x & x+y \end{vmatrix}$$

Applying $C_1 \rightarrow C_1 + C_2 - 2C_3$, we get

$$= \begin{vmatrix} 2c & c+a & a+b \\ 2r & r+p & p+q \\ 2z & z+x & x+y \end{vmatrix} = 2 \begin{vmatrix} c & c+a & a+b \\ r & r+p & p+q \\ z & z+x & x+y \end{vmatrix}$$

Now applying $C_2 \to C_2 - C_1$, we get

$$= 2 \begin{vmatrix} c & a & a+b \\ r & p & p+q \\ z & x & x+y \end{vmatrix}$$

Applying $C_3 \to C_3 - C_2$, we get

$$= 2 \begin{vmatrix} c & a & b \\ r & p & q \\ z & x & y \end{vmatrix}$$

$$= 2 \begin{vmatrix} a & b & c \\ p & q & r \\ x & y & z \end{vmatrix} \qquad \text{(by Interchanging the columns)}$$

$$= \text{R.H.S}$$

Example 20. *If x, y, z are all different and* $\begin{vmatrix} x & x^2 & 1+x^3 \\ y & y^2 & 1+y^3 \\ z & z^2 & 1+z^3 \end{vmatrix} = 0$

show that $xyz = -1$.

Solution. Given $\begin{vmatrix} x & x^2 & 1+x^3 \\ y & y^2 & 1+y^3 \\ z & z^2 & 1+z^3 \end{vmatrix} = 0$

$\Rightarrow \quad \begin{vmatrix} x & x^2 & 1 \\ y & y^2 & 1 \\ z & z^2 & 1 \end{vmatrix} + \begin{vmatrix} x & x^2 & x^3 \\ y & y^2 & y^3 \\ z & z^2 & z^3 \end{vmatrix} = 0$

Take x, y, z out from R_1, R_2 and R_3 respectively from the second determinant we get

$\Rightarrow \quad \begin{vmatrix} 1 & x & x^2 \\ 1 & y & y^2 \\ 1 & z & z^2 \end{vmatrix} + xyz \begin{vmatrix} 1 & x & x^2 \\ 1 & y & y^2 \\ 1 & z & z^2 \end{vmatrix} = 0$

$\Rightarrow \quad \begin{vmatrix} 1 & x & x^2 \\ 1 & y & y^2 \\ 1 & z & z^2 \end{vmatrix} (1 + xyz) = 0$

$\Rightarrow \quad (x - y) \cdot (y - z) \cdot (z - x) \cdot (1 + xyz) = 0$

$\Rightarrow$ $(1 + xyz) = 0$

(Because x, y, z are all distinct, so $x - y \neq 0$, $y - z \neq 0$, $z - x \neq 0$)

$\Rightarrow$ $xyz = -1$

Example 21. *Evaluate the value of x for which* $\begin{vmatrix} 4x & 6x+2 & 8x+1 \\ 6x+2 & 9x+3 & 12x \\ 8x+1 & 12x & 16x+2 \end{vmatrix} = 0$

Solution. We have $\begin{vmatrix} 4x & 6x+2 & 8x+1 \\ 6x+2 & 9x+3 & 12x \\ 8x+1 & 12x & 16x+2 \end{vmatrix} = 0$

Applying $\left(C_2 \to C_2 - \dfrac{3}{2} C_1 \right)$ and $C_3 \to C_3 - 2C_1$, we get

$$\begin{vmatrix} 4x & 2 & 1 \\ 6x+2 & 0 & -4 \\ 8x+1 & -3/2 & 0 \end{vmatrix} = 0$$

Now applying $R_3 \to R_2 + 4R_1$

$\Rightarrow$ $\begin{vmatrix} 4x & 2 & 1 \\ 22x+2 & 8 & 0 \\ 8x+1 & -3/2 & 0 \end{vmatrix} = 0$

On expanding the determinants along 3rd column, we get

$$1 \begin{vmatrix} 22x+2 & 8 \\ 8x+1 & -3/2 \end{vmatrix} = 0$$

$\Rightarrow$ $-33x - 3 - 64x - 8 = 0$

or $-97x = 11$ or $x = \dfrac{-11}{97}$

Example 22. *Without expanding show that the value of the determinant given below is zero*

$$\begin{vmatrix} \sin \alpha & \cos \alpha & \sin (\alpha + \delta) \\ \sin \beta & \cos \beta & \sin (\beta + \delta) \\ \sin \gamma & \cos \gamma & \sin (\gamma + \delta) \end{vmatrix}$$

Solution. Let $\Delta = \begin{vmatrix} \sin \alpha & \cos \alpha & \sin (\alpha + \delta) \\ \sin \beta & \cos \beta & \sin (\beta + \delta) \\ \sin \gamma & \cos \gamma & \sin (\gamma + \delta) \end{vmatrix}$

Using $\sin (A + B) = \sin A \cos B + \cos A \sin B$

$$\Delta = \begin{vmatrix} \sin\alpha & \cos\alpha & \sin\alpha\cos\delta + \cos\alpha\sin\delta \\ \sin\beta & \cos\beta & \sin\beta\cos\delta + \cos\beta\sin\delta \\ \sin\gamma & \cos\gamma & \sin\gamma\cos\delta + \cos\gamma\sin\delta \end{vmatrix}$$

$$= \begin{vmatrix} \sin\alpha & \cos\alpha & 0 \\ \sin\beta & \cos\beta & 0 \\ \sin\gamma & \cos\gamma & 0 \end{vmatrix} \qquad \text{Using } C_3 \rightarrow C_3 \ (\cos\delta)\, C_1 - (\sin\delta)C_2$$

$$= 0$$

Example 23. *Show that*

$$\begin{vmatrix} (b+c)^2 & a^2 & bc \\ (c+a)^2 & b^2 & ca \\ (a+b)^2 & c^2 & ab \end{vmatrix} = (a^2 + b^2 + c^2)(a + b + c)(b - c)(c - a)(a - b)$$

Solution. Let

$$\Delta = \begin{vmatrix} (b+c)^2 & a^2 & bc \\ (c+a)^2 & b^2 & ca \\ (a+b)^2 & c^2 & ab \end{vmatrix}$$

Applying $C_1 \rightarrow C_1 - 2C_3$, we get

$$\Delta = \begin{vmatrix} b^2 + c^2 + a^2 & a^2 & bc \\ c^2 + a^2 + b^2 & b^2 & ca \\ a^2 + b^2 + c^2 & c^2 & ab \end{vmatrix}$$

Operating $C_1 \rightarrow C_1 + C_2$, we get

$$\Delta = (a^2 + b^2 + c^2)\begin{vmatrix} 1 & a^2 & bc \\ 1 & b^2 & ca \\ 1 & c^2 & ab \end{vmatrix}$$

Operating $R_2 \rightarrow R_2 - R_1$ and $R_3 \rightarrow R_3 - R_2$

$$\Delta = (a^2 + b^2 + c^2)\begin{vmatrix} 1 & a^2 & bc \\ 0 & b^2 - a^2 & (ca - bc) \\ 0 & c^2 - a^2 & (ab - bc) \end{vmatrix}$$

$$= (a^2 + b^2 + c^2)(b - c)(c - a)\begin{vmatrix} 1 & a^2 & bc \\ 0 & b + a & -c \\ 0 & c + a & -b \end{vmatrix}$$

$R_3 \rightarrow R_3 - R_2$, we get

$$= (a^2 + b^2 + c^2)(b-c)(c-a) \begin{vmatrix} 1 & a^2 & bc \\ 0 & b+a & -c \\ 0 & c-b & c-b \end{vmatrix}$$

$$= (a^2 + b^2 + c^2)(b-a)(c-a)(c-b) \begin{vmatrix} 1 & a^2 & bc \\ 0 & b+a & -c \\ 0 & 1 & 1 \end{vmatrix}$$

Expanding along first column, we get

$$\Delta = (a^2 + b^2 + c^2)(b-a)(c-a)(c-b)(a+b+c)$$

Example 24. *Show that*

$$\begin{vmatrix} a+b & b+c & c+a \\ b+c & c+a & a+b \\ c+a & a+b & b+c \end{vmatrix} = 2 \begin{vmatrix} a & b & c \\ b & c & a \\ c & a & b \end{vmatrix}$$

Solution. Let $\quad \Delta = \begin{vmatrix} a+b & b+c & c+a \\ b+c & c+a & a+b \\ c+a & a+b & b+c \end{vmatrix}$

Applying $C_1 \rightarrow C_1 + C_2 + C_3$, we get

$$= \begin{vmatrix} 2(a+b+c) & b+c & c+a \\ 2(a+b+c) & c+a & a+b \\ 2(a+b+c) & a+b & b+c \end{vmatrix}$$

$$= 2 \begin{vmatrix} a+b+c & -a & -b \\ a+b+c & -b & -c \\ a+b+c & -c & -a \end{vmatrix} \quad \text{Applying } C_2 \rightarrow C_2 - C_1, C_3 \rightarrow C_3 - C_1$$

We get

$$= 2(-1)(-1) \begin{vmatrix} a+b+c & a & b \\ a+b+c & b & c \\ a+b+c & c & a \end{vmatrix}$$

Applying $C_1 \rightarrow C_1 - C_2 - C_3$, we get

$$= 2 \begin{vmatrix} c & a & b \\ a & b & c \\ b & c & a \end{vmatrix}$$

$$= -2 \begin{vmatrix} a & c & b \\ b & a & c \\ c & b & a \end{vmatrix} (C_1 \rightarrow C_2) = 2 \begin{vmatrix} a & b & c \\ b & c & a \\ c & a & b \end{vmatrix}$$

Example 25. *If a, b, c (all positive) are the pth, qth and rth terms respectively of a geometric progression, show that*

$$\begin{vmatrix} \log a & p & 1 \\ \log b & q & 1 \\ \log c & r & 1 \end{vmatrix} = 0$$

Solution. Consider the terms of G.P. which are $A, AR, AR^2, ...$

$$a = T_p = AR^{p-1}$$
$$b = T_q = AR^{q-1}$$
$$c = T_r = AR^{r-1}$$

Consider
$$\begin{vmatrix} \log a & p & 1 \\ \log b & q & 1 \\ \log c & r & 1 \end{vmatrix} = \begin{vmatrix} \log AR^{p-1} & p & 1 \\ \log AR^{q-1} & q & 1 \\ \log AR^{r-1} & r & 1 \end{vmatrix}$$

$$= \begin{vmatrix} \log A + (p-1)\log R & p & 1 \\ \log A + (q-1)\log R & q & 1 \\ \log A + (r-1)\log R & r & 1 \end{vmatrix}$$

$$= \begin{vmatrix} \log A & p & 1 \\ \log A & q & 1 \\ \log A & r & 1 \end{vmatrix} + \begin{vmatrix} (p-1)\log R & p & 1 \\ (q-1)\log R & q & 1 \\ (r-1)\log R & r & 1 \end{vmatrix}$$

$$= \log A \begin{vmatrix} 1 & p & 1 \\ 1 & q & 1 \\ 1 & r & 1 \end{vmatrix} + \log R \begin{vmatrix} p-1 & p & 1 \\ q-1 & q & 1 \\ r-1 & r & 1 \end{vmatrix}$$

$$= \log A \times 0 + \log R \begin{vmatrix} p & p & 1 \\ q & q & 1 \\ r & r & 1 \end{vmatrix} = 0 + \log R \times 0 = 0$$

Example 26. *Show that*

$$\begin{vmatrix} b^2 + c^2 & ab & ac \\ ba & c^2 + a^2 & bc \\ ca & cb & a^2 + b^2 \end{vmatrix} = 4a^2b^2c^2$$

Solution. Let
$$\Delta = \begin{vmatrix} b^2 + c^2 & ab & ac \\ ba & c^2 + a^2 & bc \\ ca & cb & a^2 + b^2 \end{vmatrix}$$

Multiplying R_1, R_2, R_3 by a, b, c respectively and dividing Δ by abc, we get

$$\Delta = \frac{1}{abc} \begin{vmatrix} a(b^2+c^2) & a^2 & a^2b \\ b^2a & b(c^2+a^2) & b^2c \\ c^2a & c^2b & c(a^2+b^2) \end{vmatrix}$$

Taking a, b, c common from C_1, C_2, C_3 respectively, we get

$$\Delta = \frac{abc}{abc} \begin{vmatrix} (b^2+c^2) & a^2 & a^2 \\ b^2 & c^2+a^2 & b^2 \\ c^2 & c^2 & a^2+b^2 \end{vmatrix}$$

Applying $R_1 \to R_1 + R_2 + R_3$, we get

$$= \begin{vmatrix} 2(b^2+c^2) & 2(c^2+a^2) & 2(a^2+b^2) \\ b^2 & c^2+a^2 & b^2 \\ c^2 & c^2 & a^2+b^2 \end{vmatrix}$$

Taking 2 common from R_1 we get

$$= 2 \begin{vmatrix} (b^2+c^2) & c^2+a^2 & a^2+b^2 \\ b^2 & c^2+a^2 & b^2 \\ c^2 & c^2 & a^2+b^2 \end{vmatrix}$$

Operating $R_2 \to R_2 - R_1$, $R_3 \to R_3 - R_1$, we get

$$= 2 \begin{vmatrix} b^2+c^2 & c^2+a^2 & a^2+b^2 \\ -c^2 & 0 & -a^2 \\ -b^2 & -a^2 & 0 \end{vmatrix}$$

Operating $R_1 \to R_1 + R_2 + R_3$, we get

$$\Delta = 2 \begin{vmatrix} 0 & c^2 & b^2 \\ -c^2 & 0 & -a^2 \\ -b^2 & -a^2 & 0 \end{vmatrix}$$

$$= 2[0 - c^2(0 - a^2b^2) + b^2(a^2c^2 - 0)] \qquad \text{(Expanding along } R_1)$$

$$= 2[a^2b^2c^2 + a^2b^2c^2] = 4a^2b^2c^2$$

Example 27. *Show that*

$$\begin{vmatrix} (y+z)^2 & xy & zx \\ xy & (x+z)^2 & yz \\ xz & yz & (x+y)^2 \end{vmatrix} = 2xyz\,(x+y+z)^2$$

Solution. Let

$$\Delta = \begin{vmatrix} (y+z)^2 & xy & zx \\ xy & (x+z)^2 & yz \\ xz & yz & (x+y)^2 \end{vmatrix}$$

Operating $R_1 \to xR_1$, $R_2 \to yR_2$, $R_3 \to zR_3$, we get

$$\Delta = \frac{1}{xyz} \begin{vmatrix} x(y+z)^2 & x^2y & x^2z \\ xy^2 & y(x+z)^2 & y^2z \\ xz^2 & yz^2 & z(x+y)^2 \end{vmatrix}$$

Taking x, y, z common from C_1, C_2, C_3 respectively, we get

$$\Delta = \frac{xyz}{xyz} \begin{vmatrix} (y+z)^2 & x^2 & x^2 \\ y^2 & (x+z)^2 & y^2 \\ z^2 & z^2 & (x+y)^2 \end{vmatrix}$$

$$= \begin{vmatrix} (y+z)^2 & x^2 & x^2 \\ y^2 & (x+z)^2 & y^2 \\ z^2 & z^2 & (x+y)^2 \end{vmatrix}$$

$$= \begin{vmatrix} (y+z)^2 - x^2 & 0 & x^2 \\ 0 & (z+x)^2 - y^2 & y^2 \\ z^2 - (x+y)^2 & z^2 - (x+y)^2 & (x+y)^2 \end{vmatrix}$$

Operating $C_1 \to C_1 - C_3$, $C_2 \to C_2 - C_3$, we get

$$= \begin{vmatrix} (y+z+x)(y+z-x) & 0 & x^2 \\ 0 & (z+x+y)(z+x-y) & y^2 \\ (z+x+y)(z-x-y) & (z+x+y)(z-x-y) & (x+y)^2 \end{vmatrix}$$

Taking $(x+y+z)$ common C_1 and C_2 each, we get

$$\Delta = (x+y+z)^2 \begin{vmatrix} y+z-x & 0 & x^2 \\ 0 & z+x-y & y^2 \\ z-x-y & z-x-y & (x+y)^2 \end{vmatrix}$$

Operating $R_3 \to R_3 - R_1 - R_2$, we get

$$\Delta = (x+y+z)^2 \begin{vmatrix} y+z-x & 0 & x^2 \\ 0 & z+x-y & y^2 \\ -2y & -2x & 2xy \end{vmatrix}$$

$$= (x+y+z)^2 \begin{vmatrix} y+z & x^2/y & x^2 \\ y^2/x & z+x & y^2 \\ 0 & 0 & 2xy \end{vmatrix}$$

Expanding along R_1, we get

$$\Delta = (x + y + z)^2 \cdot 2xy \begin{vmatrix} y + z & x^2/y \\ y^2/x & z + x \end{vmatrix}$$

$$= (x + y + z)^2 \cdot 2xy[(y + z)(z + x) - xy]$$
$$= (x + y + z)^2 \cdot 2xy \, (yz + z^2 + zx)$$
$$= 2xyz \, (x + y + z)^2$$

Example 28. *Solve the equation*

$$\begin{vmatrix} 3x - 8 & 3 & 3 \\ 3 & 3x - 8 & 3 \\ 3 & 3 & 3x - 8 \end{vmatrix} = 0$$

Solution. The given equation is

$$\begin{vmatrix} 3x - 8 & 3 & 3 \\ 3 & 3x - 8 & 3 \\ 3 & 3 & 3x - 8 \end{vmatrix} = 0$$

$$\Rightarrow \begin{vmatrix} 3x - 2 & 3x - 2 & 3x - 2 \\ 3 & 3x - 8 & 3 \\ 3 & 3 & 3x - 8 \end{vmatrix} = 0 \qquad \text{(By applying } R_1 \to R_1 + R_2 + R_3 \text{)}$$

$$\Rightarrow (3x - 2) \begin{vmatrix} 1 & 1 & 1 \\ 3 & 3x - 8 & 3 \\ 3 & 3 & 3x - 8 \end{vmatrix} = 0$$

$$\Rightarrow (3x - 2) \begin{vmatrix} 0 & 0 & 1 \\ 0 & 3x - 11 & 3 \\ 11 - 3x & 11 - 3x & 3x - 8 \end{vmatrix} = 0 \quad \text{(Applying } C_1 \to C_1 - C_3, C_2 \to C_2 - C_3 \text{)}$$

$$\Rightarrow (3x - 2) \times 1 \cdot \begin{vmatrix} 0 & 3x - 11 \\ 11 - 3x & 11 - 3x \end{vmatrix} = 0$$

$$(3x - 2)(3x - 11)^2 = 0$$

$$\Rightarrow \qquad x = \frac{2}{3}, \frac{11}{3}, \frac{11}{3}$$

EXERCISE 16.1

Evaluate the following determinants : (Ques. 1-7)

1. $\begin{vmatrix} \frac{1}{2} & 8 \\ 4 & 2 \end{vmatrix}$

2. $\begin{vmatrix} -2 & 3 \\ 4 & -9 \end{vmatrix}$

3. $\begin{vmatrix} \cos \theta & -\sin \theta \\ \sin \theta & \cos \theta \end{vmatrix}$

4. $\begin{vmatrix} x^2 - x + 1 & x - 1 \\ x + 1 & x + 1 \end{vmatrix}$

5. $\begin{vmatrix} 1 & 0 & 6 \\ 3 & 4 & 15 \\ 5 & 6 & 21 \end{vmatrix}$

6. $\begin{vmatrix} 23 & 12 & 11 \\ 36 & 10 & 26 \\ 63 & 26 & 37 \end{vmatrix}$

7. (a) $\begin{vmatrix} 3 & 1 & -4 \\ 3 & 2 & 5 \\ 1 & 1 & 3 \end{vmatrix}$

(b) $\begin{vmatrix} 1 & \omega & \omega^2 \\ \omega & \omega^2 & 1 \\ \omega^2 & 1 & \omega \end{vmatrix}$

Write the minor and co-factors of each element of the following determinants and also evaluate the determinant in each case : (Ques. 8-11)

8. $\begin{vmatrix} 5 & -10 \\ 0 & 3 \end{vmatrix}$

9. $\begin{vmatrix} 1 & 3 & -2 \\ 4 & -5 & 6 \\ 3 & 5 & 2 \end{vmatrix}$

10. $\begin{vmatrix} 1 & 0 & 0 \\ 0 & 1 & 0 \\ 0 & 0 & 1 \end{vmatrix}$

11. $\begin{vmatrix} 1 & 0 & 4 \\ 3 & 5 & -1 \\ 0 & 1 & 2 \end{vmatrix}$

12. Evaluate $\begin{vmatrix} x+1 & x+2 & x+4 \\ x+5 & x+6 & x+8 \\ x+7 & x+10 & x+14 \end{vmatrix}$

13. Evaluate $\begin{vmatrix} 1 & a & bc \\ 1 & b & ca \\ 1 & c & ab \end{vmatrix}$

14. Evaluate $\begin{vmatrix} x+\lambda & x & x \\ x & x+\lambda & x \\ x & x & x+\lambda \end{vmatrix}$

15. Evaluate $\begin{vmatrix} b+c & a & a \\ b & c+a & b \\ c & c & a+b \end{vmatrix}$

16. Prove that $\begin{vmatrix} 1 & x & x^2 \\ 1 & y & y^2 \\ 1 & z & z^2 \end{vmatrix} = (x-y)(y-z)(z-x)$

17. Prove that $\begin{vmatrix} -a^2 & ab & ac \\ ba & -b^2 & bc \\ ac & bc & -c^2 \end{vmatrix} = 4a^2b^2c^2$ s

18. Prove that $\begin{vmatrix} x & x^2 & yz \\ y & y^2 & zx \\ z & z^2 & xy \end{vmatrix} = (x-y)(y-$

z)(z-x)(xy+yz+zx)

19. Using properties of determinants, prove that

$$\begin{vmatrix} y+z & x & y \\ z+x & z & x \\ x+y & y & z \end{vmatrix} = (x+y+z)(x-z)^2$$

20. Using properties of determinants, prove that

$$\begin{vmatrix} a-b-c & 2a & 2a \\ 2b & b-c-a & 2b \\ 2c & 2c & c-a-b \end{vmatrix} = (a+b+c)^3$$

21. Solve the following determinants

$$\begin{vmatrix} x-2 & 2x-3 & 3x-4 \\ x-4 & 2x-9 & 3x-16 \\ x-8 & 2x-27 & 3x-64 \end{vmatrix} = 0$$

22. Using properties of determinants, prove that

$$\begin{vmatrix} 1+a^2-b^2 & 2ab & -2b \\ 2ab & 1-a^2+b^2 & 2a \\ 2b & -2a & 1-a^2-b^2 \end{vmatrix} = (1+a^2+b^2)$$

23. Prove that

$$\begin{vmatrix} x & x^2 & 1+px^2 \\ y & y^2 & 1+py^3 \\ z & z^2 & 1+pz^3 \end{vmatrix} = (1+pxyz)(x-y)(y-z)(z-x)$$

24. Prove that using properties of determinants

$$\begin{vmatrix} 3a & -a+b & -a+c \\ -b+a & 3b & -b+c \\ -c+a & -c+b & 3c \end{vmatrix} = 3(a+b+c)(ab+bc+ca)$$

25. Prove that

$$\begin{vmatrix} bc & a & a^2 \\ ca & b & b^2 \\ ab & c & c^2 \end{vmatrix} = \begin{vmatrix} 1 & a^2 & a^3 \\ 1 & b^2 & b^3 \\ 1 & c^2 & c^3 \end{vmatrix}$$

HINT TO SELECTED PROBLEMS

12. Applying $R_1 \rightarrow R_3 - R_1$ and $R_2 \rightarrow R_2 - R_1$
$\qquad C_3 \rightarrow C_3 - C_1$ and $C_2 \rightarrow C_2 - C_1$
We get the value of det $= -24$

13. Applying $R_2 \rightarrow R_2 - R_1$ and $R_3 \rightarrow R_3 - R_1$
and after expansion, we get the required result.

14. Applying $R_1 \rightarrow R_1 + R_2 + R_3$
$\qquad C_2 \rightarrow C_2 - C_1,\ C_3 \rightarrow C_3 - C_1$

15. Applying $R_1 \rightarrow R_1 + R_2 - R_3$ and expanding.

16. Applying $R_2 \rightarrow R_2 - R_1$, $R_3 \rightarrow R_3 - R_1$

17. Taking a, b, c common from first, second and third column and after then applying $R_2 \rightarrow R_2 + R_1$ and $R_3 \rightarrow R_3 + R_1$

18. Multiplying first, second and third rows of the determinant by x, y, z respectively, and then applying $C_2 \rightarrow C_2 - C_1$, $C_3 \rightarrow C_3 - C_1$

19. Applying $R_1 \rightarrow R_1 + R_2 + R_3$

    Then $C_1 \rightarrow C_1 - C_2 - C_3$

20. Applying $R_1 \rightarrow R_1 + R_2 + R_3$

    $$C_2 \rightarrow C_2 - C_1$$
    $$C_3 \rightarrow C_3 - C_1$$

22. Applying $C_1 \rightarrow C_1 - C_3$

    $$R_3 \rightarrow R_3 - R_1$$

24. Applying $C_1 \rightarrow C_1 + C_2 + C_3$

    $$R_2 \rightarrow R_2 - R_1, R_3 \rightarrow R_3 - R_1$$

ANSWERS

(1) -31 **(2)** 6   **(3)** 1   **(4)** $x^3 - x^2 + 2$ **(5)** -18

(6) 0   **(7)** (*a*) 49   (*b*) 0

(8) $M_{11} = 3, M_{12} = 0, M_{21} = -10, M_{22} = 5$
$A_{11} = 3, A_{12} = 0, A_{21} = 10, A_{22} = 5, 15$

(9) $M_{11} = -40, M_{12} = -10, M = 35, M_{21} = 16, M_{22} = 8, M_{23} = -4$
$M_{31} = 8, M_{32} = 14, M_{33} = -17$
$A_{11} = -40, A_{12} = 10, A_{13} = 35, A_{21} = -16, A_{22} = 8, A_{23} = 4$
$A_{31} = 8, A_{32} = -14, A_{33} = -17; -80$

(10) $M_{11} = 1, M_{12} = 0, M_{13} = 0, M_{21} = 0, M_{22} = 1, M_{23} = 0$
$M_{31} = 0, M_{32} = 0, M_{33} = 1$
$A_{11} = 1, A_{12} = 0, A_{13} = 0, A_{21} = 0, A_{22} = 1, A_{23} = 0$
$A_{31} = 0, A_{32} = 0, A_{33} = 1; 1$

(11) $M_{11} = 11, M_{12} = 6, M_{13} = 3, M_{21} = -4, M_{22} = 2, M_{23} = 1$
$M_{31} = -20, M_{32} = -13, M_{33} = 5$
$A_{11} = 11, A_{12} = -6, A_{13} = 3, A_{21} = 4, A_{22} = 2, A_{23} = -1$
$A_{31} = 20, A_{32} = 13, A_{33} = 5; 23$

(12) -24

(13) $(a - b)(b - c)(c - a)$

(14) $\lambda^2 (3x + \lambda)$

(15) $4abc$ **(21)** $x = 4$

CHAPTER REVIEW : A COMPETITIVE APPROACH

Selected terms and Results

TERMS

- **Determinant :** It is an arrangement of numbers in rows and columns, always in a square form and can be reduced to a single value.

- **Order of a determinant :** The determinant of a square matrix of order n is known as determinant of order n.

- **Co-factors :** If in the expression of a determinant $|a_{ij}|$ all the terms containing a_{ij} as a factor, are collected and their sum is denoted by $a_{ij}.A_{ij}$ then the factors A_{ij} is called the co-factor of the element a_{ij}.

- **Minor :** The minor $|M_{ij}|$ be the $(n-1)$ $(n-1)$ submatrix of $|a_{ij}|_{n \times n}$ obtained by deleting the ith row and jth column.

RESULTS

- The value of a determinant does not changes when a row and columns are interchanged.
- If any two rows (or columns) of a determinant are interchanged the sign of the determinant is changed.
- If two rows or column of the determinant are identical, then the value of the determinant vanishes.
- If all the elements of any row, or any column, of a determinant are multiplied by the same number then the determinant is multiplied by that number.
- If in a determinant each element in any row (or column) consists of two terms, then the determinant can be expressed as the sum of two other determinant.
- If in the determinant, the elements of a row are add m and n times the corresponding elements of the another rows (or columns), the value of the determinant does not change in particular.

$$\begin{vmatrix} a_1 + mb_1 + nc_1 & a_2 & c_1 \\ a_2 + mb_2 + nc_2 & b_2 & c_2 \\ a_3 + mb_3 + nc_3 & b_3 & c_3 \end{vmatrix} = \begin{vmatrix} a_1 & a_2 & c_1 \\ a_2 & b_2 & c_2 \\ a_3 & b_3 & c_3 \end{vmatrix}$$

Review Questions and Project Work

1. Find the value of :

$$\begin{vmatrix} 0 & 1 & \sec? \\ \tan? & -\sec? & \tan? \\ 1 & 0 & 1 \end{vmatrix}.$$

2. Find the minor and co-factors of elements of the following determinant.

$$\begin{vmatrix} 2 & -3 & 5 \\ 6 & 0 & 4 \\ 1 & 5 & -7 \end{vmatrix}$$

3. Show that

$$\begin{vmatrix} 1 & 1 & 1 \\ 1 & 1+x & 1 \\ 1 & 1 & 1+y \end{vmatrix} = xy$$

4. Prove that

$$\begin{vmatrix} a+b+c & a & b \\ c & b+c+2a & b \\ c & a & c+a+2b \end{vmatrix}$$
$$= 2(a+b+c)^3$$

5. Find the value of x if

$$\begin{vmatrix} 3+x & 5 & 2 \\ 1 & 7+x & 6 \\ 2 & 5 & 3+x \end{vmatrix} = 0$$

6. Use properties of determinants, solve the following determinant for x.

$$\begin{vmatrix} a+x & a-x & a-x \\ a-x & a+x & a-x \\ a-x & a-x & a+x \end{vmatrix} = 0$$

7. If x, y, z are all different and

$$\begin{vmatrix} x & x^2 & 1+x^3 \\ y & y^2 & 1+y^3 \\ z & z^2 & 1+z^3 \end{vmatrix} = 0$$

show that $xyz = -1$.

8. Show that

$$\begin{vmatrix} a+b & b+c & c+a \\ b+c & c+a & a+b \\ c+a & a+b & b+c \end{vmatrix} = 2 \begin{vmatrix} a & b & c \\ b & c & a \\ c & a & b \end{vmatrix}$$

Objective Type Questions

Fill in the blanks:

1. The value of $\begin{vmatrix} 5 & 2 \\ 3 & -7 \end{vmatrix} = ...$

2. The value of $\begin{vmatrix} 1 & ? \\ ? & -? \end{vmatrix} = ...$

3. A_{ij} and M_{ij} coincides if $i+j$ is ...

4. If $i+j$ is odd then $A_{ij} = ...$

True/False: *Write 'T' for true and 'F' for false statement.*

1. The value of the determinant is not changed if it is expanded along any row or column. (T/F)
2. The determinant of a square zero matrix is not necessarily zero. (T/F)
3. The value of a determinate does not change when rows and columns are interchanged. (T/F)
4. If two rows or columns of the determinant are identical then the value of the determinant is zero. (T/F)
5. If any two rows (or columns) of a determinant are interchanged, the sign of the determinant is changed. (T/F)

Multiple Choice Questions : *Choose the most appropriate one :*

1. If $\begin{vmatrix} a & a+1 & a-1 \\ -b & b+1 & b-1 \\ c & c-1 & c+1 \end{vmatrix} +$

$$\begin{vmatrix} a+1 & b+1 & c-1 \\ a-1 & b-1 & c+1 \\ (-1)^{n+2}.a & (-1)^{n+1}.b & (-1)^n.c \end{vmatrix}$$

$$= 0 \; [(b+c) \neq 0]$$

The true value of n is
(a) 0
(b) any even integer
(c) any odd integer
(d) none of these

2. If a_1, a_2, a_3 are in G.P. and $a_i > 0 \; \forall i \geq 1$ then value of

$$\begin{vmatrix} \log a_m & \log a_{m+1} & \log a_{m+2} \\ \log a_{m+3} & \log a_{m+4} & \log a_{m+5} \\ \log a_{m+6} & \log a_{m+7} & \log a_{m+8} \end{vmatrix}$$

(a) $\log a_m$
(b) $\log a_{m+2}$
(c) 0
(d) none of these

3. For positive integer x, the value of

$$\begin{vmatrix} x! & (x+1)! & (x+2)! \\ (x+1)! & (x+2)! & (x+3)! \\ (x+2)! & (x+3)! & (x+4)! \end{vmatrix} = 0$$

(a) $2x! \, (x+1)!$
(b) $2x! \, (x+1)! \, (x+2)!$
(c) $2x! \, (x+3)!$
(d) none of these

4. If $a \neq b \neq c$ then value of x satisfying

$$\begin{vmatrix} 0 & x-a & x-b \\ x+a & 0 & x-c \\ x+b & x+c & 0 \end{vmatrix} = 0 \text{ is}$$

(a) $x = 0$

(b) $x = a$

(c) $x = b$

(d) none of these

5. If $x \neq 0, y \neq 0$, then D =

$$\begin{vmatrix} 1 & 1 & 1 \\ 1 & 1+x & 1 \\ 1 & 1 & 1+y \end{vmatrix} \text{ is}$$

(a) divisible by x and y both

(b) divisible by x but not by y

(c) divisible by y and but not by x

(d) none of these

6. If $\Delta = \begin{vmatrix} 2^{r-1} & 2.3^{r-1} & 4.5^{r-1} \\ a & ß & ? \\ 2^{n-1} & 3^{n-1} & 5^{n-1} \end{vmatrix}$, then

value of $\sum_{r=1}^{n} \Delta_r$ is

(a) $\alpha + \beta + \gamma$

(b) $2\alpha + \beta + \gamma$

(c) 0

(d) none of these

7. If p, q, r are negative numbers, then

value of $\Delta = \begin{vmatrix} p & q & r \\ q & r & p \\ r & p & q \end{vmatrix}$ is

(a) > 0 (b) ≥ 0

(c) < 0 (d) none of these

8. Let $\Delta = \begin{vmatrix} 1 & \sin ? & 1 \\ -\sin ? & 1 & \sin ? \\ -1 & -\sin ? & 1 \end{vmatrix}$,

then Δ lies in the interval

(a) $[2, 4]$ (b) $[3, 4]$

(c) $[1, 4]$ (d) none of these

9. Which one of the following is a

factor of $\begin{vmatrix} a & b & c & d \\ b & c & d & a \\ c & d & a & b \\ d & a & b & c \end{vmatrix}$

(a) $a + b + c + d$ (b) $abcd$

(c) $4abc$ (d) none of these

10. Let $D = \begin{vmatrix} 1 & a & b \\ 1 & b & c \\ 1 & c & a \end{vmatrix}$ then value of

$\begin{vmatrix} a & b & c \\ b & c & a \\ 1 & 1 & 1 \end{vmatrix}$ is

(a) $-D$ (b) D

(c) 0 (d) none of these

ANSWERS

Fill in the blanks

(1) -41 (2) 1   (3) even   (4) $-M_{ij}$

True/False

(1) T   (2) F   (3) T   (4) T   (5) T

Multiple choice questions

(1) *c*   (2) *c*   (3) *b*   (4) *a*   (5) *a*

(6) *c*   (7) *a*   (8) *a*   (9) *a*   (10) *b*

Matrices

17.1 INTRODUCTION

'Matrices' is a powerful tool of modern mathematics. The study of 'Matrices' is essential in almost every important branch of science like mathematics and physics.

The word 'matrix' was used by J.J. Sylvester in 1850 and developed by 'Arthur Caylay' in 1858.

Definition

A set having mn numbers either real or complex, arranged in the form of rectangular array in which there are m rows and n columns. This rectangular arrangement is called a matrix of order $m \times n$ which is denoted by $[a_{ij}]_{m \times n}$ where $i = 1, 2, 3, ... m$ and $j = 1, 2, 3, ... n$ and a matrix of order $m \times n$ is usually written as

$$[a_{ij}]_{m \times n} = \begin{vmatrix} a_{11} & a_{12} & a_{13} & ... & a_{1n} \\ a_{21} & a_{22} & a_{23} & ... & a_{2n} \\ a_{31} & a_{32} & a_{33} & ... & a_{3n} \\ \vdots & \vdots & \vdots & ... & \vdots \\ a_{m1} & a_{m2} & a_{m3} & ... & a_{mn} \end{vmatrix}_{m \times n}$$

17.2 TYPE OF MATRICES

(*i*) **Null matrix (zero matrix) :** A matrix of order m × n is called a null matrix if it contain all mn elements zero. It is denoted by **0** and usually written as

$$\mathbf{0} = \begin{vmatrix} 0 & 0 & 0 & ... & 0 \\ 0 & 0 & 0 & ... & 0 \\ 0 & 0 & 0 & ... & 0 \\ \vdots & \vdots & \vdots & ... & \vdots \\ 0 & 0 & 0 & ... & 0 \end{vmatrix}$$

(ii) **Square matrix :** *A matrix having a number of rows equal to number of columns, is called square matrix.*

For example
$$A = \begin{bmatrix} a_{11} & a_{12} & a_{13} \\ a_{21} & a_{22} & a_{23} \\ a_{31} & a_{32} & a_{33} \end{bmatrix}_{3 \times 3}$$

This matrix A is a square matrix of order 3×3.

(iii) **Unit matrix :** *A square matrix of order $n \times n$ having all non-diagonal elements equal to zero and each of the diagonal element equal to 1 is called a unit matrix. It is denoted by I_n and is usually written as*

$$I_n = \begin{bmatrix} 1 & 0 & 0 & 0 \\ 0 & 1 & 0 & 0 \\ 0 & 0 & 1 & 0 \\ \vdots & \vdots & \vdots & \vdots \\ 0 & 0 & 0 & 1 \end{bmatrix}_{n \times m}$$

This unit matrix is also known as identity matrix.

(iv) **Row matrix :** *A matrix having only one row and n columns is called a row matrix of order $1 \times n$.*

For example $\quad A = [a_{11} a_{12} a_{13} \dots a_{1n}]_{1 \times n}.$

(v) **Column matrix :** *A matrix having m rows and only one column is called column matrix of order $m \times 1$.*

For example
$$A = \begin{bmatrix} a_{11} \\ a_{21} \\ a_{31} \\ \vdots \\ a_{m1} \end{bmatrix}_{m \times 1}$$

17.2.1. Triangular, Diagonal and Scalar Matrices

(i) **Upper triangular matrix :** *A matrix of order $n \times n$ is called an upper triangular matrix if it contains all its elements below the diagonal elements equal to zero, i.e., $A = [a_{ij}]_{n \times n}$ if $a_{ij} = 0$ for $i > j$, then A upper triangular matrix.*

For example
$$A = \begin{bmatrix} a_{11} & a_{12} & a_{13} \\ 0 & a_{22} & a_{23} \\ \vdots & \vdots & \vdots \\ 0 & 0 & a_{33} \end{bmatrix}_{3 \times 3}$$

(ii) **Lower triangular matrix :** *A matrix of order $n \times n$ is called a lower triangular matrix if it contains its all elements above the diagonal equal to zero. Suppose $A = [a_{ij}]_{m \times n}$ and if $a_{ij} = 0$ for $i < j$, then A is called lower triangular matrix.*

For example,
$$A = \begin{bmatrix} a_{11} & 0 & 0 \\ a_{21} & a_{22} & 0 \\ a_{31} & a_{32} & a_{33} \end{bmatrix}_{3 \times 3}$$

(iii) Diagonal matrix : *A matrix of order n × n is called diagonal matrix if it contains all its off-diagonal elements equal to zero. Suppose $A = [a_{ij}]_{n \times n}$ and if $a_{ij} = 0$ for all $i \neq j$, then A is called diagonal matrix, it is denoted by*

$$\text{Diag. } [a_{11}a_{12} \cdots a_{nn}].$$

(iv) Scalar matrix : *A diagonal matrix whose diagonal elements are all equal but not equal to 1, is called a scalar matrix.*

For example :
$$A = \begin{bmatrix} k & 0 & 0 \\ 0 & k & 0 \\ 0 & 0 & k \end{bmatrix}_{3 \times 3} \qquad \text{and } k \neq 1$$

(v) Idempotent matrix : *A matrix such that $A^2 = A$ is called idempotent matrix.*

For example :
$$A = \begin{bmatrix} 2 & -2 & 4 \\ -1 & 3 & 4 \\ 1 & -2 & -3 \end{bmatrix} \Rightarrow A^2 = A.A = \begin{bmatrix} 2 & -2 & 4 \\ -1 & 3 & 4 \\ 1 & -2 & -3 \end{bmatrix} = A$$

(vi) Periodic matrix : *A matrix A is called a periodic matrix of $A^{k+1} = A$, where k is a positive integer. If k is the least positive integer for which $A^{k+1} = A$, then k is said to be the period of A. If we choose k = 1, we get $A^2 = A$ and we call it to be idempotent matrix.*

(vii) Nilpotent matrix : *A matrix A is called a nilpotent matrix, if $A^k = 0$, (null matrix) where k is a positive integer. If however k is the least positive integer for which $A^k = 0$, then k is the index of the nilpotent matrix.*

For example :
$$A = \begin{bmatrix} ab & b^2 \\ -a^2 & -ab \end{bmatrix}$$

(viii) Involutary matrix : *A matrix is called a involutary matrix, if $A^2 = I$, (identity matrix).* Unit matrix is an involutary matrix.

Since $I^2 = I$ always, therefore Unit matrix is an involutary matrix.

17.3 DETERMINANT OF SQUARE MATRIX

Let A be a square matrix. Then the determinant, which is formed by the elements of a matrix A, is usually denoted by $|A|$.

For example, if
$$A = \begin{bmatrix} a_{11} & a_{12} & a_{13} \\ a_{21} & a_{22} & a_{23} \\ a_{31} & a_{32} & a_{33} \end{bmatrix}_{3 \times 3}$$

Then,
$$|A| = \begin{vmatrix} a_{11} & a_{12} & a_{13} \\ a_{21} & a_{22} & a_{23} \\ a_{31} & a_{32} & a_{33} \end{vmatrix}$$

17.4 SINGULAR AND NON-SINGULAR MATRIX

Definition : *A matrix A, whose determinant value is zero, is called singular, otherwise non-singular.*

17.5 SUB-MATRIX OF A MATRIX

Definition : *Let A be a matrix of order m × n, then a matrix which is obtained by leaving some rows and column from the given matrix A is called submatrix of matrix A.*

For example :

Let
$$A = \begin{bmatrix} a_{11} & a_{12} & a_{13} & a_{14} \\ a_{21} & a_{22} & a_{23} & a_{24} \\ a_{31} & a_{32} & a_{33} & a_{34} \end{bmatrix}_{3 \times 4}$$

Then the matrix
$$B = \begin{bmatrix} a_{11} & a_{12} & a_{13} \\ a_{31} & a_{32} & a_{33} \end{bmatrix}$$

is a submatrix of A, which is obtained by leaving second row and fourth column. If the given matrix A is a square matrix, then a square submatrix of the given matrix is called principal submatrix.

17.6 MINORS OF A MATRIX

Definition : *Let A be a matrix of order m × n, then the determinant of every square submatrix of A is called a minor of A.*

17.7 TRANSPOSE OF A MATRIX

Definition : *Consider a matrix $A = [a_{ij}]_{m \times n}$. Then a matrix which is obtained by interchanging the rows and column of A is called the transpose of A. It is denoted by A′ or A^T.*

For example :
$$A = \begin{bmatrix} a_{11} & a_{12} & a_{13} \\ a_{21} & a_{22} & a_{23} \end{bmatrix}_{2 \times 3}$$

Then the transpose of A is

$$A' = \begin{bmatrix} a_{11} & a_{21} \\ a_{12} & a_{22} \\ a_{13} & a_{23} \end{bmatrix}_{3 \times 2}$$

Remark

❖ The transpose of the transpose of matrix is the matrix itself *i.e.*, $(A')' = A$.

17.8 SYMMETRIC AND SKEW-SYMMETRIC MATRICES

Symmetric Matrix : A matrix 'A' is said to be a symmetric matrix if $A' = A$ i.e., the transpose of a matrix is equal to the matrix itself.

Skew-symmetric matrix : A matrix 'A' is said to be a skew-symmetric matrix if $A' = -A$.

For example : $\qquad A = \begin{bmatrix} 0 & 2 & 3 \\ -2 & 0 & 4 \\ -3 & -4 & 0 \end{bmatrix}$

Remark

❖ The diagonal elements of a skew symmetric matrix are all zero.

17.9 COMPLEX MATRIX

Definition : *A matrix 'A' is said to be complex matrix, if it contains some of its elements equal to complex numbers.*

For example : $\qquad A = \begin{bmatrix} 1+2i & 3i & 7 \\ -3i & 2+3i & 1+i \end{bmatrix}$

17.9.1 Conjugate of Complex Matrix

Definition : *Let A be a complex matrix, then a matrix which is obtained by replacing all the complex elements of A by their conjugate complex number, is called conjugate of a matrix and it is denoted by $\overline{A}$.*

For example : If $\qquad A = \begin{bmatrix} 1+2i & 3i & 6 \\ 7 & 2+4i & 1+i \end{bmatrix}$

Then $\qquad\qquad \overline{A} = \begin{bmatrix} 1-2i & -3i & 6 \\ 7 & 2-4i & 1-i \end{bmatrix}$

17.9.2 Transpose Conjugate of a Matrix

Definition : *The transpose of the conjugate of a matrix is called the transpose conjugate of a matrix. It is denoted by A^θ. That is*

$$A^\theta = (\overline{A})'$$

17.9.3 Hermitian and Skew-Hermitian Matrices

(i) Hermitian matrix : A matrix 'A' is said to be Hermitian if $A^\theta = A$.

For example : $\qquad A = \begin{bmatrix} 3 & 3-i \\ 3+i & 4 \end{bmatrix}$

(ii) Skew-Hermitian matrix : A matrix 'A' is said to be Skew-Hermitian if $A^\theta = -A$.

For example : $\qquad A = \begin{bmatrix} 2i & 5 \\ -5 & -i \end{bmatrix}$

Remarks

❖ The diagonal elements of a Hermitian matrix are necessarily real.
❖ The diagonal elements of a Skew-Hermitian matrix are either purely imaginary or zero.

THEOREM-1

Every square matrix can be expressed as P + iQ uniquely where P and Q are Hermitian matrices.

Proof. Let A be a square matrix. Then it can be written as

$$A = \frac{1}{2}[A + A^\theta] + i\left[\frac{1}{2i}(A - A^\theta)\right] = P + iQ$$

where $\qquad P = \frac{1}{2}(A + A^\theta), \quad Q = \frac{1}{2i}(A - A^\theta)$

We have to prove that P and Q are Hermitian matrices.

Now $\qquad P^\theta = \frac{1}{2}(A + A^\theta)^\theta = \frac{1}{2}(A^\theta + (A^\theta)^\theta) = \frac{1}{2}(A^\theta + A) = P$

$\Rightarrow$ P is Hermitian matrix.

Similarly $\qquad Q^\theta = \left[\frac{1}{2i}(A - A^\theta)\right]^\theta = -\frac{1}{2i}(A - A^\theta)^\theta = -\frac{1}{2i}(A^\theta - (A^\theta)^\theta)$

$$= -\frac{1}{2i}(A^\theta - A) = \frac{1}{2i}(A - A^\theta) = Q$$

$\Rightarrow$ Q is Hermitian matrix

UNIQUENESS: Let if possible $A = R + iS$ be another expression such that R and S are Hermitian.

$\Rightarrow$ $R^\theta = R, \ S^\theta = S$

Then $\qquad A^\theta = (R + iS)^\theta = R^\theta + (iS)^\theta = R^\theta - iS^\theta = R - iS$

$\Rightarrow \qquad A = R + iS$ and $A^\theta = R - iS$

$\Rightarrow \qquad R = \frac{1}{2}(A + A^\theta) = P$ and $S = \frac{1}{2i}(A - A^\theta) = Q$

Hence $\qquad A = R + iS = P + iQ$

$\Rightarrow$ above expression is unique.

17.10 ALGEBRA OF MATRICES

(i) Addition of Matrices : Let A and B be two matrices of $m \times n$ type. Then the sum of A and B i.e., $(A + B)$ is defined to be the matrix of the type $m \times n$ obtained by adding the corresponding elements of A and B.

Let $A = [a_{ij}]_{m \times n}$ and $B = [b_{ij}]_{m \times n}$. Then $A + B = [a_{ij} + b_{ij}]_{m \times n}$.

Thus $A + B$ is also a matrix of type $m \times n$.

Thus, if $\qquad A = \begin{bmatrix} a_{11} & a_{12} & \cdots & a_{1n} \\ a_{21} & a_{22} & \cdots & a_{2n} \\ \cdots & \cdots & \cdots & \cdots \\ a_{m1} & a_{m2} & \cdots & a_{mn} \end{bmatrix}_{m \times n}$

and
$$B = \begin{bmatrix} b_{11} & b_{12} & \cdots & b_{1n} \\ b_{21} & b_{22} & \cdots & b_{2n} \\ \cdots & \cdots & \cdots & \cdots \\ b_{m1} & b_{m2} & \cdots & b_{mn} \end{bmatrix}_{m \times n}$$

Then
$$A + B = \begin{bmatrix} a_{11}+b_{11} & a_{12}+b_{12} & \cdots & a_{1n}+b_{1n} \\ a_{21}+b_{21} & a_{22}+b_{22} & \cdots & a_{2n}+b_{2n} \\ \cdots & \cdots & \cdots & \cdots \\ a_{m1}+b_{m1} & a_{m2}+b_{m2} & \cdots & a_{mn}+b_{mn} \end{bmatrix}_{m \times n}$$

(ii) **Subtraction of two matrices :** If A and B two matrices of $m \times n$ types, then the subtraction of two matrices denoted by $A - B$ is also a matrix of $m \times n$.

Let $\quad\quad A = [a_{ij}]_{m \times n}$ and $B = [b_{ij}]_{m \times n}$

then $\quad\quad A - B = [a_{ij} - b_{ij}]_{m \times n}$.

SOLVED EXAMPLES

Example 1. *If* $A = \begin{bmatrix} 2 & 3 & -1 \\ -3 & -1 & 4 \end{bmatrix}_{2 \times 3}$ *and* $B = \begin{bmatrix} 2 & -1 & 3 \\ 7 & 2 & -1 \end{bmatrix}_{2 \times 3}$, *find* $A + B$.

Solution. We have $\quad A + B = \begin{bmatrix} 2+2 & 3-1 & -1+3 \\ -3+7 & -1+2 & 4-1 \end{bmatrix}_{2 \times 3}$

$$A + B = \begin{bmatrix} 4 & 2 & 2 \\ 4 & 1 & 3 \end{bmatrix}_{2 \times 3}$$

Example 2. *If* $A = \begin{bmatrix} 5 & -1 \\ 3 & 7 \\ 2 & 3 \end{bmatrix}_{3 \times 2}$ *and* $B = \begin{bmatrix} -3 & 3 \\ -1 & -4 \\ 1 & 1 \end{bmatrix}_{3 \times 2}$, *find* $A + B$.

Solution. We have $\quad A + B = \begin{bmatrix} 5-3 & -1+3 \\ 3-1 & 7-4 \\ 2+1 & 3+1 \end{bmatrix}_{3 \times 2} = \begin{bmatrix} 2 & 2 \\ 2 & 3 \\ 3 & 4 \end{bmatrix}_{3 \times 2}$

Example 3. *If* $A = \begin{bmatrix} a_1 & b_1 & c_1 \\ a_2 & b_2 & c_2 \end{bmatrix}_{2 \times 3}$ *and* $B = \begin{bmatrix} a_3 & b_3 & c_3 \\ a_4 & b_4 & c_4 \end{bmatrix}_{2 \times 3}$, *find* $A - B$.

Solution. We have $\quad A - B = \begin{bmatrix} a_1-a_3 & b_1-b_3 & c_1-c_3 \\ a_2-a_4 & b_2-b_4 & c_2-c_4 \end{bmatrix}_{2 \times 3}$

Example 4. *If* $A = \begin{bmatrix} 2 & 5 & -1 \\ 7 & -3 & 4 \end{bmatrix}$ *and* $B = \begin{bmatrix} -1 & 2 & -3 \\ 4 & 1 & 2 \end{bmatrix}$, *find* $A - B$.

Solution. Then $\quad A - B = \begin{bmatrix} 2+1 & 5-2 & -1+3 \\ 7-4 & -3-1 & 4-2 \end{bmatrix}_{2 \times 3} = \begin{bmatrix} 3 & 3 & 2 \\ 3 & -4 & 2 \end{bmatrix}_{2 \times 3}$

17.11   PROPERTIES OF MATRIX ADDITION

THEOREM-1

Addition of matrices is commutative, i.e., if A and B be two matrices of m × n type then
$A + B = B + A$

Proof. Let $\quad\quad A = [a_{ij}]_{m \times n}$ and $B = [b_{ij}]_{m \times n}$

then $\quad A + B = [a_{ij} + b_{ij}]_{m \times n}$

$\quad\quad\quad\quad = [b_{ij} + a_{ij}]_{m \times n}$

$\quad\quad\quad\quad = [b_{ij}]_{m \times n} + [a_{ij}]_{m \times n} \quad\quad$ [by definition of addition of two matrices]

$\quad A + B = B + A$

THEOREM-2

Addition of matrices is associative i.e. if A, B and C be three matrices of m × n type then
$$(A + B) + C = A + (B + C)$$

Proof. Let $\quad\quad A = [a_{ij}]_{m \times n}\ B = [b_{ij}]_{m \times n}$ and $C = [c_{ij}]_{m \times n}$

Then $\quad (A + B) + C = ([a_{ij}]_{m \times n} + [b_{ij}]_{m \times n}) + [c_{ij}]_{m \times n}$

$\quad\quad\quad\quad = (a_{ij} + b_{ij})_{m \times n} + [c_{ij}]_{m \times n}$

$\quad\quad\quad\quad = [(a_{ij} + b_{ij}) + c_{ij}]_{m \times n}$

$\quad\quad\quad\quad = (a_{ij})_{m \times n} + ([b_{ij} + c_{ij}])_{m \times n}$

$\quad (A + B) + C = A + (B + C)$

THEOREM-3

Existence of additive identity i.e., if $A = [a_{ij}]_{m \times n}$ is the given matrix and O be the m × n null matrix then

$$A + O = A = O + A$$

Proof. Let $\quad\quad A = [a_{ij}]_{m \times n}$ and $O = [o_{ij}]_{m \times n}$

then $\quad\quad A + O = [a_{ij}]_{m \times n} + [o]_{m \times n}$

$\quad\quad\quad\quad = [a_{ij} + o]_{m \times n} = [a_{ij}]_{m \times n} = A$

again $\quad\quad O + A = [o]_{m \times n} + [a_{ij}]_{m \times n}$

$\quad\quad\quad\quad = [o + a_{ij}]_{m \times n} = [a_{ij}]_{m \times n} = A$

Hence $\quad\quad A + O = A = O + A$

Thus the null matrix O of $m \times n$ type acts as the identity element of addition in the set of all $m \times n$ matrices.

THEOREM-4

Existence of additive inverse i.e. if $A = [a_{ij}]_{m \times n}$ be a matrix then there exist another matrix B of m × n type such that $A + B = 0 = B + A$ and the matrix B is called the additive inverse of the matrix A or the negative of A.

THEOREM-5

Cancellation law hold in case of matrix addition i.e., if A, B and C be three matrices of m × n type such that

$$A + B = A + C \quad then \ B = C$$

Proof. We have $A + B = A + C \implies -A + (A + B) = -A + (A + C)$ (adding $-A$ both sides) then, by associative law of addition, we have

$$(-A + A) + B = (-A + A) + C$$
$$O + B = O + C$$
$$B = C$$

SOLVED EXAMPLES

Example 1. *If* $A = \begin{bmatrix} 3 & 7 \\ 9 & 8 \end{bmatrix}$ *and* $B = \begin{bmatrix} -1 & 2 \\ 0 & -4 \end{bmatrix}$ *find A + B.*

Solution. we have

$$A + B = \begin{bmatrix} 3 & 7 \\ 9 & 8 \end{bmatrix} + \begin{bmatrix} -1 & 2 \\ 0 & -4 \end{bmatrix} = \begin{bmatrix} 3-1 & 7+2 \\ 9-0 & 8-4 \end{bmatrix} = \begin{bmatrix} 2 & 9 \\ 9 & 4 \end{bmatrix}$$

Example 2. *If* $A = \begin{bmatrix} 2 & 3 & 1 \\ 0 & -1 & 5 \end{bmatrix}$ *and* $B = \begin{bmatrix} 1 & 2 & -1 \\ 0 & -1 & 3 \end{bmatrix}$ *find 3A – 4B.*

Solution. We have $\quad 3A - 4B = 3\begin{bmatrix} 2 & 3 & 1 \\ 0 & -1 & 5 \end{bmatrix} - 4\begin{bmatrix} 1 & 2 & -1 \\ 0 & -1 & 3 \end{bmatrix}$

$$= \begin{bmatrix} 6 & 9 & 3 \\ 0 & -3 & 15 \end{bmatrix} - \begin{bmatrix} 4 & 8 & -4 \\ 0 & -4 & 12 \end{bmatrix}$$

$$= \begin{bmatrix} 6-4 & 9-8 & 3-(-4) \\ 0-0 & -3-(-4) & 15-12 \end{bmatrix} = \begin{bmatrix} 2 & 1 & 7 \\ 0 & 1 & 3 \end{bmatrix}$$

Example 3. *If* $A = \begin{bmatrix} 1 & 2 & -3 \\ 5 & 0 & 2 \\ 1 & -1 & 1 \end{bmatrix}$ *and* $B = \begin{bmatrix} 3 & 1 & -2 \\ 0 & 1 & 4 \\ -2 & 0 & -1 \end{bmatrix}$. *Find the matric C such that*

A + 2C = B.

Solution. Given $A + 2C = B$ or $2C = B - A$

Now $\quad 2C = \begin{bmatrix} 3 & 1 & -2 \\ 0 & 1 & 4 \\ -2 & 0 & -1 \end{bmatrix} - \begin{bmatrix} 1 & 2 & -3 \\ 5 & 0 & 2 \\ 1 & -1 & 1 \end{bmatrix}$

$$= \begin{bmatrix} 3-1 & 1-2 & -2-(-3) \\ 0-5 & 1-0 & 4-2 \\ -2-1 & 0-(-1) & -1-1 \end{bmatrix} = \begin{bmatrix} 2 & -1 & 1 \\ -5 & 1 & 2 \\ -3 & 1 & -2 \end{bmatrix}$$

$$C = \frac{1}{2}\begin{bmatrix} 2 & -1 & 1 \\ -5 & 1 & 2 \\ -3 & 1 & -2 \end{bmatrix} = \begin{bmatrix} 1 & -1/2 & 1/2 \\ -5/2 & 1/2 & 1 \\ -3/2 & 1 & -1 \end{bmatrix}$$

Example 4. *Find the additive inverse of the matrix*

$$A = \begin{bmatrix} 2 & -3 & -1 & 1 \\ 3 & -1 & 2 & 2 \\ 1 & 2 & 8 & 7 \end{bmatrix}$$

Solution. The additive inverse of matrix A is the matrix each of whose elements is the negative of the corresponding element of A. Hence, it we denote the additive inverse of A by $-A$ then we have

$$-A = \begin{bmatrix} -2 & 3 & 1 & -1 \\ -3 & 1 & -2 & -2 \\ -1 & -2 & -8 & -7 \end{bmatrix}$$

Example 5. *Solve the following equations for A and B.*

$$2A - B = \begin{bmatrix} 3 & -3 & 0 \\ 3 & 3 & 2 \end{bmatrix}, \quad 2B + A = \begin{bmatrix} 4 & 1 & 5 \\ -1 & 4 & -4 \end{bmatrix}$$

Solution. Given

$$2A - B = \begin{bmatrix} 3 & -3 & 0 \\ 3 & 3 & 2 \end{bmatrix}$$

Multiplying both sides by 2, we get

$$4A - 2B = 2\begin{bmatrix} 3 & -3 & 0 \\ 3 & 3 & 2 \end{bmatrix} = \begin{bmatrix} 6 & -6 & 0 \\ 6 & 6 & 4 \end{bmatrix} \qquad \text{...(1)}$$

Also given that

$$2B + A = \begin{bmatrix} 4 & 1 & 5 \\ -1 & 4 & -4 \end{bmatrix} \qquad \text{...(2)}$$

Adding equation (1) and (2), we get

$$5A = \begin{bmatrix} 6 & -6 & 0 \\ 6 & 6 & 4 \end{bmatrix} + \begin{bmatrix} 4 & 1 & 5 \\ -1 & 4 & -4 \end{bmatrix}$$

$$= \begin{bmatrix} 6+4 & -6+1 & 0+5 \\ 6-1 & 6+4 & 4-4 \end{bmatrix} = \begin{bmatrix} 10 & -5 & 5 \\ 5 & 10 & 0 \end{bmatrix}$$

$$\Rightarrow \qquad A = \frac{1}{5}\begin{bmatrix} 10 & -5 & 5 \\ 5 & 10 & 0 \end{bmatrix} = \begin{bmatrix} 2 & -1 & 1 \\ 1 & 2 & 2 \end{bmatrix}$$

Now again multiplying the value of A in equation (2), we get

$$2B = \begin{bmatrix} 4 & 1 & 5 \\ -1 & 4 & -4 \end{bmatrix} - \begin{bmatrix} 2 & -1 & 1 \\ 1 & 2 & 0 \end{bmatrix}$$

$$= \begin{bmatrix} 4-2 & 1-(-1) & 5-1 \\ -1-1 & 4-2 & -4-0 \end{bmatrix} = \begin{bmatrix} 2 & 2 & 4 \\ -2 & 2 & -4 \end{bmatrix}$$

or $$B = \begin{bmatrix} 1 & 1 & 2 \\ -1 & 1 & -2 \end{bmatrix}$$

Example 6. *If* $A = \begin{bmatrix} \sec^2\theta & \sin^2\theta \\ 1/3 & \cosec^2\theta \end{bmatrix}$ *and* $B = \begin{bmatrix} -\tan^2\theta & \cos^2\theta \\ 2/3 & -\cot^2\theta \end{bmatrix}$. *Find* $A + B$.

Solution. We have $A + B = \begin{bmatrix} \sec^2\theta & \sin^2\theta \\ 1/3 & \cosec^2\theta \end{bmatrix} + \begin{bmatrix} -\tan^2\theta & \cos^2\theta \\ 2/3 & -\cot^2\theta \end{bmatrix}$

$$= \begin{bmatrix} \sec^2\theta - \tan^2\theta & \sin^2\theta + \cos^2\theta \\ 1/3 + 2/3 & \cosec^2\theta - \cot^2\theta \end{bmatrix} = \begin{bmatrix} 1 & 1 \\ 1 & 1 \end{bmatrix}$$

Example 7. *If* $A = \begin{bmatrix} 0 & 1 & 2 \\ 2 & 3 & 4 \\ 4 & 5 & 6 \end{bmatrix}$ *and* $B = \begin{bmatrix} 1 & 0 & 0 \\ 0 & 1 & 0 \\ 0 & 0 & 1 \end{bmatrix}$ *then find* $3A - 4B$

Solution. We have $3A - 4B = 3\begin{bmatrix} 0 & 1 & 2 \\ 2 & 3 & 4 \\ 4 & 5 & 6 \end{bmatrix} - 4\begin{bmatrix} 1 & 0 & 0 \\ 0 & 1 & 0 \\ 0 & 0 & 1 \end{bmatrix}$

$$= \begin{bmatrix} 0 & 3 & 6 \\ 6 & 9 & 12 \\ 12 & 15 & 18 \end{bmatrix} - \begin{bmatrix} 4 & 0 & 0 \\ 0 & 4 & 0 \\ 0 & 0 & 4 \end{bmatrix}$$

$$= \begin{bmatrix} 0-4 & 3-0 & 6-0 \\ 6-0 & 9-4 & 12-0 \\ 12-0 & 15-0 & 18-4 \end{bmatrix} = \begin{bmatrix} -4 & 3 & 6 \\ 6 & 5 & 12 \\ 12 & 15 & 14 \end{bmatrix}$$

Example 8. *If* $A = \begin{bmatrix} 1 \\ -5 \\ 7 \end{bmatrix}$ *and* B $[3, 2, -2]$, *Verify that*

$(AB)^T = B^T A^T$, *where* A^T *denoted the transpose of* A.

Solution. Given that

$$A = \begin{bmatrix} 1 \\ -5 \\ 7 \end{bmatrix} \text{ and } B = [3, 2, -2]$$

then $$AB = \begin{bmatrix} 1 \\ -5 \\ 7 \end{bmatrix} . [3, 2, -2] = \begin{bmatrix} 3 & 2 & -2 \\ -15 & -10 & 10 \\ 21 & 14 & -14 \end{bmatrix}$$

Now $\qquad (AB)^T = \begin{bmatrix} 3 & -15 & 21 \\ 2 & -10 & 14 \\ -2 & 14 & -14 \end{bmatrix}$

Again $A^T = [1, -5, 7]$ and $B^T = \begin{bmatrix} 3 \\ 2 \\ -2 \end{bmatrix}$

$\therefore \qquad\qquad\qquad B^T A^T = \begin{bmatrix} 3 \\ 2 \\ -2 \end{bmatrix} \cdot [1, -5, 7]$

or $\qquad\qquad\qquad B^T A^T = \begin{bmatrix} 3 & -15 & 21 \\ 2 & -10 & 14 \\ -2 & 10 & -14 \end{bmatrix}$ $\qquad$ ...(2)

From equation (1) and (2), we conclude that
$$(AB)^T = (B^T A^T)$$

17.12 MULTIPLICATION OF MATRICES

Let $A = [a_{ij}]_{m \times n}$ and $B = [b_{ij}]_{n \times p}$ be two matrices such that the number of columns in A is equal to the number of rows in B then product of A and B denoted by AB is defined as a matrix $C = [c_{ik}]_{m \times p}$ where $c_{ik} = \Sigma a_{ij} b_{ij}$ or

The product AB is defined as the matrix whose element in the ith row and kth column is $a_{i1}b_{1k} + a_{i2}b_{2k} + a_{i3}b_{3k} + \dots + a_{in}b_{nk}$, thus we conclude that :

If A is an $m \times n$ matrix and B is an $n \times k$ matrix then the product matrix AB is an $m \times k$ matrix.

In the product AB, the matrix A is called the pre-factor and the matrix B is called the post-factor. Also we say that the matrix A has been post-multiplied by the matrix B and the matrix B has pre multiplied by the matrix A. The product in both the above cases AB and BA may or may not exist and may be equal or different. The product AB can be calculated only if the number of columns in A is equal to the number of rows in B.

Remarks

❖ If $AB = BA$, then the matrices A and B are called commutative and if $AB = -BA$ then the matrices A and B are called anticommutative.
❖ The product of two non-zero matrices may be a zero matrix.
❖ The product of matrices generally does not obey the law of cancellation.

THEOREM-1

Let A and B be symmetric matrices then AB is symmetric if and only if $AB = BA$.

Proof. It is given that A and B are symmetric matrices, therefore

$\qquad\qquad\qquad A' = A \quad B' = B$ $\qquad$ ...(1)

Let us first suppose $\qquad AB = BA$ $\qquad$ ...(2)

To prove AB is symmetric.

We know that
$$(AB)' = B'A'$$
$$= B. A \qquad \text{[using (1)]}$$
$$= AB \qquad \text{[using (2)]}$$
$$\Rightarrow \qquad (AB)' = AB$$

Hence, AB is symmetric.

Conversely, let AB be symmetric, *i.e.,*
$$(AB)' = AB$$

Consider
$$\text{L.H.S.} = (AB)'$$
$$= B'A'$$
$$= BA \qquad \text{[using (1)]}$$

Hence
$$AB = BA$$

SOLVED EXAMPLES

Example 1. *If* $A = \begin{bmatrix} 1 & 3 \\ 2 & 1 \\ 0 & 4 \end{bmatrix}$ *and* $B = \begin{bmatrix} 1 & 0 & 2 \\ 0 & 1 & 2 \\ 0 & 2 & 3 \end{bmatrix}$. *Find BA. Can we find AB?*

Solution. We cannot find AB since the number of column of A is not equal to the number of rows of B *i.e.,* column of A is 2 and rows of B is 3. They are not equal.

Since the matrix B has 3 column and matrix A has 3 rows in BA, so product BA is defined

Now
$$BA = \begin{bmatrix} 1 & 0 & 2 \\ 0 & 1 & 2 \\ 0 & 2 & 3 \end{bmatrix}_{3 \times 3} \times \begin{bmatrix} 1 & 3 \\ 2 & 1 \\ 0 & 4 \end{bmatrix}_{3 \times 2}$$

$$= \begin{bmatrix} 1.1+0.2+2.0 & 1.3+0.1+2.4 \\ 0.1+1.2+2.0 & 0.3+1.1+2.4 \\ 0.1+2.2+3.0 & 0.3+2.1+3.4 \end{bmatrix}_{3 \times 2} = \begin{bmatrix} 1 & 11 \\ 2 & 9 \\ 4 & 14 \end{bmatrix}_{3 \times 2}$$

Example 2. *If* $A = \begin{bmatrix} 1 & -2 & 3 \\ -4 & 2 & 5 \end{bmatrix}$ *and* $B \begin{bmatrix} 2 & 3 \\ 4 & 5 \\ 2 & 1 \end{bmatrix}$. *Find AB and show that*

Solution. We have
$$A = \begin{bmatrix} 1 & -2 & 3 \\ -4 & 2 & 5 \end{bmatrix} \times \begin{bmatrix} 2 & 3 \\ 4 & 5 \\ 2 & 1 \end{bmatrix}$$

$$= \begin{bmatrix} 1.2+(-2).4+3.2 & 1.3+(-2).5+3.1 \\ -4.2+2.4+5.2 & -4.3+2.5+5.1 \end{bmatrix}$$

$$= \begin{bmatrix} 0 & -4 \\ 10 & 3 \end{bmatrix}$$

Now
$$BA = \begin{bmatrix} 2 & 3 \\ 4 & 5 \\ 2 & 1 \end{bmatrix} \times \begin{bmatrix} 1 & -2 & 3 \\ -4 & 2 & 5 \end{bmatrix}$$

$$= \begin{bmatrix} 2.1+3(-4) & 2(-2)+3(2) & 2(3)+3(5) \\ 4.1+5(-4) & 4(-2)+5(2) & 4(3)+5(5) \\ 2.1+1(-4) & 2(-2)+1(2) & 2(3)+1(5) \end{bmatrix}$$

$$= \begin{bmatrix} -10 & 2 & 21 \\ -16 & 2 & 37 \\ -2 & -2 & 11 \end{bmatrix}$$

Hence, $AB \neq BA$.

Example 3. *If* $A = \begin{bmatrix} 1 & -2 & 3 \\ 2 & 3 & -1 \\ -3 & 1 & 2 \end{bmatrix}$ *and* $B = \begin{bmatrix} 1 & 0 & 2 \\ 0 & 1 & 2 \\ 1 & 2 & 0 \end{bmatrix}$. *Find AB and show that*

$AB \neq BA$.

Solution. Since A and B both are 3×3 type square matrices, therefore

$$AB = \begin{bmatrix} 1 & -2 & 3 \\ 2 & 3 & -1 \\ -3 & 1 & 2 \end{bmatrix} \times \begin{bmatrix} 1 & 0 & 2 \\ 0 & 1 & 2 \\ 1 & 2 & 0 \end{bmatrix}$$

$$= \begin{bmatrix} 1.1+(-2).0+3.1 & 1.0+(-2).1+3.2 & 1.2+(-2).2+3.0 \\ 2.1+3.0+(-1).1 & 2.0+3.1+(-1).2 & 2.2+3.2+(-1).0 \\ -3.1+1.0+2.1 & -3.0+1.1+2.2 & -3.2+1.2+2.0 \end{bmatrix}$$

$$= \begin{bmatrix} 4 & 4 & -2 \\ 1 & 1 & 10 \\ -1 & 5 & -4 \end{bmatrix}$$

Now
$$BA = \begin{bmatrix} 1 & 0 & 2 \\ 0 & 1 & 2 \\ 1 & 2 & 0 \end{bmatrix} \times \begin{bmatrix} 1 & -2 & 3 \\ 2 & 3 & -1 \\ -3 & 1 & 2 \end{bmatrix}$$

$$= \begin{bmatrix} 1.1+0.2+2.(-3) & 1.(-2)+0.3+2.1 & 1.3+0(-1)+2.2 \\ 0.1+1.2+2.(-3) & 0.(-2)+1.3+2.1 & 0.3+1.(-1)+2.2 \\ 1.1+2.2+0.(-3) & 1.(-2)+2.3+0.1 & 1.3+2.(-1)+0.2 \end{bmatrix}$$

$$= \begin{bmatrix} -5 & 0 & 7 \\ -4 & 5 & 3 \\ 5 & 4 & 1 \end{bmatrix}$$

Hence $AB \neq BA$.

Example 4. *If* $A = \begin{bmatrix} 2 & 3 & 4 \\ 1 & 2 & 3 \\ -1 & 1 & 2 \end{bmatrix}$ *and* $B = \begin{bmatrix} 1 & 3 & 0 \\ -1 & 2 & 1 \\ 0 & 0 & 2 \end{bmatrix}$, *then prove that* $AB \ne BA$.

Solution. We have

$$AB = \begin{bmatrix} 2 & 3 & 4 \\ 1 & 2 & 3 \\ -1 & 1 & 2 \end{bmatrix} \times \begin{bmatrix} 1 & 3 & 0 \\ -1 & 2 & 1 \\ 0 & 0 & 2 \end{bmatrix}$$

$$= \begin{bmatrix} 2.1+3(-1)+4.0 & 2.3+3.2+4.0 & 2.0+3.1+4.2 \\ 1.1+2(-1)+3.0 & 1.3+2.2+3.0 & 1.0+2.1+3.2 \\ (-1).1+1.(-1)+2.0 & (-1).3+1.2+2.0 & (-1).0+1.1+2.2 \end{bmatrix}$$

$$= \begin{bmatrix} -1 & 12 & 11 \\ -1 & 7 & 8 \\ -2 & -1 & 5 \end{bmatrix} \qquad \qquad ...(1)$$

Now

$$BA = \begin{bmatrix} 1 & 3 & 0 \\ -1 & 2 & 1 \\ 0 & 0 & 2 \end{bmatrix} \times \begin{bmatrix} 2 & 3 & 4 \\ 1 & 2 & 3 \\ -1 & 1 & 2 \end{bmatrix}$$

$$= \begin{bmatrix} 1.2+3.1+0.(-1) & 1.3+3.2+0.1 & 1.4+3.3+0.2 \\ (-1).2+2.1+1(-1) & (-1).3+2.2+1.1 & (-1).4+2.3+1.2 \\ 0.2+0.1+2.(-1) & 0.3+0.2+2.1 & 0.4+0.3+2.2 \end{bmatrix}$$

$$= \begin{bmatrix} 2+3+0 & 3+6+0 & 4+9+0 \\ -2+2-1 & -3+4+1 & -4+6+2 \\ 0+0-2 & 0+0+2 & 0+0+4 \end{bmatrix} = \begin{bmatrix} 5 & 9 & 13 \\ -1 & 2 & 4 \\ -2 & 2 & 4 \end{bmatrix}$$

From (1) and (2), we conclude that $AB \ne BA$

Example 5. *If* $A = \begin{bmatrix} 1 & 1 & 3 \\ 2 & 2 & 6 \\ -1 & -1 & -3 \end{bmatrix}$, *show that* $A^2 = 0$

Solution. We have

$$A^2 = A \times A = \begin{bmatrix} 1 & 1 & 3 \\ 2 & 2 & 6 \\ -1 & -1 & -3 \end{bmatrix} \times \begin{bmatrix} 1 & 1 & 3 \\ 2 & 2 & 6 \\ -1 & -1 & -3 \end{bmatrix}$$

$$= \begin{bmatrix} 1.1+1.2+3(-1) & 1.1+1.2+3(-1) & 1.3+1.6+3(-3) \\ 2.1+2.2+6(-1) & 2.1+2.2+6(-1) & 2.3+2.6+6(-3) \\ -1.2-1.2-3(-1) & -1.1-1.2-3(-1) & -1.3-1.6-3(-3) \end{bmatrix}$$

$$= \begin{bmatrix} 0 & 0 & 0 \\ 0 & 0 & 0 \\ 0 & 0 & 0 \end{bmatrix} = 0, \text{ where 0 is } 3 \times 3 \text{ null matrix.}$$

Hence, $A^2 = 0$

Example 6. *Find the product of the following matrices.*

$$A = \begin{bmatrix} 0 & c & -b \\ -c & 0 & a \\ b & -a & 0 \end{bmatrix} \text{ and } B = \begin{bmatrix} a^2 & ab & ac \\ ab & b^2 & bc \\ ac & bc & c^2 \end{bmatrix}$$

Solution. We have

$$AB = \begin{bmatrix} 0 & c & -b \\ -c & 0 & a \\ b & -a & 0 \end{bmatrix} \times \begin{bmatrix} a^2 & ab & ac \\ ab & b^2 & bc \\ ac & bc & c^2 \end{bmatrix}$$

$$= \begin{bmatrix} 0.a^2 + c.ab + (-b).ac & 0.ab + c.b^2 - b.(bc) & 0.ac + c.(bc) + (-b).c^2 \\ (-c).a^2 + 0.ab + a.ac & -c(ab) + 0.b^2 + a.(bc) & -c.(ac) + 0(bc) + a.c^2 \\ b.a^2 + (-a).ab + 0.ac & b(ab) + (-a).b^2 + 0.(bc) & b(ac) + (-a)bc + 0.c^2 \end{bmatrix}$$

$$= \begin{bmatrix} 0 & 0 & 0 \\ 0 & 0 & 0 \\ 0 & 0 & 0 \end{bmatrix}$$

Example 7. *Prove that the product of two matrices*

$$\begin{bmatrix} \cos^2 \theta & \cos \theta \sin \theta \\ \cos \theta \sin \theta & \sin^2 \theta \end{bmatrix} \text{ and } \begin{bmatrix} \cos^2 \phi & \cos \phi \sin \phi \\ \cos \phi \sin \phi & \sin^2 \phi \end{bmatrix}$$

is zero when θ *and* ϕ *differ by an odd multiple of* $\dfrac{\pi}{2}$.

Solution. The required product

$$\begin{bmatrix} \cos^2 \theta & \cos \theta \sin \theta \\ \cos \theta \sin \theta & \sin^2 \theta \end{bmatrix} \times \begin{bmatrix} \cos^2 \phi & \cos \phi \sin \phi \\ \cos \phi \sin \phi & \sin^2 \phi \end{bmatrix}$$

$$= \begin{bmatrix} \cos^2 \theta \cos^2 \phi + \cos \theta \sin \theta \cos \phi \sin \phi & \cos^2 \theta \cos \phi \sin \phi + \cos \theta \sin \theta \sin^2 \phi \\ \cos \theta \sin \theta \cos^2 \phi + \sin^2 \theta \cos \phi \sin \phi & \cos \theta \sin \theta \cos \phi \sin \phi + \sin^2 \theta \sin^2 \phi \end{bmatrix}$$

$$= \begin{bmatrix} \cos \theta \cos \phi (\cos \theta \cos \phi + \sin \theta \sin \phi) & \cos \theta \sin \phi (\cos \theta \cos \phi + \sin \theta \sin \phi) \\ \sin \theta \cos \phi (\cos \theta \cos \phi + \sin \theta \sin \phi) & \sin \theta \sin \phi (\cos \theta \cos \phi + \sin \theta \sin \phi) \end{bmatrix}$$

$$= \begin{bmatrix} \cos \theta \cos \phi \cos(\theta - \phi) & \cos \theta \sin \phi \cos(\theta - \phi) \\ \sin \theta \cos \phi \cos(\theta - \phi) & \sin \theta \sin \phi \cos(\theta - \phi) \end{bmatrix}$$

Now if $\theta - \phi =$ an odd multiple of $\dfrac{\pi}{2}$.

Then $\cos (\theta - \phi) = 0$ and consequently the above product is zero.

Example 8. *If* $A = [x, y, z]$, $B = \begin{bmatrix} a & h & g \\ h & b & f \\ g & f & c \end{bmatrix}$ *and* $C = \begin{bmatrix} x \\ y \\ z \end{bmatrix}$ *be three matrices, then find*

ABC.

Solution. We have
$$AB = [x, y, z] \times \begin{bmatrix} a & h & g \\ h & b & f \\ g & f & c \end{bmatrix}$$

$$= [x.a + y.h + z.g \quad x.h + y.b + z.f \quad x.g + y.f + z.c]$$

Now $ABC = [x.a + y.h + z.g \quad hx + yb + zf \quad xg + yf + zc] \times \begin{bmatrix} x \\ y \\ z \end{bmatrix}$

$$= [x(ax + by + gz) + y(hx + by + fz) + z(gx + fy + cz)]$$
$$= [ax^2 + by^2 + cz^2 + 2hxy + 2gzx + 2fyz]$$

Example 9. *If* $A = \begin{bmatrix} i & 0 \\ 0 & -i \end{bmatrix}$, $B = \begin{bmatrix} 0 & -1 \\ 1 & 0 \end{bmatrix}$ *and* $C = \begin{bmatrix} 0 & i \\ i & 0 \end{bmatrix}$.

Prove that $A^2 = B^2 = C^2 = -I$ *and* $AB = -C = -BA$ *where* $I = \begin{bmatrix} 1 & 0 \\ 0 & 1 \end{bmatrix}$.

Solution. We have $A^2 = \begin{bmatrix} i & 0 \\ 0 & -i \end{bmatrix} \times \begin{bmatrix} i & 0 \\ 0 & -i \end{bmatrix}$

$$= \begin{bmatrix} i.i + 0.0 & i.0 + 0(-i) \\ 0.i - i.0 & 0.0 + (-i)(-i) \end{bmatrix} = \begin{bmatrix} -1 & 0 \\ 0 & -1 \end{bmatrix} = -\begin{bmatrix} 1 & 0 \\ 0 & 1 \end{bmatrix} = -I$$

or $\qquad A^2 = -I$

Now $\qquad B = \begin{bmatrix} 0 & -1 \\ 1 & 0 \end{bmatrix} \times \begin{bmatrix} 0 & -1 \\ 1 & 0 \end{bmatrix} = \begin{bmatrix} 0.0 - 1.1 & 0(-1) + (-1).0 \\ 1.0 + 0.1 & (1)(1) + 0.0 \end{bmatrix}$

$$= \begin{bmatrix} -1 & 0 \\ 0 & -1 \end{bmatrix} = -\begin{bmatrix} 1 & 0 \\ 0 & 1 \end{bmatrix} = -I$$

or $\qquad B^2 = -I.$

Similarly we can prove that $C^2 = -I$

Hence $\qquad A^2 = B^2 = C^2 = -I$

Now $\qquad AB = \begin{bmatrix} i & 0 \\ 0 & -i \end{bmatrix}\begin{bmatrix} 0 & -1 \\ 1 & 0 \end{bmatrix} = \begin{bmatrix} i.0 + 0.1 & i(-1) + 0.0 \\ 0.0 - i(1) & 0(-1) - i.0 \end{bmatrix} \times \begin{bmatrix} 0 & -1 \\ 1 & 0 \end{bmatrix}$

$$= \begin{bmatrix} 0 & -i \\ -i & 0 \end{bmatrix} = \begin{bmatrix} 0 & i \\ i & 0 \end{bmatrix} = -C$$

and $\qquad BA = \begin{bmatrix} 0 & -1 \\ 1 & 0 \end{bmatrix} \times \begin{bmatrix} i & 0 \\ 0 & -i \end{bmatrix}$

$$= \begin{bmatrix} 0.i - 1.0 & 0.0 - (-i) \\ 1.i + 0.0 & 1.0 + 0(-i) \end{bmatrix} = \begin{bmatrix} 0 & i \\ i & 0 \end{bmatrix} = C$$

Hence $\qquad AB = -BA = -C$

Example 10. *If* $A = \begin{bmatrix} 1 & 2 & 2 \\ 2 & 1 & 2 \\ 2 & 2 & 1 \end{bmatrix}$, *show that* $A^2 - 4A - 5I = 0$

Solution. We have

$$A^2 = \begin{bmatrix} 1 & 2 & 2 \\ 2 & 1 & 2 \\ 2 & 2 & 1 \end{bmatrix} \cdot \begin{bmatrix} 1 & 2 & 2 \\ 2 & 1 & 2 \\ 2 & 2 & 1 \end{bmatrix} = \begin{bmatrix} 9 & 8 & 8 \\ 8 & 9 & 8 \\ 8 & 8 & 9 \end{bmatrix}$$

Consider $A^2 - 4A - 5I$

$$= \begin{bmatrix} 9 & 8 & 8 \\ 8 & 9 & 8 \\ 8 & 8 & 9 \end{bmatrix} - 4\begin{bmatrix} 1 & 2 & 2 \\ 2 & 1 & 2 \\ 2 & 2 & 1 \end{bmatrix} - 5\begin{bmatrix} 1 & 0 & 0 \\ 0 & 1 & 0 \\ 0 & 0 & 1 \end{bmatrix}$$

$$= \begin{bmatrix} 9 & 8 & 8 \\ 8 & 9 & 8 \\ 8 & 8 & 9 \end{bmatrix} + \begin{bmatrix} -4 & -8 & -8 \\ -8 & -4 & -8 \\ -8 & -8 & -4 \end{bmatrix} + \begin{bmatrix} -5 & 0 & 0 \\ 0 & -5 & 0 \\ 0 & 0 & -5 \end{bmatrix}$$

$$= \begin{bmatrix} 9-4-5 & 8-8+0 & 8-8+0 \\ 8-8+0 & 9-4-5 & 8-8+0 \\ 8-8+0 & 8-8+0 & 9-4-5 \end{bmatrix}$$

$$= \begin{bmatrix} 0 & 0 & 0 \\ 0 & 0 & 0 \\ 0 & 0 & 0 \end{bmatrix}$$

Hence, $A^2 - 4A - 5I = 0$

Example 11. *Find the value of x, y, z in the following equation*

$$\begin{bmatrix} 1 & 2 & 3 \\ 3 & 1 & 2 \\ 2 & 3 & 1 \end{bmatrix} \times \begin{bmatrix} x \\ y \\ z \end{bmatrix} = \begin{bmatrix} 4 & -2 \\ 0 & -6 \\ -1 & 2 \end{bmatrix} \times \begin{bmatrix} 2 \\ 1 \end{bmatrix}$$

Solution. We have $\begin{bmatrix} 1 & 2 & 3 \\ 3 & 1 & 2 \\ 2 & 3 & 1 \end{bmatrix} \times \begin{bmatrix} x \\ y \\ z \end{bmatrix} = \begin{bmatrix} 1.x + 2.y + 3.z \\ 3.x + 1.y + 2.z \\ 2.x + 3.y + 1.z \end{bmatrix}$ ...(1)

and $\begin{bmatrix} 4 & -2 \\ 0 & -6 \\ -1 & 2 \end{bmatrix} \times \begin{bmatrix} 2 \\ 1 \end{bmatrix} = \begin{bmatrix} 4.2 + (-2).1 \\ 0.2 + (-6).1 \\ -1.2 + 2.1 \end{bmatrix} = \begin{bmatrix} 6 \\ -6 \\ 0 \end{bmatrix}$ ...(2)

With the help of (1) and (2) the given equation reduces to

$$\begin{bmatrix} x + 2y + 3z \\ 3x + y + 2z \\ 2x + 3y + z \end{bmatrix} = \begin{bmatrix} 6 \\ -6 \\ 0 \end{bmatrix}$$

On comparing the corresponding elements on both sides, we get

$$x + 2y + 3z = 6$$
$$3x + y + 2z = -6$$

and $$2x + 3y + z = 0$$

Solving these, we get

$$\left. \begin{aligned} x &= -4 \\ y &= 2 \\ z &= 2 \end{aligned} \right\}$$

Example 12. If $A = \begin{bmatrix} 4 & 2 \\ -1 & 1 \end{bmatrix}$ *find* $(A - 2I)(A - 3I)$.

Solution. We have

$$A - 2I = \begin{bmatrix} 4 & 2 \\ -1 & 1 \end{bmatrix} - 2\begin{bmatrix} 1 & 0 \\ 0 & 1 \end{bmatrix}$$

$$= \begin{bmatrix} 4 & 2 \\ -1 & 1 \end{bmatrix} - \begin{bmatrix} 2 & 0 \\ 0 & 2 \end{bmatrix} + \begin{bmatrix} 2 & 2 \\ -1 & -1 \end{bmatrix}$$

Also

$$A - 3I = \begin{bmatrix} 4 & 2 \\ -1 & 1 \end{bmatrix} - 3\begin{bmatrix} 1 & 0 \\ 0 & 1 \end{bmatrix} = \begin{bmatrix} 4 & 2 \\ -1 & 1 \end{bmatrix} - \begin{bmatrix} 3 & 0 \\ 0 & 3 \end{bmatrix} = \begin{bmatrix} 1 & 2 \\ -1 & -2 \end{bmatrix}$$

$$\therefore \quad (A - 2I)(A - 3I) = \begin{bmatrix} 2 & 2 \\ -1 & -1 \end{bmatrix} \begin{bmatrix} 1 & 2 \\ -1 & -2 \end{bmatrix} = \begin{bmatrix} 0 & 0 \\ 0 & 0 \end{bmatrix} = 0$$

Hence $\quad (A - 2I)(A - 3I) = 0$

Example 13. If $I = \begin{bmatrix} 1 & 0 \\ 0 & 1 \end{bmatrix}$, $C = \begin{bmatrix} 0 & 1 \\ 0 & 0 \end{bmatrix}$ *show that* $(aI + bC)^3 = a^3 I + 3a^2 bC$.

Solution. We have

$$aI + bC = a\begin{bmatrix} 1 & 0 \\ 0 & 1 \end{bmatrix} + b\begin{bmatrix} 0 & 1 \\ 0 & 0 \end{bmatrix}$$

$$= \begin{bmatrix} a & 0 \\ 0 & a \end{bmatrix} + \begin{bmatrix} 0 & b \\ 0 & 0 \end{bmatrix} = \begin{bmatrix} a & b \\ 0 & a \end{bmatrix}$$

$$\therefore \quad (aI + bC)^2 = \begin{bmatrix} a & b \\ 0 & a \end{bmatrix} \begin{bmatrix} a & b \\ 0 & a \end{bmatrix} = \begin{bmatrix} a^2 & 2ab \\ 0 & a^2 \end{bmatrix}$$

$$\therefore \quad (aI + bC)^3 = \begin{bmatrix} a^2 & 2ab \\ 0 & a^2 \end{bmatrix} \begin{bmatrix} a & b \\ 0 & a \end{bmatrix} = \begin{bmatrix} a^3 & 3a^2 b \\ 0 & a^3 \end{bmatrix}$$

Now $\qquad a^3I + 3a^2bC = a^3\begin{bmatrix} 1 & 0 \\ 0 & 1 \end{bmatrix} + 3a^2b\begin{bmatrix} 0 & 1 \\ 0 & 0 \end{bmatrix}$

$$= \begin{bmatrix} a^3 & 0 \\ 0 & a^3 \end{bmatrix} + \begin{bmatrix} 0 & 3a^2b \\ 0 & 0 \end{bmatrix} = \begin{bmatrix} a^3 & 3a^2b \\ 0 & a^3 \end{bmatrix}$$

Hence $\qquad (aI + bC)^3 = a^3I + 3a^2bC$

Example 14. *If* $A = \begin{bmatrix} 1 & 1 & -1 \\ 2 & 0 & 3 \\ 3 & -1 & 2 \end{bmatrix}, B = \begin{bmatrix} 1 & 3 \\ 0 & 2 \\ -1 & 4 \end{bmatrix}$ *and* $C = \begin{bmatrix} 1 & 2 & 3 & -4 \\ 2 & 0 & -2 & 1 \end{bmatrix}$.

Find A (BC) and (AB) C and hence show that A (BC) = (AB) C

Solution. We have

$$AB = \begin{bmatrix} 1 & 1 & -1 \\ 2 & 0 & 3 \\ 3 & -1 & 2 \end{bmatrix}\begin{bmatrix} 1 & 3 \\ 0 & 2 \\ -1 & 4 \end{bmatrix}$$

$$= \begin{bmatrix} 1.1+1.0+(-1).(-1) & 1.3+1.2-1.4 \\ 2.1+0.0+3(-1) & 2.3+0.2+3.4 \\ 3.1-1.0+2.(-1) & 3.3-1.2+2.4 \end{bmatrix}$$

$$= \begin{bmatrix} 2 & 1 \\ -1 & 18 \\ 1 & 15 \end{bmatrix} \qquad \qquad \qquad \dots(1)$$

$$BC = \begin{bmatrix} 1 & 3 \\ 0 & 2 \\ -1 & 4 \end{bmatrix}\begin{bmatrix} 1 & 2 & 3 & -4 \\ 2 & 0 & -2 & 1 \end{bmatrix}$$

$$= \begin{bmatrix} 1.2+3.2 & 1.2+3.0 & 1.3+3.(-2) & 1.(-4)+3.1 \\ 0.1+2.2 & 0.2+2.0 & 0.3+2.(-2) & 0.(-4)+2.1 \\ -1.1+4.2 & -1.2+4.0 & -1.3+4.(-2) & -1.(-4)+4.1 \end{bmatrix}$$

$$= \begin{bmatrix} 7 & 2 & -3 & -1 \\ 4 & 0 & -4 & 2 \\ 7 & -2 & -11 & 8 \end{bmatrix} \qquad \qquad \dots(2)$$

Now $\qquad (AB)\,C = \begin{bmatrix} 2 & 1 \\ -1 & 18 \\ 1 & 15 \end{bmatrix}\begin{bmatrix} 1 & 2 & 3 & -4 \\ 2 & 0 & -2 & 1 \end{bmatrix}$

$$= \begin{bmatrix} 2.1+1.2 & 2.2+1.0 & 2.3+1.(-2) & 2.(-4)+1.1 \\ -1.1+18.2 & -1.2+18.0 & -1.3+18.(-2) & -1.(-4)+18.1 \\ -1.1+15.2 & 1.2+15.0 & 1.3+15.(-2) & 1.(-4)+15.1 \end{bmatrix}$$

$$= \begin{bmatrix} 4 & 4 & 4 & -7 \\ 35 & -2 & -39 & 22 \\ 31 & 2 & -27 & 11 \end{bmatrix} \qquad \qquad ...(3)$$

And

$$A\,(BC) = \begin{bmatrix} 1 & 1 & -1 \\ 2 & 0 & 3 \\ 3 & -1 & 2 \end{bmatrix}\begin{bmatrix} 7 & 2 & -3 & -1 \\ 4 & 0 & -4 & 2 \\ 7 & -2 & -11 & 8 \end{bmatrix}$$

$$= \begin{bmatrix} 7+4-1.7 & 2.1+1.0+.(-1)(-2) & 1(-3)+1.(-4)-1.(-11) & 1(-1)+1(2)-1.8 \\ 14+0.4+3.7 & 2.2+0.0+3.(-2) & 2.(-3)+0.(-4)+3.(-11) & 2.(-1)+0.2+3.8 \\ 3.7-1.4+2.7 & 3.2-1.0+2.(-2) & 3.(-1)+1(-4)+2(-11) & 3(-1)-1(2)+2.8 \end{bmatrix}$$

$$= \begin{bmatrix} 4 & 4 & 4 & -7 \\ 35 & -2 & -39 & 22 \\ 31 & 2 & -27 & 11 \end{bmatrix} \qquad \qquad ...(4)$$

Hence, from equation (3) and (4), we get
$$(AB)\,C = A\,(BC)$$

Example 15. *If* $A = \begin{bmatrix} 3 & -4 \\ 1 & -1 \end{bmatrix}$ *show that* $A^k = \begin{bmatrix} 1+2k & -4k \\ k & 1-2k \end{bmatrix}$ *where* k *is any positive integer.*

Solution. We have

$$A^2 = \begin{bmatrix} 3 & -4 \\ 1 & -1 \end{bmatrix}\begin{bmatrix} 3 & -4 \\ 1 & -1 \end{bmatrix} = \begin{bmatrix} 5 & -8 \\ 2 & -3 \end{bmatrix} = \begin{bmatrix} 1+2.2 & -4.2 \\ 2 & 1-2.2 \end{bmatrix} \qquad ...(1)$$

$$A^3 = \begin{bmatrix} 5 & -8 \\ 2 & -3 \end{bmatrix}\begin{bmatrix} 3 & -4 \\ 1 & -1 \end{bmatrix} = \begin{bmatrix} 7 & -12 \\ 3 & -5 \end{bmatrix} = \begin{bmatrix} 1+2.3 & -4.3 \\ 3 & 1-2.3 \end{bmatrix} \qquad ...(2)$$

Thus the result is true for $k = 2, 3$.

Now assume that result is true for integer k i.e., $A^k = \begin{bmatrix} 1+2k & -4k \\ k & 1-2k \end{bmatrix}$ then

$$A^{k+1} = \begin{bmatrix} 1+2k & -4k \\ k & 1-2k \end{bmatrix}\begin{bmatrix} 3 & -4 \\ 1 & -1 \end{bmatrix}$$

$$= \begin{bmatrix} (1+2k).3-4k.1 & (1+2k).(-4)+(-4k)(-1) \\ k.3+(1-2k).1 & k.(-4)+(1-2k).(-1) \end{bmatrix} = \begin{bmatrix} 3+2k & -4(k+1) \\ k+1 & -1-2k \end{bmatrix}$$

$$= \begin{bmatrix} 1+2(k+1) & -4(k+1) \\ k+1 & 1-2(k+1) \end{bmatrix}$$

Hence, the result is true for A^k. Then it is also true for A^{k+1}. Hence by induction the required result follows.

Example 16. If $A_\alpha = \begin{bmatrix} \cos \alpha & \sin \alpha \\ -\sin \alpha & \cos \alpha \end{bmatrix}$ then show that

$$A_\alpha^n = \begin{bmatrix} \cos n\alpha & \sin n\alpha \\ -\sin n\alpha & \cos n\alpha \end{bmatrix}$$

Solution. Let $\qquad A_\alpha = \begin{bmatrix} \cos \alpha & \sin \alpha \\ -\sin \alpha & \cos \alpha \end{bmatrix}$

$$A_\alpha^2 = A_\alpha \cdot A_\alpha = \begin{bmatrix} \cos \alpha & \sin \alpha \\ -\sin \alpha & \cos \alpha \end{bmatrix} \begin{bmatrix} \cos \alpha & \sin \alpha \\ -\sin \alpha & \cos \alpha \end{bmatrix}$$

$$= \begin{bmatrix} \cos \alpha \cos \alpha - \sin \alpha \sin \alpha & \cos \alpha \sin \alpha + \sin \alpha \cos \alpha \\ -\sin \alpha \cos \alpha - \sin \alpha \cos \alpha & -\sin \alpha \sin \alpha + \cos \alpha \cos \alpha \end{bmatrix}$$

$$= \begin{bmatrix} \cos(\alpha + \alpha) & \sin(\alpha + \alpha) \\ -\sin(\alpha + \alpha) & \cos(\alpha + \alpha) \end{bmatrix}$$

$\therefore \qquad A_\alpha^2 = \begin{bmatrix} \cos 2\alpha & \sin 2\alpha \\ -\sin 2\alpha & \cos 2\alpha \end{bmatrix}$ \qquad\qquad ...(1)

In the equation (1) and the given value of A_α, let us assume that the

$$A_\alpha^n = \begin{bmatrix} \cos n\alpha & \sin n\alpha \\ -\sin n\alpha & \cos n\alpha \end{bmatrix}$$

Now to prove $\qquad A_\alpha^{n+1} = A_\alpha^n \cdot A_\alpha$ \qquad\qquad ...(2)

$$\text{RHS} = \begin{bmatrix} \cos n\alpha & \sin n\alpha \\ -\sin n\alpha & \cos n\alpha \end{bmatrix} \begin{bmatrix} \cos \alpha & \sin \alpha \\ -\sin \alpha & \cos \alpha \end{bmatrix}$$

$$= \begin{bmatrix} \cos n\alpha \cos \alpha - \sin n\alpha \sin \alpha & \cos n\alpha . \sin \alpha + \sin n\alpha \cos \alpha \\ -\sin n\alpha \cos \alpha - \cos n\alpha \sin \alpha & -\sin n\alpha \sin \alpha + \cos n\alpha . \cos \alpha \end{bmatrix}$$

$$= \begin{bmatrix} \cos(n+1)\alpha & \sin(n+1)\alpha \\ -\sin(n+1)\alpha & \cos(n+1)\alpha \end{bmatrix}$$

i.e., equation (2) holds for $n + 1$ if it is true for n. Hence by mathematical induction, we have

$$A_\alpha^n = \begin{bmatrix} \cos n\alpha & \sin n\alpha \\ -\sin n\alpha & \cos n\alpha \end{bmatrix}$$

Example 17. Evaluate $A^3 - 2A^2 + A - I$ if

$$A = \begin{bmatrix} 1 & 3 & 2 \\ 2 & 0 & 3 \\ 1 & -1 & 1 \end{bmatrix} \text{ where } I = \begin{bmatrix} 1 & 0 & 0 \\ 0 & 1 & 0 \\ 0 & 0 & 1 \end{bmatrix}$$

Solution. We have

$$A^2 = \begin{bmatrix} 1 & 3 & 2 \\ 2 & 0 & 3 \\ 1 & -1 & 1 \end{bmatrix} \begin{bmatrix} 1 & 3 & 2 \\ 2 & 0 & 3 \\ 1 & -1 & 1 \end{bmatrix}$$

$$= \begin{bmatrix} 1.1+3.2+2.1 & 1.3+3.0+2.(-1) & 1.2+3.3+2.1 \\ 2.1+0.2+3.1 & 2.3+0.0+3.(-1) & 2.2+0.3+3.1 \\ 1.1-1.2+1.1 & 1.3-1.0+1.(-1) & 1.2-1.3+1.1 \end{bmatrix} = \begin{bmatrix} 9 & 1 & 13 \\ 5 & 3 & 7 \\ 0 & 2 & 0 \end{bmatrix}$$

$$A^3 = \begin{bmatrix} 9 & 1 & 13 \\ 5 & 3 & 7 \\ 0 & 2 & 0 \end{bmatrix} \begin{bmatrix} 1 & 3 & 2 \\ 2 & 0 & 3 \\ 1 & -1 & 1 \end{bmatrix}$$

$$= \begin{bmatrix} 9.1+1.2+3.1 & 9.3+10+13.(-1) & 9.2+1.3+13.1 \\ 5.1+3.2+7.1 & 5.3+3.0+7.(-1) & 5.2+3.3+7.1 \\ 0.1+2.2+0.1 & 0.3+2.0+0.(-1) & 0.2+2.3+0.1 \end{bmatrix}$$

$$= \begin{bmatrix} 24 & 14 & 34 \\ 18 & 8 & 26 \\ 4 & 0 & 6 \end{bmatrix}$$

Hence, $A^3 - 2A^2 + A - I$

$$= \begin{bmatrix} 24 & 14 & 34 \\ 18 & 8 & 26 \\ 4 & 0 & 6 \end{bmatrix} - 2\begin{bmatrix} 9 & 1 & 13 \\ 5 & 3 & 7 \\ 0 & 2 & 0 \end{bmatrix} + \begin{bmatrix} 1 & 3 & 2 \\ 2 & 0 & 3 \\ 1 & -1 & 1 \end{bmatrix} - \begin{bmatrix} 1 & 0 & 0 \\ 0 & 1 & 0 \\ 0 & 0 & 1 \end{bmatrix}$$

$$= \begin{bmatrix} 24 & 14 & 34 \\ 18 & 8 & 26 \\ 4 & 0 & 6 \end{bmatrix} + \begin{bmatrix} -18 & -2 & -26 \\ -10 & -6 & -14 \\ 0 & -4 & 0 \end{bmatrix} + \begin{bmatrix} 1 & 3 & 2 \\ 2 & 0 & 3 \\ 1 & -1 & 1 \end{bmatrix} - \begin{bmatrix} 1 & 0 & 0 \\ 0 & 1 & 0 \\ 0 & 0 & 1 \end{bmatrix}$$

$$= \begin{bmatrix} 24-18+1-1 & 14-2+3+0 & 34-26+2+0 \\ 18-10+2+0 & 8-6+0-1 & 26+4+3+0 \\ 4+0+1+0 & 0-4-1+0 & 6+0+1-1 \end{bmatrix}$$

$$= \begin{bmatrix} 6 & 15 & 10 \\ 10 & -1 & 15 \\ 5 & -5 & 6 \end{bmatrix}$$

Example 18. *If* $A = \begin{bmatrix} 1 & 2 \\ 3 & 4 \\ 5 & 6 \end{bmatrix}$ *and* $B = \begin{bmatrix} -3 & -2 \\ 1 & -5 \\ 4 & 3 \end{bmatrix}$. *Find* $D = \begin{bmatrix} p & q \\ r & s \\ t & u \end{bmatrix}$ *such that* $A + B - D = 0$.

Solution. We have $A + B - D = 0$ or $D = A + B$

or

$$D = \begin{bmatrix} 1 & 2 \\ 3 & 4 \\ 5 & 6 \end{bmatrix} + \begin{bmatrix} -3 & -2 \\ 1 & -5 \\ 4 & 3 \end{bmatrix} = \begin{bmatrix} 1-3 & 2-2 \\ 3+1 & 4-5 \\ 5+4 & 6+3 \end{bmatrix}$$

$$D = \begin{bmatrix} -2 & 0 \\ 4 & -1 \\ 9 & 9 \end{bmatrix} \text{ or } \begin{bmatrix} p & q \\ r & s \\ t & u \end{bmatrix} = \begin{bmatrix} -2 & 0 \\ 4 & -1 \\ 9 & 9 \end{bmatrix}$$

Hence $p = -2, q = -1, r = 4, s = -1, t = 9, u = 9$.

Example 19. *If* $A = \begin{bmatrix} 0 & 1 \\ -1 & 0 \end{bmatrix}$. *Find number a, b, so that* $(aI + bA)^2 = A$.

Solution. We have

$$aI + bA = a\begin{bmatrix} 1 & 0 \\ 0 & 1 \end{bmatrix} + b\begin{bmatrix} 0 & 1 \\ -1 & 0 \end{bmatrix}$$

$$= \begin{bmatrix} a & 0 \\ 0 & a \end{bmatrix} + \begin{bmatrix} 0 & b \\ -b & 0 \end{bmatrix} = \begin{bmatrix} a & b \\ -b & a \end{bmatrix}$$

$$\therefore \qquad (aI + bA)^2 = \begin{bmatrix} a & b \\ -b & a \end{bmatrix} \times \begin{bmatrix} a & b \\ -b & a \end{bmatrix}$$

$$= \begin{bmatrix} a^2 - b^2 & ab + ba \\ -ab - ab & -b^2 + a^2 \end{bmatrix}$$

$$= \begin{bmatrix} a^2 - b^2 & 2ab \\ -2ab & a^2 - b^2 \end{bmatrix}$$

$\therefore$ If $(aI + bA)^2 = A$, then, we have

$$\begin{bmatrix} a^2 - b^2 & 2ab \\ -2ab & a^2 - b^2 \end{bmatrix} = \begin{bmatrix} 0 & 1 \\ -1 & 0 \end{bmatrix}$$

Equating the corresponding elements, we have

$$a^2 - b^2 = 0, \quad 2ab = 1$$

$$\Rightarrow \qquad a = b = 1/\sqrt{2}.$$

Example 20. *Show that* $E^2F + F^2E = E$ *where* $E = \begin{bmatrix} 0 & 0 & 1 \\ 0 & 0 & 1 \\ 0 & 0 & 0 \end{bmatrix}$ $F = \begin{bmatrix} 1 & 0 & 0 \\ 0 & 1 & 0 \\ 0 & 0 & 1 \end{bmatrix}$.

Solution. We have

$$E^2 = \begin{bmatrix} 0 & 0 & 1 \\ 0 & 0 & 1 \\ 0 & 0 & 0 \end{bmatrix} \times \begin{bmatrix} 0 & 0 & 1 \\ 0 & 0 & 1 \\ 0 & 0 & 0 \end{bmatrix} = \begin{bmatrix} 0 & 0 & 0 \\ 0 & 0 & 0 \\ 0 & 0 & 0 \end{bmatrix}$$

Now

$$E^2F = \begin{bmatrix} 0 & 0 & 0 \\ 0 & 0 & 0 \\ 0 & 0 & 0 \end{bmatrix} \times \begin{bmatrix} 1 & 0 & 0 \\ 0 & 1 & 0 \\ 0 & 0 & 1 \end{bmatrix} = \begin{bmatrix} 0 & 0 & 0 \\ 0 & 0 & 0 \\ 0 & 0 & 0 \end{bmatrix} \qquad ...(1)$$

Again

$$F^2 = \begin{bmatrix} 1 & 0 & 0 \\ 0 & 1 & 0 \\ 0 & 0 & 1 \end{bmatrix} \times \begin{bmatrix} 1 & 0 & 0 \\ 0 & 1 & 0 \\ 0 & 0 & 1 \end{bmatrix} = \begin{bmatrix} 1 & 0 & 0 \\ 0 & 1 & 0 \\ 0 & 0 & 1 \end{bmatrix}$$

and

$$F^2E = \begin{bmatrix} 1 & 0 & 0 \\ 0 & 1 & 0 \\ 0 & 0 & 1 \end{bmatrix} \times \begin{bmatrix} 0 & 0 & 1 \\ 0 & 0 & 1 \\ 0 & 0 & 0 \end{bmatrix} = \begin{bmatrix} 0 & 0 & 1 \\ 0 & 0 & 1 \\ 0 & 0 & 0 \end{bmatrix} \qquad ...(2)$$

∴ From (1) and (2) we conclude that

$$E^2F + F^2E = \begin{bmatrix} 0 & 0 & 0 \\ 0 & 0 & 0 \\ 0 & 0 & 0 \end{bmatrix} + \begin{bmatrix} 0 & 0 & 1 \\ 0 & 0 & 1 \\ 0 & 0 & 0 \end{bmatrix} = \begin{bmatrix} 0 & 0 & 1 \\ 0 & 0 & 1 \\ 0 & 0 & 0 \end{bmatrix} = E.$$

Hence $E^2F + F^2E = E$.

Example 21. $A = \begin{bmatrix} 1 & -1 \\ 2 & -1 \end{bmatrix}$, $B = \begin{bmatrix} a & -1 \\ b & -1 \end{bmatrix}$ *and* $(A + B)^2 = A^2 + B^2$ *find a and b.*

Solution. We have

$$A^2 = \begin{bmatrix} 1 & -1 \\ 2 & -1 \end{bmatrix} \times \begin{bmatrix} 1 & -1 \\ 2 & -1 \end{bmatrix} = \begin{bmatrix} 1-2 & -1+1 \\ 2-2 & -2+1 \end{bmatrix} = \begin{bmatrix} -1 & 0 \\ 0 & -1 \end{bmatrix}$$

and $$B^2 = \begin{bmatrix} a & 1 \\ b & -1 \end{bmatrix} \times \begin{bmatrix} a & 1 \\ b & -1 \end{bmatrix} = \begin{bmatrix} a^2+b & a-1 \\ ab-b & b+1 \end{bmatrix}$$

∴ $$A^2 + B^2 = \begin{bmatrix} -1 & 0 \\ 0 & -1 \end{bmatrix} + \begin{bmatrix} a^2+b & a-1 \\ ab-b & b+1 \end{bmatrix} = \begin{bmatrix} a^2+b-1 & a-1 \\ ab-b & b \end{bmatrix} \qquad \ldots(1)$$

Also $$A + B = \begin{bmatrix} 1 & -1 \\ 2 & -1 \end{bmatrix} + \begin{bmatrix} a & 1 \\ b & -1 \end{bmatrix} = \begin{bmatrix} 1+a & -1+1 \\ 2+b & -1-1 \end{bmatrix} = \begin{bmatrix} 1+a & 0 \\ 2+b & -2 \end{bmatrix}$$

∴ $$(A + B)^2 = \begin{bmatrix} 1+a & 0 \\ 2+b & -2 \end{bmatrix} \times \begin{bmatrix} 1+a & 0 \\ 2+b & -2 \end{bmatrix} = \begin{bmatrix} (1+a)^2 +0 & 0+0 \\ (2+b)(1+a)-2(2+b) & 0+4 \end{bmatrix}$$

$$= \begin{bmatrix} (1+\alpha)^2 & 0 \\ (2+\beta)(\alpha-1) & 4 \end{bmatrix} \qquad \ldots(2)$$

Now given that $(A + B)^2 = A^2 + B^2$.

Hence from (1) and (2), we get

$$\begin{bmatrix} (1+a)^2 & 0 \\ (2+b)(a-1) & 4 \end{bmatrix} = \begin{bmatrix} a^2+b-1 & a-1 \\ ab-b & b \end{bmatrix}$$

or $a - 1 = 0$ and $b = 4$.

Hence $a = 1$, $b = 4$.

Example 22. *If A, B are square matrices, such that* $A^2 = A$, $B^2 = B$ *show that* $(AB)^2 = AB$ *if A, B commute.*

Solution. Consider $$(AB)^2 = (AB)(AB) = A (BA) B$$

$$= A (AB) B \qquad (\because AB = BA)$$
$$= (AA)(BB)$$
$$= A^2B^2$$
$$= AB \qquad (\because A^2= A, B^2 = B)$$

Example 23. If $A_\alpha = \begin{bmatrix} \cos \alpha & -\sin \alpha \\ \sin \alpha & \cos \alpha \end{bmatrix}$ *then show that*

(i) $A_\alpha A_{-\alpha} = I$ (ii) $(A_\alpha A_\beta) = A_{(\alpha + \beta)}$

Solution. We have

$$A_\alpha = \begin{bmatrix} \cos \alpha & -\sin \alpha \\ \sin \alpha & \cos \alpha \end{bmatrix}$$

$$\Rightarrow \quad A_{-\alpha} = \begin{bmatrix} \cos(-\alpha) & -\sin(-\alpha) \\ \sin(-\alpha) & \cos(-\alpha) \end{bmatrix} = \begin{bmatrix} \cos \alpha & \sin \alpha \\ -\sin \alpha & \cos \alpha \end{bmatrix}$$

(i) $\quad A_\alpha A_{-\alpha} = \begin{bmatrix} \cos \alpha & -\sin \alpha \\ \sin \alpha & \cos \alpha \end{bmatrix}\begin{bmatrix} \cos \alpha & \sin \alpha \\ -\sin \alpha & \cos \alpha \end{bmatrix}$

$$= \begin{bmatrix} \cos^2 \alpha + \sin^2 \alpha & \sin \alpha \cos \alpha - \cos \alpha \sin \alpha \\ \sin \alpha \cos \alpha - \sin \alpha \cos \alpha & \sin^2 \alpha + \cos^2 \alpha \end{bmatrix}$$

$$= \begin{bmatrix} 1 & 0 \\ 0 & 1 \end{bmatrix} = 1$$

(ii) $\quad A_\alpha \cdot A_\beta = \begin{bmatrix} \cos \alpha & -\sin \alpha \\ \sin \alpha & \cos \alpha \end{bmatrix}\begin{bmatrix} \cos \beta & -\sin \beta \\ \sin \beta & \cos \beta \end{bmatrix}$

$$= \begin{bmatrix} \cos \alpha \cos \beta - \sin \alpha \sin \beta & \cos \alpha \sin \beta + \sin \alpha \cos \beta \\ \sin \alpha \cos \beta + \cos \alpha \sin \beta & -\sin \alpha \sin \beta + \cos \alpha \cos \beta \end{bmatrix}$$

$$= \begin{bmatrix} \cos(\alpha + \beta) & \sin(\alpha + \beta) \\ \sin(\alpha + \beta) & \cos(\alpha + \beta) \end{bmatrix}$$

$$= A_{(\alpha + \beta)}$$

Example 24. If $A = \begin{bmatrix} 3 & -3 & 4 \\ 2 & -3 & 4 \\ 0 & -1 & 1 \end{bmatrix}$ *prove that* $A^3 = A^{-1}$

Solution. We have

$$A^2 = \begin{bmatrix} 3 & -3 & 4 \\ 2 & -3 & 4 \\ 0 & -1 & 1 \end{bmatrix} \cdot \begin{bmatrix} 3 & -3 & 4 \\ 2 & -3 & 4 \\ 0 & -1 & 1 \end{bmatrix} = \begin{bmatrix} 3 & -4 & 4 \\ 0 & -1 & 0 \\ -2 & 2 & -3 \end{bmatrix}$$

$$\Rightarrow \quad A^4 = A^2 \cdot A^2 = \begin{bmatrix} 3 & -4 & 4 \\ 0 & -1 & 0 \\ -2 & 2 & -3 \end{bmatrix} \cdot \begin{bmatrix} 3 & -4 & 4 \\ 0 & -1 & 0 \\ -2 & 2 & -3 \end{bmatrix}$$

$$= \begin{bmatrix} 1 & 0 & 0 \\ 0 & 1 & 0 \\ 0 & 0 & 1 \end{bmatrix} = I$$

$$A^4 = I$$

Multiplying by A^{-1} on both sides, we get

$$A^3 = A^{-1}$$

EXERCISE 17.1

1. If $A = \begin{bmatrix} 2 & 3 & 1 \\ 0 & -1 & 5 \end{bmatrix}$ and $B = \begin{bmatrix} 1 & 2 & -1 \\ 0 & -1 & 3 \end{bmatrix}$. Find $2A - 3B$.

2. If $A = \begin{bmatrix} 1 & 2 & 3 \\ 0 & 5 & 7 \\ 6 & 8 & 9 \end{bmatrix}$ and $B = \begin{bmatrix} 2 & 0 & 3 \\ 3 & 0 & 5 \\ 5 & 7 & 0 \end{bmatrix}$. Find $3A - 2B$.

3. If $A = \begin{bmatrix} 1 & 2 \\ 3 & 0 \\ 4 & 1 \end{bmatrix}$ and $B = \begin{bmatrix} 0 & 1 & 0 \\ 0 & 2 & 1 \\ 2 & 3 & 0 \end{bmatrix}$. Find BA.

4. Find the product of the matrices $A = \begin{bmatrix} 4 & 2 & -1 & 2 \\ 3 & -7 & 1 & -8 \\ 2 & 4 & -3 & 1 \end{bmatrix}$ and $B = \begin{bmatrix} 2 & 3 \\ -3 & 0 \\ 1 & 5 \\ 3 & 1 \end{bmatrix}$

5. If $A = \begin{bmatrix} 1 & -3 & 2 \\ 2 & 1 & -3 \\ 4 & -3 & -1 \end{bmatrix}$, $B = \begin{bmatrix} 1 & 4 & 1 & 0 \\ 2 & 1 & 1 & 1 \\ 1 & -2 & 1 & 2 \end{bmatrix}$ and $C = \begin{bmatrix} 2 & 1 & -1 & -2 \\ 3 & -2 & -1 & -1 \\ 2 & -5 & -1 & 0 \end{bmatrix}$.

   Show that $AB = AC$.

6. If $A = \begin{bmatrix} 1 & -1 \\ 1 & 1 \end{bmatrix}$ show that $A^2 = 2A$ and $A^3 = 4A$.

7. If $A = \begin{bmatrix} 1 & 2 \\ 3 & 4 \end{bmatrix}$, $B = \begin{bmatrix} 2 & 1 \\ 4 & 2 \end{bmatrix}$, $C = \begin{bmatrix} 5 & 1 \\ 7 & 4 \end{bmatrix}$, show that $A(B + C) = AB + AC$.

8. If $I = \begin{bmatrix} 1 & 0 \\ 0 & 1 \end{bmatrix}$ and $E = \begin{bmatrix} 0 & 1 \\ 0 & 0 \end{bmatrix}$, prove that $(aI + bE)^3 = a^3I + 3a^2bE$.

9. If $A = \begin{bmatrix} 2 & 0 & 1 \\ 2 & 1 & 3 \\ 1 & -1 & 0 \end{bmatrix}$. Find $A^2 - 5A + 6I$.

10. If $A = \begin{bmatrix} \cosh u & \sinh u \\ \sinh u & \cosh u \end{bmatrix}$ then show that $A^n = \begin{bmatrix} \cosh nu & \sinh nu \\ \sinh nu & \cosh nu \end{bmatrix}$, where n in any

    positive integer.

11. Show that multiplication of matrices is not commutative.

HINT TO SELECTED PROBLEMS

10. Let
$$A = \begin{bmatrix} \cosh u & \sinh u \\ \sinh u & \cosh u \end{bmatrix}$$

$$A^2 = A.\,A = \begin{bmatrix} \cosh u & \sinh u \\ \sinh u & \cosh u \end{bmatrix} \begin{bmatrix} \cosh u & \sinh u \\ \sinh u & \cosh u \end{bmatrix} = \begin{bmatrix} \cosh 2u & \sinh 2u \\ \sinh 2u & \cosh 2u \end{bmatrix} \quad ...(1)$$

Let us assume that the

$$A^n = \begin{bmatrix} \cosh n\alpha & \sinh n\alpha \\ \sinh n\alpha & \cosh n\alpha \end{bmatrix}$$

Now to show $A^{n+1} = A^n.A$ ...(2)

$$= \begin{bmatrix} \cosh n\alpha & \sinh n\alpha \\ \sinh n\alpha & \cosh n\alpha \end{bmatrix} \begin{bmatrix} \cosh \alpha & \sinh \alpha \\ \sinh \alpha & \cosh \alpha \end{bmatrix}$$

$$= \begin{bmatrix} \cosh (n+1)\alpha & \sinh (n+1)\alpha \\ \sinh (n+1)\alpha & \cosh (n+1)\alpha \end{bmatrix}$$

i.e., equation (2) holds for $n + 1$ if it true for n.
Hence, by mathematical induction, we have

$$A^n = \begin{bmatrix} \cosh n\alpha & \sinh n\alpha \\ \sinh n\alpha & \cosh n\alpha \end{bmatrix}$$

11. Let us take
$$A = \begin{bmatrix} 1 & -2 & 3 \\ 2 & 3 & -1 \\ -3 & 1 & 2 \end{bmatrix} \text{ and } B = \begin{bmatrix} 1 & 0 & 2 \\ 0 & 1 & 2 \\ 1 & 2 & 0 \end{bmatrix}$$

$$AB = \begin{bmatrix} 1 & -2 & 3 \\ 2 & 3 & -1 \\ -3 & 1 & 2 \end{bmatrix} \begin{bmatrix} 1 & 0 & 2 \\ 0 & 1 & 2 \\ 1 & 2 & 0 \end{bmatrix}$$

$$= \begin{bmatrix} 1.1+(-2).0+3.1 & 1.0+(-2).1+3.2 & 1.2+(-2).2+3.0 \\ 2.1+3.0+(-1).1 & 2.0+3.1+(-1).2 & 2.2+3.2+(-1).0 \\ -3.1+1.0+2.1 & -3.0+1.1+2.2 & -3.2+1.2+2.0 \end{bmatrix}$$

$$= \begin{bmatrix} 4 & 4 & -2 \\ 1 & 1 & 10 \\ -1 & 5 & -4 \end{bmatrix}$$

Now
$$BA = \begin{bmatrix} 1 & 0 & 2 \\ 0 & 1 & 2 \\ 1 & 2 & 0 \end{bmatrix} \times \begin{bmatrix} 1 & -2 & 3 \\ 2 & 3 & -1 \\ -3 & 1 & 2 \end{bmatrix}$$

$$= \begin{bmatrix} -5 & 0 & 7 \\ -4 & 5 & 3 \\ 5 & 4 & 1 \end{bmatrix}$$

Hence $AB \neq BA$

Clearly, the multiplication of matrix is not commutative.

ANSWERS

(1) $\begin{bmatrix} 1 & 0 & 5 \\ 0 & 1 & 1 \end{bmatrix}$

(2) $\begin{bmatrix} -1 & 6 & 3 \\ -6 & 15 & 11 \\ 8 & 10 & 27 \end{bmatrix}$

(3) $\begin{bmatrix} 3 & 0 \\ 10 & 1 \\ 11 & 4 \end{bmatrix}$

(4) $AB = \begin{bmatrix} 7 & 9 \\ 4 & 6 \\ -8 & -8 \end{bmatrix}$, BA is underfind

(9) $\begin{bmatrix} 1 & -1 & -3 \\ -1 & -1 & -10 \\ -5 & 4 & 4 \end{bmatrix}$

17.13 ADJOINT OF A MATRIX

Let $A = [a_{ij}]_{n \times n}$ be a square matrix of order $n \times n$. Then the adjoint of A is a matrix of the same order $n \times n$ which is obtained by the transposing of a matrix whose elements are the cofactor of the elements of A in the determinant A. That is if $B = [A_{ij}]_{n \times n}$, where A_{ij} are the cofactors of the elements a_{ij} in the determinant $|A|$, then B' is called the adjoint of A. It is denoted by adj A.

Remark

❖ Sometimes the adjoint of a matrix is also called the adjugate of the matrix.

THEOREM-1

If A is a square matrix, then
$$A.(adj\ A) = (adj\ A).\ A = |A|.\ I$$
where I is the unit matrix of the same order as A.

Proof. We have the $(i, j)^{th}$ element of the product $A\ (adj.\ A)$ = product of the ith row of A and column of *adj A*

$$= a_{1i}A_{j1} + a_{2i}A_{j2} + ... + a_{ni}A_{jn}.$$

$$= \begin{cases} 0 \text{ when } i \neq j \\ |A| \text{ when } i = j. \end{cases}$$

Thus, in the product, only the diagonal element exist and each is equal to $|A|$ while all other elements are zero, so that

$$A.\ (adj\ A) = \begin{bmatrix} |A| & 0 & 0 & ... & 0 \\ 0 & |A| & 0 & ... & 0 \\ 0 & 0 & |A| & ... & 0 \\ ... & ... & ... & ... & ... \\ 0 & 0 & 0 & ... & |A| \end{bmatrix}$$

or $\qquad A.\,(\text{adj }A) = |A| \begin{bmatrix} 1 & 0 & \ldots & 0 \\ 0 & 1 & \ldots & 0 \\ \ldots & \ldots & \ldots & \ldots \\ 0 & 0 & \ldots & 1 \end{bmatrix} = |A|.\,I.$

Similarly, $\qquad\qquad\qquad\qquad (\text{adj }A).\,A = |A|.\,I$

Hence the theorem is proved.

SOLVED EXAMPLES

Example 1. *Find the adjoint of the matrix.*

$$A = \begin{bmatrix} 1 & 2 & 4 \\ 5 & 7 & 8 \\ 9 & 10 & 12 \end{bmatrix}$$

Solution. For the given matrix A, we have

$A_{11} = \begin{bmatrix} 7 & 8 \\ 10 & 12 \end{bmatrix} = 4; \qquad A_{12} = -\begin{bmatrix} 5 & 8 \\ 9 & 12 \end{bmatrix} = 12; \qquad A_{13} = \begin{bmatrix} 5 & 7 \\ 9 & 10 \end{bmatrix} = -13;$

$A_{21} = -\begin{bmatrix} 2 & 4 \\ 10 & 12 \end{bmatrix} = 16; \qquad A_{22} = \begin{bmatrix} 1 & 4 \\ 9 & 12 \end{bmatrix} = -24; \qquad A_{23} = -\begin{bmatrix} 1 & 2 \\ 9 & 12 \end{bmatrix} = 8;$

$A_{31} = \begin{bmatrix} 2 & 4 \\ 7 & 8 \end{bmatrix} = -12; \qquad A_{32} = -\begin{bmatrix} 2 & 4 \\ 7 & 8 \end{bmatrix} = 12; \qquad A_{33} = \begin{bmatrix} 1 & 2 \\ 5 & 7 \end{bmatrix} = -3;$

Therefore the matrix B formed by the cofactors of the elements of $|A|$ is :

$$B = \begin{bmatrix} 4 & 12 & -13 \\ 16 & -24 & 8 \\ -12 & 12 & -3 \end{bmatrix}$$

Now $\qquad\qquad$ adj A = transpose of the matrix B

$$= \begin{bmatrix} 4 & 12 & -12 \\ 12 & -24 & 12 \\ -13 & 8 & -3 \end{bmatrix}$$

17.14 INVERSE OR RECIPROCAL OF MATRIX

Let A be a square matrix of order $n \times n$ and there exists a square matrix of the same order such that $AB = BA = I$, where I is a unit matrix of order $n \times n$. Then, the matrix B is called the inverse of a matrix A.

Remarks

❖ A matrix 'A' is invertible if it is non-singular.

❖ $A^{-1} = \dfrac{adj.A}{|A|}$, $|A| \neq 0$

THEOREM-1

The inverse of a matrix is unique.

Proof. If possible, let B and C be two inverses of the matrix A, then by definition

$$AB = BA = I \qquad \qquad ...(1)$$

and $\qquad \qquad AC = CA = I \qquad \qquad ...(2)$

From (1) and (2), we get $AB = CA$, each being equal to I

or $\qquad B(AB) = B(AC) \qquad$ or $\quad (BA)B = (BA)C$

or $\qquad IB = IC$ from (1) or $\quad B = C.$

Hence, the inverse of a matrix, if exists is unique.

THEOREM-2

A square matrix A has an inverse if and only if A is non-singular.

Proof. The condition is necessary : Let B be the inverse of the matrix A, then $AB = I.$

$$|A||B| = |I| = I$$

Therefore, $\qquad \qquad |A| \neq 0.$

The condition is sufficient : Let $|A| \neq 0$, we assume that

$$B = \frac{adj\, A}{|A|}$$

$$AB = A\left(\frac{adj\, A}{|A|}\right)$$

$$= \frac{1}{|A|}(A\, adj\, A) = \frac{|A|I}{|A|} = I$$

Similarly $\qquad \qquad BA = I$

$\therefore \qquad \qquad AB = BA = I$

Hence, A has an inverse.

THEOREM-3

If A and B are two non-singular matrices of the same order, then AB is also non-singular and $AB)^{-1} = B^{-1}A^{-1}.$

Proof. Let A^{-1} and B^{-1} exist. Since A and B are non-singular, therefore,

$$(AB)(B^{-1}A^{-1}) = A(BB^{-1})A^{-1} \qquad \text{(By associative law)}$$

$$= AIA^{-1} \qquad \text{(By inverse property)}$$

$$= AA^{-1} = I$$

Similarly, $\qquad (B^{-1}A^{-1})(AB) = I$

$$(B^{-1}A^{-1})(AB) = (AB)(B^{-1}A^{-1}) = I$$

i.e., $B^{-1}A^{-1}$ is the inverse of AB or $(AB)^{-1} = B^{-1}A^{-1}$ and as such AB is also non-singular.

THEOREM-4

If A is a non-singular matrix, then $(A^{-1})^{-1} = A$.

Proof. Let A^{-1} be the given matrix instead of A, then

$$(A^{-1})A = A(A^{-1}) = I$$

This show that A is the unique inverse of A^{-1}.

i.e., $$A = (A^{-1})^{-1}$$

THEOREM-5

The inverse of the transpose of a matrix A is the transpose of the inverse of A, i.e., $(A')^{-1} = (A^{-1})'$.

Proof. We have $$AA^{-1} = I = A^{-1}A$$

$\therefore$ $$(AA^{-1})' = I' = (A^{-1}A)' \text{ or } (A^{-1})'A' = I = A'(A^{-1})'$$

Hence $(A^{-1})'$ is the inverse of A'

i.e., $$(A')^{-1} = (A^{-1})'$$

17.15 ORTHOGONAL AND UNITARY MATRICES

17.15.1 Orthogonal Matrix :

Definition : *A matrix A is called an orthogonal matrix. If $AA' = I = A'A$, where I is an identity matrix.*

THEOREM-1

If A is an orthogonal matrix, then A^{-1} is also an orthogonal matrix.

Proof. By definition, if A is orthogonal, then

$$AA' = A'A = I$$

or $(AA')^{-1} = (A'A)^{-1} = I^{-1}$ or $(A')^{-1}A^{-1} = A^{-1}(A')^{-1} = I$ $\qquad [\because I^{-1} = I]$

or $\qquad (A^{-1})'A^{-1} = A^{-1}(A^{-1})' = I$.

Hence A^{-1} is orthogonal.

i.e., inverse of an orthogonal matrix is also orthogonal.

THEOREM-2

If A and B are n-square orthogonal matrices then AB and BA are orthogonal matrices.

Proof. Since A and B are orthogonal matrices, we have

$$AA' = I \quad \text{and} \quad BB' = I \qquad\qquad ...(1$$

Now $\qquad (AB)(AB)' = (AB)(B'A')$

$\qquad\qquad\qquad = A(BB')A'$ $\qquad\qquad$ by associative law

$\qquad\qquad\qquad = AIA'$ $\qquad\qquad$ since $BB' = I = AA$

Hence AB is an orthogonal matrix. Similarly we can show that BA is also orthogonal.

17.15.2 Unitary Matrix :

Definition : *A square matrix A is said to be unitary if $A^\theta A = I$, where I is an identity matrix and A^θ is the transposed conjugate of A; the elements of A are complex numbers.*

Since $A^\theta = |\bar{A}|$ and $|AA^\theta| = |A||A^\theta|$, therefore if $AA^\theta = I$, we have $|A||\bar{A}| = I$.

Thus determinant of a unitary matrix is of unit modulus.

For a matrix to be unitary, it must be non-singular.

Hence $\qquad AA^\theta = I \implies A^\theta A = I$

i.e., $\qquad AA^\theta = I = A^\theta A$.

SOLVED EXAMPLES

Example 1. *Find the adjoint of the matrix $A = \begin{bmatrix} 1 & 2 \\ 3 & -5 \end{bmatrix}$ and verify the theorem*

$$A(adj\, A) = (adj\, A)\, A = |A|\, I.$$

Solution. We have $\qquad |A| = \begin{vmatrix} 1 & 2 \\ 3 & -5 \end{vmatrix} = 1(-5) - (3)2 = -5 - 6 = -11$

The cofactors of the elements of the first row of $|A|$ are $-5, -3$ respectively. The cofactors of the elements of second row of $|A|$ are $-2, 1$ respectively. Therefore, the matrix B formed by the cofactors of the elements of $|A|$ is

$$B = \begin{bmatrix} -5 & -3 \\ -2 & 1 \end{bmatrix}$$

$\therefore$ adj A = transpose of the matrix $\quad B = \begin{bmatrix} -5 & -2 \\ -3 & 1 \end{bmatrix}$

Now $\qquad A\,(\text{adj}\, A) = \begin{bmatrix} 1 & 2 \\ 3 & -5 \end{bmatrix} \begin{bmatrix} -5 & -2 \\ -3 & 1 \end{bmatrix}$

$$= \begin{bmatrix} -5-6 & -2+2 \\ -15+15 & -6-5 \end{bmatrix} = \begin{bmatrix} -11 & 0 \\ 0 & -11 \end{bmatrix}$$

$$= (-11)\begin{bmatrix} 1 & 0 \\ 0 & 1 \end{bmatrix} = |A|\, I$$

Also $\qquad (\text{adj}\, A) = \begin{bmatrix} -5 & -2 \\ -3 & 1 \end{bmatrix} \begin{bmatrix} 1 & 2 \\ 3 & -5 \end{bmatrix}$

$$= \begin{bmatrix} -5-6 & -10+10 \\ -3+3 & -6-5 \end{bmatrix}$$

$$= \begin{bmatrix} -11 & 0 \\ 0 & -11 \end{bmatrix} = (-11)\begin{bmatrix} 1 & 0 \\ 0 & 1 \end{bmatrix}$$

Hence $\quad A\,(\text{adj}\, A) = (\text{adj}\, A)\, A = |A|\, I.$

Example 2. *Find the adjoint of the matrix A, where : $A = \begin{bmatrix} 3 & -3 & 4 \\ 2 & -3 & 4 \\ 0 & -1 & 1 \end{bmatrix}$*

Solution. For the given matrix A, we have

$$A_{11} = \begin{vmatrix} -3 & 4 \\ -1 & 1 \end{vmatrix} = 1, \qquad A_{12} = - \begin{vmatrix} 2 & 4 \\ 0 & 1 \end{vmatrix} = -2$$

$$A_{13} = \begin{vmatrix} 2 & -3 \\ 0 & -1 \end{vmatrix} = -2, \qquad A_{21} = - \begin{vmatrix} -3 & 4 \\ -1 & 1 \end{vmatrix} = -1$$

$$A_{22} = \begin{vmatrix} 3 & 4 \\ 0 & 1 \end{vmatrix} = 3, \qquad A_{23} = - \begin{vmatrix} 3 & -3 \\ 0 & -1 \end{vmatrix} = 3$$

$$A_{31} = \begin{vmatrix} -3 & 4 \\ -3 & 4 \end{vmatrix} = 0, \qquad A_{32} = - \begin{vmatrix} 3 & 4 \\ 2 & 4 \end{vmatrix} = -4$$

$$A_{33} = \begin{vmatrix} 3 & -3 \\ 2 & -3 \end{vmatrix} = -3$$

Therefore the matrix B formed by the cofactors of the elements of A is

$$B = \begin{bmatrix} 1 & -2 & -2 \\ -1 & 3 & 3 \\ 0 & -4 & -3 \end{bmatrix}$$

$$\text{Adj } A = B^T = \begin{bmatrix} 1 & -1 & 0 \\ -2 & 3 & -4 \\ -2 & 3 & -3 \end{bmatrix}$$

Example 3. *If* $A = \begin{bmatrix} 1 & 2 & 3 \\ 0 & 5 & 0 \\ 2 & 4 & 3 \end{bmatrix}$, *find* $A^2 - 2A + \text{Adj } A$

Solution. We have

$$A^2 = \begin{bmatrix} 1 & 2 & 3 \\ 0 & 5 & 0 \\ 2 & 4 & 3 \end{bmatrix} \times \begin{bmatrix} 1 & 2 & 3 \\ 0 & 5 & 0 \\ 2 & 4 & 3 \end{bmatrix}$$

$$= \begin{bmatrix} 1+0+6 & 2+10+12 & 3+0+9 \\ 0+0+0 & 0+25+0 & 0+0+0 \\ 2+0+6 & 4+20+12 & 6+0+9 \end{bmatrix}$$

$$= \begin{bmatrix} 7 & 24 & 12 \\ 0 & 25 & 0 \\ 8 & 36 & 15 \end{bmatrix} \qquad \qquad \text{...(1)}$$

Also

$$A_{11} = \begin{vmatrix} 5 & 0 \\ 4 & 3 \end{vmatrix} = 15, \qquad A_{12} = - \begin{vmatrix} 0 & 0 \\ 2 & 3 \end{vmatrix} = 0$$

$$A_{13} = \begin{vmatrix} 0 & 5 \\ 2 & 4 \end{vmatrix} = -10, \qquad A_{21} = - \begin{vmatrix} 2 & 3 \\ 4 & 3 \end{vmatrix} = -6$$

$$A_{22} = \begin{vmatrix} 1 & 3 \\ 2 & 3 \end{vmatrix} = -3, \qquad A_{23} = - \begin{vmatrix} 1 & 2 \\ 2 & 4 \end{vmatrix} = 0$$

$$A_{31} = \begin{vmatrix} 2 & 3 \\ 5 & 0 \end{vmatrix} = -15, \qquad A_{32} = - \begin{vmatrix} 1 & 3 \\ 0 & 0 \end{vmatrix} = 0$$

$$A_{33} = \begin{vmatrix} 1 & 2 \\ 2 & 5 \end{vmatrix} = 5$$

$$\therefore \qquad B = \begin{bmatrix} 15 & 0 & -10 \\ 6 & -3 & 0 \\ -15 & 0 & 5 \end{bmatrix}$$

$$\therefore \qquad \text{Adj } A = B' = \begin{bmatrix} 15 & 6 & -15 \\ 0 & -3 & 0 \\ -10 & 0 & 5 \end{bmatrix} \qquad \qquad \text{...(2)}$$

$$\therefore \quad A^2 - 2A + \text{Adj } A = \begin{bmatrix} 7 & 24 & 12 \\ 0 & 25 & 0 \\ 8 & 36 & 15 \end{bmatrix} - 2\begin{bmatrix} 1 & 2 & 3 \\ 0 & 5 & 0 \\ 2 & 4 & 3 \end{bmatrix} + \begin{bmatrix} 15 & 6 & -15 \\ 0 & -3 & 0 \\ -10 & 0 & 5 \end{bmatrix}$$

(Using (1) and (2))

$$= \begin{bmatrix} 7 & 24 & 12 \\ 0 & 25 & 0 \\ 8 & 36 & 15 \end{bmatrix} - \begin{bmatrix} 2 & 4 & 6 \\ 0 & 10 & 0 \\ 4 & 8 & 6 \end{bmatrix} + \begin{bmatrix} 15 & 6 & -15 \\ 0 & -3 & 0 \\ -10 & 0 & 5 \end{bmatrix}$$

$$= \begin{bmatrix} 7-2+15 & 24-4+6 & 12-6-15 \\ 0-0-0 & 25-10-3 & 0-0-0 \\ 8-4-10 & 36-8+0 & 15-6+5 \end{bmatrix}$$

$$= \begin{bmatrix} 20 & 26 & -9 \\ 0 & 12 & 0 \\ -6 & 28 & 14 \end{bmatrix}$$

Example 4. *Find the inverse of* $A = \begin{bmatrix} 1 & 2 & 3 \\ 2 & 4 & 5 \\ 3 & 5 & 6 \end{bmatrix}$

Solution. For the given matrix A, we have

Also $\qquad A_{11} = \begin{vmatrix} 4 & 5 \\ 5 & 6 \end{vmatrix} = -1, \qquad A_{12} = - \begin{vmatrix} 2 & 5 \\ 3 & 6 \end{vmatrix} = 3$

$$A_{13} = \begin{vmatrix} 2 & 4 \\ 3 & 5 \end{vmatrix} = -2, \qquad A_{21} = - \begin{vmatrix} 2 & 3 \\ 5 & 6 \end{vmatrix} = 3$$

$$A_{22} = \begin{vmatrix} 1 & 3 \\ 3 & 6 \end{vmatrix} = -3, \qquad A_{23} = - \begin{vmatrix} 1 & 2 \\ 3 & 5 \end{vmatrix} = -1$$

$$A_{31} = \begin{vmatrix} 2 & 3 \\ 4 & 5 \end{vmatrix} = -2, \qquad A_{32} = -\begin{vmatrix} 1 & 3 \\ 2 & 5 \end{vmatrix} = 1$$

$$A_{33} = \begin{vmatrix} 1 & 2 \\ 2 & 4 \end{vmatrix} = 0$$

$$\therefore \qquad B = \begin{bmatrix} -1 & 3 & -2 \\ 3 & -3 & 1 \\ -2 & 1 & 0 \end{bmatrix}$$

$$\therefore \qquad \text{Adj } A = C' = \begin{bmatrix} -1 & 3 & -2 \\ 3 & -3 & 1 \\ -1 & 1 & 0 \end{bmatrix}$$

Also
$$|A| = \begin{vmatrix} 1 & 2 & 3 \\ 2 & 4 & 5 \\ 3 & 5 & 6 \end{vmatrix}$$
$$= 1(24 - 25) - 2(12 - 15) + 3(10 - 12)$$
$$= -1 + 6 - 6$$
$$= -1$$

Now $\quad A^{-1} = \dfrac{\text{adj } A}{|A|} = \dfrac{1}{-1} \begin{bmatrix} -1 & 3 & -2 \\ 3 & -3 & 1 \\ -2 & 1 & 0 \end{bmatrix} = \begin{bmatrix} 1 & -3 & 2 \\ -3 & 3 & -1 \\ 2 & -1 & 0 \end{bmatrix}$

Example 5. *Find the inverse of the following matrix*

$$A = \begin{bmatrix} 1 & 3 & 3 \\ 1 & 4 & 3 \\ 1 & 3 & 4 \end{bmatrix}$$

Solution. We have

$$|A| = \begin{vmatrix} 1 & 3 & 3 \\ 1 & 4 & 3 \\ 1 & 3 & 4 \end{vmatrix}$$
$$= 1\,(4 \times 4 - 3 \times 3) - 3\,(4 \times 1 - 3 \times 1) + 3\,(1 \times 3 - 4 \times 1)$$
$$= 1\,(16 - 9) - 3(4 - 3) + 3\,(3 - 4)$$
$$= 7 - 3 - 3 = 1 \neq 0$$

Now the cofactors of the elements of the first row of $|A|$ are

$$\begin{vmatrix} 4 & 3 \\ 3 & 4 \end{vmatrix}, \; -\begin{vmatrix} 1 & 3 \\ 1 & 4 \end{vmatrix}, \; \begin{vmatrix} 1 & 4 \\ 1 & 3 \end{vmatrix}, \text{ i.e., } 7, -1, -1 \text{ respectively.}$$

The cofactors of the elements of the second row of $|A|$ are

$$-\begin{vmatrix} 3 & 3 \\ 3 & 4 \end{vmatrix}, \; \begin{vmatrix} 1 & 3 \\ 1 & 4 \end{vmatrix}, \; -\begin{vmatrix} 1 & 3 \\ 1 & 3 \end{vmatrix}, \text{ i.e., } -3, 1, 0 \text{ respectively.}$$

Similarly, the cofactors of the elements of the third row $|A|$ are

$$\begin{vmatrix} 3 & 3 \\ 4 & 3 \end{vmatrix}, -\begin{vmatrix} 1 & 3 \\ 1 & 3 \end{vmatrix}, \begin{vmatrix} 1 & 3 \\ 1 & 4 \end{vmatrix}, i.e., -3, 1, 0 \text{ respectively.}$$

Hence, the matrix of cofactors

$$\begin{bmatrix} 7 & -1 & -1 \\ -3 & 1 & 0 \\ -3 & 0 & 1 \end{bmatrix}$$

$$\Rightarrow \qquad Adj\ (A) = \begin{bmatrix} 7 & -3 & -3 \\ -1 & 1 & 0 \\ -1 & 0 & 1 \end{bmatrix}$$

$$\Rightarrow \qquad A^{-1} = \frac{Adj(A)}{|A|} = \begin{bmatrix} 7 & -3 & -3 \\ -1 & 1 & 0 \\ -1 & 0 & 1 \end{bmatrix}$$

Example 6. *Given that* $A = \begin{bmatrix} 1 & 2 & 1 \\ 3 & 2 & 3 \\ 1 & 1 & 2 \end{bmatrix}$, *compute :* (i) *det A* (ii) *Adj A* (iii) A^{-1}.

Solution. We have

(i) $$A = \begin{bmatrix} 1 & 2 & 1 \\ 3 & 2 & 3 \\ 1 & 1 & 2 \end{bmatrix}$$

$$= 1\ (4-3) + 2\ (6-3) + 1\ (3-2)$$
$$= 1 - 6 + 1$$
$$= -4.$$

(ii) Now the cofactors of the elements of the first row of $|A|$ are

$$\begin{vmatrix} 2 & 3 \\ 1. & 2 \end{vmatrix}, \begin{vmatrix} 3 & 3 \\ 1 & 2 \end{vmatrix}, \begin{vmatrix} 3 & 2 \\ 2 & 1 \end{vmatrix} \ i.e.,\ 1, -3, 1 \text{ respectively.}$$

The cofactors of the elements of the second row of $|A|$ are

$$-\begin{vmatrix} 2 & 1 \\ 1 & 2 \end{vmatrix}, \begin{vmatrix} 1 & 1 \\ 1 & 2 \end{vmatrix}, -\begin{vmatrix} 1 & 2 \\ 1 & 1 \end{vmatrix} \ i.e.,\ -3, 1, 1 \text{ respectively.}$$

The cofactors of the elements of the third row of $|A|$ are

$$\begin{vmatrix} 2 & 1 \\ 2 & 2 \end{vmatrix}, -\begin{vmatrix} 1 & 1 \\ 3 & 3 \end{vmatrix}, \begin{vmatrix} 1 & 2 \\ 3 & 2 \end{vmatrix} \ i.e.,\ 4, 0, -4 \text{ respectively.}$$

$$\therefore \qquad B = \begin{bmatrix} 1 & -3 & 1 \\ -3 & 1 & 1 \\ 4 & 0 & -4 \end{bmatrix}$$

$$\therefore \qquad Adj\ A = B' = \begin{bmatrix} 1 & -3 & 4 \\ -3 & 1 & 0 \\ 1 & 1 & -4 \end{bmatrix}\ 2$$

(*iii*) The inverse of A = $\dfrac{\text{Adj } A}{|A|}$ = $-\dfrac{1}{4}\begin{bmatrix} 1 & -3 & 4 \\ -3 & 1 & 0 \\ 1 & 1 & -4 \end{bmatrix}$ = $\begin{bmatrix} -\dfrac{1}{4} & \dfrac{3}{4} & -1 \\ \dfrac{3}{4} & -\dfrac{1}{4} & 0 \\ -\dfrac{1}{4} & -\dfrac{1}{4} & 1 \end{bmatrix}$

Example 7. If $A = \begin{bmatrix} 0 & 0 & 1 \\ 0 & 1 & 0 \\ 1 & 0 & 0 \end{bmatrix}$, *then show that* $A^{-1} = A$

Solution. We have $|A| = \begin{vmatrix} 0 & 0 & 1 \\ 0 & 1 & 0 \\ 1 & 0 & 0 \end{vmatrix} = \begin{vmatrix} 0 & 1 \\ 1 & 0 \end{vmatrix} = -1.$

Now for the given matrix A, we have

Also $A_{11} = \begin{vmatrix} 1 & 0 \\ 0 & 0 \end{vmatrix} = 0,$ $A_{12} = -\begin{vmatrix} 0 & 0 \\ 1 & 0 \end{vmatrix} = 0$

$A_{13} = \begin{vmatrix} 0 & 1 \\ 1 & 0 \end{vmatrix} = -1,$ $A_{21} = -\begin{vmatrix} 0 & 1 \\ 0 & 0 \end{vmatrix} = 0$

$A_{22} = \begin{vmatrix} 0 & 1 \\ 1 & 0 \end{vmatrix} = -1,$ $A_{23} = -\begin{vmatrix} 0 & 0 \\ 1 & 0 \end{vmatrix} = 0$

$A_{31} = \begin{vmatrix} 0 & 1 \\ 1 & 0 \end{vmatrix} = -1,$ $A_{32} = -\begin{vmatrix} 0 & 1 \\ 0 & 0 \end{vmatrix} = 0$

$A_{33} = \begin{vmatrix} 0 & 0 \\ 0 & 1 \end{vmatrix} = 0$

Therefore, the matrix B formed by the cofactor of the elements of $|A|$ is

$$B = \begin{bmatrix} 0 & 0 & -1 \\ 0 & -1 & 0 \\ -1 & 0 & 0 \end{bmatrix}$$

Adj $A = B' =$ transpose of the matrix B

$$= \begin{bmatrix} 0 & 0 & -1 \\ 0 & -1 & 0 \\ -1 & 0 & 0 \end{bmatrix}$$

∴ $A^{-1} = \dfrac{\text{Adj } A}{|A|} = \dfrac{1}{(-1)} \cdot \begin{bmatrix} 0 & 0 & -1 \\ 0 & -1 & 0 \\ -1 & 0 & 0 \end{bmatrix} = \begin{bmatrix} 0 & 0 & 1 \\ 0 & 1 & 0 \\ 1 & 0 & 0 \end{bmatrix} = A$

Example 8. *Find the inverse of the matrix*

$$A = \begin{bmatrix} \cos\alpha & -\sin\alpha & 1 \\ \sin\alpha & \cos\alpha & 0 \\ 0 & 0 & 1 \end{bmatrix}$$

Solution. We have $|A| = \begin{bmatrix} \cos\alpha & -\sin\alpha & 1 \\ \sin\alpha & \cos\alpha & 0 \\ 0 & 0 & 1 \end{bmatrix} = 1 \begin{vmatrix} \cos\alpha & -\sin\alpha \\ \sin\alpha & \cos\alpha \end{vmatrix}$

On expanding the determinant along the third row, we get

$$\cos^2\alpha + \sin^2\alpha = 1.$$

Since $|A| \neq 0$, therefore A^{-1} exist.

Now for the given matrix, we have

Also $\quad A_{11} = \begin{vmatrix} \cos\alpha & 0 \\ 0 & 1 \end{vmatrix} = \cos\alpha, \quad A_{12} = -\begin{vmatrix} \sin\alpha & 0 \\ 0 & 1 \end{vmatrix} = -\sin\alpha$

$A_{13} = \begin{vmatrix} \sin\alpha & \cos\alpha \\ 0 & 0 \end{vmatrix} = 0, \quad A_{21} = -\begin{vmatrix} -\sin\alpha & 0 \\ 0 & 1 \end{vmatrix} = \sin\alpha$

$A_{22} = \begin{vmatrix} \cos\alpha & 0 \\ 0 & 1 \end{vmatrix} = \cos\alpha, \quad A_{23} = -\begin{vmatrix} \cos\alpha & -\sin\alpha \\ 0 & 0 \end{vmatrix} = 0$

$A_{31} = \begin{vmatrix} -\sin\alpha & 0 \\ \cos\alpha & 0 \end{vmatrix} = 0, \quad A_{32} = -\begin{vmatrix} \cos\alpha & 0 \\ \sin\alpha & 0 \end{vmatrix} = 0$

$A_{33} = \begin{vmatrix} \cos\alpha & -\sin\alpha \\ \sin\alpha & \cos\alpha \end{vmatrix} = 0$

$\therefore \qquad\qquad B = \begin{bmatrix} \cos\alpha & -\sin\alpha & 0 \\ \sin\alpha & \cos\alpha & 0 \\ 0 & 0 & 1 \end{bmatrix}$

$\therefore$ Adj A = transpose of the matrix B

$$= \begin{bmatrix} \cos\alpha & \sin\alpha & 0 \\ -\sin\alpha & \cos\alpha & 0 \\ 0 & 0 & 1 \end{bmatrix}$$

Now $A^{-1} = \dfrac{1}{|A|}$ Adj A and here $|A| = 1$

$\therefore \qquad\qquad A^{-1} = \begin{bmatrix} \cos\alpha & \sin\alpha & 0 \\ -\sin\alpha & \cos\alpha & 0 \\ 0 & 0 & 1 \end{bmatrix}$

Example 9. *Show that*

$$A = \begin{bmatrix} 5 & 3 \\ -1 & -2 \end{bmatrix}$$

satisfies the equation $A^2 - 3A - 7I = 0$. *Also find* A^{-1}.

Solution. We have

$$A^2 = A.\,A = \begin{bmatrix} 5 & 3 \\ -1 & -2 \end{bmatrix} \cdot \begin{bmatrix} 5 & 3 \\ -1 & -2 \end{bmatrix}$$

$$= \begin{bmatrix} 5 \cdot 5 + 3 \cdot (-1) & 5 \cdot 3 + 3 \cdot (-2) \\ (-1) \cdot 5 + (-2) \cdot (-1) & (-1) \cdot 3 + (-2)(-2) \end{bmatrix}$$

$$= \begin{bmatrix} 22 & 9 \\ -3 & 1 \end{bmatrix}$$

$\therefore \qquad A^2 - 3A - 7I = \begin{bmatrix} 22 & 9 \\ -3 & 1 \end{bmatrix} - 3 \begin{bmatrix} 5 & 3 \\ -1 & -2 \end{bmatrix} - 7 \begin{bmatrix} 1 & 0 \\ 0 & 1 \end{bmatrix}$

$$= \begin{bmatrix} 22 - 15 - 7 & 9 - 9 + 0 \\ -3 + 3 - 0 & 1 + 6 - 7 \end{bmatrix}$$

$$= \begin{bmatrix} 0 & 0 \\ 0 & 0 \end{bmatrix}$$

$\Rightarrow \qquad A^2 - 3A - 7I = 0$

Again $\qquad |A| = \begin{vmatrix} 5 & 3 \\ -1 & -2 \end{vmatrix} = -10 + 3 = -7 \neq 0$

so A^{-1} exist.

For the given matrix, we have

$$A_{11} = -2, \ A_{12} = 1, \ A_{21} = -3, \ A_{22} = 5$$

Therefore the matrix B formed by the cofactor of the element $|A|$ is

$$B = \begin{bmatrix} -2 & 1 \\ -3 & 5 \end{bmatrix}$$

Adj A = Transpose of $B = \begin{bmatrix} -2 & -3 \\ 1 & 5 \end{bmatrix}$

Hence $\qquad A^{-1} = \dfrac{\text{Adj } A}{|A|} = \dfrac{1}{(-7)} \begin{bmatrix} -2 & -3 \\ 1 & 5 \end{bmatrix}$

$\therefore \qquad A^{-1} = \begin{bmatrix} 2/7 & 3/7 \\ 1/7 & 5/7 \end{bmatrix}$

Example 10. *Find the inverse of matrix* $A = \begin{bmatrix} i & -1 & 2i \\ 2 & 0 & 2 \\ -1 & 0 & 1 \end{bmatrix}$

Solution. We have $\qquad |A| = \begin{bmatrix} i & -1 & 2i \\ 2 & 0 & 2 \\ -1 & 0 & 1 \end{bmatrix}$

$$= i(0 - 0) + 1(2 + 2) + 2i(0 - 0) = 4 \neq 0$$

Now the matrix A, we have

$$A_{11} = \begin{vmatrix} 0 & 2 \\ 0 & 1 \end{vmatrix} = 0, \qquad A_{12} = - \begin{vmatrix} 2 & 2 \\ -1 & 1 \end{vmatrix} = -4$$

$$A_{13} = \begin{vmatrix} 2 & 0 \\ -1 & 0 \end{vmatrix} = 0, \qquad A_{21} = - \begin{vmatrix} -1 & 2i \\ 0 & 1 \end{vmatrix} = 1$$

$$A_{22} = \begin{vmatrix} i & 2i \\ -1 & 1 \end{vmatrix} = 3i, \qquad A_{23} = - \begin{vmatrix} i & -1 \\ -1 & 0 \end{vmatrix} = 1$$

$$A_{31} = \begin{vmatrix} -1 & 2i \\ 0 & 2 \end{vmatrix} = -2, \qquad A_{32} = - \begin{vmatrix} i & 2i \\ 2 & 2 \end{vmatrix} = 2i$$

$$A_{33} = \begin{vmatrix} i & -1 \\ 2 & 0 \end{vmatrix} = 0$$

$\therefore$

$$B = \begin{bmatrix} 0 & -4 & 0 \\ 1 & 3i & 1 \\ -2 & 2i & 2 \end{bmatrix}$$

$\therefore$
$$\text{Adj } A = \text{transpose of the matrix } B$$

$$= \begin{bmatrix} 0 & 1 & -2 \\ -4 & 3i & 2i \\ 0 & 1 & 2 \end{bmatrix}$$

$\therefore$

$$A^{-1} = \frac{\text{Adj } A}{|A|} = \frac{1}{4} \begin{bmatrix} 0 & 1 & -2 \\ -4 & 3i & 2i \\ 0 & 1 & 2 \end{bmatrix} = \begin{bmatrix} 0 & \dfrac{1}{4} & -\dfrac{1}{2} \\ -1 & \dfrac{3}{4}i & \dfrac{1}{2}i \\ 0 & \dfrac{1}{4} & \dfrac{1}{2} \end{bmatrix}$$

Example 11. *Show that the matrix* $A = \dfrac{1}{3} \begin{bmatrix} 1 & 2 & 2 \\ 2 & 1 & -2 \\ -2 & 2 & -1 \end{bmatrix}$ *is orthogonal.*

Solution. We have
$$A = \frac{1}{3} \begin{bmatrix} 1 & 2 & 2 \\ 2 & 1 & -2 \\ -2 & 2 & -1 \end{bmatrix}$$

Then
$$A' = \frac{1}{3} \begin{bmatrix} 1 & 2 & -2 \\ 2 & 1 & 2 \\ 2 & -2 & -1 \end{bmatrix}$$

We have
$$AA' = \frac{1}{3} \times \frac{1}{3} \begin{bmatrix} 1 & 2 & 2 \\ 2 & 1 & -2 \\ -2 & 2 & -1 \end{bmatrix} \begin{bmatrix} 1 & 2 & -2 \\ 2 & 1 & 2 \\ 2 & -2 & -1 \end{bmatrix}$$

$$= \frac{1}{9}\begin{bmatrix} 9 & 0 & 0 \\ 0 & 9 & 0 \\ 0 & 0 & 9 \end{bmatrix} = \frac{1}{9} \times 9\begin{bmatrix} 1 & 0 & 0 \\ 0 & 1 & 0 \\ 0 & 0 & 1 \end{bmatrix} = \begin{bmatrix} 1 & 0 & 0 \\ 0 & 1 & 0 \\ 0 & 0 & 1 \end{bmatrix} = I.$$

Hence, by the definition of orthogonality, the matrix A is orthogonal.

Example 12. *Show that the matrix* $\begin{bmatrix} \cos\theta & \sin\theta \\ -\sin\theta & \cos\theta \end{bmatrix}$ *is orthogonal*

Solution. Let $\quad A = \begin{bmatrix} \cos\theta & \sin\theta \\ -\sin\theta & \cos\theta \end{bmatrix}$

Then $\quad\quad\quad A' = \begin{bmatrix} \cos\theta & -\sin\theta \\ \sin\theta & \cos\theta \end{bmatrix}$

We have $\quad\quad AA' = \begin{bmatrix} \cos\theta & \sin\theta \\ -\sin\theta & \cos\theta \end{bmatrix}\begin{bmatrix} \cos\theta & -\sin\theta \\ \sin\theta & \cos\theta \end{bmatrix}$

$$= \begin{bmatrix} \cos^2\theta + \sin^2\theta & \cos\theta\sin\theta - \sin\theta\cos\theta \\ \sin\theta\cos\theta - \cos\theta\sin\theta & \sin^2\theta + \cos^2\theta \end{bmatrix}$$

$$= \begin{bmatrix} 1 & 0 \\ 0 & 1 \end{bmatrix} = I.$$

Hence, the condition of orthogonality is satisfied. Therefore, the given matrix is orthogonal.

Example 13. *Prove that the matrix* $\begin{bmatrix} \dfrac{1+i}{2} & \dfrac{-1+i}{2} \\ \dfrac{1+i}{2} & \dfrac{1-i}{2} \end{bmatrix}$ *is unitary.*

Solution. Let $\quad\quad\quad A = \begin{bmatrix} \dfrac{1+i}{2} & \dfrac{-1+i}{2} \\ \dfrac{1+i}{2} & \dfrac{1-i}{2} \end{bmatrix}$

Then $\quad\quad\quad\quad\quad A' = \begin{bmatrix} \dfrac{1+i}{2} & \dfrac{1+i}{2} \\ \dfrac{-1+i}{2} & \dfrac{1-i}{2} \end{bmatrix}$

$\therefore \quad\quad A^\theta = (\bar{A}') = \begin{bmatrix} \dfrac{1-i}{2} & \dfrac{1-i}{2} \\ \dfrac{-1-i}{2} & \dfrac{1+i}{2} \end{bmatrix}$

Now $\quad A^\theta A = \begin{bmatrix} \dfrac{1-i}{2} & \dfrac{1-i}{2} \\ \dfrac{-1-i}{2} & \dfrac{1+i}{2} \end{bmatrix}\begin{bmatrix} \dfrac{1+i}{2} & \dfrac{-1+i}{2} \\ \dfrac{1+i}{2} & \dfrac{1-i}{2} \end{bmatrix}$

$$= \begin{bmatrix} \frac{1}{4}(1-i^2)+\frac{1}{4}(1-i^2) & -\frac{1}{4}(1-i^2)+\frac{1}{4}(1-i)^2 \\ -\frac{1}{4}(1+i)^2 +\frac{1}{4}(1+i^2) & \frac{1}{4}(1-i^2)+\frac{1}{4}(1-i^2) \end{bmatrix} = \begin{bmatrix} 1 & 0 \\ 0 & 1 \end{bmatrix} = I.$$

Hence, the matrix A is unitary.

EXERCISE 17.2

1. Find the adjoint of matrix $A = \begin{bmatrix} 1 & 0 & 2 \\ 2 & 1 & 0 \\ 3 & 2 & 1 \end{bmatrix}$

2. Find the adjoint of the matrix $A = \begin{bmatrix} 1 & 1 & 1 \\ 1 & 2 & -3 \\ 2 & -1 & 3 \end{bmatrix}$

3. Find the inverse of the matrix $A = \begin{bmatrix} 5 & 2 \\ 7 & 3 \end{bmatrix}$

4. Find the inverse of the matrix $A = \begin{bmatrix} 1 & 0 & 0 \\ 1 & 1 & 0 \\ 1 & 0 & 1 \end{bmatrix}$

5. Find the inverse of the matrix $\begin{bmatrix} 2 & -2 & 4 \\ 2 & 3 & 2 \\ -1 & 1 & -1 \end{bmatrix}$

6. Show that the matrix $\begin{bmatrix} \frac{1}{3} & \frac{2}{3} & \frac{2}{3} \\ \frac{2}{3} & \frac{1}{3} & -\frac{2}{3} \\ -\frac{2}{3} & \frac{2}{3} & -\frac{1}{3} \end{bmatrix}$ is orthogonal.

7. Show that the matrix $A = \frac{1}{\sqrt{2}}\begin{bmatrix} 1 & i \\ -i & -1 \end{bmatrix}$ is unitary.

8. Find the inverse of the matrix $A = \begin{bmatrix} 2 & 1 & 2 \\ 2 & 2 & 1 \\ 1 & 1 & 1 \end{bmatrix}$

9. Find the inverse of the matrix A, where

$$A = \begin{bmatrix} 3 & -3 & 4 \\ 2 & -3 & 4 \\ 0 & -1 & 1 \end{bmatrix}$$

10. Find the inverse of the matrix A, where

$$A = \begin{bmatrix} 1 & 2 & 3 \\ 4 & 5 & 6 \\ 6 & 7 & 9 \end{bmatrix}$$

11. Find the inverse of $A = \begin{bmatrix} 3 & -10 & -1 \\ -2 & 8 & 2 \\ 2 & -4 & -2 \end{bmatrix}$

ANSWERS

(1) $\begin{bmatrix} 1 & 4 & -2 \\ -2 & 5 & 4 \\ 1 & 2 & 1 \end{bmatrix}$ **(2)** $\begin{bmatrix} 3 & -4 & -5 \\ -9 & 1 & 4 \\ -5 & 3 & 1 \end{bmatrix}$ **(3)** $\begin{bmatrix} 3 & -2 \\ -7 & 5 \end{bmatrix}$ **(4)** $\begin{bmatrix} 1 & 0 & 0 \\ -1 & 1 & 0 \\ -1 & 0 & 1 \end{bmatrix}$

(5) $\dfrac{1}{10}\begin{bmatrix} -5 & 2 & -16 \\ 0 & 2 & 4 \\ 5 & 0 & 10 \end{bmatrix}$ **(8)** $\begin{bmatrix} 1 & 1 & -3 \\ -1 & 0 & 2 \\ 0 & -1 & 2 \end{bmatrix}$ **(9)** $\begin{bmatrix} 1 & -1 & 0 \\ -2 & 3 & -4 \\ -2 & 3 & -3 \end{bmatrix}$

(10) $\begin{bmatrix} -1 & -1 & 1 \\ 0 & 3 & -2 \\ 2/3 & -5/3 & 1 \end{bmatrix}$ **(11)** $\begin{bmatrix} 1/2 & 1 & 3/4 \\ 0 & 1/4 & 1/4 \\ 1/2 & 1/2 & -1/4 \end{bmatrix}$

17.16   RANK OF A MATRIX

Definition : *A positive integer r is said to be the rank of matrix A if it contains at least one square submatrix of order $r \times r$, whose determinant is non-zero while any square submatrix of A of order $(r + 1) \times (r + 1)$ or greater is singular i.e., having determinant zero. The rank of a matrix A is denoted by $\rho(A)$.*

It is obvious that the rank r of a matrix of order $m \times n$ may at most be equal to the smaller of the numbers m and n, but it may be less.

If the rank of a square matrix A of order $n \times n$ is r and $r < n$, then the matrix A is said to be *singular*, on the other hand if $r = n$, then the matrix is said to be *non-singular*.

Remarks

❖   If the rank of a matrix is zero, then matrix is a null matrix.
❖   The rank of every non-zero matrix must be greater than or equal to 1.
❖   The rank of a unit matrix is equal to the order to the unit matrix.

17.17   ECHELON FORM OF A MATRIX

Definition : *A matrix A is said to be in Echelon form if it satisfies following conditions :*

(i) *Every row of A has all its entries zero which occurs below the every row having a non-zero entry.*

(ii) *The numbers of zeros before the first non-zero entry in the same row is less than the number of zeros in the next row.*

Remark

❖ The rank of a matrix is equal to the number of non-zero rows in **Echelon form** of the given matrix.

For example :

Let A = $\begin{bmatrix} 0 & 2 & 3 & 5 \\ 0 & 0 & 3 & 2 \\ 0 & 0 & 0 & 0 \end{bmatrix}$

This matrix A is in Echelon form and it has two non-zero rows since rank of A is equal to the number of non-zero rows. Hence rank of $A = 2$.

THEOREM-1

The rank of the transpose of a matrix is the same as that of the original matrix.

Proof. Let us suppose A is any matrix and A' is its transpose and let rank of $A = r$. This implies that A contains at least one r-rowed square matrix whose determinant is non-zero, let it be B. Obviously B' is a submatrix of A' but we know that det $B' = $ det B and since det $B \neq 0 \Rightarrow$ det $B' = 0$. Thus the rank of $A' \geq r$. Now if A contains a $(r + 1)$-rowed square submatrix C, then det $C = 0$ because rank of $A = r$. Obviously C' is a submatrix of A' and det $C' = $ det $C = 0$, if follows that A' does not contain $(r + 1)$-rowed square submatrix with non-zero determinant. Hence rank of $A' \leq r$ and consequently we obtained rank of $A' = $ rank of A.

17.18   ELEMENTARY TRANSFORMATIONS OF A MATRIX

Definition : *A transformation is said to be elementary if it is one of the followings :*
  (i) *Interchanging of any two rows (or columns).*
 (ii) *Multiplying any row (or column) by any non-zero number.*
(iii) *Addition of any row to k times the other row, where k is any non-zero number.*

Remarks

❖ If the elementary transformation (or E-transformation) is performed on rows, then it is called by row-transformation.
❖ If the E-transformation is performed on column, it is called column-transformation.

17.19   ELEMENTARY MATRICES

Definition : *A matrix which is obtained by a single E-transformation is called an elementary matrix.* For example

$\begin{bmatrix} 0 & 0 & 1 \\ 0 & 1 & 0 \\ 1 & 0 & 0 \end{bmatrix}, \begin{bmatrix} 1 & 0 & 0 \\ 0 & 1 & 0 \\ 0 & 0 & 1 \end{bmatrix}$ etc.

$$A = \begin{bmatrix} 3 & -10 & -1 \\ -2 & 8 & 2 \\ 2 & -4 & -2 \end{bmatrix}$$

Here first E-matrix is obtained from I_3 by interchanging C_1 and C_3 columns and the second E-matrix is obtained by $R_1 \rightarrow R_1 + 2R_2$

Remarks

❖ All the elementary matrices are non-singular.

❖ Each elementary matrix possesses its inverse.

17.20 INVARIANCE OF RANK UNDER E-TRANSFORMATION

THEOREM-1

Elementary transformation (E-transformation) do not change the rank of matrix.

Proof. Since we know that E-transformations are of three types. Therefore, we shall provide this theorem for following three cases :

Case I. *Interchanging the rows (or columns) does not change the rank.*

Let A be a matrix of order $m \times n$ of rank r and let B be a matrix obtained from A by interchanging the rows R_i and R_j *i.e.*, by E-transformations $R_i \leftrightarrow R_j$. Let the rank of B be s. Then we shall prove $r = s$.

Since rank of $A = r$. This implies A contains at least one. r-rowed square submatrix with non-zero determinant let it be R *i.e.*, det $R \neq 0$. Let us suppose S be the r-rowed square submatrix of B having the same rows as are in R though these rows may be in different positions. Then either

$$\det S = \det R \quad \text{or} \quad \det S = -\det R$$

But $$\det R \neq 0 \implies \det S \neq 0$$

∴ $$\text{Rank of } B \geq r \implies s \geq r. \qquad \qquad ...(1)$$

Further since the matrix A can also be obtained from B by E-transformation $R_i \leftrightarrow R_j$. Then we have

$$r \geq s \qquad \qquad ...(2)$$

Hence from (1) and (2), we conclude that $r = s$.

Case II. *Multiplication of the elements of a row by a non-zero number does not change the rank.*

Let A be a matrix of order $m \times n$ of rank r and let B be a matrix which obtained from A by E-transformation $R_i \rightarrow kR_j$ where $k \neq 0$ and let rank of B be s. Therefore we shall prove that $s = r$. Suppose B_0 is an $(r + 1)$-rowed square submatrix of B, then there exists A_0 of $(r + 1)$-rowed square submatrix of A such that either

$$\det B_0 = \det A_0$$

or $$\det B_0 = k\det A_0$$

But rank of $A = r$, this means that every $(r + 1)$-rowed square submatrix of A has zero determinant.

$$\therefore \qquad\qquad |A_0| = 0$$

$$\Rightarrow \qquad\qquad |B_0| = 0$$

$\Rightarrow$ Every $(r + 1)$-rowed square submatrix will have zero determinant

$\Rightarrow$ Rank of B can exceed the rank of A

$$\Rightarrow \qquad\qquad s \le r. \qquad\qquad\qquad ...(1)$$

Further since the matrix A can also be obtained from B by E-transformation $R_i \to \left(\dfrac{1}{k}\right) R_j$. Thus we have

$$r \le s. \qquad\qquad ...(2)$$

Hence from (1) and (2), we conclude that

$$r = s.$$

Case III. *Addition of any row to the product of any number k and other row does not change the rank.*

Let the rank of matrix A of order $m \times n$ be r and let B is obtained by the E-transformation $R_i \to R_i + kR_j$ and let rank of B be s. Then we shall prove $s = r$.

Now if B_0 is an $(r + 1)$-rowed square submatrix of B, there exists uniquely A_0 an $(r + 1)$-rowed square submatrix of A.

Since we know that any E-transformation does not change the determinant value. Therefore if no row of A_0 is a part of ith row of A, or if two rows of A_0 are the parts of the ith and j^{th} rows of A, then $\det B_0 = \det A_0$.

But the rank of $\qquad\qquad A = r \Rightarrow \det A_0 = 0 \Rightarrow B_0 = 0.$

Now suppose if a row of A_0 is a part of ith row of A and no row is a part of jth row, then

$$\det B_0 = \det A_0 + k \det C_0$$

where C_0 is an $(r + 1)$-rowed square submatrix which is obtained from A_0 by E-transformation $R_i \to R_i + kR_j$.

Clearly, all the $(r + 1)$ rows of C_0 are exactly same as the rows of some $(r + 1)$-rowed square submatrix of A, though in some different position. Therefore $\det C_0$ is ± 1 times det of some $(r + 1)$-rowed square submatrix A. But the rank of A is r. This implies every $(r + 1)$-rowed square submatrix will have zero determinant.

$$\therefore \qquad \det A_0 = 0, \det C_0 = 0 \Rightarrow \det B_0 = 0$$

Hence rank of B can not exceed the rank of A

$$\therefore \qquad\qquad s \le r. \qquad\qquad\qquad ...(1)$$

Further since A can also be obtained from B by E-transformation $R_i \to R_i - kR_j$, therefore we have

$$r \le s. \qquad\qquad ...(2)$$

From (1) and (2), we conclude that

$$r = s.$$

Remarks

❖ The rank of a matrix does not change by a series of E-transformation.

❖ The rank of a matrix does not change by a column-transformation.

17.21   NORMAL FORM

Definition : *If a matrix is reduced to the form* $\begin{pmatrix} I_r & 0 \\ 0 & 0 \end{pmatrix}$. *Then this form is called normal form of the given matrix.*

THEOREM-1

Every matrix of order $m \times n$ of rank r can be reduced to the form $\begin{pmatrix} I_r & 0 \\ 0 & 0 \end{pmatrix}$ *by a finite number of E-transformations, where I_r is the unit matrix of order $r \times r$.*

Proof. Let $A = [a_{ij}]_{m \times n}$ be a matrix of order $m \times n$ and of rank r. If A is zero matrix, then its rank is zero and thus A can be written as $\begin{pmatrix} I_r & 0 \\ 0 & 0 \end{pmatrix}$.

Let us suppose A is a non-zero matrix it means that it has at least one of its element non-zero. Let this non-zero element be $a_{ij} = k \neq 0$.

Let B be a matrix which is obtained from A by E-transformations $R_1 \leftrightarrow R_i$ and $C_1 \leftrightarrow C_i$ and whose leading element is k. Again using the E-transformation $R_1 \to \frac{1}{k} R_1$ on B and we get a matrix C whose leading element becomes 1. Let this matrix C be

$$C = \begin{bmatrix} 1 & C_{12} & C_{13} & \cdots & C_{1n} \\ C_{21} & C_{22} & C_{23} & \cdots & C_{2n} \\ C_{31} & C_{32} & C_{33} & \cdots & C_{3n} \\ \cdots & \cdots & \cdots & \cdots & \cdots \\ C_{m1} & C_{m2} & C_{m3} & \cdots & C_{mn} \end{bmatrix}_{m \times n}$$

Now, substracting first column after multiplying by suitable number from remaining columns of C and subtracting first row after multiplying by suitable number from remaining rows of C. We, therefore, obtain a matrix D whose elements of the first row and first column are zero except the leading element. Let D be given as

$$D = \begin{bmatrix} 1 & 0 & 0 & \cdots & \cdots & 0 \\ 0 & & & & & \\ 0 & & & & & \\ 1 & & & A_1 & & \\ 1 & & & & & \\ 0 & & & & & \end{bmatrix}_{m \times n}$$

where A_1 is a matrix of order $(m-1) \times (n-1)$.

If this matrix A_1 is non-zero matrix, then we shall apply above process on A_1. Since we know that E-transformation will not effect the first row and first column of D, so that we shall apply E-transformation on D and there no need to take A_1 separately. Continuing this process finitely we obtain a matrix M such that

$$M = \begin{pmatrix} I_k & O \\ O & O \end{pmatrix}$$

This implies that matrix M has a rank k. But M is obtained from A by a finite number of E-transformations and we know that E-transformations do not change the rank, therefore k must be equal to r. Hence the matrix A of order $m \times n$ of rank r can be reduced to the form $\begin{pmatrix} I_r & O \\ O & O \end{pmatrix}$ by a finite number of E-transformations.

Remark

❖ The form $\begin{pmatrix} I_r & O \\ O & O \end{pmatrix}$ of A is also called first canonical form.

Corollary 1. *The rank of matrix of order $m \times n$ is r if it can be reduced to $\begin{pmatrix} I_r & O \\ O & O \end{pmatrix}$ by a finite number of E-transformations.*

Corollary 2. *If A is a matrix of order $m \times n$ of rank r, then there exist non-singular matrices P and Q such that*

$$PAQ = \begin{pmatrix} I_r & O \\ O & O \end{pmatrix}.$$

17.22 EQUIVALENCE OF MATRICES

Definition : *Let A be a matrix of order $m \times n$. If a matrix B of order $m \times n$ is obtained from A by a finite number of E-transformations, then A is called **equivalent** to B. It is denoted by $A \sim B$ (Read as A is equivalent to B).*

THEOREM-1

The relation "$\sim$" in the set of all $m \times n$ matrices is an equivalence relation.
Proof.
(*i*) **Reflexivity.** If A is a matrix of order $m \times n$ then A is equivalent to A i.e., $A \sim A$.

(*ii*) **Symmetry.** Let A and B be two matrix of order $m \times n$ and $A \sim B$. This implies if B is obtained from A by a finite number of E-transformation, then A can also be obtained from B by a finite E-transformations. Hence $A \sim A$.

(*iii*) **Transitivity.** Let A, B, C be three matrices of order $m \times n$ and $A \sim B$, $B \sim C$. This implies that if B is obtained from A by a finite number of E-transformations and C is obtained from B by a finite number of E-transformations, then C can also be obtained from A by a finite number of E-transformations. Hence $A \sim C$.

Hence the relation "$\sim$" is an equivalence relation.

Remarks

❖ An equivalence relation is a relation which is reflexive, symmetric and transitive.

❖ Two equivalent matrices have the same rank.

❖ Two matrices of same order and of same rank are always equivalent.

17.23   ROW-TRANSFORMATIONS (EMPLOYMENT)

THEOREM-1

Let A be a matrix of order m × n of rank r. Then there exists a non-singular matrix P such that

$$PA = \begin{bmatrix} G \\ O \end{bmatrix}$$

where G is a matrix of order r × n of rank r and O is zero matrix of order (m − r) × n.

Proof. Since we know that if A is a matrix of order $m \times n$ and of rank r, then there exists non-singular matrices P and Q such that

$$PAQ = \begin{pmatrix} I_r & O \\ O & O \end{pmatrix} \qquad \qquad ...(i)$$

Further since we know that every non-singular matrix can be expressed as the product of elementary matrices.

So let $\qquad \qquad Q = Q_1 Q_2 Q_3 \cdots Q_k$

where $Q_1, Q_2, \cdots Q_k$ are elementary matrices. Substitute the value of Q in (1), we get

$$PAQ_1 Q_2 Q_3 \cdots Q_k = \begin{pmatrix} I_r & O \\ O & O \end{pmatrix} \qquad \qquad ...(ii)$$

Since we know that every E-column transformation of a matrix is equivalent to post multiplication with corresponding E-matrix and no column transformation can effect the last $(m - r)$ rows of the R.H.S. of (ii). Thus post multiplying the L.H.S. of (ii) by E-matrics $Q_k^{-1}, Q_k^{-1} \cdots Q_2^{-1}, Q_1^{-1}$, we get

$$PA = \begin{bmatrix} G \\ O \end{bmatrix}$$

and the rank of PA is same as that of A because E-transformation do not change the rank.

Thus the rank of $\begin{bmatrix} G \\ O \end{bmatrix}$ is r and hence rank of G is also γ, because G has r rows.

17.24. COLUMN-TRANSFORMATIONS (EMPLOYMENT)

THEOREM-1

Let A be a matrix of order m × n and of rank r. Then there exists a non-singular matrix Q such that
$$AQ = [H \; O]$$
where H is a matrix of order m × r of rank r and O is a zero matrix of order m × (n − r).

Proof. Same as above

17.25   RANK OF PRODUCT OF MATRICES

THEOREM-1

The rank of a product of two matrices cannot exceed the rank of either matrix.

Proof. Let A and B be two matrices of order $m \times n$ and $n \times p$ respectively. Let r_1 and r_2

be the ranks of A and B respectively and let r be the rank of AB. We shall prove that $r \leq r_1$ and $r \leq r_2$.

Since the rank of A is r_1, then there exist a non-singular matrix P such that

$$PA = \begin{bmatrix} G \\ O \end{bmatrix} \qquad \qquad \text{...}(i)$$

where G is a matrix of order $r_1 \times n$ of rank r_1 and O is a zero matrix of order $(m - r_1) \times n$. Now post multiplying the both sides of (i) by B, we get

$$PAB = \begin{bmatrix} G \\ O \end{bmatrix} B. \qquad \qquad \text{...}(ii)$$

Since we know that rank of (PAB) = rank of (AB)

$\therefore$ rank of $(PAB) = r$ [$\because$ rank of $(AB) = r$]

or   rank of $\begin{bmatrix} G \\ O \end{bmatrix} = r$ ...(iii)

Since rank of G is r_1 so it has only r_1 non zero rows, therefore the matrix.

$$\begin{bmatrix} G \\ O \end{bmatrix} B$$

cannot have more than r_1 non-zero rows. Thus we have

$$\text{rank of } \begin{bmatrix} G \\ O \end{bmatrix} B \leq r_1$$

or $r \leq r_1$ [form (iii)]

or   Rank of (AB) = Rank of A.   ...(iv)

Further since we have rank (AB) = rank $(AB)'$ and $(AB)' = B'A'$.

$\therefore$ rank of (AB) = rank of $(B'A') \leq$ rank of B' using (iv)

or   rank of $(AB) \leq$ rank of B' = rank of B

or   rank of $(AB) \leq$ rank of B

or $r \leq r_2.$

Remark

❖ The rank of matrix does change by pre (post) multiplication with a non-singular matrix.

SOLVED EXAMPLES

Example 1. *Determine the rank of the following matrices :*

$$(i) \begin{bmatrix} 1 & 2 & 3 & 4 \\ 2 & 4 & 6 & 8 \\ 3 & 6 & 9 & 12 \end{bmatrix} \qquad (ii) \begin{bmatrix} 1 & 2 & 3 \\ 3 & 4 & 5 \\ 4 & 5 & 6 \end{bmatrix}.$$

Solution. (i) The square submatrices of the given matrix are

$$A_1 = \begin{bmatrix} 1 & 2 & 3 \\ 2 & 4 & 6 \\ 3 & 6 & 9 \end{bmatrix}, A_2 = \begin{bmatrix} 1 & 2 & 4 \\ 2 & 4 & 8 \\ 3 & 6 & 12 \end{bmatrix}, A_3 = \begin{bmatrix} 1 & 3 & 4 \\ 2 & 6 & 8 \\ 3 & 9 & 12 \end{bmatrix}, A_4 = \begin{bmatrix} 2 & 3 & 4 \\ 4 & 6 & 8 \\ 6 & 9 & 12 \end{bmatrix}$$

$$\det A_1 = 1\ (36 - 36) + 2\ (18 - 18) + 3\ (12 - 12) = 0$$
$$\det A_2 = 1\ (48 - 48) + 2\ (24 - 24) + 4\ (12 - 12) = 0$$
$$\det A_3 = 1\ (72 - 72) + 3\ (24 - 24) + 4\ (18 - 18) = 0$$
$$\det A_4 = 2\ (72 - 72) + 3\ (48 - 48) + 4\ (36 - 36) = 0$$

Therefore, determinant of all square submatrices of the given matrix of order 3×3 are zero so the rank of the given matrix is less than 3. Now the square submatrices of the given matrix of order 2×2 are

$$\begin{bmatrix} 1 & 2 \\ 2 & 4 \end{bmatrix}, \begin{bmatrix} 1 & 3 \\ 2 & 6 \end{bmatrix}, \begin{bmatrix} 1 & 4 \\ 2 & 8 \end{bmatrix}, \begin{bmatrix} 2 & 3 \\ 4 & 6 \end{bmatrix}, \begin{bmatrix} 2 & 4 \\ 4 & 8 \end{bmatrix}, \begin{bmatrix} 2 & 4 \\ 3 & 8 \end{bmatrix}$$

$$\begin{bmatrix} 2 & 6 \\ 3 & 9 \end{bmatrix}, \begin{bmatrix} 2 & 8 \\ 3 & 12 \end{bmatrix}, \begin{bmatrix} 4 & 6 \\ 6 & 9 \end{bmatrix}, \begin{bmatrix} 4 & 8 \\ 6 & 12 \end{bmatrix}, \begin{bmatrix} 6 & 8 \\ 9 & 12 \end{bmatrix}, \begin{bmatrix} 1 & 2 \\ 3 & 6 \end{bmatrix}, \begin{bmatrix} 1 & 3 \\ 3 & 9 \end{bmatrix}$$

$$\begin{bmatrix} 1 & 4 \\ 3 & 12 \end{bmatrix}, \begin{bmatrix} 2 & 3 \\ 6 & 9 \end{bmatrix}, \begin{bmatrix} 2 & 4 \\ 6 & 12 \end{bmatrix}, \begin{bmatrix} 3 & 4 \\ 9 & 12 \end{bmatrix}$$

Obviously, the determinant of all square submatrices of order 2×2 are zero. Thus the rank of the given matrix is less than 2. Since the given matrix is non-zero matrix. Hence, the rank of the given matrix is 1.

(*ii*)
$$A = \begin{bmatrix} 1 & 2 & 3 \\ 3 & 4 & 5 \\ 4 & 5 & 6 \end{bmatrix}$$

$$\det A = 1\ (24 - 25) + 2\ (20 - 18) + 3\ (15 - 16)$$
$$= -1 + 4 - 3 = -2 + 2 = 0$$

Therefore the rank $A \neq 3$.

Now the square submatrices of A of order 2×2 are

$$A_1 = \begin{bmatrix} 1 & 2 \\ 3 & 4 \end{bmatrix}, A_2 = \begin{bmatrix} 1 & 3 \\ 3 & 5 \end{bmatrix} \text{ etc.}$$

$$\det A_1 = 4 - 6 = -2 \neq 0.$$

Hence the rank $A = 2$.

Example 2. If $A = \begin{bmatrix} 0 & 1 & 0 & 0 \\ 0 & 0 & 1 & 0 \\ 0 & 0 & 0 & 1 \\ 0 & 0 & 0 & 0 \end{bmatrix}$, *find the rank of A and* A^2.

Solution. Since the matrix A is in *Echelon form* and there are three non-zero rows. Therefore, rank of A is equal to the number of non-zero rows. Hence, rank of $A = 3$.

Next find A^2

$$A^2 = \begin{bmatrix} 0 & 1 & 0 & 0 \\ 0 & 0 & 1 & 0 \\ 0 & 0 & 0 & 1 \\ 0 & 0 & 0 & 0 \end{bmatrix} \cdot \begin{bmatrix} 0 & 1 & 0 & 0 \\ 0 & 0 & 1 & 0 \\ 0 & 0 & 0 & 1 \\ 0 & 0 & 0 & 0 \end{bmatrix} = \begin{bmatrix} 0 & 0 & 1 & 0 \\ 0 & 0 & 0 & 1 \\ 0 & 0 & 0 & 0 \\ 0 & 0 & 0 & 0 \end{bmatrix}$$

Obviously, A^2 is an *Echelon form* and having two non-zero rows. Hence the rank of $A^2 = 2$.

Example 3. *Use E-transformation to reduce the following matrix A to triangular form and hence find the rank of A.*

$$A = \begin{bmatrix} 8 & 1 & 3 & 6 \\ 0 & 3 & 2 & 2 \\ -8 & -1 & -3 & 4 \end{bmatrix}$$

Solution. Since we have

$$A = \begin{bmatrix} 8 & 1 & 3 & 6 \\ 0 & 3 & 2 & 2 \\ -8 & -1 & -3 & 4 \end{bmatrix}$$

$$\sim \begin{bmatrix} 1 & 1 & 3 & 6 \\ 0 & 3 & 2 & 2 \\ -1 & -1 & -3 & 4 \end{bmatrix} \text{by } C_1 \to \frac{1}{8}C_1 \sim \begin{bmatrix} 1 & 1 & 3 & 6 \\ 0 & 3 & 2 & 2 \\ 0 & 0 & 0 & 10 \end{bmatrix} \text{by } R_3 \to R_3 + R_1.$$

This matrix is a triangular matrix (Echelon form) and it contains three non-zero rows. Hence, the rank of $A = 3$.

Example 4. *Reduce the matrix* $A = \begin{bmatrix} 1 & 1 & 3 & 6 \\ 0 & 3 & 2 & 2 \\ 0 & 0 & 0 & 10 \end{bmatrix}$ *to the normal form* $\begin{pmatrix} I_r & O \\ O & O \end{pmatrix}$ *and hence, find the rank of A.*

Solution. Since the given matrix is

$$A = \begin{bmatrix} 1 & -1 & 2 & -3 \\ 4 & 1 & 0 & 2 \\ 0 & 3 & 0 & 4 \\ 0 & 1 & 0 & 2 \end{bmatrix}$$

performing $C_2 \to C_2 + C_1, C_3 \to C_3 - 2C_1, C_4 \to C_4 + 3C_1$

$$\sim \begin{bmatrix} 1 & 0 & 0 & 0 \\ 4 & 5 & -8 & 14 \\ 0 & 3 & 0 & 4 \\ 0 & 1 & 0 & 2 \end{bmatrix}$$

performing $R_2 \to R_2 - 4R_1$

$$\sim \begin{bmatrix} 1 & 0 & 0 & 0 \\ 0 & 5 & -8 & 14 \\ 0 & 3 & 0 & 4 \\ 0 & 1 & 0 & 2 \end{bmatrix}$$

performing $R_2 \longleftrightarrow R_4$

$$\sim \begin{bmatrix} 1 & 0 & 0 & 0 \\ 0 & 1 & 0 & 2 \\ 0 & 3 & 0 & 4 \\ 0 & 5 & -8 & 14 \end{bmatrix}$$

performing $C_4 \to C_4 - 2C_2$ $\sim \begin{bmatrix} 1 & 0 & 0 & 0 \\ 0 & 1 & 0 & 0 \\ 0 & 3 & 0 & -2 \\ 0 & 5 & -8 & 4 \end{bmatrix}$

performing $R_3 \to R_3 - 3R_2,\ R_4 \to R_4 - 5R_2$

$\sim \begin{bmatrix} 1 & 0 & 0 & 0 \\ 0 & 1 & 0 & 0 \\ 0 & 0 & 0 & -2 \\ 0 & 0 & -8 & 4 \end{bmatrix}$

performing $C_3 \leftrightarrow C_4$ $\sim \begin{bmatrix} 1 & 0 & 0 & 0 \\ 0 & 1 & 0 & 0 \\ 0 & 0 & -2 & 0 \\ 0 & 0 & 4 & -8 \end{bmatrix}$

performing $R_3 \to -\dfrac{1}{2}R_3$ $\sim \begin{bmatrix} 1 & 0 & 0 & 0 \\ 0 & 1 & 0 & 0 \\ 0 & 0 & 1 & 0 \\ 0 & 0 & 4 & -8 \end{bmatrix}$

performing $R_4 \to R_4 - 4R_3$ $\sim \begin{bmatrix} 1 & 0 & 0 & 0 \\ 0 & 1 & 0 & 0 \\ 0 & 0 & 1 & 0 \\ 0 & 0 & 0 & -8 \end{bmatrix}$

performing $R_4 \to -\dfrac{1}{8}R_4$ $\sim \begin{bmatrix} 1 & 0 & 0 & 0 \\ 0 & 1 & 0 & 0 \\ 0 & 0 & 1 & 0 \\ 0 & 0 & 0 & 1 \end{bmatrix}$

$\therefore$ $A \sim I_4$

Hence the rank of A = 4.

Example 5. *Find two non-singular matrices P and Q such that PAQ is in the normal form where*

$$A = \begin{bmatrix} 1 & 1 & 1 \\ 1 & -1 & -1 \\ 3 & 1 & 1 \end{bmatrix}.$$

Solution. Since we have

$$A = I_3\, A\, I_3$$

i.e.,

$$\begin{bmatrix} 1 & 1 & 1 \\ 1 & -1 & -1 \\ 3 & 1 & 1 \end{bmatrix} = \begin{bmatrix} 1 & 0 & 0 \\ 0 & 1 & 0 \\ 0 & 0 & 1 \end{bmatrix} A \begin{bmatrix} 1 & 0 & 0 \\ 0 & 1 & 0 \\ 0 & 0 & 1 \end{bmatrix}. \qquad \ldots(1)$$

Now applying E-transformation on the matrix A on the L.H.S. of (1) until A reduced to the normal form. In this process we apply E-row transformation to pre-factor R_3 of R.H.S. of (1) and E-column transformation to post-factor I_3 of R.H.S. of (1). Now performing $R_2 \rightarrow R_2 - R_1,\ R_3 \rightarrow R_3 - 3R_1$, we get

$$\begin{bmatrix} 1 & 1 & 1 \\ 0 & -2 & -2 \\ 0 & -2 & -2 \end{bmatrix} = \begin{bmatrix} 1 & 0 & 0 \\ -1 & 1 & 0 \\ -3 & 0 & 1 \end{bmatrix} A \begin{bmatrix} 1 & 0 & 0 \\ 0 & 1 & 0 \\ 0 & 0 & 1 \end{bmatrix}$$

performing $C_2 \rightarrow C_2 - C_1,\ C_3 \rightarrow C_3 - C_1$

$$\begin{bmatrix} 1 & 0 & 0 \\ 0 & -2 & -2 \\ 0 & -2 & -2 \end{bmatrix} = \begin{bmatrix} 1 & 0 & 0 \\ -1 & 1 & 0 \\ -3 & 0 & 1 \end{bmatrix} A \begin{bmatrix} 1 & -1 & -1 \\ 0 & 1 & 0 \\ 0 & 0 & 1 \end{bmatrix}$$

performing $R_2 \rightarrow -\dfrac{1}{2} R_1$

$$\begin{bmatrix} 1 & 0 & 0 \\ 0 & 1 & 1 \\ 0 & -2 & -2 \end{bmatrix} = \begin{bmatrix} 1 & 0 & 0 \\ \dfrac{1}{2} & -\dfrac{1}{2} & 0 \\ -3 & 0 & 1 \end{bmatrix} A \begin{bmatrix} 1 & -1 & -1 \\ 0 & 1 & 0 \\ 0 & 0 & 1 \end{bmatrix}$$

performing $R_3 \rightarrow R_3 + 2R_2$

$$\begin{bmatrix} 1 & 0 & 0 \\ 0 & 1 & 1 \\ 0 & 0 & 0 \end{bmatrix} = \begin{bmatrix} 1 & 0 & 0 \\ \dfrac{1}{2} & -\dfrac{1}{2} & 0 \\ -2 & -1 & 1 \end{bmatrix} A \begin{bmatrix} 1 & -1 & -1 \\ 0 & 1 & 0 \\ 0 & 0 & 1 \end{bmatrix}$$

performing $C_3 \rightarrow C_3 - C_2$

$$\begin{bmatrix} 1 & 0 & 0 \\ 0 & 1 & 0 \\ 0 & 0 & 0 \end{bmatrix} = \begin{bmatrix} 1 & 0 & 0 \\ \dfrac{1}{2} & -\dfrac{1}{2} & 0 \\ -2 & -1 & 1 \end{bmatrix} A \begin{bmatrix} 1 & -1 & 0 \\ 0 & 1 & -1 \\ 0 & 0 & 1 \end{bmatrix}$$

$$\therefore \qquad \begin{pmatrix} I_2 & O \\ O & O \end{pmatrix} = PAQ$$

where $P = \begin{bmatrix} 1 & 0 & 0 \\ \dfrac{1}{2} & -\dfrac{1}{2} & 0 \\ -2 & -1 & 1 \end{bmatrix},\ Q = \begin{bmatrix} 1 & -1 & 0 \\ 0 & 1 & -1 \\ 0 & 0 & 1 \end{bmatrix}$

Hence rank of $A = 2$.

Example 6. *With the help of E-transformation, find the rank of the following matrix :*

$$\begin{bmatrix} 1 & 1 & 2 & 3 \\ 1 & 3 & 0 & 3 \\ 1 & -2 & -3 & -3 \\ 1 & 1 & 2 & 3 \end{bmatrix}$$

Solution. Let $A = \begin{bmatrix} 1 & 1 & 2 & 3 \\ 1 & 3 & 0 & 3 \\ 1 & -2 & -3 & -3 \\ 1 & 1 & 2 & 3 \end{bmatrix}$

performing $R_2 \to R_2 - R_1$, $R_3 \to R_3 - R_1$, $R_4 \to R_4 - R_1$, we get

$$\sim \begin{bmatrix} 1 & 1 & 2 & 3 \\ 0 & 2 & -2 & 0 \\ 0 & -3 & -5 & -6 \\ 0 & 0 & 0 & 0 \end{bmatrix}$$

performing $R_2 \to \dfrac{1}{2} R_2$

$$\sim \begin{bmatrix} 1 & 1 & 2 & 3 \\ 0 & 1 & -1 & 0 \\ 0 & -3 & -5 & -6 \\ 0 & 0 & 0 & 0 \end{bmatrix}$$

performing $R_3 \to R_2 + 3R_2$

$$\sim \begin{bmatrix} 1 & 1 & 2 & 3 \\ 0 & 1 & -1 & 0 \\ 0 & -3 & -5 & -6 \\ 0 & 0 & 0 & 0 \end{bmatrix}$$

This is an *Echelon form* and having three non-zero rows. Hence, the rank of the give matrix = 3.

Example 7. *Find the rank of matrix* $A = \begin{bmatrix} 1 & 1 & 1 \\ a & b & c \\ a^3 & b^3 & c^3 \end{bmatrix}$ *a, b, c being all real numbers.*

Solution. Let

$$|A| = \begin{bmatrix} 1 & 1 & 1 \\ a & b & c \\ a^3 & b^3 & c^3 \end{bmatrix} \quad \begin{matrix} C_2 \to C_2 - C_1 \\ C_3 \to C_3 - C_1 \end{matrix}$$

$$= \begin{bmatrix} 1 & 0 & 0 \\ a & b-a & c-a \\ a^3 & b^3 - a^3 & c^3 - a^3 \end{bmatrix}$$

$$= \begin{vmatrix} b-a & c-a \\ b^3 - a^3 & c^3 - a^3 \end{vmatrix}$$

$$= (b-a)(c-a) \begin{vmatrix} 1 & 1 \\ b^2 + ab + a^2 & c^2 + ca + a^2 \end{vmatrix} \qquad C_2 \to C_2 - C_1$$

$$= (b-a)(c-a) \begin{vmatrix} 1 & 0 \\ b^2 + ab + a^2 & c^2 + ca - b^2 - ab \end{vmatrix}$$

$$= (b-a)(c-a)[(c^2 + ca - b^2 - ab) - 0]$$

$$= (b-a)(c-a)[(c^2 - b^2) + a(c - b)]$$

$$= (b-a)(c-a)[(c-b)(c+b+a)]$$

$\therefore \qquad |A| = (a-b)(b-c)(c-a)(a+b+c).$

Now, following cases arise.

Case I. Let $a = b = c$, then

$$A = \begin{bmatrix} 1 & 1 & 1 \\ a & a & a \\ a^3 & a^3 & a^3 \end{bmatrix}$$

Therefore all minors of order 3 and 2 of A are zero. Also as no element of A is zero, so A has non-zero minors of order 1.

Hence rank $(A) = 1$.

Case II. Let $a = b \neq c$, then

$$|A| = \begin{bmatrix} 1 & 1 & 1 \\ a & a & c \\ a^3 & a^3 & a^3 \end{bmatrix} = 0 \text{ as } C_1 \text{ and } C_2 \text{ are identical.}$$

Also A have a minor of order 2 *viz.*

$$\begin{vmatrix} 1 & 1 \\ a & c \end{vmatrix} = c - a \neq 0$$

Hence, rank $(A) = 2$.

Similarly, we can discuss the cases $b = c \neq a$ and $c = a \neq b$.

Case III. Let a, b, c be all different such that $a + b + c = 0$

By above discussion $|A| = 0$.

Also, A has a minor of order 2

$$\begin{vmatrix} 1 & 1 \\ a & b \end{vmatrix} = b - a \neq 0 \qquad\qquad (\because a \neq b)$$

Case IV. Let a, b, c be all different but $a + b + c \neq 0$. In this case from (case I), it is evident that $|A| \neq 0$ *i.e.*, A has a non-zero minor of order 3.

Also A has no minor of order greater than 3.

Hence, $\qquad r(A) = 3$

EXERCISE 17.3

Determine the rank of the following matrices :

1. $\begin{bmatrix} 1 & 2 & 3 \\ 2 & 1 & 0 \\ 0 & 1 & 2 \end{bmatrix}$

2. $\begin{bmatrix} 1 & 1 & 1 & 1 \\ 1 & 1 & 1 & 1 \\ 1 & 1 & 1 & 1 \\ 1 & 1 & 1 & 1 \end{bmatrix}$

3. $\begin{bmatrix} 1 & 2 & 3 \\ 3 & 4 & 6 \\ 4 & 5 & 6 \end{bmatrix}$

4. $\begin{bmatrix} 1 & 2 & -7 & 5 \\ 0 & 5 & 0 & 8 \\ 0 & 0 & 0 & -8 \end{bmatrix}$

5. $\begin{bmatrix} 2 & -1 & 3 & 4 \\ 0 & 3 & 4 & 1 \\ 2 & 3 & 7 & 5 \\ 2 & 5 & 11 & 6 \end{bmatrix}$

6. $\begin{bmatrix} -2 & -1 & -3 & -1 \\ 1 & 2 & 3 & -1 \\ 1 & 0 & 1 & 1 \\ 0 & 1 & 1 & -1 \end{bmatrix}$

7. $\begin{bmatrix} 2 & 3 & -1 & -1 \\ 1 & -1 & -2 & -4 \\ 3 & 1 & 3 & -2 \\ 6 & 3 & 0 & -7 \end{bmatrix}$

8. $\begin{bmatrix} 1 & a & b & 0 \\ 0 & c & d & 1 \\ 1 & a & b & 0 \\ 0 & c & d & 1 \end{bmatrix}$

9. $\begin{bmatrix} 3 & -2 & 0 & -1 \\ 0 & 2 & 2 & 1 \\ 1 & -2 & -3 & 2 \\ 0 & 1 & 2 & 1 \end{bmatrix}$

10. $\begin{bmatrix} 1 & -1 & 3 & 6 \\ 1 & 3 & -3 & -4 \\ 5 & 3 & 3 & 11 \end{bmatrix}$

11. $\begin{bmatrix} 1 & -3 & 4 & 7 \\ 9 & 1 & 2 & 0 \end{bmatrix}$

12. $\begin{bmatrix} 1 & 0 & 2 & 1 \\ 0 & 1 & -2 & 1 \\ 1 & -1 & 4 & 0 \\ -2 & 2 & 8 & 0 \end{bmatrix}$

13. $\begin{bmatrix} 6 & 1 & 3 & 8 \\ 4 & 2 & 6 & -1 \\ 10 & 3 & 9 & 7 \\ 16 & 4 & 12 & 15 \end{bmatrix}$

14. $\begin{bmatrix} 8 & 0 & 0 & 1 \\ 1 & 0 & 8 & 1 \\ 0 & 0 & 1 & 8 \\ 0 & 1 & 1 & 8 \end{bmatrix}$

15. (a) $\begin{bmatrix} 1 & 2 & -1 & 3 \\ 4 & 1 & 2 & 1 \\ 3 & -1 & 1 & 2 \\ 1 & 2 & 0 & 1 \end{bmatrix}$

(b) $\begin{bmatrix} 1 & 2 & 3 & 0 \\ 2 & 4 & 3 & 2 \\ 3 & 2 & 1 & 3 \\ 6 & 8 & 7 & 5 \end{bmatrix}$

16. Reduce the following matrix to its Echelon form and find its rank :

(a) $\begin{bmatrix} 1 & 3 & 4 & 5 \\ 3 & 9 & 12 & 9 \\ -1 & -3 & -4 & -3 \end{bmatrix}$

(b) $\begin{bmatrix} 5 & 3 & 14 & 4 \\ 0 & 1 & 2 & 1 \\ 1 & -1 & 2 & 0 \end{bmatrix}$

17. Reduce the following matrix to normal form and find its rank :

(a) $\begin{bmatrix} 0 & 1 & 3 & -1 \\ 1 & 0 & 1 & 1 \\ 3 & 1 & 0 & 2 \\ 1 & 1 & -2 & 0 \end{bmatrix}$

(b) $\begin{bmatrix} 1 & 2 & 3 \\ 2 & 4 & 7 \\ 3 & 6 & 10 \end{bmatrix}$

18. Reduce the following matrix to normal form and find its rank :

$$\begin{bmatrix} 2 & -2 & 0 & 6 \\ 4 & 2 & 0 & 2 \\ 1 & -1 & 0 & 3 \\ 1 & -2 & 1 & 2 \end{bmatrix}$$

19. Find the rank of A, B, $A + B$ and $\dot{A}B$, where

$$A = \begin{bmatrix} 1 & 1 & -1 \\ 2 & -3 & 4 \\ 3 & -2 & 3 \end{bmatrix}, \quad B = \begin{bmatrix} -1 & -2 & -1 \\ 6 & 12 & 6 \\ 5 & 10 & 5 \end{bmatrix}$$

20. If A and B are two equivalent matrices, then show that rank A = rank B.

21. Change the following matrix A into normal form and find its rank

$$(i)\ A = \begin{bmatrix} 1 & 2 & -1 & 4 \\ 2 & 4 & 3 & 5 \\ -1 & -2 & 6 & -7 \end{bmatrix} \qquad (ii)\ \begin{bmatrix} 1 & 2 & 3 & 1 \\ 2 & 4 & 6 & 2 \\ 1 & 2 & 3 & 2 \end{bmatrix}$$

HINT TO SELECTED PROBLEMS

1. $A = \begin{bmatrix} 1 & 2 & 3 \\ 2 & 1 & 0 \\ 0 & 1 & 2 \end{bmatrix}$

$\therefore$ det $A = 1\,(2 - 0) - 2\,(4 - 0) + 3\,(2 - 0) = 2 - 8 + 6 = 0$

$\therefore$ Rank $A < 3$.

Now the square submatrices of the given matrix are

$$A_1 = \begin{bmatrix} 1 & 2 \\ 2 & 1 \end{bmatrix}, \ A_2 = \begin{bmatrix} 2 & 3 \\ 1 & 0 \end{bmatrix} \text{ etc.}$$

and   det $A_1 = 1 - 4 = -3 \neq 0$

Hence Rank $A = 2$.

2. $\begin{bmatrix} 1 & 1 & 1 & 1 \\ 1 & 1 & 1 & 1 \\ 1 & 1 & 1 & 1 \\ 1 & 1 & 1 & 1 \end{bmatrix}$

Obviously det $A = 0 \Rightarrow$ Rank $A < 4$

Also determinant values of all square sub matrices order 3×3 and 2×2 are zero. Thus Rank < 2. Since A is not a null matrix. Hence Rank $(A) = 1$.

3. $A = \begin{bmatrix} 1 & 2 & 3 \\ 3 & 4 & 5 \\ 4 & 5 & 6 \end{bmatrix}$

Now   det $A = 1\,(24 - 25) - 2\,(18 - 20) + 3\,(15 - 16)$

$= -1 + 4 - 3$

$= 0$

$\therefore$ Rank $(A) < 3$.

The square submatrices of A are $A_1 = \begin{bmatrix} 1 & 2 \\ 3 & 4 \end{bmatrix}, A_2 = \begin{bmatrix} 2 & 3 \\ 4 & 5 \end{bmatrix}$ etc.

and $$\det (A_1) = 4 - 6 = -2 \neq 0$$
$$\text{Rank } (A) = 2$$

4. Let $A = \begin{bmatrix} 1 & 2 & -7 & 5 \\ 0 & 5 & 0 & 8 \\ 0 & 0 & 0 & -8 \end{bmatrix}$

Since the matrix A is of Echelon form and contains three non-zero rows. Hence Rank of A in 3.

5. $\begin{bmatrix} 2 & -1 & 3 & 4 \\ 0 & 3 & 4 & 1 \\ 2 & 3 & 7 & 5 \\ 2 & 5 & 11 & 6 \end{bmatrix}$

Performing $R_3 \rightarrow R_3 - R_1$, $R_4 \rightarrow R_4 - R_1$

$$\sim \begin{bmatrix} 2 & -1 & 3 & 4 \\ 0 & 3 & 4 & 1 \\ 0 & 4 & 4 & 1 \\ 0 & 6 & 8 & 2 \end{bmatrix}$$

Performing $R_2 \rightarrow \dfrac{1}{3} R_2$

$$\sim \begin{bmatrix} 2 & -1 & 3 & 4 \\ 0 & 1 & 4/3 & 1/3 \\ 0 & 4 & 4 & 1 \\ 0 & 6 & 8 & 2 \end{bmatrix}$$

Performing $R_3 \rightarrow R_3 - 4R_2$, $R_4 \rightarrow R_4 - 6R_1$

$$A \sim \begin{bmatrix} 2 & -1 & 3 & 4 \\ 0 & 1 & 4/3 & 1/3 \\ 0 & 0 & -4/3 & -1/3 \\ 0 & 0 & 0 & 0 \end{bmatrix}$$

Thus A is converted into Echelon form, having 3 non-zero rows. Hence Rank of $A = 3$.

6. Let $A = \begin{bmatrix} -2 & -1 & -3 & -1 \\ 1 & 2 & 3 & -1 \\ 1 & 0 & 1 & 1 \\ 0 & 1 & 1 & -1 \end{bmatrix}$

Then use the following steps :

(1) Performing $R_2 \leftrightarrow R_1$

(2) Performing $R_2 \rightarrow R_2 - 2R_1$, $R_3 \rightarrow R_3 - R_1$

(3) Performing $R_4 \leftrightarrow R_2$

(4) Performing $R_3 \rightarrow R_3 - 2R_2$, $R_4 \rightarrow R_4 - 3R_1$, we get

$$\sim \begin{bmatrix} 1 & 2 & 3 & -1 \\ 0 & 1 & 1 & -1 \\ 0 & 0 & 0 & 0 \\ 0 & 0 & 0 & 0 \end{bmatrix}$$

Then A is converted into Echelon form having two non-zero rows.

Hence Rank of $A = 2$

7. $A = \begin{bmatrix} 2 & 3 & -1 & -1 \\ 1 & -1 & -2 & -4 \\ 3 & 1 & 3 & -2 \\ 6 & 3 & 0 & -7 \end{bmatrix}$

Then using the following steps :

(1) Performing $R_2 \leftrightarrow R_1$

(2) Performing $R_2 \rightarrow R_2 - 2R_1$, $R_3 \rightarrow R_3 - 3R_1$, $R_4 \rightarrow R_4 - 6R_1$

(3) Performing $R_2 \rightarrow \dfrac{1}{5} R_2$

(4) Performing $R_3 \rightarrow R_3 - 4R_2$, $R_4 \rightarrow R_4 - 9R_2$

(5) Finally performing $R_4 \rightarrow R_4 - R_3$

$$A \sim \begin{bmatrix} 1 & -1 & -2 & -4 \\ 0 & 1 & 3/5 & 7/5 \\ 0 & 0 & 33/5 & 22/5 \\ 0 & 0 & 0 & 0 \end{bmatrix}$$

Then A is converted into Echelon form having 3 non-zero rows.

Hence Rank of $A = 3$.

8. $A = \begin{bmatrix} 1 & a & b & 0 \\ 0 & c & d & 1 \\ 1 & a & b & 0 \\ 0 & c & d & 1 \end{bmatrix}$

Here $R_1 = R_3$ and $R_2 = R_4$

Let $A = 0 \Rightarrow$ Rank of $A < 4$.

Also, all the square submatrices of order 3×3 having two rows or two columns identical so that their determinant value will zero. Thus rank of $A < 3$. Further the square submatrices of order 2×2 are

$$A_1 = \begin{bmatrix} 1 & a \\ 0 & 0 \end{bmatrix}, A_2 = \begin{bmatrix} a & b \\ c & d \end{bmatrix} \text{ etc.}$$

and det $A_1 = c - 0 = c \neq 0$

Hence Rank of $A = 2$.

10. Let us assume $\quad A = \begin{bmatrix} 2 & -2 & 0 & 6 \\ 4 & 2 & 0 & 2 \\ 1 & -1 & 0 & 3 \\ 1 & -2 & 1 & 2 \end{bmatrix}$.

Performing $R_3 \leftrightarrow R_1$

$$\begin{bmatrix} 1 & -1 & 0 & 3 \\ 4 & 2 & 0 & 2 \\ 1 & -1 & 0 & 3 \\ 1 & -2 & 1 & 2 \end{bmatrix}$$

Then, using the following steps :

(1)   Performing $R_2 \to R_2 - 4R_1,\ R_3 \to R_3 - 2R_1,\ R_4 \to R_4 - R_1$
(2)   Performing $R_4 \leftrightarrow R_2$
(3)   Performing $C_2 \to C_2 + C_1,\ C_4 \to C_4 - 3C_1$
(4)   Performing $R_4 \to R_4 + 6R_2$
(5)   Performing $C_3 \to C_3 + C_2,\ C_4 \to C_4 - C_2$
(6)   Performing $R_3 \leftrightarrow R_4$

(7)   Performing $R_3 \to \dfrac{1}{6} R_3$

(8)   Performing $C_4 \to C_4 + \dfrac{8}{3} C_3$

(9)   Finally performing $R_2 \to (-1)R_1,$ we get $A \sim \begin{bmatrix} 1 & 0 & 0 & 0 \\ 0 & 1 & 0 & 0 \\ 0 & 0 & 1 & 0 \\ 0 & 0 & 0 & 0 \end{bmatrix}$

or $\qquad\qquad\qquad A \sim \begin{pmatrix} I_3 & 0 \\ 0 & 0 \end{pmatrix}$

Then A is reduced to normal form and hence the rank of $A = 3$.

19. Rank of A :

$$\det A = 1\,(-9+8) - 1\,(6-12) - 1\,(-4+9) = -1 + 6 - 5 = 0$$

$\Rightarrow$ Rank of $(A) < 3$.

The square submatrices of A are

$$A_1 = \begin{bmatrix} 1 & 1 \\ 2 & -3 \end{bmatrix}, A_2 = \begin{bmatrix} 1 & -1 \\ -3 & 4 \end{bmatrix} \text{ etc.}$$

and $\qquad\qquad \det A_1 = -3 - 2 = -5 \neq 0 \ \Rightarrow\ $ Rank $(A) = 2$.

Hence $\qquad\qquad$ Rank $A = 2$

Now $\qquad\qquad\qquad B = \begin{bmatrix} -1 & -2 & -1 \\ 6 & 12 & 6 \\ 5 & 10 & 5 \end{bmatrix}$

Performing $R_2 \to R_2 + 6R_1,\ R_3 \to R_3 + 5R_1$

$$\sim \begin{bmatrix} -1 & -2 & -1 \\ 0 & 0 & 0 \\ 0 & 0 & 0 \end{bmatrix}$$

$\therefore$ B is reduced to Echelon form, which is having one non-zero row.

Hence, Rank of $B = 1$.

Further

$$A + B = \begin{bmatrix} 1 & 1 & -1 \\ 2 & -3 & 4 \\ 3 & -2 & 3 \end{bmatrix} + \begin{bmatrix} -1 & -2 & -1 \\ 6 & 12 & 6 \\ 5' & 10 & 5 \end{bmatrix} = \begin{bmatrix} 0 & -1 & -2 \\ 8 & 9 & 10 \\ 8 & 8 & 8 \end{bmatrix}$$

$\therefore$

$$\det (A + B) = -1\,(80 - 64) - 2\,(64 - 72)$$
$$= -16 + 16 = 0$$

Rank of $(A + B) < 3$

Now the square submatrices of $(A + B)$ are

$$(A + B) = \begin{bmatrix} 0 & -1 \\ 8 & 9 \end{bmatrix}; \begin{bmatrix} 0 & -2 \\ 8 & 10 \end{bmatrix} \text{etc.}$$

$$\det (A + B) = 0 + 8 = 8 \neq 0$$

$\Rightarrow$ Rank of $(A + B) = 2$.

Next, $AB = \begin{bmatrix} 1 & 1 & -1 \\ 2 & -3 & 4 \\ 3 & -2 & 3 \end{bmatrix} \begin{bmatrix} -1 & -2 & -1 \\ 6 & 12 & 6 \\ 5 & 10 & 5 \end{bmatrix} = \begin{bmatrix} 0 & 0 & 0 \\ 0 & 0 & 0 \\ 0 & 0 & 0 \end{bmatrix}$

$\therefore$ AB is Null matrix, then Rank of $(AB) = 0$

20. Since A and B are equivalent. Then by definition B is obtained by a finite chain E-transformation applied on 'A' and vise-versa and further since we know that E-transformation do not change the rank. Hence

$$\text{Rank } A = \text{Rank } B$$

21. Since we have

$$A = \begin{bmatrix} 1 & 2 & -1 & 4 \\ 2 & 4 & 3 & 5 \\ -1 & -2 & 6 & -7 \end{bmatrix}$$

Performing $R_2 \to R_2 - 2R_1, R_3 \to R_3 + R_1$

$$A \sim \begin{bmatrix} 1 & 2 & -1 & 4 \\ 0 & 0 & 5 & -3 \\ 0 & 0 & 5 & -3 \end{bmatrix}$$

Performing $C_2 \to C_2 - 2C_1, C_3 \to C_3 + C_1, C_4 \to C_4 - 4C_1$

$$\sim \begin{bmatrix} 1 & 0 & 0 & 0 \\ 0 & 0 & 5 & -3 \\ 0 & 0 & 5 & -3 \end{bmatrix}$$

Performing $R_3 \rightarrow R_3 - R_2$

$$A \sim \begin{bmatrix} 1 & 0 & 0 & 0 \\ 0 & 0 & 5 & -3 \\ 0 & 0 & 0 & 0 \end{bmatrix}$$

Performing $R_2 \rightarrow \dfrac{1}{5} R_2$

$$\sim \begin{bmatrix} 1 & 0 & 0 & 0 \\ 0 & 0 & 1 & -3/5 \\ 0 & 0 & 0 & 0 \end{bmatrix}$$

Performing $C_4 \rightarrow C_4 + \dfrac{3}{5} C_3$

$$A \sim \begin{bmatrix} 1 & 0 & 0 & 0 \\ 0 & 1 & 0 & 0 \\ 0 & 0 & 0 & 0 \end{bmatrix}$$

Performing $C_2 \leftrightarrow C_3$

or

$$A \sim \begin{pmatrix} I_2 & 0 \\ 0 & 0 \end{pmatrix}$$

Thus A is reduced to Normal form and hence rank of $A = 2$.

ANSWERS

(1) 2   **(2)** 1   **(3)** 2   **(4)** 3   **(5)** 3

(6) 2   **(7)** 3   **(8)** 2   **(9)** 4   **(10)** 3

(11) 2   **(12)** 3   **(13)** 2   **(14)** 4   **(15)** (a) 3   (b) 3

(16) (a) 2 (b) 3   **(17)** (a) 3   (b) 2   **(18)** 3

(19) Rank $A = 2$, Rank $B = 1$, Rank $(A + B) =$ Rank $/AB = 0$

(21) (i) 2 (ii) 1.

17.26   LINEAR EQUATIONS

In this section we shall study two types of linear equations :

(1) Homogeneous linear equation

(2) Non-Homogeneous linear equation.

17.26.1 Homogeneous Linear Equations

Let us consider a system of linear homogeneous equation as follows :

$$\left.\begin{array}{l} a_{11}x_1 + a_{12}x_2 + \ldots + a_{1n}x_n = 0 \\ a_{21}x_2 + a_{22}x_2 + \ldots + a_{2n}x_n = 0 \\ \ldots\ldots\ldots\ldots\ldots\ldots\ldots\ldots\ldots\ldots\ldots \\ a_{m1}x_1 + a_{m2}x_2 + \ldots + a_{mn}x_n = 0 \end{array}\right\} \qquad \ldots(1)$$

These are m equation in n unknown. Any set of numbers $x_1, x_2, \ldots x_n$ that satisfies all the equation (1) is called a solution of (1).

17.26.2 Trivial Solution

The solution $x_1 = 0$, $x_2 = 0$, ... $x_n = 0$ of the equation (1) are called trivial solution.

17.26.3 Non-Trivial Solution

Any solution other than trivial, if exist, is called non-trivial solution of equations (1). Let the coefficient matrix be

$$A = \begin{bmatrix} a_{11} & a_{12} & \cdots & a_{1n} \\ a_{21} & a_{22} & \cdots & a_{2n} \\ \cdots & \cdots & \cdots & \cdots \\ a_{m1} & a_{m2} & \cdots & a_{mn} \end{bmatrix}_{m \times n} \quad \text{and} \quad X = \begin{bmatrix} x_1 \\ x_2 \\ x_3 \\ \cdots \\ x_n \end{bmatrix}_{n \times 1}, \quad 0 = \begin{bmatrix} 0 \\ 0 \\ 0 \\ \cdots \\ \cdots \\ 0 \end{bmatrix}_{m \times 1}$$

Then the equation (1) can also be written as
$$AX = 0 \qquad \qquad \text{...(2)}$$
This equation (2) is called a matrix equation.

THEOREM-1

If X_1 and X_2 are two non-trivial solutions of (2), then $k_1 X_1 + k_2 X_2$ is also a solution of (2), where k_1 and k_2 are any arbitrary numbers.

Proof. Since the equation (2) is $AX = 0$

and $\quad AX_1 = 0$, $AX_2 = 0$ are given.

Now consider $A(k_1 X_1 + k_2 X_2) = k_1 (AX_1) + k_2 (AX_2) = k_1(0) + k_2(0)$.

Hence, $k_1 X_1 + k_2 X_2$ is the solution of (2).

17.27 NATURE OF THE SOLUTION OF EQUATION AX = 0

Since $AX = 0$ is a matrix equation of a system of m homogeneous linear equations in n unknowns and A is a coefficients matrix of order $m \times n$. Let the rank of A be r. Then obviously r can not be greater than n. So, that either r is n or is less than n. Therefore there are following cases :

Case I : If $r = n$, then the equation $AX = 0$ will have no linearly independent solution. So, in this case only trivial solution will exist.

Case II : If $r < n$, then there will be $(n - r)$ linearly independent solution of $AX = 0$ and thus, in this case we shall have infinite solutions.

Case III : Suppose the number of equation are less than number of unknowns i.e., $m < n$ and since $r \le m$, then obviously $r < n$ thus, in this case a non-zero solution will exist. Therefore the equation $AX = 0$ will have infinite solution.

SOLVED EXAMPLES

Example 1. *Find the solution of the following system of linear homogeneous equations.*
$$2x_1 - x_2 + x_3 = 0$$
$$3x_1 + 2x_2 + x_3 = 0$$
$$x_1 - 3x_2 + 5x_3 = 0$$

Solution. The given equation in matrix form $AX = B$ can be written as

$$\begin{bmatrix} 2 & -1 & 1 \\ 3 & 2 & 1 \\ 1 & -1 & 5 \end{bmatrix} \begin{bmatrix} x_1 \\ x_2 \\ x_3 \end{bmatrix} = \begin{bmatrix} 0 \\ 0 \\ 0 \end{bmatrix}$$

Therefore,
$$A = \begin{bmatrix} 2 & -1 & 1 \\ 3 & 2 & 1 \\ 1 & -3 & 5 \end{bmatrix}$$

Performing $R_2 \rightarrow 2R_2 - 3R_1$, $R_3 \rightarrow 2R_3 - R_1$

$$\begin{bmatrix} 2 & -1 & 1 \\ 0 & 7 & -1 \\ 0 & -5 & 9 \end{bmatrix}$$

Now, $R_3 \rightarrow 7R_3 + 5R_2$

$$\Rightarrow \qquad \begin{bmatrix} 2 & -1 & 1 \\ 0 & 7 & -1 \\ 0 & 0 & 58 \end{bmatrix} \begin{bmatrix} x_1 \\ x_2 \\ x_3 \end{bmatrix} = \begin{bmatrix} 0 \\ 0 \\ 0 \end{bmatrix}$$

Hence rank of $A = 3$, *i.e.*, $r = 3$.

So trivial solution exists which is given by

$$2x_1 - x_2 + x_3 = 0$$
$$7x_2 - x_3 = 0$$
$$58x_3 = 0$$
$$2x_1 = 0 \quad \Rightarrow \quad x_1 = 0$$
$$7x_2 = 0 \quad \Rightarrow \quad x_2 = 0$$
$$x_3 = \frac{0}{58} = 0$$
$$x_1 = 0, \ x_2 = 0 \ \ x_3 = 0$$

Also, the rank of A is 3 and equal to the number of unknowns x_1, x_2 and x_3. Hence, $x_1 = x_2 = x_3 = 0$.

Example 2. *Find the solution of the following system of linear homogeneous equations :*

$$x + y + z = 0$$
$$2x - y - 3z = 0$$
$$3x - 5y + 4z = 0$$
$$x + 17y + 4z = 0$$

Solution. The coefficient matrix is given by

$$A = \begin{bmatrix} 1 & 1 & 1 \\ 2 & -1 & -3 \\ 3 & -5 & 4 \\ 1 & 17 & 4 \end{bmatrix}$$

First reduce A into Echelon form

Performing $R_2 \to R_2 - 2R_1$, $R_3 \to R_4 - 3R_1$, $R_4 \to R_4 - R_1$

$$\sim \begin{bmatrix} 1 & 1 & 1 \\ 0 & -3 & -5 \\ 0 & -8 & 1 \\ 0 & 16 & 3 \end{bmatrix}$$

Performing $R_2 \to \dfrac{1}{3} R_2$, we get

$$\sim \begin{bmatrix} 1 & 1 & 1 \\ 0 & 1 & \dfrac{5}{3} \\ 0 & -8 & 1 \\ 0 & 16 & 3 \end{bmatrix}$$

Performing $R_3 \to R_3 + 8R_2$, $R_4 \to R_4 - 16R_4$, we get

$$\sim \begin{bmatrix} 1 & 1 & 1 \\ 0 & 1 & \dfrac{5}{3} \\ 0 & 0 & \dfrac{43}{3} \\ 0 & 0 & \dfrac{71}{3} \end{bmatrix}$$

Performing $R_3 \to \dfrac{3}{43} R_3$, we get

$$\sim \begin{bmatrix} 1 & 1 & 1 \\ 0 & 1 & \dfrac{5}{3} \\ 0 & 0 & 1 \\ 0 & 0 & -\dfrac{71}{3} \end{bmatrix}$$

Performing $R_4 \to R_4 + \dfrac{71}{3} R_3$

$$\sim \begin{bmatrix} 1 & 1 & 1 \\ 0 & 1 & \dfrac{5}{3} \\ 0 & 0 & 1 \\ 0 & 0 & 0 \end{bmatrix}$$

This is an Echelon form and having three non-zero rows so A has the rank 3.

Since there are 3 number of unknown, hence a trivial solution exists *i.e.*, $x = 0$, $y = 0$, $z = 0$ is the only solution.

17.28   **NON-HOMOGENEOUS EQUATIONS**

Let us consider a system of equation which are non-homogeneous as follows :

$$\left.\begin{array}{l} a_{11}x_1 + a_{12}x_2 + ... + a_{1n}x_n = b_1 \\ a_{21}x_2 + a_{22}x_2 + ... + a_{2n}x_n = b_2 \\ .... \\ a_{m1}x_1 + a_{m2}x_2 + ... + a_{mn}x_n = b_n \end{array}\right\} \qquad ...(1)$$

There are m equations in n unknown. Let

$$A = \begin{bmatrix} a_{11} & a_{12} & ... & a_{1n} \\ a_{21} & a_{22} & ... & a_{2n} \\ ... & ... & ... & ... \\ a_{m1} & a_{m2} & ... & a_{mn} \end{bmatrix}_{m \times n} \qquad X = \begin{bmatrix} x_1 \\ x_2 \\ ... \\ x_n \end{bmatrix}_{n \times 1} , \quad B = \begin{bmatrix} b_1 \\ b_2 \\ ... \\ b_n \end{bmatrix}_{m \times 1}$$

Then the system of equations (1) can also be written as

$$AX = B \qquad ...(2)$$

This equation is called a matrix equation. If $x_1, x_2, ... x_n$ simultaneously satisfy the equation (2), then $(x_1, x_2, ... x_n)$ is called the solution of (2).

17.28.1 Consistancy and inconsistancy : When there will exist one or more than one solution of the equation $AX = B$. Then the equations are said to be consistant otherwise said to be inconsistent.

17.28.2 Augmented matrix : The matrix of the type

$$[A \mid B] = \begin{bmatrix} a_{11} & a_{12} & ... & a_{1n} & b_1 \\ a_{21} & a_{22} & ... & a_{2n} & b_2 \\ ... & ... & ... & ... & ... \\ a_{m1} & a_{m2} & ... & a_{mn} & b_m \end{bmatrix}$$

is called the augmented matrix of the equations.

17.29   **CONDITION FOR CONSISTENCY**

THEOREM-1

The equation $AX = B$ is consistent if and only if the rank of A and the rank of the augmented matrix $[A \mid B]$ are same.

Proof. Since the equation is $AX = B$ $\qquad ...(1)$

The matrix A can be written as $A = [C_1, C_2, ... C_n]$

where $C_1, C_2 ... C_n$ are column vectors. Then the equation (1) can be written as

$$[C_1, C_2, ... C_n] \begin{bmatrix} x_1 \\ x_2 \\ ... \\ x_n \end{bmatrix} = B$$

or $$x_1C_1 + x_2C_2 + ... + x_nC_n = B \qquad ...(2)$$

Suppose the rank of A is r, then A has r linearly independent columns. Let these columns be $C_1, C_2, ... C_r$ and these $C_1, C_2, ... C_r$ are linearly independent and remaining $(n - r)$ columns are in linear combination of $C_1, C_2, ... C_r$.

Necessary condition : Suppose the equation are consistent, there must exist $k_1 k_2, ... k_r$ such that

$$k_1C_1 + k_2C_2 + ... + k_nC_n = B \qquad ...(3)$$

But $C_{r+1}, C_{r+2}, ..., C_n$ is linear combination of $C_1, C_2, ... C_r$, then from (2) it is obvious that B is also a linear combination of $C_1, C_2, ... C_r$ and thus $[A/B]$ has the rank r. Hence, the rank of A is same as the rank of $[A/B]$.

Sufficient condition : Suppose rank A = rank $[A/B]$ = r. This implies that $[A/B]$ has r linearly independent columns. But $C_1, C_2, ... C_r$ of $[A/B]$ are already linearly independent.

Thus B can be expressed as

$$B = k_1C_1 + k_2C_2 + ... + k_rC_r \qquad ...(4)$$

where $k_1, k_2, ... k_r$ are scalers.

Now, equation (4) becomes

$$B = k_1C_1 + k_2C_2 + ... + k_rC_r + 0.\,C_{r+1} + ... + 0.\,C_n \qquad ...(5)$$

Comparing (2) and (5), we get

$$x_1 = k_1, x_2 = k_2, ... x_r = k_r, x_{r+1} = 0, ... x_n = 0$$

and these values of $x_1, x_2, ... x_n$ are the solution of $AX = B$. Hence, the equations are consistent.

17.30 CONDITIONS FOR A SYSTEM OF n-EQUATIONS IN n-UNKNOWNS TO HAVE A UNIQUE SOLUTION

THEOREM-1

If A be an n-rowed non-singular matrix, X be an n × 1 matrix, B be an n × 1 matrix, then the system of equation AX = B has a unique solution.

Proof. If A be an n-rowed non-singular matrix, the ranks of matrices A and $[A/B]$ both is n. Therefore the system of equations $AX = B$ is consistant.

Multiplying both sides of $AX = B$ by A^{-1}, we have
$$A^{-1}AX = A^{-1}B. \text{ or } IX = A^{-1}B.$$

or $X = A^{-1}B$ is a solution of the equation $AX = B$.

To show that the solution is unique, let us suppose that X_1 and X_2 be two solutions of $AX = B$.

Then $$AX_1 = B, AX_2 = B \implies AX_1 = AX_2$$
$$\implies \qquad A^{-1}AX_1 = A^{-1}AX_2$$
$$\implies \qquad IX_1 = IX_2 \implies X_1 = X_2.$$

Hence, the solution is unique.

Remarks

❖ If rank A < rank of $[A/B]$, then there is no solution.
❖ If $r = n$, then there will be a unique solution.

❖ If $r < n$, then $(n - r)$ variables can be assigned arbitrary values. Thus there will be infinite solution and $(n - r + 1)$ solution will be linearly independent.

❖ If $m < n$ and $r \leq m < n$, then equation will have infinite solutions.

SOLVED EXAMPLES

Example 1. *Express the following system of equations in matrix form*
$$9x + 7y + 3z = 6;\ 5x + y + 4z = 1;\ 6x + 8y + 2z = 4.$$

Solution. The given equations are
$$9x + 7y + 3z = 6$$
$$5x + y + 4z = 1$$
$$6x + 8y + 2z = 6$$

∴   The required matrix form of these equations is given by $AX = B$

where $\quad A = \begin{bmatrix} 9 & 7 & 3 \\ 5 & 1 & 4 \\ 6 & 8 & 2 \end{bmatrix};\ X = \begin{bmatrix} x \\ y \\ z \end{bmatrix}$ and $B = \begin{bmatrix} 6 \\ 1 \\ 4 \end{bmatrix}.$

Example 2. *Solve by the matrix method*
$$x + y + z = 6,\ x - y + z = 2,\ 2x + y - z = 1.$$

Solution. The given equations are
$$x + y + z = 6$$
$$x - y + z = 2$$
$$2x + y - z = 1$$

Let $\quad A = \begin{bmatrix} 1 & 1 & 1 \\ 1 & -1 & 1 \\ 2 & 1 & -1 \end{bmatrix},\ B = \begin{bmatrix} 6 \\ 2 \\ 1 \end{bmatrix}$

and assume that there exists a matrix $X = \begin{bmatrix} x \\ y \\ z \end{bmatrix}$ such that $AX = B$.

Then $\quad \begin{bmatrix} 1 & 1 & 1 \\ 1 & -1 & 1 \\ 2 & 1 & -1 \end{bmatrix} \begin{bmatrix} x \\ y \\ z \end{bmatrix} = \begin{bmatrix} 6 \\ 2 \\ 1 \end{bmatrix}$

Performing $R_2 \rightarrow R_2 - R_1$ and $R_3 \rightarrow R_3 - 2R_1$,

we get $\quad \begin{bmatrix} 1 & 1 & 1 \\ 0 & -2 & 0 \\ 0 & -1 & -3 \end{bmatrix} \begin{bmatrix} x \\ y \\ z \end{bmatrix} = \begin{bmatrix} 6 \\ -4 \\ -11 \end{bmatrix}$

Performing $R_1 \rightarrow R_1 + R_3$

$\Rightarrow \quad \begin{bmatrix} 1 & 0 & -2 \\ 0 & -2 & 0 \\ 0 & -1 & -3 \end{bmatrix} \begin{bmatrix} x \\ y \\ z \end{bmatrix} = \begin{bmatrix} -5 \\ -4 \\ -11 \end{bmatrix}$

$\Rightarrow \quad x - 2z = -5,\ -2y = -4,\ -y - 3z = 11$

$\Rightarrow \qquad\qquad x - 2z = -5, \; y = 2, \; 2 + 3z = 11$

$\Rightarrow \qquad\qquad x = 2z - 5, \; y = 2, \; z = 3$

$\Rightarrow \qquad\qquad x = 1, \; y = 2, \; z = 3.$

Example 3. *Solve the following equations by matrix method :*

$$x + y + z = 9$$
$$2x + 5y + 7z = 52$$
$$2x + y - z = 0$$

Solution. The given equation can be written as

$$\begin{bmatrix} 1 & 1 & 1 \\ 2 & 5 & 7 \\ 2 & 1 & -1 \end{bmatrix} \begin{bmatrix} x \\ y \\ z \end{bmatrix} = \begin{bmatrix} 9 \\ 52 \\ 0 \end{bmatrix}$$

i.e., $\qquad\qquad AX = B$.

∴ Augmented matrix is

$$[A \,|\, B] = \begin{bmatrix} 1 & 1 & 1 & : & 9 \\ 2 & 5 & 7 & : & 52 \\ 2 & 1 & -1 & : & 0 \end{bmatrix}$$

Performing $R_2 \to R_2 - 2R_1$ and $R_3 \to R_3 - 2R_1$, we get

$$[A \,|\, B] \sim \begin{bmatrix} 1 & 1 & 1 \\ 0 & 3 & 5 \\ 0 & -1 & -3 \end{bmatrix} \begin{bmatrix} x \\ y \\ z \end{bmatrix} = \begin{bmatrix} 9 \\ 34 \\ -18 \end{bmatrix}$$

Performing $R_2 \to R_1 + 3R_3$

$$\Rightarrow \qquad\qquad \sim \begin{bmatrix} 1 & 1 & 1 \\ 0 & 0 & -4 \\ 0 & -1 & -3 \end{bmatrix} \begin{bmatrix} x \\ y \\ z \end{bmatrix} = \begin{bmatrix} 9 \\ -20 \\ -18 \end{bmatrix}$$

The matrix equation is equivalent to the equations

$$x + y + z = 9$$
$$- 4z = - 20$$
$$- y - 3z = - 18$$

which gives $x = 1, y = 3, z = 5$.

Example 4. *Solve the following equations by matrix method*

$$x - 2y + 3z = 6$$
$$3x + y - 4z = - 7$$
$$5x - 3y + 2z = 5$$

Solution. The given equation can be written as

$$\begin{bmatrix} 1 & -2 & 3 \\ 3 & 1 & -4 \\ 5 & -3 & 2 \end{bmatrix} \begin{bmatrix} x \\ y \\ z \end{bmatrix} = \begin{bmatrix} 6 \\ -7 \\ 5 \end{bmatrix}$$

i.e., $AX = B$

∴ Augmented matrix is

$$[A \mid B] = \begin{bmatrix} 1 & -2 & 3 & : & 6 \\ 3 & 1 & -4 & : & -7 \\ 5 & -3 & 2 & : & 5 \end{bmatrix}$$

Performing $R_2 \to R_2 - 3R_1$ and $R_3 \to R_3 - 5R_1$, we get

$$[A \mid B] \sim \begin{bmatrix} 1 & -2 & 3 & : & 6 \\ 0 & 7 & -13 & : & -25 \\ 0 & 7 & -13 & : & -25 \end{bmatrix}$$

Performing $R_3 \to R_3 - R_2$

$$\sim \begin{bmatrix} 1 & -2 & 3 & : & 6 \\ 0 & 7 & -13 & : & -25 \\ 0 & 0 & 0 & : & 0 \end{bmatrix}$$

This is an Echelon form and having two non-zero rows and rank A = rank $[A \mid B]$ = 2. Thus, the equation are consistant.

$$\begin{bmatrix} 1 & -2 & 3 \\ 0 & 7 & -13 \\ 0 & 0 & 0 \end{bmatrix} \begin{bmatrix} x \\ y \\ z \end{bmatrix} = \begin{bmatrix} 6 \\ -25 \\ 0 \end{bmatrix}$$

i.e., $x - 2y + 3z = 6$

$$7y - 13z = -25$$

Let $z = c$ then

$$y = -\frac{25}{7} + \frac{13}{7}c \quad \text{and} \quad x = -\frac{8}{7} + \frac{5}{7}c.$$

Hence, the solution is

$$x = -\frac{8}{7} + \frac{5}{7}c, \, y = -\frac{25}{7} + \frac{13}{7}c, \, z = c, \text{where } c \text{ is an arbitrary constant.}$$

Example 5. *Solve the following equations by matrix method :*

$$2x + 3y + z = 9$$
$$x + 2y + 3z = 6$$
$$3x + y + 2z = 8$$

Solution. The given equation can be written as

$$\begin{bmatrix} 2 & 3 & 1 \\ 1 & 2 & 3 \\ 3 & 1 & 2 \end{bmatrix} \begin{bmatrix} x \\ y \\ z \end{bmatrix} = \begin{bmatrix} 9 \\ 6 \\ 8 \end{bmatrix}$$

i.e., $AX = B$

∴ Augmented matrix is $$[A \mid B] = \begin{bmatrix} 2 & 3 & 1 & : & 9 \\ 1 & 2 & 3 & : & 6 \\ 3 & 1 & 2 & : & 8 \end{bmatrix}$$

Performing $R_1 \rightarrow R_1 - 2R_1$ and $R_3 \rightarrow R_3 - 3R_2$, we get

$$\sim \begin{bmatrix} 0 & -1 & -5 & : & -3 \\ 1 & 2 & 3 & : & 6 \\ 0 & -5 & -7 & : & -10 \end{bmatrix}$$

Performing $R_3 \rightarrow R_3 - 5R_1$, we get

$$\sim \begin{bmatrix} 0 & -1 & -5 & : & -3 \\ 1 & 2 & 3 & : & 6 \\ 0 & 0 & 18 & : & 5 \end{bmatrix}$$

which gives
$$-y - z = -3$$
$$x + 2y + 3z = 6$$
$$18z = 5$$

Hence, the solution is
$$x = \frac{35}{18}, y = \frac{29}{18}, z = \frac{5}{18}.$$

Example 6. *Show that the equation*
$$x + 2y - z = 3, 3x - y + 2z = 1, \ 2x - 2y + 3z = 2, \ x - y + z = -1$$
are consistant and solve them.

Solution. The given equation can be written as

$$\begin{bmatrix} 1 & 2 & -1 \\ 3 & -1 & 2 \\ 2 & -2 & 3 \\ 1 & -1 & 1 \end{bmatrix} \begin{bmatrix} x \\ y \\ z \end{bmatrix} = \begin{bmatrix} 3 \\ 1 \\ 2 \\ -1 \end{bmatrix}$$

i.e., $\qquad\qquad AX = B$

Therefore augmented matrix is

$$[A|B] = \begin{bmatrix} 1 & 2 & -1 & : & 3 \\ 3 & -1 & 2 & : & 1 \\ 2 & -2 & 3 & : & 2 \\ 1 & -1 & 1 & : & -1 \end{bmatrix}$$

Performing $R_2 \rightarrow R_1 - 3R_1, R_3 \rightarrow R_3 - 2R_1, R_4 \rightarrow R_4 - R_1$, we get

$$[A|B] \sim \begin{bmatrix} 1 & 2 & -1 & : & 3 \\ 0 & -7 & 5 & : & -8 \\ 0 & -6 & 5 & : & -4 \\ 0 & -3 & 2 & : & -4 \end{bmatrix}$$

Performing $R_2 \rightarrow R_2 - R_3$

$$\sim \begin{bmatrix} 1 & 2 & -1 & : & 3 \\ 0 & -1 & 0 & : & -4 \\ 0 & -6 & 5 & : & -4 \\ 0 & -3 & 2 & : & -4 \end{bmatrix}$$

Performing $R_3 \to R_3 - 6R_2$, $R_4 \to R_4 - 3R_2$, we get

$$\sim \begin{bmatrix} 1 & 2 & -1 & : & 3 \\ 0 & -1 & 0 & : & -4 \\ 0 & 0 & 5 & : & 20 \\ 0 & 0 & 2 & : & 8 \end{bmatrix}$$

Performing $R_3 \to \dfrac{1}{5} R_3$, $R_4 \to \dfrac{1}{2} R_4$

$$\sim \begin{bmatrix} 1 & 2 & -1 & : & 3 \\ 0 & -1 & 0 & : & -3 \\ 0 & 0 & 1 & : & 4 \\ 0 & 0 & 1 & : & 4 \end{bmatrix}$$

Performing $R_4 \to R_1 - R_3$

$$\sim \begin{bmatrix} 1 & 2 & -1 & : & 3 \\ 0 & -1 & 0 & : & -4 \\ 0 & 0 & 1 & : & 4 \\ 0 & 0 & 0 & : & 0 \end{bmatrix}$$

This is an Echelon form and having three non-zero rows. Thus rank A = rank of $[A \mid B]$ = 3. Therefore, the equations are consistent

and

$$\begin{bmatrix} 1 & 2 & -1 \\ 0 & -1 & 0 \\ 0 & 0 & 1 \\ 0 & 0 & 0 \end{bmatrix} \begin{bmatrix} x \\ y \\ z \end{bmatrix} = \begin{bmatrix} 3 \\ -3 \\ 4 \\ 0 \end{bmatrix}$$

$\therefore \qquad x - 2y - z = 3, -y = -4, z = 4.$

Hence, the solution is $x = -1, y = 4, z = 4$.

Example 7. *If the system of following equations is consistent then find the solution :*

$$x + y + 4z = 6,$$
$$3x - 2y - 2z = 9,$$
$$5x + y + 2z = 13$$

Solution. The given equation can be written as

$$\begin{bmatrix} 1 & 1 & 4 \\ 3 & 2 & -2 \\ 5 & 1 & 2 \end{bmatrix} \begin{bmatrix} x \\ y \\ z \end{bmatrix} = \begin{bmatrix} 6 \\ 9 \\ 13 \end{bmatrix}$$

i.e., $\qquad\qquad\qquad AX = B$

$\therefore$ augmented matrix is

$$[A \mid B] = \begin{bmatrix} 1 & 1 & 4 & : & 6 \\ 3 & 2 & -2 & : & 9 \\ 5 & 1 & 2 & : & 13 \end{bmatrix}$$

Performing $R_2 \rightarrow R_2 - 3R_1, R_3 \rightarrow R_3 - R_1$

$$\sim \begin{bmatrix} 1 & 1 & 4 & : & 6 \\ 0 & -1 & -14 & : & -9 \\ 4 & 0 & -2 & : & 7 \end{bmatrix}$$

Performing $R_3 \rightarrow R_3 - 4R_1$, we get

$$\sim \begin{bmatrix} 1 & 1 & 4 & : & 6 \\ 0 & -1 & -14 & : & -9 \\ 0 & -4 & -18 & : & -17 \end{bmatrix}$$

Performing $R_3 \rightarrow R_3 - 4R_2$, we get

$$\sim \begin{bmatrix} 1 & 1 & 4 & : & 6 \\ 0 & -1 & -14 & : & -9 \\ 0 & 0 & 38 & : & 19 \end{bmatrix}$$

Performing $R_3 \rightarrow \dfrac{1}{38} R_3$, we get

$$\sim \begin{bmatrix} 1 & 1 & 4 & : & 6 \\ 0 & -1 & -14 & : & -9 \\ 0 & 0 & 1 & : & 1/2 \end{bmatrix}$$

Performing $R_1 \rightarrow R_1 - 4R_3, R_2 \rightarrow R_2 + 14R_3$, we get

$$\sim \begin{bmatrix} 1 & 0 & 0 & : & 4 \\ 0 & -1 & 0 & : & -2 \\ 0 & 0 & 1 & : & 1/2 \end{bmatrix}$$

Performing $R_1 \rightarrow R_1 + R_2$, we get

$$\sim \begin{bmatrix} 1 & 0 & 0 & : & 2 \\ 0 & -1 & 0 & : & -2 \\ 0 & 0 & 1 & : & 1/2 \end{bmatrix}$$

Performing $R_2 \rightarrow - R_2$, we get

$$\sim \begin{bmatrix} 1 & 0 & 0 & : & 2 \\ 0 & 1 & 0 & : & 2 \\ 0 & 0 & 1 & : & 1/2 \end{bmatrix}$$

This is an Echelon form and having three non-zero rows hence rank A = rank of $[A \mid B] = 3$.

Thus, the system of equations are consistent

$$\therefore \quad \begin{bmatrix} 1 & 0 & 0 \\ 0 & 1 & 0 \\ 0 & 0 & 1 \end{bmatrix} \begin{bmatrix} x \\ y \\ z \end{bmatrix} = \begin{bmatrix} 2 \\ 2 \\ 1/2 \end{bmatrix}$$

Hence, the solution is $x = 2, y = 2, z = 1/2$.

Example 8. *Show that the equations*

$$x + y + z = -3$$
$$3x + y - 2z = -2$$
$$2x + 4y + 7z = 7$$

are not consistent.

Solution. The given equations can be written as

$$\begin{bmatrix} 1 & 1 & 1 \\ 3 & 1 & -2 \\ 2 & 4 & 7 \end{bmatrix} \begin{bmatrix} x \\ y \\ z \end{bmatrix} = \begin{bmatrix} -3 \\ -2 \\ 7 \end{bmatrix}$$

i.e., $AX = B$

$\therefore$ Augmented matrix is

$$[A \mid B] = \begin{bmatrix} 1 & 1 & 1 & : & -3 \\ 3 & 1 & -2 & : & -2 \\ 2 & 4 & 7 & : & 7 \end{bmatrix}$$

Performing $R_2 \to R_2 - 3R_1$, $R_3 \to R_3 - 2R_1$, we get

$$[A \mid B] \sim \begin{bmatrix} 1 & 1 & 1 & : & -3 \\ 0 & -2 & -5 & : & 7 \\ 0 & 2 & 5 & : & 13 \end{bmatrix}$$

Performing $R_3 \to R_2 + R_3$

$$\sim \begin{bmatrix} 1 & 1 & 1 & : & -3 \\ 0 & -2 & -5 & : & 7 \\ 0 & 0 & 0 & : & 20 \end{bmatrix}$$

This is an Echelon form and having three non-zero rows therefore the rank $[A \mid B] = 3$ and we see that

$$A \sim \begin{bmatrix} 1 & 1 & 1 \\ 0 & -2 & -5 \\ 0 & 0 & 0 \end{bmatrix}$$

Obviously rank $A = 2$.

Since, rank $A \neq$ rank $[A \mid B]$

Therefore, the given equations are not consistant.

Example 9. *Investigate for what value of λ, μ the simultaneous equations*

$$x + y + z = 6$$
$$x + 2y + 3z = 10$$
$$x + 2y + \lambda z = \mu$$

have : (i) no solution (ii) a unique solution (iii) an infinite solutions.

Solution. The given equation can be written as

$$\begin{bmatrix} 1 & 1 & 1 \\ 1 & 2 & 3 \\ 1 & 2 & \lambda \end{bmatrix} \begin{bmatrix} x \\ y \\ z \end{bmatrix} = \begin{bmatrix} 6 \\ 10 \\ \mu \end{bmatrix}$$

i.e., $\qquad\qquad\qquad\quad AX = B$

Therefore, augmented matrix is

$$[A \,|\, B] = \begin{bmatrix} 1 & 1 & 1 & : & 6 \\ 1 & 2 & 3 & : & 10 \\ 1 & 2 & \lambda & : & \mu \end{bmatrix}$$

Performing $R_2 \to R_2 - R_1, R_3 \to R_3 - R_1$, we get

$$\sim \begin{bmatrix} 0 & 1 & 1 & : & 6 \\ 0 & 1 & 2 & : & 4 \\ 0 & 1 & \lambda-1 & : & \mu-6 \end{bmatrix}$$

Performing $R_3 \to R_3 - R_2$

$$\sim \begin{bmatrix} 1 & 1 & 1 & : & 6 \\ 0 & 1 & 2 & : & 4 \\ 0 & 0 & \lambda-3 & : & \mu-10 \end{bmatrix}$$

If $\lambda = 3$, the rank A = rank $[A \,|\, B]$ = 3. Thus in this case a unique solution exists. If $\lambda = 3$ and $\mu \neq 10$ then rank $A \neq$ rank $[A \,|\, B]$ is 3. Thus rank $A \neq$ rank $[A \,|\, B]$. Hence, in this case equations are inconsistent.

If $\lambda = 3$, and $\mu = 10$, then rank A = rank = $[A \,|\, B]$ = 2. Thus, in this case infinite solutions exist.

EXERCISE 17.4

1. Use matrix method to solve the equations
   $$2x - y + 3z = 9, x + y + z = 6, \ x - y + z = 2.$$

2. Use matrix method to solve the equations
   $$x + 2y + z = 2, 3x + 5y + 5z = 4, \ 2x + 4y + 3z = 3.$$

3. Show that the equations are consistent and hence solve them
   $$x - 3y - 8z + 10 = 0, 3x + y - 4z = 0, \ 2x + 5y + 6z - 13 = 0$$

Solve the following equations by matrix method :

4. $5x + 3y + 7z = 4, 3x + 26y - 2z = 9, \ 7x + 2y + 10z = 5$

5. $5x - 6y + 4z = 15, 7x + 4y - 3z = 19, \ 2x + y + 6z = 46$

6. $2x - y + 3z = 8, -x + 2y + z = 4, \ 3x + y - 4z = 0$

7. Show that the following equations are not consistent.
   $$2x - y + z = 4, 3x - y + z = 6, \ 4x - y + 2z = 7, \ -x + y - z = 9$$

8. Prove that the following system of equations have a unique solution
   $$5x + 3y + 14z = 4, y + 2z = 1, \ x - y + 2z = 0$$

9. Use matrix method, solve $3x + y + 2z = 3,\ 2x - 3y - z = -3,\ x + 2y + z = 4$

HINT TO SELECTED PROBLEMS

1. Here the given system of equation can be written in the form of
$$AX = B$$

$$\Rightarrow \quad \begin{bmatrix} 2 & -1 & 3 \\ 1 & 1 & 1 \\ 1 & -1 & 1 \end{bmatrix} \begin{bmatrix} x \\ y \\ z \end{bmatrix} = \begin{bmatrix} 9 \\ 6 \\ 2 \end{bmatrix}$$

Consider the augmented matrix

$$[A \,|\, B] = \begin{bmatrix} 2 & -1 & 3 & : & 9 \\ 1 & 1 & 1 & : & 6 \\ 1 & -1 & 1 & : & 2 \end{bmatrix} \qquad \text{(by } R_1 \leftrightarrow R_2)$$

We get
$$\sim \begin{bmatrix} 1 & 1 & 1 & : & 6 \\ 2 & -1 & 3 & : & 9 \\ 1 & -1 & 1 & : & 2 \end{bmatrix}$$

Now $R_2 \to R_2 - 2R_1,\ R_3 \to R_3 - R_1$

$$[A \,|\, B] = \begin{bmatrix} 1 & 1 & 1 & : & 6 \\ 0 & -3 & 1 & : & -3 \\ 0 & -2 & 0 & : & -4 \end{bmatrix}$$

$$\sim \begin{bmatrix} 1 & 1 & 1 & : & 6 \\ 0 & -3 & 1 & : & -3 \\ 0 & -2 & -2/3 & : & -2 \end{bmatrix} \qquad \left(\text{by } R_3 \to R_3 - \frac{2}{3} R_2 \right)$$

$\therefore$ Rank $[A \,|\, B]$ = The number of non-zero rows in Echelon form = 3

Similarly, we may get the rank of A, which is also equal to 3.

Now, Since the Rank of $[A \,|\, B]$ = Rank of $[A]$, therefore the given system of equations is consistent.

Also, the rank of A is equal to number of unknowns therefore, the given system of equations has a unique solution. The given system of equations can be reduces to

$$\begin{bmatrix} 1 & 1 & 1 \\ 0 & -3 & 1 \\ 0 & 0 & -2/3 \end{bmatrix} \begin{bmatrix} x \\ y \\ z \end{bmatrix} = \begin{bmatrix} 6 \\ -3 \\ -2 \end{bmatrix}$$

which gives
$$x + y + z = 6$$
$$-3y + z = -3$$
$$-\frac{2}{3} z = -2$$

$\Rightarrow \quad z = 3,\ y = 2,\ x = 1.$

Hence $x = 1,\ y = 2,\ z = 3$, is the required solution of the given system of equations.

3. Here, the given system of equations can be written in the form of
$$AX = B$$

such that
$$\begin{bmatrix} 1 & -3 & -8 \\ 3 & 1 & -4 \\ 2 & 5 & -6 \end{bmatrix} \begin{bmatrix} x \\ y \\ z \end{bmatrix} = \begin{bmatrix} -10 \\ 0 \\ 13 \end{bmatrix}$$

Now consider the augmented matrix.

$$[A\,|\,B] = \begin{bmatrix} 1 & -3 & -8 & : & -10 \\ 3 & 1 & -4 & : & 0 \\ 2 & 5 & 6 & : & 13 \end{bmatrix}$$

Now $R_2 \rightarrow R_2 - 3R_1, R_3 \rightarrow R_3 - 2R_1$

We have
$$\sim \begin{bmatrix} 1 & -3 & -8 & : & -10 \\ 0 & 10 & 20 & : & 30 \\ 0 & 11 & 22 & : & 33 \end{bmatrix}$$

$$\sim \begin{bmatrix} 1 & -3 & -8 & : & -10 \\ 0 & 1 & 2 & : & 3 \\ 0 & 1 & 2 & : & 3 \end{bmatrix} \quad \left(\text{by } R_2 \rightarrow \frac{1}{10}R_2, R_3 \rightarrow \frac{1}{11}R_3 \right)$$

Now $R_3 \rightarrow R_3 - R_2$

We get
$$\sim \begin{bmatrix} 1 & -3 & -8 & : & -10 \\ 0 & 1 & 2 & : & 3 \\ 0 & 0 & 0 & : & 0 \end{bmatrix}$$

The rank of $[A\,|\,B]$ = Number of non-zero rows = 2
Here we observe that

$$\text{Rank } A = \text{Rank } [A\,|\,B]$$

Therefore, the given system of equations is consistent.

Now, Rank $(A) = 2$, which is less than the number of unknown.

$\Rightarrow$ Given system of equations having infinite number of solutions. Now the given system of equation reduces to

$$\begin{bmatrix} 1 & -3 & -8 \\ 0 & 1 & 1 \\ 0 & 0 & 0 \end{bmatrix} \begin{bmatrix} x \\ y \\ z \end{bmatrix} = \begin{bmatrix} -10 \\ 3 \\ 0 \end{bmatrix}$$

which gives
$$x - 3y - 8z = -10$$
$$y + 2z = 3$$

Let $z = c$ then $y = 3 - 2c$ and $x = 2c - 1$.

Hence the solution of the given system of equations is $x = 2c - 1$, $y = 3 - 2c$ and $z = c$.

7. Consider the augmented matrix of the given system

$$[A\,|\,B] = \begin{bmatrix} 2 & -1 & 1 & : & 4 \\ 3 & -1 & 1 & : & 6 \\ 4 & -1 & 2 & : & 7 \\ -1 & 1 & -1 & : & 9 \end{bmatrix}$$

$$\sim \begin{bmatrix} -1 & 1 & -1 & : & 9 \\ 3 & -1 & 1 & : & 6 \\ 4 & -1 & 2 & : & 7 \\ 2 & -1 & 1 & : & 4 \end{bmatrix} \qquad \text{(by } R_1 \leftrightarrow R_4)$$

Now $R_2 \to R_2 + 3R_1$, $R_3 \to R_3 + 4R_1$, $R_4 \to R_4 + 2R_1$

We get

$$\sim \begin{bmatrix} -1 & 1 & -1 & : & 9 \\ 0 & 2 & -2 & : & 33 \\ 0 & 3 & -2 & : & 43 \\ 0 & 1 & -1 & : & 22 \end{bmatrix}$$

$$\sim \begin{bmatrix} -1 & 1 & -1 & : & 9 \\ 0 & 1 & -1 & : & 22 \\ 0 & 2 & -2 & : & 33 \\ 0 & 3 & -2 & : & 43 \end{bmatrix} \qquad \text{(by } R_2 \leftrightarrow R_4)$$

$$\sim \begin{bmatrix} -1 & 1 & -1 & : & 9 \\ 0 & 1 & -1 & : & 22 \\ 0 & 0 & 0 & : & -11 \\ 0 & 0 & 1 & : & 3 \end{bmatrix} \quad \text{(by } R_3 \to 2R_2, R_4 \to R_4 - 3R_2)$$

$$\sim \begin{bmatrix} -1 & 1 & -1 & : & 9 \\ 0 & 1 & -1 & : & 22 \\ 0 & 0 & 1 & : & 23 \\ 0 & 0 & 0 & : & -11 \end{bmatrix}$$

This is Echelon form.

Here, we observe that

Rank of $[A \mid B] = 4$ and Rank of $A = 3$

Here, the given system of equations is inconsistent.

8. The given system of equations can be written as

$$AX = B$$

$$\Rightarrow \begin{bmatrix} 1 & 3 & -8 \\ 0 & 1 & -4 \\ 1 & -1 & -6 \end{bmatrix} \begin{bmatrix} x \\ y \\ z \end{bmatrix} = \begin{bmatrix} 4 \\ 1 \\ 0 \end{bmatrix}$$

Consider the augmented matrix

$$[A \mid B] = \begin{bmatrix} 1 & 3 & 14 & : & 4 \\ 0 & 1 & 2 & : & 1 \\ 1 & -1 & 2 & : & 0 \end{bmatrix}$$

Now $R_3 \to R_3 - R_1$

We get

$$\sim \begin{bmatrix} 1 & 3 & 14 & : & 4 \\ 0 & 1 & 2 & : & 1 \\ 0 & -4 & -12 & : & -4 \end{bmatrix}$$

$$\sim \begin{bmatrix} 1 & 3 & 14 & : & 4 \\ 0 & 1 & 2 & : & 1 \\ 0 & 0 & -4 & : & 0 \end{bmatrix} \qquad \text{(by } R_3 \to R_3 + 4R_1\text{)}$$

Which is Echelon form

$$\text{Rank of } [A \,|\, B] = 3 = \text{Rank of } A$$

$\Rightarrow$ System is inconsistent and have a unique solution such that

$$x + 3y + 14z = 4$$
$$y + 2z = 1$$
$$-4z = 0$$
$$z = 0$$
$$y = 1, \quad \text{and} \quad x = 1$$

Hence the given system of equations have a unique solution given by

$$x = 1, \, y = 1, \quad \text{and} \quad z = 0$$

ANSWERS

(1) $x = 1, y = 2, z = 3$

(2) $x = 3, y = 0, z = -1$

(3) Constistent : $x = 2c - 1, y = 3 - 2c, \, z = c$

(4) $x = \dfrac{7}{11}, y = \dfrac{3}{11}, z = 0$

(5) $x = 3, \, y = 4, z = 6$

(6) $x = 2, \, y = 2, z = 2.$

(9) $x = 1, \, y = 2, z = -1$

17.31 CRAMER'S RULE

Consider the system of linear equations

$$a_1 x + b_1 y + c_1 z = d_1$$
$$a_2 x + b_2 y + c_2 z = d_2$$
$$a_3 x + b_3 y + c_3 z = d_3 \qquad \qquad ...(1)$$

We define Δ = determinant of coefficients

$$= \begin{bmatrix} a_1 & b_1 & c_1 \\ a_2 & b_2 & c_2 \\ a_3 & b_3 & c_3 \end{bmatrix}$$

Now we define Δ_x which is obtained by suppressing the column of coefficients of x and replacing it by the column of constant terms d_1, d_2, d_3 on right hand side

$$\therefore \qquad \Delta_x = \begin{bmatrix} d_1 & b_1 & c_1 \\ d_2 & b_2 & c_2 \\ d_3 & b_3 & c_3 \end{bmatrix}$$

Similarly, we obtained

$$\Delta_y = \begin{bmatrix} a_1 & d_1 & c_1 \\ a_2 & d_2 & c_2 \\ a_3 & d_3 & c_3 \end{bmatrix} \quad \text{and } \Delta_z = \begin{bmatrix} a_1 & b_1 & d_1 \\ a_2 & b_2 & d_2 \\ a_3 & b_3 & d_3 \end{bmatrix}$$

Now

Case I. If $\Delta \neq 0$ solution of system (1) is given by

$$x = \frac{\Delta_x}{\Delta}, y = \frac{\Delta_y}{\Delta}, z = \frac{\Delta_z}{\Delta}$$

and system is called consistent.

Case II. $\Delta = 0$ but at least one of $\Delta_x, \Delta_y, \Delta_z \neq 0$, then the system does not possess any common solution and so system is called inconsistent.

Case III. $\Delta = 0$ also $\Delta_x = \Delta_y = \Delta_z = 0$, and at least one cofactor of $\Delta \neq 0$, then system has infinitely many solutions and the system then be solved by elimination method.

Elimination of one unknown from three equations gives any one equation in two unknowns therefore two unknowns can be found in terms of the other we give this unknown an arbitrary value.

If $\Delta = \Delta_x = \Delta_y = \Delta_z = 0$ and all cofactors of $\Delta, \Delta_x, \Delta_y$ and Δ_z are zero then system is equivalent to only one equation in three unknowns and then we give any two unknowns arbitrary values to find the remaining unknown in terms of three constants.

SOLVED EXAMPLES

Example 1. *Using the Cramer's rule, solve the following equations*
$$x + y - 4 = 0, 2x - 3y - 3 = 0$$

Solution. The given equation is

$$x + y - 4 = 0 \qquad \qquad ...(1)$$
$$2x - 3y - 3 = 0 \qquad \qquad ...(2)$$

Here
$$\Delta = \begin{vmatrix} 1 & 1 \\ 2 & -3 \end{vmatrix} = -5 \neq 0$$

$$\Delta_x = \begin{vmatrix} 4 & 1 \\ 3 & -3 \end{vmatrix} = -15$$

$$\Delta_y = \begin{vmatrix} 1 & 4 \\ 2 & 3 \end{vmatrix} = -5$$

$\therefore$ By Cramer's rule

$$x = \frac{\Delta_x}{\Delta} = 3, y = \frac{\Delta_y}{\Delta} = 1.$$

Example 2. *Show that the system of equations $x + y - 2 = 0, 2x + 3y - 5 = 0, 4x - y - 3 = 0$ is consistent. Find the solution using Cramer's rule.*

Solution. The given system of equations is

$$x + y - 2 = 0 \qquad \qquad ...(1)$$
$$2x + 3y - 5 = 0 \qquad \qquad ...(2)$$
$$4x - y - 3 = 0 \qquad \qquad ...(3)$$

is consistent (*i.e.*, have common solution), if the determinant

$$\therefore \qquad \Delta^* = \begin{vmatrix} 1 & 1 & -2 \\ 2 & 3 & -5 \\ 4 & -1 & -3 \end{vmatrix} = 0 \ i.e., \ \begin{vmatrix} 1 & 0 & 0 \\ 2 & 1 & -1 \\ 4 & -5 & 5 \end{vmatrix} = 0$$

$\therefore \quad \Delta^* = 5 - 5 = 0$, hence the system is consistent, so it is sufficient to solve any two equations by Cramer's rule

Let us consider equation (1) and (2)

$$\Delta = \begin{vmatrix} 1 & 1 \\ 2 & 3 \end{vmatrix} = 1 \, (\neq 0)$$

$$\Delta_x = \begin{vmatrix} 2 & 1 \\ 5 & 3 \end{vmatrix} = 6 - 5 = 1$$

$$\Delta_y = \begin{vmatrix} 1 & 2 \\ 2 & 5 \end{vmatrix} = 5 - 4 = 1$$

$$x = \frac{\Delta_x}{\Delta} = 1, y = \frac{\Delta_y}{\Delta} = 1.$$

Hence the required solution is given by $x = y = 1$.

Example 3. *Solve the following by Cramer's rule*

$$x + y + z = 6$$
$$x - y + z = 2$$
$$3x + 2y - 4z = -5$$

Solution. We have $\Delta = \begin{vmatrix} 1 & 1 & 1 \\ 1 & -1 & 1 \\ 3 & 2 & -4 \end{vmatrix} = \begin{vmatrix} 1 & 0 & 0 \\ 1 & -2 & 0 \\ 3 & -1 & 7 \end{vmatrix} = 14 \neq 0$

$$\Delta_x = \begin{vmatrix} 6 & 1 & 1 \\ 2 & -1 & 1 \\ -5 & 2 & 4 \end{vmatrix} = \begin{vmatrix} 6 & 1 & 1 \\ -4 & -2 & 0 \\ 19 & 6 & 0 \end{vmatrix} = 14$$

$$\Delta_y = \begin{vmatrix} 1 & 6 & 1 \\ 1 & 2 & 1 \\ 3 & -5 & -4 \end{vmatrix} = \begin{vmatrix} 1 & 6 & 1 \\ 0 & -4 & 0 \\ 0 & -23 & -7 \end{vmatrix} = 28$$

$$\Delta_z = \begin{vmatrix} 1 & 1 & 6 \\ 1 & -1 & 2 \\ 3 & 2 & -5 \end{vmatrix} = \begin{vmatrix} 1 & 1 & 6 \\ 0 & -2 & -4 \\ 0 & -1 & -23 \end{vmatrix} = 42$$

Hence, by Cramer's rule

$$x = \frac{\Delta_x}{\Delta} = 1, \ \frac{\Delta_y}{\Delta} = 2 = y, z = \frac{\Delta_z}{\Delta} = 3.$$

Hence, the solution is given by $x = 1, y = 2, z = 3$.

Example 4. *Find the value of λ for which the system of equations $x + y - 2z = 0, 2x - 3y + z = 0, x - 5y + 4z = \lambda$ are consistent and find the solutions for all such values of λ.*

Solution. The given system of equations is

$$x - 5y + 4z = \lambda \qquad \qquad ...(1)$$
$$x + y - 2z = 0 \qquad \qquad ...(2)$$
$$2x - 3y + z = 0 \qquad \qquad ...(3)$$

$$\Delta = \begin{vmatrix} 1 & -5 & 4 \\ 1 & 1 & -2 \\ 2 & -3 & 1 \end{vmatrix} = \begin{vmatrix} 1 & -5 & 4 \\ 0 & 6 & -6 \\ 0 & 7 & -7 \end{vmatrix} = 0$$

Hence, the system is consistent only when

$$\Delta_x = \Delta_y = \Delta_z = 0$$

Now $\quad \Delta_x = \begin{vmatrix} \lambda & -5 & 4 \\ 0 & 1 & -2 \\ 0 & -3 & 1 \end{vmatrix} = -5\lambda = 0 \implies \lambda = 0$

For $\lambda = 0$, clearly $\Delta_y = \Delta_z = 0$.

$\therefore$ System is consistent if $\lambda = 0$, then no eliminating x from (1), (2) and (1), (3), we have

$$6y - 6z = 0 \qquad y - z = 0$$
and $\qquad\quad 7y - 7z = 0 \quad$ or $\qquad y = z$

Let $y = z = k \in$ R, then from (1), we have $x = 5k - 4k = k$.

Hence, solution is given by $x = y = z = k \in$ R.

Example 5. *Solve the equations by Cramer's rule*

$$\frac{4}{x+5} + \frac{3}{y+7} = -1$$

$$\frac{6}{x+5} - \frac{6}{y+7} = -5$$

Solution. The given system of equations is

$$\frac{4}{x+5} + \frac{3}{y+7} = -1$$

$$\frac{6}{x+5} - \frac{6}{y+7} = -5$$

Now putting $\dfrac{1}{x+5} = a, \quad \dfrac{1}{y+7} = b$ the equations becomes

$$4a + 3b = -1$$
$$6a - 6b = -5$$

$$\Delta = \begin{vmatrix} 4 & 3 \\ 6 & -6 \end{vmatrix} = -42 \neq 0$$

$$\Delta_a = \begin{vmatrix} -1 & 3 \\ -5 & -6 \end{vmatrix} = 21, \quad \Delta_b = \begin{vmatrix} 4 & -1 \\ 6 & -5 \end{vmatrix} = -14$$

So by Cramer's rule

$$a = \frac{\Delta_a}{\Delta} = \frac{21}{-42} = -\frac{1}{2}, \quad b = \frac{\Delta_b}{\Delta} = \frac{-14}{-42} = \frac{1}{3}$$

$\therefore \qquad x + 5 = -2, \quad y + 7 = 3$

or $\qquad x = -7, \quad y = -4$.

Hence, the solution is $x = -7, \; y = -4$.

Example 6. *Using Cramer's rule solve the following system of linear equations :*

$$x + 2y + 3z = 6$$
$$2x + 4y + z = 17$$
$$3x + 2y + 9z = 2$$

Solution. We have

$$\Delta = \begin{vmatrix} 1 & 2 & 3 \\ 2 & 4 & 1 \\ 3 & 2 & 9 \end{vmatrix} = -20$$

$$\Delta_x = \begin{vmatrix} 6 & 2 & 3 \\ 17 & 4 & 1 \\ 2 & 2 & 9 \end{vmatrix} = -20$$

$$\Delta_y = \begin{vmatrix} 1 & 6 & 3 \\ 2 & 17 & 1 \\ 3 & 2 & 9 \end{vmatrix} = -80$$

$$\Delta_z = \begin{vmatrix} 1 & 2 & 6 \\ 2 & 4 & 17 \\ 3 & 2 & 2 \end{vmatrix} = 20$$

Then by Cramer's rule, we have

$$x = \frac{\Delta_x}{\Delta} = \frac{-20}{-20} = 1$$

$$y = \frac{\Delta_y}{\Delta} = \frac{-80}{-20} = 4$$

$$z = \frac{\Delta_z}{\Delta} = \frac{20}{-20} = -1.$$

EXERCISE 17.5

1. Using Cramer's rule, solve the following equations :

$$x + y + z = 6$$
$$2x + y - z = 1$$
$$x + y - 2z = -3$$

2. Find the value of k if the following equations are consistent :

$$x + y - 3 = 0$$
$$(1 + k)x + (2 + k)y - 8 = 0$$
$$x - (1 + k)y + (2 + k) = 0$$

3. Find the value of k if the system of equations :

$$(k + 1)^3 x + (k + 2)^3 y = (k + 3)^3$$
$$(k + 1) x + (k + 2) y = (k + 3)$$
$$x + y = 1; \text{ is consistent.}$$

4. If the system of equations :

$$x + 2y = 5, \quad 2x - y = 5, \quad x + 3y = 6 \text{ is consistent, solve it.}$$

5. Solve the following by Cramer's rule :

$$x + y + z = 11$$
$$2x - 6y - z = 0$$
$$3x + 4y + 2z = 0$$

6. Show that the system of equations

$$3x - y + 4z = 3$$
$$x + 2y - 3z = -2$$
$$6x + 5y + \lambda z = -3$$

has at least one solution for any real numbers λ. Find the set of solutions if $\lambda = -5$.

7. Using Cramer's rule to solve the following system of linear equations :

$$2x - 3y + z = 7$$
$$2x + y + z = 1$$
$$4y + 3z = -11$$

8. Using Cramer's rule, solve the following equations :

$$x + y + z = 1$$
$$3x + 5y + 6z = 4$$
$$9x + 2y - 36z = 17$$

9. Solve the following equations by Cramer's rule :

$$x + 9y - z = 4$$
$$2x + 7y + 3z = 7$$
$$3x + 10y + 4z = 9$$

10. Solve the Cramer's rule

$$2x + 3y - 3z = 0$$
$$5x - 2y + 2z = 19$$
$$x + 7y - 5z = 5$$

ANSWERS

(1) $x = 1, y = 2, z = 3$ **(2)** $k = 1$ or $-5/3$ **(3)** $k = -2$

(5) $x = -8, y = -7, z = 26$ **(7)** $x = 1, y = -2, z = -1$

(8) $x = \dfrac{1}{3}, y = 1, z = \dfrac{-1}{3}$ **(9)** $x = -3, y = 1, z = 2$

17.32   GAUSS ELIMINATION METHOD

Let us consider a system of linear equations

$$AX = B \qquad \qquad \text{...(1)}$$

assuming $\det A \neq 0$, equation (1) has the following form

$$\left. \begin{aligned} a_{11}x_1 + a_{12}x_2 + ... + a_{1n}x_n &= b_1 \\ a_{21}x_1 + a_{22}x_2 + ... + a_{2n}x_n &= b_2 \\ \cdots \quad \cdots \quad \cdots \quad \cdots \quad \cdots \quad \cdots \\ a_{n1}x_1 + a_{n2}x_2 + ... + a_{nn}x_n &= b_n \end{aligned} \right\} \qquad \text{...(2)}$$

Assuming $a_{11} \neq 0$ and divide first equation by a_{11} and then we substract this equation multiplies by $a_{21}, a_{31}, ... a_{n1}$ from second, third, nth equation, we get

$$x_1 + a_{12}'x_2 + \dots + a_{1n}'x_n = b_1' \\ a_{22}'x_2 + \dots + a_{2n}'x_n = b_2' \\ \dots \quad \dots \quad \dots \quad \dots \quad \dots \quad \dots \\ a_{n2}'x_2 + \dots + a_{n2}'x_n = b_n' \quad\Bigg\} \quad\dots(3)$$

Now, we divide second equation of (3) by a_{22}' and substract after multiplied by a_{32}', $a_{42}', \dots a_n^2$ from (3), (4) ... nth equation of (3), we get

$$x_1 + a_{12}'x_2 + \dots + a_{1n}'x_n = b_1' \\ x_2 + a_{23}''x_3 + \dots + a_{2n}''x_n = b_2'' \\ a_{33}''x_3 + \dots + a_{3n}''x_n = b_3'' \\ \dots \quad \dots \quad \dots \quad \dots \quad \dots \\ a_{3n}''x_3 + \dots + a_{nn}''x_n = b_n'' \quad\Bigg\} \quad\dots(4)$$

Continuing in this way, we get

$$x_1 + c_{12}x_2 + c_{13}x_3 + \dots + x_{1n}x_n = d_1 \\ x_2 + c_{23}x_3 + \dots + c_{2n}x_n = d_2 \\ \dots \quad \dots \quad \dots \quad \dots \quad \dots \quad \dots \\ a_{nn}x_n = d_n \quad\Bigg\} \quad\dots(5)$$

this is a form of upper triangular system.

From back substitution we can find the solution of system of given equations.

Let us consider

$$a_{11}x_1 + a_{12}x_2 + a_{13}x_3 = b_1 \\ a_{21}x_1 + a_{22}x_2 + a_{23}x_3 = b_2 \\ a_{31}x_1 + a_{32}x_2 + a_{33}x_3 = b_3 \quad\Bigg\} \quad\dots(6)$$

Step I. First, eliminate x_1 from (2) and (3) assuming $a_{11} \neq 0$. Now dividing (1) by a_{11} and then substract from (2) and (3) after multiplied by a_{21} and a_{31} respectively, we get

$$x_1 + a_{12}'x_2 + a_{13}'x_3 = b_1' \\ a_{22}'x_2 + a_{23}'x_3 = b_2' \\ a_{32}'x_2 + a_{33}'x_3 = b_3' \quad\Bigg\} \quad\dots(7)$$

where $\quad a_{12}' = \dfrac{a_{12}}{a_{11}} \qquad a_{13}' = \dfrac{a_{13}}{a_{11}} \qquad a_{22}' = a_{22} - a_{21}a_{12}'$

$$a_{23}' = a_{23} - a_{21}a_{13}', \quad a_{32}' = a_{32} - a_{31}a_{12}' \quad a_{33}' = a_{33} - a_{31}a_{13}'$$

$$b_1' = \dfrac{b_1}{a_{11}} \qquad b_2' = b_2 - a_{21}b_1' \qquad b_3' = b_3 - a_{31}b_1'$$

Step II. Now eliminating x_2 from third equation in (7).

Again assuming $a_{22}' \neq 0$, dividing equation (2) in (7) by a_{22}' and then substract from third equation after multiplied by a_{32}', we get

$$x_1 + a_{12}'x_2 + a_{13}'x_3 = b_1' \\ x_2 + a_{23}''x_3 = b_2'' \\ a_{33}''x_3 = b_3'' \quad\Bigg\} \quad\dots(8)$$

where $a_{23}'' = \dfrac{a_{23}'}{a_{22}'},$ $a_{33}'' = a_{33}' - a_{32}'a_{23}''$

$b_2'' = \dfrac{b_2'}{a_{22}'}$ $b_3'' = b_3' - a_{32}'b_2''$

Step III. Evaluating x_1, x_2 and x_3 from (8) by back substitution.

Remarks

❖ The coefficient a_{11}, a_{22}' and a_{33}'' are called pivots.

❖ This method will fail it any one of pivots a_{11}, a_{22}' and a_{33}'' becomes zero. In such cases we rewrite the equations in a different order so that pivots are non-zero.

❖ From each step of procedure, the largest coefficient of x is closen on pivot elements.

SOLVED EXAMPLES

Example 1. *Solve by Gauss's elimination method of the following :*
$$6x + 3y + 2z = 6,$$
$$6x + 4y + 3z = 0$$
$$20x + 15y + 12z = 0$$

Solution. Here pivot elements is 6. Now divide first equation by 6, we get

$$x + \frac{1}{2}y + \frac{1}{3}z = 1 \qquad\qquad\qquad ...(1)$$

Now eliminating x from second and third equation with the help of (1), such that

$$y + z = -6 \qquad\qquad\qquad ...(2)$$

$$5y + \frac{16}{3}z = -20 \qquad\qquad\qquad ...(3)$$

Now eliminating y from (3) with the help of (2), we get

$$\left(\frac{16}{3} - 5\right)z = -20 + 30$$

$$\frac{1}{3}z = 10, \quad z = 30$$

Substituting the value of z in (2) we get
$$y = -6 - 30 = -36$$
and again substituting the values of y and z into (1), we get

$$x + \frac{1}{2}(-36) + \frac{1}{3}(30) = 1$$

$$x - 18 + 10 = 1, \ x = 9$$

Hence the solution of equations are
$$x = 9, y = -36, z = 30.$$

Example 2. *By Gauss's elimination method, solve*
$$5x - y - 2z = 142$$
$$x - 3y - z = -30$$
$$2x - y - 3z = -50$$

Solution. The largest coefficient in the first equation is 5 which is pivot element so divide first equation by 5, we get

$$x - \frac{1}{5}y - \frac{2}{5}z = \frac{1+2}{5} \qquad \qquad ...(1)$$

Now eliminating x from second and third equation with the help of (1), we get

$$-\frac{14}{5}y - \frac{3}{5}z = -\frac{292}{5} \qquad \qquad ...(2)$$

$$-\frac{3}{5}y - \frac{11}{5}z = -\frac{309}{5} \qquad \qquad ...(3)$$

Eliminating y from (2) and (3), we get

$$-\frac{145}{5}z = -\frac{3450}{5}$$

$$z = \frac{3450}{145} = 23 \cdot 79$$

Substituting the value of z into (3), we get

$$-\frac{3}{5}y - \frac{11}{5}(23 \cdot 79) = -\frac{309}{5}$$

$$-\frac{3}{5}y = -\frac{309}{5} + \frac{11(23 \cdot 79)}{5}$$

$$-3y = -309 + 11 (23 \cdot 79)$$

$$= -309 + 261 \cdot 69$$

$$-3y = -47 \cdot 31$$

$$y = 15 \cdot 77$$

Substituting the values of y and z into (1), we get

$$x - \frac{1}{5}(15 \cdot 77) - \frac{2}{5}(23 \cdot 79) = \frac{142}{5}$$

$$x = \frac{142}{5} + \frac{15 \cdot 77}{5} + \frac{2(23 \cdot 79)}{5}$$

$$= \frac{205 \cdot 35}{5}$$

$$x = 41 \cdot 07$$

Hence, the solution is given by

$$x = 41 \cdot 07, y = 15 \cdot 77, z = 23 \cdot 79.$$

Example 3. *Using Gauss's elimination method, solve*

$$2x_1 + 4x_2 + x_3 = 3$$
$$3x_1 + 2x_2 - 2x_3 = 2$$
$$x_1 - x_2 + x_3 = 6$$

Solution. Dividing first equation by 2, we get

$$x_1 + 2x_2 + \frac{1}{2}x_3 = \frac{3}{2} \qquad \qquad ...(1)$$

Multiplying (1) by (3) and substract from (2) and also from (3) equation, we get

$$4x_2 + \frac{7}{2}x_3 = \frac{5}{2} \qquad \ldots(2)$$

$$-3x_2 + \frac{1}{2}x_3 = \frac{9}{2} \qquad \ldots(3)$$

Now dividing (2) by 4 and substract after multiplied by -3 form (3), we get

$$25x_3 = 51$$

$$x_3 = \frac{51}{25} = 2 \cdot 04$$

Substituting the value of x_3 into (2), we get

$$4x_2 + \frac{7}{2}(2 \cdot 04) = \frac{5}{2}$$

$$4x_2 = \frac{5}{2} - \frac{7(2 \cdot 04)}{2}$$

$$x_2 = -1 \cdot 16$$

Now substituting the values of x_2 and x_3 into (1), we get

$$x_1 + 2(-1 \cdot 16) + \frac{1}{2}(2 \cdot 04) = \frac{3}{2}$$

$$x_1 = \frac{3}{2} + 2(1 \cdot 16) - \frac{1}{2}(2 \cdot 04)$$

$$= \frac{5.6}{2}$$

$$x_1 = 2 \cdot 8$$

Hence, the solution is given by

$$x_1 = 2 \cdot 8, \, x_2 = -1 \cdot 16, \, x_3 = 2 \cdot 04.$$

EXERCISE 17.6

1. Solve by the Gauss elimination method :
$$2x + y + 4z = 12$$
$$8x - 3y + 2z = 23$$
$$4x + 11y - z = 33$$

2. Apply Gauss-elimination method to solve the equations :
$$x + 4y - z = -5$$
$$x + y - 6z = -12$$
$$3x - y - z = 4$$

3. Solve the following system by Gauss elimination method :
$$2x + y + z = 10$$
$$3x + 2y + 3z = 18$$
$$x + 4y + 9z = 16$$

4. By Gauss elimination method, solve
$$4x + 11y - z = 33$$
$$x + y + 4z = 12$$
$$8x - 3y + 2z = 20$$

5. Solve by the Gauss elimination method :
$$x + 2y + z = 3$$
$$2x + 3y + 3z = 10$$
$$3x - y + 2z = 13$$

6. Solve the equations by Gauss-elimination method :
$$x_1 + x_2 + 3x_3 = 4$$
$$3x_1 + x_2 - 3x_3 = 4$$
$$2x_1 - 3x_2 - 5x_3 = -5$$

7. Solve the equations by Gauss-elimination method :
$$2x_1 + x_2 + 4x_3 = 12$$
$$8x_1 - 3x_2 + 2x_3 = 20$$
$$4x_1 + 11x_2 - x_3 = 33$$

8. Solve the following equations by Gauss-elimination method :
$$x_1 + x_2 + x_3 = 10$$
$$2x_1 + x_2 + 2x_3 = 17$$
$$3x_1 + 2x_2 + x_3 = 17$$

9. Solve by Gauss elimination method :
$$2x + 3y - z = 5$$
$$4x + 4y - 3z = 3$$
$$2x - 3y + 2z = 2$$

10. Solve by Gauss elimination method :
$$2x_1 + 4x_2 + x_3 = 2$$
$$3x_1 + 2x_2 - 2x_3 = -2$$
$$x_1 - x_2 + x_3 = 6$$

11. Solve the equations by Gauss elimination method :
$$2x + y + z = 10$$
$$3x + 2y + 3z = 18$$
$$x + 4y + 9z = 16$$

ANSWERS

(1) $x = \dfrac{47}{14}, y = \dfrac{13}{7}, z = \dfrac{6}{7}$ (2) $x = \dfrac{117}{71}, y = \dfrac{81}{71}, z = \dfrac{148}{71}$

(3) $x = 7, y = -9, z = 5.$ (4) $x = 2.856, y = 2.121, z = 1.756$

(5) $x = 2, y = -1, z = 3.$ (6) $x_1 = 1.13, x_2 = 1.79, x_3 = 0.36$

(7) $x_1 = 3, x_2 = 2, x_3 = 1.$ (8) $x_1 = 2, x_2 = 3, x_3 = 5$

(9) $x = 1, y = 2, z = 3.$ (10) $x_1 = 2, x_2 = -1.2, x_3 = 2.8$

(11) $x = 7, y = -9, z = 5.$

CHAPTER REVIEW : A COMPETITIVE APPROACH

Selected terms and Results

TERMS

- **Matrix :** A set having mn numbers either real or complex arranged in the form of rectangular array in which there are m rows and n columns. This arrangement is called matrix.

- **Null matrix :** A matrix of order $m \times n$ is called a null matrix if it contains all mn elements zero.

- **Square matrix :** A matrix having a number of rows equal to number of columns is called square matrix.

- **Unit matrix :** A square matrix of order $n \times n$ having all non-diagonal elements equal to zero and each of the diagonal element equal to 1 is called a unit matrix.

- **Row matrix :** A matrix having only one row and n columns is called a row matrix of order $1 \times n$.

- **Column matrix :** A matrix having m rows and only one column is called column matrix of order $m \times 1$.

- **Upper Triangular matrix :** A matrix of order $n \times n$ is called an upper triangular matrix if it contains its all elements above the diagonal elements equal to zero.

- **Lower Triangular matrix :** A matrix of order $n \times n$ is called a lower triangular matrix if it contains its all elements below the diagonal element equal to zero.

- **Diagonal matrix :** A matrix of order $n \times n$ is called a diagonal matrix if it contains all its off-diagonal element equal to zero.

- **Scalar matrix :** A diagonal matrix whose diagonal elements are all equal but not equal to 1 is called a scalar matrix.

- **Periodic matrix :** A matrix A is called periodic if $A^{k+1} = A$ for some k.

- **Nilpotent matrix :** A matrix A is called nilpotent if $A^k = 0$.

- **Involuntry matrix :** A matrix A is called an involuntry matrix if $A^2 = 0$.

- **Singular and non-singular matrix :** A matrix A whose determinant value is zero is called singular otherwise non-singular.

- **Transpose of a matrix :** The matrix which is obtained by interchanging the rows and columns of A is called transpose A' of A.

- **Symmetric matrix :** A matrix A is called symmetric if $A' = A$

- **Skew-symmetric matrix :** A matrix A is called skew symmetric if $A' = A$

- **Adjoint of a matrix :** The transpose of the matrix of cofactors is called adjoint.

- **Inverse of a matrix :** Let A be a square matrix of order $n \times n$ and there exists a square matrix of the same order such that $AB = BA = I$, the unit matrix. Then B is called the inverse of A.

RESULTS

- There is a lot of difference between a matrix and a determinant. The value of a determinant is a number while matrix is not a number.

- The matrix addition is commutative and associative.

- Matrix multiplication is not commutative.

- The transpose of the product of two matrices is the product in reverse order of their transpose.

- The necessary and sufficient condition for square matrix to possess an inverse is that it must be non-singular.

- Inverse of any square matrix if exists is unique.
- A positive integer r is said to be the rank of a non-zero matrix A if

  (*i*) there exists at least one minor in A of order r which is not zero.

  (*ii*) every minor in A of order greater than r is zero.
- A matrix obtained from a unit matrix by subjecting it to any of the elementary transformation is called elementry matrix.
- In Echelon form :

  (*i*) All the non-zero rows of A, if any proceed the zero rows.

  (*ii*) The number of zero proceeding the first non-zero element in a row is less than the number of such zero in the succeeding row.

  (*iii*) The first non-zero element in a row is unity.
- To obtain the rank of a matrix we reduce it to upper triangular matrix and we count the number of non-zero rows, the rank of the matrix is equal to the number of non-zero rows.
- A system of equations is said to be consistent if they have one or more solutions.

Review Questions and Project Work

1. Define the following :
   - (*i*) Null matrix
   - (*ii*) Unit matrix
   - (*iii*) Diagonal matrix
   - (*iv*) Singular and non-singular matrix
   - (*v*) Symmetric and skewsymmetric matrix
   - (*vi*) Adjoint of a matrix
   - (*vii*) Orthogonal matrix
   - (*viiii*) Unitary matrix

2. If $A = \begin{bmatrix} 1 \\ -5 \\ 7 \end{bmatrix}$ and $B = \begin{bmatrix} 3 & 2 & -2 \end{bmatrix}$.

Verify that $(AB)^T = B^T A^T$.

3. If A and B are symmetric matrices, show that AB is symmetric if and only if $AB = BA$.

4. If $A = \begin{bmatrix} 3 & -4 \\ 1 & -1 \end{bmatrix}$ show that $A^k = \begin{bmatrix} 1+2k & -4k \\ k & 1-2k \end{bmatrix}$, where k is any positive integer.

5. If $A_\alpha = \begin{bmatrix} \cos a & \sin a \\ -\sin a & \cos a \end{bmatrix}$, then show that $A_a^n = \begin{bmatrix} \cos na & \sin na \\ -\sin na & \cos na \end{bmatrix}$

6. If A, B are square matrices such that $A^2 = A . B^2 = B$ show that $(AB)^2 = AB$ if $AB = BA$.

7. If $A_\alpha = \begin{bmatrix} \cos a & -\sin a \\ \sin a & \cos a \end{bmatrix}$ then show that

   (*i*) $A_\alpha A_{-\alpha} = I$ (*ii*) $A_\alpha . A_\beta = A_{\alpha+\beta}$

8. If $A = \begin{bmatrix} 3 & -3 & 4 \\ 2 & -3 & 4 \\ 0 & 1 & 1 \end{bmatrix}$, show that $A^3 = A^{-1}$.

9. Show that matrix multiplication is not commutative.

10. Show that a square matrix A has an inverse if and only if A is non-singular.

11. If A and B are non-singular matrices of same order, then show that $A \cdot B$ is also non-singular and $(AB)^{-1} = B^{-1}A^{-1}$.

12. Find the adjoint of $A = \begin{bmatrix} 3 & -3 & 4 \\ 2 & -3 & 4 \\ 0 & -1 & 1 \end{bmatrix}$

13. Find the inverse of the following matrix $A = \begin{bmatrix} 1 & 3 & 3 \\ 1 & 4 & 3 \\ 1 & 3 & 4 \end{bmatrix}$.

14. Show that $A = \begin{bmatrix} 5 & 3 \\ -1 & -2 \end{bmatrix}$ satisfies the

equation

$A^2 - 3A - 7I = 0$. Also find A^{-1}.

15. Show that the matrix $A =$

$\dfrac{1}{3} \begin{bmatrix} 1 & 2 & 2 \\ 2 & 1 & -2 \\ -2 & 2 & -1 \end{bmatrix}$ is orthogonal.

16. Find the inverse of the following matrix :

(*i*) $\begin{bmatrix} 2 & 1 & 2 \\ 2 & 2 & 1 \\ 1 & 1 & 1 \end{bmatrix}$

(*ii*) $\begin{bmatrix} 3 & -3 & 4 \\ 2 & -3 & 4 \\ 0 & -1 & 1 \end{bmatrix}$

(*iii*) $\begin{bmatrix} 1 & 2 & 3 \\ 4 & 5 & 6 \\ 6 & 7 & 9 \end{bmatrix}$

(*iv*) $\begin{bmatrix} 3 & -10 & 1 \\ -2 & 8 & 2 \\ -2 & -4 & -2 \end{bmatrix}$

17. Find the rank of the following matrices :

(*i*) $\begin{bmatrix} 1 & 1 & 1 \\ a & b & c \\ a^3 & b^3 & c^3 \end{bmatrix}$

(*ii*) $\begin{bmatrix} 1 & 2 & 3 & 0 \\ 2 & 4 & 3 & 2 \\ 3 & 2 & 1 & 3 \\ 6 & 8 & 7 & 5 \end{bmatrix}$

(*iii*) $\begin{bmatrix} 5 & 3 & 14 & 4 \\ 0 & 1 & 2 & 1 \\ 1 & -1 & 2 & 0 \end{bmatrix}$

(*iv*) $\begin{bmatrix} 1 & 2 & 3 \\ 2 & 4 & 7 \\ 3 & 6 & 10 \end{bmatrix}$

(*v*) $\begin{bmatrix} 1 & 2 & 3 & 1 \\ 2 & 4 & 6 & 2 \\ 1 & 2 & 3 & 2 \end{bmatrix}$

18. Investigate for what value of λ, μ the simultaneous equations

$$x + y + z = 6$$
$$x + 2y + 3z = 10$$
$$x + 2y + \lambda = \mu$$

have (*i*) no solution (*ii*) a unique solution (*iii*) an infinite solution

19. Use matrix method solve

$$3x + y + 2z = 3$$
$$2x - 3y - z = -3$$
$$x + 2y + z = 4$$

20. Solve the following by Cramer's rule:

(*i*) $x + y + z = 6$
         $x - y + z = 2$
         $3x + 2y + 4z = -5$

(*ii*) $x + 2y + 3z = 6$
         $2x + 4y + z = 17$
         $3x + 2y + 9z = 2$

(*iii*) $2x - 3y + z = 7$
         $2x + y + z = 1$
         $4y + 3z = -11$

Objective Type Questions

Fill in the blanks:

1. The matrix $AX = 0$ is a system of linear .... equations.

2. If the rank of A is r, then the number of linearly independent solutions of m homogeneous equation in n variable is ....

3. If X_1 and X_2 are the solutions of $AX = 0$ then .... is a solution of $AX = 0$

4. If the rank of A is equal to the number

of unknown in $AX = 0$ then there exists only ....

5. If the rank of A is less than number of unknown in $AX = 0$ then there are ....

6. The matrix equation $AX = B$ is system of linear ....

7. The equation $AX = B$ is consistent if the rank A ....

8. The rank $A < $ rank $(A|B)$ then the equation $AX = B$ is ....

9. If rank $A = $ rank $(A|B)$ and is also equal to the number of unknown, then there will be ....

10. If rank $A = $ rank $(A|B) = r$ and $r < n$ then there will only .... linearly independent solution of $AX = B$.

11. If the rank of square matrix is not equal to its order then the matrix is ....

12. The rank of every non-zero matrix is always greater then or equal to ....

13. The rank of I_n is .... where I is a unit matrix of order n.

14. The rank of a matrix is less than equal to r, if all $(r + 1)$ rowed minors of the matrix ....

15. The rank of A and A' are ....

16. Non-square matrix has .... inverse.

17. A matrix A is said to be singular if $|A| = $ ....

18. A matrix is said to be .... if it is square and non-singular

19. If $|A| \neq 0$ then matrix is said to be ....

20. The inverse of a matrix, if exist is ....

21. If A and B are two non-singular

matrices of the same order then AB is ....

22. $(AB)^{-1} = $ ....

23. The transpose of the matrix of cofactors is known as ....

24. The necessary and sufficient condition that a square matrix may possess an inverse is that it be ....

25. The inverse of the inverse of a matrix A is equal to ....

True/False: *Write 'T' for true and 'F' for false statement.*

1. Every square matrix possess inverse (T/F)

2. Every non-singular matrix possess inverse. (T/F)

3. If A, B are any two $n \times n$ matrices such that $BA = 0$, then at least one of them is non-singular. (T/F)

4. The inverse of an orthogonal matrix is not necessarily orthogonal (T/F)

5. Adj $(AB) = $ Adj $(A).$Adj(B). (T/F)

6. The inverse of matrix A exist if A is singular (T/F)

7. The rank of zero matrix is 1 (T/F)

8. The rank of I_4 is 4. (T/F)

9. The rank of a matrix, if it reduced to a Echelon form is equal to the number of non-zero rows in Echelon form. (T/F)

10. If $A = [adj]_{n \times m}$ and $[A] = 0$ then the rank of $A \geq n$.

Multiple Choice Questions : *Choose the most appropriate one :*

1. If the rank of $A = \begin{bmatrix} a_{11} & a_{12} \\ a_{21} & a_{22} \end{bmatrix}$ is 2 then

rank of $\begin{bmatrix} a_{11} & a_{12} \\ a_{21} & a_{22} \end{bmatrix}$ is :

(a) 3   (b) 2
(c) 1   (d) none of these

2. If A is a null matrix, then its rank is

: (a) 0
(b) 1
(c) 2
(d) none of these

3. The rank of I_6 is
   (a) 3
   (b) 4
   (c) 6
   (d) none of these

4. If A and B are equivalent and rank $(A) = r$ then rank of B is
   (a) $r - 1$
   (b) r
   (c) $r + 1$
   (d) none of these

5. If $A = \begin{bmatrix} 1 & 1 & 1 \\ 1 & 1 & 1 \\ 1 & 1 & 1 \end{bmatrix}$ then the rank of A^2 is
   (a) 3
   (b) 2
   (c) 1
   (d) none of these

6. If $A = \begin{bmatrix} 1 \\ 1 \\ 1 \end{bmatrix}$, $B = [2 \quad 3 \quad 4]$ then
   rank $(AB) =$
   (a) 2
   (b) 3
   (c) 1
   (d) none of these

7. If the rank of a matrix $\geq r$ then there is at least one r-rowed minor of the matrix whose rank is
   (a) 1
   (b) 0

(c) 2
(d) r

8. There are m equations in n variable then the order of coefficient matrix is
   (a) $n \times m$
   (b) $m \times n$
   (c) $m \times m$
   (d) $n \times n$

9. The matrix equation $AX = 0$ represent
   (a) non-homogeneous linear equations
   (b) homogeneous linear equations
   (c) homogeneous non-linear equation
   (d) none of these

10. The equations $AX = B$ are linear and
    (a) homogeneous
    (b) non-homogeneous
    (c) may be homogeneous
    (d) none of these

11. If X_1 and X_2 are the solution of $AX = 0$ then which one is also the solution of $AX = 0$
    (a) $X_1^2 + X_2^2$
    (b) $X_1 + X_2$
    (c) $(X_1 + X_2)^2$
    (d) none of these

12. If the equations $AX = B$ are consistent and rank of $(A \mid B) = 4$ then rank of A is
    (a) 4
    (b) 8
    (c) 3
    (d) none of these

ANSWERS

Fill in the blanks

(1) homogeneous
(2) $n - r$
(3) $C_1X_2 + C_2X_2$
(4) zero solution
(5) infinite solution
(6) non-homogeneous equation
(7) rank $(A \mid B)$
(8) inconsistent
(9) zero solution
(10) $n - r$
(11) singular
(12) 1
(13) n
(14) vanish
(15) same
(16) no
(17) 0
(18) invertible
(19) non-singular
(20) unique
(21) non-singular
(22) $B^{-1}A^{-1}$
(23) adjoint
(24) non-singular
(25) A itself

True/False

(1) F   (2) T   (3) T   (4) F   (5) T
(6) F   (7) F   (8) T   (9) T   (10) F

Multiple choice questions

(1) b (2) a (3) c (4) b (5) c (6) c
(7) b (8) b (9) b (10) b (11) b (12) a

Index